T0213258

Lecture Notes in Artificial Intelligence 9920

Subseries of Lecture Notes in Computer Science

More information about this series at http://www.springer.com/series/1244

Víctor Flores · Fernando Gomide
Andrzej Janusz · Claudio Meneses
Duoqian Miao · Georg Peters
Dominik Ślęzak · Guoyin Wang
Richard Weber · Yiyu Yao (Eds.)

Rough Sets

International Joint Conference, IJCRS 2016
Santiago de Chile, Chile, October 7–11, 2016
Proceedings

Springer

Editors
Víctor Flores
Universidad Católica del Norte
Antofagasta
Chile

Fernando Gomide
University of Campinas
Campinas, SP
Brazil

Andrzej Janusz
University of Warsaw
Warsaw
Poland

Claudio Meneses
Universidad del Católica del Norte
Antofagasta
Chile

Duoqian Miao
Tongji University
Shanghai
China

Georg Peters
University of Applied Sciences
Munich
Germany

Dominik Ślęzak
University of Warsaw and Infobright
Warsaw
Poland

Guoyin Wang
Chongqing University of Posts
    and Telecommunications
Chongqing
China

Richard Weber
Universidad de Chile
Santiago
Chile

Yiyu Yao
University of Regina
Regina, SK
Canada

ISSN 0302-9743   ISSN 1611-3349   (electronic)
Lecture Notes in Artificial Intelligence
ISBN 978-3-319-47159-4   ISBN 978-3-319-47160-0   (eBook)
DOI 10.1007/978-3-319-47160-0

Library of Congress Control Number: 2016952879

LNCS Sublibrary: SL7 – Artificial Intelligence

Printed on acid-free paper

This Springer imprint is published by Springer Nature
The registered company is Springer International Publishing AG
The registered company address is: Gewerbestrasse 11, 6330 Cham, Switzerland

Preface

This book contains regular and workshop papers selected for presentation at the 2016 International Joint Conference on Rough Sets (IJCRS 2016) held at Universidad de Chile, Santiago de Chile, during October 7–11, 2016.

IJCRS 2016 merged four main areas referring to major topics of rough set conferences held so far: rough sets and data science (in relation to RSCTC series organized since 1998), rough sets and granular computing (in relation to RSFDGrC series organized since 1999), rough sets and knowledge technology (in relation to RSKT series organized since 2006), and rough sets and intelligent systems (in relation to RSEISP series organized since 2007). It followed the success of the Joint Rough Set Symposiums (currently called International Joint Conferences on Rough Sets) established in Toronto, Canada (2007) and then continued in Chengdu, China (2012), Halifax, Canada (2013), Granada and Madrid, Spain (2014), and Tianjin, China (2015), where the new acronym – IJCRS – was used for the first time. Its goal was to attract experts from academia and industry from all over the world, including those working in various fields related to theoretical foundations and practical applications of rough sets, those working in other fields, wishing to discuss their results and experiences with the rough set community, as well as those dealing with real-world problems, wishing to discuss them with others and to look for new inspirations.

IJCRS 2016 comprised a vital mix of regular presentations and plenary sessions. The conference opening anniversary talk and the special plenary memorial session were dedicated to the seminal achievements of Zdzisław I. Pawlak (1926–2006) – a Polish mathematician and computer scientist, the founder of rough sets (1982), who also contributed to the design of the first Polish computer (1950), introduced a new approach to random number generation (1953), introduced a positional numeral system with base -2, introduced a generalized class of reverse Polish notation languages, proposed a new formal model of a digital machine, created the first mathematical model of DNA (1965), and proposed a new, very well-received mathematical model of conflict analysis (1984). The conference program also included 12 other keynotes and plenary talks, two tutorials, the 4th International Workshop on Three-way Decisions, Uncertainty, and Granular Computing (TWDUG), and the annual meeting of the International Rough Set Society (IRSS) at which its newly elected officers (for the period 2016–2018) and newly appointed fellows and senior members were welcomed.

IJCRS 2016 attracted 109 submissions (not including invited and special memorial session contributions), which underwent a rigorous reviewing process. Each accepted full-length paper was evaluated by three to five experts on average. In the present volume, 47 regular and workshop submissions are published as full-length papers. Moreover, 27 papers are published in the form of extended abstracts in additional conference materials. All full-length papers were gathered into nine sections that reflect some of the main trends in rough set research and illustrate how rough sets can co-exist with other approaches. Section 1 includes full-length papers prepared by keynote speakers, tutorial speakers, and IRSS fellows invited to deliver plenary talks at IJCRS 2016. Sections 2 and 3 contain papers showing how rough sets relate to the concepts of approximation, granulation, non-determinism, and incompleteness. Section 4 gathers full-length papers accepted to the TWDUG workshop. Section 5 contains both rough-set-related as well as not-rough-set-related papers on fuzziness and similarity in knowledge representation. Finally, Sections 6–9 correspond to the topics of machine learning and decision making, ranking and clustering, derivation and application of rule-based classifiers, as well as various rough-set-related aspects of working with feature subsets in knowledge discovery. We would like to thank all authors for contributing to the conference, as well as all Program Committee members and external reviewers for their hard work and very insightful comments.

The conference would not have been successful without support received from distinguished individuals and organizations. We express our gratefulness to the IJCRS 2016 honorary chairs, Andrzej Skowron and Bo Zhang, for their great leadership. We thank Davide Ciucci, Pablo A. Estévez, Jerzy W. Grzymała-Busse, Qinghua Hu, Xiaohua Tony Hu, Masahiro Inuiguchi, Pawan Lingras, Ernestina Menasalvas, Marco Orellana, Sankar K. Pal, Lech T. Polkowski, Roman Słowiński, and Shusaku Tsumoto for delivering excellent keynote and plenary talks. We thank Davide Ciucci, Salvatore Greco, Jouni Järvinen, Tianrui Li, Wojciech Moczulski, Hung Son Nguyen, Piotr Przystałka, Marek Sikora, Andrzej Skowron, and Radosław Zimroz for preparing tutorial materials. We would also like to thank Jerzy Błaszczyński, Yasuo Kudo, Dun Liu, Jaime Pavlich, Diego Urrutia, and Juan D. Velásquez, who supported the conference as tutorial, workshop, and publicity chairs. We are grateful to Soledad Arriagada, Juan Bekios, Karla Jaramillo, Aurora Radich, and all other representatives of Universidad de Chile and Universidad Católica del Norte who were involved in the conference organization. We acknowledge Davide Ciucci, Chris Cornelis, Marcin Szeląg, and Marcin Szczuka for their additional significant help at various stages of the conference publicity and material preparation. We would like to thank our sponsors, Springer, IRSS, and Data Mining Services Ltd., for their strategic and financial support. IJCRS 2016 was partially funded by the Complex Engineering Systems Institute, ISCI (ICM-FIC: P05-004-F, CONICYT: FB0816). We also acknowledge that we used EasyChair to conduct the paper-reviewing process.

October 2016

Víctor Flores
Fernando Gomide
Andrzej Janusz
Claudio Meneses
Duoqian Miao
Georg Peters
Dominik Ślęzak
Guoyin Wang
Richard Weber
Yiyu Yao

IJCRS 2016 Organization

Honorary Chairs

Andrzej Skowron   University of Warsaw, Poland
Bo Zhang   Tsinghua University, China

Conference Chairs

Dominik Ślęzak   University of Warsaw and Infobright, Poland
Richard Weber   Universidad de Chile, Chile

Steering Committee

Claudio Meneses   Universidad Católica del Norte, Chile
Duoqian Miao   Tongji University, China
Georg Peters   Munich University of Applied Sciences, Germany
Yiyu Yao   University of Regina, Canada

Program Chairs

Víctor Flores   Universidad Católica del Norte, Chile
Fernando Gomide   Universidade Estadual de Campinas, Brazil
Andrzej Janusz   University of Warsaw, Poland
Guoyin Wang   Chongqing University of Posts and
                         Telecommunications, China

Workshop and Tutorial Chairs

Dun Liu   Southwest Jiaotong University, China
Diego Urrutia   Universidad Católica del Norte, Chile

Publicity Chairs

Jerzy Błaszczyński   Poznań University of Technology, Poland
Yasuo Kudo   Muroran Institute of Technology, Japan
Jaime Pavlich   Pontificia Universidad Javeriana, Colombia
Juan D. Velásquez   Universidad de Chile, Chile

Local Organizing Committee

Soledad Arriagada	Universidad de Chile, Chile
Juan Bekios	Universidad Católica del Norte, Chile
Karla Jaramillo	Universidad de Chile, Chile
Aurora Radich	Universidad de Chile, Chile

IJCRS 2016 Program Committee

Piotr Artiemjew	University of Warmia and Mazury, Poland
Nouman Azam	FAST-NU, Pakistan
Ahmad Taher Azar	Benha University, Egypt
Cesar Azurdia	Universidad de Chile, Chile
Philippe Balbiani	Inria de Toulouse, France
Mohua Banerjee	IIT Kanpur, India
Alan Barton	Carleton University, Canada
Juan Bekios	Universidad Católica del Norte, Chile
Rafael Bello	Universidad Central de Las Villas, Cuba
Jerzy Błaszczyński	Poznań University of Technology, Poland
Nizar Bouguila	Concordia University, Canada
Mihir Chakraborty	Jadavpur University, India
Shampa Chakraverty	Netaji Subhas Institute of Technology, India
Chien-Chung Chan	University of Akron, USA
Hongmei Chen	Southwest Jiaotong University, China
Mu-Chen Chen	National Chiao Tung University, Taiwan
Davide Ciucci	University of Milano-Bicocca, Italy
Chris Cornelis	Ghent University, Belgium
Zoltán Ernő Csajbók	University of Debrecen, Hungary
Jianhua Dai	Tianjin University, China
Martine De Cock	University of Washington Tacoma, USA
Ali Dehghanfirouzabadi	Universidad de Santiago de Chile, Chile
Dayong Deng	Zhejiang Normal University, China
Thierry Denoeux	Université de Technologie de Compiègne, France
Lipika Dey	TCS Innovation Lab Delhi, India
Fernando Diaz	University of Valladolid, Spain
Didier Dubois	IRIT/RPDMP, France
Ivo Düntsch	Brock University, Canada
Zied Elouedi	Institut Suprieur de Gestion de Tunis, Tunisia
G. Ganesan	Adikavi Nannaya University, India
Yang Gao	Nanjing University, China
Anna Gomolińska	University of Białystok, Poland
Salvatore Greco	University of Catania, Italy
Jerzy Grzymała-Busse	University of Kansas, USA
Shen-Ming Gu	Zhejiang Ocean University, China
Christopher Henry	University of Winnipeg, Canada
Chris Hinde	Loughborough University, UK

Qinghua Hu	Tianjin University, China
Dmitry Ignatov	National Research University Higher School of Economics, Russia
Masahiro Inuiguchi	Osaka University, Japan
Ryszard Janicki	McMaster University, Canada
Jouni Järvinen	TXODDS and Lappeenranta University of Technology, Finland
Richard Jensen	Aberystwyth University, UK
Xiuyi Jia	Nanjing University of Science and Technology, China
Md. Aquil Khan	IIT Indore, India
Yoo-Sung Kim	Inha University, Korea
Beata Konikowska	Institute of Computer Science PAS, Poland
Jacek Koronacki	Institute of Computer Science PAS, Poland
Bożena Kostek	Gdańsk Univerity of Technology, Poland
Marzena Kryszkiewicz	Warsaw University of Technology, Poland
Yasuo Kudo	Muroran Institute of Technology, Japan
Yoshifumi Kusunoki	Osaka University, Japan
Sergei O. Kuznetsov	National Research University Higher School of Economics, Russia
Carson K. Leung	University of Manitoba, Canada
Rory A. Lewis	University of Colorado Colorado Springs, USA
Huaxiong Li	Nanjing University, China
Tianrui Li	Southwest Jiaotong University, China
Decui Liang	University of Electronic Science and Technology of China, China
Jiye Liang	Shanxi University, China
Churn-Jung Liau	Academia Sinica, Taiwan
Pawan Lingras	Saint Mary's University, Canada
Caihui Liu	Gannan Normal University, China
Dun Liu	Southwest Jiaotong University, China
Guilong Liu	Beijing Language and Culture University, China
Pradipta Maji	Indian Statistical Institute, India
A. Mani	Calcutta University, India
Victor Marek	University of Kentucky, USA
Nikolaos Matsatsinis	Technical University of Crete, Greece
Jesús Medina-Moreno	University of Cádiz, Spain
Ernestina Menasalvas	Technical University of Madrid, Spain
Jose Merigo	Universidad de Chile, Chile
Ju-Sheng Mi	Xi'an Jiaotong University, China
Tamás Mihálydeák	University of Debrecen, Hungary
Fan Min	Southwest Petroleum University, China
Boris Mirkin	National Research University Higher School of Economics, Russia
Sushmita Mitra	Indian Statistical Institute, India
Sadaaki Miyamoto	University of Tsukuba, Japan
Mikhail Ju. Moshkov	KAUST, Saudi Arabia

Michinori Nakata	Josai International University, Japan
Amedeo Napoli	LORIA Nancy, France
Maria C. Nicoletti	FACCAMP and UFSCar, Brazil
Sergey Nikolenko	Steklov Mathematical Institute, Russia
Vilém Novák	University of Ostrava, IRAFM, Czech Republic
Hannu Nurmi	University of Turku, Finland
Piero Pagliani	Research Group on Knowledge and Communication Models, Italy
Krzysztof Pancerz	University of Rzeszów, Poland
Andrei Paun	University of Bucharest, Romania
Witold Pedrycz	University of Alberta, Canada
James F. Peters	University of Manitoba, Canada
Frederick Petry	Naval Research Lab, USA
Jonas Poelmans	Katholieke Universiteit Leuven, Belgium
Lech T. Polkowski	Polish-Japanese Academy of Information Technology, Poland
Henri Prade	IRIT-CNRS, France
Jin Qian	Jiangsu University of Technology, China
Anna Maria Radzikowska	Warsaw University of Technology, Poland
Elisabeth Rakus-Andersson	Blekinge Institute of Technology, Sweden
Sheela Ramanna	University of Winnipeg, Canada
C. Raghavendra Rao	University of Hyderabad, India
Zbigniew Raś	University of North Carolina Charlotte, USA
Leszek Rutkowski	Technical University of Częstochowa, Poland
Henryk Rybiński	Warsaw University of Technology, Poland
Hiroshi Sakai	Kyushu Institute of Technology, Japan
Gerald Schaefer	Loughborough University, UK
Lin Shang	Nanjing University, China
B. Uma Shankar	Indian Statistical Institute, India
Zhongzhi Shi	Chinese Academy of Sciences, China
Marek Sikora	Silesian University of Technology, Poland
Andrzej Skowron	University of Warsaw, Poland
Roman Słowiński	Poznań University of Technology, Poland
Jerzy Stefanowski	Poznań University of Technology, Poland
John G. Stell	University of Leeds, UK
Jarosław Stepaniuk	Białystok University of Technology, Poland
Zbigniew Suraj	University of Rzeszów, Poland
Piotr Synak	Infobright, Poland
Andrzej Szałas	University of Warsaw, Poland
Marcin Szczuka	University of Warsaw, Poland
Domenico Talia	University of Calabria, Italy
Raúl M. del Toro	Technical University of Madrid, Spain
B.K. Tripathy	VIT University Vellore, India
Li-Shiang Tsay	North Carolina A&T State University, USA
Diego Urrutia	Universidad Católica del Norte, Chile
Changzhong Wang	Bohai University, China

Hai Wang	Saint Mary's University, Canada
Hui Wang	University of Ulster, UK
Xin Wang	University of Calgary, Canada
Arkadiusz Wojna	Infobright, Poland
Marcin Wolski	Maria Curie-Skłodowska University, Poland
Wei-Zhi Wu	Zhejiang Ocean University, China
Xindong Wu	University of Vermont, USA
Weihua Xu	Chongqing University of Posts and Telecommunications, China
Zhan-Ao Xue	Henan Normal University, China
Yan Yang	Southwest Jiaotong University, China
JingTao Yao	University of Regina, Canada
Dongyi Ye	Fuzhou University, China
Hong Yu	Chongqing University of Posts and Telecommunications, China
Xiaodong Yue	Shanghai University, China
Sławomir Zadrożny	Systems Research Institute PAS, Poland
Bo Zhang	Tsinghua University, China
Hongyun Zhang	Tongji University, China
Nan Zhang	Yantai University, China
Qinghua Zhang	Chongqing University of Posts and Telecommunications, China
Yan-Ping Zhang	Anhui University, China
Shu Zhao	University of California Berkeley, USA
Ning Zhong	Maebashi Institute of Technology, Japan
Bing Zhou	Sam Houston State University, USA
Wojciech Ziarko	University of Regina, Canada

IJCRS 2016 External Reviewers

María José Benítez Caballero
Manuel De Buenaga
Katarzyna Borowska
Maria Eugenia Cornejo Piñero
Golnoosh Farnadi
Brian Keith
Yonghwa Kim
Vojtěch Molek
Eloisa Ramírez Poussa
Yanyan Yang

Contents

Fuzziness and Similarity in Knowledge Representation

Machine Learning and Decision Making

Ranking and Clustering

Derivation and Application of Rules and Trees

Derivation and Application of Feature Subsets

Keynotes, Tutorials and Expert Papers

Advances in Rough and Soft Clustering: Meta-Clustering, Dynamic Clustering, Data-Stream Clustering

Pawan Lingras[✉] and Matt Triff

Mathematics and Computing Science, Saint Mary's University, Halifax, Canada
pawan@cs.smu.ca, matt.triff@gmail.com

Abstract. Over the last five decades, clustering has established itself as a primary unsupervised learning technique. In most major data mining projects clustering can serve as a first step in understanding the available data. Clustering is used for creating meaningful profiles of entities in an application. It can also be used to compress the dataset into more manageable granules. The initial methods of crisp clustering objects represented using numeric attributes have evolved to address the demands of the real-world. These extensions include the use of soft computing techniques such as fuzzy and rough set theory, the use of centroids and medoids for computational efficiency, modes to accommodate categorical attributes, dynamic and stream clustering for managing continuous accumulation of data, and meta-clustering for correlating parallel clustering processes. This paper uses applications in engineering, web usage, retail, finance, and social networks to illustrate some of the recent advances in clustering and their role in improved profiling, as well as augmenting prediction, classification, association mining, dimensionality reduction, and optimization tasks.

Keywords: Clustering · Rough sets · Fuzzy sets · Finance · Retail · Social networks · Web usage · Engineering · Meta-clustering · Dynamic clustering

1 Introduction

Clustering is one of the most versatile data mining techniques. Since it is an unsupervised learning technique, it can be part of the initial pattern analysis in a dataset. Clustering can also be used at different stages of a knowledge discovery process. The objective of this paper is to use real world applications in domains ranging from retail, mobile/social networks, finance, web usage, and engineering to demonstrate how clustering can play an important role in data mining. Researchers have proposed a number of extensions to the original crisp clustering techniques. The applications described in this paper describe how these extensions improve the unsupervised learning process in the real-world.

The paper first illustrates how the available raw data with limited input from domain experts can initiate a knowledge discovery process. We will see

© Springer International Publishing AG 2016
V. Flores et al. (Eds.): IJCRS 2016, LNAI 9920, pp. 3–22, 2016.
DOI: 10.1007/978-3-319-47160-0_1

why the initial crisp clustering algorithms are not able to model clustering in real-world applications. Fuzzy and rough set theories are shown to provide better alternatives for some of the real-world applications. Fuzzy clustering provides a degree of membership to the clusters, but does not provide obvious boundaries between clusters. Rough clustering can provide a happy medium between the fuzzy and crisp clustering. Rough clustering can also complement fuzzy clustering to provide descriptive memberships and identifiable rough boundaries of clusters. The paper also describes how one can derive well delineated rough clusters from a fuzzy clustering scheme.

Clustering technology continues to evolve to respond to new challenges. We will discuss an emerging area of meta-clustering that use hierarchical, network, and temporal relationships between objects for parallel clustering processes that feed knowledge to each other, creating semantically enhanced meta-clustering schemes. We will briefly review dynamic, incremental, and decremental clustering algorithms the have been developed to address continuous accumulation of data. These techniques reorganize the clustering schemes by adding new clusters, deleting obsolete clusters, and merging clusters that start to converge. The discussion will also include the need for efficient handling of high velocity datastreams. The versatility of clustering is further demonstrated by showing its usage for improving the quality of other data mining techniques. For example, grouping similar patterns can improve the quality of prediction techniques. Clustering can also be used to summarize the results of other data mining techniques, such as evolutionary optimization. Finally, we will discuss how clustering can provide an alternative, or supplementary, classification or association mining technique. The objective of the paper is not to provide a comprehensive review of clustering research, but to demonstrate its pivotal role in real-world data mining applications.

2 Crisp Clustering

In this section, we will look at a web usage mining application of a popular clustering algorithm called k-means [6]. The k-means algorithm identifies the centroids (means) of the clusters in a dataset. It begins with random centroids. Objects are assigned to the closest centroid. The centroid of the objects assigned to different clusters is recalculated. The process continues until the centroids converge.

Yelp.com is an online review and recommendation community. Yelp was founded in 2004, is available in 32 countries worldwide, and currently has over 100 million unique monthly visitors. Yelp provides value to consumers by allowing users to research written reviews, ratings, business details such as business hours and whether or not a business has free WiFi, as well as pictures posted by other users of the business and its products. Yelp also provides a social platform for its users, allowing them to create events, lists of recommended businesses to share and comment on, and to message and become friends with other users.

In Spring 2013, Yelp released a large set of data, covering the entire Phoenix Metropolitan Area (PMA) as part of the Yelp Dataset Challenge. The Yelp

Dataset Challenge was open-ended and aimed at finding innovative uses for the data Yelp collects. Yelp posed potential questions to answer, such as "What time of day is a restaurant busy, based on its reviews?", "What makes a review funny or cool?", "Which business is likely to be reviewed next by a user?", and more. Yelp encouraged the submission in any form that entrants felt conveyed the appeal of their project, which would later be judged for one of ten cash prizes. The data covering the PMA came as four separate files, one each for businesses, check-ins, users, and reviews. Business information includes each businesses unique ID, name, neighborhoods that they are located within, full localized address, city, state, latitude, longitude, average star rating out of five (rounded to half stars) from reviewers, categories, and a variable set for whether or not the business is still active. Reviews contained the business ID of the business being reviewed, the ID of the reviewer writing the review, the number of stars the reviewer gave the business out of five (rounded to the half star), the text of the written review, the date the review was given, and the number of votes other users have given the review, in the categories of "Funny", "Cool" and "Useful". Reviewer data contained the unique user ID, first name, number of reviews they have given, average stars rated (as a floating point average of all the reviews they have made), and the total number of votes their reviews have received for the three categories previously mentioned. Finally, check-in data contained information of which business the check-in data related to, and the total number of people who had checked-in to the business on the mobile Yelp app or webpage for each hour of each day of the week (168 categories, or 24 categories for each day of the week).

As a first data mining activity, we can group the reviewers and businesses based on the number of reviews from different categories (*, **, ***, ****, *****), as well as votes received by the reviewer. Representing the objects in the dataset is one of the most important aspects of clustering. A reviewer is represented by $sr_j = (total, *, **, ***, ****, *****, votes)$, where total is the number of reviews, * is the number of one star reviews, ** means the number of two star reviews, and so on. A business is represented by $sb_i = (total, *, **, ***, ****, *****)$. Note that there are no votes in the representation of a business.

Table 1. Centroids from crisp clustering of business data

Cluster ID	Total	*	**	***	****	*****	Size
sbc_1	5.75	4.99	2.18	2.74	3.92	86.14	2073
sbc_2	5.29	55.80	11.44	7.80	11.13	13.82	1221
sbc_3	11.10	6.45	7.10	11.70	59.22	15.50	2212
sbc_4	12.42	12.61	20.05	34.21	22.61	10.50	2301
sbc_5	13.64	7.04	5.83	9.24	31.04	46.83	2782
sbc_6	101.39	5.92	9.43	16.04	37.45	31.13	844
sbc_7	334.85	3.18	6.43	13.64	38.59	38.13	104

Table 1 shows the results of crisp clustering applied to the information about the businesses. We can describe the resulting profiles of business clusters as:

sbc_1 **Sparsely but very well rated** - Fewest number of reviews, mostly five stars.

sbc_2 **Sparsely and lowly rated** - Fewest number of reviews, mostly one and two stars.

sbc_3 **Well rated** - Modest number of reviews, mostly five and four stars.

sbc_4 **Ambivalently rated** - Modest number of evenly spread reviews.

sbc_5 **Reasonably rated** - Modest number of reviews, mostly four and five stars.

sbc_6 **Well rated** - Large number of reviews, mostly four and five stars with noticeable three stars.

sbc_7 **Reasonably rated** - Largest number of reviews, mostly four and five stars with noticeable three stars.

Table 2. Centroids from crisp clustering of reviewer data

Cluster ID	Total	*	**	***	****	*****	Votes	Size
src_2	1.60	87.33	0.69	2.29	0.85	8.82	2.39	5154
src_5	2.47	1.55	1.22	3.06	3.65	90.50	3.06	16569
src_6	2.94	5.64	65.53	6.94	10.10	11.77	4.18	4581
src_4	4.75	3.43	2.70	8.49	69.60	15.75	6.26	14321
src_3	22.19	2.63	4.34	63.06	17.01	12.94	85.01	3171
src_1	183.04	4.50	9.43	23.14	39.48	23.41	2278.42	75
src_6	442.5	2.27	5.65	26.29	46.04	19.73	13073.5	2

Table 2 shows the results of crisp clustering applied to the information about the reviewers. We can describe the resulting profiles of reviewer clusters as:

src_1 **Infrequent and hard** -Very few and mostly one and two star reviews.

src_2 **Infrequent and soft** - Very few and mostly five and four star reviews.

src_3 **Infrequent and very soft** - Very few and almost exclusively five star reviews.

src_4 **Infrequent and balanced** - Very few and mostly five star reviews, with noticeable two and three stars reviews as well.

src_5 **Somewhat prolific and balanced** - Modest number of reviews and votes, mostly four, five, and three stars.

src_6 **Prolific and balanced** - Large number of reviews and votes are mostly four, three, and five stars.

src_7 **Extremely prolific and balanced** - This group of two is essentially an outlier with a large number of reviews and votes, and these users should be treated separately as prolific reviewers.

More details of the experiments can be found in [13].

Table 3. Centroids from fuzzy clustering of business data

Cluster ID	Total	*	**	***	****	*****	Size
$sbcf_1$	5.47	1.83	0.92	1.13	2.36	93.73	1799
$sbcf_2$	8.77	36.66	14.52	16.19	17.77	14.84	1948
$sbcf_3$	10.74	6.88	4.89	7.67	29.05	51.47	2246
$sbcf_4$	12.15	7.74	7.88	14.49	49.96	19.89	2229
$sbcf_5$	13.43	12.90	14.26	25.06	31.29	16.46	2115
$sbcf_6$	76.85	6.87	9.73	16.40	37.63	29.34	1032
$sbcf_7$	262.98	3.83	6.96	14.23	38.72	36.25	168

Table 4. Centroids from fuzzy clustering of reviewer data

Cluster ID	Total	*	**	***	****	*****	Votes	Size
$srcf_6$	1.46	97.77	0.40	0.36	0.57	0.88	3.79	4079
$srcf_4$	1.52	0.34	0.24	0.26	0.39	98.75	2.20	13842
$srcf_3$	1.71	0.33	0.44	0.75	97.72	0.74	3.13	6740
$srcf_2$	1.83	0.76	95.69	0.79	1.36	1.37	5.82	2274
$srcf_5$	2.12	0.76	0.91	94.47	2.09	1.74	6.40	2223
$srcf_6$	6.14	4.23	4.41	5.89	40.15	45.30	10.70	7032
$srcf_1$	11.09	8.99	11.82	18.24	33.88	27.04	22.64	7683

3   Fuzzy Clustering

Conventional clustering assigns various objects to precisely one cluster. A fuzzy generalization of the clustering, Fuzzy C-means, uses a fuzzy membership function to describe the degree of membership (ranging from 0 to 1) of an object to a given cluster. There is a stipulation that the sum of fuzzy memberships of an object to all the clusters must be equal to 1. The algorithm was first proposed by Dunn in 1973 [4].

Table 3 shows the results of applying fuzzy clustering to the businesses in the yelp.com dataset. One of the major advantages of the fuzzy clustering is the fact that the businesses can belong to multiple clusters. Another interesting feature of fuzzy clustering is that the resulting centroids tend to be less extreme, better separated, and the cluster sizes are more uniformly distributed. The overall clustering profiles do match the crisp clustering and are given below:

$sbcf_1$ **Sparsely but very well rated** - Fewest number of reviews, mostly five stars.

$sbcf_2$ **Sparsely and lowly rated** - Few reviews, majority are one and two stars.

$sbcf_3$ **Well rated** - Modest number of reviews, mostly five and four stars.

$sbcf_4$ **Reasonably rated** - Modest number of reviews, mostly four and five stars.

$sbcf_5$ **Ambivalently rated** - Modest number of evenly spread reviews.

$sbcf_6$ **Reasonably rated** - Large number of reviews, mostly four and five stars with noticeable three stars.

$sbcf_7$ **Reasonably rated** - Largest number of reviews, mostly four and five stars with noticeable three stars.

Table 4 shows the results of fuzzy clustering applied to the information about the reviewers. The moderating effect of fuzzy C-means is more pronounced for the reviewer dataset. The last two crisp clusters src_6 and src_7 consisted of a total of 77 reviewers with extremely high values for total reviews and votes. The corresponding fuzzy clusters, $srcf_6$ and $srcf_7$, have more moderate centroids and represent more than 14000 reviewers. The outlying reviewers have been essentially absorbed in the crisp cluster src_5. These moderate profiles are possible because these reviewers can belong to multiple clusters. Due to the more pronounced effect of fuzzy clustering, the reviewer fuzzy profiles are somewhat different from their crisp counter-parts and can be described as:

$srcf_1$ **Infrequent and hardest** - Very few and mostly one star reviews.

$srcf_2$ **Infrequent and soft** - Very few and mostly five star reviews.

$srcf_3$ **Infrequent and very soft** - Very few and almost exclusively five star reviews.

$srcf_4$ **Infrequent and hard** - Very few and mostly two star reviews.

$srcf_5$ **Infrequent middle of the road** - Very few and mostly three star reviews.

$srcf_6$ **Frequent and somewhat soft** - Modest number of reviews and votes.

$srcf_7$ **Prolific and balanced** - Large number of evenly spread reviews.

In summary, the comparison of crisp and fuzzy profiles shows that the fuzzy clustering allows objects to belong to multiple clusters. This assignment to multiple clusters leads to centroids that are less extreme and well separated. Moreover, the objects are more uniformly distributed among all the clusters. More details of the experiments can be found in [13].

4   Rough Clustering

Fuzzy C-means makes it possible to assign an object to multiple clusters with different degrees of membership. These memberships can be too descriptive for most users. Moreover, one cannot easily identify the cluster boundaries. The rough K-means [14] algorithm and its various extensions [15, 16] have been found to be effective in creating well delineated lower and upper bounds of clusters. A comparative study of crisp, rough and evolutionary clustering depicts how rough clustering outperforms crisp clustering [8]. Peters et al. [17] provide a good comparison of rough clustering and other conventional clustering algorithms. Rough set clustering has also been reformulated in the context of related theories such as the interval set clustering formulation by Yao et al. [24]. Yu et al. [25, 26] proposed clustering of incomplete data based on a new and dynamic extension of rough set theory called three-way decision theory. This section presents an engineering application to demonstrate the effectiveness of rough clustering.

Seasonal and permanent traffic counters scattered across a highway network are the major sources of traffic data. These traffic counters measure the traffic volume – the number of vehicles that have passed through a particular section of a lane or highway in a given time period. Traffic volumes can be expressed in terms of hourly or daily traffic. More sophisticated traffic counters record additional information such as the speed, length and weight of the vehicle. Highway agencies generally have records from traffic counters collected over a number of years. In addition to obtaining data from traffic counters, traffic engineers also conduct occasional surveys of road users to get more information.

The permanent traffic counter (PTC) sites are grouped to form various road classes. These classes are used to develop guidelines for the construction, maintenance and upgrading of highway sections. In one commonly used system, roads are classified on the basis of trip purpose and trip length characteristics [19]. Examples of resulting classes are commuter, business, long distance, and recreational highways. The trip purpose provides information about the road users, an important criterion in a variety of traffic engineering analyses. Trip purpose information can be obtained directly from the road users, but since all users cannot be surveyed, traffic engineers study various traffic patterns obtained from seasonal and permanent traffic counters and sample surveys of a few road users.

The present study is based on a sample of 264 monthly traffic patterns - variation of monthly average daily traffic volume in a given year - recorded between 1987 and 1991 on Alberta highways. The distribution of PTCs in various regions are determined based on the traffic flow through the provincial highway networks. The patterns obtained from these PTCs represent traffic from all major regions in the province. The hypothetical classification scheme consisted of three classes:

1. Commuter/business,
2. Long distance, and
3. Recreational.

The rough set classification scheme was expected to specify lower and upper bounds of these classes.

The resulting rough set classification schemes were subjectively compared with the conventional classification scheme. The upper and lower approximations of the commuter/business, long distance, and recreational classes were also checked against the geography of Alberta highway networks. More details of the experiment can be found in [9].

Figure 1 shows the monthly patterns for the lower approximations of the three groups: commuter/business, long distance, and recreational. The average pattern for the lower approximation of the commuter/business class has the least variation over the year. The recreational class, conversely, has the most variation. The variation for the long distance class is less than the recreational, but more than the commuter/business class. Figure 2 shows one of the highway sections near counter number C013201 that may have been commuter/business or long

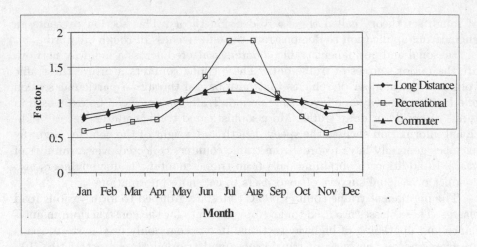

Fig. 1. Monthly patterns for the lower approximations

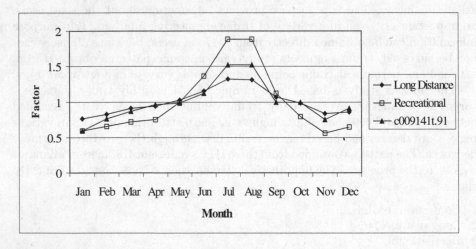

Fig. 2. Monthly pattern that may be long distance or recreational

distance in 1985. It is clear that the monthly pattern for the highway section falls in between the two classes. The counter C013201 is located on Highway 13, 20 km west of the Alberta-Saskatchewan border. It is an alternate route for travel from the city of Saskatoon and surrounding townships to townships surrounding the city of Edmonton. A similar observation can be made in Fig. 3 for highway section C009141 that may have been long distance or recreational in 1991. The counter C009141 is located on Highway 9, 141 km west of the Alberta-Saskatchewan border. The traffic on that particular road seems to have higher seasonal variation than a long distance road. Rough set representation of clusters enables us to identify such intermediate patterns.

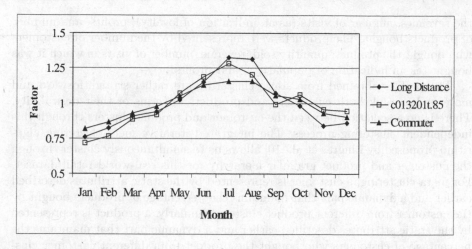

Fig. 3. Monthly pattern that may be commuter/business or long distance

5   Meta-Clustering

This section describes a novel set of integrated secondary data mining approaches to clustering. The techniques presented in this paper enhance the conventional clustering techniques for hierarchical, network, and temporal data.

5.1   Hierarchical Meta-Clustering

In granular computing, a granule represents an object associated with a set of information. For example, a customer with certain purchasing patterns could represent an information granule. A granule can include a collection of finer granules. For example, a customer granule could include many visits, which are finer granules. A visit, in turn, can include the purchase of a number of products, which are even finer granules. This results in a hierarchy of customers-visits-products. Profiles of customers created by clustering should also include the profiles of visits that these customers make. The profiling of visits should in turn include profiles of customers. Similarly, profiles of products should be both influenced by and should influence profiles of customers and visits. We describe an iterative clustering technique that iterates back and forth through a granular hierarchy to obtain a stable set of profiles of objects at all levels of the hierarchy. This section reports experiments with a real-world dataset consisting of all the purchases made by customers from a small retail chain.

The data spans three years from 2005–2007, consisting of 15,341 customers and 8,987 products. Many of the customers do not come often enough and buy enough products to provide meaningful iterative clustering. Similarly, a number of products are not brought frequently enough and not by a diverse customer population. Therefore, we chose to restrict our analysis to the top 1000 customers and the top 500 products, based on their revenue. Customers were represented by the revenues, number of visits (as an indication of loyalty), profits, and number of products bought. The products were represented by the number of customers who bought the product, quantity sold, revenue, number of visits in which it was bought (as an indication of popularity), and profits.

The profiles obtained from static clustering are rather straightforward and more or less rank both customers and products in terms of their desirability. There is no association between the customer and product clusters through this independent clustering process. The integrated iterative meta-clustering algorithm proposed by Lingras et al. [10] allows us to simultaneously cluster through the customer and product granular hierarchy for this real-world retail dataset. For meta-clustering, a customer is represented by the static attributes described earlier and a dynamic part that maintains the percentage of products bought by the customer from different product clusters. Similarly, a product is represented by the static attributes described earlier and a dynamic part that maintains the percentage of customers who bought the product from different customer clusters. Meta-clustering begins with static clustering of customers and products. The results from static clustering are used to create the dynamic parts of the representations of products and customers. The augmented representations are used to re-cluster both products and customers. The process is repeated until the dynamic representations converge.

The meta-centroids of the customer clusters are shown in Table 5 and the meta-centroids of the product clusters are shown in Table 6. The last column in each table shows the size of each cluster.

The resulting customer profiles are now more refined as they use association with the profiles of the products that the customers buy. We can describe these enhanced profiles as follows:

Table 5. Centroids from iterative meta-clustering of real customer data

	Revenue	Visits	Profits	Products	pc_1	pc_2	pc_3	Size
	sc_1	sc_2	sc_3	sc_4				
cc_1	0.48	0.56	0.81	0.59	0.77	0.61	0.66	11
cc_2	0.25	0.17	0.68	0.20	0.46	0.20	0.28	99
cc_3	0.07	0.05	0.56	0.07	0.11	0.08	0.10	603
cc_4	0.11	0.08	0.58	0.09	0.34	0.10	0.15	287

Table 6. Centroids from iterative meta-clustering of real product data

	NumCustomers	Quantity	Revenue	Visits	Profits	cc_1	cc_2	cc_3	cc_4	Size
	sp_1	sp_2	sp_3	sp_4	sp_5					
pc_1	0.15	0.08	0.07	0.09	0.06	0.15	0.22	0.15	0.37	166
pc_2	0.52	0.46	0.49	0.54	0.39	0.55	0.61	0.48	0.66	11
pc_3	0.07	0.03	0.04	0.03	0.04	0.07	0.11	0.07	0.14	323

cc_1 Highest spending, profitable and most loyal customers who buy more or less equally from all product groups.

cc_2 Moderately spending, profitable and moderately loyal customers who seem to favour the second most desirable group of products, given by pc_1.

cc_3 Along with cc_4, these customer contribute least to the store's business. The distinguishing feature for this cluster is the fact that they buy uniformly few products from all the three clusters.

cc_4 While comparable in contributions to cc_3, these customers seem to favour the second most desirable group of products, given by pc_1.

The association of product information with customer clusters is inversely applicable to the product profiles, which are refined using the profiles of the customers who buy these products. These augmented product profiles can be described as:

pc_1 Moderate revenue, profit and moderately popular products that are modestly preferred by all customers. There is a slightly higher preference by customers from the third ranked group of customers, in cluster cc_4.

pc_2 Highest revenue/profit, and most popular products that are favoured highly by customers from all the groups.

pc_3 Least contributing products who seem to have similarly low patronage from customers across the spectrum.

5.2   Network Meta-Clustering

Interdependencies between objects can also be observed in a networked environment, where objects such as phone users are connected to other phone users. In such a case, the profile of a phone user should include the profiles of other users created by the same clustering process. These dependencies are applicable to any social network. This section presents a recursive clustering technique for such networked environment using a data set provided by Eagle [5].

The objective of the present study is to use recursive clustering to converge to a set of user profiles. The data set comprises of 182,208 phone calls data collected from about 102 users over a period of nine months. The following variables were used to represent a phone call:

1. Average duration of phone calls
2. Average number of weekend/weekday
3. Average number of daytime/night-time
4. Average number of outgoing/incoming
5. Average number of SMS
6. Average number of voice calls
7. Average number of long duration calls

The clustering results can be analyzed in two parts - static and dynamic. The static part corresponds to the clustering analysis based on the static part of the data as described earlier. The dynamic part of a user maintains the percentage of users from each cluster that were called by this user. Meta-clustering begins with static clustering of users. The results from static clustering are used to create the dynamic parts for the users. The augmented representations are used to re-cluster the users. The process is repeated until the dynamic representations converge.

Profile of Cluster 1: These users make low number of calls, low average duration calls, low weekend calls, highest day time calls, highest outgoing calls, least SMS calls, highest voice calls and the fewest long duration calls.

Profile of Cluster 2: This cluster is made up of phone numbers which make the highest number of calls, low average duration of calls, low weekend calls, least day time calls, low number of outgoing calls, moderate SMS calls and moderate number of voice calls.

Profile of Cluster 3: This cluster is made up of phone numbers that make the least number of calls, low average duration of calls, low weekend calls, moderate day time calls, moderate outgoing calls, moderate SMS calls and high number of voice calls.

Profile of Cluster 4: This cluster is made up of phone numbers that make moderate number of calls, least average duration of calls, high number of weekend calls, moderate day time calls, least outgoing calls, highest SMS calls, least voice calls and the least number of long duration calls.

Profile of Cluster 5: This cluster is made up of phone numbers that made moderate number of calls, highest average duration of calls, high weekend calls, low daytime calls, high outgoing calls, low SMS calls, high voice calls and the highest number of long distance calls.

Table 7. The cluster centers corresponding to the dynamic part

Cluster number	$m_{j,1}^i$	$m_{j,2}^i$	$m_{j,3}^i$	$m_{j,4}^i$	$m_{j,5}^i$
1	0.008	0.017	0.047	0.002	0.010
2	0.005	0.010	0.022	0.000	0.005
3	0.011	0.001	0.012	0.001	0.005
4	0.023	0.008	0.011	0.000	0.009
5	0.003	0.000	0.000	0.000	0.001

The cluster centers for the dynamic part in the Table 7 show some differentiation between the clusters. Each row gives us the cluster center for each cluster and each column gives us a dimension.

Since the rows are actually cluster centers, each of the values in the row is the value of the cluster center along a particular dimension. If a cluster center of cluster l is high with respect to a particular column k, it means that phone numbers belonging to cluster l frequently communicate with the phone numbers of cluster k. We can extend this concept to mention that for cluster l, if the values along most dimensions are high (i.e. most of the values are high along the row), then cluster l is a very social cluster and it is in contact with most of the clusters.

The popularity values (column wise values) are indicative of how important a dimension is for the cluster centers for all the clusters. As seen in the Table 7, the column of cluster 2 has high values for clusters 1, 2 and 4. However, all values along the column 1 are high. This means that cluster 1 is a very popular and important cluster for all other clusters.

Please note that sociability and popularity are independent of each other. Hence, it is possible that a cluster l is social with cluster k but is not popular with cluster k. The results of the clustering as shown in Table 7 in terms of sociability and popularity are summarized below.

- Cluster 1 is the most popular cluster. Cluster 1 is also a very social cluster as its row-wise values are high.
- Cluster 2 is very popular among the phone numbers from the clusters 1, 2 and 4. Cluster 2 is also fairly social except with phone numbers from cluster 4.
- Cluster 3 is very popular with all the clusters except cluster 5. Cluster 3 is also social with all clusters.
- Cluster 4 is the least popular cluster. Also, cluster 4 phone numbers are social with all the clusters except their own cluster. These phone numbers indicate people who are very selective of the people they communicate with.
- Cluster 5 phone numbers are the least social phone numbers. They socialize only with themselves and with phone numbers from cluster 1. But they are popular phone numbers and all clusters communicate with them.

From the above observations, we can conclude that while certain phone users tend to concentrate their destination numbers to a particular group of people (who fall within the same cluster because of their inherent calling behavior), others are more diversely networked. We can also distinguish between their sociability and popularity characteristics which can help to build a sophisticated model of the social network represented by the data set.

5.3 Temporal Meta-Clustering

The recursive clustering developed for a networked environment is also extended to temporal databases, where profiles of daily patterns of a quantity should include profiles of previous daily patterns, and in some cases profiles of future daily patterns. For example, let us assume that we need to profile a stock based on its volatility. The volatility in a stock price today should take into account volatility of stock prices in the immediate past and immediate future. One can look at such a daily pattern as an object that is connected to the past and future daily patterns.

Volatility of financial data series is an important indicator used by traders. The fluctuation in prices creates trading opportunities. Volatility is a measure for variation of price of a financial instrument over time. Distribution of prices during the day can provide an elaborate description of price fluctuations. A trader finds a daily price pattern interesting when it is volatile. The higher the fluctuations in prices, the more volatile the pattern. The Black Scholes index is a popular way to quantify volatility of a pattern [3]. We can segment daily patterns based on the values of the Black Scholes index. This segmentation is essentially a clustering of one dimensional representation (Black Scholes index) of the daily pattern. Black Scholes index is a single concise index to identify volatility in a daily pattern. However, a complete distribution of prices during the day can provide a more elaborate information on the volatility during the day. While a distribution consisting of frequency of different prices is not a concise description for a single day, it can be a very useful representation of daily patterns for clustering based on volatility. Lingras and Haider [11] described how to create a rough ensemble of clustering using both of these representations. In this paper, we only use the daily price distribution to demonstrate the recursive temporal meta-clustering. However, the proposed approach can use either of the two representations, or even an ensemble of the two clustering methods.

Following [11], we use five percentile values; 10 %, 25 %, 50 %, 75 % and 90 % to represent the price distribution. 10 % of the prices are below the 10th percentile value, 25 % of the prices are below the 25th percentile value and so on. Our data set contains average prices at 10 min intervals for 223 instruments transacted on 121 days, comprising a total of 27,012 records. Each daily pattern has 39 intervals. This data set is used to create a five dimensional pattern, which represents 10, 25, 50, 75 and 90 percentile values of the prices. The prices are normalized by the opening price so that a commodity selling for $100 has the same pattern as the one that is selling for $10. Afterwards, the natural logarithms of the five percentiles are calculated.

As before, the clustering results can be analyzed in two parts - static and dynamic. The static part corresponds to the clustering analysis based on the static part of the data as described earlier. The dynamic part of a stock is the volatility cluster the stock belonged to for the previous ten days. Meta-clustering begins with static clustering of the stocks. The results from static clustering are used to create the dynamic parts for the stocks. The augmented representations are used to re-cluster the stocks. The process is repeated until the dynamic representations converge as shown in Table 8.

Table 8. Final ranked centers for percentile data

Rank	Cluster	p25	p50	p75	p90	d_{m-9}	d_{m-8}	d_{m-7}	d_{m-6}	d_{m-5}	d_{m-4}	d_{m-3}	d_{m-2}	d_{m-1}	d_m
1	C4	0.17	0.37	0.58	0.78	1.04	1.02	1.05	1.04	1.03	1.04	1.03	1.03	1.06	1.10
2	C3	0.20	0.42	0.66	0.87	1.27	1.29	1.46	1.67	1.97	2.35	2.70	2.98	3.01	2.93
3	C2	0.20	0.43	0.68	0.90	3.31	3.17	2.71	2.35	1.95	1.56	1.33	1.21	1.20	1.23
4	C5	0.25	0.52	0.80	1.04	2.92	3.05	3.18	3.31	3.39	3.34	3.19	2.99	2.86	2.71
5	C1	0.91	1.92	2.72	3.24	1.27	1.21	1.25	1.22	1.23	1.19	1.19	1.17	1.16	1.21

The static and dynamic profiles for the days of all the financial instruments (stocks) provide us the meta-profiles of each cluster as follows:

Cluster C4 (Rank 1) - least volatile The stocks in this cluster are not volatile today nor have they shown any volatility for last two weeks (10 trading days).

Cluster C3 (Rank 2) - low volatility today, but volatile over last week The stocks in this cluster are not volatile today. However, they were volatile last week (5 trading days). The volatility in these stocks may be subsiding and it may be relatively safer to sell them.

Cluster C2 (Rank 3) - moderate volatility today and last week, but volatile two weeks ago The stocks in this cluster are somewhat volatile today. They have not shown much volatility last week (5 trading days) either. However, they were quite active two weeks ago. The volatility in these stocks has definitely subsided and it may be better to sell them as they are unlikely to rise in the next little while.

Cluster C5 (Rank 4) - moderate volatility today, but volatile for last two weeks The stocks in this cluster are not very volatile today. However, they were volatile over the last two weeks (10 trading days). The volatility in these stocks seems to have come to a screeching halt. It may be good idea to study the news on these stocks and trade accordingly.

Cluster C1 (Rank 5) - high volatility today, but was not volatile for last two weeks The stocks in this cluster are attracting the interest of the traders. They may be in early phase of activity and potential buying opportunities.

The cluster profiles described above put the volatility in a historical perspective and may allow traders to look at stocks differently, leading to a more informed decision. More details of the meta-clustering experiments can be found in [12].

6 Other Extensions and Applications of Clustering

Clustering continues to evolve to address more practical research challenges. This section provides a brief overview of various challenges addressed by a number of researchers.

6.1 Use of Medoids and Modes as an Alternative to Centroids

The K-means algorithm [6] continues to be the basis of many of the clustering efforts, such as fuzzy C-means and rough K-means. However, the use of centroids can sometimes be a limiting factor in the application of rough set theory. For example, the size of the search space while using Genetic Algorithms to evolve a clustering scheme can be daunting if we use centroids in a high dimensional space. In such cases, using a medoid (an object in the dataset that is closest to the centroid) can improve the computational efficiency, as shown by Joshi and Lingras [8]. The centroid used in the K-means algorithm is geared towards numerical attributes. In a number of real-world applications we need to use categorical attributes. In such cases modes (values that appear most frequently in a cluster) have been used successfully [1, 2].

6.2   Dynamic and Stream Clustering in Big Data

All the examples discussed earlier in this paper use a static dataset. These profiles can be used for analysis. However, for real-time analysis of a changing environment, clustering algorithms need to be modified. Peters et al. [18] used supermarket data to show how the changing purchasing patterns can be modeled using rough set based dynamic clustering. Such a dynamic clustering involves creation of new clusters, deletion of obsolete clusters, and merging of clusters that are becoming indistinguishable. Ammar et al. [1, 2] used a variation of dynamic clustering, called incremental and decremental clustering, to adapt to the changing nature of the data using possibilistic soft clustering. The approach was shown to be useful even for standard data mining datasets. Another incremental clustering approach based on three-way decision theory was reported by Yu et al. [26].

The problem of continuous influx of data has further intensified with the emergence of new high volume and high velocity datastreams such as Internet of Things (IoT) or wearable devices. These devices produce a large amount of continuously streaming data. In addition to the problems of emergence, obsolescence, and merging of clusters, the data stream clustering algorithms need to be able to manage clustering in a single pass, i.e. they can look at a data point only once before making a decision about its membership. Silva et al. [20] provide a comprehensive overview of data stream clustering.

A commercial application of rough set based clustering can be found in Infobright's database software [23], where the incoming rows are grouped into packrows labeled by statistical summaries for faster querying purposes. This is an excellent practical example of how rough clustering can be used for compressing data in real-world big data processing. Such packrows can be created in a natural order according to how rows have been loaded into the database. Infobright's software uses a more intelligent approach where the original ordering of the stream of rows is dynamically modified in order to produce packrows with better quality statistical summaries [22]. This intelligent approach could be viewed as a practical implementation of rough stream clustering, because some rows drop into the outlying packrows and, as their summaries are less precise, they are accessed by more queries on average.

6.3   Augmenting Other Data Mining Tasks Using Clustering

Clustering can also be used to improve results of other data mining tasks such as prediction, classification, and association mining.

Zhang et al. [27] found that variations in sales patterns of different products make it difficult to obtain a generic time series prediction model that fits best to an entire data set. They proposed time series clustering to analyze the data set to identify local groups of products that exhibit typical seasonal sales patterns, or stability in sales patterns for a certain period. For such local groups depending upon seasonality or stability, they recommended better inventory forecasting strategies [27].

Joshi and Lingras [8] show how clustering can be used to identify various English alphabets - a task that is normally performed through classification. Researchers have also proposed semi-supervised learning methods that combine the strengths of clustering with traditional classification models.

The meta-clustering of products and customers discussed earlier, and described in detail by Lingras et al. [10], provides a list of products that are typically bought by a group of similar customers. This information can be used to augment a product recommender system. If some of the customers are not buying products that others in their group of buying, the system can recommend purchase of these products.

Clustering can also be used to reduce the number of possibilities created by optimization algorithms such as genetic algorithms. The sophisticated optimization algorithms create solutions that are vectors of real numbers. However, some of the solutions vary by an insignificantly small amount such as 2.056 in one solution could be 2.055987 in a different solution. Rounding off of results does not always work when we are considering vectors of real numbers as can be seen in Fig. 4. Clustering can be used to find a smaller number of distinct vectors. For example, we could use the centroids of the four clusters in Fig. 4 as four distinct optimization solutions.

Another interesting application of rough sets in big data can be found in the attribute clustering proposal by Janusz and Slezak [7] to reduce high dimensional

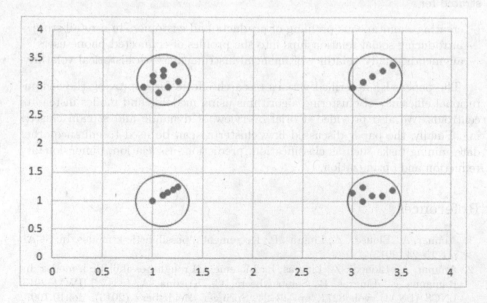

Fig. 4. Optimization solutions that can be grouped into four clusters

space, such as gene expression datasets [21], into a more manageable number of dimensions for the application of other data mining tasks. They combined attribute clustering with attribute reduction (feature selection) methods. The resulting clusters of attributes were derived in order to run attribute reduction algorithms on their representatives, instead of the full attribute space. These clusters of attributes were heuristically evaluated and were found to be interchangeable in decision models.

It should be noted here that clustering does not replace the classical data mining techniques, but can enhance their results.

7   Summary and Conclusions

This paper described the role of clustering in real-world data mining and how the algorithmic and theoretical development in clustering has responded to new challenges.

We looked at the use of conventional clustering for analysis of web usage on yelp.com by creating profiles of businesses and reviewers using the K-means algorithm.

We then studied the fuzzy theoretic extension, called the fuzzy C-means algorithm, that provided more moderate cluster centroids and an ability for objects to belong to multiple clusters.

Another soft computing extension based on rough set theory called rough K-means was used to show how clustering of highway sections can be made more meaningful with the help of lower and upper bounds.

The usefulness of three types of meta-clustering techniques were demonstrated for

- creating simultaneous profiling of products and customers in a retail store
- introducing social relationships into the profiles of connected phone users
- augmenting daily volatility of financial instruments with historical volatility.

The paper also described various research efforts for improving the computational efficiency of clustering algorithms using medoids and modes instead of centroids. We also provided a brief overview of dynamic and stream clustering. Finally, the paper discussed how clustering can be used to enhance other data mining tasks such as classification, prediction, association, dimensionality reduction and optimization.

References

1. Ammar, A., Elouedi, Z., Lingras, P.: Decremental possibilistic k-modes. In: SCAI, pp. 15–24 (2013)
2. Ammar, A., Elouedi, Z., Lingras, P.: Incremental rough possibilistic k-modes. In: Ramanna, S., Lingras, P., Sombattheera, C., Krishna, A. (eds.) MIWAI 2013. LNCS (LNAI), vol. 8271, pp. 13–24. Springer, Heidelberg (2013). doi:10.1007/978-3-642-44949-9_2

3. Black, F., Scholes, M.: The pricing of options and corporate liabilities. J. Polit. Econ. **81**, 637–654 (1973)
4. Dunn, J.C.: A fuzzy relative of the isodata process and its use in detecting compact well-separated clusters. J. Cybern. **3**, 32–57 (1973)
5. Eagle, N.: The reality mining data (2010). http://eprom.mit.edu/data/ RealityMining_ReadMe.pdf
6. Hartigan, J.A., Wong, M.A.: Algorithm as 136: a k-means clustering algorithm. J. R. Stat. Soc. Ser. C (Appl. Stat.) **28**(1), 100–108 (1979). http://www.jstor.org/stable/2346830
7. Janusz, A., Slezak, D.: Rough set methods for attribute clustering and selection. Appl. Artif. Intell. **28**(3), 220–242 (2014)
8. Joshi, M., Lingras, P.: Evolutionary and iterative crisp and rough clustering ii: experiments. In: Chaudhury, S., Mitra, S., Murthy, C.A., Sastry, P.S., Pal, S.K. (eds.) PReMI 2009. LNCS, vol. 5909, pp. 621–627. Springer, Heidelberg (2009). doi:10.1007/978-3-642-11164-8_101
9. Lingras, P.: Unsupervised rough set classification using gas. J. Intell. Inf. Syst. **16**(3), 215–228 (2001)
10. Lingras, P., Elagamy, A., Ammar, A., Elouedi, Z.: Iterative meta-clustering through granular hierarchy of supermarket customers and products. Inf. Sci. **257**, 14–31 (2013)
11. Lingras, P., Haider, F.: Rough ensemble clustering. In: Intelligent Data Analysis, Special Issue on Business Analytics in Finance and Industry (2014)
12. Lingras, P., Haider, F., Triff, M.: Granular meta-clustering based on hierarchical, network, and temporal connections. Granular Comput. **1**(1), 71–92 (2016)
13. Lingras, P., Triff, M.: Fuzzy and crisp recursive profiling of online reviewers and businesses. IEEE Trans. Fuzzy Syst. **23**(4), 1242–1258 (2015)
14. Lingras, P., West, C.: Interval set clustering of web users with rough k-means. J. Intell. Inf. Syst. **23**(1), 5–16 (2004)
15. Mitra, S.: An evolutionary rough partitive clustering. Pattern Recogn. Lett. **25**(12), 1439–1449 (2004)
16. Peters, G.: Some refinements of rough k-means clustering. Pattern Recogn. **39**(8), 1481–1491 (2006)
17. Peters, G., Crespo, F., Lingras, P., Weber, R.: Soft clustering-fuzzy and rough approaches and their extensions and derivatives. Int. J. Approximate Reasoning **54**(2), 307–322 (2013)
18. Peters, G., Weber, R., Nowatzke, R.: Dynamic rough clustering and its applications. Appl. Soft Comput. **12**(10), 3193–3207 (2012)
19. Sharma, S.C., Werner, A.: Improved method of grouping provincewide permanent traffic counters. Transp. Res. Rec. **815**, 13–18 (1981)
20. Silva, J.A., Faria, E.R., Barros, R.C., Hruschka, E.R., de Carvalho, A.C., Gama, J.: Data stream clustering: a survey. ACM Comput. Surv. (CSUR) **46**(1), 13 (2013)
21. Slezak, D.: Rough sets and few-objects-many-attributes problem: the case study of analysis of gene expression data sets. In: Frontiers in the Convergence of Bioscience and Information Technologies, FBIT 2007, pp. 437–442. IEEE (2007)
22. Ślęzak, D., Kowalski, M.: Intelligent data granulation on load: improving info-bright's knowledge grid. In: Lee, Y., Kim, T., Fang, W., Ślęzak, D. (eds.) FGIT 2009. LNCS, vol. 5899, pp. 12–25. Springer, Heidelberg (2009). doi:10.1007/ 978-3-642-10509-8_3
23. Ślęzak, D., Synak, P., Wojna, A., Wróblewski, J.: Two database related interpretations of rough approximations: data organization and query execution. Fundamenta Informaticae **127**(1–4), 445–459 (2013)

24. Yao, Y., Lingras, P., Wang, R., Miao, D.: Interval set cluster analysis: a re-formulation. In: Sakai, H., Chakraborty, M.K., Hassanien, A.E., Ślęzak, D., Zhu, W. (eds.) RSFDGrC 2009. LNCS (LNAI), vol. 5908, pp. 398–405. Springer, Heidelberg (2009). doi:10.1007/978-3-642-10646-0_48

25. Yu, H., Su, T., Zeng, X.: A three-way decisions clustering algorithm for incomplete data. In: Miao, D., Pedrycz, W., Ślęzak, D., Peters, G., Hu, Q., Wang, R. (eds.) RSKT 2014. LNCS (LNAI), vol. 8818, pp. 765–776. Springer, Heidelberg (2014). doi:10.1007/978-3-319-11740-9_70

26. Yu, H., Zhang, C., Wang, G.: A tree-based incremental overlapping clustering method using the three-way decision theory. Knowl.-Based Syst. **91**, 189–203 (2016)

27. Zhang, P., Joshi, M., Lingras, P.: Use of stability and seasonality analysis for optimal inventory prediction models. J. Intell. Syst. **20**(2), 147–166 (2011)

Clinical Narrative Analytics Challenges

Ernestina Menasalvas$^{(\boxtimes)}$, Alejandro Rodriguez-Gonzalez, Roberto Costumero,
Hector Ambit, and Consuelo Gonzalo

Centro de Tecnología Biomédica, Universidad Politécnica de Madrid, Madrid, Spain
{ernestina.menasalvas,alejandro.rodriguezg,roberto.costumero,
hector.ambit}@upm.es, chelo@fi.upm.es

Abstract. Precision medicine or evidence based medicine is based on the extraction of knowledge from medical records to provide individuals with the appropriate treatment in the appropriate moment according to the patient features. Despite the efforts of using clinical narratives for clinical decision support, many challenges have to be faced still today such as multilinguarity, diversity of terms and formats in different services, acronyms, negation, to name but a few. The same problems exist when one wants to analyze narratives in literature whose analysis would provide physicians and researchers with highlights. In this talk we will analyze challenges, solutions and open problems and will analyze several frameworks and tools that are able to perform NLP over free text to extract medical entities by means of Named Entity Recognition process. We will also analyze a framework we have developed to extract and validate medical terms. In particular we present two uses cases: (i) medical entities extraction of a set of infectious diseases description texts provided by MedlinePlus and (ii) scales of stroke identification in clinical narratives written in Spanish.

Keywords: Clinical narratives · Natural language processing

1 Introduction

Electronic Health Records (EHR) and their use in medical institutions are becoming more and more popular and its adoption has been increased during the last years [1].

Physicians complain of having tools to store information but no tools to extract information out of these records. This is a consequence of the unstructured nature of the information contained in the EHRs and consequently still remains a difficult task to perform Query and Answering processes in an accurate way [2].

Clinical narratives lack structure or they have an structure that depends on the hospital or even service of the hospital, contain abbreviations, numbers and they are written in the language of the country. Besides, concepts frequently appear in clinical notes as hypothetical, negated, or expressing temporal relationships and all these issues have to be identify to properly understand and relate concepts in EHRs. Thus traditional NLP process has to be enhaced with

© Springer International Publishing AG 2016
V. Flores et al. (Eds.): IJCRS 2016, LNAI 9920, pp. 23–32, 2016.
DOI: 10.1007/978-3-319-47160-0_2

new modules. In particular disambiguation of acronyms is required as the same acronym can have multiple meanings depending on the context.

Multilinguarity affects the development of these systems. Most of the existing solutions are for medical text written in English. In [3] a scheme in which emotion recognition from text through classification with the rough set theory and the support vector machines (SVMs) is proposed for Chinesse language. The experiment results showed that rough set theory and SVMs method are effective in emotion recognition. In [4] a rough set-based semi-naive Bayesian classification method is applied to dependency parsing. The rough set-based classifier is embedded with Nivre deterministic parsing algorithm to conduct dependency parsing task on a Chinese corpus showing the method has a good performance on this task.

One paramount step of the text analysis is the Named Entity Recognition that relies on ontologies. Unified Medical Language System (UMLS) [5] is composed from different ontologies and databases and are organized so every common concept contains a unique identifier. Even that translations of UMLS to different languages are available, these translations only contain a small amount of terms. On the other hand, enrichement of these vocabularies is required in order to add those terms that are specific for a disease, treatment or speciality. H2A [6] is a system composed of several software components to process clinical narratives written in Spanish.

In this paper we present two case studies about the application of Natural Language Processing methods and models. In the first case study we have applied two well-known NLP tools named MetaMap and Apache cTAKES to extract medical diagnosis terms (symptoms, signs, laboratory procedures and tests and diagnostic procedures) as an extension of a previous work [7] to compare the accuracy of both methods. The tools were applied against a set of 30 infectious diseases provided by Medline Plus. In the second case study we have applied our framework for NLP with text written in Spanish to discover scales of stroke in clinical narratives. The rest of the paper is organized as follows: In Sect. 2 the state of the art concerning NLP tools in the health sector is reviewed. The challenges of the NLP process in which we focus are briefly described in Sect. 3 while in Sect. 4 the two case studies are presented: Subsect. 4.1 presents the application of a NER application to extract concepts of infectious diseases and in Subsect. 4.2 the detection of stroke scales on Spanish narratives is presented. To conclude Sect. 5 discusses the achievements so far and presents the outlook of the future developments.

2   Background

Application of Natural Language Processing techniques to extract information from Electronic Health Records has been extensively studied as in the last decade, although most solutions are English-centric.

Electronic Health Record (EHR) contain valuable clinical information expressed in narrative form. This information is nowadays stored in digital form, but the content still lacks from structure, typos are common, etc.

The analysis of different uses of information extraction from textual documents in EHR has been analyzed in [8]. According to that publication, this extraction poses new challenges due to the problems mentioned before. The growth in the use of EHRs has generated a significant development in Medical Language Processing systems (MLP), information extraction techniques and applications [8–23].

A medical text processor is described in Friedman et al. which processes radiology reports [16]. Clinical documents are analyzed in order to transform them into terms pertaining to a controlled vocabulary. The MedLEE system is presented in [13]. MedLEE was developed to extract structures and to encode clinical information from textual patient reports. The first version of MedLEE was evaluated in chest radiology reports. MedLEE was extended to work on mammography reports and discharge summaries [14], electrocardiography, echocardiography and pathology reports [15]. The performance of MedLEE using differents lexicons (LUMLS, M-CUR, M+UMLS) was evaluated in [19].

Patient discharge summaries (PDS) were processed using MENELAS [23] to extract information from them. MENELAS can analyze reports in French, English and Dutch. cTAKES, a clinical Text Analysis and Knowledge Extraction System is introduced in [22]. cTAKES is an open-source NLP system that uses rule-based and machine learning techniques to process and extract information to support clinical research. The cTAKES components are sentence boundary detectors, tokenizers, normalizers, Part-of-Speech (PoS) taggers, shallow parsers and Named Entity Recognition (NER) annotators. HITEx (Health Information Text Extraction) [24], an open-source application based on the GATE framework, were developed to solve common problems in medical domains such as diagnosis extraction, discharge medications extraction and smoking status extraction. HITEx has been also used in [25] to extract the main diagnosis from a set of 150 discharge summaries. Co-morbidity and smoking status showed a positive performance.

MedTAS/P [10] is a system based on the Unstructured Information Management Architecture (UIMA) [26] open source framework that uses NLP techniques, machine learning and rules to map free-text pathology reports automatically into concepts represented by CDKRM (Cancer Disease Knowledge Representation Model) for storing cancer characteristics and their relationships. Fiszman et al. [12] introduced Sym Text, a NLP tool to extract relevant clinical information from radiology (Ventilation/Perfusion lung scan) reports. To evaluate the use of current NLP techniques in an automatic knowledge acquisition domain, a system is introduced in Taboada et al. [27]. The system reuses OpenNLP, Stanford parsers, SemRep and UMLS NormalizeString service as building blocks. Using an ontology, clinical practice guidelines documents are enriched. In Thomas et al. [28], an NLP program to identify patients with prostate cancer and to retrieve pathological information from their EMR is evaluated. The results show that NLP can do it accurately.

Some systems have been developed [29,30] to process clinical text in German. An approach called Left-Associative Grammar (LAG) was used in MediTas [30], to parse summary sections of cytopathological findings reports for a Medical Text Analysis System for German. For the German SNOMED II version another Natural Language Processing (NLP) parser is presented in [29]. The parser divides a medical term into fragments which might contain other SNOMED terms.

There are nearly 500 million Spanish speakers worldwide, however, tools to extract medical information from Spanish EHR are practically non existent. Savana Médica [31] is one of the solutions that are starting to be present in the Spanish medical environment.

The introduction of the TIDA architecture (currently renamed to H2A: Human Health Analytics) proposed for a medical decision support system was done in [32,33]. This architecture constitutes a software that analyzes text, images and the structure data from the patient in order to give the doctors answers to complex questions. Previous works on the analysis of negation detection in Spanish for medical documents has been introduced in [34] and analysis on the creation of models for performing NLP in the medical domain has been explained in [35].

3 Challenges of Extracting Valuable Information from EHR and Medical Texts

3.1 Traditional NLP Pipeline

The NLP process is a pipeline (see Fig. 1) that detects sentences, tokens, Part-of-Speech (PoS), phrases and parse tree and is able to find entities such as locations, people, or in the case of healthcare, drugs, diseases or parts of the body. The main modules of the process are the following:

– Part-of-Speech (PoS) tagging and Parsing for Spanish EHRs. NLP techniques applied to PoS and Shallow Parsing to train models are mainly based on supervised learning and that, at least for English, is a solved problem. Semi-supervised learning is also used to bootstrap the creation of corpora that is used to train the models, especially in the cases of specialized corpora as the process of annotation is very time-consuming. Unsupervised learning is currently trending as a problem to solve so annotation can be skipped. These models trained have been developed and used in different frameworks on the medical domain as seen in Sect. 2. The main challenge has to deal with the interoperability using different frameworks and most of them are English centric and some of them are proprietary. Generally, the improvement in those models depends on the corpus used to train them; and the lack of these annotated corpora in the clinical domain which are accessible, specially in languages such as Spanish, is a challenge yet to be solved.
– Named Entity Recognition (NER) is responsible of extracting the entities that are relevant in a domain and getting the relationships among them. NERs typically rely on the use of ontologies and dictionaries to detect, structure and analyse the data contained in a particular domain. UMLS is the most frequently used thesaurus in the health sector however the translation of UMLS

Fig. 1. Named Entity Recognition (NER) pipeline for clinical domain

to other languages rather than English, is not complete what clearly decreases the power of tools using them.

3.2 Discovering the Meaning of Numbers and Metrics

If one observes a clinical narrative, it is easy to spot many features that are particular for this kind of texts. In particular, an interesting problem is that of numbers that appear in the text which typically are followed by some metrics. This can be the case for a treatment (eg. ibuprofen 2 cp/d), laboratory tests (eg. glucose 140 mg), or blood pressure measure (eg. 140/92 mmHg). Dates is another typical feature in clinical narratives which can be absolute (eg. MRI on 14/02/2016) or relative (eg. patient suffered from headache two days ago). But numbers can also make reference to the status of the patient regarding a disease (eg. the cancer has spread and the patient is on stadium IV). In Sect. 4.2 we present an application of H2A in which the NER module developed is able to find numbers and metrics in particular to detect the scales that are reported for patients suffering a stroke.

3.3 Identifying Diagnosis Terms and Elements

The identification of diagnosis elements in medical texts is a crucial task. It is mainly used for the development of medical diagnosis systems, since nowadays it is very difficult to find open databases with information regarding the symptoms, signs or procedures to diagnose a disease (also known as diagnosis criterion; see DCM model [36]). Other relevant uses can be found in the construction of human symptoms disease networks [37], a challenging linea of research where this information is very important.

4 Materials and Methods

In what follows we present results of the experiments conducted with our framework applied for two different problems: (i) find names and symptoms of infectious diseases in English text and (ii) find the scale that is reported for a patient suffering a stroke.

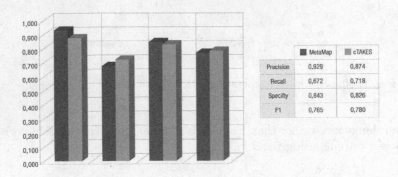

	MetaMap	cTAKES
Precision	0,929	0,874
Recall	0,672	0,718
Specifity	0,843	0,826
F1	0,765	0,780

Fig. 2. Comparison of results of MetaMap and cTakes

4.1 Extracting Clinical Terms from Medical Texts

As an extension of a previous work [7] we present the resutls of using a Apache cTAKES to retrieve clinical terms from MedlinePlus texts. We have used the same set of 30 infectious diseases. The idea of this use case is to show a comparative in accuracy regarding the extraction of generalist medical terms that only affect to terms used in the diagnosis context. As has been outlined in [7] only those semantic types that belong to the classes related with diagnostic elements were used to filter. The experiment was performed using our framework in which Apache cTAKES is used as NER. We have manually analyzed the results and made a comparison between MetaMap ([7]) and Apache cTAKES.

As it can be seen in Fig. 2, the mean results are quite similar between the executions using MetaMap or cTAKES. Precision is higher on MetaMap, while recall is higher in cTAKES. Specificity and F1 score are roughly the same, the former being higher in MetaMap, while the latter is higher in cTAKES. The main differences are found in the analysis of individual diseases. cTAKES typically performs better on laboratory or test results or locating rare symptoms, but increases in most cases the number of false positives, incorrectly annotating several elements as findings. In this case, it could also be relevant that the number of true negatives is higher because the NLP process annotates more elements, but the validation usually classifies them correctly. The tools analyzed have a good performance in general in terms of NER process. However, the general behaviour could be improved by adding specific vocabularies of acronyms or complex terms to the validation terms. Future work will be focused on the analysis of further NLP tools as well the creation of hybrid approaches where more than one NLP tool could be applied to capture the knowledge within the texts.

4.2 Stroke Scales Detection in Clinical Narratives in Spanish

Neurologists have defined different protocols to be able to determine the severity of their patients when they are affected by a stroke. This is reflected on the report as values of scale of the stroke. There are different scales: Barthel, Modified Rankin Scale, National Institute of Health Stroke Scale (NIHSS), Canadian

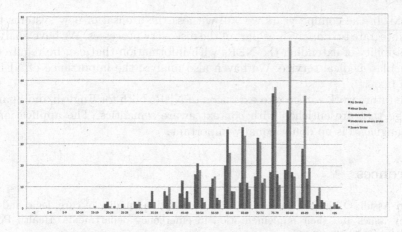

Fig. 3. Severity of the patients extracted from clinical notes

Neurological Scale, ... In order to detect these scales in the narrative we have improved the NLP process in particular enhancing the NER for Spanish narratives. In particular the following functionalities have been added:

– Detection of different possible metrics, such as the detection of the doses of a particular drug, the doses that the patient must take of it, or the different values that are indicative of different values in a laboratory test.
– Detection of contextual temporal information, retrieving absolute and relative dates that allow the automatic correlation of events in the order that happened and causation could be analyzed after when associating these dates to particular entities.
– Detection of particular scales or indexes that indicate the severity of the patient or a grade in a particular condition, like the values of the different stroke scales for patients, or the stadium of the cancer patients.

As an example of use case, EHR can be automatically analyzed to detect stroke scales such as: NIHSS, Modified Rankin Scale, Canadian Neurological Scale or the Barthel Index. These scales are typically used to determine the severity of a patient related to a stroke or the functional and neurological status of the patient after suffering from that condition. These scales are automatically extracted from the free-text written by doctors so their value and an interpretation can be given.

Figure 3 shows a possible application generating graphical distributions by age and severity of patients records according to their NIHSS values to analyze a given population.

5   Conclusions and Future Work

Clinical narratives analysis lays on the root of personalized medicine. In this paper some of the challenges that narrative analytics has to face have been reviewed. In the coming years we will witness the arousal of new systems that would integrate text analysis as part of the discovery processes. We have analyzed the possibility of extending the NERs with information that can be required in a particular medical service. We haven also analyse the importance of NER in english texts.

The future work will go into extension of NERs both for English and Spanish language that are enriched with context aware semantics. The application to other languages is no doubt equally important.

References

1. Ben-Assuli, O.: Electronic health records, adoption, quality of care, legal and privacy issues and their implementation in emergency departments. Health Policy **119**(3), 287–297 (2015)
2. Hanauer, D.A., Mei, Q., Law, J., Khanna, R., Zheng, K.: Supporting information retrieval from electronic health records: a report of university of michigans nine-year experience in developing and using the electronic medical record search engine (EMERSE). J. Biomed. Inf. **55**, 290–300 (2015)
3. Teng, Z., Ren, F., Kuroiwa, S.: Emotion recognition from text based on the rough set theory and the support vector machines. In: 2007 International Conference on Natural Language Processing and Knowledge Engineering, pp. 36–41. IEEE (2007)
4. Ji, Y., Shang, L., Dai, X., Ma, R.: Apply a rough set-based classifier to dependency parsing. In: Wang, G., Li, T., Grzymala-Busse, J.W., Miao, D., Skowron, A., Yao, Y. (eds.) RSKT 2008. LNCS (LNAI), vol. 5009, pp. 97–105. Springer, Heidelberg (2008). doi:10.1007/978-3-540-79721-0_18
5. Humphreys, B.L., Lindberg, D.A.: The UMLS project: making the conceptual connection between users and the information they need. Bull. Med. Libr. Assoc. **81**(2), 170 (1993)
6. Rodriguez, A., Gonzalo, C., Menasalvas, E., Costumero, R., Ambit, H.: H2a - human health analytics: a natural language processing system for electronic health records. In: Proceedings of the AMIA Symposium. IJCRS-Chile (2016, to appear)
7. Rodríguez-González, A., Martínez-Romero, M., Costumero, R., Wilkinson, M.D., Menasalvas-Ruiz, E.: Diagnostic knowledge extraction from medlineplus: an application for infectious diseases. In: Overbeek, R., Rocha, M.P., Fdez-Riverola, F., Paz, J.F. (eds.) 9th International Conference on Practical Applications of Computational Biology and Bioinformatics. AISC, vol. 375, pp. 79–87. Springer, Heidelberg (2015). doi:10.1007/978-3-319-19776-0_9
8. Meystre, S.M., Savova, G.K., Kipper-Schuler, K.C., Hurdle, J.F., et al.: Extracting information from textual documents in the electronic health record: a review of recent research. Yearb Med. Inform. **35**, 128–144 (2008)
9. Christensen, L.M., Haug, P.J., Fiszman, M.: Mplus: a probabilistic medical language understanding system. In: Proceedings of the ACL 2002 Workshop on Natural Language Processing in the Biomedical Domain, vol. 3, pp. 29–36. Association for Computational Linguistics (2002)
10. Coden, A., Savova, G.K., Sominsky, I.L., Tanenblatt, M.A., Masanz, J.J., Schuler, K., Cooper, J.W., Guan, W., de Groen, P.C.: Automatically extracting cancer disease characteristics from pathology reports into a disease knowledge representation model. J. Biomed. Inf. **42**(5), 937–949 (2009)

11. Doan, S., Mike Conway, T., Phuong, M., Ohno-Machado, L.: Natural language processing in biomedicine: a unified system architecture overview. arXiv preprint arXiv:1401.0569 (2014)
12. Fiszman, M., Haug, P.J., Frederick, P.R.: Automatic extraction of pioped interpretations from ventilation/perfusion lung scan reports. In: Proceedings of the AMIA Symposium, pp. 860–864 (1998)
13. Friedman, C., Hripcsak, G., DuMouchel, W., Johnson, S.B., Clayton, P.D.: Natural language processing in an operational clinical information system. Nat. Lang. Eng. 1(01), 83–108 (1995)
14. Friedman, C.: Towards a comprehensive medical language processing system: methods and issues. In: Proceedings of the AMIA Annual Fall Symposium, p. 595. American Medical Informatics Association (1997)
15. Friedman, C.: A broad-coverage natural language processing system. In: Proceedings of the AMIA Symposium, p. 270. American Medical Informatics Association (2000)
16. Friedman, C., Alderson, P.O., Austin, J.H., Cimino, J.J., Johnson, S.B.: A general natural-language text processor for clinical radiology. J. Am. Med. Inf. Assoc. 1(2), 161–174 (1994)
17. Friedman, C., Hripcsak, G.: Natural language processing and its future in medicine. Acad. Med. 74(8), 890–895 (1999)
18. Friedman, C., Knirsch, C., Shagina, L., Hripcsak, G.: Automating a severity score guideline for community-acquired pneumonia employing medical language processing of discharge summaries. In: Proceedings of the AMIA Symposium, p. 256. American Medical Informatics Association (1999)
19. Friedman, C., Liu, H., Shagina, L., Johnson, S., Hripcsak, G.: Evaluating the UMLS as a source of lexical knowledge for medical language processing. In: Proceedings of the AMIA Symposium, p. 189. American Medical Informatics Association (2001)
20. Goryachev, S., Sordo, M., Zeng, Q.T.: A suite of natural language processing tools developed for the I2B2 project. In: AMIA Annual Symposium Proceedings, vol. 2006, p. 931. American Medical Informatics Association (2006)
21. Hripcsak, G., Austin, J.H.M., Alderson, P.O., Friedman, C.: Use of natural language processing to translate clinical information from a database of 889,921 chest radiographic reports 1. Radiology 224(1), 157–163 (2002)
22. Savova, G.K., Masanz, J.J., Ogren, P.V., Zheng, J., Sohn, S., Kipper-Schuler, K.C., Chute, C.G.: Mayo clinical text analysis and knowledge extraction system (cTAKES): architecture, component evaluation and applications. J. Am. Med. Inf. Assoc. 17(5), 507–513 (2010)
23. Zweigenbaum, P.: Menelas: an access system for medical records using natural language. Comput. Method Prog. Biomed. 45(1), 117–120 (1994)
24. Goryachev, S.: Hitex manual. https://www.i2b2.org/software/projects/hitex/hitex_manual.html
25. Zeng, Q.T., Goryachev, S., Weiss, S., Sordo, M., Murphy, S.N., Lazarus, R.: Extracting principal diagnosis, co-morbidity and smoking status for asthma research: evaluation of a natural language processing system. BMC Med. Inf. Decis. Making 6(1), 30 (2006)
26. Ferrucci, D., Lally, A.: UIMA: an architectural approach to unstructured information processing in the corporate research environment. Nat. Lang. Eng. 10(3–4), 327–348 (2004)
27. Taboada, M., Meizoso, M., Martínez, D., Riaño, D., Alonso, A.: Combining open-source natural language processing tools to parse clinical practice guidelines. Expert Syst. 30(1), 3–11 (2013)

28. Thomas, A.A., Zheng, C., Jung, H., Chang, A., Kim, B., Gelfond, J., Slezak, J., Porter, K., Jacobsen, S.J., Chien, G.W.: Extracting data from electronic medical records: validation of a natural language processing program to assess prostate biopsy results. World J. Urology **32**(1), 99–103 (2014)

29. Hohnloser, J.H., Holzer, M., Fischer, M.R., Ingenerf, J., Günther-Sutherland, A.: Natural language processing, automatic snomed-encoding of free text: An analysis of free text data from a routine electronic patient record application with a parsing tool using the german snomed ii. In: Proceedings of the AMIA Annual Fall Symposium, p. 856. American Medical Informatics Association (1996)

30. Pietrzyk, P.M.: A medical text analysis system for german-syntax analysis. Method Inf. Med. **30**(4), 275–283 (1991)

31. Savana Médica: Savana médica (2015)

32. Costumero, R., Gonzalo, C., Menasalvas, E.: TIDA: a spanish EHR semantic search engine. In: Saez-Rodriguez, J., Rocha, M.P., Fdez-Riverola, F., De Paz, J.F., Santana, L.F. (eds.) PACBB 2014. AISP, vol. 294, pp. 235–242. Springer, Heildelberg (2014)

33. Costumero, R., Garcia-Pedrero, A., Sánchez, I., Gonzalo, C., Menasalvas, E.: 1 electronic health records analytics: natural language processing and image annotation. In: Big Data and Applications, p. 1 (2014)

34. Costumero, R., Lopez, F., Gonzalo-Martín, C., Millan, M., Menasalvas, E.: An approach to detect negation on medical documents in Spanish. In: Ślęzak, D., Tan, A.H., Peters, J.F., Schwabe, L. (eds.) BIH 2014. LNCS (LNAI), vol. 8609, pp. 366–375. Springer, Heidelberg (2014). doi:10.1007/978-3-319-09891-3_34

35. Costumero, R., García-Pedrero, Á., Gonzalo-Martín, C., Menasalvas, E., Millan, S.: Text analysis and information extraction from Spanish written documents. In: Ślęzak, D., Tan, A.-H., Peters, J.F., Schwabe, L. (eds.) BIH 2014. LNCS (LNAI), vol. 8609, pp. 188–197. Springer, Heidelberg (2014). doi:10.1007/978-3-319-09891-3_18

36. Rodríguez-González, A., Alor-Hernández, G.: An approach for solving multi-level diagnosis in high sensitivity medical diagnosis systems through the application of semantic technologies. Comput. Biol. Med. **43**(1), 51–62 (2013)

37. Zhou, X., Menche, J., Barabási, A.-L., Sharma, A.: Human symptoms-disease network. Nat. Commun. **5** (2014)

Modern ICT and Mechatronic Systems
in Contemporary Mining Industry

Wojciech Moczulski[1,2]([✉]), Piotr Przystałka[1], Marek Sikora[3,4],
and Radosław Zimroz[5,6]

[1] Institute of Fundamentals of Machinery Design, Silesian University of Technology,
Konarskiego 18A, 44-100 Gliwice, Poland
{wojciech.moczulski,piotr.przystalka}@polsl.pl
[2] SkyTech Research Sp. z o.o., Konarskiego 18C, 44-100 Gliwice, Poland
[3] Institute of Informatics, Silesian University of Technology,
Akademicka 12, 44-100 Gliwice, Poland
marek.sikora@polsl.pl
[4] Institute of Innovative Technology EMAG, Leopolda 31, 40-189 Katowice, Poland
[5] Faculty of Geoengineering, Mining and Geology, Technical University of Wrocław,
Na Grobli 15, 50-421 Wrocław, Poland
radoslaw.zimroz@pwr.wroc.pl
[6] KGHM Cuprum Sp. z o.o. CBR, Sikorskiego 2-8, 53-659 Wrocław, Poland

Abstract. The paper deals with modern ICT techniques and systems,
and mechatronic systems for mining industry, with particular attention
paid to results achieved by the authors and their research groups. IT systems concern process and machinery monitoring, fault detection and iso-
lation of processes and machinery, and assessment of risk and hazards in
mining industry. Furthermore, innovative applications of AI methods are
addressed, including pattern recognition and interpretation for process
control, classification of seismic events, estimating loads of conveyors,
and the others. Special attention is paid to applications of mechatronic
solutions, such as: unmanned working machinery and longwalls in coal
mines, and specialised robots for basic work. Mobile robots for inspecting areas of mines affected by catastrophes are presented, too. Moreover,
recent communication solutions for collision avoidance, localisation of
mining machinery, and wireless transmission are addressed. The paper
concludes with most likely development of ICT and mechatronic systems
for mining industry.

Keywords: ICT in mining industry · Risk and hazards assessment ·
Mechatronic working systems for mines · Robotized inspection

The paper includes some results achieved during research carried out in the framework of research projects: *Disesor* partially financed by Polish National Centre for
Research and Development under grant No. PBS2/B9/20/2013; *TeleRescuer* partially financed by the Research Fund for Coal and Steel under grant No. RFCR-CT-
2014-00002 and by the Polish Ministry for Science and High Education.

V. Flores et al. (Eds.): IJCRS 2016, LNAI 9920, pp. 33–42, 2016.
DOI: 10.1007/978-3-319-47160-0_3

1   Motivation

Mining industry is experiencing technological and organizational revolution in recent years. There are several pan-European initiatives supporting such changes (European Innovation Partnership on Raw Materials, European Insitute of Innovation and Technology - Raw Materials Branch, EURobotics with Topic Group "Robotics in Mining", etc.) [11–13]. It motivates us to highlight current trends in mining industry and present several research activities exploiting very advanced technologies (rarely recognized as mining applications). Depending on type of mined raw materials, used technology, environmental hazard, etc., mining companies might face very different problems. For the needs of this paper we will use several perspectives. The first critical feature is type of a mine (opencast, underground). Next critical issue is energetic or non-energetic type of materials. The first one is mainly related to hard coal mined in general in underground mines with explosive atmosphere. The second one, at least in Poland, is associated with underground mines and deeply located copper ore deposits. A deep underground mine means hard rock with extremely harsh environmental conditions, natural hazard (seismic risk) and relatively poor automation of mining operations (room and pillar technology, no mechanical excavation in contrast to longwall systems in underground hard coal mines). Relatively easier case could be noticed in open cast mines (lignite brown coal or non-energetic raw materials including aggregates). There are examples of nearly fully automatic mines with central room control, autonomous machines, analytic centre, etc. Regardless of mentioned issues in all cases motivation to introduce advanced technology to the mining industry is very similar. To be competitive enough, a mining company should be economically effective, safe and environment-friendly. Nowadays, mining companies are great examples of advanced technology users. Big mining companies like Codelco, Rio Tinto, KGHM, etc., use advanced IT technologies (monitoring of objects and processes, data modelling for processes optimization, decision support systems, reporting tools etc.), advanced process control systems, robots for inspections or partially robotized mining operation (drilling/bolting, transport, etc.) and many other smart solutions. In this paper we address modern ICT systems and mechatronic (and especially robotic) systems. The discussion illustrated by selected real world examples will cover underground hard coal mines, opencast lignite mines and deep underground copper ore mines.

2   IT Systems in Mining Industry

The diversity of IT systems in the mining industry is vast and covers many areas of current activity of the company. Some of them are very general (electronic document circulation, purchasing, finance management, etc.) and have been implemented in the past [4,9,16]. However, specific solutions dedicated to production monitoring management in mining are non-trivial and their successful application is challenging for many companies. The most prominent examples are SCADA systems used to monitor condition of machines, process, environment, seismic activity and many others.

The authors are especially involved in developing methods and systems for fault detection of machinery and for assessment of risk and hazards. As an example, the *Disesor* system is presented which is a decision support tool for fault diagnosis, hazard prediction and analysis in mining industry [29]. The Polish consortium leaded by the EMAG Institute of Innovative Technology together with the Silesian University of Technology, Warsaw University and Sevitel Sp. z o.o. has developed the system. *Disesor* is a shell system that can be filled in with data and facts concerning a given problem domain. It can deal with fault diagnosis in different plants (processes or machinery), and with hazard prediction and analyses of such phenomena as e.g. methane hazard in coal mines, or rock burst hazard in underground mines. The architecture of the system consists of: a Data Repository, Data Preparation and Cleaning module, Prediction Module, Analytical Module and Expert System Module. The core of Analytical, Prediction and Expert System modules is based on *RapidMiner* platform [15,26]. The goal of the Data Preparation and Cleaning Module, which is referred further as ETL2, is to integrate the data stored in data warehouse and process it to the form acceptable by the methods creating prediction and classification models. In other words the ETL2 module prepares training sets. The Prediction Module is aimed to apply classification and prediction models created in the Analytical Module for a given time horizon and frequency of the values measured by the chosen sensors. This module also tracks the trends in the incoming measurements. Created predictive models are adapted to the analysed process on the basis of the incoming data stream and the models learnt on historical data (within the Analytical Module). The module provides interfaces that enable the choice of quality indices and their thresholds that ensure the minimal prediction quality. If the quality of predictions meets the conditions set by the user, the predictions will be treated as values provided by a *virtual sensor*. They can be further utilised e.g. by the Expert System but also can be presented to a dispatcher of a monitoring system. The Expert System module is aimed to perform on-line and off-line fault diagnosis of machinery and other technical equipment. It is also capable of supervising processes and supporting the dispatcher or expert by decision-making with respect to both technical condition of the equipment and improper development of the process. The Expert System allows reasoning by means of multi-domain knowledge representations and multi-inference engines. The inference process is performed by means of classical inference based on Boolean logic or fuzzy inference system as well as probabilistic inference with the use of belief networks. The Analytical Module is aimed to perform analysis of historical data (off-line) and to report identified significant dependencies and trends. The results generated by this module are stored in the repository only when accepted by the user. Therefore, this module supports the user in decision-making of what is interesting from monitoring and prediction point of view. It also provides additional information that can be utilised to enrich the knowledge base of the Expert System or that can be utilised to comparative analysis. The module supports identification of changes and trends in the monitored processes and tools and it also enables to compare the operator's and dispatcher's work.

The more detailed description of the system can be found in [21,29]. Examples of applications of this system to practical mining problems will be presented in the tutorial. Moreover, the authors also view *Disesor* as a tool that can be used for a wider spectrum of applications, and therefore it is considered to apply this software for solving similar problems/tasks [19].

Other examples concern brown coal mines, where giant bucket wheel excavators and belt conveyors systems are of special importance to the operation of a mine. Researchers from Wrocław University of Technology have developed a monitoring and diagnostic system with original decision making scheme for the case of nonstationary operating conditions (variable load/speed) [1,8,35]. This system includes instrumentation technology, signal processing, diagnostic procedures and maintenance strategies for drive units (gears, bearings) as well as for belts with steel cords. Unique, original solutions have been also developed for complex multistage planetary gearbox.

The next application concerns a belt conveyor network, which is distributed over large area and can include up to 100 conveyors with many components to monitor. Implementation of online monitoring system on each of them might be considered as expensive. Thus, another solution for underground mine has been developed. It assumes that data will be acquired on demand by a portable system. The central part of this solution is related to CMMS system for simple support of maintenance stuff. The personnel needs to be informed when and what should be checked. Measurements require minimum skills, data processing is fast and decision is available immediately. A key issue is long term monitoring of objects behaviour and extracting appropriate knowledge from population of machines of the same type but with different condition [31].

The last example is related to load-haul-dump (LHD) machines: loaders, trucks, drilling and bolting machines. The most difficult issues here are: complexity of the machine (more than 70 variables are monitored), its mobility, extremely harsh environment (these machines are working in mining face area), and number of machines in operation. KGHM Cuprum has been a partner in the SYNAPSA project focused on monitoring LHD machines. It has been implemented in the deep underground mines in KGHM "Polish Copper" S.A. Again, holistic view on instrumentation (on-board monitoring system, wireless communication, fiber optics network, data warehouse) and advanced analytics and reporting tools were developed in this project, which is described e.g. in [36].

3   Artificial Intelligence Applications

Artificial Intelligence (AI) methods and applications have become ever more and more popular in the mining industry. One of the most frequently applied are pattern processing, analysis, recognition and interpretation methods. As a pattern we understand an ordered set of data, including images collected by video and infrared camera systems. There are many applications of pattern recognition (PR) techniques for detecting faults basing on process data and residual signals.

One prominent example of using AI is diagnostics of machinery. Based on measured data (vibration, temperature, pressure, etc.) one needs to make a decision regarding machine condition. In the most of cases it is two-class classification (good/bad condition). Such an approach requires appropriate training data sets that cover both good and bad condition cases. Unfortunately, mining machines are often very specialised, designed on demand, for the particular mine (see for example a bucket wheel excavator). No diagnostic tests are allowed (introducing artificial damages to learn about behaviour of machine in bad condition). In such a case, the so called *one-class classification* could help [2,27]. Mining machines are working in time varying conditions, including also transients. It means that apart from classical diagnostic features there is a need to use descriptors of operational conditions (speed or load values). It leads to data fusion, multidimensional, multivariate data analysis. Again, fundamental problems arise (are data representative, are they redundant, can dimensionality be reduced, are there feaures dependent linearly or nonlinearly, etc.). Some of these issues have been discussed in [2,3,5].

PR can be also applied to detect and classify seismic events. Monitoring of seismic events is operating 24 h 7d. Till now from the mining perspective the energy and localisation of a seismic event is automatically evaluated by the monitoring system. By appropriate parameterisation of a seismic signal it might be possible to recognise automatically the character of a seismic event. It has been shown by Xu *et al.* [33] that spectral representation of different types of seismic events is significantly different and can be the basis for classification. Such a classification for copper ore mine was discussed by Sokołowski *et al.* [30].

Very important AI applications concern data mining (DM) in databases collected by multiple systems employed in the mining industry. Data carries very useful information of relations between process parameters, the state of the plant (object or process) and outputs. There are multiple systems developed to acquire knowledge from data in an automatic way, with the *Disesor* system [29] as a prominent example. But knowledge engineering is not limited to DM applications only. For example, the *Disesor* system provides a bunch of tools for acquiring knowledge from domain experts, which can then be implemented in a dedicated Expert System capable of supporting the user (operator, diagnoser, manager) in taking decisions concerning further development of the process under control [29]. It is well-grounded that expert systems can be effectively applied in many areas such as medicine, education, entertainment, risk management and fault diagnosis [7,23]. The present activity of scientists and engineers confirms that expert systems also play very important role in the field of mining engineering. In [34] the authors present an expert system for supervising workstations of coal mines. This system applied production rules to express knowledge, whereas the inference process was realized by means of forward as well as backward reasoning. A safety management system for coal mine was suggested in [32]. The authors created the information processing system that was based on web site technology. This system is used in Zibo mining industry group and Xu Chang coal plants in China. The authors of [14] developed an expert system for assessment and optimization of coal mines in terms of their eco-efficiency. This was not a typical expert system, but rather a set of software tools for managing a mining company. The more detailed and deeper survey on expert systems for aiding the mining industry may be found for instance in [6].

Another area of application of AI (in particular machine learning – ML) in mines is hazard assessment and prediction. For example the *Disesor* system enables predictions of two types of hazards: methane and seismic ones [29]. In case of methane hazard the system predicts a maximum methane concentration at a longwall end area. The prediction horizon is medium-term (several minutes). An accurate medium-term prediction of maximum methane concentration would let the mine dispatcher monitor labour security factors in a more efficient way, which would result in reduction in the number of costly automatic power switch-offs caused by overrun of the admissible methane concentration level. One of the main tasks of coal mine geophysical stations is to determine the current state of seismic hazard (particularly, hazard of high-energy destructive tremor which may result in a rockburst) in underground mining places. Rockbursts, as phenomena related with mining seismicity, pose a serious hazard to miners and can destroy longwalls and the equipment. The *Disesor* system predicts the value of the seismic energy which will be emitted within the longwall area. Higher ($> 5 \cdot 10^5$ J) energy values are treated as potentially dangerous situations.

In both the cases ML methods are used to construct predictive models. Due to the imbalanced nature of the data (dangerous situations are rare) the problem of constructing good predictive models characterised by good sensitivity and specificity is difficult [18]. For this reason there were organized two international data mining competitions [17] in order to develop the methodology for predictive models suitable for the problems mentioned above. The best models are based on an advanced data pre-processing and the use of ensemble classification paradigm. The best methods were implemented within the *Disesor* system. The implementation includes: tuning the model to conditions existing in the specific longwall and monitoring the quality of predictions (by setting minimum thresholds related to wrong decisions) and concept drift identification.

Bearing in mind the current state of the art in this subject it can be concluded that recent challenges for mining companies cause that advanced expert systems are viewed as indispensable decision support tools which are being more often involved in their activities. The authors of the paper still see the need for development of more advanced expert system shells that can be successfully applied in the mining industry [25]. Their studies focus on the practical applications of such software tools for solving different problems such as fault diagnosis of belt conveyors [20] or longwall shearers [28], hazard prediction [21], etc.

4   Mechatronic Systems and Robotics

Much effort has been made since decades to release human miners from carrying out very heavy and unsafe work. In the paper we address the most impressive systems such as unmanned working machinery and longwalls. There are two different approaches to releasing human operators from direct operation of machinery working in hazardous environment. The first one consists in equipping the machinery with the option of *remote control*. The operator can remain in a safe place (and in the best solution need not to go down to the underground part of the mine) and control the operation of the machinery basing on information provided by video and other sensory systems [10]. Moreover, an expert system can support him by taking decisions. The more advanced technology consists in completely autonomous operation of the machinery, which requires significantly higher intelligence of the system. Autonomous and remote controlled systems are also offered for mining trucks, bulldozers, drills, and shovels [22].

Very broad research and applications concern copper ore mining industry, where its good communication infrastructure facilitates remote operation of machinery.

To make a next step in robotisation one should consider specialized autonomous robots for basic work including automated drilling machines, automated machinery for loading blasting holes, autonomous LHD (loaders and trucks), robotized arms for oversized copper ore lumps crushing, etc. Currently in KGHM "Polish Copper" S.A. such projects are in progress and results are expected in near future. Also some works related to inspection robots (flying, walking, etc.) are considered for spatially distributed infrastructure inspection. As mentioned earlier, it is not reasonable to monitor online everything in a large scale mine. To minimise humans effort one might exploit mobile robots (in teleoperation mode or fully autonomous). Since it is very difficult to develop a machine that deserves absolute autonomy despite instant situation in an operating scene, a very innovative approach of *virtual teleportation of the operator to the operational scene* is developed and implemented (the key know-how of SkyTech Research Sp. z o.o.).

Other group of mechatronic systems is devoted to robotized inspection. Such systems can release human inspectors from very hazardous and troublesome work. An example of this can be inspection of an area of coal mine affected by a catastrophe such as fire, explosion of gases, etc. An exemplary system called *TeleRescuer* [24] replaces human rescuers in inspections of roadways closed by a dam, since it is capable of operating in explosive atmosphere and in increased temperatures. The system is composed of two main parts: a mobile robot satisfying ATEX M1 requirements, and an innovative Human-Machine Interface (HMI) that can evoke virtual teleportation of the operator to the scene where just the robot operates, but the rescuer could not be there due to unacceptable hazard to his life. Although the main operation mode is remote control, the system is capable of operating autonomously in case of losing communication with the control station located in a fresh-air base. This research is carried out by a consortium where Silesian University of Technology (Poland) acts as the leader, and the partners are: VSB – Technical University of Ostrava (Czech Republic), the University Carlos III of Madrid (Spain), SkyTech Research Sp. z o.o. (Poland), Simmersion GmbH (Austria), and KOPEX S.A. (Poland). More details of this system will be presented in the tutorial.

5   Sensing and Communication

Collision avoidance and proximity detection systems become ever more and more popular in the mining industry. The reason is that mobile mining machinery becomes bigger and operates with higher speeds of its organs, while there remain significant dead sight areas where the operator is unable to detect other machines or even humans. The collision avoidance systems are based on RFID, radar, vision or ultrasonic systems. Their operation is affected by harsh environment conditions such as high relative humidity, dust or coal culm.

Other essential issue concerns communication. If one considers coal mines, the coal itself that constitutes walls and ceilings of a roadway affects propagation of radio waves. Nonetheless, countable wireless communication systems are developed, capable of working in underground coal mines. The complexity of the system depends on data transfer rate required to transmit measurement results and images. Recently wireless systems capable of operating in coal mines take advantage of a network of repeaters based on motes – simple communication modules that can organize themselves in an 'ad hoc' network. Such a system is developed in the framework of *TeleRescuer* project playing the role of a backup communication system for the robot.

6   Conclusions

In the paper some issues concerning ICT and mechatronic systems that are used nowadays in mining industry have been discussed. Development of such systems can have great influence on the further evolution of this industry. If one considers increasing the productivity by simultaneous reduction of costs and hazards, then intensive automation with the goal to develop autonomous machines that might totally replace human operators is the answer. Nevertheless, it is impossible to isolate completely the human operator from supervising the system in some rare, yet very critical situations when automatic control cannot assure flawless operation. To this end, the innovative *virtual teleportation technology* could help by allowing remote intervention of the experienced operator who personally could assess the situation and find solution to the problem unsolvable by the autonomous system itself. The next step could be to include a *self-learning functionality* to the system to allow collecting new skills that could be used in the future to operate when facing problems too difficult to solve until now. Such ambitious goals can be undertaken by international consortia and require both the R&D and implementation work. Research centres and universities can play very important role in this process by developing new methods and demonstrators. Very close collaboration with the industry would be of great importance.

References

1. Bartelmus, W., Zimroz, R.: Vibration condition monitoring of planetary gearbox under varying external load. Mech. Syst. Signal Proc. **23**(1), 246–257 (2009). Special Issue: Non-linear Structural Dynamics

2. Bartkowiak, A., Zimroz, R.: Outliers analysis and one class classification approach for planetary gearbox diagnosis. J. Phys. Conf. Ser. **305**(1), 012031 (2011)
3. Bartkowiak, A., Zimroz, R.: Dimensionality reduction via variables selection - linear and nonlinear approaches with application to vibration-based condition monitoring of planetary gearbox. Appl. Acoust. **77**, 169–177 (2014)
4. Brzychczy, E.: The intelligent computer-aided support in designing mining operations at underground hard coal mines. In: 23th World Mining Congress, August 11–15 2013, Montreal (2013)
5. Cempel, C.: Multidimensional condition monitoring of mechanical systems in operation. Mech. Syst. Signal Process. **17**(6), 1291–1303 (2003)
6. Chekushina, E.V., Vorobev, A.E., Chekushina, T.V.: Use of expert systems in the mining. Middle-East J. Sci. Res. **18**(1), 1–3 (2013)
7. Cholewa, W.: Expert systems in technical diagnostics. In: Korbicz, J., Kowalczuk, Z., Kościelny, J., Cholewa, W. (eds.) Fault Diagnosis, pp. 591–631. Springer, Heidelberg (2004)
8. Cioch, W., Knapik, O., Leśkow, J.: Finding a frequency signature for a cyclostationary signal with applications to wheel bearing diagnostics. Mech. Syst. Signal Process. **1**(38), 55–64 (2013)
9. COIG. http://www.coig.pl/en/mining, (as of Aug. 2016)
10. CSIRO. Mining safety and automation. http://www.csiro.au/en/Research/EF/Areas/Coal-mining/Mining-safety-and-automation, (as of Aug. 2016)
11. EC-Europa. https://ec.europa.eu/growth/tools-databases/eip-rawmaterials/en/content/strategic-implementation-plan-sip-0, (as of Aug. 2016)
12. EIT. http://eitrawmaterials.eu/, (as of Aug. 2016)
13. EU-Robotics. https://eu-robotics.net/, (as of Aug. 2016)
14. Golak, S., Wieczorek, T.: Koncepcja system ekspertowego do oceny i poprawy ekoefektywności kopalń (in Polish). Studia Informatica **116**(2), 213–222 (2014)
15. Hofmann, M., Klinkenberg, R.: RapidMiner: Data Mining Use Cases and Business Analytics Applications. Chapman & Hall/CRC, Boca Raton (2013)
16. IBM. http://www-935.ibm.com/industries/metalsmining/, (as of Aug. 2016)
17. Janusz, A., Sikora, M., Wróbel, L., Stawicki, S., Grzegorowski, M., Wojtas, P., Slezak, D.: Mining data from coal mines: IJCRS'15 data challenge. In: Yao, Y. (ed.) RSFDGrC 2015. LNCS, vol. 9437, pp. 429–438. Springer, Heidelberg (2015). doi:10.1007/978-3-319-25783-9_38
18. Kabiesz, J.: Effect of the form of data on the quality of mine tremors hazard forecasting using neural networks. Geotech. Geol. Eng. **24**(5), 1131–1147 (2006)
19. Kadlec, P., Gabrys, B., Strandt, S.: Data-driven soft sensors in the process industry. Comput. Chem. Eng. **33**(4), 795–814 (2009)
20. Kalisch, M., Przystałka, P., Timofiejczuk, A.: A concept of meta-learning schemes for context-based fault diagnosis. In: XV International Technical Systems Degradation Conference, TSD International Conference, Liptovsky Mikulas, 30 March – 2 April 2016, pp. 113–114 (2016)
21. Kozielski, M., Sikora, M., Wróbel, Ł.: *Disesor* - decision support system for mining industry. In: Ganzha, M., Maciaszek, L., Paprzycki, M. (eds.) Proceedings of the 2015 Federated Conference on Computer Science and Information Systems, vol. 5 of Annals of Computer Science and Information Systems, pp. 67–74. IEEE (2015)
22. Leica Geosystems. Autonomous and remote controlled mining. http://mining.leica-geosystems.com/news/all-news/autonomous-and-remote-controlled-mining, (as of Aug. 2016)
23. Liebowitz, J.: The Handbook of Applied Expert Systems. CRC Press LLC, Boca Raton (1997)

24. Moczulski, W., Cyran, K., Januszka, M., Novak, P., Timofiejczuk, A.: *Telerescuer* - an innovative robotized system for supporting mining rescuers by inspecting road-ways affected by catastrophes. In: 24th World Mining Congress (2016)
25. Przystałka, P., Moczulski, W., Timofiejczuk, A., Kalisch, M., Sikora, M.: Development of Expert System Shell for Coal Mining Industry. Springer, Heidelberg (2016)
26. RapidMiner: *RapidMiner* software website, August 2016. https://rapidminer.com/
27. Sáez, J.A., Krawczyk, B., Woźniak, M.: Analyzing the oversampling of different classes, types of examples in multi-class imbalanced datasets. Pattern Recogn. **57**(C), 164–178 (2016)
28. Sikora, M., Moczulski, W., Timofiejczuk, A., Przystałka, P., Ślęzak, D.: DIS-ESOR: An integrated shell decision support system for systems of monitoring processes, equipment and hazards. In: Mechanizacja, automatyzacja i robotyzacja w górnictwie, pp. 39–47 (2015)
29. Sikora, M., Przystałka, P., (eds.): Zintegrowany, szkieletowy system wspomagania decyzji dla systemów monitorowania procesów, urządzeń i zagrożeń (in Polish). Publishing House of the Institute for Sustainable Technologies, National Research Institute, Radom, Poland (2016) (in print)
30. Sokołowski, J., Obuchowski, J., Madziarz, M., Wyłomańska, A., Zimroz, R.: Features based on instantaneous frequency for seismic signals clustering. J. VibroEng. **18**(3), 1654–1667 (2016)
31. Stefaniak, P., Zimroz, R., Bartelmus, W., Hardygóra, M.: Computerised decision-making support system based on data fusion for machinery systems management and maintenance. Appl. Mech. Mater. **683**, 108–113 (2014)
32. Wang, C., Wang, Z.: Design and implementation of safety expert information management system of coal mine based on fault tree. J. Softw. **5**(10), 1114–1120 (2010)
33. Xu, X., Dou, L., Lu, C., Zhang, Y.: Frequency spectrum analysis on micro-seismic signal of rock bursts induced by dynamic disturbance. Min. Sci. Technol. **20**(5), 682–685 (2010)
34. Yingxu, Q., Hongguo, Y.: Design and application of expert system for coal mine safety. In: Second IITA International Conference on Geoscience and Remote Sensing (2010)
35. Zimroz, R., Hardygóra, M., Błażej, R.: Maintenance of belt conveyor systems in Poland - an overview. In: Proceedings of the 12th International Symposium Continuous Surface Mining - Aachen, pp. 21–30. Springer, Heidelberg (2015)
36. Zimroz, R., Wodecki, J., Król, R., Andrzejewski, M., Śliwiński, P., Stefaniak, P.: Self-propelled Mining Machine Monitoring System - Data Validation, Processing and Analysis. Springer, Heidelberg (2014)

Rough Sets of Zdzisław Pawlak Give New Life to Old Concepts. A Personal View ....

Lech T. Polkowski[⊠]

Polish-Japanese Academy of Information Technology,
Koszykowa str. 86, 02-008 Warsaw, Poland
polkow@pjwstk.edu.pl

Abstract. Zdzisław Pawlak influenced our thinking about uncertainty by borrowing the idea of approximation from geometry and topology and carrying those ideas into the realm of knowledge engineering. In this way, simple and already much worn out mathematical notions, gained a new life given to them by new notions of decision rules and algorithms, complexity problems, and problems of optimization of relations and rules. In his work, the author would like to present his personal remembrances of how his work was influenced by Zdzisław Pawlak interlaced with discussions of highlights of research done in enliving classical concepts in new frameworks, and next, he will go to more recent results that stem from those foundations, mostly on applications of rough mereology in behavioral robotics and classifier synthesis via granular computing.

Keywords: Rough sets · Rough mereology · Granular computing · Betweenness · Mobile robot navigation · Kernel and residuum in data

1 Meeting Professor Pawlak First Time: First Problems

It was in the year 1992 and the person who contacted us was Professor Helena Rasiowa, the eminent world–renowned logician. Zdzisław asked me to create a topological theory of rough set spaces: He was eager to introduce into rough sets the classical structures; some logic and algebra already were therein. The finite case was well recognized so I followed an advice by Stan Ulam:*'if you want to discuss a finite case, go first to the infinite one'*, I considered information systems with countably infinitely many attributes. Let me sum up the essential results which were warmly welcomed by Zdzisław.

1.1 Rough Set Topology: A Context and Basic Notions

Assume given a set[1] (a *universe*) U of *objects* along with a sequence $A = \{a_n : n = 1, 2, \ldots\}$ of *attributes*;[2] without loss of generality, we may assume that

---

L.T. Polkowski—An invited Fellow IRSS talk.

[1] Results on topology of rough sets can be best found in author's [4].

[2] The pair IS = (U,A) will be called an *information system*; each $a_n \in A$ maps U into a set V of *possible values*.

© Springer International Publishing AG 2016
V. Flores et al. (Eds.): IJCRS 2016, LNAI 9920, pp. 43–53, 2016.
DOI: 10.1007/978-3-319-47160-0_4

$Ind_n \subseteq Ind_{n+1}$ for each n, where $Ind_n = \{(u,v) : u, v \in U, a_n(u) = a_n(v)\}$. Letting $Ind = \bigcap_n Ind_n$, we may assume that the family $\{Ind_n : n = 1, 2, \ldots\}$ *separates objects*, i.e., for each pair $u \neq v$, there is a class $P \in U/Ind_n$ for some n such that $u \in P, v \notin P$, otherwise we would pass to the quotient universe U/Ind. We endow U with some topologies.

1.2　Topologies Π_n, the Topology Π_0 and Exact and Rough Sets

For each n, the topology Π_n is defined as the partition topology obtained by taking as open sets unions of families of classes of the relation Ind_n. The topology Π_0 is the union of topologies Π_n for $n = 1, 2, \ldots$. We apply the topology Π_0 to the task of discerning among subsets of the universe U^3:

$$\text{A set } Z \subseteq U \text{ is} \Pi_0\text{-exact if } Cl_{\Pi_0} Z = Int_{\Pi_0} Z \text{ else } Z \text{ is } \Pi_0\text{-rough.} \qquad (1)$$

1.3　The Space of Π_0-rough Sets is Metrizable

Each Π_0-rough set can be represented as a pair (Q, T) where $Q = Cl_{\Pi_0} X, T = U \setminus Int_{\Pi_0} X$ for some $X \subseteq U$. The pair (Q, T) has to satisfy the conditions: 1. $U = Q \cup T$. 2. $Q \cap T \neq \emptyset$. 3. If $\{x\}$ is a Π_0-open singleton then $x \notin Q \cap T$. We define a metric d_n as[4]

$$d_n(u, v) = 1 \text{ in case } [u]_n \neq [v]_n \text{ else } d_n(u, v) = 0. \qquad (2)$$

and the metric d:

$$d(u, v) = \sum_n 10^{-n} \cdot d_n(u, v). \qquad (3)$$

Theorem 1. *Metric topology of d is Π_0.*

We employ the notion of the *Hausdorff metric* and apply it to pairs (Q, T) satisfying 1–3 above, i.e., representing Π_0-rough sets. For pairs $(Q_1, T_1), (Q_2, T_2)$, we let

$$D((Q_1, T_1), (Q_2, T_2)) = max\{d_H(Q_1, Q_2), d_H(T_1, T_2)\} \qquad (4)$$

and

$$D^*((Q_1, T_1), (Q_2, T_2)) = max\{d_H(Q_1, Q_2), d_H(T_1, T_2), d_H(Q_1 \cap Q_2, T_1 \cap T_2)\}, \qquad (5)$$

where $d_H(A, B) = max\{max_{x \in A} dist(x, B), max_{y \in B} dist(y, A)\}$ is the Hausdorff metric on closed sets[5]. The main result is

Theorem 2. *If each descending sequence $\{[u_n]_n : n = 1, 2, \ldots\}$ of classes of relations Ind_n has a non–empty intersection, then each D^*–fundamental sequence of Π_0–rough sets converges in the metric D to a Π_0–rough set. If, in addition, each relation Ind_n has a finite number of classes, then the space of Π_0–rough sets is compact in the metric D.*

---

[3] Cl_τ is the closure operator and Int_τ is the interior operator with respect to a topology τ.

[4] $[u]_n$ is the Ind_n-class of u.

[5] $dist(x, A) = min_{y \in A} d(x, y)$.

1.4   The Space of Almost Π_0-rough Sets is Metric Complete

In notation of preceding sections, it may happen that a set X is Π_n-rough for each n but it is Π_0-exact. We call such sets *almost rough sets*. We denote those sets as Π_ω-rough. Each set X of them, is represented in the form of a sequence of pairs $(Q_n, T_n) : n = 1, 2, \ldots$ such that for each n, 1. $Q_n = Cl_{\Pi_n} X, T_n = U \setminus Int_{\Pi_n} X$. 2. $Q_n \cap T_n \neq \emptyset$. 3. $Q_n \cup T_n = U$. 4. $Q_n \cap T_n$ contains no singleton $\{x\}$ with $\{x\}$ Π_n-open. To introduce a metric into the space of Π_ω-rough sets, we apply again the Hausdorff metric but in a modified way: for each n, we let $d_{H,n}$ to be the Hausdorff metric on Π_n-closed sets, and for representations (Q_n, T_n) and $(Q_n^*, T_n^*)_n$ of Π_ω-rough sets X, Y, respectively, we define the metric D' as:

$$D'(X, Y) = \sum_n 10^{-n} \cdot max\{d_{H,n}(Q_n, Q_n^*), d_{H,n}(T_n, T_n^*)\}. \tag{6}$$

It turns out that

Theorem 3. *The space of Π_ω-rough sets endowed with the metric D' is complete, i.e., each D'-fundamental sequence of Π_ω-rough sets converges to a Π_ω-rough set.*

Apart from theoretical value of these results, there was an applicational tint in them.

1.5   Approximate Collage Theorem

Consider an Euclidean space E^n along with an information system $(E^n, A = \{a_k : k = 1, 2, \ldots\})$, each attribute a_k inducing the partition P_k of E^n into cubes of the form $\prod_{i=1}^n [m_i + \frac{j_i}{2^k}, m_i + \frac{j_i+1}{2^k})$, where m_i runs over integers and $j_i \in [0, 2^k-1]$ is an integer. Hence, $P_{k+1} \subseteq P_k$, each k. We consider *fractal objects*, i.e., systems of the form $[(C_1, C_2, \ldots, C_p), f, c]$, where each C_i is a compact set and f is an affine contracting mapping on E^n with a contraction coefficient $c \in (0, 1)$. The resulting fractal is the limit of the sequence $(F_n)_n$ of compacta, where 1. $F_0 = \bigcup_{i=1}^p C_i$. 2. $F_{n+1} = f(F_n)$. In this context, fractals are classical examples of Π_0-rough sets. Assume we perceive fractals through their approximations by consecutive grids P_k, so each F_n is viewed on as its *upper approximations* $a_k^+ F_n$ for each k[6]. As $diam(P_k) \to_{k \to \infty} 0$, it is evident that the symmetric difference $F \triangle F_n$ becomes arbitrarily close to the symmetric difference $a_k^+ F \triangle a_k^+ F_n$. Hence, in order to approximate F with F_n it suffices to approximate $a_k^+ F$ with $a_k^+ F_n$. The question poses itself: what is the least k which guarantees for a given ε, that if $a_k^+ F_n = a_k^+ F$ then $d_H(F, F_n) \leq \varepsilon$. We consider the metric D on fractals and their approximations. We had proposed a counterpart to Collage Theorem, by replacing fractals F_n by their grid approximations[7].

---

[6] This theorem comes from the chapter by the author in [3].

[7] The upper approximation of a set $X \subseteq U$ with respect to a partition P on U is $\bigcup\{q \in P : q \cap X \neq \emptyset\}$.

Theorem 4 *(Approximate Collage Theorem). Assume a fractal F generated by the system $(F_0 = \bigcup_{i=1}^{p} C_i, f, c)$ in the space of Π_0-rough sets with the metric D. In order to satisfy the requirement $d_H(F, F_n) \leq \varepsilon$, it is sufficient to satisfy the requirement $a_{k_0}^+ F_n = a_{k_0}^+ F$ with $k_0 = \lceil \frac{1}{2} - log_2\varepsilon \rceil$ and $n \geq \lceil \frac{log_2[2^{-k_0+\frac{1}{2}} \cdot K^{-1} \cdot (1-c)]}{log_2 c} \rceil$, where $K = d_H(F_0, F_1)$.*

2 Mereology and Rough Mereology

It was a characteristic feature of Professor Pawlak that He had a great interest in theoretical questions. He remembered how He browsed through volumes in the Library at Mathematical Institute of the Polish Academy of Sciences. No doubt that the emergence of rough set theory owes much to those excursions into philosophical writings of Frege, Russell and others. At one time, Zdzisław mentioned some fascicles of the works of Stanisław Leśniewski, the creator of the first formal theory of Mereology. Zdzisław was greatly interested in various formalizations of the idea of a concept and in particular in possible relations between Mereology and Rough Sets. From our analysis of the two theories Rough Mereology emerged.

2.1 Basic Mereology

The primitive notion is here that of a *part*. The relation of *being a part of*, denoted $prt(u,v)$, is defined on a universe U by requirements: 1. $prt(u,u)$ holds for no u. 2. $prt(u,v)$ and $prt(v,w)$ imply $prt(u,w)$: $prt(u,v)$ means that u is a *proper part* of v. To account for *improper parts*, i.e., *wholes* the notion of *an ingredient, element, ingr* for short, was proposed which is $prt \cup$ '=', i.e., $ingr(u,v)$ if and only if $prt(u,v)$ or $u = v$. Ingredients are essential in mereological reasoning by the Leśniewski Inference Rule (LIR for short):[8]

LIR: For $u, v \in U$, if for each w such that $ingr(w,u)$, there exist t, q such that $ingr(t,w), ingr(t,q), ingr(q,v)$, then $ingr(u,v)$.

Ingredients are instrumental in forming individuals–classes of individuals: for each non-void property $\mathcal{C}$ of individuals in U, there exists a unique individual, the class of $\mathcal{C}$, $Cls\mathcal{C}$ in symbols, defined by requirements: 1. If u satisfies $\mathcal{C}$ then $ingr(u, Cls\mathcal{C})$. 2. For each u with $ingr(u, Cls\mathcal{C})$, there exist t, q such that $ingr(t,u), ingr(t,q)$ and q satisfies $\mathcal{C}$. Classes are instrumental in our definition of granules. The favorite example of Leśniewski was the chessboard as the class of white and black squares.

2.2 Rough Mereology

The basic notion of a part to a degree is rendered as the relation $\mu(u,v,r) \subseteq U^2 \times [0,1]$, read as '*u is a part of v to a degree of at least r*' which is defined by

---

[8] To acquaint oneself with this theory it is best to read Lesniewski [2]. This is a rendering by E. Luschei of the original work *Foundations of Set Theory*. Polish Scientific Circle. Moscow 1916.

requirements: 1. $\mu(u, v, 1)$ if and only if $ingr(u, v)$. 2. If $\mu(u, v, 1)$ and $\mu(w, u, r)$ then $\mu(w, v, r)$. 3. If $\mu(u, v, r)$ and $s < r$ then $\mu(u, v, s)$. The relation μ was termed by us a *rough inclusion*. Relation of rough mereology to rough set theory becomes clear when we realize that the latter is about concepts and their approximations and that the containment relation is a particular case of the part relation, hence approximations upper and lower are classes of indiscernibility classes which are ingredients or, respectively, parts to a positive degree of a concept. Rough inclusions in information systems are usually defined in the attribute–value format, examples are for instance given by t–norms. It is well–known that Archimedean t-norms, the Łukasiewicz t–norm $L(x, y) = max\{0, x + y - 1\}$ and the Menger (product) t–norm $P(x, y) = x \cdot y$, allow the representation of the form $T(x, y) = g(f(x) + f(y))$, where $f : [0, 1] \rightarrow [0, 1]$ is a decreasing continuous function with $f(1) = 0$ and g is the pseudo–inverse to f. For an information system $IS = (U, A)$, the *discernibility set* $Dis(u, v)$ equals $A \setminus Ind(u, v)$[9].

Theorem 5. *For an Archimedean t–norm $T(x, y) = g(f(x) + f(y))$, the relation $\mu_T(u, v, r)$ if and only if $g(\frac{card(Dis(u,v))}{card(A)}) \geq r$ is a rough inclusion on the universe U.*

As an example, we define the *Łukasiewicz rough inclusion* μ_L as $\mu_L(u, v, r)$ if and only if $g(\frac{card(Dis(u,v))}{card(A)}) \geq r$. As in case of Łukasiewicz rough inclusion, $g(x) = 1 - x$, we have $\mu_L(u, v, r)$ if and only if $\frac{card(Ind(u,v))}{card(A)} \geq r$: a fuzzified indiscernibility. We recall that each t–norm T defines the *residual implication* $\rightarrow_T$ via the equivalence $x \rightarrow_T y \geq r$ if and only if $T(x, r) \leq y$.

Theorem 6. *Let $\rightarrow_T$ be a residual implication and $f : U \rightarrow [0, 1]$ an embedding of U into the unit interval. Then $\mu(u, v, r)$ if and only if $f(u) \rightarrow_T f(v) \geq r$ is a rough inclusion.*

We have therefore a collection of rough inclusions to be selected.

3   Rough Mereology in Behavioral Robotics

Autonomous robots are one of the best examples for the notion of an intelligent agent. Problems of their navigation in environments with obstacles are basic in behavioral robotics. We recall here an approach based on rough mereology[10].

3.1   Betweenness Relation in Navigating of Teams of Intelligent Agents

Betweenness relation is one of primitive, apart from equidistance, relations adopted by Alfred Tarski in His axiomatization of plane geometry. This relation was generalized by Johan van Bentham in the form of the relation $B(x, y, z)$,

---

[9] Please see relevant chapters in Polkowski [5].
[10] Please see Polkowski L., Osmialowski P. [8].

x, y, z points in an Euclidean space of a finite dimension (it reads: 'x *is between* y *and* z'), with a metric d, in the form:

$$B(x,y,z) \text{ if and only if for each } q \neq x : \; d(x,y) < d(q,y) \text{ or } d(x,z) < d(q,z). \tag{7}$$

Rough mereology offers a quasi–distance function:

$$\kappa(x,y) = min\{sup_r\mu(x,y,r), sup_s\mu(y,x,s)\}. \tag{8}$$

We apply in definition of $\kappa(x,y)$ the rough inclusion $\mu(a,b,r)$, where a,b are bounded measurable sets in the plane,

$$\mu(a,b,r) \text{ if and only if } \frac{area(a \cap b)}{area(a)} \geq r. \tag{9}$$

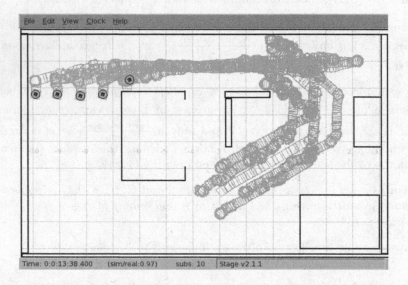

Fig. 1. Trails of robots moving in the line formation through the passage.

Consider autonomous robots in the plane as embodiments of intelligent agents. We model robots as rectangles (in fact squares) regularly placed, i.e., with edges parallel to coordinate axes. For such robots denoted a, b, c,... , the betweenness relation can be expressed as follows, see [8]:

Theorem 7. *Robot a is between robots b and c, i.e. $B(a,b,c)$ holds true, with respect to betweenness defined in (7), distance defined in (8) and the rough inclusion defined in (9) if and only if $a \subseteq ext(b,c)$, where $ext(b,c)$ is the extent of b and c, i.e., the minimal rectangle containing b and c.*

For a team of robots $\mathcal{T} = \{a_1, a_2, \ldots, a_m\}$, a *formation* on $\mathcal{T}$ is a relation B on $\mathcal{T}$. Figure 1 shows a team of robots in Line formation mediating a bottleneck passage after which they return to the Cross formation.

4. Granular Computing

The last of rough mereology applications Zdzisław could be acquainted with is a theory of granular computing presented first at GrC 2005 at Tsinghua University in Beijing, China. Given a rough inclusion μ on a universe U of an information system (U, A), define a *granule* $g_\mu(u, r)$ about $u \in U$ of the radius r as $g_\mu(u, r) = Cls\{v \in U : \mu(v, u, r)\}$. For practical reasons, we compute granules as sets $\{v \in U : \mu(v, u, r)\}$. The class and the set coincide for many rough inclusions, cf. [5][11].

4.1 Granular Classifers: Synthesis via Rough Inclusions

We assume that we are given a decision system $DS = (U, A, d)$ from which a classifier is to be constructed; on the universe U, a rough inclusion μ is given, and a radius $r \in [0, 1]$ is chosen. We can find granules $g_\mu(u, r)$ for all $u \in U$, and make them into the set $G(\mu, r)$. From this set, a covering $Cov(\mu, r)$ of the universe U can be selected by means of a chosen strategy $\mathcal{G}$, i.e.,[12]

$$Cov(\mu, r) = \mathcal{G}(G(\mu, r)). \tag{10}$$

We intend that $Cov(\mu, r)$ becomes a new universe of the decision system whose name will be the *granular reflection* of the original decision system. It remains to define new attributes for this decision system. Each granule g in $Cov(\mu, r)$ is a collection of objects; attributes in the set $A \cup \{d\}$ can be factored through the granule g by means of a chosen strategy $\mathcal{S}$, usually the majority vote, i.e., for each attribute $a \in A \cup \{d\}$, the new factored attribute $\overline{a}$ is defined by means of the formula

$$\overline{a}(g) = \mathcal{S}(\{a(v) : ingr(v, g_\mu(u, r))\}) \tag{11}$$

In effect, a new decision system $(Cov(\mu, r), \{\overline{a} : a \in A\}, \overline{d})$ is defined. The thing[13] $\overline{v_g}$ with the *information set* $Inf(\overline{v_g})$ defined as[14]

$$Inf(\overline{v_g}) = \{(\overline{a}, \overline{a}(g)) : a \in A \cup \{d\}\} \tag{12}$$

is called the *granular reflection of g*. We consider a standard data set *the Australian Credit Data Set* from Repository at UC Irvine and we collect the best results for this data set by various rough set based methods in the table of Fig. 2. In Fig. 3, we give for this data set the results of exhaustive classifier on granular structures: meanings of symbols are r = granule radius, tst = test set size, trn = train set size, rulex = rule number, aex = accuracy, cex = coverage[15].

---

[11] Please consult Polkowski [5] Ch. 9 and Polkowski, Artiemjew [6].

[12] An information system $IS = (U, A)$ augmented by a new attribute $d : U \rightarrow V$, the *decision*, is called the *decision system* $DS = (U, A, d)$.

[13] The philosophical term 'thing' is reserved for beings of virtual character possibly not present in the given information/decision system.

[14] In a decision system (U, A, d), for $u \in U$, the information set of u is $Inf(u) = \{(a, a(u)) : a \in A \cup \{d\}\}$.

[15] MI is the *Michalski index*. $MI = \frac{1}{2} \cdot aex + \frac{1}{4} \cdot aex^2 + \frac{1}{2} \cdot cex - \frac{1}{4} \cdot aex \cdot cex$.

source	method	accuracy	coverage	MI
Bazan	$SNAPM(0.9)$	$error = 0.130$	$--$	$--$
Nguyen SH	$simple.templates$	0.929	0.623	0.847
Nguyen SH	$general.templates$	0.886	0.905	0.891
Nguyen SH	$tolerance.gen.templ.$	0.875	1.0	0.891
Wroblewski	$adaptive.classifier$	0.863	$-$	$--$

Fig. 2. Best results for Australian credit by some rough set based algorithms

r	tst	trn	$rulex$	aex	cex	MI
nil	345	345	5597	0.872	0.994	0.907
0.0	345	1	0	0.0	0.0	0.0
0.0714286	345	1	0	0.0	0.0	0.0
0.142857	345	2	0	0.0	0.0	0.0
0.214286	345	3	7	0.641	1.0	0.762
0.285714	345	4	10	0.812	1.0	0.867
0.357143	345	8	23	0.786	1.0	0.849
0.428571	345	20	96	0.791	1.0	0.850
0.5	345	51	293	0.838	1.0	0.915
0.571429	345	105	933	0.855	1.0	0.896
0.642857	345	205	3157	0.867	1.0	0.904
0.714286	345	309	5271	0.875	1.0	0.891
0.785714	345	340	5563	0.870	1.0	0.890
0.857143	345	340	5574	0.864	1.0	0.902
0.928571	345	342	5595	0.867	1.0	0.904

Fig. 3. Australian credit granulated

We can compare results: for template based methods, the best MI is 0.891, for exhaustive classifier (r = nil) MI is equal to 0.907 and for granular reflections, the best MI value is 0.915 with few other values exceeding 0.900. What seems worthy of a moment's reflection is the number of rules in the classifier. Whereas for the exhaustive classifier (r = nil) in non–granular case, the number of rules is equal to 5597, in granular case the number of rules can be surprisingly small with a good MI value, e.g., at $r = 0.5$, the number of rules is 293, i.e., 5 percent of the exhaustive classifier size, with the best MI of all of 0.915. This compression of classifier seems to be the most impressive feature of granular classifiers.

5   Betweenness Revisited in Data Sets

We can use in a given information set $IS = (U, A)$, the Łukasiewicz rough inclusion μ_L in order to obtain the mereological distance κ of (8) and the *generalized betweenness* relation GB (read: 'u is between $v_1, v_2, ..., v_k$')[16]:

---

[16] A detailed account please find in Polkowski, Nowak [7].

$GB(u, v_1, v_2, ..., v_k)$ if for each $v \neq u$, there is v_i such that $\kappa(u, v_i) \geq \kappa(u, v)$.

$$(13)$$

One proves cf. [7] that betweenness GB can be expressed as a convex combination:

Theorem 8. $GB(u, v_1, v_2, ..., v_k)$ *if and only if* $Inf(u) = \bigcup_{i=1}^{k} C_i$, *where* $C_i \subseteq Inf(v_i)$ *for* $i = 1, 2, ..., k$ *and* $C_i \cap C_j = \emptyset$ *for each pair* $i \neq j$.

In order to remove ambiguity in representing u, we introduce the notion of a *neighborhood* $N(u)$ over a set of *neighbors* $\{v_1, v_2, \ldots, v_k\}$ as the structure of the form:

$$< (v_1, C_1 \subseteq Ind(u, v_1), q(v_1)), \ldots, (v_k, C_k \subseteq Ind(u, v_k), q(v_k)) > \qquad (14)$$

with neighbors $v_1, v_2, \ldots, v_k$ ordered in the descending order of the factor q, where $q_i = \frac{card(C_i)}{card(A)}$. Clearly, $\sum_{i=1}^{k} q_i = 1$ and $q_i > 0$ for each $i \leq k$.

5.1 Dual Indiscernibility Matrix, Kernel and Residuum

Dual indiscernibility matrix DIM, for short, is defined as the matrix $M(U, A) = [m_{a,v}]$ where $a \in A, v$ a value of a and $m_{a,v} = \{u \in U : a(u) = v\}$ for each pair a, v. The *residuum* of the information system (U, A), *Res* in symbols, is the set $\{u \in U:$ there exists a pair(a, v) with $m_{a,v} = \{u\}\}$. The difference $U \setminus Res$ is the *kernel*, *Ker* in symbols. Clearly, $U = Ker \cup Res$, $Ker \cap Res = \emptyset$. The rationale behind those notions is that Ker consists of objects mutually exchangeable so averaged decisions on neighbors should transfer to test objects, while Res consists of objects with outliers which may serve as antennae catching test objects. It is interesting to see how those subsets do in tasks of classification into decision classes. Figure 4 shows results of applying C4.5 and k-NN to whole data set, Ker and Res for a few data sets from UC Irvine Repository. Results are very satisfying in terms of accuracy and size of data sets. Please observe that, for data considered, sets Ker and Res as a rule yield better of results for C4.5 and k-NN on the whole set[17].

5.2 A Novel Approach: Partial Approximation of Data Objects

The Pair classifier approaches a test object with inductively selected pairs of neighbors of training objects covering it partly[18].

Induction is driven by degree of covering from maximal down to the threshold number of steps. Successive pairs are indexed with *level* L. Objects in pairs up to a given level are pooled and they vote for decision value by majority voting.

---

[17] In order to split the data set into parts of which one is GB-self-contained and the other GB-vacuous, we propose the DIM matrix.

[18] A relaxed idea of convex combinations of objects lies in approximating only parts of data objects with training objects, see Artiemjew, Nowak, Polkowski [1].

database	set tested	accuracy of C4.5	accuracy of k-NN	number of samples
adult	whole set	$.857 \pm .003$	$.837 \pm .003$	39074.0
	Ker	$.853 \pm .004$	$.835 \pm .003$	22366.0
	Res	$.849 \pm .003$	$.833 \pm .003$	16708.0
PID	whole set	$.733 \pm .027$	$.723 \pm .021$	614.4
	Ker	$.704 \pm .037$	$.711 \pm .032$	212.9
	Res	$.724 \pm .035$	$.745 \pm .030$	401.5
fertility	whole set	$.852 \pm .073$	$.866 \pm .060$	80.0
diagnosis	Ker	$.846 \pm .075$	$.880 \pm .064$	71.6
	Res	$.852 \pm .068$	$.880 \pm .064$	8.4
german	whole set	$.713 \pm .023$	$.732 \pm .025$	800.0
credit	Ker	$.671 \pm .045$	$.714 \pm .038$	98.9
	Res	$.712 \pm .023$	$.726 \pm .030$	701.1
heart	whole set	$.750 \pm .054$	$.825 \pm .048$	216.0
disease	Ker	$.742 \pm .061$	$.822 \pm .051$	109.2
	Res	$.767 \pm .054$	$.827 \pm .041$	106.8

Fig. 4. Classification results

database	kNN	Bayes	Pair–best	Pair-0
Adult	.841	.864	.853L1	.823
Australian	.855	.843	.859L4,5	.859
Diabetes	.631	.652	.721L0	.710
German credit	.730	.704	.722L1	.721
Heart disease	.837	.829	.822L1	.800
Hepatitis	.890	.845	.892L0	.831
Congressional voting	.938	.927	.928L0	.928
Mushroom	1.0	.910	1.0L0	1.0
Nursery	.578	.869	.845L0	.845
Soybean large	.928	.690	.910L0	.910

Fig. 5. Pair classifier

Figure 5 shows results in comparison to k-NN and Bayes classifiers. The symbol Lx denotes the level of covering, Pair-0 is the simple pair classifier with approximations by the best pair and Pair–best denotes the best result over levels studied.

6   Conclusions

The paper presents some results along two threads: along one thread results are highlighted obtained by following Zdzisław Pawlak's 'research requests' and the other thread illustrates results obtained in classical settings by considering new contexts of knowledge engineering created by vision of Zdzisław Pawlak. Further work will focus on rational search for small decision-representative subsets of data with Big Data on mind and rough set based Approximate Ontology in biological and medical data.

Acknowledgements. This is in remembrance of Prof. Zdzisław Pawlak on the 10th anniversary of His demise. To organizers of IJCRS 2016 Prof. Richard Weber and Dr. Dominik Ślęzak DSc my thanks go for the invitation to talk henceforth the occasion to remember. To referees my thanks go for their useful remarks.

References

1. Artiemjew, P., Nowak, B., Polkowski, L.: A classifier based on the dual indiscernibility matrix. In: Dregvaite, G., Damasevicius, R. (eds.) Forthcoming in Communications in Computer and Information Science, Proceedings ICIST 2016, CCIS639, pp. 1–12. Springer (2016). doi:10.1007/978-3-319-46254-7_30
2. Lesniewski, S.: On the foundations of mathematics. Topoi **2**, 7–52 (1982)
3. Polkowski, L.: Approximate mathematical morphology. In: Pal, S.K., Skowron, A. (eds.) Rough Fuzzy Hybridization, pp. 151–162. Springer, Singapore (1999)
4. Polkowski, L.T.: Rough Sets. Mathematical Foundations. Springer, Physica, Heidelberg (2002)
5. Polkowski, L.T.: Approximate Reasoning by Parts. An Introduction to Rough Mereology. Springer, Switzerland (2011)
6. Polkowski, L., Artiemjew, P.: Granular Computing in Decision Approximation. An Application of Rough Mereology. Springer, Switzerland (2015)
7. Polkowski, L., Nowak, B.: Betweenness, Łukasiewicz rough inclusion, Euclidean representations in information systems, hyper-granules, conflict resolution. Fundamenta Informaticae **147**(2-3) (2016)
8. Polkowski, L., Osmialowski, P.: Navigation for mobile autonomous robots and their formations. An application of spatial reasoning induced from rough mereological geometry. In: Barrera, A. (ed.) Mobile Robots Navigation, pp. 329–354. Intech, Zagreb (2010)

Rough Set Approaches to Imprecise Modeling

Masahiro Inuiguchi[✉]

Graduate School of Engineering Science, Osaka University, Toyonaka,
Osaka 560-8531, Japan
inuiguti@sys.es.osaka-u.ac.jp

Abstract. In the classical rough set approaches, lower approximations
of single decision classes have been mainly treated. Based on those
approximations, attribute reduction and rule induction have been developed. In this paper, from the authors' recent studies, we demonstrate
that various analyses are conceivable by treating lower approximations
of unions of multiple decision classes.

Keywords: Attribute reduction · Attribute importance · Imprecise
rules · MLEM2

1 Introduction

Rough set theory [1,2] provides useful tools for reasoning from data. Attribute
reduction and rule induction are well developed techniques based on rough set
theory. They are applied to various fields including data analysis, signal processing, knowledge discovery, machine learning, artificial intelligence, medical informatics, decision analysis, granular computing, Kansei engineering, and so forth.
In the approach, the lower approximation (a set of objects whose classification is
consistent in all given data) and upper approximation (a set of possible members
in view of given data) are calculated for each decision class. The lower approximation of each decision class has been majorly used for obtaining attribute
reduction and rule induction, so far. However, as is known in the literature [3,4],
the replacement of the lower approximation with the upper approximation provides a different aspect of the analysis.

In this paper, we assume that a decision table with multiple decision classes
(more than two decision classes) is given. From the author's recent study on
rough set and imprecise modeling, we show several results obtained by using
the lower approximations of unions of k decision classes instead of lower approximations of single decision classes. This approach can be seen as a rough set
approach to imprecise modeling because it provides the analysis based on the
preservation of imprecise classification, i.e., correct classification up to k possible
decision classes. After a brief introduction of the classical rough set approaches,
we describe the following recent results obtained by the replacement of the lower
approximation of each decision class with that of each union of k decision classes:

This work was partially supported by JSPS KAKENHI Grant Number 26350423.

V. Flores et al. (Eds.): IJCRS 2016, LNAI 9920, pp. 54–64, 2016.
DOI: 10.1007/978-3-319-47160-0_5

(1) In Subsect. 3.2, the attribute reduction based on lower approximations of unions of k decision classes provides an intermediate between two extreme attribute reductions using lower and upper approximations of single decision classes. The two extremes are obtained by special parameter settings of k.

(2) In Sect. 4, it shows that the evaluation of attribute importance changes drastically by the selection of parameter k. It implies that the attribute importance cannot be evaluated univocally.

(3) In Subsect. 5.2, the classifier with rules induced for unions of k decision classes achieves a better performance than the classifier with rules induced for single decision classes.

2   A Brief Review of Rough Sets

The classical rough sets are defined under an equivalence relation which is often called an indiscernibility relation. In this paper, we restrict ourselves to discussions of the classical rough sets under decision tables. A decision table is characterized by four-tuple $\mathcal{I} = \langle U, C \cup \{d\}, V, \rho \rangle$, where U is a finite set of objects, C is a finite set of condition attributes, d is a decision attribute, $V = \bigcup_{a \in C \cup \{d\}} V_a$ and V_a is a domain of the attribute a, and $\rho : U \times C \cup \{d\} \to V$ is an information function such that $\rho(x, a) \in V_a$ for every $a \in C \cup \{d\}, x \in U$.

Given a set of attributes $A \subseteq C \cup \{d\}$, we can define an equivalence relation I_A referred to as an indiscernibility relation by $I_A = \{(x, y) \in U \times U \mid \rho(x, a) = \rho(y, a), \forall a \in A\}$. From I_A, we have an equivalence class, $[x]_A = \{y \in U \mid (y, x) \in I_A\}$. When $A = \{d\}$, we define

$$\mathcal{D} = \{D_j, \ j = 1, 2, \ldots, p\} = \{[x]_{\{d\}} \mid x \in U\}, \ D_i \neq D_j \ (i \neq j). \qquad (1)$$

D_j is called a 'decision class'. There exists a unique $v_j \in V_d$ such that $\rho(x, d) = v_j$ for each $x \in D_j$, i.e., $D_j = \{x \in U \mid \rho(x, d) = v_j\}$. Moreover, since $D_i \cap D_j = \emptyset$ $(i \neq j)$ and $\bigcup \mathcal{D} = U$ hold, $\mathcal{D}$ forms a partition.

For a set of condition attributes $A \subseteq C$, the lower and upper approximations of an object set $X \subseteq U$ are defined as follows:

$$A_*(X) = \{x \mid [x]_A \subseteq X\}, \ \ A^*(X) = \{x \mid [x]_A \cap X \neq \emptyset\}. \qquad (2)$$

A pair $(A_*(X), A^*(X))$ is called a rough set of X. The boundary region of X is defined by $BN_A(X) = A^*(X) - A_*(X)$. Since $[x]_A$ can be seen as a set of objects indiscernible from $x \in U$ in view of condition attributes in $A, A_*(X)$ is interpreted as a collection of objects whose membership to X is noncontradictive in view of condition attributes in A. $BN_A(X)$ is interpreted as a collection of objects whose membership to X is doubtful in view of condition attributes in A. $A^*(X)$ is interpreted as a collection of possible members. For $x \in U$, the generalized decision class $\partial_A(x)$ of x with respect to a condition attribute set $A \subseteq C$ is defined by $\partial_A(x) = \{\rho(y, d) \mid y \in [x]_A\}$ (see [3,5,6]).

Let $X, Y \subseteq U$. We have the following properties:

$$A_*(X) \subseteq X \subseteq A^*(X), \tag{3}$$

$$A \subseteq B \Rightarrow A_*(X) \subseteq B_*(X), \ A^*(X) \supseteq B^*(X), \tag{4}$$

$$A_*(X \cap Y) = A_*(X) \cap A_*(Y), \ A^*(X \cup Y) = A^*(X) \cup A^*(Y), \tag{5}$$

$$A_*(X \cup Y) \supseteq A_*(X) \cup A_*(Y), \ A^*(X \cap Y) \subseteq A^*(X) \cap A^*(Y), \tag{6}$$

$$BN_A(X) = A^*(X) \cap A^*(U - X), \tag{7}$$

$$A_*(X) = X - BN_A(X), \tag{8}$$

$$A^*(X) = X \cup BN_A(X) = U - A_*(U - X), \tag{9}$$

$$A_*(X) = A^*(X) - A^*(U - X) = U - A^*(U - X). \tag{10}$$

3 Attribute Reduction

3.1 Previous Approaches

Attribute reduction is one of the major topics in rough set approaches. It indicates minimally necessary attributes to classify objects without the deterioration of classification accuracy, and reveals important attributes. A set of minimally necessary attributes is called a reduct. In the classical rough set analysis, reducts preserving lower approximations are frequently used. Namely, a set of condition attributes, $A \subseteq C$ is called a reduct if and only if it satisfies (L1) $A_*(D_j) = C_*(D_j), j = 1, 2, \ldots, p$ and (L2) $\not\exists a \in A, (A - \{a\})_*(D_j) = C_*(D_j), j = 1, 2, \ldots, p$. Since we discuss several kinds of reducts, we call this reduct, a 'reduct preserving lower approximations' or an 'L-reduct' for short. Let $\mathcal{R}^L$ be a set of L-reducts. Then $\bigcap \mathcal{R}^L$ is called the 'core preserving lower approximation' or the 'L-core'. Attributes in the L-core are important because we cannot preserve all lower approximations of decision classes without any of them.

We consider reducts preserving upper approximation or equivalently, preserving boundary regions [3, 4]. A set of condition attributes, $A \subseteq C$ is called a 'reduct preserving upper approximations' or a 'U-reduct' for short if and only if it satisfies (U1) $A^*(D_j) = C^*(D_j), j = 1, 2, \ldots, p$ and (U2) $\not\exists a \in A, (A - \{a\})^*(D_j) = C^*(D_j), j = 1, 2, \ldots, p$. On the other hand, a set of condition attributes, $A \subseteq C$ is called a 'reduct preserving boundary regions' or a 'B-reduct' for short if and only if it satisfies (B1) $BN_A(D_j) = BN_C(D_j), j = 1, 2, \ldots, p$ and (B2) $\not\exists a \in A, BN_{(A-\{a\})}(D_j) = BN_C(D_j), j = 1, 2, \ldots, p$.

For those reducts, we have (R1) A U-reduct is also a B-reduct and vice versa, (R2) There exists an L-reduct A for a U-reduct B such that $B \supseteq A$ and (R3) There exists an L-reduct A for a B-reduct B such that $B \supseteq A$. Those relations can be proved easily from (5), (9) and (10). Since B-reduct is equivalent to U-reduct, we describe only U-reduct in what follows. Let $\mathcal{R}^U$ be a set of U-reducts. Then $\bigcap \mathcal{R}^U$ is called the 'core preserving upper approximation' or the 'U-core'. Attributes in the U-core are important because we cannot preserve all upper approximations of decision classes without any of them.

To obtain a part or all of reducts, many approaches have been proposed in the literature [2,7]. Among them, we mention an approach based on a discernibility matrix [5,7]. In this approach, we construct a Boolean function which characterizes the preservation of the lower approximations to obtain L-reducts. Each L-reduct is obtained as a prime implicant of the Boolean function. For the detailed discussion of the discernibility matrix for L-reducts, see reference [5,7].

3.2   Refinement of Attribute Reduction

Consider a cover $\mathcal{F}_k = \{D_{i_1} \cup D_{i_2} \cup \cdots \cup D_{i_k} \mid 1 \leq i_1 < i_2 < \cdots < i_k \leq p\}$ for $k \in \{1, 2, \ldots, p-1\}$. A condition attribute set A is called an $\mathcal{F}_k$-reduct if and only if (F1(k)) $A_*(F) = C_*(F)$ for all $F \in \mathcal{F}_k$ and (F2(k)) $\nexists a \in A, (A - \{a\})_*(F) = C_*(F)$ for all $F \in \mathcal{F}_k$.

From (9) and (10), we know that an $\mathcal{F}_k$-reduct A is a minimal set such that $A^*(F) = C^*(F)$ for all $F \in \mathcal{F}_{p-k}$. Moreover, from (5), an $\mathcal{F}_k$-reduct A satisfies (F$_l$1) for all $l \leq k$ and therefore, from (9) and (10), it satisfies $A^*(F) = C^*(F)$ for all $F \in \mathcal{F}_{p-l}$ and for all $l \leq k$. Note that $\mathcal{F}_1$-reducts are equivalent to L-reducts and $\mathcal{F}_{p-1}$-reducts are equivalent to U-reducts. From this observation the strong-weak relations among $\mathcal{F}_k$-reducts for $1 \leq k \leq p - 1$ can be depicted as in Fig. 1. The reducts located on the upper side of Fig. 1 are strong, i.e., the condition to be the upper reduct is stronger than the lower. On the contrary, the reducts located on the lower side of Fig. 1 are weak, i.e., the condition to be the lower reduct is weaker than the upper. Therefore, for any reduct A located on the upper side, there exists a reduct B located on the lower side such that $B \subseteq A$.

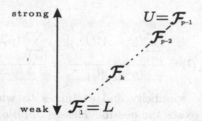

Fig. 1. The strong-weak relation among reducts

Let $\mathcal{R}(k)$ be a set of $\mathcal{F}_k$-reducts. Then $\bigcap \mathcal{R}(k)$ is called the '$\mathcal{F}_k$-core'. Attributes in the $\mathcal{F}_k$-core are important because we cannot preserve $C_*(F)$ for all $F \in \mathcal{F}_k$ without any of them.

As all L-reducts can be calculated using a discernibility matrix [5,7], all $\mathcal{F}_k$-reducts for $1 \leq k \leq p - 1$ can be calculated by a discernibility matrix. The (i, j)-component $\mathcal{D}_{ij}^k$ of the discernibility matrix $\mathcal{D}^k$ for calculating $\mathcal{F}_k$-reducts is obtained as the following set of attributes:

$$\mathcal{D}_{ij}^k = \begin{cases} \{a \in C \mid \rho(x_i, a) \neq \rho(x_j, a)\}, & \text{if } \partial_C(x_i) \neq \partial_C(x_j) \text{ and } |\partial_C(x_i)| \leq k, \\ C, & \text{otherwise.} \end{cases}$$

$$(11)$$

Then all $\mathcal{F}_k$-reducts are obtained as prime implicants of a Boolean function,

$$f^k = \bigwedge_{i,j:x_i,x_j \in U} \bigvee \mathcal{D}_{ij}^k, \tag{12}$$

where we regard $a \in \mathcal{D}_{ij}^k$ as a statement that 'the reduct includes a'. The computational complexity is NP-hard as in the classical decision matrix method [7].

Note that $\mathcal{D}_{ij}^k$ can be obtained from $\mathcal{D}_{ij}^l$ with $l > k$ by exchanging all components of i-th row such that $|\partial_C(x_i)| > k$ with C. Then, once $\mathcal{D}_{ij}^{p-1}$ is obtained, the other decision matrices can be obtained easily.

4   Attribute Importance

Corresponding to L-reducts, the following set function μ^Q is considered:

$$\mu^Q(A) = \gamma_A(\mathcal{D}) = \frac{\sum_{i=1}^{p} |A_*(D_i)|}{|U|}, \tag{13}$$

where $\gamma_A(\mathcal{D})$ is called a 'quality of approximation' of partition $\mathcal{D}$ and evaluates to what extent the set of condition attributes clearly classifies the objects into decision classes. L-reducts can be defined by μ^Q as follows: $A \subseteq C$ is an L-reduct if and only if A satisfies (L1) $\mu^Q(A) = \mu^Q(C)$ and (L2) $\nexists a \in A, \mu^Q(A - \{a\}) = \mu^Q(C)$.

Similarly, corresponding to U-reducts, the following set functions μ^{sp} is considered (see [4]):

$$\mu^{sp}(A) = \sigma_A(\mathcal{D}) = \frac{\sum_{i=1}^{p} |U - A^*(D_i)|}{(p-1)|U|} = \frac{\sum_{x \in U} (p - |\partial_A(x)|)}{(p-1)|U|}, \tag{14}$$

μ^{sp} shows the degree of specificity and evaluates to what extent the set of condition attributes decreases the possible classes of objects. U-reducts can be defined by μ^{sp} as follows: $A \subseteq C$ is a U-reduct if and only if A satisfies (U1) $\mu^{sp}(A) = \mu^{sp}(C)$ and (U2) $\nexists a \in A, \mu^{sp}(A - \{a\}) = \mu^{sp}(C)$.

We can regard those set functions as characteristic functions of cooperative game theory. Applying the Sharpley value and Harsanyi dividend (called also, Möbius transform) defined respectively by

$$I_\mu^S(A) = \sum_{K \subseteq C-A} \frac{(|C| - |K| - |A|)!|K|!}{(|C| - |A| + 1)!} \sum_{L \subseteq A} (-1)^{|A|-|L|} \mu(K \cup L), \tag{15}$$

$$m_\mu(A) = \sum_{B \subseteq A} (-1)^{|A-B|} \mu(B), \tag{16}$$

we may evaluate the attribute importance as well as interaction among attributes.

Attribute importance can be considered with respect to $\mathcal{F}_k$. We define the following simple set functions:

$$\mu_k^{\mathrm{re}}(A) = \begin{cases} 1 \text{ if } A_*(F) = C_*(F), \ \forall F \in \mathcal{F}_k, \\ 0 \text{ otherwise,} \end{cases} \quad A \subseteq C. \tag{17}$$

This set function shows simply whether A satisfies the requirement (F1(k)) of $\mathcal{F}_k$-reduct or not. By the definition, a minimal set $A \subseteq C$ such that $\mu_k^{\mathrm{re}}(A) = 1$ is an $\mathcal{F}_k$-reduct.

Moreover, we can define the following set function showing the constrained specificity (see [8,9]):

$$\mu_k^{\mathrm{sp}}(A) = \frac{\displaystyle\sum_{\substack{i_1,\ldots,i_k \in \{1,\ldots,p\} \\ i_j \neq i_l}} |A_*(D_{i_1} \cup \cdots \cup D_{i_k})|}{\dbinom{p-1}{k-1}|U|} = \frac{\displaystyle\sum_{i:|\partial_A(x_i)| \leq k} \dbinom{p - |\partial_A(x_i)|}{k - |\partial_A(x_i)|}}{\dbinom{p-1}{k-1}|U|}. \tag{18}$$

The constrained specificity implies that only objects $x \in U$ such that $|\partial_A(x)| \leq k$ are taken into consideration in μ_k^{sp} while all objects are considered in μ^{sp}. Note that we have $\mu_k^{\mathrm{sp}} = \mu^{\mathrm{Q}}$ when $k = 1$ and $\mu_k^{\mathrm{sp}} = \mu^{\mathrm{sp}}$ when $k = p - 1$. As μ_k^{sp} corresponds to $\mathcal{F}_k$-reducts, $\mathcal{F}_k$-reducts are characterized by (F1(k)′) $\mu_k^{\mathrm{sp}}(A) = \mu_k^{\mathrm{sp}}(C)$ and (F2(k)′) $\nexists a \in A, \mu_k^{\mathrm{sp}}(A - \{a\}) = \mu_k^{\mathrm{sp}}(C)$. We note that μ_k^{sp} does not be influenced by $\partial_C(x), x \in U$ but only by $\partial_A(x), x \in U$. We have other set functions corresponding to $\mathcal{F}_k$-reducts (see [8,9]) but we consider only μ_k^{re} and μ_k^{sp}.

The next example shows that the attribute importance is very different by k and the meaning of set function.

Example 1. Consider a decision table shown in Table 1. The decision table is given in a profile-wise way. There are four decision attribute values $1, 2, 3$ and 4. In column d of Table 1, a frequency distribution of objects sharing a common profile (condition attribute values) is given. For example, $(1, 0, 1, 0)$ on row w_1 implies that there are two objects taking 1 for a_1, 1 for a_2, 1 for a_3 and 1 for a_4 and one of them takes 1 for d while the other takes 3 for d. Similarly, $(2, 0, 0, 0)$ on row w_3 implies that there are two objects taking 2 for a_1, 2 for a_2, 3 for a_3, 1 for a_4 and 1 for d. We obtain $\{a_2\}$ and $\{a_1, a_3\}$ as $\mathcal{F}_1$-reducts, $\{a_1, a_3\}$ and $\{a_1, a_2, a_4\}$ as $\mathcal{F}_2$-reducts, and $\{a_1, a_2, a_4\}$ and $\{a_1, a_3, a_4\}$ as $\mathcal{F}_3$-reducts. We have $\{a_1\}$ as $\mathcal{F}_2$-core and $\{a_1, a_4\}$ as $\mathcal{F}_3$-core. We have no $\mathcal{F}_1$-core, i.e., $\mathcal{F}_1$-core is the empty set. Measures μ_k^{re} and μ_k^{sp} as well as their Shapley interaction

Table 1. A decision table

Profile	a_1	a_2	a_3	a_4	d	Profile	a_1	a_2	a_3	a_4	d	Profile	a_1	a_2	a_3	a_4	d
w_1	1	1	1	1	$(1,0,1,0)$	w_4	3	3	4	2	$(0,1,0,0)$	w_7	2	4	4	2	$(0,0,2,0)$
w_2	1	1	2	2	$(1,0,0,1)$	w_5	4	1	2	2	$(0,1,1,0)$	w_8	4	1	5	5	$(0,1,1,1)$
w_3	2	2	3	1	$(2,0,0,0)$	w_6	2	5	3	2	$(1,0,0,0)$	w_9	4	1	5	4	$(1,0,1,1)$

Table 2. Measures, Shapley interaction indices and Harsanyi dividends ($\beta = \mathrm{re}, \mathrm{sp}$)

A	μ_1^β	$I^S_{\mu_1^\beta}$	$m_{\mu_1^\beta}$	μ_2^β	$I^S_{\mu_2^\beta}$	$m_{\mu_2^\beta}$	μ_3^β	$I^S_{\mu_3^\beta}$	$m_{\mu_3^\beta}$
a_1	0	0.1667	0	0	0.5833	0	0	0.4167	0
a_2	1	0.6667	1	0	0.0833	0	0	0.0833	0
a_3	0	0.1667	0	0	0.25	0	0	0.0833	0
a_4	0	0	0	0	0.0833	0	0	0.4167	0
a_1a_2	1	−0.5	0	0	0.1667	0	0	0.1667	0
a_1a_3	1	0.5	1	1	0.6667	1	0	0.1667	0
a_1a_4	0	0	0	0	0.1667	0	0	0.6667	0
a_2a_3	1	−0.5	0	0	−0.3333	0	0	−0.3333	0
a_2a_4	1	0	0	0	0.1667	0	0	0.1667	0
a_3a_4	0	0	0	0	−0.3333	0	0	0.1667	0
$a_1a_2a_3$	1	−1	−1	1	−0.5	0	0	−0.5	0
$a_1a_2a_4$	1	0	0	1	0.5	1	1	0.5	1
$a_1a_3a_4$	1	0	0	1	−0.5	0	1	0.5	1
$a_2a_3a_4$	1	0	0	0	−0.5	0	0	−0.5	0
C	1	0	0	1	−1	−1	1	−1	−1
a_1	0.0556	0.0556	0.0556	0.1481	0.1204	0.1481	0.3148	0.2099	0.3148
a_2	0.3333	0.1759	0.3333	0.3333	0.1512	0.3333	0.3333	0.1265	0.3333
a_3	0.1667	0.0926	0.1667	0.2593	0.1265	0.2593	0.3519	0.1481	0.3519
a_4	0	0.0093	0	0.0741	0.0463	0.0741	0.2593	0.1821	0.2593
a_1a_2	0.3333	−0.1296	−0.0556	0.3333	−0.1358	−0.1481	0.4074	−0.1420	−0.2407
a_1a_3	0.3333	0.0370	0.1111	0.4444	0.0123	0.0370	0.5556	−0.0494	−0.1111
a_1a_4	0.1667	0.0370	0.1111	0.3333	0.0494	0.1111	0.6111	0.0247	0.0370
a_2a_3	0.3333	−0.1852	−0.1667	0.3704	−0.1728	−0.2222	0.4074	−0.1605	−0.2778
a_2a_4	0.3333	−0.0185	0.0000	0.3704	−0.0247	−0.0370	0.5185	−0.0309	−0.0741
a_3a_4	0.1667	−0.0185	0	0.2593	−0.0988	−0.0741	0.4630	−0.1420	−0.1481
$a_1a_2a_3$	0.3333	−0.0556	−0.1111	0.4444	0.0556	0.0370	0.5556	0.1667	0.1852
$a_1a_2a_4$	0.3333	−0.0556	−0.1111	0.4444	−0.0185	−0.0370	0.6667	0.0185	0.0370
$a_1a_3a_4$	0.3333	−0.0556	−0.1111	0.4444	−0.0926	−0.1111	0.6667	−0.0556	−0.0370
$a_2a_3a_4$	0.3333	0.0556	0	0.3704	0.0556	0.0370	0.5185	0.0556	0.0741
C	0.3333	0.1111	0.1111	0.4444	0.0370	0.0370	0.6667	−0.0370	−0.0370

indices and Harsanyi dividends are shown in Table 2. The values in the upper part of Table 2 are those when $\beta = \mathrm{re}$ and the values in the lower part of Table 2 are those when $\beta = \mathrm{sp}$.

As shown in Table 2, the attribute importance is very different by k and by the meaning of set function. The detailed descriptions are found in [8,9]. This result shows that we should decide the significance of narrowing down the possible classes for an object and what information is meaningful when the attribute importance is considered. By the setting of information gain the attribute importance is very different.

5 Rule Induction

5.1 The Conventional Approach with Precise Rules

The other major topic in rough set approaches is the minimal rule induction, i.e., inducing rules inferring the membership to D_j with minimal conditions which can differ members of $C_*(D_j)$ from non-members, are investigated well. In this paper, we use minimal rule induction algorithms proposed in the field of rough sets, i.e., LEM2 and MLEM2 algorithms [10]. By those algorithms, we obtain minimal set of rules with minimal conditions which can explain all objects in lower approximations of X under a given decision table. LEM2 algorithm and MLEM2 algorithm [10] are different in their forms of condition parts of rules: by LEM2 algorithm, we obtain rules of the form of "if $f(u, a_1) = v_1, f(u, a_2) = v_2, \ldots$ and $f(u, a_s) = v_s$ then $u \in X$", while by MLEM2 algorithm, we obtain rules of the form of "if $v_1^L \leq f(u, a_1) \leq v_1^R, v_2^L \leq f(u, a_2) \leq v_2^R, \ldots$ and $v_s^L \leq f(u, a_s) \leq v_s^R$ then $u \in X$". Namely, MLEM2 algorithm is a generalized version of LEM2 algorithm to cope with numerical/ordinal condition attributes. For each decision class D_i we induce rules inferring the membership of D_i. Using all those rules, we build a classifier system by applying the idea of LERS [10].

5.2 Classification with Imprecise Rules

In the same way as the induction method for rules about D_i, we can induce rules about a union of D_i's (see [11–13]). Namely, LEM2-based algorithms can be applied to the induction of rules about a union of D_i's (imprecise rules). We note that imprecise rules can be induced when the number of classes is larger than two, i.e., $p > 2$. Moreover, in the same way, we can build a classifier by induced rules about $\bigcup_{j \in \{i_1, i_2, \ldots, i_l\}} D_j$. A rule about a union of l classes in its conclusion is called an l-imprecise rule. The classification of a new object u under rules about unions of D_i's is done by the following procedure:

1. When u matches to at least one of the conditions of the rule, we calculate

$$\hat{S}(D_i) = \sum_{\substack{\text{matching rule } r \\ \text{for } Z \supseteq D_i}} Stren(r) \times Spec(r), \tag{19}$$

where r is called a *matching rule* if the condition part of r is satisfied with u. The strength $Stren(r)$ is the total number of objects in the given dataset correctly classified by rule r. The specificity $Spec(r)$ is the total number of condition attributes in the condition part of rule r. Z is a variable showing a union of classes. For convenience, when there is no matching rules about $Z \supseteq D_i$, we define $\hat{S}(D_i) = 0$. If there exists D_j such that $\hat{S}(D_j) > 0$, the class D_i with the largest $\hat{S}(D_i)$ is selected. If a tie occurs, class D_i with smallest index i is selected from tied classes.

2. When u does not match totally to any rule, for each D_i, we calculate

$$\hat{M}(D_i) = \sum_{\substack{\text{partially matching} \\ \text{rules } r \text{ for } Z \supseteq D_i}} Mat\_f(r) \times Stren(r) \times Spec(r), \qquad (20)$$

where r is called a *partially matching rule* if a part of the premise of r is satisfied. The matching factor $Mat\_f(r)$ is the ratio of the number of matched conditions of rule r to the total number of conditions of rule r. Then the class D_i with the largest $\hat{M}(D_i)$ is selected. If a tie occurs, class D_i with smallest index i is selected from tied classes.

We note that this classification method reduced to the conventional one when Z is a decision class D_i.

Table 3. Classification accuracies of the classifiers with imprecise rules

$A(l)$	No. rules	Accuracy (%)	$A(l)$	No. rules	Accuracy (%)
C(1)	57.22 ± 1.74	98.67 ± 0.97	E(1)	35.89 ± 2.03	75.52 ± 6.21
C(2)	128.02 ± 3.16	$98.96^{**} \pm 0.73$	E(2)	220.67 ± 8.93	$83.20^{**} \pm 5.66$
C(3)	69.55 ± 1.37	$99.68^{**} \pm 0.49$	E(3)	565.67 ± 21.48	$84.66^{**} \pm 5.64$
D(1)	12.09 ± 1.27	92.32 ± 4.42	E(4)	781.36 ± 28.42	$84.87^{**} \pm 5.71$
D(2)	61.32 ± 4.07	$94.58^{**} \pm 3.59$	E(5)	617.06 ± 23.06	$83.74^{**} \pm 6.26$
D(3)	103.58 ± 6.11	$96.03^{**} \pm 3.26$	E(6)	269.27 ± 10.50	$82.56^{**} \pm 6.26$
D(4)	77.28 ± 4.45	$95.58^{**} \pm 3.69$	E(7)	54.09 ± 2.86	78.38 ± 6.70
D(5)	23.84 ± 1.81	$91.87^{*} \pm 4.75$	Z(1)	9.67 ± 0.55	95.84 ± 6.63
G(1)	25.38 ± 1.50	63.34 ± 10.18	Z(2)	48.54 ± 2.10	95.55 ± 7.15
G(2)	111.40 ± 4.33	$72.57^{**} \pm 8.81$	Z(3)	105.37 ± 4.25	$96.74^{*} \pm 5.45$
G(3)	178.35 ± 5.41	$73.44^{**} \pm 9.19$	Z(4)	113.78 ± 3.74	$96.84^{*} \pm 5.22$
G(4)	130.14 ± 4.96	$71.16^{*} \pm 9.91$	Z(5)	66.76 ± 2.69	$97.24^{**} \pm 5.07$
G(5)	39.59 ± 2.18	$65.04^{**} \pm 9.96$	Z(6)	17.72 ± 0.66	96.05 ± 6.51

We examined the classification accuracy of the classifier with imprecise rules for several datasets: car ($|U| = 1728, |C| = 6, |V_d| = 4$), dermatology ($|U| = 358, |C| = 34, |V_d| = 6$), ecoli ($|U| = 336, |C| = 7, |V_d| = 8$), glass ($|U| = 214, |C| = 9, |V_d| = 6$) and zoo ($|U| = 101, |C| = 16, |V_d| = 7$) obtained from UCI machine learning repository [14]. The results are shown in Table 3. In columns of '$A(l)$', the name of dataset and the number l of decision classes to be combined are indicated by the initial letter and by the number in the parentheses, respectively. The data shown in other columns are obtained by 10 times run of 10-fold cross validation. The results of the conventional approach with precise rules are shown in the rows with '(1)' ($l = 1$). Each entry in Table 3 shows the average ave and the standard deviation dev in the form of $ave \pm dev$. Asterisk $*$

and two asterisks ∗∗ implies the significant differences in the paired t-test with significance levels $\alpha = 0.05$ and $\alpha = 0.01$, respectively. As shown in Table 3, the classification accuracy is improved by using imprecise rules. However the number of rules are increased. It can affect the interpretability of results as well as the computational time. The reduction of number of rules is investigated in [12,13].

6   Concluding Remarks

In this paper, we described the rough set approaches to decision tables based on the lower approximations of unions of k decision classes instead of lower approximations of single decision classes. We demonstrated that significantly different results are obtained by the selection of k. In attribute reduction and importance, the selection of k depends on to what extent of imprecision is meaningful/allowable in object classification. On the other hand, in rule induction, k can be selected about $p/2$, because of the classification accuracy of the classifier.

The imprecise rules can be applied to privacy preservation [15] when the publication of rules is requested. The applications and improvements of the proposed approaches as well as the comparison with other multi-class rule mining methods are future topics.

References

1. Pawlak, Z.: Rough sets. Int. J. Comput. Inf. Sci. **11**(5), 341–356 (1982)
2. Pawlak, Z.: Rough Sets: Theoretical Aspects of Reasoning About Data. Kluwer Academic Publishing, Dordrecht (1991)
3. Ślęzak, D.: Various approaches to reasoning with frequency based decision reducts: a survey. In: Polkowski, L., Tsumoto, S., Lin, T.Y. (eds.) Rough Set Methods and Applications, pp. 235–285. Physica, Heidelberg (2000)
4. Inuiguchi, M., Tsurumi, M.: Measures based on upper approximations of rough sets for analysis of attribute importance and interaction. Int. J. Innov. Comput. Inf. Control **2**(1), 1–12 (2006)
5. Pawlak, Z., Skowron, A.: Rough sets and boolean reasoning. Inf. Sci. **177**(1), 41–73 (2007)
6. Ślęzak, D.: On generalized decision functions: reducts, networks and ensembles. In: Yao, Y., Hu, Q., Yu, H., Grzymala-Busse, J.W. (eds.) RSFDGrC 2015. LNCS (LNAI), vol. 9437, pp. 13–23. Springer, Heidelberg (2015). doi:10.1007/ 978-3-319-25783-9_2
7. Skowron, A., Rauser, C.M.: The discernibility matrix and function in information systems. In: Słowiński, R. (ed.) Intelligent Decision Support: Handbook of Application and Advances of Rough Set Theory. Theory and Decision Library, vol. 11, pp. 331–362. Kluwer Academic Publishers, Dordrecht (1992)
8. Inuiguchi, M.: Attribute importance degrees corresponding to several kinds of attribute reduction in the setting of the classical rough sets. In: Torra, V., Dahlbom, A., Narukawa, Y. (eds.) Fuzzy Sets, Rough Sets, Multisets and Clustering. Springer (in press)
9. Inuiguchi, M.: Variety of rough set based attribute importance. In: USB Proceedings of SCIS-ISIS, pp. 548–551 (2016)

10. Grzymala-Busse, J.W.: MLEM2 - Discretization during rule induction. In: Kłopotek, M.A., Wierzchoń, S.T., Trojanowski, K. (eds.) IIPWM 2003. AISC, vol. 22. Springer, Heidelberg (2003)
11. Inuiguchi, M., Hamakawa, T.: The utilities of imprecise rules and redundant rules for classifiers. In: Huynh, V.N., Denoeux, T., Tran, D.H., Le, A.C., Pham, S.B. (eds.) KSE 2013. AISC, vol. 245, pp. 45–56. Springer, Heidelberg (2014). doi:10. 1007/978-3-319-02821-7_6
12. Hamakawa, T., Inuiguchi, M.: On the utility of imprecise rules induced by MLEM2 in classification. In: Kudo Y., Tsumoto, S. (eds.) Proceedings of 2014 IEEE International Conference on Granular Computing (GrC), pp. 76–81 (2014)
13. Inuiguchi, M., Hamakawa, T., Ubukata, S.: Utilization of imprecise rules induced by MLEM2 algorithm. In: Proceedings of the 10th Workshop on Uncertainty Processing (WUPES 2015), pp. 73–84 (2015)
14. UCI Machine Learning Repository. http://archive.ics.uci.edu/ml/
15. Inuiguchi, M., Hamakawa, T., Ubukata, S.: Imprecise rules for data privacy. In: Ciucci, D., Wang, G., Mitra, S., Wu, W.-Z. (eds.) RSKT 2015. LNCS (LNAI), vol. 9436, pp. 129–139. Springer, Heidelberg (2015). doi:10.1007/978-3-319-25754-9_12

Rule Set Complexity for Incomplete Data Sets with Many Attribute-Concept Values and "Do Not Care" Conditions

Patrick G. Clark[1], Cheng Gao[1], and Jerzy W. Grzymala-Busse[1,2(⊠)]

[1] Department of Electrical Engineering and Computer Science,
University of Kansas, Lawrence, KS 66045, USA
patrick.g.clark@gmail.com, {cheng.gao,jerzy}@ku.edu
[2] Department of Expert Systems and Artificial Intelligence,
University of Information Technology and Management, 35-225 Rzeszow, Poland

Abstract. In this paper we present results of novel experiments conducted on 12 data sets with many missing attribute values interpreted as attribute-concept values and "do not care" conditions. In our experiments complexity of rule sets, in terms of the number of rules and the total number of conditions induced from such data, are evaluated. The simpler rule sets are considered better. Our first objective was to check which interpretation of missing attribute values should be used to induce simpler rule sets. There is some evidence that the "do not care" conditions are better. Our secondary objective was to test which of the three probabilistic approximations: singleton, subset or concept, used for rule induction should be used to induce simpler rule sets. The best choice is the subset probabilistic approximation and the singleton probabilistic approximation is the worst choice.

Keywords: Incomplete data · Attribute-concept values · "Do not care" conditions · Probabilistic approximations · MLEM2 rule induction algorithm

1 Introduction

In this paper data sets with missing attribute values are mined using probabilistic approximations. The probabilistic approximation, with a probability α, is an extension of a standard approximation, a basic idea of rough set theory. If $\alpha = 1$, the probabilistic approximation becomes the lower approximation, for very small and positive α, the probabilistic approximation is identical with the upper approximation. The idea of the probabilistic approximation was introduced in [20] and further developed in [19, 22–24].

Data sets with missing attribute values need special kinds of approximations, called singleton, subset and concept [12, 13]. Such approximations were generalized to singleton, subset and concept probabilistic approximations in [15]. The

© Springer International Publishing AG 2016
V. Flores et al. (Eds.): IJCRS 2016, LNAI 9920, pp. 65–74, 2016.
DOI: 10.1007/978-3-319-47160-0_6

first experiments on probabilistic approximations were presented in [1]. In experiments reported in this paper, we used all three kinds of probabilistic approximations: singleton, subset and concept.

In this paper, missing attribute values may be interpreted in two different ways, as attribute-concept values or as "do not care" conditions. Attribute-concept values, introduced in [14], are typical values for a given concept. For example, if the concept is a set of people sick with flu, and a value of the attribute *Temperature* is missing for some person who is sick with flu, using this interpretation, we would consider typical values of *Temperature* for other people sick with flu, such as *high* and *very high*. A "do not care" condition is interpreted as if the original attribute value was irrelevant, we may replace it by any existing attribute value [8,17,21].

The first experiments on data sets with missing attribute values interpreted as lost values and "do not care" conditions, with 35 % of missing attribute values, were reported in [7]. Research on data with missing attribute values interpreted as attribute-concept values and "do not care" conditions was presented in [2–6]. In [6] two imputation methods for missing attribute values were compared with rough-set approaches based on two interpretations of missing attribute values, as lost values and "do not care" conditions, combined with using singleton, subset and concept probabilistic approximations. It was shown that the rough-set approaches were better than imputation for five out of six data sets. The smallest error rate was associated with data sets with lost values. In [3] experiments were related to the error rate computed by ten-fold cross validation for mining data sets with attribute-concept values and "do not care" conditions using only three probabilistic approximations: lower, middle (with $\alpha = 0.5$) and upper. Results were not conclusive, in four cases attribute-concept values were better, while in two cases "do not care" conditions were better, in remaining 18 cases differences between the two were statistically insignificant. In [4] the error rate was evaluated for data sets with many missing attribute-concept values and "do not care" conditions. In two cases "do not care" conditions were better, in one case attribute-concept values were better, in remaining three cases differences were statistically insignificant.

With inconclusive results of experiments on the error rate, the question is which interpretation of missing attribute values is associated with smaller complexity of rule sets. In [2], experiments on complexity of rule sets induced from data sets with attribute-concept values and "do not care" conditions using lower, middle and upper approximations were presented. For half of the cases the number of rules was smaller for attribute-concept values, similarly for the total number of rule conditions. Results on the choice of the best type of probabilistic approximation (singleton, subset or concept) were inconclusive. In [5] experiments were also focused on complexity of rules sets, this time for data sets with 35 % of attribute-concept values and "do not care" conditions. For 13 combinations (out of 24) the attribute-concept values were associated with simpler rules, for five combinations "do not care" conditions were better, similarly for the total number of rule conditions.

The difference in performance between the two interpretations of missing attribute values, as attribute-concept values or "do not care" conditions, is more clear for data sets with many missing attribute values. Results of this paper are more conclusive than in our previous research.

Thus, our first objective was to check which interpretation of missing attribute values should be used to induce simpler rule sets, in terms of the number of rules and total number of rule conditions, from data sets with many attribute-concept values and "do not care" conditions, using the Modified Learning from Examples Module version 2 (MLEM2) system for rule induction [11]. There is some evidence that the "do not care" conditions are better. Our secondary objective was to test which of the three probabilistic approximations: singleton, subset or concept, used for rule induction should be used to induce simpler rule sets. The best choice is the subset probabilistic approximation and the singleton probabilistic approximation is the worst choice.

2   Incomplete Data

In this paper the input data sets are in the form of a *decision table*. A decision table has rows representing *cases* and columns defining *variables* with the set of all cases denoted by U. The dependent variable d is called the *decision* and the independent variables are labeled *attributes*. The set of all attributes will be denoted by A. Additionally, the value for a specific case x and attribute a is denoted by $a(x)$.

There are multiple ways to represent missing attribute values, however in this paper we distinguish them with two interpretations. The first, attribute-concept values, are identified using $-$ and the second, denoted by $*$ are "do not care" conditions.

One of the most important ideas of rough set theory [18] is an indiscernibility relation, defined for complete data sets. Let B be a nonempty subset of A. The indiscernibility relation $R(B)$ is a relation on U defined for $x, y \in U$ as follows:

$$(x, y) \in R(B) \text{ if and only if } \forall a \in B \ (a(x) = a(y)).$$

The indiscernibility relation $R(B)$ is an equivalence relation. Equivalence classes of $R(B)$ are called *elementary sets* of B and are denoted by $[x]_B$. A subset of U is called *B-definable* if it is a union of elementary sets of B.

The set X of all cases defined by the same value of the decision d is called a *concept*. The largest B-definable set contained in X is called the *B-lower approximation* of X, denoted by $\underline{appr}_B(X)$, and defined as follows

$$\cup\{[x]_B \mid [x]_B \subseteq X\},$$

while the smallest B-definable set containing X, denoted by $\overline{appr}_B(X)$ is called the *B-upper approximation* of X, and is defined as follows

$$\cup\{[x]_B \mid [x]_B \cap X \neq \emptyset\}.$$

For a variable a and its value v, (a, v) is called a variable-value pair. When considering a complete data set, the *block* of (a, v), denoted by $[(a, v)]$, is the set $\{x \in U \mid a(x) = v\}$ [9]. However, when representing missing information and incomplete data sets, the definition of a block of an attribute-value pair is modified in the following way.

- For an attribute a, where there exists a case x such that $a(x) = -$, the case x should be included in blocks $[(a, v)]$ for all specified values $v \in V(x, a)$ of attribute a, where

$$V(x, a) = \{a(y) \mid a(y) \text{ is specified, } y \in U, \ d(y) = d(x)\},$$

- For an attribute a, where there exists a case x such that $a(x) = *$, the case x should be included in blocks $[(a, v)]$ for all specified values v of the attribute a.

For a case $x \in U$ and $B \subseteq A$, the *characteristic set* $K_B(x)$ is defined as the intersection of the sets $K(x, a)$, for all $a \in B$, where the set $K(x, a)$ is defined in the following way.

- If $a(x)$ is specified, then $K(x, a)$ is the block $[(a, a(x))]$ of attribute a and its value $a(x)$,
- If $a(x) = -$, then the corresponding set $K(x, a)$ is equal to the union of all blocks of attribute-value pairs (a, v), where $v \in V(x, a)$ if $V(x, a)$ is nonempty. If $V(x, a)$ is empty, $K(x, a) = U$,
- If $a(x) = *$, then the set $K(x, a) = U$, where U is the set of all cases.

3   Lower and Upper Approximations

We quote some definitions from [16]. Let X be a subset of U and let B be a subset of the set A of all attributes. The B-*singleton lower approximation* of X, denoted by $\underline{appr}_B^{singleton}(X)$, is defined as follows

$$\{x \mid x \in U, K_B(x) \subseteq X\}.$$

The B-*singleton upper approximation* of X, denoted by $\overline{appr}_B^{singleton}(X)$, is defined as follows

$$\{x \mid x \in U, K_B(x) \cap X \neq \emptyset\}.$$

The B-*subset lower approximation* of X, denoted by $\underline{appr}_B^{subset}(X)$, is defined as follows

$$\cup \{K_B(x) \mid x \in U, K_B(x) \subseteq X\}.$$

The B-*subset upper approximation* of X, denoted by $\overline{appr}_B^{subset}(X)$, is defined as follows

$$\cup \{K_B(x) \mid x \in U, K_B(x) \cap X \neq \emptyset\}.$$

The B-*concept lower approximation* of X, denoted by $\underline{appr}_B^{concept}(X)$, is defined as follows

$$\cup \{K_B(x) \mid x \in X, K_B(x) \subseteq X\}.$$

The B-*concept upper approximation* of X, denoted by $\overline{appr}_B^{concept}(X)$, is defined as follows

$$\cup\{K_B(x) \mid x \in X, K_B(x) \cap X \neq \emptyset\} = \cup\{K_B(x) \mid x \in X\}.$$

4 Probabilistic Approximations

The B-singleton probabilistic approximation of X with the threshold α, $0 < \alpha \leq 1$, denoted by $appr_{\alpha,B}^{singleton}(X)$, is defined as follows

$$\{x \mid x \in U,\ Pr(X|K_B(x)) \geq \alpha\},$$

where $Pr(X|K_B(x)) = \frac{|X \cap K_B(x)|}{|K_B(x)|}$ is the conditional probability of X given $K_B(x)$.

A B-subset probabilistic approximation of the set X with the threshold α, $0 < \alpha \leq 1$, denoted by $appr_{\alpha,B}^{subset}(X)$, is defined as follows

$$\cup\{K_B(x) \mid x \in U,\ Pr(X|K_B(x)) \geq \alpha\}.$$

A B-concept probabilistic approximation of the set X with the threshold α, $0 < \alpha \leq 1$, denoted by $appr_{\alpha,B}^{concept}(X)$, is defined as follows

$$\cup\{K_B(x) \mid x \in X,\ Pr(X|K_B(x)) \geq \alpha\}.$$

In general, all three probabilistic approximations are distinct, even for the same value of the parameter α. Additionally, if for a given set X a probabilistic approximation $appr_\beta(X)$ is not listed, then $appr_\beta(X)$ is equal to the closest probabilistic approximation $appr_\alpha(X)$ of the same type with α larger than or equal to β.

If a characteristic relation $R(B)$ is an equivalence relation, all three types of probabilistic approximation: singleton, subset and concept are reduced to the same probabilistic approximation.

5 Experiments

Our experimental data sets are based on six data sets available from the University of California at Irvine *Machine Learning Repository*. Basic information about these data sets are presented in Table 1.

Incomplete data sets were produced from the base data by creating a set of templates. To create the templates, existing specified attribute values are replaced at 5 % increments with a corresponding *attribute-concept* value. So the

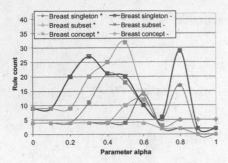

Fig. 1. Number of rules for the *breast cancer* data set

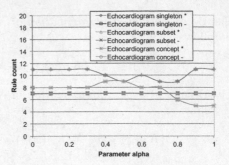

Fig. 2. Number of rules for the *echocardiogram* data set

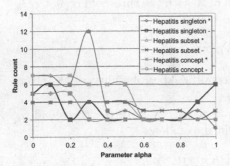

Fig. 3. Number of rules for the *hepatitis* data set

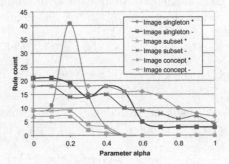

Fig. 4. Number of rules for the *image segmentation* data set

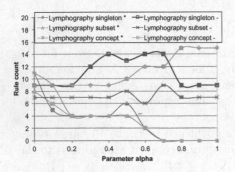

Fig. 5. Number of rules for the *lymphography* data set

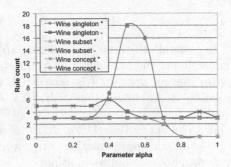

Fig. 6. Number of rules for the *wine recognition* data set

template creation begins with no missing values, then 5 % of the values are randomly replaced with *attribute-concept* values, then an additional 5 % are randomly replaced. The process continues with the data set until at least one row of the decision table attribute values are all missing values. Three attempts were made to randomly replace specified values with missing values where either a

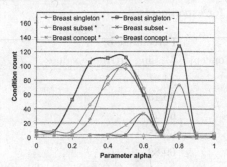

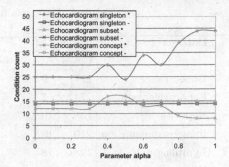

Fig. 7. Total number of conditions for the *breast cancer* data set

Fig. 8. Total number of conditions for the *echocardiogram* data set

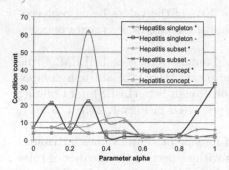

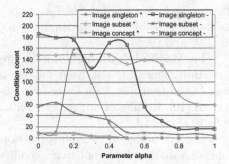

Fig. 9. Total number of conditions for the *hepatitis* data set

Fig. 10. Total number of conditions for the *image segmentation* data set

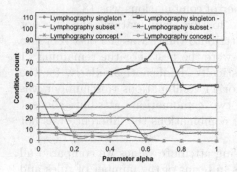

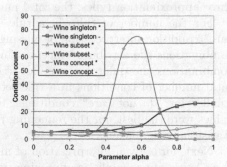

Fig. 11. Total number of conditions for the *lymphography* data set

Fig. 12. Total number of conditions for the *wine recognition* data set

new data set with an extra 5 % is created or the process stops. To produce the "do not care" condition data sets, the same templates are used, replacing − with ∗.

In this paper, data sets with many missing attribute values are studied. We chose the maximum number of missing values that could be synthesized and for

Table 1. Data sets used for experiments

Data set	Number of			Percentage of
	Cases	Attributes	Concepts	Missing attribute values
Breast cancer	277	9	2	44.81
Echocardiogram	74	7	2	40.15
Hepatitis	155	19	2	60.27
Image segmentation	210	19	7	69.85
Lymphography	148	18	4	69.89
Wine recognition	178	13	3	64.65

this research, has been defined as more than 40 % of the values being replaced. As shown in Table 1, the maximum percentage of missing values ranges between 40.15 % and 69.89 %.

The Modified Learning from Examples Module version 2 (MLEM2) rule induction algorithm was used for our experiments [11]. MLEM2 is a component of the Learning from Examples based on Rough Sets (LERS) data mining system [10]. Results of our experiments are presented in Figs. 1, 2, 3, 4, 5, 6, 7, 8, 9, 10, 11 and 12.

First we compared two interpretations of missing attribute values, attribute-concept values and "do not care" conditions with respect to the number of rules in a rule set. For every data set type, separately for singleton, subset and concept probabilistic approximations, the Wilcoxon matched-pairs signed rank test was used with a 5 % level of significance two-tailed test. With six data set types and three approximation types, the total number of combinations was 18.

For the number of rules in a rule set, for five combinations the "do not care" condition interpretation of missing attribute values was the best. For two combinations the attribute-concept values were the best. For the remaining 11 combinations the difference was not statistically significant. Similarly, for the total number of conditions in a rule set, for 11 combinations this number was smaller for "do not care" conditions, for two combinations attribute-concept values were the best, for the remaining five combinations the difference was not statistically significant.

Next, for a given interpretation of missing attribute values we compared all three types of probabilistic approximations in terms of the number of rules and the total number of conditions in a rule set using multiple comparisons based on Friedman's nonparametric test. Here, with six types of data sets and two interpretations of missing attribute values, the total number of combinations was 12. For the number of rules, the smallest number was associated with the subset probabilistic approximations for three combinations, with one tie between subset and concept probabilistic approximations. For remaining combinations the difference was not statistically significant. The singleton probabilistic approximation was never a winner. For the total number of rule conditions, the smallest number was also associated with the subset probabilistic approximations for six combinations, with one tie between subset and concept probabilistic approximations. For remaining combinations the difference was not statistically significant. Again, the singleton probabilistic approximation was never a winner.

6   Conclusions

As follows from our experiments, there is some evidence that the number of rules and the total number of conditions are smaller for "do not care" conditions than for attribute-concept values. Additionally, the best probabilistic approximation that should be used for rule induction from data with many attribute-concept values and "do not care" conditions is the subset probabilistic approximation. On the other hand, the singleton probabilistic approximation is the worst.

References

1. Clark, P.G., Grzymala-Busse, J.W.: Experiments on probabilistic approximations. In: Proceedings of the 2011 IEEE International Conference on Granular Computing, pp. 144–149 (2011)
2. Clark, P.G., Grzymala-Busse, J.W.: Complexity of rule sets induced from incomplete data sets with attribute-concept values and and "do not care" conditions. In: Proceedings of the Third International Conference on Data Management Technologies and Applications, pp. 56–63 (2014)
3. Clark, P.G., Grzymala-Busse, J.W.: Mining incomplete data with attribute-concept values and "do not care" conditions. In: Polycarpou, M., de Carvalho, A.C.P.L.F., Pan, J.-S., Woźniak, M., Quintian, H., Corchado, E. (eds.) HAIS 2014. LNCS (LNAI), vol. 8480, pp. 156–167. Springer, Heidelberg (2014). doi:10.1007/978-3-319-07617-1_14
4. Clark, P.G., Grzymala-Busse, J.W.: Mining incomplete data with many attribute-concept values and do not care conditions. In: Proceedings of the IEEE International Conference on Big Data, pp. 1597–1602 (2015)
5. Clark, P.G., Grzymala-Busse, J.W.: On the number of rules and conditions in mining data with attribute-concept values and "do not care" conditions. In: Kryszkiewicz, M., Bandyopadhyay, S., Rybinski, H., Pal, S.K. (eds.) PReMI 2015. LNCS, vol. 9124, pp. 13–22. Springer, Heidelberg (2015). doi:10.1007/978-3-319-19941-2_2
6. Clark, P.G., Grzymala-Busse, J.W., Kuehnhausen, M.: Mining incomplete data with many missing attribute values. a comparison of probabilistic and rough set approaches. In: Proceedings of the Second International Conference on Intelligent Systems and Applications, pp. 12–17 (2013)
7. Clark, P.G., Grzymala-Busse, J.W., Rzasa, W.: Mining incomplete data with singleton, subset and concept approximations. Inf. Sci. **280**, 368–384 (2014)
8. Grzymala-Busse, J.W.: On the unknown attribute values in learning from examples. In: Proceedings of the 6th International Symposium on Methodologies for Intelligent Systems, pp. 368–377 (1991)
9. Grzymala-Busse, J.W.: LERS-a system for learning from examples based on rough sets. In: Slowinski, R. (ed.) Intelligent Decision Support. Handbook of Applications and Advances of the Rough Set Theory, pp. 3–18. Kluwer Academic Publishers, Dordrecht (1992)

10. Grzymala-Busse, J.W.: A new version of the rule induction system LERS. Fundamenta Informaticae **31**, 27–39 (1997)
11. Grzymala-Busse, J.W.: MLEM2: a new algorithm for rule induction from imperfect data. In: Proceedings of the 9th International Conference on Information Processing and Management of Uncertainty in Knowledge-Based Systems, pp. 243–250 (2002)
12. Grzymala-Busse, J.W.: Rough set strategies to data with missing attribute values. In: Notes of the Workshop on Foundations and New Directions of Data Mining, in Conjunction with the Third International Conference on Data Mining, pp. 56–63 (2003)
13. Grzymala-Busse, J.W.: Data with missing attribute values: generalization of indiscernibility relation and rule induction. Trans. Rough Sets **1**, 78–95 (2004)
14. Grzymala-Busse, J.W.: Three approaches to missing attribute values—a rough set perspective. In: Proceedings of the Workshop on Foundation of Data Mining, in Conjunction with the Fourth IEEE International Conference on Data Mining, pp. 55–62 (2004)
15. Grzymala-Busse, J.W.: Generalized parameterized approximations. In: Yao, J.T., Ramanna, S., Wang, G., Suraj, Z. (eds.) RSKT 2011. LNCS (LNAI), vol. 6954, pp. 136–145. Springer, Heidelberg (2011). doi:10.1007/978-3-642-24425-4_20
16. Grzymala-Busse, J.W., Rzasa, W.: Definability and other properties of approximations for generalized indiscernibility relations. Trans. Rough Sets **11**, 14–39 (2010)
17. Kryszkiewicz, M.: Rules in incomplete information systems. Inf. Sci. **113**(3–4), 271–292 (1999)
18. Pawlak, Z.: Rough sets. Int. J. Comput. Inform. Sci. **11**, 341–356 (1982)
19. Pawlak, Z., Skowron, A.: Rough sets: some extensions. Inf. Sci. **177**, 28–40 (2007)
20. Pawlak, Z., Wong, S.K.M., Ziarko, W.: Rough sets: probabilistic versus deterministic approach. Int. J. Man Mach. Stud. **29**, 81–95 (1988)
21. Stefanowski, J., Tsoukias, A.: Incomplete information tables and rough classification. Comput. Intell. **17**(3), 545–566 (2001)
22. Yao, Y.Y.: Probabilistic rough set approximations. Int. J. Approximate Reasoning **49**, 255–271 (2008)
23. Yao, Y.Y., Wong, S.K.M.: A decision theoretic framework for approximate concepts. Int. J. Man Mach. Stud. **37**, 793–809 (1992)
24. Ziarko, W.: Probabilistic approach to rough sets. Int. J. Approximate Reasoning **49**, 272–284 (2008)

Rough Sets, Approximation and Granulation

Rough Sets Are ε-Accessible

Mohamed Quafafou[⊠]

Aix-Marseille University, Marseille, France
mohamed.quafafou@univ-amu.fr
http://www.quafafou.com

Abstract. This paper focuses on the relationship between perceptions and sets considering that perceptions are not only imprecise or doubtful, but they are also multiple. Accessible sets are developed according to this view, where sets representation is a central problem depending not only on features of its objects, but also on their perceptions. The accessibility notion is related to the perception and can be summarized as follows "to be accessible is to be perceived", which is more weak than the Berkeley's idealism. In this context, we revisit Rough sets showing that: (1) the Pawlak's perception of sets can be written using only two perceivers, which are respectively pessimistic and optimistic, and (2) Rough sets are ε-accessible. Moreover, we introduce a rough set computational theory of perception, denoted π-RST and discuss the perception dynamic problem laying its foundation on social interaction between perceivers, granularity and preference.

Keywords: Rough sets · Perception · Accessibility · Hypergraph · Preference

1 Introduction

We consider here the classical set theory, denoted ZF Theory, which is defined by Georges Cantor and axiomatized by Zermelo and Fraenkel [8]. Let U be a set of objects called the universe and a fundamental binary relation, denoted $\in$, which is defined between an object x and a set $X \subset U$, where $x \in X$ expresses that x is a member (or element) of X. Thus, the characteristic (membership) function of a set X, denoted 1_X, can take on only two values 0 and 1, and consequently, $1_X(x) = 1$ or 0 according as x does or does not belong to X. However, several classes of objects encountred in the real world reveal the fallacy of this assumption because such objects have not precise criteria. Hence the need to replace the boolean membership with a continum of grades of membership [1]. Using fuzzy sets, L.A. Zadeh has introduced, in his paper [2], a computational theory of perception considering that perceptions are intrinsically imprecise and stressed the need of "a methodology in which the objects of computation are perceptions - perceptions of time, distance, form, direction, color, shape, truth, likelihood, intent, and other attributes of physical and mental objects". More recently, Z. Pawlak introduces Rough sets to express vagueness based on sets

© Springer International Publishing AG 2016
V. Flores et al. (Eds.): IJCRS 2016, LNAI 9920, pp. 77–86, 2016.
DOI: 10.1007/978-3-319-47160-0_7

boundary regions [3,4]. Next, a methodology of Perception Based Computing (PBC) was introduced by A. Jankowski and A. Skowron in [5,6] considering a rough-granular point of view. We continue this effort making a bridge between perception and set theories, introducing Accessible sets, where the accessibility is related to the perception and can be summarized as follows "to be accessible is to be perceived". This perception is more weak than Berkeley's idealism, where objects are nothing more than our experiences of them, i.e. "to be is to be perceived". The epistemology is the study of the theory of knowledge. It is among the most important areas of philosophy and addresses multiple questions including the following: What is knowledge?, From where do we get our knowledge?, How are our beliefs justified?, etc. In this paper, we mainly focus on the issue of "How do we perceive the world around us?". According to perceptions theories, the basic view is known as Naive realism, where we directly perceive the world as it is. The perception [7] is a passive process as we simply receive information about the world through our senses: "objects have the properties that they appear to us to have. An alternative view, developed by John Locke, is the Representative realism, where we are actively involved in perception. In fact, there are Primary Qualities, which objects have independent of any observer, and Secondary Qualities, which objects only have because they are perceived. Finally, the Idealism defended by George Berkeley who is persuaded by the thought that we have direct access only to our experiences of the world, and not to the world itself: to be is to be perceived. Thus, objects are nothing more than our experiences of them and God is constantly perceiving everything. Is there any relation between sets and perceptions theories ? At least both involve objects of the universe and our primary goal is to bridge these two research fields to formally define a class of sets depending on their perception. To achieve this goal, *we do not consider that perceptions are only imprecise and vague, but they are also multiple.*

Throughout this paper, we introduce basic notions defining the perception of sets and how to compute with perceptions using the hypergraph theory. We give a brief introduction to accessible set theory and underline its connection with the epistemology and more particularly with perception theories. Next, we revisit Rough sets showing that the Pawlak's perception of sets can be written using only two perceivers, which are respectively optimistic and optimistic. In this context, rough sets are ε-accessible. After that, we introduce a rough set computational theory of perception, denoted π-RST and discuss the perception dynamic problems by founding it on social interaction between perceivers.

2   Accessible Sets: A Brief Introduction

2.1   Basic Notions

Let U be the universe of objects, $I \subseteq \mathbb{N}$ a set of perceivers, and f_i the elementary perception function of the perceiver i that permits him to perceive a set $X \subseteq U$ as $f_i(X) \in 2^U$. Consequently, the perceiver i represents, or perceives, a set X as $f_i(X)$ and the image of f_i, denoted $<_i U$, depends on its nature and can be equal to 2^U. However, it is not always the case, especially, when the perception function is imprecise, vague, etc. The universe of perceived objects and concepts, considering the set I of perceivers, denoted $<U$ is equal to $\{f_i(X) : i \in I, X \in 2^U\}$. On the basis the elementary perception, we introduce a ternary relation depending on objects and subsets of the universe U, and their perceivers.

Definition 1 (Ternary relation $\in_i$). *The perceiver i perceives that $x \in U$ is a member of the set $X \in 2^U$ if,*

$$x \in_i X \Leftrightarrow x \in f_i(X). \tag{1}$$

where, $x \in_i X$ means "x is perceived, by the perceiver i, to be a member of X". Thus, a given set X is defined by several characteristic functions, hence twice as many perceivers: $\chi_X : U \times I \to \{0, 1\}$, where

$$\chi_X(x, i) = \begin{cases} 1 & \text{if} \quad x \in_i X \\ 0 & \text{else} \end{cases}$$

As we have said before, we assume that the perception is multiple and consequently any concept is represented by a family of sets rather than only one. This perception multiplicity is defined by the function $\mathbb{Q}_I$ as follows: $\mathbb{Q}_I(X) = (f_i(X))_{i \in I}$, where $X \in 2^U$. In this particular context, the perception space is the pair (U, I), where each perceiver i has his own perception function f_i and his knowledge base K_i representing his knowledge, preferences, interests, etc. The main question is "to be perceived as", instead of "to be", a member of a set. The accessible notion depends on the nature of perceptions.

Definition 2 (Accessible set). *Let I be a set of perceivers, we say that $X \in 2^U$ is accessible, denoted $\mathbb{Q}_I \Vdash X$, iff:*

$$\forall i \in I, f_i(X) = X. \tag{2}$$

In the case where a set X is not accessible, it may be partially accessible, ϵ-Accessible, or not accessible at all.

Definition 3 (ϵ-Accessible Set). *Let U be the universe of objects, I a set of perceivers, $X \in 2^U$ is said to be ϵ-Accessible iff:*

$$\exists J \subset I, \forall i \in J, f_i(X) = X; \tag{3}$$

where $\epsilon = \frac{|J|}{|I|} \in [0, 1[$.

2.2 Perceptions and Perceivers

In this section, we outline some of the features in order to go beyond the single elementary perception.

Definition 4 (Core and Support). *The perception function of a given set X is characterized by its Core $\kappa(X) = \cap\{f_i(X) : i \in I\}$ and its support $\sigma(X) = \cup\{f_i(X) : i \in I\}$.*

Unlike elementary perceptions, shared perceptions are alternative representations of a set X taking into account its perception by the different perceivers.

Definition 5 (Minimal shared perception). *Let U be the universe of objects, $X \subset U$, I the index set of perceivers, f_i the elementary perception of the perceiver i and $\mathbb{Q}_I(X) = (f_i(X))_{i \in I}$ is the perception of X. The set of minimal shared perception of X, denoted $\widehat{\mathbb{Q}}_I(X)$, is defined as follows:*

$$\widehat{X} \in \widehat{\mathbb{Q}}_I(X) \Leftrightarrow (\forall i \in I, \widehat{X} \cap f_i(X) \neq \emptyset) \wedge (\forall Y \subset \widehat{X}, \exists i \in I, Y \cap f_i(X) = \emptyset) . \quad (4)$$

The set $\widehat{X}$ is an elementary shared perception and $\widehat{\mathbb{Q}}_I(X)$ is an antichain in the lattice $(2^{<U}, \subset)$.

Definition 6 (Consistent shared perceptions). *The space of consistent shared perceptions considering the set of perceivers I, is the interval $[\widehat{\mathbb{Q}}_I(X), \cup \{Y \in \widehat{\mathbb{Q}}_I(X)\}]$*

This interval define the space of consistent perceptions considering all perceivers. We distinguish the following five main categories of perceivers according how they perceive all the universe.

Definition 7 (Perceivers categories).

- *Perfect $\mathbb{Q}$: $f_i \in \mathbb{Q} \Leftrightarrow \forall X \in 2^U, f_i(X) = X$*
- *Pessimistic $\mathbb{Q}^{\downarrow}$: $f_i \in \mathbb{Q}^{\downarrow} \Leftrightarrow \forall X \in 2^U, f_i(X) \subset X$*
- *Optimistic $\mathbb{Q}^{\uparrow}$: $f_i \in \mathbb{Q}^{\uparrow} \Leftrightarrow \forall X \in 2^U, X \subset f_i(X)$*
- *Partial $\mathbb{Q}^{\neq}$: $f_i \in \mathbb{Q}^{\neq} \Leftrightarrow \forall X \in 2^U, f_i(X) \cap X \neq \emptyset$ and $f_i(X) \cap -X \neq \emptyset$*
- *Ignorant $\mathbb{Q}^{\neg}$: $f_i \in \mathbb{Q}^{\neg} \Leftrightarrow \forall X \in 2^U, f_i(X) \cap X = \emptyset$*

All these perceptions are more or less consistent with the perceived set X, except the Ignorant perception class as $f_i(X) \cap X = \emptyset$. Thus, the ignorant perceiver is actually an erroneous perceiver and what he perceives is wrong.

2.3   Computing with Perceptions

In this section, we represent the perception of a set X by the hypergraph $\mathcal{H}_I(X) = (\mathcal{V}_I(X), \mathcal{E}_I(X))$, where the set of its nodes is $\mathcal{V}_I(X) = \cup_{i \in I}\{f_i(X)\}$ and $\mathcal{E}_I(X) = \{f_i(X) : i \in I\}$ is the set of its hyperedges.

Proposition 1 (Minimal shared perception). *Let $X \subset U$ a set of objects, I a finite subset of $\mathbb{N}$ and $F = \{f_i : i \in I\}$ a set of perceivers. The perception function of X, i.e. $\mathbb{Q}_I(X)$ is represented by the hypergraph $\mathcal{H}_I(X) = (\mathcal{V}_I(X), \mathcal{E}_I(X))$, than the set of its minimal transverses, denoted $MinTr(\mathcal{H}_I(X)$, corresponds to the set of minimal shared perception: $\widehat{\mathbb{Q}}_I(X) = MinTr(\mathcal{H}_I(X))$.*

In fact, according to the Definition 5, a traversal intersect the perception of each perceiver and it is minimal. In contrast to elementary perceptions, each shared perception contains at least on member of each elementary perception and consequently represents all perceivers. However, the number of minimal transverslas of the hypergraph $\mathcal{H}_I(X)$ may be very high, so it is more convienient to compute a minimal set of shared perceptions which allow to generate all shared perceptions. These minimal perceptions are called shared perceptions generator. To achieve this goal, the authors of [15] introduce the notion of generalized node in order to reduce the large number of intermediate partial transverses, which are computed during the minimal transverse search - and, therefore, improve the total running time of the algorithm and reduce its storage requirements.

Definition 8 (Generalized node). *Let $\mathcal{H}_I$ be a hypergraph on $\mathcal{V}_I$. The set $X \subseteq \mathcal{V}_I$ is a generalized node of $\mathcal{H}_I$ if all nodes in X belong in exactly the same hyperedges of $\mathcal{H}_I$.*

This kind of nodes have been used successively in [15] to perform efficient computation of minimal transverse. They are also used here to compute the shared perception generators.

3   Rough Set Theory Revisited

Let us now focus on rough sets interpreting than according to the perception of sets we introduced above.

Definition 9 (Pawlak's perception function). *Let $X \subset U$, $I = \{1, 2\}$ two perceivers such that $f_1 = \underline{f}$ and $f_2 = \overline{f}$, which are defined as follows: $\underline{f}(X) = \underline{X}$ and $\overline{f}(X) = \overline{X}$*

According to this Pawlakian perception, denoted $\mathbb{Q}_p$, we have $\mathbb{Q}_p(X) = (\underline{f}(X), \overline{f}(X))$, and an object $x \in U$ is perceived to be certainty a member of X if it is perceived by $\underline{f}$ to be a member of X: $x \underline{\in} X \Leftrightarrow x \in \underline{f}(X)$. Similarly, x is perceived to be possibly a member of a set X if it perceived by $\overline{f}$ in X: $x \overline{\in} X \Leftrightarrow x \in \overline{f}(X)$.

Proposition 2 (Pawlak's perception of objects and sets). *Let $X \subset U$, $(\underline{f}, \overline{f})$ be the two perceivers, where $\underline{f}(X) = \underline{X}$ and $\overline{f}(X) = \overline{X}$, so,*

- *$|I| = 2$*
- *A rough set is not accessible: $\forall\, X \in 2^U, (\underline{X} \neq \overline{X}) \Rightarrow \neg(\mathbb{Q}_p \Vdash X)$*

(U, I) is Pawlak's perception space, which is defined as follows:

- U is the universe of objects,
- I the set of Pawlakian's perceivers reduced only to two perceivers. The elementary perception of perceivers are $\underline{f}$ and $\overline{f}$. Moreover, each perceiver has a specific knowledge base $\underline{K} = (\underline{K}_u, \underline{K}_p)$ and $\overline{K} = (\overline{K}_u, \overline{K}_p)$, respectively, where $\underline{K}_u = \{\underline{\in}\}$, $\overline{K}_u = \{\overline{\in}\}$ are perception biases as the two Pawlakian perceivers are characterized by a doubt that makes the boundary of a rough set non empty. Furthermore, $\underline{K}_p = \overline{K}_p = \{$ *information system* $\}$ as the two perceivers observe the same objects, which are described by features and we have additional information representing by different instances (information table).

Please remark that $\prec_p U = 2^U$ as both the certain and possible perception return classical (crisp) sets. Furthermore, $\forall X \in 2^U, (\underline{X} \neq \overline{X}) \Rightarrow \neg(\mathbb{Q}_p \Vdash X)$, which means that rough sets are not accessible and $\mathbb{Q}_p(X) = (\underline{X}, \overline{X})$ as $\underline{f}(X) = \underline{X}$ and $\overline{f}(X) = \overline{X}$.

Properties of Pawlak's perception $\mathbb{Q}_p$ are:
$\forall X, Y \in 2^U$, (see Definition 2)

(1) the perception of $\underline{f}$ is more specific, or precise, than that of $\overline{f}$, which is more general, $\underline{f}(X) \subseteq \overline{f}(X)$:
$\underline{X} \subseteq \overline{X}$
(2) all objects and only crisp sets, which are not rough, are accessible, $\forall x \in U \; \mathbb{Q}_p \Vdash x$, and $\forall X \in 2^U, (\underline{X} \neq \overline{X}) \Rightarrow \neg(\mathbb{Q}_p \Vdash X)$.
(3) the perception of the intersection of two sets is included into the intersection of their perception, $\mathbb{Q}_p(X \cap Y) \subseteq \mathbb{Q}_p(X) \cap \mathbb{Q}_p(Y)$:
$\overline{f}(X \cap Y) \subseteq \overline{f}(X) \cap \overline{f}(Y)$ and $\underline{f}(X) \cap \underline{f}(Y) = \underline{f}(X \cap Y)$
(4) the perception of the union of two sets contains the intersection of their perception, $\mathbb{Q}_p(X) \cup \mathbb{Q}_p(Y) \subseteq \mathbb{Q}_p(X \cup Y)$:
$\overline{f}(X \cup Y) = \overline{f}(X) \cup \overline{f}(Y)$ and $\underline{f}(X) \cup \underline{f}(Y) \subseteq \underline{f}(X \cup Y)$
(5) the perception of the complement of a set is equal the complement of its perception, $\mathbb{Q}_p(-X) = -\mathbb{Q}_p(X)$:
$\underline{f}(-X) = -\overline{f}(X)$ and $\overline{f}(-X) = -\underline{f}(X)$
(6) the perception is monotonous, $X \subseteq Y \Rightarrow \mathbb{Q}_p(X) \subseteq \mathbb{Q}_p(Y)$:
if $X \subseteq Y$ than $\underline{f}(X) \subseteq \underline{f}(Y)$ and $\overline{f}(X) \subseteq \overline{f}(Y)$

Proposition 3 (Core and Support). *The perception of a given set X is characterized by its Core $\kappa(X) = \underline{f}(X)$ and its support $\sigma(X) = \overline{f}(X)$.*

Proposition 4 (Minimal shared perceptions). *Let U be the universe of objects, $X \subset U$, I the set of perceivers, f_i the elementary perception of the perceiver i and $\mathbb{Q}_p(X) = (\underline{f}(X), \overline{f}(X))$ is the perception of X. The set of certain minimal shared perception of is $\widehat{\mathbb{Q}}_p(X) = \{\{x\} : x \in \underline{f}(X)\}$*

Proposition 5 (Consistent shared perceptions). *The space of consistent shared perceptions considering the set of perceivers I, is the interval $[\widehat{\mathbb{Q}}_p(X), \overline{f}(X)]$*

4 π-RST: A Rough Computational Theory of Perceptions

Information granules are collections of objects, which are grouped together according to their similarity, indistinguishability, coherency, etc. Furthermore, granular computing is an approach that takes advantage of this concept using different granularity levels in designing intelligent systems [10]. Rough set theory explicit such granules and represent them as the result of partitioning of the universe according to a given equivalence relation. In this context, such granules used to define two approximation operators that allow to recognize if a set a crip, well defined, or not, i.e. rough. Furthermore, a rough set approach to computation was introduced in [9] allowing Calculi of Granules. Also, information granulation provided the basis for the development of α-Rough Set Theory, denoted α-RST [12], considering a family of equivalence relations instead of the alone relation in the rough set theory. For example, the familly of equivalence relations $(R_i)_{1 \leq i \leq 3}$ leading to different partitions of the universe U into equivalence classes, which are more or less coarse, see Fig. 1. The usefulness of this generalization, and consequently of information granulation was shown for different problems, for example, Feature selection [12], Concept learning [13], Linguistic negation modelling [14], etc.

In Sect. 3, we have defined the Pawlak perception considering only two perceivers and we would like to extend it here by introducing more perceivers. One way to achieve this goal is to perceive objects and concepts through different granules. For this reason, we consider the family of relations $(R_i)_{1 \leq i \leq |\ I\ |}$, which allows the construction granular perceivers.

Definition 10 (Granular perceivers). *Let $I \subseteq \mathbb{N}$, $(R_i)_{1 \leq i \leq |\ I\ |}$ be a family of equivalence relations, each relation generates two Pawlakian perceivers $\underline{f_i}$ and $\overline{f_i}$ defined as follows:*

$$\forall X \in 2^U, \underline{f_i}(X) = \underline{R_i}(X) \ and \ \forall X \in 2^U, \overline{f_i}(X) = \overline{R_i}(X) \tag{5}$$

and U/R_i are sets of equivalence classes.

The relation R_i of the family are more or less coarse. A relation R_i is finer than R_j if every equivalence class of R_i is a subset of an equivalence class of R_j, and thus every equivalence class of R_j is a union of equivalence classes of R_i. The relation R_3 is finer than R_2, which in turn is finer than R_1, see Fig. 1. Pursuant to the Definition 10, the use of a family of $|\ I\ |$ equivalence relations leads to the generation of $2 \times |\ I\ |$ granular perceivers. In the simplest terms, each relation R_k is represented by a matrix, where each equivalence classe is defined by the line i and the column j is denoted $[i \cdot j]_{R_k}$ and $\forall x, y \in U, x, y \in [i \cdot j]_{R_k} \Leftrightarrow x R_k y$. Thus, the sets of granules defined by the three equivalence relations are: $U/R_1 = \{[i \cdot j]_{R_1} : 1 \leq i \leq 3, 1 \leq j \leq 4\}$, $U/R_2 = \{[i \cdot j]_{R_2} : 1 \leq i \leq 6, 1 \leq j \leq 8\}$ and $U/R_3 = \{[i \cdot j]_{R_3} : 1 \leq i \leq 12, 1 \leq j \leq 16\}$. In the rough set theory, such granules are the basis for the lower and upper approximations of sets, which are more or less precise according to the finesse of equivalence relations - the more we use finer equivalence relations, the more approximations are precise. Figure 1 includes three equivalence relations, R_1, R_2, and R_3 and shows the approximation of a set. Each equivalence class has a square shape, where objects of white cells, see Fig. 1 (left), do not belong to the set, those of grey-white cells belong certainly to the set representing the perception of $\underline{f_1}$, whereas grey cells representing the boundary of the set and their union with grey-white cells define the perception of $\overline{f_1}$. The same is valid for perceivers f_2 (middle) and f_3 (right). These granules represent different partitions of the same space, for example $[2 \cdot 2]_{R_1} = \cup \{[i \cdot j]_{R_2} : 3 \leq i, j \leq 4\}$.

Perception are naturally dynamic and changes over time for different reasons and their dynamic play a crucial role for intelligent systems and human behaviors. Such reasoning is often referred as adaptive process over time and we focus

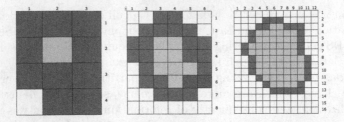

Fig. 1. Perceiving the world through coarse or fine granules: R_1 (left), R_2 (middle) and R_3 (right)

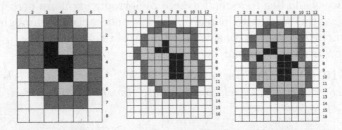

Fig. 2. Perception dynamic through social interaction

here on the adaptation of perceptions. We want just to put forward the benefit of interaction between perceivers allowing to one among them to compare his perception with others, and consequently adapts his perception. Thus, We introduce two operators, which are Expansion ($\oplus : 2^U \rightarrow 2^U$) and Contraction ($\ominus : 2^U \rightarrow 2^U$), such that $\forall X \in 2^U, X \subset \oplus(X)$ and $\ominus(X) \subset X$. These operators define how to update, i.e.; expand or contract, the perception of a given perceiver of X, i.e. $f_i(X)$ when comparing his perception with the perceiver f_j:

$$\oplus(f_i(X)) = f_i(X) \cup \{[x]_{R_j} : x \in f_j(X) \wedge \mathcal{P}^+(f_i, [x]_{R_j})\}$$
$$\ominus(f_i(X)) = f_i(X) - \{[x]_{R_j} : x \in f_j(X) \wedge \mathcal{P}^-(f_i, [x]_{R_j})\}$$

We could go on about the dynamic of perceptions and the semantic definition these two operators but this is not the goal of this paper. Let us just explain the idea behind these operators considering the three equivalence relations and the perception of the perceiver f_1, see Fig. 2.

Consider for example $\mathcal{P}^+ = \{[x]_{R_j} : Perf([x]_{R_j}, f_i)\}$ the set of classes defined by R_j, which are dominated by elements preferred by the perceiver i, whereas $\mathcal{P}^- = \{[x]_{R_j} : \neg Perf([x]_{R_j}, f_i)\}$ contains granules dominated by non-preferred objects. In this case, $\oplus(f_i(X)) = f_i(X) \cup \mathcal{P}^+$ and $\ominus(f_i(X)) = f_i(X) - \mathcal{P}^-$. The Fig. 2 shows the adaptation of the perception of the perceiver f_1 comparing successively with f_2 and f_3. He starts from his initial perception Fig. 1 (left) to update it constructing a more refeined perception Fig. 2 (right). The perceiver f_1 starts comparing his perception with the one of f_2 and remarks that granules $\overline{[5 \cdot 4]}_{R_2}$ are dominated by objects he prefers, while it is the reverse case for the granule $[3 \cdot 4]_{R_2}$. So, he updates his perception consequently:

$$f_1(X) = \oplus(f_1(X)) = f_1(X) \cup \{[5 \cdot 4]_{R_2}\}$$
$$\underline{f_1}(\overline{X}) = \ominus(\underline{f_1}(\overline{X})) = \underline{f_1}(\overline{X}) - \{[3 \cdot 4]_{R_2}\}$$

The result is shown in Fig. 2 (left), black cells represents the new perception of f_1. Next he compare his updated perception with the perception of $f_3(X)$ and remarks that $[7 \cdot 4]_{R_3}, [8 \cdot 9]_{R_3}$ are dominated by his preferred objects, whereas it is the contrary for granules $[5 \cdot 5]_{R_3}, [7 \cdot 5]_{R_3}, [7 \cdot 6]_{R_3}$. He updates his perception accordingly:

$$f_1(X) = \oplus(f_1(X)) = f_1(X) \cup \{[7 \cdot 4]_{R_3}, [8 \cdot 9]_{R_3}\}$$
$$\underline{f_1}(\overline{X}) = \ominus(\underline{f_1}(\overline{X})) = \underline{f_1}(\overline{X}) - \{[5 \cdot 5]_{R_3}, [7 \cdot 5]_{R_3}, [7 \cdot 6]_{R_3}\}$$

Consequently, the final perception $f_1(X)$, Fig. 2 (right) is more precise than initial perception Fig. 1 (left). I would like to conclude by stressing that the adaptation process is nor linear, nor monotonous. It is a more complex process resulting from the interaction between perceivers, their relative influence, their experience with objects of the universe, etc. Unfortunately, the limited size of the paper does not allow us to go any further with the dynamic of perceptions and their relationship with social interactions.

5 Conclusion

In this paper we propose a conceptual set framework based on a perception theory where the main question is not "is an object belong to a given set?", but "is it perceived to belong to it?". Doing so, a new line of research that make a bridge between perception and set theories is introduced, i.e. accessible sets, where the accessibility is related to the perception and can be summarized as follows "to be accessible is to be perceived". This perception is more weak than Berkeley's idealism, where objects are nothing more than our experiences of them, i.e. "to be is to be perceived". Accessible sets are defined not only by objects they contain, but, they are also determined by how they are perceived. Using this new framework, we have analyzed the Pawlak's perception of sets showing that rough sets are only ε-Accessible. Finally, our proposal can also be seen as an attempt to define a computational theory of perceptions.

References

1. Zadeh, L.A.: Fuzzy sets. Inf. Control **8**, 338–353 (1965)
2. Zadeh, L.A.: A new direction in AI - toward a computational theory of perceptions. AI Mag. **22**(1), 73–84 (2001)
3. Pawlak, Z.: Rough sets. Int. J. Comput. Inform. Sci. **11**, 341–356 (1982)
4. Pawlak, Z.: Rough Sets: Theoretical Aspect of Reasoning about Data. Kluwer Academic Pubilishers, Dordrecht (1991)
5. Skowron, A., Wasilewski, P.: An introduction to perception based computing. In: Kim, T., Lee, Y., Kang, B.-H., Ślezak, D. (eds.) FGIT 2010. LNCS, vol. 6485, pp. 12–25. Springer, Heidelberg (2010). doi:10.1007/978-3-642-17569-5_2

6. Jankowski, A., Skowron, A.: Toward perception based computing: a rough-granular perspective. In: Zhong, N., Liu, J., Yao, Y., Wu, J., Lu, S., Li, K. (eds.) Web Intelligence Meets Brain Informatics. LNCS (LNAI), vol. 4845, pp. 122–142. Springer, Heidelberg (2007)
7. Hatfield, G.: Perception & Cognition. Oxford University Press, Oxford (2009)
8. Enderton, H.B.: Elements of Set Theory. Academic Press, New York (1977)
9. Polkowski, L., Skowron, A.: Rough mereological calculi of granules: a rough set approach to computation. Comput. Intell. **17**(3), 472–492 (2001)
10. Bargiela, A., Pedrycz, W.: Granular Computing: An Introduction. Kluwer Academic Publishers, Dordrecht (2003)
11. Quafafou, M.: α-RST: a generalization of rough set theory. Inf. Sci. **124**, 301–316 (2000)
12. Quafafou, M., Boussouf, M.: Generalized rough sets based feature selection. Intell. Data Anal. **4**(1), 3–17 (2000)
13. Dubois, V., Quafafou, M.: Concept learning with approximation: rough version spaces. In: Alpigini, J.J., Peters, J.F., Skowron, A., Zhong, N. (eds.) RSCTC 2002. LNCS (LNAI), vol. 2475, pp. 239–246. Springer, Heidelberg (2002)
14. Pacholczyk, D., Quafafou, M., Garcia, L.: Optimistic vs. pessimistic interpretation of linguistic negation. In: Scott, D. (ed.) AIMSA 2002. LNCS (LNAI), vol. 2443, p. 132. Springer, Heidelberg (2002)
15. Kavvadias, D.J., Stavropoulos, E.C.: An efficient algorithm for the transversal hypergraph generation. J. Graph Algorithms Appl. **9**(3), 239–264 (2005)

Refinements of Orthopairs and IUML-algebras

Stefano Aguzzoli[1], Stefania Boffa[2(✉)], Davide Ciucci[3], and Brunella Gerla[2]

[1] Università di Milano, Milano, Italy
[2] Università dell' Insubria, Varese, Italy
sboffa@uninsubria.it
[3] Università di Milano–Bicocca, Milano, Italy

Abstract. In this paper we consider sequences of orthopairs given by refinement sequences of partitions of a finite universe. While operations among orthopairs can be fruitfully interpreted by connectives of three-valued logics, we describe operations among sequences of orthopairs by means of the logic IUML of idempotent uninorms having an involutive negation.

Keywords: Orthopairs · Forests · IUML-algebras

1 Introduction

An *orthopair* is a pair of disjoint subsets of a universe U. Despite their simplicity, orthopairs arise in several situations of knowledge representation and granular computing [5,6]. They are commonly used to model uncertainty and to deal with approximation of sets. In particular, any rough approximation of a set determines an orthopair. Indeed, given a partition P of a universe U, every subset X of U determines the orthopair $(\mathcal{L}_P(X), \mathcal{E}_P(X))$, where $\mathcal{L}_P(X)$ is the *lower approximation* of X, i.e., the union of the blocks of P included in X, and $\mathcal{E}_P(X)$ is the *impossibility domain* or *exterior region*, namely the union of blocks of P with no elements in common with X [5].

Several kinds of operations have been considered among rough sets [7], corresponding to connectives in three-valued logics. Logical approaches to some of these connectives have been given, such as Łukasiewicz, Nilpotent Minimum, Nelson and Gödel connectives [2,3,11]. In this paper we focus on Sobociński conjunction and the related algebraic structures called IUML-algebras.

The Sobociński conjunction $*$ is defined on $\{0, 1/2, 1\}$ by $\min(x, y)$ for $x \leq 1 - y$ and $\max(x, y)$ otherwise [12]. Such operation (like all three-valued operations) can be defined also on orthopairs on a universe U [7], assigning 1 and 0 to elements in the first and second component of the orthopair respectively, and $1/2$ to elements of U not belonging to any of the components. In this way we get an operation between orthopairs defined as follows:

$$(A, B) * (C, D) = (((A \backslash D) \cup (C \backslash B)), B \cup D).$$

The set $\{0, 1/2, 1\}$ with the Sobociński conjunction is an example of three-valued IUML-algebra (see Definition 1). We are interested in establishing a relationship

© Springer International Publishing AG 2016
V. Flores et al. (Eds.): IJCRS 2016, LNAI 9920, pp. 87–96, 2016.
DOI: 10.1007/978-3-319-47160-0_8

between sequences of successive refinements of orthopairs (over a finite universe) and (not necessarily three-valued) IUML-algebras. Such result is achieved through the dual representation of finite IUML-algebras as finite forests [1]. This is the more innovative part of our work, which spans through algebraic logic, dynamics in rough sets and partial approximation spaces.

More in detail, we consider refinements of partitions and the related orthopairs. Refinements naturally arise in knowledge representation and in the rough set framework, where a refinement corresponds to a finer partition of the universe. A refinement sequence of the universe U is a sequence $\mathcal{P} = (P_0, \ldots, P_n)$ of partitions of subsets of U such that every block of P_i is contained in a block of P_{i-1} for each i from 1 to n. We notice that in our approach we will deal with refinements built on partial partitions, that is partitions that do not cover all the universe [8,9]. As an example, refinement sequences can be used for ontology construction through partitions given by an increasing number of attributes.

Example 1. A refinement sequence can represent classifications in which we want to better specify some classes while ignoring others: suppose to start from animals first classified as *Vertebrata* (first partition). But you are really interested in *Amphibia* and *Mammalia*, that do not form a partition of *Vertebrata* (second partial partition). Then you want to refine such a classification by considering two groups of *Amphibia* (*Anura* and *Caudata*) and three groups of *Mammalia* (*Marsupialia, Cetacea* and *Felidae*) (third partial partition) and further, in the group of *Cetacea* you are interested in *Odontoceti* and *Mysticeti* (fourth partial partition). In this case any set of individuals can be approximated by four orthopairs corresponding to the four partial partitions.

Another paradigmatic example is the temporal evolution, specifically, when new attributes are added to an information table.

Example 2. Let us imagine that a clothing store sells 21 pieces of garment, labeled $c_1 \ldots c_{21}$. The table below shows the available information on the sold items.

	c_1	c_2	c_3	c_4	c_5	c_6	c_7	c_8	c_9	c_{10}	c_{11}	c_{12}	c_{13}	c_{14}	c_{15}	c_{16}	c_{17}	c_{18}	c_{19}	c_{20}	c_{21}
Man	Y	Y	Y	Y	Y	Y	Y	Y	Y	Y	×	N	N	N	N	N	N	N	N	N	N
30 %	Y	Y	Y	Y	Y	×	N	N	N	N	×	Y	Y	Y	Y	N	N	N	N	N	×
50 %	Y	Y	×	N	N	×	Y	Y	N	N	×	Y	Y	N	N	Y	Y	×	N	N	×

Every row divides the items into two groups: according to wearer gender (2nd row), and according to discount rates (3rd row—30% and 4th row—50 %)[1]. Moreover, we consider the attributes *Man*, 30 % and 50 % respectively in the instants of time t_0, t_1 and t_2, where $t_0 < t_1 < t_2$. Note that some of the elements lacks some of information (× in the table); for example, c_{11} is not classified in any group since it is not for sale, and c_6 has been sold already and needs no discount

---

[1] For the sake of space, rows and columns are switched with respect to the standard convention of representing attributes as columns and rows as objects.

rate information. Therefore, at each instant t_i, we obtain a partial partition P_i from the attributes available at time t_i: $P_0 = \{\{c_1, \ldots, c_{10}\}, \{c_{12}, \ldots, c_{21}\}\}$, $P_1 = \{\{c_1, \ldots, c_5\}, \{c_7, \ldots, c_{10}\}, \{c_{12}, \ldots, c_{15}\}, \{c_{16}, \ldots, c_{20}\}\}$, $P_2 = \{\{c_1, c_2\}, \{c_4, c_5\}, \{c_7, c_8\}, \{c_9, c_{10}\}, \{c_{12}, c_{13}\}, \{c_{14}, c_{15}\}, \{c_{16}, c_{17}\}, \{c_{19}, c_{20}\}\}$.

Our goal is to show that not only three-valued IUML-algebras correspond to orthopairs, but each sequence of successive refinements of orthopairs over finite universe can be represented by a (not necessarily three-valued) finite IUML-algebra.

The paper is organized as follows: Sect. 2 recalls the basic notions of IUML-algebras and the correspondence with forests. Sections 3 and 4 contain the main results of the work: in Sect. 3 we define sequences of orthopairs and put them in correspondence with pairs of disjoint upsets of a forest. In Sect. 4 we finally show how to equip the set of sequences of orthopairs with a structure of IUML-algebra.

2 IUML-algebras

This section describes some fundamental notions on IUML-algebras and finite forests. We refer the reader to [1,10] for an exhaustive treatment of the subject.

Let $(P, \leq)$ be an ordered set and $X \subseteq P$. X is an *upset* if, whenever $x \in X$, $y \in P$ and $x \leq y$, we have $y \in X$. Given $X \subseteq P$ and $x \in P$, we define $\uparrow X = \{y \in P : (\exists x \in X)\, y \geq x)\}$ and $\uparrow x = \{y \in P : y \geq x\}$. A *forest* is a partially ordered set $(F, \leq)$ such that for every $x \in F$ the set $\{y \in F \mid y \leq x\}$ is totally ordered. A map $f : F \to G$ between forests is *open* if, for $a \in G$ and $b \in F$, whenever $a \leq f(b)$ there exists $c \in F$ with $c \leq b$ such that $f(c) = a$. Equivalently, open maps carry upsets to upsets.

Definition 1. A *idempotent uninorm mingle logic algebra (IUML-algebra)* [10] is a idempotent commutative bounded residuated lattice $A = (A, \wedge, \vee, *, \rightarrow, \perp, \top, e)$, satisfying $e \leq (x \to y) \vee (y \to x)$ and $(x \to e) \to e = x$ for every $x, y \in A$.

In any IUML-algebra, if we define the unary operation $\neg$ as $\neg x = x \to e$ then $\neg\neg x = x$ ($\neg$ is involutive) and $x \to y = \neg(x * \neg y)$.

Example 3. On the interval $[0, 1]$ consider the following operations:

$$x * y = \begin{cases} \min(x, y) & \text{if } x \leq 1 - y \\ \max(x, y) & \text{otherwise.} \end{cases} \qquad x \to y = \begin{cases} \max(1 - x, y) & \text{if } x \leq y \\ \min(1 - x, y) & \text{otherwise.} \end{cases}$$

Then the structure $([0, 1], \wedge, \vee, *, \to, 0, 1, 1/2)$ is an IUML-algebra. Note that the set $\{0, 1/2, 1\}$ equipped with the restriction of $*$ and $\to$ is again an IUML-algebra, with $*$ being Sobociński conjunction.

In [1] a dual categorical equivalence is described between finite forests F with order preserving open maps and finite IUML-algebras with homomorphisms. For any finite forest F denote by $SP(F)$ the set of pairs of disjoint upsets of F and define the following operations: if $X = (X^1, X^2)$ and $Y = (Y^1, Y^2) \in SP(F)$ are pairs of disjoint upsets of F, we set:

- $(X^1, X^2) \sqcap (Y^1, Y^2) = (X^1 \cap Y^1, X^2 \cup Y^2)$ and $(X^1, X^2) \sqcup (Y^1, Y^2) = (X^1 \cup Y^1, X^2 \cap Y^2)$;
- $(X^1, X^2) * (Y^1, Y^2) = ((X^1 \cap Y^1) \cup (X \diamond Y), (X^2 \cup Y^2) \backslash (X \diamond Y))$ where, for each $U = (U^1, U^2), V = (V^1, V^2) \in SP(F)$, letting $U^0 = F \backslash (U^1 \cup U^2)$, we set $U \diamond V = \uparrow ((U^0 \cap V^1) \cup (V^0 \cap U^1))$;
- $(X^1, X^2) \rightarrow (Y^1, Y^2) = \neg((X^1, X^2) * (Y^2, Y^1))$ where $\neg(X^1, X^2) = (X^2, X^1)$.

Theorem 1 [1]. *For every finite forest* F, $(SP(F), \sqcap, \sqcup, *, \rightarrow, (\emptyset, F), (F, \emptyset), (\emptyset, \emptyset))$ *is an IUML-algebra. Vice-versa, each finite IUML-algebra is isomorphic with* $SP(F)$ *for some finite forest* F.

3   Sequences of Orthopairs and Forests

Definition 2. A *partial partition* of U is a partition of a subset of U. A sequence $P_0, \ldots, P_n$ of partial partitions of U is a *refinement sequence* if each element of P_i is contained in an element of P_{i-1}, for $i = 1, \ldots, n$.

Example 4. If $U = \{a, b, c, d, e, f, g, h, i, j\}$ then $P_0 = \{\{a, b, c, d, e\}, \{f, g, h, i\}\}$, $P_1 = \{\{a, b\} \{c, d\}, \{f, g\}, \{h, i\}\}$ is a refinement sequence of partial partitions of U.

From now on, we do not consider partitions that contain singletons, that is blocks with only one element. This constraint is necessary in order to prove the desired results, see Proposition 2. Let us fix à refinement sequence $P = P_0, \ldots, P_n$ of partial partitions of U. For any $X \subseteq U$ and for every $i = 0, \ldots, n$ we consider the *orthopair* $(\mathcal{L}_i(X), \mathcal{E}_i(X))$ determined by P_i. To every refinement sequence we assign a set of sequences of orthopairs.

Definition 3. Let $\mathcal{P} = (P_0, \ldots, P_n)$ be a refinement sequence of U and let $X \subseteq U$. Then we denote by $\mathcal{O}_\mathcal{P}(X)$ the sequence of orthopairs

$$((\mathcal{L}_0(X), \mathcal{E}_0(X)), \ldots, (\mathcal{L}_n(X), \mathcal{E}_n(X))).$$

Example 5. Given $U = \{a, b, c, d, e, f, g, h, i, j\}$, $X = \{a, b, c, d, e\}$ and the following refinement sequence of partial partitions of U: $P_0 = \{\{a, b, c, d, e, f, g, h, i, j\}\}$, $P_1 = \{\{a, b, c, d\}, \{e, f, g, h, i\}\}$, $P_2 = \{\{a, b\}, \{c, d\}, \{e, f\}, \{g, h\}\}$, then the sequence of orthopairs of X is $\mathcal{O}(X) = ((\emptyset, \emptyset), (\{a, b, c, d\}, \emptyset), (\{a, b, c, d\}, \{g, h\}))$.

Definition 4. Let $\mathcal{P}$ be a refinement sequence of partial partitions of U. We associate with $\mathcal{P}$ a forest $(F_\mathcal{P}, \leq_{F_\mathcal{P}})$, where:

1. $F_\mathcal{P} = \bigcup_{i=0}^n P_i$ (the set of nodes is the set of all subsets of U belonging to the partitions $P_0, \ldots, P_n$), and
2. for $N, M \in F_\mathcal{P}$, $N \leq_{F_\mathcal{P}} M$ if and only if there exists $i \in \{0, \ldots, n-1\}$ such that $N \in P_i$, $M \in P_{i+1}$ and $M \subseteq N$ (the partial order relation is the reverse inclusion).

Example 6. The forest associated with the refinement sequence (P_0, P_1) of partitions of $\{a, b, c, d, e, f, g, h, i, j\}$, where $P_0 = \{\{a, b, c, d\}, \{e, f, g, h, i\}\}$ and $P_1 = \{\{a, b\}, \{c, d\}, \{e, f\}, \{g, h\}\}$, is shown in the following figure:

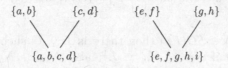

For any $X \subseteq U$ the sequence $\mathcal{O}_\mathcal{P}(X)$ of orthopairs with respect to $\mathcal{P}$ determines two subsets of the forest $F_\mathcal{P}$, obtained by considering the blocks contained in $\mathcal{L}_i(X)$ and the blocks contained in $\mathcal{E}_i(X)$. This observation leads to the following definition.

Definition 5. For every refinement sequence $\mathcal{P} = (P_0, \ldots, P_n)$ of U and any $X \subseteq U$ we let $(X^1_\mathcal{P}, X^2_\mathcal{P})$ be such that $X^1_\mathcal{P} = \{N \in F_\mathcal{P} : N \subseteq X\}$ and $X^2_\mathcal{P} = \{N \in F_\mathcal{P} : N \cap X = \emptyset\}$. Let $SO(F_\mathcal{P})$ be the set $\{(X^1_\mathcal{P}, X^2_\mathcal{P}) \mid X \subseteq U\}$.

Example 7. Given U and $\mathcal{P}$ of Example 6, if $X = \{a, b, e, g\}$ then $X^1_\mathcal{P} = \{\{a, b\}\}$ and $X^2_\mathcal{P} = \{\{c, d\}\}$.

We write (X^1, X^2) instead of $(X^1_\mathcal{P}, X^2_\mathcal{P})$, when $\mathcal{P}$ is clear from the context.

Theorem 2. *Given a set U and a refinement sequence $\mathcal{P}$ of U, the map*

$$h : \mathcal{O}_\mathcal{P}(X) \in \{\mathcal{O}_\mathcal{P}(X) \mid X \subseteq U\} \mapsto (X^1, X^2) \in SO(F_\mathcal{P})$$

is a bijection.

Proof. First of all we prove that h is well defined and injective, that is $\mathcal{O}_\mathcal{P}(X) = \mathcal{O}_\mathcal{P}(Y)$ if and only if $(X^1, X^2) = (Y^1, Y^2)$.

($\Rightarrow$). We note that $N \in X^1$ if and only if $N \in P_i$ and $N \subseteq X$ for some $i \in \{0, \ldots, n\}$, namely $N \in P_i$ and $N \subseteq \mathcal{L}_i(X)$. Consequently $N \in Y^1$, since $\mathcal{L}_i(X) = \mathcal{L}_i(Y)$. Dually $N \in X^2$ if and only if $N \in Y^2$, since $\mathcal{E}_i(X) = \mathcal{E}_i(Y)$ for each $i \in \{0, \ldots, n\}$.

($\Leftarrow$). Let $i \in \{0, \ldots, n\}$. $n \in \mathcal{L}_i(X)$ if and only if there is $N \in F_\mathcal{P}$ such that $n \in N$ and $N \subseteq X$. For hypothesis $N \subseteq Y$. Then $n \in \mathcal{L}_i(Y)$. Dually we can prove that $\mathcal{E}_i(X) = \mathcal{E}_i(Y)$ for each $i \in \{0, \ldots, n\}$, since $X^2 = Y^2$.

Trivially every pair (X^1, X^2) of $SO(F_\mathcal{P})$ is image of $\mathcal{O}_\mathcal{P}(X)$, hence h is a bijection. $\quad\square$

Proposition 1. *Let $\mathcal{P}$ be a sequence of partitions of U. Then $X^1_\mathcal{P}$ and $X^2_\mathcal{P}$ are disjoint upsets of $F_\mathcal{P}$ for each $X \subseteq U$.*

Proof. $X^1_\mathcal{P}$ and $X^2_\mathcal{P}$ are disjoint by definition. If $N \in X^1_\mathcal{P}$ and $N \leq_{F_\mathcal{P}} M$, then by definition $M \subseteq N \subseteq X$ hence $M \subseteq X$ and $M \in X^1_\mathcal{P}$. Analogously, if $N \in X^2_\mathcal{P}$ and $N \leq_{F_\mathcal{P}} y$ then $M \subseteq N$ and $N \cap X = \emptyset$, hence $M \cap X = \emptyset$ and $M \in X^2_\mathcal{P}$. $\quad\square$

We recall that $SP(F_\mathcal{P})$ is the set of all pairs of disjoint upsets of $F_\mathcal{P}$. Note that $SO(F_\mathcal{P}) \subseteq SP(F_\mathcal{P})$, but the opposite does not always hold.

Proposition 2. *Let $\mathcal{P}$ be a refinement sequence of partitions of U and $(L^1, L^2) \in SP(F_{\mathcal{P}})$. Then $(L^1, L^2) \in SO(F_{\mathcal{P}})$ if and only if for each $N \in F_{\mathcal{P}} \backslash L^1$ there is no subset $\{N_1, \ldots, N_m\}$ of L^1 such that $N = N_1 \cup \ldots \cup N_m$ and for each $M \in F_{\mathcal{P}} \backslash L^2$ there is no subset $\{M_1, \ldots, M_m\}$ of L^2 such that $M = M_1 \cup \ldots \cup M_m$.*

Proof. ($\Rightarrow$) If $(L^1, L^2) \in SO(F_{\mathcal{P}})$ then there is $L \subseteq U$ such that $L^1 = \{M \in F_{\mathcal{P}} \mid M \subseteq L\}$. Then if $N_1, \ldots, N_m \in L^1$, also $N_1, \ldots, N_m \subseteq L$ and hence $N = N_1 \cup \cdots \cup N_m \subseteq L$ and $N \in L^1$. The case for L^2 is analogous.

($\Leftarrow$) Let $(L^1, L^2) \in SF(F_{\mathcal{P}})$ and let $H = \bigcup_{N \in L^1} N$ and $K = \bigcup_{N \in L^2} N$. Let $\mathcal{M} := \{M \in F_{\mathcal{P}} \backslash (L^1 \cup L^2) \mid M \cap H = \emptyset\}$, pick elements x_M belonging to $M \in \mathcal{M}$ such that $x_M \notin K$ and $x_M = x_N$ for each $N \in \mathcal{M}$ such that $N \cap M \neq \emptyset$. Then we set $X = H \cup \{x_M \mid M \in \mathcal{M}\}$. The proof follows from the fact that $(L^1, L^2) = h(\mathcal{O}_{\mathcal{P}}(X))$, since every block contains at least two elements and hence nodes $M \in \mathcal{M}$ are neither contained in, nor disjoint with X. $\qquad\square$

Corollary 1. *Let $\mathcal{P}$ be a refinement sequence of partitions of U. Then $SO(F_{\mathcal{P}}) = SP(F_{\mathcal{P}})$ if and only if every node of $F_{\mathcal{P}}$ strictly contains the union of its successors.*

Example 8. We consider the universe and the refinement sequence of Example 6. We have $SO(F_{\mathcal{P}}) \subset SP(F_{\mathcal{P}})$, indeed the pair $(\{\{a, b\}, \{c, d\}, \{e, f\}, \{g, h\}, \{e, f, g, h, i\}\}, \emptyset)$ does not belong to $SO(F_{\mathcal{P}})$, since there is no subset of U that contains the two sets $\{a, b\}$ and $\{c, d\}$, but it does not contain their union.

On the other hand, if we consider the universe $U = \{a, b, c, d, e, f, g, h, i, j, k\}$ and the refinement sequence $\mathcal{P}$, where $P_0 = \{\{a, b, c, d, e, f, g, h, i, j, k\}\}$, $P_1 = \{\{a, b, c, d, e\}, \{f, g, h, i, j\}\}$, and $P_2 = \{\{a, b\}, \{c, d\}, \{f, g\}, \{h, i\}\}$, then we have $SO(F_{\mathcal{P}}) = SP(F_{\mathcal{P}})$.

Definition 6. *Let $\mathcal{P}$ be a refinement sequence of partial partitions of the universe U. Then if (L^1, L^2) and (M^1, M^2) are in $SP(F_{\mathcal{P}})$, we set $(L^1, L^2) \simeq_{SP} (M^1, M^2)$ if and only if $\bigcup_{N \in L^1} N = \bigcup_{N \in M^1} N$ and $\bigcup_{N \in L^2} N = \bigcup_{N \in M^2} N$.*

The relation $\simeq_{SP}$ is an equivalence relation in $SP(F_{\mathcal{P}})$.

Theorem 3. *Let $CP(F_{\mathcal{P}}) = \{(L^1, L^2) \in SP(F_{\mathcal{P}}) \mid (L^1, L^2) \simeq_{SP} (X^1, X^2), \text{ with } (X^1, X^2) \in SO(F_{\mathcal{P}})\}$. Then $CP(F_{\mathcal{P}}) = SP(F_{\mathcal{P}})$, hence $CP(F_{\mathcal{P}})$ is the support of an IUML-algebra.*

Proof. Clearly every pair in $CP(F_{\mathcal{P}})$ belongs to $SP(F_{\mathcal{P}})$. In order to prove the opposite inclusion, let $(L^1, L^2) \in SP(F_{\mathcal{P}})$ and consider, for $i = 1, 2$, the sets $K^i = \{N \in F_{\mathcal{P}} : N = N_1 \cup \ldots \cup N_m, \text{ where } N_1, \ldots N_m \in L^i\}$. Then, by Proposition 2, $(L^1 \cup K^1, L^2 \cup K^2)$ belongs to $SO(F_{\mathcal{P}})$ and $(L^1 \cup K^1, L^2 \cup K^2) \simeq_{SP} (L^1, L^2)$, hence $(L^1, L^2) \in CP(F_{\mathcal{P}})$.

Example 9. With $U = \{a, b, c, d\}$ and $\mathcal{P} = (P_0 = \{\{a, b, c, d\}\}, P_1 = \{\{a, b\}, \{c, d\}\})$, the IUML-algebra $Sub(F_{\mathcal{P}})$ has the following Hasse diagram:

$$(\{\{a, b, c, d\}, \{a, b\}, \{c, d\}\}, \emptyset)$$
$$|$$
$$(\{\{a, b\}, \{c, d\}\}, \emptyset)$$
$$\diagup \quad \diagdown$$
$$(\{\{a, b\}\}, \emptyset) \quad (\{\{c, d\}\}, \emptyset)$$
$$\diagup \quad \diagdown \quad \diagup \quad \diagdown$$
$$(\{a, b\}, \{c, d\}) \quad (\emptyset, \emptyset) \quad (\{c, d\}, \{a, b\})$$
$$\diagdown \quad \diagup \quad \diagdown \quad \diagup$$
$$(\emptyset, \{c, d\}) \quad (\emptyset, \{a, b\})$$
$$\diagdown \quad \diagup$$
$$(\emptyset, \{\{a, b\}, \{c, d\}\})$$
$$|$$
$$(\emptyset, \{\{a, b, c, d\}, \{a, b\}, \{c, d\}\}$$

Example 10. We consider the universe and the refinement sequence described in Example 2 of Sect. 1. Then, from the table showed in the example, we build a forest where every level represents a partial partition of the universe of codes:

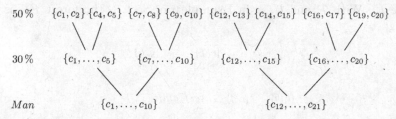

A sequence of orthopairs and a pair of disjoint upsets of the forest are assigned to every subset of $\{c_1, \ldots, c_{21}\}$. For example, if $X = \{c_{15}, c_{16}, c_{17}, c_{18}, c_{19}, c_{20}\}$, then we have the following sequence of orthopairs of X:

$$\mathcal{O}(X) = \begin{pmatrix} (\emptyset, \{c_1, \ldots, c_{10}\})_{t_0} \\ (\{c_{16}, \ldots, c_{20}\}, \{c_1, \ldots, c_5, c_7, \ldots, c_{10}\})_{t_1} \\ (\{c_{16}, c_{17}, c_{19}, c_{20}\}, \{c_1, c_2, c_4, c_5, c_7, c_8, c_9, c_{10}, c_{12}, c_{13}\})_{t_2} \end{pmatrix}.$$

We observe that the orthopair at instant t_i contains more information than the orthopair at instant t_{i-1}, for $i \in \{1, 2\}$, but some elements no longer appear since they are no more classified (see the element c_3 that appears in the orthopair at instant t_1 and does not appear at instant t_2). This is the consequence of having a sequence of partitions that can lose objects when refined.

The sequence of orthopairs can be equivalently described by the following pair of disjoint upsets of our forest: $(\{\{c_{16}, c_{17}\}, \{c_{19}, c_{20}\}, \{c_{16}, \ldots, c_{20}\}\}, \{\{c_1, c_2\}, \{c_4, c_5\}, \{c_7, c_8\}, \{c_9, c_{10}\}, \{c_{12}, c_{13}\}, \{c_1, \ldots, c_5\}, \{c_7, \ldots, c_{10}\}, \{c_1, \ldots, c_{10}\}\})$.

4   Sequences of Orthopairs as IUML-algebras

In this section, given a refinement sequence $\mathcal{P}$ of the universe U, we provide $SO(F_{\mathcal{P}})$ with a structure of IUML-algebra, and we also find its dual forest that in general will be different from $F_\mathcal{P}$. Indeed, we build a forest $F_{\mathcal{P}'}$ assigned to a new refinement sequence $\mathcal{P}'$ of U, by removing from $F_\mathcal{P}$ all nodes equal to the union of their successors (cfr. [4]).

Let $\mathcal{P} = (P_0, \ldots, P_n)$ be a refinement sequence of partial partitions of U. Then we build a refinement sequence $\mathcal{P}' = (P_0', \ldots, P_m')$ (with $m \leq n$) of partial partitions of U determined by the following conditions:

- $P_m' = P_n$,
- for every $i \in \{0, \ldots, n-1\}$ and $N \in P_i$, if there are no $N_1, \ldots, N_l \in P_{i+1}'$ such that $N = N_1 \cup \ldots \cup N_l$ then $N \in P_i'$, otherwise $N \notin P_i'$ but $N_j \in P_i'$ for each $j = 1, \ldots, l$.

Example 11. Consider U and $\mathcal{P}$ of Example 6 of Sect. 3. We obtain $\mathcal{P}' = (P_0', P_1')$, where $P_0' = \{\{a, b\}, \{c, d\}, \{e, f, g, h, i\}\}$, and $P_1' = \{\{a, b\}, \{c, d\}, \{e, f\}, \{g, h\}\}$.

When we build $\mathcal{P}'$ from P, we associate with $F_\mathcal{P}$ the forest $F_{\mathcal{P}'}$.

Proposition 3. *Let $N \in F_\mathcal{P}$. Then $N \in F_{\mathcal{P}'}$ if and only if there is no subset $\{N_1, \ldots, N_l\}$ of nodes of $F_\mathcal{P}$ such that $N = N_1 \cup \ldots \cup N_l$.*

Therefore every node of $F_{\mathcal{P}'}$ is not equal to the union of its sons.

Example 12. With $\mathcal{P}$ and U as in Example 6 of Sect. 3, the forest $F_{\mathcal{P}'}$ is the following:

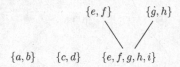

Proposition 4. *Let $\mathcal{P} = (P_0, \ldots, P_n)$ be a refinement sequence of partitions of U. Then $\mathcal{P}'$ consists of the only partition P_n if and only if P_i is a partition of U, for each $i \in \{0, \ldots, n\}$ and in this case $F_{\mathcal{P}'}$ is the subforest of $F_\mathcal{P}$ made of all leaves of $F_\mathcal{P}$.*

Proposition 5. *Let $\mathcal{P} = (P_0, \ldots, P_n)$ be a refinement sequence of partitions of U. Then $\mathcal{P} = \mathcal{P}'$ if and only if it does not exist a subset P_i that is a partition of some element of P_{i-1} for each $i \in \{1, \ldots, n\}$.*

On the other hand, $\mathcal{P} = \mathcal{P}'$ implies that $F_\mathcal{P} = F_{\mathcal{P}'}$.

Example 13. Let $U = \{a, b, c, d, e, f, g, h, i, j, k\}$ be a universe and $\mathcal{P} = (P_0, P_1, P_2)$ a refinement sequence of U such that $P_0 = \{\{a, b, c, d, e, f, g, h, i, j, k\}\}$, $P_1 = \{\{a, b, c, d, e\}, \{f, g, h, i, j\}\}$, and $P_2 = \{\{a, b\}, \{c, d\}, \{f, g\}, \{h, i\}\}$. So $\mathcal{P} = \mathcal{P}'$ and $F_\mathcal{P} = F_{\mathcal{P}'}$.

Theorem 4. *Let $\mathcal{P}$ be a refinement sequence of partitions of U and consider the map*

$$f : SO(F_\mathcal{P}) \longrightarrow SP(F_{\mathcal{P}'})$$

where $f((X_\mathcal{P}^1, X_\mathcal{P}^2)) = (X_{\mathcal{P}'}^1, X_{\mathcal{P}'}^2)$, for $(X_\mathcal{P}^1, X_\mathcal{P}^2) \in SO(F_\mathcal{P})$. Then f is bijective.

Proof. Let $X, Y \subseteq U$ such that $(X_{\mathcal{P}'}^1, X_{\mathcal{P}'}^2) = (Y_{\mathcal{P}'}^1, Y_{\mathcal{P}'}^2)$. Then $X_{\mathcal{P}'}^1 = Y_{\mathcal{P}'}^1$ and $X_{\mathcal{P}'}^2 = Y_{\mathcal{P}'}^2$. So $X_{\mathcal{P}}^1 = Y_{\mathcal{P}}^1$, since $X_{\mathcal{P}}^1$ and $Y_{\mathcal{P}}^1$ are obtained adding respectively to $X_{\mathcal{P}'}^1$ and $Y_{\mathcal{P}'}^1$ the same nodes, namely all the nodes that belong to $F_{\mathcal{P}}$ and that are union of some nodes of $X_{\mathcal{P}'}^1$. For the same reason $X_{\mathcal{P}}^2 = Y_{\mathcal{P}}^2$. Consequently f is an injective function. Moreover f is a surjection, since given $X \subseteq U$, then $(X_{\mathcal{P}'}^1, X_{\mathcal{P}'}^2)$ corresponds to the pair of $SO(F_{\mathcal{P}})$ generated by X. $\qquad\square$

Now using f and the operations on $SP(F_{\mathcal{P}'})$ as in Theorem 1, we introduce the following operations in $SO(F_{\mathcal{P}})$:

- $(X_{\mathcal{P}}^1, X_{\mathcal{P}}^2) \cap_{SO} (Y_{\mathcal{P}}^1, Y_{\mathcal{P}}^2) := f^{-1}((X_{\mathcal{P}'}^1, X_{\mathcal{P}'}^2) \sqcap (Y_{\mathcal{P}'}^1, Y_{\mathcal{P}'}^2))$,
- $(X_{\mathcal{P}}^1, X_{\mathcal{P}}^2) \cup_{SO} (Y_{\mathcal{P}}^1, Y_{\mathcal{P}}^2) := f^{-1}((X_{\mathcal{P}'}^1, X_{\mathcal{P}'}^2) \sqcup (Y_{\mathcal{P}'}^1, Y_{\mathcal{P}'}^2))$,
- $(X_{\mathcal{P}}^1, X_{\mathcal{P}}^2) *_{SO} (Y_{\mathcal{P}}^1, Y_{\mathcal{P}}^2) := f^{-1}((X_{\mathcal{P}'}^1, X_{\mathcal{P}'}^2) * (Y_{\mathcal{P}'}^1, Y_{\mathcal{P}'}^2))$.
- $(X_{\mathcal{P}}^1, X_{\mathcal{P}}^2) \rightarrow_{SO} (Y_{\mathcal{P}}^1, Y_{\mathcal{P}}^2) := f^{-1}((X_{\mathcal{P}'}^1, X_{\mathcal{P}'}^2) \rightarrow (Y_{\mathcal{P}'}^1, Y_{\mathcal{P}'}^2))$.

Then the following trivially holds:

Theorem 5. *Let $\mathcal{P}$ be a refinement sequence of partial partitions of U. Then $L(F_{\mathcal{P}}) = (SO(F_{\mathcal{P}}), \cap_{SO}, \cup_{SO}, *_{SO}, \rightarrow_{SO}, (\emptyset, F_{\mathcal{P}}), (F_{\mathcal{P}}, \emptyset), (\emptyset, \emptyset))$ is an IUML-algebra.*

Example 14. Given U and $\mathcal{P}$ of the Example 9, then the Hasse diagram of the IUML-algebra $L(F_{\mathcal{P}})$ is the following:

Theorem 6. *Given an universe U and a refinement sequence $\mathcal{P} = (P_0, \ldots, P_n)$, the structure of IUML-algebra on $SO(F_{\mathcal{P}})$ induces on sequences of orthopairs the operation*

$$\mathcal{O}_{\mathcal{P}}(X) \odot_{SO} \mathcal{O}_{\mathcal{P}}(Y) = h^{-1}(h(\mathcal{O}_{\mathcal{P}}(X)) *_{SO} h(\mathcal{O}_{\mathcal{P}}(Y)))$$

(for every $X, Y \subseteq U$) that is equal in turns to the sequence of orthopairs $((A_0, B_0), \ldots, (A_n, B_n))$ defined as follows: for each $i = 1, \ldots, n$, we firstly set

$$(A_i', B_i') = (\mathcal{L}_i(X), \mathcal{E}_i(X)) * (\mathcal{L}_i(Y), \mathcal{E}_i(Y))$$

(where $*$ is the Sobociński conjunction) and then $A_0 = A_0'$ and, for $i > 0$, $A_{i+1} = A_{i+1}' \cup \{N \in P_{i+1} \mid N \subseteq A_i\}$, while $B_i = B_i' \backslash A_i$.

In other words, the operation $\odot_{SO}$ maps each pair of sequences of orthopairs to the sequence of orthopairs given by applying the Sobociński conjunction between orthopairs relative to same partition and then closing with respect to the inclusion in the first component.

5   Conclusions and Further Works

In this paper, we studied the refinement sequences obtained by partial partitions and we proved that, starting from Sobociński conjunction between orthopairs, sequences of orthopairs can be equipped with a structure of IUML-algebra. The converse direction, that consists in associating with every finite IUML-algebra a universe and a sequence of orthopairs on it, is rather straightforward. In future works we plan to provide a natural interpretation of the operation defined in Theorem 6. Further, we shall generalize other three-valued operations between orthopairs to sequences of orthopairs and study the obtained algebraic structures. Another direction to investigate is to widen the applicability of this approach to refinements based on (partial) coverings instead of partitions.

References

1. Aguzzoli, S., Flaminio, T., Marchioni, E.: Finite forests. Their algebras and logics (Submitted)
2. Banerjee, M., Chakraborty, K.: Algebras from rough sets. In: Pal, S., Skowron, A., Polkowski, L. (eds.) Rough-Neural Computing, pp. 157–188. Springer, Heidelberg (2004)
3. Bianchi, M.: A temporal semantics for nilpotent minimum logic. Int. J. Approx. Reason. **55**(1, part 4), 391–401 (2014)
4. Calegari, S., Ciucci, D.: Granular computing applied to ontologies. Int. J. Approx. Reasoning **51**(4), 391–409 (2010)
5. Ciucci, D.: Orthopairs: a simple and widely used way to model uncertainty. Fundamenta Informaticae **108**(3–4), 287–304 (2011)
6. Ciucci, D.: Orthopairs and granular computing. Granular Comput. **1**(3), 159–170 (2016)
7. Ciucci, D., Dubois, D.: Three-valued logics, uncertainty management and rough sets. In: Peters, J.F., Skowron, A. (eds.) Transactions on Rough Sets XVII. LNCS, vol. 8375, pp. 1–32. Springer, Heidelberg (2014)
8. Ciucci, D., Mihálydeák, T., Csajbók, Z.E.: On definability and approximations in partial approximation spaces. In: Miao, D., Pedrycz, W., Slezak, D., Peters, G., Hu, Q., Wang, R. (eds.) RSKT 2014. LNCS, vol. 8818, pp. 15–26. Springer, Heidelberg (2014)
9. Csajbók, Z.E.: Approximation of sets based on partial covering. In: Peters, J.F., Skowron, A., Ramanna, S., Suraj, Z., Wang, X. (eds.) Transactions on Rough Sets XVI. LNCS, vol. 7736, pp. 144–220. Springer, Heidelberg (2013)
10. Metcalfe, G., Montagna, F.: Substructural fuzzy logics. J. Symb. Logic **72**(3), 834–864 (2007)
11. Pagliani, P.: Rough set theory and logic-algebraic structures. In: Orłowska, E. (ed.) Incomplete Information: Rough Set Analysis, vol. 13, pp. 109–190. Springer, Heidelberg (1998)
12. Sobociński, B.: Axiomatization of a partial system of three-value calculus of propositions. J. Comput. Syst. **1**, 23–55 (1952)

Rough Approximations Induced
by Orthocomplementations
in Formal Contexts

Tong-Jun Li[1,2](✉), Wei-Zhi Wu[1,2], and Shen-Ming Gu[1,2]

[1] School of Mathematics, Physics and Information Science,
Zhejiang Ocean University, Zhoushan 316022, Zhejiang, China
[2] Key Laboratory of Oceanographic Big Data Mining and Application
of Zhejiang Province, Zhejiang Ocean University,
Zhoushan 316022, Zhejiang, China
{litj,wuwz,gsm}@zjou.edu.cn

Abstract. Formal contexts is a common framework for rough set theory and formal concept analysis, and some rough set models in formal contexts have been proposed. In this paper, based on the theory of abstract approximation spaces presented by Cattaneo [1], a Brouwer orthocomplementation on the set of objects of a formal context is presented, as a result, a pair of new lower and upper rough approximation operators is introduced. Comparison between the new approximation operators and the existing approximation operators is made, and two necessary and sufficient conditions about equivalence of the operators are obtained. Relationships and algebraic structures among the definable subsets of these approximation operators are investigated.

Keywords: Rough sets · Formal contexts · Orthocomplemetations · Quasi BZ lattices

1   Introduction

Rough set theory is an approach to approximate concepts, therein unknown (imprecise, vague, unclassified, approximable) concepts are approximated by two known (precise, crisp, classified) concepts which are called lower and upper approximations respectively. There are various forms of known concepts in application data. Originally, the known concepts in Pawlak rough set model [10] are described by an equivalence relation derived by information systems. Since inception, various approaches of concept description were introduced, for example, probabilistic rough sets [17], similarity binary relation based rough sets [11], covering rough sets [3,18], etc. The lower and upper approximations of unknown concepts are proposed by using various approaches.

Formal contexts can be viewed as special information systems. They are the base of formal concept analysis [12]. Rough sets can also be applied to formal contexts, and the approaches of formal concept analysis and rough sets

© Springer International Publishing AG 2016
V. Flores et al. (Eds.): IJCRS 2016, LNAI 9920, pp. 97–106, 2016.
DOI: 10.1007/978-3-319-47160-0_9

are complementary. On the one hand, approaches of rough sets can be introduced into concept lattices, so concept approximations of formal concepts were put forward [5,9]. On the other hand, technology of formal concept analysis can also be taken into rough sets, consequently different types of concept lattices were proposed [4,15,16].

As for rough sets constructed on formal contexts, in the literature [6], a covering and three binary relations on sets of objects were defined, and by means of the approaches of covering rough sets and binary relation based rough sets, four types of rough approximation operators were proposed. In [1], Cattaneo presented the notion of abstract approximation space, and given a method of constructing rough sets on quasi BZ posets or BZ posets. In this paper, we implement concretely the approaches of Cattaneo to define new rough approximation operators in formal contexts, and discuss some related issues. The remainder of the paper is organized as follows: Sect. 2 reviews some basic notions and knowledge related to the work. Section 3 presents a Brouwer orthocomplementation on the power set of objects, by which two pairs of rough approximation operators are defined, the definable sets and their algebraic properties are investigated. Section 4 presents a summary of conclusions.

2 Preliminaries

In this section, rough approximations based on binary relations and induced by formal contexts are reviewed respectively.

2.1 Rough Approximations Based on Binary Relations

Let U be a finite and nonempty set called the universe of discourse. The class of all subsets of U will be denoted by $\mathcal{P}(X)$. The complement of a subset A in U will be denoted by A^c, that is, $A^c = \{x \in U | x \notin A\}$.

Let U and W be two finite and nonempty universes of discourse. A *binary crisp relation* (*binary relation in short*) R from U to W is a subset of $U \times W$, we also denote $(x, y) \in R$ by xRy and call x a *predecessor* of y, and y a *successor* of x. If $U = W$, R is called a binary relation on U. The relation R on U is said to be *reflexive* if $(x, x) \in R, \forall x \in U$; R is said to be *symmetric* if $(x, y) \in R \Rightarrow (y, x) \in R, \forall x, y \in U$. If R is reflexive and symmetric, then R is said to be a *tolerance relation* on U.

For a binary relation R from U to W, we will write R^{-1} to denote the *inverse relation* of R, that is, $R^{-1} = \{(x, y) \in W \times U | (y, x) \in R\}$. For each $x \in U$, the *successor neighborhood* of x with respect to (w.r.t.) R will be defined by $R(x) = \{y \in W | (x, y) \in R\}$.

Let U and W be two finite nonempty universes of discourse, and R a binary relation from U to W. The triple (U, W, R) is called a *generalized crisp approximation space* in [13]. For $X \in \mathcal{P}(W)$, *the lower and upper rough approximations of X w.r.t.* (U, W, R), denoted by $\underline{R}(X)$ and $\overline{R}(X)$ respectively, are defined by

$$\underline{R}(X) = \{x \in U | R(x) \subseteq X\}, \quad \overline{R}(X) = \{x \in U | R(x) \cap X \neq \emptyset\}.$$

The rough approximation operators $\underline{R}$ and $\overline{R}$ satisfy the following properties:
$\forall X, Y \in \mathcal{F}(W)$,

(L1) $\underline{R}(X) = (\overline{R}(X^c))^c$,   (U1) $\overline{R}(X) = (\underline{R}(X^c))^c$;
(L2) $\underline{R}(W) = U$,   (U2) $\overline{R}(\emptyset) = \emptyset$;
(L3) $\underline{R}(X \cap Y) = \underline{R}(X) \cap \underline{R}(Y)$, (U3) $\overline{R}(X \cup Y) = \overline{R}(X) \cup \overline{R}(X)$;
(L4) $X \subseteq Y \Rightarrow \underline{R}(X) \subseteq \underline{R}(Y)$,   (U4) $X \subseteq Y \Rightarrow \overline{R}(X) \subseteq \overline{R}(Y)$.

The above properties are fundamental for $\underline{R}$ and $\overline{R}$. In particular, Properties (L1) and (U1) show that $\underline{R}$ and $\overline{R}$ are dual to each other. Based on a variety of approximation operators, certain types of binary relations on U, say, reflexive, symmetric, transitive, and Euclidean relation, can be characterized [13,14].

Let R be an arbitrary binary relation on U, and $\underline{R}$ and $\overline{R}$ the lower and upper generalized rough approximation operators. Then

(1) R is reflexive $\iff$ (L5) $\underline{R}(X) \subseteq X$, $\forall X \in \mathcal{P}(U)$,
$\iff$ (U5) $X \subseteq \overline{R}(X)$, $\forall X \in \mathcal{P}(U)$,
(2) R is symmetric $\iff$ (L6) $\overline{R}(\underline{R}(X)) \subseteq X$, $\forall X \in \mathcal{P}(U)$,
$\iff$ (U6) $X \subseteq \underline{R}(\overline{R}(X))$, $\forall X \in \mathcal{P}(U)$.

2.2 Rough Approximations Induced in Formal Contexts

A *formal context* is a tripe (U, A, I), where U and A are two nonempty and finite sets, the elements of U and A are respectively called objects, and attributes or properties, thus U and A are also called set of objects and set of attributes respectively, and I is a binary relation from U to A, where $(x, a) \in I$ means that object x has attribute a.

For a formal context (U, A, I), the formal context (U, A, I^c) determined by the complement of I is called the complement context of (U, A, I), where $I^c = U \times A - I$.

Example 1. Table 1 shows a formal context $T = (U, A, I)$, where $U = \{1, 2, 3, 4, 5, 6, 7\}$ and $A = \{a, b, c, d, e, f\}$. In this table, for example, the object 4 has the properties a, c and d. The property b is possessed by the objects 2 and 5.

In [6], four types of rough approximations on formal contexts are introduced, two of them will be used in the following, which are reviewed as follows: Let (U, A, I) be a formal context.

Model 1 (Based on a Covering). Assume that (U, A, I) is regular, that is, $I(x) \neq \emptyset$ and $I(x) \neq A$ for all $x \in U$; $I^{-1}(a) \neq \emptyset$ and $I^{-1}(a) \neq U$ for all $y \in A$. The family $C_I = \{I^{-1}(a) | a \in A\}$ of subsets of objects is a covering of U, that is, $\bigcup_{a \in A} I^{-1}(a) = U$. Then the pair (U, C_I) is a covering approximation space, the first type of rough approximations on (U, A, I) are covering-based rough approximations defined as follows: for any $X \subseteq U$,

$$\underline{C_I}(X) = \bigcup \{Y \in C_I | Y \subseteq X\}, \quad \overline{C_I}(X) = \bigcap \{Y \in C_{I^c} | X \subseteq Y\}. \tag{1}$$

Table 1. A formal context $T = (U, A, I)$

U	a	b	c	d	e	f
1	1	0	0	1	0	0
2	0	1	0	0	1	0
3	1	0	0	0	0	1
4	1	0	1	1	0	0
5	0	1	1	0	0	0
6	0	0	0	0	1	0
7	0	0	0	0	0	1

A formal context (U, A, I) can be viewed as a generalized approximation space, and the rough approximations defined by (1) can be represented by the lower and upper approximations w.r.t. (U, A, I) as follows: For any $X \subseteq U$,

$$\underline{C_I}(X) = \overline{I}(\underline{I^{-1}}(X)), \quad \overline{C_I}(X) = \underline{I}(\overline{I^{-1}}(X)).$$

Model 2 (Based on a tolerance relation). A tolerance relation S_I on U can be defined by $S_I = \{(x, y) \in U \times U | \exists a \in A(xIa \wedge yIa)\}$, then the second type of rough approximations on (U, A, I) are the rough approximations with respect to (U, S_I), that is, for any $X \subseteq U$,

$$\underline{S_I}(X) = \{x \in U | S_I(x) \subseteq X\}, \ \overline{S_I}(X) = \{x \in U | S_I(x) \cap X \neq \emptyset\}.$$

Similarly, $\underline{S_I}$ and $\overline{S_I}$ can also be expressed by the generalized rough approximation operators w.r.t. (U, A, I), that is, for any $X \subseteq U$,

$$\underline{S_I}(X) = \underline{I}(\underline{I^{-1}}(X)), \quad \overline{S_I}(X) = \overline{I}(\overline{I^{-1}}(X)). \tag{2}$$

The following relation can be seen in [6]: for any $X \subseteq U$,

$$\underline{S_I}(X) \subseteq \underline{C_I}(X) \subseteq X \subseteq \overline{C_I}(X) \subseteq \overline{S_I}(X). \tag{3}$$

3   Rough Approximations in Formal Contexts

In [1], Cattaneo introduced the notion of abstract approximation space, and by means of the theory of quasi BZ posets [2] a method for constructing rough approximation operators was proposed. In this section, we will present two types of rough approximations in the framework of formal contexts by using the method proposed by Cattaneo.

The following description will be carried out in a common formal context (U, A, I).

3.1   Brouwer Orthocomplementation Induced by Formal Contexts

A preclusivity (or distinguishability) relation $\# \subseteq U \times U$ on the set U of objects can be induced as follows:

$$\# = \{(x, y) \in U \times U | I(x) \cap I(y) = \emptyset\}.$$

Then the relation $\#$ is irreflexive ($x\#y$ implies $x \neq y$) and symmetric. For any $x, y \in U$, $x\#y$ means that x and y are distinguishable, which holds if and only if x and y have no the same attributes.

The preclusive orthocomplementation of any subset X of the universe U is defined as

$$X^{\#} = \{x \in U | \forall y \in X(x\#y)\}.$$

Facts 1 ([1]). The operator $\# : \mathcal{P}(U) \to \mathcal{P}(U)$ is a Brouwer orthocomplementation mapping [1], i.e. for which the following holds whatever $X, Y \subseteq U$,

(1) $X \subseteq X^{\#\#}$;
(2) $X \subseteq Y$ implies $Y^{\#} \subseteq X^{\#}$;
(3) $X \cap X^{\#} = \emptyset$.

From the definition of the relation $\#$, it can be seen that $\#$ is just the complement relation of S_I, that is, $\# = (S_I)^c$. Thus orthocomplementation $\#$ on $\mathcal{P}(U)$ is closely related to the approximation operators based on the binary relation S_I.

Proposition 1. $X^{\#c} = \overline{S(I)}(X)$ for all $X \in \mathcal{P}(U)$.

3.2   Rough Approximations Induced by Brouwer Orthocomplementation

The standard set theoretic complement c on $\mathcal{P}(U)$ is a Kleene orthocomplementation [1] and $\#$ defined on $\mathcal{P}(U)$ is a Brouwer orthocomplementation, by Property (U5) and Proposition 1 we know that c and $\#$ satisfy the weak interconnection rule:

$$X^{\#} \subseteq X^c, \forall X \in \mathcal{P}(U).$$

Since $(\mathcal{P}(U), \cup, \cap)$ is an atomic distributive complete lattice (w.r.t. set theoretic union $\cup$ and intersection $\cap$), so the structure $(\mathcal{P}(U), \cup, \cap, c, \#, \emptyset, U)$ is a quasi BZ lattice [1].

Using the method introduced by Cattaneo in [1], we can define two pairs of lower and upper rough approximation operators. For any $X \in \mathcal{P}(U)$, from Condition (win) it follows that $X^{\#\#} \subseteq X^{\#c}$, by Facts 1 (1) we get

$$X \subseteq X^{\#\#} \subseteq X^{\#c}.$$

Therefore we can define $X^{\#\#}$ and $X^{\#c}$ as two upper approximations of $X \subseteq U$, and re-denoted by $\mathcal{C}(X)$ and $\mu(X)$ respectively, that is, for any $X \in \mathcal{P}(U)$,

$$\mathcal{C}(X) = X^{\#\#}, \mu(X) = X^{\#c}.$$

In order to define two corresponding lower approximation operators, we denote the dual operator of $\#$ w.r.t. the complement operator c by $\&$, i.e. $\& = c\#c$.

Facts 2 ([1]). The operator $\&$ is an anti-Brouwer orthocomplementation w.r.t. $\#$, i.e. whatever $X, Y \in \mathcal{P}(U)$:

(1) $X^{\&\&} \subseteq X$;
(2) $X \subseteq Y$ implies $Y^{\&} \subseteq X^{\&}$;
(3) $X \cup X^{\&} = U$.

The dual operator of $\#c$ w.r.t. the complement operator c is $c(\#c)c$, which obviously equals $c\#$. Therefore, for any $X \subseteq U$, with respect to the upper approximation $\mu(X)$, a lower approximation of X is defined as $X^{c\#}$, and re-denoted by $\nu(X)$.

Similarly, as the dual operator of $\#\#$ is $c(\#\#)c$ and $c(\#\#)c = (c\#c)(c\#c) = \&\&$, for any $X \subseteq U$, w.r.t. the upper approximation $\mathcal{C}(X)$, a lower approximation of X is defined as $X^{\&\&}$, and re-denoted by $\mathcal{O}(X)$.

As a result, two pairs of dual rough approximation operators are obtained, that is, for any $X \in \mathcal{P}(U)$,

$$\nu(X) = X^{c\#}, \ \mu(X) = X^{\#c};$$
$$\mathcal{O}(X) = X^{\&\&}, \mathcal{C}(X) = X^{\#\#}.$$

Based on Proposition 1 the following conclusions can be proved easily.

Theorem 1. *For any $X \in \mathcal{P}(U)$, we have*

(1) $\nu(X) = \underline{S_I}(X), \mu(X) = \overline{S_I}(X)$;
(2) $\mathcal{O}(X) = \overline{S_I}(\underline{S_I}(X)), \mathcal{C}(X) = \underline{S_I}(\overline{S_I}(X))$.

From Theorem 1 it is clear that the operators ν and μ coincide with the existing operators $\underline{S_I}$ and $\overline{S_I}$ respectively, thus they are rough approximation operators based on a tolerance relation basically. As for the operators $\mathcal{O}$ and $\mathcal{C}$, based on Theroem 1 and Eq. (2) we have the following conclusions.

Corollary 1. *For any $X \in \mathcal{P}(U)$, we have*

$$\mathcal{O}(X) = \overline{I}(\overline{I^{-1}}(\underline{I}(\underline{I^{-1}}(X)))), \mathcal{C}(X) = \underline{I}(\underline{I^{-1}}(\overline{I}(\overline{I^{-1}}(X)))).$$

Since S_I is a tolerance relation on U, so the successor neighborhoods of all the objects of U form a covering of U, denote it by $\mathcal{C}_{S_I}$, then $\mathcal{C}_{S_I} = \{S_I(x) | x \in U\}$.

Proposition 2. *For any $X \in \mathcal{P}(U)$, we have*

$$\mathcal{O}(X) = \underline{\mathcal{C}_{S_I}}(X), \ \mathcal{C}(X) = \overline{\mathcal{C}_{S_I}}(X).$$

Therefore, $\mathcal{O}$ and $\mathcal{C}$ are rough approximation operators based on coverings. For the coverings $\mathcal{C}_I$ and $\mathcal{C}_{S_I}$, it is easy to prove that for any $x \in U$, $S_I(x) = \bigcup\{I^{-1}(a) | xIa\}$, that is, $\mathcal{C}_I$ is finer than $\mathcal{C}_{S_I}$ [7]. According to Theorem 3 in [7] and by Properties (L5) and (U5), and (L6) and (U6), the following conclusions can be obtained.

Theorem 2. *For any $X \in \mathcal{P}(U)$, we have*

$$\nu(X) \subseteq \mathcal{O}(X) \subseteq \underline{\mathcal{C}_I}(X) \subseteq X \subseteq \overline{\mathcal{C}_I}(X) \subseteq \mathcal{C}(X) \subseteq \mu(X). \tag{4}$$

Theorem 2 shows that $(\mathcal{O}(X), \mathcal{C}(X))$ captures less information about any subset $X \subseteq U$ than $(\underline{\mathcal{C}_I}(X), \overline{\mathcal{C}_I}(X))$, and more information than $(\nu(X), \mu(X))$.

The example below shows that $(\mathcal{O}, \mathcal{C})$ is different from $(\underline{\mathcal{C}_I}, \overline{\mathcal{C}_I})$ and (ν, μ).

Example 2. Let (U, A, I) is the formal context in Example 1 and $X = \{1, 2, 4, 5, 7\}$. According to the definitions of the approximation operators in (4), we have $\mathcal{O}(X) = \{2, 4, 5\}$, $\underline{\mathcal{C}_I}(X) = \{1, 2, 4, 5\}$ and $\nu(X) = \{5\}$. Thus, $\mathcal{O}(X)$ does not equal $\underline{\mathcal{C}_I}$ and $\nu(X)$. By the duality of the lower and upper approximation operators, we know that $\mathcal{C}(X)$ does not equal $\overline{\mathcal{C}_I}$ and $\mu(X)$ either.

The following theorem shows that under some conditions, $(\mathcal{O}, \mathcal{C})$ may be equal to $(\underline{\mathcal{C}_I}, \overline{\mathcal{C}_I})$ or (ν, μ).

Theorem 3. *For the operators $(\mathcal{O}$ and $\mathcal{C})$, following statements hold:*

(1) $\mathcal{C}(X) = \mu(X)$ *(or, $\mathcal{O}(X) = \nu(X)$) for all $X \in \mathcal{P}(U)$ if and only if (i) the family $\{S_I(x) | x \in U\}$ is a partition of U.*
(2) $\mathcal{C}(X) = \overline{\mathcal{C}_I}(X)$ *(or, $\mathcal{O}(X) = \underline{\mathcal{C}_I}(X)$) for all $X \in \mathcal{P}(U)$ if and only if (ii) for any $a \in A$, there exists a $x \in U$ with $I^{-1}(a) = S_I(x)$.*

It should be noted that the conditions (i) and (ii) in Theorem 3 are not equivalent.

The equation $\mathcal{C}(X) = \mu(X)$ is tantamount to the following interconnection rule:

$$(\text{in}) \ X^{\#\#} = X^{\#c}, \forall X \in \mathcal{P}(U),$$

which is stronger than (win). From [1] it follows that under the condition (i) in Theorem 3, the quasi BZ lattice $(\mathcal{P}(U), \cup, \cap, ^c, ^\#, \emptyset, U)$ turns into a BZ lattice.

3.3 Algebraic Properties of Rough Approximations

Referring to the properties of the rough approximations based on coverings [7], we can give properties of $\mathcal{O}$ and $\mathcal{C}$.

Theorem 4. *Let $X, Y \subseteq U$. Then*

(6) $\mathcal{O}(U) = U, \mathcal{C}(\emptyset) = \emptyset$;
(7) $X \subseteq Y, \mathcal{O}(X) \subseteq \mathcal{O}(Y), \mathcal{C}(X) \subseteq \mathcal{C}(Y)$;
(8) $\mathcal{O}(X) \subseteq X, X \subseteq \mathcal{C}(Y)$;
(9) $\mathcal{O}(X) \subseteq \mathcal{O}\mathcal{O}(X)), \mathcal{C}(\mathcal{C}(X)) \subseteq \mathcal{C}(X)$.

Definition 1. *For any $X \in \mathcal{P}(U)$,*

(1) X *is said to be lower definable w.r.t. (ν, μ) if $\nu(X) = X$, and the set of all lower definable subsets is denoted by Σ_{rl}^{ν};*

(2) X is said to be *upper definable* w.r.t. (ν, μ) if $\mu(X) = X$, and the set of all upper definable subsets is denoted by Σ_{ru}^{μ};

(3) X is said to be *rough definable* w.r.t. (ν, μ) if X is lower definable and upper definable w.r.t. (ν, μ), denote the set of all rough definable subsets as $\Sigma_r^{(\nu,\mu)}$, then $\Sigma_r^{(\nu,\mu)} = \Sigma_{rl}^{\nu} \cap \Sigma_{ru}^{\mu}$;

(4) X is said to be *open* w.r.t. $(\underline{\mathcal{C}_I}, \overline{\mathcal{C}_I})$ if $\underline{\mathcal{C}_I} = X$, and the set of all open subsets is denoted by $\Sigma_o^{\mathcal{C}_I}$;

(5) X is said to be *closed* w.r.t. $(\underline{\mathcal{C}_I}, \overline{\mathcal{C}_I})$ if $\overline{\mathcal{C}_I}(X) = X$, and the set of all closed subsets is denoted by $\Sigma_c^{\mathcal{C}_I}$;

(6) X is said to be *clopen* w.r.t. $(\underline{\mathcal{C}_I}, \overline{\mathcal{C}_I})$ if X is open and closed w.r.t. $(\underline{\mathcal{C}_I}, \overline{\mathcal{C}_I})$, denote the set of all clopen subsets as $\Sigma_{co}^{\mathcal{C}_I}$, then $\Sigma_{co}^{\mathcal{C}_I} = \Sigma_o^{\mathcal{C}_I} \cap \Sigma_c^{\mathcal{C}_I}$;

(7) X is said to be *open* w.r.t. $(\mathcal{O}, \mathcal{C})$ if $\mathcal{O}(X) = X$, and the set of all open subsets is denoted by $\Sigma_o^{\mathcal{C}_{SI}}$;

(8) X is said to be *closed* w.r.t. $(\mathcal{O}, \mathcal{C})$ if $\mathcal{C}(X) = X$, and the set of all closed subsets is denoted by $\Sigma_c^{\mathcal{C}_{SI}}$;

(9) X is said to be *clopen* w.r.t. $(\mathcal{O}, \mathcal{C})$ if X is open and closed w.r.t. $(\mathcal{O}, \mathcal{C})$, denote the set of all clopen subsets as $\Sigma_{co}^{\mathcal{C}_{SI}}$, then $\Sigma_{co}^{\mathcal{C}_{SI}} = \Sigma_o^{\mathcal{C}_{SI}} \cap \Sigma_c^{\mathcal{C}_{SI}}$.

As ν and μ, $\mathcal{O}$ and $\mathcal{C}$, and $\underline{\mathcal{C}_I}$ and $\overline{\mathcal{C}_I}$ are dual to each other, respectively, we have $\Sigma_{rl}^{\nu} = (\Sigma_{ru}^{\nu})^c$, $\Sigma_o^{\mathcal{C}_{SI}} = (\Sigma_c^{\mathcal{C}_{SI}})^c$, and $\Sigma_o^{\mathcal{C}_I} = (\Sigma_c^{\mathcal{C}_I})^c$, where for a family $\mathcal{S}$ of subsets of U, $\mathcal{S}^c$ denotes a family with $\mathcal{S}^c = \{X \subseteq U | X^c \in \mathcal{S}\}$.

According to the chain of inclusions (4) and by Definition 1, we get the following theorem.

Theorem 5. $\Sigma_{rl}^{\nu} \subseteq \Sigma_o^{\mathcal{C}_{SI}} \subseteq \Sigma_o^{\mathcal{C}_I}$, $\Sigma_{ru}^{\mu} \subseteq \Sigma_c^{\mathcal{C}_{SI}} \subseteq \Sigma_c^{\mathcal{C}_I}$, so $\Sigma_r^{(\nu,\mu)} \subseteq \Sigma_{co}^{\mathcal{C}_{SI}} \subseteq \Sigma_{co}^{\mathcal{C}_I}$.

From Theorem 3 it follows that if the condition (i) holds then $\Sigma_r^{(\nu,\mu)} = \Sigma_o^{\mathcal{C}_{SI}} = \Sigma_c^{\mathcal{C}_{SI}} = \Sigma_{co}^{\mathcal{C}_{SI}}$, and if the condition (ii) holds then $\Sigma_o^{\mathcal{C}_{SI}} = \Sigma_o^{\mathcal{C}_I}$, $\Sigma_c^{\mathcal{C}_{SI}} = \Sigma_c^{\mathcal{C}_I}$, and $\Sigma_{co}^{\mathcal{C}_{SI}} = \Sigma_{co}^{\mathcal{C}_I}$. Furthermore, by Theorem 3.16 of [8], if $\mathcal{C}_I$ is a partition of U then all the sets of definable subsets become the same one.

Theorem 6. $\mathcal{O}(X), \mu(X) \in \Sigma_o^{\mathcal{C}_{SI}}$, $\mathcal{C}(X), \nu(X) \in \Sigma_c^{\mathcal{C}_{SI}}$, $\forall X \in \mathcal{P}(U)$.

As for the definable subsets w.r.t. (ν, μ), the following conclusion holds.

Proposition 3. For any $X \in \mathcal{P}(U)$, X is lower definable w.r.t. (ν, μ) if and only if it is upper definable w.r.t. (ν, μ).

By Proposition 3 we have $\Sigma_{rl}^{\nu} = \Sigma_{ru}^{\mu} = \Sigma_r^{(\nu,\mu)}$. It can be verified that $\emptyset$ and U belong to all the sets of definable subsets of Theorem 5.

Theorem 7. $\Sigma_o^{\mathcal{C}_{SI}}, \Sigma_c^{\mathcal{C}_{SI}}, \Sigma_r^{(\nu,\mu)}, \Sigma_o^{\mathcal{C}_I}$ and $\Sigma_c^{\mathcal{C}_I}$ are all complete lattice under the set inclusion relation $\subseteq$, and following statements hold:

(1) For $(\Sigma_o^{\mathcal{C}_{SI}}, \subseteq)$, denote the join and meet by $\vee_o^{\mathcal{C}_{SI}}$ and $\wedge_o^{\mathcal{C}_{SI}}$ respectively, then $\vee_o^{\mathcal{C}_{SI}}$ coincides with the set theoretic union $\cup$, and $X \wedge_o^{\mathcal{C}_{SI}} Y = \mathcal{O}(X \cap Y)$, $\forall X, Y \in \Sigma_o^{\mathcal{C}_{SI}}$;

(2) *For* $(\Sigma_c^{\mathcal{C}_{SI}}, \subseteq)$, *denote the join and meet by* $\vee_c^{\mathcal{C}_{SI}}$ *and* $\wedge_c^{\mathcal{C}_{SI}}$ *respectively, then* $\wedge_c^{\mathcal{C}_{SI}}$ *coincides with the set theoretic intersection* $\cap$*, and* $X \vee_c^{\mathcal{C}_{SI}} Y = \mathcal{C}(X \cup Y), \forall X, Y \in \Sigma_c^{\mathcal{C}_{SI}}$.

(3) *For* $(\Sigma_r^{(\nu,\mu)}, \subseteq)$, *denote the join and meet by* $\vee_r^{(\nu,\mu)}$ *and* $\wedge_r^{(\nu,\mu)}$ *respectively, then* $\vee_r^{(\nu,\mu)}$ *and* $\wedge_r^{(\nu,\mu)}$ *coincide with the set theoretic union* $\cup$ *and intersection* $\cap$ *respectively;*

(4) *For* $(\Sigma_o^{\mathcal{C}_I}, \subseteq)$, *denote the join and meet by* $\vee_o^{\mathcal{C}_I}$ *and* $\wedge_o^{\mathcal{C}_I}$ *respectively, then* $\vee_o^{\mathcal{C}_I}$ *coincides with the set theoretic union* $\cup$*, and* $X \wedge_o^{\mathcal{C}_I} Y = \underline{\mathcal{C}_I}(X \cap Y), \forall X, Y \in \Sigma_o^{\mathcal{C}_I}$;

(5) *For* $(\Sigma_c^{\mathcal{C}_I}, \subseteq)$, *denote the join and meet by* $\vee_c^{\mathcal{C}_I}$ *and* $\wedge_c^{\mathcal{C}_I}$ *respectively, then* $\wedge_c^{\mathcal{C}_I}$ *coincides with the set theoretic intersection* $\cap$*, and* $X \vee_c^{\mathcal{C}_I} Y = \overline{\mathcal{C}_I}(X \cup Y), \forall X, Y \in \Sigma_c^{\mathcal{C}_I}$.

Based on Theorem 7 we can obtain the following conclusions.

Corollary 2. $\Sigma_{co}^{\mathcal{C}_{SI}}$, $\Sigma_r^{(\nu,\mu)}$, *and* $\Sigma_{co}^{\mathcal{C}_I}$ *are complete lattices under the set inclusion relation* $\subseteq$*, and all the join and meet operations coincide with set theoretic union and intersection respectively.*

4   Summaries

In the framework of formal contexts, many rough approximation operations have been proposed. In order to obtain some different rough approximation operators in a formal context (U, A, I), the method used by Cattaneo in [1] is adopted in this paper, then a pair of new lower and upper rough approximation operators on the power set $\mathcal{P}(U)$ is introduced, i.e. $(\mathcal{O}, \mathcal{C})$, which locates between two pairs of the existing rough approximation operators, i.e. (ν, μ) and $(\underline{\mathcal{C}_I}, \overline{\mathcal{C}_I})$. Some conditions are obtained, under which the new operators and the near operators are equal. After definable subsets of these operators are presented, relationships and algebra properties of the definable subsets are discussed.

It is well known that attribute reduction of formal contexts is a key issue in formal concept analysis, so with respect to the new operators $\mathcal{O}$ and $\mathcal{C}$, we will study corresponding attribute reduction of formal contexts.

Acknowledgements. This work was supported by grants from the National Natural Science Foundation of China (Nos. 11071284, 61075120, 61272021, 61202206) and the Zhejiang Provincial Natural Science Foundation of China (Nos. LY14F030001, LZ12F03002, LY12F02021).

References

1. Cattaneo, G.: Abstract approximation spaces for rough theories. In: Polkowski, S. (ed.) Rough Set and Kownoledge Discovery 1. Studies in Fuzziness and Soft Computing, vol. 18, pp. 59–98. Physica-Verlag, Heidelberg (1998)

2. Cattaneo, G., Marino, G.: Non-usual orthocomplementations onpartially ordered-sets and fuzziness. Fuzzy Sets Syst. **25**(1), 107–123 (1988)
3. Bonikowski, Z., Bryniarski, E., Wybraniec, U.: Extensions and intentions in the rough set theory. Inf. Sci. **107**, 149–167 (1998)
4. Gediga, G., Duntsch, I.: Modal-style operators in qualitative data analysis. In: Proceedings of the 2002 IEEE International Conference on Data Mining, pp. 155–162 (2002)
5. Kent, R.E.: Rough concept analysis. In: Ziarko, W.P. (ed.) Rough Sets, Fuzzy Sets and Knowledge Discovery, pp. 248–255. Springer, London (1993)
6. Li, T.J., Zhang, W.X.: Rough approximations in formal contexts. In: Proceedings of the Fourth International Conference on Machine Learning and Cybernetics, ICMLC 2005, Guangzhou, pp. 18–21 (2005)
7. Li, T.-J.: Rough approximation operators in covering approximation spaces. In: Greco, S., Hata, Y., Hirano, S., Inuiguchi, M., Miyamoto, S., Nguyen, H.S., Słowiński, R. (eds.) RSCTC 2006. LNCS (LNAI), vol. 4259, pp. 174–182. Springer, Heidelberg (2006)
8. Li, T.J.: Rough approximation operators on two univeses of discourse and their fuzzy extensions. Fuzzy Sets Syst. **159**, 3033–3050 (2008)
9. Pagliani, P.: From concept lattices to approximation spaces: algebraic structures of some spaces of partial objects. Fundamenta Informaticae **18**(1), 1–25 (1993)
10. Pawlak, Z.: Rough sets. Int. J. Comput. Inf. Sci. **11**, 341–356 (1982)
11. Slowinski, R., Vanderpooten, D.: A generalized definition of rough approximations based on similarity. IEEE Trans. Knowl. Data Eng. **12**, 331–336 (2000)
12. Wille, R.: Restructuring lattice theory: an approach based on hierarchies of concepts. In: Rival, I. (ed.) Ordered Sets, pp. 445–470. Reidel, Dordrecht, Boston (1982)
13. Wu, W.Z., Zhang, W.X.: Constructive and axiomatic approaches of fuzzy approximation operators. Inf. Sci. **159**, 233–254 (2004)
14. Yao, Y.Y.: Constructive and algebraic methods of the theory of rough sets. Inf. Sci. **109**, 21–47 (1998)
15. Yao, Y.: A comparative study of formal concept analysis and rough set theory in data analysis. In: Tsumoto, S., Słowiński, R., Komorowski, J., Grzymała-Busse, J.W. (eds.) RSCTC 2004. LNCS (LNAI), vol. 3066, pp. 59–68. Springer, Heidelberg (2004)
16. Yao, Y.Y.: Concept lattices in rough set theory. In: Proceedings of 23rd International Meeting of the North American Fuzzy Information Processing Society, pp. 796–801 (2004)
17. Yao, Y.Y.: The superiority of three-way decisions in probabilistic rough set models. Inf. Sci. **181**, 1080–1096 (2011)
18. Zhu, W., Wang, F.Y.: Reduction and axiomization of covering generalized rough sets. Inf. Sci. **152**, 217–230 (2003)

On Optimal Approximations of Arbitrary Relations by Partial Orders

Ryszard Janicki[✉]

Department of Computing and Software, McMaster University,
Hamilton, ON L8S 4K1, Canada
janicki@mcmaster.ca

Abstract. The problem of optimal quantitative approximation of an arbitrary binary relation by a partial order is discussed and some solution is provided. It is shown that even for a very simple quantitative measure the problem is NP-hard. Some quantitative metrics are also applied for known property-driven approximations by partial orders.

1 Introduction

A motivation for this kind of work has been clearly described in [9].

"Consider the following problem: we have a set of data that have been obtained in an empirical manner. From the nature of the problem we know that the set should be partially ordered, but because the data are empirical it is not. In a general case, this relation may be arbitrary. What is the best partially ordered approximation of an arbitrary relation and how this approximation can be computed?"

This paper provides an orthogonal approach to that of [9], however it can be read independently. In [9] property-driven partial order approximations of an arbitrary binary relation were provided and discussed in both the classical algebraic model and the Rough Set settings [14,15]. No quantitative metrics were used in [9].

In this paper we propose two simple metrics for measuring similarity and difference between relations, and a definition of *optimal* approximation. We also provide some justification of both metrics and the definition.

In [9,10], a special attention is paid to two partially ordered approximations of R, denoted by $(R^\bullet)^+$ and $(R^+)^\bullet$ for a given relation R. Using graph terminology, $R^\bullet$ is derived from R by erasing all arcs from all strongly connected components (or equivalently, removing all arcs from all cycles). The relation R^+ is a transitive closure of R. The relation $(R^+)^\bullet$ is a classical approximation, first proposed by Schröder in 1895 [16], which is often regarded as 'the' partially ordered approximation. We will show that with respect to our metrics, $(R^\bullet)^+$ is better approximation of R than Schröder's $(R^+)^\bullet$.

In memory of Prof. Zdzisław Pawlak.

This research has partially been supported by a Discovery NSERC grant of Canada.

V. Flores et al. (Eds.): IJCRS 2016, LNAI 9920, pp. 107–119, 2016.
DOI: 10.1007/978-3-319-47160-0_10

We will also show that finding quantitative optimal approximation, with respect to simple metrics proposed in this paper, is NP-hard.

Finally we will argue that quantitative optimal approximations (with any reasonable metrics) are somehow inconsistent with property-driven approximations of [9,10].

2   Relations, Directed Graphs and Partial Orders

In this section we recall some fairly known concepts and results that will be used later in this paper [1,4].

Let X be a set. *We assume all sets considered in this paper are finite.* Note that every relation $R \subseteq X \times X$ can be interpreted as a *directed graph* $G_R = (V, E)$ where $V = X$ is the set of vertices and $E = R$ is the set of edges (c.f. [1]).

A relation $< \in X \times X$ is a *(sharp) partial order* if it is irreflexive and transitive, i.e. if $\neg(a < a)$ and $a < b < c \implies a < c$, for all $a, b, c \in X$.

We write $a \sim_< b$ if $\neg(a < b) \wedge \neg(b < a)$, that is if a and b are either *distinctly incomparable* (w.r.t. $<$) or *identical* elements. We also write

$$a \equiv_< b \iff (\{x \mid a < x\} = \{x \mid b < x\} \wedge \{x \mid x < a\} = \{x \mid x < b\}).$$

The relation $\equiv_<$ is an *equivalence relation* (i.e. it is reflexive, symmetric and transitive) and it is called *the equivalence with respect to* $<$, since if $a \equiv_< b$, there is nothing in $<$ that can distinguish between a and b (c.f. [4]). We always have $a \equiv_< b \implies a \sim_< b$.

Let $\mathbb{PO}(X)$ denote the set on all partial orders included in $X \times X$.

For every relation $R \subseteq X \times X$, the relation $R^+ = \bigcup_{i=1}^{\infty} R^i$ is called the *transitive closure* of R, the relation $R^{-1} = \{(b, a) \mid (a, b) \in R\}$ is called the *inverse* of R, and a relation R is *acyclic* if and only if $\neg x R^+ x$ for all $x \in X$. In graph terminology, if R is acyclic then G_R is DAG (*Directed Acyclic Graph*), while if for all $x \in X$ we have $x R^+ x$ then the graph G_R is *strongly connected*.

Let R be a relation and let $a \in X$. We define

$$Ra = \{x \mid xRa\}, R^\circ a = Ra \cup \{a\} \text{ and } aR = \{x \mid aRx\}, aR^\circ = aR \cup \{a\}.$$

Now, for every relation R we can define the relations R^{cyc}, $R^\bullet$, R^{C} and $\equiv_R$ as follows

- $aR^{cyc}b \iff aR^+b \wedge bR^+a$,
- $aR^\bullet b \iff aRb \wedge \neg(aR^{cyc}b)$, i.e. $R^\bullet = R \backslash R^{cyc}$.
- $aR^{\mathsf{C}}b \iff bR^\circ \subset aR^\circ \wedge R^\circ a \subset R^\circ b$,
- $a \equiv_R b \iff aR = bR \wedge Ra = Rb$.

In [9,10], the relation $R^\bullet$ is called an *acyclic refinement* of R and the relation R^{C} is called an *inclusion property kernel* of R. In graph terminology, if $aR^{cyc}b$ then a and b are *strongly connected* in G_R, and the graph $G_{R^\bullet} = (X, R^\bullet)$ has been derived from $G_R = (X, R)$ by deleting all edges from all *strongly connected components* of G_R. The relation $\equiv_R$ is an extension of $\equiv_<$ for arbitrary relation R proposed in [9]. As for partial orders, if $a \equiv_R b$, there is nothing in R that can distinguish between a and b (with respect to R).

Corollary 1. *1. $R^\bullet \subseteq R$, $R^\bullet$ is acyclic (i.e. also irreflexive), and $aR^\bullet b \iff aRb \wedge \neg(bR^+a)$.*
2. $R^c \subseteq R$ and R^c is a partial order,
3. If R is a partial order then $R = R^+ = R^\bullet = R^c$. □

Let $|X| = n$. Given R, the complexity of calculating R^+ is $O(n^3)$ (Floyd-Warshall Algorithm). Calculating $R^\bullet$ comprises of finding all strongly connected components of G_R (Tarjan Algorithm can be used) and then deleting all edges from all strongly connected components so the time complexity is $O(|X|+|R|) = O(n^2)$ (c.f. [1]).

3 Problems with Optimal Approximation

Let R and S be two relations on X and $G_R = (X, R)$, $G_S = (X, S)$ their appropriate graph representations. Without loosing any generality we may assume that both R and S are irreflexive, i.e. $(a, a) \notin R \cup S$ for any $a \in X$. How can we measure a difference or similarity between R and S? One possibility is just to count common edges of the graphs G_R and G_S, which leads to

$$sim(R, S) = |R \cap S|.$$

We will call $sim(R, S)$ an *absolute similarity* between relations R and S. We added *absolute* to distinguish it from *similarity* as formally defined for instance in [11, 19].

The other possibility is to count the edges that were removed from R and to the number of edges that were added to R to get S. In this case we can define:

$$dist(R, S) = |R \backslash S| + |S \backslash R| = |R \cup S| - |R \cap S|.$$

We will call $dist(R, S)$ an *absolute distance* between relations R and S.

When we scale both *sim* and *dist* to $[0, 1]$, we get well known and popular *Jaccard similarity* and *Jaccard distance* [7]: $sim_J(R, S) = \frac{|R \cap S|}{|R \cup S|}$ and $dist_J(R, S) = \frac{|R \cup S| - |R \cap S|}{|R \cup S|} = 1 - sim_J(R, S)$. However in this paper we will use unscaled measures *sim* and *dist* instead of Jaccard indexes. While $sim_J(R, S) + dist_J(R, S) = 1$, there is no that type of relationship that involves *only* $sim(R, S)$ and $dist(R, S)$. In our approach the relation S is a partial order and we will show that $sim(R, S)$ measures different aspects of approximation than $dist(R, S)$. Moreover, Jaccard indexes are meaningless when $S = \emptyset$, and $\emptyset$ is a valid and useful partial order.

If S is interpreted as some approximation of R, we may use:

$$dist_{sim}(R, S) = |R| - sim(R, S)$$

as a measure of closeness of S to R. If $R = S$ then $dist_{sim}(R, S) = 0$.

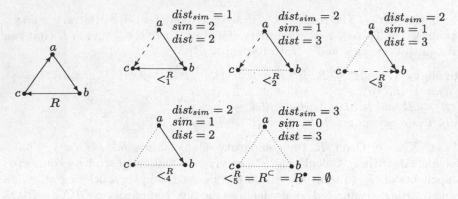

Fig. 1. An example of a relation R and its all potential partial order approximations (up to isomorphism). Dashed edges are added, dotted lines represent incomparability. For example $R \setminus <_1^R = \{(c,a)\}$ and $<_1^R \setminus R = \{(a,c)\}$, so $dist(R, <_1^R) = 2$.

Let $R \subseteq X \times X$ be an arbitrary relation. It is temped to say that a partial order $<^R$ on X is the best partial order approximation of R if $sim(R, <^R)$ is *maximal* for all partial orders on X, and/or if $dist(R, <^R)$ is *minimal* for all partial orders on X. However such straightforward approach may lead to unexpected and maybe undesired results.

Consider the relation R from Fig. 1. There are five non-isomorphic partial orders on the three elements set $\{a, b, c\}$, and they are named $<_i^R$, $i = 1, \dots, 5$, in Fig. 1. If only values of *sim* and *dist* are taken into account, a partial order $<_1^R$ (or *any order isomorphic to it*) is an optimal or best approximation, as for all partial orders $<$ on $\{a, b, c\}$, $sim(R, <) \leq 2$, $dist(R, <) \geq 2$, and $sim(R, <_1^R) = 2$, $dist(R, <_1^R) = 2$. The order $<_1^R$ had been obtained by 'flipping' edge (c, a) of G_R, two additional isomorphic orders can be obtained by 'flipping' edges (a, b) and (b, c) respectively. In all three cases the values of *sim* and *dist* are the same as for $<_1^R$. But why have we chosen 'flipping' (c, a)? Why not (a, b) or (b, c)? Are we allowed to flip at all without seriously alternating input data, especially if choice of what to flip appears to be random?

In decision and ranking theory, where outcomes are expected to be specialized partial orders, cycles in input relation R are usually interpreted as *indifference* or *incomparability* [5,8]. With this interpretation, $<_5^R$ would be considered as the *only* acceptable partially ordered approximation of R, but $sim(R, <_5^R) = 0$ and $dist(R, <_5^R) = 3$, so according to the values of *sim* and *dist*, the partial order $<_5^R$ is the *worst* partial order approximation.

Consider now the relation Q from Fig. 2, which is acyclic but not a partial order. It can be shown by inspection that for all partial orders over the set $\{a, b, c, d\}$ the value $dist(Q, <_1^Q) = 1$ is minimal. The partial order $<_1^Q$ resulted from deleting the edge (b, c) from the graph G_Q. But why should we delete (a, b), i.e. make a and b incomparable (recall that Q represents empirical data but the nature of problem demands that Q should be a partial order)? The relation Q is acyclic but it lacks transitivity, which most likely results from the fact

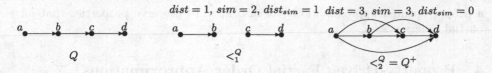

Fig. 2. A relation Q and its two potential partial order approximation. Note that $Q^C = Q^{cyc} = \emptyset$ and $Q^\bullet = Q$. The picture describing Q is *not* a Hasse diagram, it describes the full relation Q, so Q is *not* a partial order!

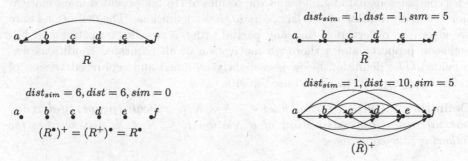

Fig. 3. A relation R, one of its optimal acyclic approximation $\widehat{R}$, and its two potential partial order approximations: $(R^\bullet)^+$ and $(\widehat{R})^+$.

that the empirical data represented by Q are incomplete, or because providing explicit transitivity was considered unneeded. In this case the most natural and proper way to transform Q into an appropriate partial order is to compute Q^+, the transitive closure of Q. We have $dist(Q, <_2^Q) = 3 > 1 = dist(Q, <_1^Q)$, but on the other hand $sim(Q, <_2^Q) = 3 > 2 = sim(Q, <_1^Q)$ and additionally $dist_{sim}(Q, <_2^Q) = 0$. So, which approximation is better, $<_1^Q$ or $<_2^Q$?

The situation that, for some S_1 and S_2, $sim(R, S_1) > sim(R, S_2)$, so S_1 is a better approximation of R with respect to $sim(\ldots)$, and $dist(R, S_2) < dist(R, S_1)$, so S_2 is a better approximation of R with respect to $dist(\ldots)$, occurs quite often. Consider the relation R from Fig. 3. The graph G_R is strongly connected and $\widehat{R}$ is one of the optimal acyclic approximations of R, while $(R^\bullet)^+$ and $(\widehat{R})^+$ are two partial orders that can be regarded as partial order approximations of R. We have $sim(R, (\widehat{R})^+) > sim(R, (R^\bullet)^+)$ while $dist(R, (R^\bullet)^+) < dist(R, (\widehat{R})^+)$. The relation $(\widehat{R})^+$ has the same problem as the relation $<_1^R$ of Fig. 1, $\widehat{R}$ is one of six different optimal acyclic approximations, it resulted by removing the edge (f, a). But why this edge, not any other?

Consider $Q = \widehat{R}$, where both R and $\widehat{R}$ are these from Fig. 3. In this case $Q^\bullet = Q$, $Q^+ = (Q^\bullet)^+ = (Q^+)^\bullet$ and $Q^+ = (\widehat{R})^+$. Moreover, $sim(Q, Q^+) = 5$, $dist_{sim}(Q, Q^+) = 0$ and $dist(Q, Q^+) = 9$. Since $dist_{sim}(Q, Q^+) = 0$, Q^+ is the best partial order approximation of Q with respect to measure $sim(\ldots)$.

These results indicate that using only $sim(R, S)$ and/or $dist(R, S)$ (or any particular numerical measure in fact) is not sufficient when we are looking for a proper optimal approximation. We also have to preserve properties that all partial orders posses.

4   Property-Driven Partial Order Approximations

In [9] the definitions of property driven partial order approximation and weak partial order approximation were presented and discussed. Ranking process, pairwise comparisons paradigm [8] and the results of [18,20] provided main motivation for interpretation of partial orders in these definitions. The definitions were based on the observation that any partial order is acyclic, transitive and has inclusion property and a thorough motivation of all requested conditions was provided. The definition below is a slightly modified and rephrased version of definitions proposed in [9] and used in [10].

Definition 1 ([9]). *A partial order $< \subseteq X \times X$ is a **weak (property-driven)** **partial order approximation** of a relation $R \subseteq X \times X$ if it satisfies the following four conditions:*

1. $a < b \implies aR^+b,$ 3. $aR^\bullet b \implies a < b$
2. $aR^C b \implies a < b,$ 4. $a \equiv_R b \implies a \equiv_< b.$

*A weak partial order approximation $<$ is a **(property-driven) partial order approximation** if additionally*

5. $a < b \implies \neg aR^{cyc}b$ *(or, equivalently $a < b \implies \neg bR^+a$).*   □

In [9,10] the conditions (2) and (3) where represented by one stronger condition, namely, $aR^C b \wedge aR^\bullet b \implies a < b$, which we now believe should be replaced by two separate implications. Definition 1 was motivated by the following intuitions [10]. Since R^+ is the smallest transitive relation containing R, and due to informational noise, imprecision, randomness, etc., some parts of R might be missing, it is reasonable to assume that R^+ is the upper bound of $<$, so condition (1). Conditions (2) and (3) define lower bounds. The greatest partial order included in R usually does not exist, but when R is interpreted as an estimation of a ranking, these lower bounds appear to be reasonable (c.f. [10]). Condition (4) ensures preservation of the equivalence with respect to R. Condition (5) says that if $aR^{cyc}b$ then usually a and b are incomparable. If R is interpreted as an estimation of a ranking, then in most cases $aR^{cyc}b$ is interpreted that a and b are indifferent [5]. Similar interpretations exist in concurrency theory [13].

The following result characterizing property-driven partial order approximations has been proven in [9] (they hold for new version of Definition 1 too).

Theorem 1 ([9]).

1. The relations $(R^\bullet)^C$, $(R^\bullet)^+$, $(R^+)^\bullet$ are (property-driven) partial order approximations of R.
2. The relation R^C is a weak (property-driven) partial order approximation of R.
3. $(R^\bullet)^C \subseteq (R^\bullet)^+ \subseteq (R^+)^\bullet$ and $(R^\bullet)^C \subseteq R^C \subseteq R$. $\qquad\qquad\qquad$ □

For the examples from Fig. 1, the partial orders $<_i^R$, $i = 1, 2, 3$ do not satisfy the condition (1) of Definition 1. We have $R^+ = R$ and for example $(a, c) \in <_1^R \setminus R^+$. The partial orders $<_i^R$, $i = 1, \ldots, 4$ do not also satisfy the condition (5) of Definition 1. For example $b <_1^R a$ and bR^+a. From Fig. 1, only $<_5^R$ satisfies the condition (5) of Definition 1. The condition (3) implies that if R is acyclic, i.e. G_R is DAG, then $R \subseteq <$. The partial order $<_1^Q$ from Fig. 2 does not satisfy the condition (3), as $(b, c) \in Q = Q^\bullet$ but $(b, c) \notin <_1^Q$. The partial order $<_2^Q$ from Fig. 2 satisfies all five conditions of Definition 1.

All these examples indicate that *property-driven* partial order approximations do not fit well to numerical estimations given by functions $sim(\ldots)$ and $dist(\ldots)$ proposed in previous section. However, we may use quantitative estimations for established property-driven approximations as $(R^\bullet)^+$ and $(R^+)^\bullet$.

5 Quantitative Properties of $(R^\bullet)^+$ and $(R^+)^\bullet$

In this section we will apply measures $sim(\ldots), dist(\ldots)$ and Jaccard index $sim_J(\ldots)$ to property-driven approximations $(R^\bullet)^+$ and $(R^+)^\bullet$. We will start with characterization of their intersections with a given relation R.

Lemma 1. *For every relation $R \subseteq X \times X$, we have:* $R \cap (R^\bullet)^+ = R \cap (R^+)^\bullet = R^\bullet$.

Proof. Since $(R^\bullet)^+ \subseteq (R^+)^\bullet$, then $R \cap (R^\bullet)^+ \subseteq R \cap (R^+)^\bullet$. Assume $(a, b) \in R \cap (R^+)^\bullet$. From the definition of acyclic refinement '•', we have

$$(a, b) \in (R^+)^\bullet \iff (a, b) \in R^+ \land (b, a) \notin R^+.$$

Hence: $(a, b) \in R \cap (R^+)^\bullet \iff (a, b) \in R \land (a, b) \in R^+ \land (b, a) \notin R^+ \iff (a, b) \in R \land (b, a) \notin R^+ \iff (a, b) \in R^\bullet$. Hence $R \cap (R^\bullet)^+ \subseteq R \cap (R^+)^\bullet = R^\bullet$. On the other hand $R^\bullet \subseteq R$ and obviously $R^\bullet \subseteq (R^\bullet)^+$, so $R \cap (R^\bullet)^+ = R \cap (R^+)^\bullet = R^\bullet$. $\qquad\qquad$ □

We may now formulate the main result of this section.

Proposition 1. *For every relation R, we have:*

1. $sim(R, (R^\bullet)^+) = sim(R, (R^+)^\bullet) = |R^\bullet|$,
2. $dist(R, (R^\bullet)^+) \leq dist(R, (R^+)^\bullet)$,
3. $sim_J(R, (R^\bullet)^+) \geq sim_J(R, (R^+)^\bullet)$.

Proof. (1) A consequence of Lemma 1.

(2) Since $(R^\bullet)^+ \subseteq (R^+)^\bullet$, then $|R \cup (R^\bullet)^+| \leq |R \cup (R^+)^\bullet|$. By 1, $|R \cap (R^\bullet)^+| = |R \cap (R^+)^\bullet|$. Hence $dist(R, (R^\bullet)^+) = |R \cup (R^\bullet)^+| - |R^\bullet| \leq |R \cup (R^+)^\bullet| - |R^\bullet| = dist(R, (R^+)^\bullet)$.

(3) Since $|R \cap (R^\bullet)^+| = |R \cap (R^+)^\bullet|$ and $(R^\bullet)^+ \subseteq (R^+)^\bullet$, then

$$sim_J(R, (R^\bullet)^+) = \frac{|R \cap (R^\bullet)^+|}{|R \cup (R^\bullet)^+|} \geq \frac{|R \cap (R^+)^\bullet|}{|R \cup (R^+)^\bullet|} = sim_J(R, (R^+)^\bullet). \qquad \square$$

It appears that with respect to numerical similarity and distance measures, including Jaccard index (and all indexes consistent with Jaccard index, see [11]), the relation $(R^\bullet)^+$ seems to be better approximation than Schröder's $(R^+)^\bullet$.

While time complexity of calculating $(R^\bullet)^+$ and $(R^+)^\bullet$ is $O(n^3)$ for both (as calculating transitive closure is a dominating factor in both cases), practical time complexity is always smaller for $(R^\bullet)^+$ as $|R^\bullet| \leq |R|$.

6   Approximations Based on Absolute Similarity and Distance

In Sect. 3 we have discussed problems related to quantitative optimal approximation. In this section some solution, based on analysis from previous sections, is proposed. This section also contains the main results of this paper. We start with a definition of an *optimal simple partial order approximation* of a given relation R.

Definition 2. *For every relation R on X, a partial order $R^\oplus$ on X is an* **optimal simple partial order approximation** *of R if the following conditions are satisfied:*

1. *$R^\bullet \subseteq R^\oplus$,*
2. *$sim(R, R^\oplus) = \max\{sim(R, <) \mid < \in \mathbb{PO}(X)\}$,*
3. *$dist(R, R^\oplus) = \min\{dist(R, <) \mid sim(R, <) = sim(R, R^\oplus)\}$.* $\qquad \square$

Condition (1) defines a lower bound. The relation $R^\bullet$ is R with all cycles removed and is considered as a necessary part of $R^\oplus$. Lack of acyclicity is considered bigger problem than lack of transitivity. The latter could be intensional (c.f. Hasse diagrams, dependency graphs, etc. [3,5,13]), the former is often a serious error [5,8]. Conditions (2) and (3) of Definition 2 capture this asymmetry by making absolute similarity the dominant measure and absolute distance the secondary measure. We call this approximation '*simple*' as most of the properties from Definition 1 is no longer required. They are just too restrictive for quantitative optimization.

We will consider two distinct cases:

Case 1. R is acyclic, i.e. $R = R^\bullet$.
Case 2. R contains a cycle, i.e. $R^{cyc} \neq \emptyset$.

The case 1 is simple, one just has to use transitive closure.

Proposition 2. *If R is acyclic, then $R^{\oplus} = R^{+}$.*

Proof. For any partial order $<$ containing R the condition (2) of Definition 2 is satisfied as then $dist_{sim}(R, <) = 0$. Since R^{+} is the smallest partial order containing R (c.f. [4]), then the condition (3) of Definition 2 is satisfied too. $\square$

The second case involves removing cycles and it is much more complex. For every relation $R \subseteq X \times X$, let

$$\mathbb{MDAG}(R) = \{\widehat{R} \mid sim(R, \widehat{R}) = \max\{sim(R, S) \mid S \subseteq R \land S^{cyc} = \emptyset\}\}.$$

The elements of $\mathbb{MDAG}(R)$ are maximal (with respect to number of arcs) directed acyclic graphs included in R, and for every $\widehat{R}$, $R\backslash\widehat{R}$ is a *minimum feedback arc set* (c.f. [1,12,17]), which is NP-complete.

Theorem 2 (Karp 1972 [12]). *Minimum feedback arc set problem is NP-complete.* $\square$

The next theorem is the main result of this section.

Theorem 3. *If R contains a cycle, i.e. $R^{cyc} \neq \emptyset$ then:*

1. $R^{\oplus} = \widehat{R}^{+}$, *where $\widehat{R}$ is **some** relation from $\mathbb{MDAG}(R)$.*
2. *Finding an optimal simple partial order approximation $R^{\oplus}$ is NP-hard.*
3. *There are R such that for some $\tilde{R} \in \mathbb{MDAG}(R)$, $\tilde{R}^{+} \neq R^{\oplus}$.*

Proof. (1) In general $\widehat{R}$ is not unique. By Proposition 2, $\widehat{R}^{+}$ is the optimal simple partial order order approximation of $\widehat{R}$. Clearly $\widehat{R} = R \cap \widehat{R}^{+}$.

We will show that $sim(R, \widehat{R}^{+}) = \max\{sim(R, <) \mid < \in \mathbb{PO}(X)\}$. Suppose that there is $S \in \mathbb{PO}(X)$ such that $|R \cap S| > |R \cap \widehat{R}^{+}|$. But this means that $R \cap S \subseteq R$, $(R \cap S)^{cyc} = \emptyset$ and $sim(R, R \cap S) > sim(R, \widehat{R})$, a contradiction as $\widehat{R}$ is a maximal acyclic approximation of R. Hence $\widehat{R}^{+}$ is an optimal simple partial order approximation of R. Note that we only have proven that there is some $\widehat{R} \in \mathbb{MDAG}(R)$ such that $R^{\oplus} = \widehat{R}^{+}$. It does not have to be true for all members of $\mathbb{MDAG}(R)$.

(2) Suppose $R^{\oplus}$ is known. Define $R' = R \cap R^{\oplus}$. We will show that $R' = \widehat{R}$, i.e. $sim(R, R') = \max\{sim(R, S) \mid S \subseteq R \land S^{cyc} = \emptyset\}$. Suppose there exist S such that $S \subseteq R \land S^{cyc} = \emptyset$ and $|R \cap S| > |R \cap R'|$. But $R \cap R' = R \cap R^{\oplus}$, so we have $|R \cap S^{+}| \geq |R \cap S| > |R \cap R'| = |R \cap R^{\oplus}|$, i.e. $|R \cap S^{+}| > |R \cap R^{\oplus}|$, so $R^{\oplus}$ is not an optimal simple partial order approximation. Hence $R' = \widehat{R}$, and $\widehat{R} = R \cap R^{\oplus}$. We can derive $\widehat{R} = R \cap R^{\oplus}$ from R and $R^{\oplus}$ in $O(n^2)$ where $n = |X|$, which means that finding $\widehat{R}$ is polynomially reduced to finding $R^{\oplus}$. But $R\backslash\widehat{R}$ is *minimum feedback arc set*, and the minimum feedback arc set problem is one of the first problems proven NP-complete (Karp 1972 [12]).

(3) Consider R, $\widehat{R}$ and $\tilde{R}$ from Fig. 4. We have $\widehat{R}, \tilde{R} \in \mathbb{MDAG}(R)$, $\widehat{R}^{+} = R^{\oplus}$ but $\tilde{R}^{+} \neq R^{\oplus}$. $\square$

If $R^{cyc} \neq \emptyset$, feasible construction of $R^{\oplus}$ is problematic. Not only finding any element of $\mathbb{MDAG}(R)$ is NP-complete, but only for *some* $\widehat{R} \in \mathbb{MDAG}(R)$ we have $R^{\oplus} = \widehat{R}^{+}$.

However, if some suboptimal solution is acceptable, there are many efficient approximation, heuristic algorithms, or exact, but feasible, that can be used for $\tilde{R} \in \mathbb{MDAG}(R)$ [2,6,17], and then we can calculate $\tilde{R}^{+}$ in $O(|X|^{3})$ time.

For our case studies, we have: in Fig. 1: $R^{\oplus} = <_{1}^{R}$; in Fig. 2: $Q^{\oplus} = <_{2}^{Q}$, in Fig. 3: $R^{\oplus} = \widehat{R}^{+}$, and in Fig. 4 $R^{\oplus}$ is $\widehat{R}^{+}$.

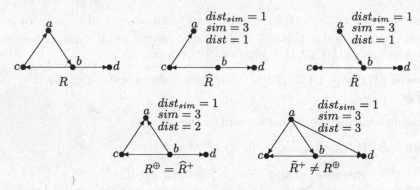

Fig. 4. An example of R, where $\widehat{R}, \tilde{R} \in \mathbb{MDAG}(R)$, and $\widehat{R}^{+} = R^{\oplus}$ but $\tilde{R}^{+} \neq R^{\oplus}$.

The result below shows the relationship between $R^{\oplus}$ and property-driven partial order approximations.

Proposition 3. *1. If $R^{cyc} \neq \emptyset$ then:*
(a) $R^{\oplus} \backslash R^{+} \neq \emptyset$, so the condition (1) of Definition 1 is never satisfied,
(b) $R^{\oplus} \cap R^{cyc} \neq \emptyset$, so the condition (5) of Definition 1 is never satisfied.
2. $R^{C} \subseteq R^{\oplus}$, so the condition (2) of Definition 1 is satisfied.
3. $(R^{\bullet})^{+} \subseteq R^{\oplus}$ and $R^{\bullet} \subseteq R^{\oplus}$, so the condition (3) of Definition 1 is also satisfied.

Proof. This very simple proof is left to a reader.   □

For the examples from Figs. 1, 2, 3 and 4, the condition (4) of Definition 1 is satisfied, however for the relations R and $\widehat{R}$ from Fig. 5, we have $b \equiv_{R} d$ but $\neg(b \equiv_{\widehat{R}} d)$, and it can easily be shown $\neg(b \equiv_{R^{\oplus}} d)$, so for this case the condition (4) of Definition 1 is not satisfied.

7   Final Comment

In many applications of partial orders, full transitivity is never or seldom explicitly used. Quite often use of acyclic relations that uniquely represent partial

Fig. 5. An example where $\neg(\equiv_R \subseteq \equiv_{\widehat{R}})$. We have $b \equiv_R d$ but $\neg(b \equiv_{\widehat{R}} d)$.

orders, as Hasse diagrams, dependency graphs, etc., is sufficient and more efficient (c.f. [3,5,13]). If $(a,b) \in R, (b,c) \in R$ but $(a,c) \notin R, (c,a) \notin R$, and R
is interpreted as partial ordering, then either the relationship between a and c
was analyzed and declared that a and c are incomparable, and then we have
some inconsistency; or the relationship between a and c was just not analyzed as
transitivity of R was implicitly assumed. Cycles, on the other hand, are always
a result of errors or data inconsistency. Even if the case $(a,b) \in R, (b,c) \in R$ but
$(a,c) \notin R, (c,a) \notin R$ is the result of errors or inconsistencies, this case appears
to be less serious problem than the case $(a,b) \in R, (b,c) \in R$ and $(c,a) \in R^+$.

In our model the measure $sim(\ldots)$ punishes more for having cycles, while
$dist(\ldots)$ for not being transitive. Since we believe that cycles are more problematic than lack of transitivity (that might be intentional, just the result of specific
procedure of an experiment), we emphasize $sim(\ldots)$ in Definition 2, and do not
use measures like Jaccard index [7] (or other used in [11]), where a differentiation
between $sim(\ldots)$ and $dist(\ldots)$ is also impossible).

We have shown that the quantitative approach to partial order approximations of arbitrary relation is somehow inconsistent with the property-driven
approach presented in [9,10], however when quantitative measures are applied to
property-driven approximations $(R^\bullet)^+$ and $(R^+)^\bullet$, the relation $(R^\bullet)^+$ is better
approximation of R than $(R^+)^\bullet$.

In Rough sets setting, that is not discussed due to space limit, we can define
the set of relations $\mathbb{MDAG}(R)$ as an outcome of applying some α-lower approximation (in a sense of [9,10]) to R. The main difference is that in this case
the result is not exactly one α-lower approximation, as in [9,10], but the set
$\mathbb{MDAG}(R)$. Then we can apply transitive closure as a particular α-upper approximation as in [9,10].

In the approach taken in this paper, for every relation R and its approximation S, we consider three distinct cases:

1. $(a,b) \in R$ but $(a,b) \notin S \wedge (b,a) \notin S$. In this case, to transform R into S, we
   just remove (a,b), and the cost of this operation is assumed to be *one*.
2. $(a,b) \notin R \wedge (b,a) \notin R$ but $(a,b) \in S$. In this case, to transform R into S, we
   just add (a,b), and the cost of this operation is again *one*.
3. $(a,b) \notin R \wedge (b,a) \in R$ but $(a,b) \in S \wedge (b,a) \notin R$. Now, to transform R into
   S, we add (a,b) *and* remove (b,c), so the cost of this operation is *two*.

Such approach makes formulas for $sim(...)$ and $dist(...)$ very simple and easy to handle. However, some may argue that for the case (3), which is often called 'flipping', the cost also should be *one* not *two*.

How the results of this paper would change if we assume that the cost of 'flipping' is *one*, instead of *two*? We guess that not much but do not have any result yet.

Acknowledgment. The author gratefully acknowledges the anonymous referees, whose comments significantly contributed to the final version of this paper. George Karakostas is thanked for a hint that helped to prove Theorem 3 and Ian Munro for influential comments on the nature of 'flipping'. The problem itself has been first discussed with late Zdzisław Pawlak in late nineties during one of his visits to McMaster.

References

1. Cormen, T.H., Leiserson, C.E., Rivest, D.L., Stein, C.: Introduction to Algorithms. MIT Press, Cambridge (2001)
2. Eades, P., Lin, X., Smyth, W.F.: A fast and effective heuristic for the feedback arc set problem. Inf. Process. Lett. **47**, 319–323 (1993)
3. Diekert, V., Rozenberg, G. (eds.): The Book of Traces. World Scientific, Singapore (1995)
4. Fishburn, P.C.: Interval Orders and Interval Graphs. Wiley, New York (1985)
5. French, S.: Decision Theory. Ellis Horwood, New York (1986)
6. Hassin, R., Rubinstein, S.: Approximations for the maximum acyclic subgraph problem. Inf. Process. Lett. **51**(3), 133–140 (1994)
7. Jaccard, P.: Étude comparative de la distribution florale dans une portion des Alpes et des Jura. Bulletin de la Société Vaudoise des Sciences Naturalles **37**, 547–549 (1901)
8. Janicki, R.: Pairwise comparisons based non-numerical ranking. Fundamenta Informaticae **94**(2), 197–217 (2009)
9. Janicki, R.: Approximations of arbitrary binary relations by partial orders: classical and rough set models. Trans. Rough Sets **13**, 17–38 (2011)
10. Janicki, R.: Property-driven rough sets approximations of relations. In: Skowron, A., Suraj, Z. (eds.) Rough Sets and Intelligent Systems - Professor Zdzisław Pawlak in Memoriam. Intelligent Systems Reference Library, vol. 42, pp. 333–357. Springer, Heidelberg (2013)
11. Janicki, R., Lenarčič, A.: Optimal approximations with rough sets and similarities in measure spaces. Int. J. Approx. Reason. **71**, 1–14 (2016)
12. Karp, M.: Reducibility among combinatorial problems. In: Miller, R.E., Thatcher, J.W. (eds.) Complexity of Computer Computations, pp. 85–103. Plenum, New York (1972)
13. Kleijn, J., Koutny, M.: Formal languages and concurrent behaviour. Stud. Comput. Intell. **113**, 125–182 (2008)
14. Pawlak, Z.: Rough sets. Int. J. Comput. Inform. Sci. **34**, 557–590 (1982)
15. Pawlak, Z.: Rough Sets. Kluwer, Dordrecht (1991)
16. Schröder, E.: Algebra der Logik. Teuber, Leipzig (1895)
17. Skiena, S.: The Algorithm Design Manual. Springer, London (2010)
18. Skowron, A., Stepaniuk, J.: Tolerence approximation spaces. Fundamenta Informaticae **27**, 245–253 (1996)

19. Tversky, A.: Features of similarity. Psychol. Rev. **84**(4), 327–352 (1977)
20. Yao, Y.Y., Wang, T.: On rough relations: an alternative formulation. In: Zhong, N., Skowron, A., Ohsuga, S. (eds.) RSFDGrC 1999. LNCS (LNAI), vol. 1711, pp. 82–90. Springer, Heidelberg (1999). doi:10.1007/978-3-540-48061-7_12

On Approximation of Relations by Generalized Closures and Generalized Kernels

Agnieszka D. Bogobowicz[1] and Ryszard Janicki[2(✉)]

[1] Polish-Japanese Institute of Information Technology, 02-008 Warsaw, Poland
abogobow@pjwstk.edu.pl
[2] Department of Computing and Software, McMaster University,
Hamilton, ON L8S 4K1, Canada
janicki@mcmaster.ca

Abstract. Various concepts of *closures* and *kernels* are introduced and discussed in the context of approximation of arbitrary relations by relations with specific properties.

1 Introduction

Relations are universal and simple tools used for modelling various properties of both data and systems. Usually the nature of data and systems enforces some special properties of relations, for example being a partial order or equivalence relation, etc. On the other hand when relations are created from empirical data they may not have the desired properties. *What is the "best" approximation that has the desired structure and properties and how it can be computed?* Note that we use the word "best", not "optimal" as our judgment may not be based on quantitative metrics.

For the approximation of arbitrary relations by partial orders this problem was discussed and some solutions were proposed in [6] (within both the standard theory of relations [14] and Rough Sets paradigm [12,13]). More general approach, with arbitrary binary relations instead of partial orders, and in terms of Rough Sets, has been proposed in [7]. The solutions discussed in both [6] and [7] are called "property-driven" and *do not use* any quantitative metrics. They enforce the desired properties of relations by applying some sequences of appropriate lower and upper approximations. These approximations, called α-lower and α-upper approximations, where α is a formula defining some relational property as for example transitivity, are extensions of classical rough set approximations [12,13].

Quantitative metrics and concepts of "optimal" approximations were introduced in [8,9]. In the standard Rough Sets model, every set X has two approximations, *lower* approximation $\mathbf{A}(X)$ and *upper* approximation $\overline{\mathbf{A}}(X)$. In [9], the third approximation, an *optimal* approximation (with respect to a class of similarity measures consistent with Marczewski-Steinhaus index [3,11]), has been

Partially supported by NSERC grant of Canada.

V. Flores et al. (Eds.): IJCRS 2016, LNAI 9920, pp. 120–130, 2016.
DOI: 10.1007/978-3-319-47160-0_11

introduced. Optimal approximations may not be unique, but for each optimal approximation O of X, $O \in OPT(X)$, we have $\underline{\mathbf{A}}(X) \subseteq O \subseteq \overline{\mathbf{A}}(X)$. Time complexity of finding an optimal approximation is the same as time complexities for $\underline{\mathbf{A}}(X)$ and $\overline{\mathbf{A}}(X)$, i.e. $O(|U|^2)$, where U is the universe containing X (cf. [9,15]). Optimal approximations of arbitrary relations by partial orders have been proposed and analyzed in [8]. It turns out the problem is NP-hard even for very simple metrics close to Jaccard index [3,5].

In this paper we will expand some ideas of [7]. In particular, standard concepts of *closure* and *kernel* (cf. [2,10,14]) will be generalized and used as approximation tools. No particular quantitative metrics will be used, however the use of such metrics in framework of 'property-driven' models will be discussed. Due to page limits, most proofs are omitted, for some sketches are presented.

2 Relations, Closures and Kernels

In this section we recall some fairly known concepts and results that will be used in the following sections [2,10,13,14].

Let X be a set, any $R \subseteq X \times X$ then R is called a *binary relation* (on X). We often will write aRb to denote $(a,b) \in R$.

A relation $\equiv$ is an equivalence relation   iff   it is reflexive, symmetric and transitive, i.e. $x \equiv x, x \equiv y \Rightarrow y \equiv x$, and $x \equiv y \equiv z \Rightarrow x \equiv z$, for all $x, y, z \in X$. For every equivalence relation $\equiv$ on X and every $x \in X$, the set $[x]_\equiv = \{y \mid x \equiv y\}$ denotes an *equivalence class* containing the element x.

The set of all equivalence classes of an equivalence relation $\equiv$ is denoted as $X/_\equiv$, and it is a *partition* of X, i.e. the sets from $X/_\equiv$ are disjoint and cover the whole X.

Definition 1 (Closure and Kernel). *Let X be a set and let $\mathcal{P}_\alpha \subseteq 2^X$ be a family of sets. The family $\mathcal{P}_\alpha$ is interpreted as a family of sets having a required property α.*

For every $S \subseteq X$, we define

$$\mathsf{C}_\mathcal{P}(S) = \bigcap_{S \subseteq Q \wedge Q \in \mathcal{P}_\alpha} Q \quad and \quad \mathsf{K}_\mathcal{P}(S) = \bigcup_{Q \subseteq S \wedge Q \in \mathcal{P}_\alpha} Q.$$

*1. If $\mathsf{C}_\mathcal{P}(S) \in \mathcal{P}_\alpha$, then $\mathsf{C}_\mathcal{P}(S)$ is an α-**closure** of S.*
*2. If $\mathsf{K}_\mathcal{P}(S) \in \mathcal{P}_\alpha$, then $\mathsf{K}_\mathcal{P}(S)$ is an α-**kernel** of S.*   □

If $\mathsf{C}_\mathcal{P}(S) \in \mathcal{P}_\alpha$, then it is the least superset of S that belongs to $\mathcal{P}_\alpha$, if $\mathsf{K}_\mathcal{P}(S) \in \mathcal{P}_\alpha$, is the greatest subset of S that belongs to $\mathcal{P}_\alpha$.

The popular closures of relations as transitive closure, reflexive closure, etc., are simple property set closures.

Proposition 1 (Explicit Expressions for Closures [14] and Kernels [7] of Binary Relations). *Let R be a relation on X.*

1. $R^{ref} = R \cup id$ *is the reflexive closure of R.*
2. $R^{\overline{sym}} = R \cup R^{-1}$ *is the symmetric closure of R.*
3. $R^{+} = \bigcup_{i=1}^{\infty} R^i$ *is the transitive closure of R.*
4. $R^{*} = \bigcup_{i=0}^{\infty} R^i$ *is the reflexive-transitive closure of R.*
5. $R^{\underline{sym}} = \{(a,b) \mid (a,b) \in R \wedge (b,a) \in R\}$ *is the symmetric kernel of R.* □

The closures and kernels have various important applications in many parts of mathematics, cf. [2,10,14], and *Pawlak's Rough Sets Lower and Upper Approximations* [12,13] can be considered as one of the most important. Let U be a finite set, called *universe*, $E \subseteq U \times U$ be an equivalence relation on U, and let $\mathcal{P}_E$ be the set of all unions of equivalence classes from U/E, i.e., $\mathcal{P}_E = \{X \mid X = \bigcup_{x \in X} [x]_E\}$. Then the Rough Sets *Lower Approximation* is just $\underline{\mathbf{A}}(S) = \mathsf{K}_{\mathcal{P}_E}(S)$, and the Rough Sets *Upper Approximation* is just $\overline{\mathbf{A}}(S) = \mathsf{C}_{\mathcal{P}_E}(S)$.

Unfortunately for the purpose of approximating arbitrary relations by relations with specific properties, standard closures and kernels are of limited use. For properties more complex than these from Proposition 1, very often $\mathbb{C}_{\mathcal{P}} = \emptyset$ or $\mathbb{K}_{\mathcal{P}} = \emptyset$, or both are empty. For example the property of partial ordering is closed under intersection, but if R contains a cycle, then there is no partial order that contains R. Even decomposing more complex properties into simpler one does not help much. For example a partial order approximation of a given relation R can be obtained by applying the transitive closure to R and then removing all cycles, or removing all cycles first and then applying transitive closure (the results are usually different, see [6] for details). While transitive closure is a simple property set closure, the operation of removing all cycles is neither a simple property set closure nor a simple property set kernel. If a relation contains a cycle than an acyclic superset does not exist (which rules out simple closure) and the property of acyclicity is not closed under set union (which rules out simple property kernel). It turns out very few complex properties are closed on either intersection or union (cf. [6,7,16]). For instance, transitivity is not closed under union and having a cycle is not closed under intersection. Some properties, like "having exactly one cycle" are preserved by neither union nor intersection.

To deal with these and similar problems, a new kind of approximations has been proposed in [7].

3   Generalized Closures and Kernels for Relations

In this section we generalize some ideas of [7] and formulate them is more general setting with Rough Sets being just a special case.

Let α be any predicate that describe some *property of binary relations*. As an example we can take

$$\alpha = [\forall x, y, z \in X.\ \neg(xRx) \wedge (xRyRz \Rightarrow xRz)]$$

i.e. a definition of (sharp) partial order.

Let X be a finite set and let

$$Rel_\alpha = \{R \mid R \subseteq X \times X \text{ and } R \text{ satisfies the property } \alpha\}.$$

If α is a predicate defined above, then Rel_α is the set of all (sharp) partial orders on X.

Let Prop denote the set of predicates such that $\alpha \in$ Prop $\implies Rel_\alpha \neq \emptyset \wedge Rel_\alpha \neq \{\emptyset\}$.

Note that we allow the case $\alpha \in$ Prop and $\emptyset \in Rel_\alpha$. The restrictions of Prop are merely for technical reasons, to avoid considering pathological cases in each result and proof. For more details the reader is referred to [7].

We start with adapting well known concepts of lower and upper bounds, and minimal and maximal elements for our purposes.

Definition 2. *Let $R \subseteq X \times X$ be a non-empty relation and $\alpha \in$ Prop. We say that:*

1. *R has α-**lower bound** $\iff \exists Q \in Rel_\alpha. Q \subseteq R$,*
2. *R has α-**upper bound** $\iff \exists Q \in Rel_\alpha. R \subseteq Q$.*

We also define

3. *$\mathbf{lb}_\alpha(R) = \{Q \mid Q \in Rel_\alpha \wedge Q \subseteq R\}$, the set of all α-lower bounds of R, and*
4. *$\mathbf{ub}_\alpha(R) = \{Q \mid Q \in Rel_\alpha \wedge R \subseteq Q\}$, the set of all α-upper bounds of R.* □

Both $\mathbf{lb}_\alpha(R)$ and $\mathbf{ub}_\alpha(R)$ always exist but they might be empty.

Definition 3. *For every family of relations $\mathcal{F} \subseteq 2^{X \times X}$, we define*

1. *$\max(\mathcal{F}) = \{R \mid \forall Q \in \mathcal{F}. R \subseteq Q \Rightarrow R = Q\}$, the set of all maximal elements of $\mathcal{F}$,*
2. *$\min(\mathcal{F}) = \{R \mid \forall Q \in \mathcal{F}. Q \subseteq R \Rightarrow R = Q\}$, the set of all minimal elements of $\mathcal{F}$.* □

We will now provide the main concept of this paper. Intersection is the *greatest lower bound* and union is the *least upper bound* of a given family of sets. We will generalize α-kernels and α-closures by replacing the *greatest lower bound* by the *set* of all *maximal lower bounds* and the *least upper bound* by the *set* of all *minimal upper bounds*.

Definition 4. *Let $R \subseteq X \times X$ and $\alpha \in$ Prop. Define the sets:*

$$\mathbb{K}_\alpha(R) = \max(\mathbf{lb}_\alpha(R)) \quad and \quad \mathbb{C}_\alpha(R) = \min(\mathbf{ub}_\alpha(R)).$$

1. *The set $\mathbb{K}_\alpha(R)$ is called the **generalized α-kernel** of R.*
2. *The set $\mathbb{C}_\alpha(R)$ is called the **generalized α-closure** of R.* □

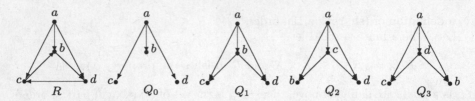

Fig. 1. For the relation R above and $\alpha = [\forall a. \, \neg(aQa) \wedge \forall a, b, c. \, aQb \wedge bQa \Rightarrow aQc]$ (partial ordering), we have $\mathbf{K}_\alpha(R) = Q_0, \mathbb{K}_\alpha(R) = \max(\mathbf{lb}_\alpha(R)) = \{Q_1, Q_2, Q_3\}$. Neither $\mathbf{C}_\alpha(R)$ nor $\mathbb{C}_\alpha(R)$ exist as $\mathbf{ub}_\alpha(R) = \emptyset$. The notions $\mathbf{K}_\alpha(R)$ and $\mathbf{C}_\alpha(R)$ are defined and discussed in Sect. 5

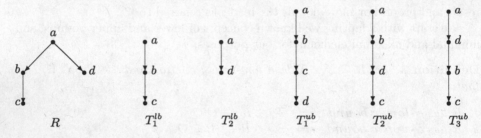

Fig. 2. The relation R is a partial order, all the others are total orders (represented as Hasse Diagrams [14]). If $\beta = [(\forall a. \, \neg(aQa)) \wedge (\forall a, b, c. \, aQb \wedge bQa \Rightarrow aQc) \wedge (\forall a, b. \, aQb \vee bQa)]$ (total ordering), then $\mathbb{K}_\beta(R) = \max(\mathbf{lb}_\beta(R)) = \{T_1^{lb}, T_2^{lb}\}$, and $\mathbb{C}_\beta(R) = \min(\mathbf{ub}_\beta(R)) = \{T_1^{ub}, T_2^{ub}, T_3^{ub}\}$

Definition 4 is illustrated in Figs. 1 and 2. Both $\mathbb{K}_\alpha(R)$ and $\mathbb{C}_\alpha(R)$ always exist, they might be empty, however they do not require α being preserved by intersection or union, as it is implicitly assumed for standard kernels and closures.

For a set of relations $\mathcal{R} \subseteq 2^{X \times X}$, we define $\mathbb{K}_\alpha(\mathcal{R})$ and $\mathbb{C}_\alpha(\mathcal{R})$ standardly as

$$\mathbb{K}_\alpha(\mathcal{R}) = \bigcup_{R \in \mathcal{R}} R \quad \text{and} \quad \mathbb{C}_\alpha(\mathcal{R}) = \bigcup_{R \in \mathcal{R}} R$$

We can now discuss properties of compositions as $\mathbb{K}_\alpha(\mathbb{K}_\beta(R)), \mathbb{C}_\alpha(\mathbb{K}_\alpha(R))$, etc.

The below two results characterize operational and compositional properties of generalized α-kernels and generalized α-closures. They resemble the properties of Pawlak's lower and upper approximations [12,13], but, since the concept used are more general, are weaker.

Proposition 2. *If $R, Q \subseteq X \times X$ have α-lower bound then:*

1. *If $R \in Rel_\alpha$ then $\mathbb{K}_\alpha(R) = \{R\}$.*
2. *$R \subseteq Q \implies \forall R' \in \mathbb{K}_\alpha(R) \, \exists Q' \in \mathbb{K}_\alpha(Q). \, R' \subseteq Q'$.*
3. *$\forall Q \in \mathbb{K}_\alpha(R). \, Q \subseteq R$.*
4. *$\mathbb{K}_\alpha(R) = \mathbb{K}_\alpha(\mathbb{K}_\alpha(R))$.*
5. *if R has α-upper bound then $\mathbb{C}_\alpha(R) = \mathbb{K}_\alpha(\mathbb{C}_\alpha(R))$.* $\qquad\qquad\square$

Proposition 3. *If $R, Q \subseteq X \times X$ have α-lower bound then:*

1. *If $R \in Rel_\alpha$ then $\mathbb{C}_\alpha(R) = \{R\}$.*
2. *$R \subseteq Q \implies \forall R' \in \mathbb{C}_\alpha(R) \; \exists Q' \in \mathbb{C}_\alpha(Q). \; R' \subseteq Q'$.*
3. *$\forall Q \in \mathbb{C}_\alpha(R). \; R \subseteq Q$,*
4. *$\mathbb{C}_\alpha(R) = \mathbb{C}_\alpha(\mathbb{C}_\alpha(R))$,*
5. *if R has α-upper bound then $\mathbb{K}_\alpha(R) = \mathbb{C}_\alpha(\mathbb{K}_\alpha(R))$.* □

Most of the interesting properties are conjunctions of simpler predicates. For example, a binary relation can be made a partial order by applying transitive closure first and making the outcome acyclic later, or in the opposite order (see [6,7]); or a relation can be made an equivalence relation by applying reflexive, symmetric and transitive closures in this order (see [7]).

However composing generalized kernels and closures is tricky. If R has a property β different from α, neither the elements of $\mathbb{K}_\alpha(R)$ nor the elements of $\mathbb{C}_\alpha(R)$ may satisfy β. For example if R is transitive, its symmetric closure is symmetric, but may not be transitive any longer [14]. Hence such compositions can be used only in some, although frequent, circumstances.

Definition 5. *Let $\alpha, \beta \in \mathsf{Prop}$.*

1. *We say that a property α $\mathbb{K}$-**preserves** a property β iff for every $R \in Rel_\beta$, if R has α-lower bound then $\mathbb{K}_\alpha(R) \subseteq Rel_\beta$.*
2. *We say that a property α $\mathbb{C}$-**preserves** a property β iff for every $R \in Rel_\beta$, if R has α-upper bound then $\mathbb{C}_\alpha(R) \subseteq Rel_\beta$.* □

The compositions $\mathbb{K}_\alpha(\mathbb{K}_\beta(R))$ and $\mathbb{K}_\alpha(\mathbb{C}_\beta(R))$ are *well defined* only if $\alpha\mathbb{K}$-preserves β; and the compositions $\mathbb{C}_\alpha(\mathbb{K}_\beta(R))$ and $\mathbb{C}_\alpha(\mathbb{C}_\beta(R))$ are well defined only if α $\mathbb{C}$-preserves β.

When writing appropriate compositions we will assume that they are well defined.

In general there are no specific relationships between $\mathbb{K}_\alpha(\mathbb{K}_\beta(R))$ and $\mathbb{K}_\beta(\mathbb{K}_\alpha(R))$, or between $\mathbb{C}_\alpha(\mathbb{C}_\beta(R))$ and $\mathbb{C}_\beta(\mathbb{C}_\alpha(R))$. However, we can show a simple (and expected) relationship between $\mathbb{K}_\alpha(\mathbb{C}_\beta(R))$ and $\mathbb{C}_\beta(\mathbb{K}_\alpha(R))$.

Proposition 4. *Let $R \subseteq X \times X, \alpha, \beta \in \mathsf{Prop}$, and*

– *R has α-upper bound and β-lower bound,*
– *each $S \in \mathbb{C}_\alpha(R)$ has β-lower bound, and each $S \in \mathbb{K}_\beta(R)$ has α-upper bound,*

then: $\forall S \in \mathbb{C}_\alpha(\mathbb{K}_\beta(R)) \; \exists Q \in \mathbb{K}_\beta(\mathbb{C}_\alpha(R)). \; S \subseteq Q$.

Proof (sketch). Since $\forall R' \in \mathbb{C}_\alpha(R). \; R \subseteq R'$, then $\forall T \in \mathbb{K}_\beta(R) \; \exists T' \in \mathbb{K}_\beta(\mathbb{C}_\alpha(R)). \; T \subseteq T'$, and by the same reason, $\forall S \in \mathbb{C}_\alpha(\mathbb{K}_\beta(R)) \; \exists Q \in \mathbb{C}_\alpha(\mathbb{K}_\beta(\mathbb{C}_\alpha(R))). \; S \subseteq Q$. Since $\beta\mathbb{K}$-preserves α, $\mathbb{K}_\beta(\mathbb{C}_\alpha(R)) \subseteq Rel_\alpha$, so $\mathbb{C}_\alpha(\mathbb{K}_\beta(\mathbb{C}_\alpha(R))) = \mathbb{K}_\beta(\mathbb{C}_\alpha(R))$. □

4   Mixing Generalized Kernels and Closures

For complex predicates α, generalized α-kernels and α-closures suffer from the same problems as discussed at the end of Sect. 2 for standard kernels and closures, quite often $\mathbb{K}_\alpha(R)$ and/or $\mathbb{C}_\alpha(R)$ are empty, or their elements are 'far away' from R (cf. [6,7]). However most of the complex properties of relations α, can be presented as a conjunction of simpler properties, so we usually have: $\alpha = \alpha_1 \wedge \ldots \wedge \alpha_k$, where α_i are relatively simple predicates. For example *partial ordering = irreflexivity $\wedge$ transitivity* (or equivalently *partial ordering = acyclity $\wedge$ transitivity*), and *equivalence = reflexivity $\wedge$ symmetry $\wedge$ transitivity*. More precisely, *partial ordering* $= \alpha_1 \wedge \alpha_2$ where $\alpha_1 = \forall x \in X.\ \neg(xRx)$ and $\alpha_2 = \forall x, y, z \in X.\ xRy \wedge yRz \Rightarrow xRz$, and *equivalence* $= \beta_1 \wedge \beta_2 \wedge \beta_3$, where $\beta_1 = \forall x \in X.\ xRx, \beta_2 = \forall x, y \in X.\ xRy \Leftrightarrow yRx$, and $\beta_3 = \forall x, y, z \in X.\ xRy \wedge yRz \Rightarrow xRz$.

This suggests using an appropriate sequence of α_i-kernels and α_j-closures instead of single α-kernel or/and α-closure (cf. [6,7]). A property in the form $\alpha = \alpha_1 \wedge \ldots \wedge \alpha_k$ is called *compositional* [7].

The following result shows the relationships between $\mathbb{K}_{\alpha \wedge \beta}(R)$ and $\mathbb{K}_\alpha(\mathbb{K}_\beta(R))$, and between $\mathbb{C}_{\alpha \wedge \beta}(R)$ and $\mathbb{C}_\alpha(\mathbb{C}_\beta(R))$.

Proposition 5. *Assume that α, β belong to* Prop.

1. *If R has β-lower bound and $(\alpha \wedge \beta)$-lower bound, and each $Q \in \mathbb{K}_\beta(R)$ has α-lower bound, then:* $\forall S \in \mathbb{K}_{\alpha \wedge \beta}(R)\ \exists T \in \mathbb{K}_\alpha(\mathbb{K}_\beta(R)).\ S \subseteq T \subseteq R$.
2. *If R has β-upper bound and $(\alpha \wedge \beta)$-upper bound, and each $Q \in \mathbb{C}_\beta(R)$ has α-upper bound, then:* $\forall T \in \mathbb{C}_\alpha(\mathbb{C}_\beta(R))\ \exists S \in \mathbb{C}_{\alpha \wedge \beta}(R).\ R \subseteq T \subseteq S$.

Proof (sketch of (1)). Let $S \in \mathbb{K}_{\alpha \wedge \beta}(R)$. Clearly $\mathbf{lb}_{\alpha \wedge \beta}(R) \subseteq \mathbf{lb}_\beta(R)$, so $\max(\mathbf{lb}_{\alpha \wedge \beta}(R)) \subseteq \mathbf{lb}_\beta(R)$. Hence $\exists S_\beta \in \mathbb{K}_\beta(R)$ such that $S \subseteq S_\beta$. Since $S \in \mathbb{K}_{\alpha \wedge \beta}(R)$, then $S \in Rel_\alpha$, so, by Proposition 2(2), $\exists T \in \mathbb{K}_\alpha(S_\beta).\ S \subseteq T$. But $T \in \mathbb{K}_\alpha(\mathbb{K}_\beta(R))$ and, Proposition 2(3), $T \subseteq R$. $\square$

Proposition 5 suggests an important technique for the design of approximation schema. It says in principle that using a complex predicate as a property usually results in a *worse* approximation than when the property is decomposed into simpler ones, and then we approximate a given relation over all these simpler properties.

We will adopt the following convention $\mathbb{P}_\alpha^{(0)}(R) = \mathbb{K}_\alpha(R)$ and $\mathbb{P}_\alpha^{(1)}(R) = \mathbb{C}_\alpha(R)$.

Let $\alpha = \alpha_1 \wedge \ldots \wedge \alpha_k$ be a compositional property, $i_1, \ldots, i_k$ a sequence with $i_j \in \{0, 1\}$, and $R \subseteq X \times X$. Define a sequence s as $s = (\alpha_1, i_1), \ldots, (\alpha_k, i_k)$. In [7], a sequence like s is called an *approximation schedule*.

We will consider the following composition of kernels and closures.

Definition 6. *Let $R \subseteq X \times X, \alpha = \alpha_1 \wedge \ldots \wedge \alpha_k$ and $s = (\alpha_1, i_1), \ldots, (\alpha_k, i_k)$. We define a set of* **composed approximations** *of R as:*

$$Comp_{(s)}(R) = \mathbb{P}_{\alpha_1}^{(i_1)}(\ldots \mathbb{P}_{\alpha_{k-1}}^{(i_{k-1})}(\mathbb{P}_{\alpha_k}^{(i_k)}(R))\ldots),$$

where every $S \in \mathbb{P}_{\alpha_j}^{(i_j)}(\ldots \mathbb{P}_{\alpha_{k-1}}^{(i_{k-1})}(\mathbb{P}_{\alpha_k}^{(i_k)}(R))\ldots)$ *has* α_j-*lower bound if* $i_j = 0$ *or has* α_j-*upper bound if* $i_j = 1$, *for* $j = 1, \ldots, k$.   □

Since to have a composition well defined, the properties α_i appearing in the sequence s must satisfy Definition 5, $Comp_{(s)}(R)$ may not exist or it may be empty. However it it is not empty, every relation $Q \in Comp_{(s)}(R)$ can be considered as a *property-driven* approximation of $R \subseteq X \times X$.

This approach seems to be especially useful when we want to use some metrics to find 'optimal' approximation. Suppose that for a given property α we have a procedure that can provide, for any relation $R \subseteq X \times X$, a set *optimal* approximations $Opt_\alpha(R)$. We assume that we have some *similarity* measure between relations $sim_\alpha(R, S)$, and $Opt_\alpha(R) = \{S \mid sim_\alpha(R, S)$ is maximal$\}$.

While there are plenty of universal, convenient and useful similarity measures for sets [3,5,9,11], some of them as Marczewski-Steinhaus Index [11] are very well suited for Rough Sets [4,9], for the relations we have an opposite situation. As it was argued in [8] for the case of approximation for partial orders (and also implicitly in [7]), an universal similarity measure for relations probably does not exists, each particular property α most likely requires some special similarity measure, and for complex α it is difficult to figure out how to measure such similarity, and make it consistent with property-driven approach. This task is apparently easier for simple properties α.

Hence we propose the following procedure for finding 'optimal' approximation of R for a given compositional property $\alpha = \alpha_1 \wedge \ldots \wedge \alpha_k$.

Definition 7. *Let* $R \subseteq X \times X, \alpha \in$ Prop *and* $sim_\alpha(\ ,\ )$ *be a given similarity for relations w.r.t. property* α. *Define the sets:*

$$\mathbb{K}_\alpha^{opt}(R) = \{S \mid S \in \mathbb{K}_\alpha(R) \wedge sim_\alpha(R, S) = maximum\} \ and$$
$$\mathbb{C}_\alpha^{opt}(R) = \{S \mid S \in \mathbb{C}_\alpha(R) \wedge sim_\alpha(R, S) = maximum\}.$$

1. *The set* $\mathbb{K}_\alpha^{opt}(R)$ *is called the* **generalized optimal** α-**kernel** *of* R.
2. *The set* $\mathbb{C}_\alpha^{opt}(R)$ *is called the* **generalized optimal** α-**closure** *of* R.   □

Clearly $\mathbb{K}_\alpha^{opt}(R) \subseteq \mathbb{K}_\alpha(R)$ and $\mathbb{C}_\alpha^{opt}(R) \subseteq \mathbb{C}_\alpha(R)$, for every $R \subseteq X \times X$. We may now apply the ideas of Definition 6 to $\mathbb{K}^{opt}$ and $\mathbb{C}^{opt}$. As before, we will adopt the convention $\mathbb{P}_\alpha^{opt(0)}(R) = \mathbb{K}_\alpha^{opt}(R)$ and $\mathbb{P}_\alpha^{opt(1)}(R) = \mathbb{C}_\alpha^{opt}(R)$.

Definition 8. *Let* $R \subseteq X \times X, \alpha = \alpha_1 \wedge \ldots \wedge \alpha_k$ *and* $s = (\alpha_1, i_1), \ldots, (\alpha_k, i_k)$. *We define a set of* **optimal composed approximations** *of* R *as*

$$Comp_{(s)}^{opt}(R) = \mathbb{P}_{\alpha_1}^{opt(i_1)}(\ldots \mathbb{P}_{\alpha_{k-1}}^{opt(i_{k-1})}(\mathbb{P}_{\alpha_k}^{opt(i_k)}(R))\ldots),$$

where every $S \in \mathbb{P}_{\alpha_j}^{opt(i_j)}(\ldots \mathbb{P}_{\alpha_{k-1}}^{opt(i_{k-1})}(\mathbb{P}_{\alpha_k}^{opt(i_k)}(R))\ldots)$ *has* α_j-*lower bound if* $i_j = 0$ *or has* α_j-*upper bound if* $i_j = 1$, *for* $j = 1, \ldots, k$.   □

Similarly as $Comp_{(s)}(R)$, for some s the set $Comp_{(s)}^{opt}(R)$ may not exist or be empty, and, if it exists $Comp_{(s)}^{opt}(R) \subseteq Comp_{(s)}(R)$. Clearly the result depends on particular $sim_{\alpha_i}, i = 1, \ldots, k$, but as the results of [8] suggest, this could be a more feasible approach than trying to compute $Opt_\alpha(R)$ directly. Finding appropriate sim_{α_i} could be problematic and in many cases most likely NP-complete [8]. In some cases, NP-hard general analytic solutions can be regarded as property-constraints and approximation methods based on algebraic properties of a group derived for physical phenomena can be used [1].

5   Unique Approximations

Both $\mathbb{K}_\alpha(R)$ and $\mathbb{C}_\alpha(R)$ are sets, and having a *set* of 'best' lower or upper approximations is a mixed blessing, often a nuisance, often we would prefer to have just one lower or upper approximation (cf. [8]). This is the idea explored in [7] and earlier in [12,13]. However some further restrictions are needed.

Definition 9. *1. A property α is* **closed under union** *if*

$$X, Y \in Rel_\alpha \implies X \cup Y \in Rel_\alpha.$$

2. A property α is **closed under intersection** *if $X, Y \in Rel_\alpha \implies X \cap Y \in Rel_\alpha$.*

3. Let Prop$^{\cup\cap}$ *denote the set of all α that are closed under union* **or** *intersection.*   □

The following model has been proposed and explored in [6,7] in Rough Sets framework. We will present this model here in a slightly more general setting. In principle we have replaced sets of maximal lower bounds with their intersection and sets of minimal upper bounds with their union.

Definition 10 ([6,7]). *Let $R \subseteq X \times X$ and let $\alpha \in$ Prop$^{\cup\cap}$.*

1. If R has α-lower bound then we define its **closed α-generalized kernel** *as:*

$$\mathbf{K}_\alpha(R) = \bigcap\{X \mid X \in \max(\mathbf{lb}_\alpha(R))\}.$$

2. If R has α-upper bound then we define its closed α-generalized **closure** *as:*

$$\mathbf{C}_\alpha(R) = \bigcup\{X \mid X \in \min(\mathbf{ub}_\alpha(R))\}.$$ □

The simple relationship between $\mathbb{K}_\alpha(R), \mathbb{C}_\alpha(R)$ and $\mathbf{K}_\alpha(R), \mathbf{C}_\alpha(R)$ is given below.

Corollary 1. *Let $R \subseteq X \times X$ and let $\alpha \in$ Prop$^{\cup\cap}$.*

1. If R has α-lower bound then there exist $Q \in \mathbb{K}_\alpha(R). \mathbf{K}_\alpha(R) \subseteq Q$.
2. If R has α-upper bound then there exist $Q \in \mathbb{C}_\alpha(R). Q \subseteq \mathbf{C}_\alpha(R)$.   □

Counterparts of Propositions 2 and 3, i.e. operational and compositional properties, are much more elaborate in this case.

Proposition 6 ([7]). *If $R, Q \subseteq X \times X$ have α-lower bound and $R', Q' \subseteq X \times X$ have α-upper bound then:*

1. $R \subseteq Q \implies \mathbf{K}_\alpha(R) \subseteq \mathbf{K}_\alpha(Q), R' \subseteq Q' \implies \mathbf{C}_\alpha(R') \subseteq \mathbf{C}_\alpha(Q')$
2. $\mathbf{K}_\alpha(R) \subseteq R, R' \subseteq \mathbf{C}_\alpha(R')$,
3. $\mathbf{K}_\alpha(R) = \mathbf{K}_\alpha(\mathbf{K}_\alpha(R)), \mathbf{C}_\alpha(R') = \mathbf{C}_\alpha(\mathbf{C}_\alpha(R'))$
4. $\mathbf{K}_\alpha(R \cap Q) = \mathbf{K}_\alpha(\mathbf{K}_\alpha(R) \cap \mathbf{K}_\alpha(Q)), \mathbf{C}_\alpha(R' \cup Q') = \mathbf{C}_\alpha(\mathbf{C}_\alpha(R') \cup \mathbf{C}_\alpha(Q'))$,
5. *if α is closed under intersection then $\mathbf{K}_\alpha(R \cap Q) = \mathbf{K}_\alpha(R) \cap \mathbf{K}_\alpha(Q)$, while if α is closed under union then $\mathbf{C}_\alpha(R \cup Q) = \mathbf{C}_\alpha(R) \cup \mathbf{C}_\alpha(Q)$,*
6. *if R has α-upper bound then $\mathbf{C}_\alpha(R) = \mathbf{K}_\alpha(\mathbf{C}_\alpha(R))$, while if R has α-lower bound then $\mathbf{K}_\alpha(R) = \mathbf{C}_\alpha(\mathbf{K}_\alpha(R))$.* $\qquad\square$

If α is closed under union than we can provide simpler formula for $\mathbf{K}_\alpha(A)$ and if α is closed under intersection than we we can provide simpler formula for $\mathbf{C}_\alpha(A)$.

Proposition 7 ([7]).

1. *If α is closed under union and R has α-lower bound, then*

$$\mathbf{K}_\alpha(R) = \bigcup\{S \mid S \in \mathbf{lb}_\alpha(R)\} = \bigcup\{S \mid S \subseteq R \wedge S \in Rel_\alpha\}.$$

2. *If α is closed under intersection and A has α-upper bound, then*

$$\mathbf{C}_\alpha(A) = \bigcap\{X \mid X \in \mathbf{ub}_\alpha(A)\} = \bigcap\{X \mid A \subseteq X \wedge X \in \mathcal{P}_\alpha\}. \qquad\square$$

The right formulas of Proposition 7 have the same pattern as appropriate lower and upper Rough Sets approximations [12,13], which means that our model is one of many extensions of the original Pawlak's ideas [12,13]. Detailed properties of $\mathbf{K}_\alpha(A)$ and $\mathbf{C}_\alpha(A)$ have been discussed in [6,7].

6 Final Comment

We have presented a fairly general framework for finding 'best' approximations of arbitrary relations by relations with specific properties defined by a given predicate α. First we introduced generalized α-kernels and generalized α-closures, $\mathbb{K}_\alpha$ and $\mathbb{C}_\alpha$, and show their basic properties. For complex α we suggest representing them as a conjunction of simpler ones, $\alpha = \alpha_1 \ldots \alpha_k$, and if possible, use an appropriate sequence of α_i-kernels and α_j-closures to calculate a set of potential approximations. No specific quantitative metrics were used however use of abstract similarity measures was allowed in the proposed approximation procedure. Due to space restrictions only two specific examples were presented, however some of the examples from [7,8] also illustrate issues discussed in this paper.

Acknowledgment. The authors gratefully acknowledge four anonymous referees, whose comments significantly contributed to the final version of this paper. This work was done during the first author visit to the Department of Computing and Software, McMaster University.

References

1. Bogobowicz, A.D.: Non-newtonian creep into two-dimensional cavity of near-rectangular shape. J. Appl. Mech. ASME **63**(4), 1047–1051 (1996)
2. Cohn, P.M.: Universal Algebra. Harper and Row, New York (1965)
3. Deza, M.M., Deza, E.: Encyclopedia of Distances. Springer, Berlin (2012)
4. Gomolinska, A.: On certain rough inclusion functions. Trans. Rough Sets **9**, 35–55 (2008)
5. Jaccard, P.: Étude comparative de la distribution florale dans une portion des Alpes et des Jura. Bulletin de la Société Vaudoise des Sciences Naturalles **37**, 547–549 (1901)
6. Janicki, R.: Approximations of arbitrary binary relations by partial orders: classical and rough set models. Trans. Rough Sets **13**, 17–38 (2011)
7. Janicki, R.: Property-driven rough sets approximations of relations. In: Skowron, A., Suraj, Z. (eds.) Rough Sets, Intelligent Systems - Professor Zdzisław Pawlak in Memoriam. Intelligent Systems Reference Library, vol. 42, pp. 333–357. Springer, Heidelberg (2013)
8. Janicki, R.: On optimal approximations of arbitrary relations by partial orders. In: Flores, V., et al. (eds.) Rough Sets. LNCS, vol. 9920, pp. 107–119. Springer, Heidelberg (2016)
9. Janicki, R., Lenarčič, A.: Optimal approximations with rough sets and similarities in measure spaces. Int. J. Approx. Reason. **71**, 1–14 (2016)
10. Lang, S.: Algebraic Structures. Addison-Wesley, Reading (1968)
11. Marczewski, E., Steinhaus, H.: On a certain distance of sets and corresponding distance of functions. Colloquium Mathematicum **4**, 319–327 (1958)
12. Pawlak, Z.: Rought sets. Int. J. Comput. Inform. Sci. **34**, 557–590 (1982)
13. Pawlak, Z.: Rough Sets. Kluwer, Dordrecht (1991)
14. Rosen, K.H.: Discrete Mathematics and Its Applications. McGraw-Hill, New York (1999)
15. Saquer, J., Deogun, J.S.: Concept approximations based on rough sets and similarity measures. Int. J. Appl. Math. Comput. Sci. **11**(3), 655–674 (2001)
16. Yao, Y.Y., Wang, T.: On rough relations: an alternative formulation. In: Zhong, N., Skowron, A., Ohsuga, S. (eds.) RSFDGrC 1999. LNCS (LNAI), vol. 1711, pp. 82–90. Springer, Heidelberg (1999). doi:10.1007/978-3-540-48061-7_12

Ordered Information Systems and Graph Granulation

John G. Stell[(⊠)]

School of Computing, University of Leeds, Leeds, U.K.
j.g.stell@leeds.ac.uk

Abstract. The concept of an Information System, as used in Rough Set theory, is extended to the case of a partially ordered universe equipped with a set of order preserving attributes. These information systems give rise to partitions of the universe where the set of equivalence classes is partially ordered. Such ordered partitions correspond to relations on the universe which are reflexive and transitive. This correspondence allows the definition of approximation operators for an ordered information system by using the concepts of opening and closing from mathematical morphology. A special case of partial orders are graphs and hypergraphs and these provide motivation for the need to consider approximations on partial orders.

Keywords: Ordered information system · Graph granulation · Graph partitioning

1 Introduction

From one perspective the theory of rough sets allows us to move between two levels of detail. Elements at the more detailed level are grouped together and these granules become elements at a more abstract (less detailed) level of detail. The process of granulation, that is the process of forming the granules, can be parameterized by a relation on a set. In the classic case described by Pawlak [8] the relation is an equivalence relation, but arbitrary relations give rise to several different granulations as described by Yao [18]. Stell [14] showed how these could be generalized to the case of a relation on a hypergraph, as opposed to the relations on a set considered in [18], by using operations of erosion and dilation from mathematical morphology. However, the treatment in [14] did not consider how to connect relations on a hypergraph with partitions of the underlying set of edges and nodes. The present paper is also related to the work of Lin [7] on granular computing and neighbourhood systems, since neighbourhood systems correspond to arbitrary relations. However, neighbourhood systems are structures on sets whereas here the more general case of partially ordered sets is studied.

The paper start by reviewing how hypergraphs can be seen as partial orders. This leads to a motivating example of a graph granulation which prompts an

© Springer International Publishing AG 2016
V. Flores et al. (Eds.): IJCRS 2016, LNAI 9920, pp. 131–142, 2016.
DOI: 10.1007/978-3-319-47160-0_12

examination of ordered information systems. This shows how the partitions induced by such structures have corresponding reflexive and transitive relations. Using these relations we use the well-known operations of opening and closing, but novelly in the general case of a partial order rather than a set, to obtain appropriate approximation operators.

2   Background on Graphs and Hypergraphs

In this section we start by outlining the approach to graphs and hypergraphs that we use.

2.1   Graphs and Hypergraphs

We work with graphs which are undirected and which may have multiple edges between nodes as well as multiple loops on nodes. In a graph each edge is incident with one or two nodes, but consideration of binary relations on graphs leads naturally to hypergraphs as we will see later. In a hypergraph [1] there are edges and nodes, but each edge may be incident with any number of nodes. In our work we require that edges are incident with a non-zero number of nodes. One formalization of these structures is to have two disjoint sets for the nodes and for the edges. We use an alternative approach with a single set consisting of all the node and edges together with an incidence relation which expresses which edges are incident with which nodes. This has been used in [13,14] and is based on using a similar approach to graphs in [3]. Relations appear in two ways in the paper: every hypergraph is treated as a set of edges and nodes equipped with an incidence relation on this set, and an indiscernibility relation on the hypergraphs is a further binary relation on the set of edges and nodes subject to appropriate constraints.

Definition 1. *A **hypergraph** consists of a set U and a reflexive incidence relation $H \subseteq U \times U$ such that for all $u, v, w \in U$, if $(u, v) \in H$ and $(v, w) \in H$ then $u = v$ or $v = w$.*

*Given a hypergraph (U, H), an element $u \in U$ is an **edge** if there is some $v \in U$ where $(u, v) \in H$ and $u \neq v$. An element which is not an edge is a **node**.*

It is straightforward to check that in a hypergraph (U, H) the incidence relation H will be transitive as well as reflexive so that it is a preorder and, in fact, a partial order too. Hypergraphs defined in this way may have edges that are incident with arbitrary non-empty sets of nodes and not just with one or two nodes as in the case of a graph. Graphs arise as a special case of hypergraphs as in the next definition.

Definition 2. *A **graph** is a hypergraph (U, H) which satisfies the constraint that for every $u \in U$ the set $\{v \in U \mid (u, v) \in H \text{ and } u \neq v\}$ has at most two elements.*

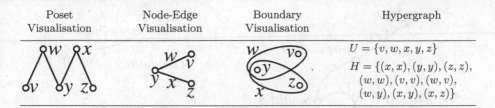

| Poset Visualisation | Node-Edge Visualisation | Boundary Visualisation | Hypergraph |

Fig. 1. Three ways of visualising a graph

We use the terminology 'subgraph' for the structural parts of arbitrary hypergraphs.

Definition 3. *A **subgraph** of a hypergraph (U, H) is defined as a subset $K \subseteq U$ for which $k \in K$ and $(k, u) \in H$ imply $u \in K$.*

Figure 1 shows three different ways in which graphs may be visualised. The example shows a graph with two edges and three nodes. The node-edge visualisation is the most familiar depiction, with each edge drawn as line between two nodes. A useful alternative to this, and the only viable possibility once we are dealing with hypergraphs which are not graphs is the *Boundary Visualisation*. In this each edge is shown as a boundary enclosing all the nodes with which it is incident.

3   Granules in Graphs and Hypergraphs

Figure 2 provides a motivating example of a graph and its abstraction to a less detailed view. The left hand diagram provides the more detailed view. It shows an imaginary transport network of rail lines in a city. There are nine labelled stations: *West*, *Mid*, *C1*, *C2*, *C3*, *North*, *South*, *SouthEast*, and *SouthWest*. Five lines are labelled: *a* and *b* as well as *p*, *q*, and *r*.

Now consider the representation on the right-hand side. This provides a less detailed view. Some of the stations are unchanged from the more detailed view. The three stations *C1*, *C2* and *C3* together with the three lines joining them have become a single node labelled *Centre*. This node does not represent a single station at the more detailed level; it represents a subgraph consisting of three nodes and three edges. The line labelled *westline* consists of two lines *a* and *b* together with the intermediate station *Mid*. Here the subgraph consisting of nodes *West*, *North*, *Mid* and edges *a* and *b* has become a subgraph consisting of one edge and two nodes.

The more detailed representation is a graph, with undirected edges which connect pairs of nodes. However, the less detailed representation is a *hypergraph* having two edges, one of which is incident with three nodes, *North*, *Centre* and *South*. This edge, labelled *circleline*, stands for a granule at the more detailed level consisting of the *SouthWest* and *SouthEast* nodes together with the edges connecting them to *South*, to *North* and to *C1*, *C2*, *C3*.

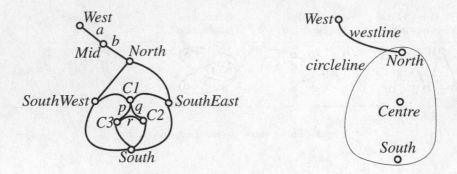

Fig. 2. Motivating example of coalescing edges and nodes

The clustering of separate entities at the more detailed level into a single entity at the less detailed level can be understood as the action of an attribute which assigns values in the lower level to entities at the higher level. To explain how this works in the case of a partially ordered set of entities, it is necessary next to introduce the concept of an ordered information system.

4   Ordered Information Systems

In this section we consider a generalization of the notion of Information System in which the universe is not merely a set but carries a partial order. In this setting the attributes defined on the universe are monotone, or order-preserving, functions to partially ordered sets of values. We see how this gives rise to a partition of the universe, as in the well-known set-based case, but with the additional structure of a partial order on the equivalence classes. In order to understand the appropriate way to define upper and lower approximations we then need to connect these partially ordered partitions with relations on the partially ordered universe, which we do in Theorem 2 below. Although our motivation used graphs and hypergraphs, we shall see that it is the partial order that provides the essential structure.

4.1   Information Systems on a Partially Ordered Universe

We recall the set case following the terminology of [9]. An **Information System**, $\mathcal{A} = (U, A)$, consists of a set U and a set A of functions called **attributes** defined on U where $\alpha : U \rightarrow V_\alpha$ for each $\alpha \in A$. Each subset $B \subseteq A$ gives rise to an **indiscernibility relation** IND_B where $(u_1, u_2) \in IND_B$ iff $\alpha(u_1) = \alpha(u_2)$ for all $\alpha \in B$. The relation IND_B is an equivalence relation on U, and we denote the equivalence class of u by $[u]_B$ or just $[u]$ where no ambiguity arises.

Each subset $X \subseteq U$ has an **upper approximation**, $\overline{B}(X)$, and a **lower approximation**, $\underline{B}(X)$, with respect to a given $B \subseteq A$ where,

$$\overline{B}(X) = \{u \in U \mid \exists w((w, u) \in IND_B \text{ and } w \in X)\}$$
$$\underline{B}(X) = \{u \in U \mid \forall w((u, w) \in IND_B \text{ implies } w \in X)\}$$

Now we generalize this to the case of a partially ordered set (U, H).

Definition 4. *An **Ordered Information System**, $\mathcal{A} = (U, H, A)$, consists of set U and partial order H on U, and a set, A, of order preserving functions called **ordered attributes** defined on (U, H) where $\alpha : (U, H) \to (V_\alpha, K_\alpha)$.*

An ordered information system gives rise to an **indiscernibility relation** IND_B as before where $(u_1, u_2) \in IND_B$ iff $\alpha(u_1) = \alpha(u_2)$ for all $\alpha \in B$. The equivalence classes are partially ordered if we define $[u_1] \leq [u_2]$ iff $\forall \alpha \in B \,(\alpha u_1 \, K_\alpha \, \alpha u_2)$, This defined ordering necessarily satisfies

$$u_1 \, H \, u_2 \text{ implies } [u_1] \leq [u_2]. \tag{1}$$

As an example we can consider the change of level of detail in Fig. 2 as an Ordered Information System. For simplicity, take just the part of the hypergraph at the detailed level consisiting of nodes $\{West, Mid, North\}$ and of edges $\{a, b\}$. The set U is then $\{West, Mid, North, a, b\}$, and the partial order H relates every edge in U to its two incident nodes. Thus $(a, West) \in H$ and $(a, Mid) \in H$ and so on. In addition H is defined to be reflexive. In this simple example there is just one attribute α where $V_\alpha = \{West, westline, North\}$ with the partial order K_α containing $(westline, North)$ and $(westline, West)$ as well as the identity pairs $(westline, westline)$ etc. The attribute α assigns the value $westline$ to both a and b as well as to Mid because this is the less-detailed feature to which these entities belong, as well as satisfying $\alpha(West) = West$ and $\alpha(North) = North$.

In a partially ordered universe approximations need to respect structure. To do this we need to make use of definitions of approximations based on relations. So next we introduce relations on partial orders.

4.2 Relations on Partial Orders

A relation on a graph, a special case of Definition 5, is a relation on the set of all edges and nodes of the graph which in addition respects the incidence structure. The consequences of this definition have been explored in more detail in [15,16], but all works in the more general setting of a partially ordered set (U, H).

Here we only need a few properties of these relations, which we set out in Theorem 1; proofs can be found in the above references. In the definition we write composition of relations as ; and we take $R; S$ to mean composition in the following order

$$R; S = \{(u, w) \in U \times U \mid \exists v \in U \, ((u, v) \in R \text{ and } (v, w) \in S)\}.$$

Definition 5. *A relation $R \subseteq U \times U$ is **stable** for a partial order H on U if $H; R; H \subseteq R$.*

The significance of Definition 5 is that stable relations on a hypergraph (U, H) correspond to union-preserving functions on the lattice of subgraphs [15]. Arbitrary relations correspond to union-preserving functions on the lattice of subsets, but here subgraphs are approximated.

Theorem 1 (Stell [15]). *Let R and S be stable relations on (U, H). We use $R \,;\, S$, to denote the composition, of R and S as relations on U.*

1. *$R \,;\, S$ is stable.*
2. *The equality relation $I \subseteq U \times U$ need not be stable, but H is a stable relation and satisfies $H \,;\, R = R = R \,;\, H$.*
3. *Neither the converse $\check{R}$ nor the complement $\overline{R}$ need be stable, but stable relations are closed under the converse complement operation $\frown R = \overline{\check{R}} = \check{\overline{R}}$ and this satisfies $\frown\frown R = R$.*

Definition 6. *A relation R on (U, H) is **reflexive** if $H \subseteq R$, and is **transitive** if $R \,;\, R \subseteq R$.*

We note in passing that the notion of symmetry for such relations is not straightforward on account of the lack of an involutory converse operation [15]. In fact there are several different ways of defining symmetry but these are outside the scope of the present paper.

It can be seen that any relation on U (without any assumption of stability) which satisfies the reflexivity and transitivity conditions must actually be stable. This is because we would have $H \,;\, R \,;\, H \subseteq R \,;\, R \,;\, R \subseteq R$.

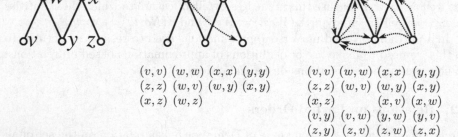

$$(v,v) \ (w,w) \ (x,x) \ (y,y)$$
$$(z,z) \ (w,v) \ (w,y) \ (x,y)$$
$$(x,z) \ (w,z)$$

$$(v,v) \ (w,w) \ (x,x) \ (y,y)$$
$$(z,z) \ (w,v) \ (w,y) \ (x,y)$$
$$(x,z) \qquad\quad (x,v) \ (x,w)$$
$$(v,y) \ (v,w) \ (y,w) \ (y,v)$$
$$(z,y) \ (z,v) \ (z,w) \ (z,x)$$

Fig. 3. Graph from Fig. 1 with two examples of reflexive transitive stable relations

Figure 3 provides two examples of stable relations on the partial order, which is also more specifically a graph, from Fig. 1. These two relations, which are described in detail as subsets of $U \times U$ in the figure, are reflexive which allows the convenient visualisation shown. In a reflexive relation we only need show those arrows (ordered pairs) which are present in addition to H. These added arrows are shown as dashed lines with arrow heads in the figure.

4.3   Correspondence Between Partitions and Relations on Partial Orders

The well-known correspondence between partitions of a set and equivalence relations on a set allows us to switch between two ways of thinking about indiscernibility. The consideration of relations on sets that need not be equivalence relations has often been used in rough set theory [11,12,18], but the use of relations on hypergraphs, or more generally on partial orders has been relatively unexplored. The approach in [14] deals with several approximation operators in terms of relations on hypergraphs, but did not consider what connection there might be between partitions (in some generalized sense) of a hypergraph and relations on the hypergraph with properties analogous to reflexivity, transitivity and symmetry. As already noted, symmetry will not be considered in this paper, so we will just deal with reflexive and transitive relations in the sense of Definition 6.

Definition 7. *For any relation R on U and any $u \in U$, define the u-**dilate** of R to be $uR = \{v \in U \mid u\,R\,v\}$. More generally, for any $X \subseteq U$, we will use XR to denote $\{v \in U \mid \exists x\,(x\,R\,v\ and\ x \in X)\}$.*

The terminology 'dilate' is used as in mathematical morphology [2,10] uR is the dilation by R of the set $\{u\}$ usually denoted $\{u\} \oplus R$. In the case that R is an equivalence relation on U, the u-dilates are just the equivalence classes. For more general R, two dilates can overlap without being equal and this can be seen in the examples in Fig. 4.

Lemma 1. *Let R be a reflexive and transitive relation on U. Then for any $u, v \in U$ the following three statements are equivalent:*

$$v \in uR, \qquad\qquad u\,R\,v, \qquad\qquad vR \subseteq uR.$$

In generalizing the notion of a partition from a set to a hypergraph, and more generally to a partial order, it is clear that if the blocks of the partition are disjoint then they will not, except in trivial cases, respect the additional structure of the set. In the case of a graph, if we require that blocks are disjoint and that in addition they are always subgraphs then a connected graph will only have a single partition consisting of just one block. In the case of a partial order requiring blocks to both be disjoint and to be downward closed sets leads to a generalized form of this limitation.

Thus, we expect a general partition of a partial order either to have blocks that overlap or to have blocks that need not be downward closed sets. But a good notion of partition should also be connected with relations on the partial order and should be capable of supporting approximation operators with good properties. The following result demonstrates how these requirements are connected.

Theorem 2. *Let U be a set and H a partial order on U. The following three sets of structures are in bijective correspondence with each other.*

1. *Relations R on U such that $H \subseteq R$ and $R\,;R \subseteq R$.*
2. *Partitions of U equipped with a partial order, P, on the set of equivalence classes such that*

$$\forall u, v \in U\ (u\,H\,v \Rightarrow [u]\,P\,[v]), \tag{2}$$

where $[u]$ denotes the equivalence class of u.

3. *Sets $\mathcal{C}$ of subsets of U such that if we define for $u \in U$*

$$\ulcorner u \urcorner^{\mathcal{C}} = \bigcap \{B \in \mathcal{C} \mid u \in B\}$$

then for every $u \in U$ we have $\ulcorner u \urcorner^{\mathcal{C}} \in \mathcal{C}$ and if $u \, H \, v$ then $\ulcorner v \urcorner^{\mathcal{C}} \subseteq \ulcorner u \urcorner^{\mathcal{C}}$.

Proof. We show first that *1* corresponds with *2*.

Given a relation R as in *1*, define a relation $\equiv_R$ by $x \equiv_R y$ if $x \, R \, y$ and $y \, R \, x$. It is straightforward to check that this is an equivalence relation. From R we also define a relation P_R on the equivalence classes of $\equiv_R$ by $[x] \, P_R \, [y]$ if $x \, R \, y$. To check this is well-defined we need to check that if $x \equiv_R x'$ and $y \equiv_R y'$ then $x \, R \, y$ iff $x' \, R \, y'$, but this is routine. It is also clear that P_R satisfies the property stated of P in (2).

In the other direction, given an equivalence relation $\equiv$ and a partial order P satisfying (2), define a relation, $S_{\bar{P}}^{\equiv}$, on U by $x \, S_{\bar{P}}^{\equiv} \, y$ iff $[x] \, P \, [y]$. We can check that $S_{\bar{P}}^{\equiv}$ contains H and is transitive.

To justify that we have a bijection, suppose first that we have a relation R as in *1*. We need to show that $S_{P_R}^{\equiv_R} = R$. Secondly, given an equivalence relation E and a partial order Q satisfying (2), we need to show that $\equiv_{S_Q^E} = E$ and that $P_{S_Q^E} = Q$. These are both routine calculations from the definitions.

We now show that *1* corresponds with *3*.

Given a relation R as in *1*, define a set of subsets of U by $\mathcal{B}_R = \{uR \subseteq U \mid u \in U\}$. The key observation here is that $\ulcorner u \urcorner^{\mathcal{B}_R} = uR$. To justify this, note that for any $w \in U$ the condition $u \, R \, w$ is equivalent, since R is reflexive and transitive, to

$$\forall v \, (v \, R \, u \Rightarrow v \, R \, w). \tag{3}$$

Now (3) is equivalent to $w \in \bigcap \{vR \subseteq U \mid u \in vR\}$ and hence to $w \in \ulcorner u \urcorner^{\mathcal{B}_R}$.

To map a set of subsets $\mathcal{C}$ as in *3* to a relation as in *1* we define the relation $T_{\mathcal{C}}$ by

$$u \, T_{\mathcal{C}} \, v \quad \text{iff} \quad \ulcorner v \urcorner^{\mathcal{C}} \subseteq \ulcorner u \urcorner^{\mathcal{C}}.$$

Finally, to show these two constructions provide a bijection, we have to check that given any relation R as in *1*, and any set of subsets $\mathcal{C}$ as in *3*, that $T_{\mathcal{B}_R} = R$ and that $\mathcal{B}_{T_{\mathcal{C}}} = \mathcal{C}$. These are routine calculations from the definitions. $\square$

The above result shows how reflexive and transitive relations on a partial order are equivalent to two ways of weakening the usual notion of a partition on a set. Clearly, in the special case that H is the identity relation on U and in addition that R is symmetric, the structures in parts 2 and 3 both reduce to an ordinary partition on U.

Besides equivalence relations and partitions of a set U, functions defined on U provide another way of performing granulation. This too generalizes to the partial order case. We omit the proof as it uses similar techniques to that of Theorem 2.

Theorem 3. *Let (U, H) and (V, K) be posets and φ a function from U to V such that for all $u_1, u_2 \in U$ we have $u_1 \, H \, u_2$ implies $\varphi u_1 \, K \, \varphi u_2$. Then the relation $\mathcal{R}_\varphi$ on U defined by $u_1 \, \mathcal{R}_\varphi \, u_2$ iff $\varphi u_1 \, K \, \varphi u_2$ is transitive and contains H.* $\qquad\square$

Figure 5 illustrates Theorem 5 by showing how the relations used in Fig. 4 have corresponding order-preserving functions defined on the underlying partial orders.

4.4 Two Kinds of Granulation

In a set the notion of granulation, that is the formation of granules, involves grouping or clustering together subsets of the elements. This happens too with graphs. The lower example in Fig. 4 shows a case where one edge and one node form one cluster and the remaining two nodes and edge form another cluster. However, granulating a partial order, such as a graph, is not just a matter of clustering elements together. Such clustering by itself only yields a discrete set and cannot produce non-trivial partial orders. The second component to granulation on a partial order is the provision of a partial order on the clusters of elements. Theorem 2 shows that reflexive and transitive relations on partial orders correspond to granulations consisting of the formation of clusters together with ordering the clusters.

The upper example in Fig. 4 shows that even if the formation of clusters gives only singleton clusters we can still have an ordering on these clusters which strictly extends the original ordering. The cluster ordering is shown by bold arrows in the figure.

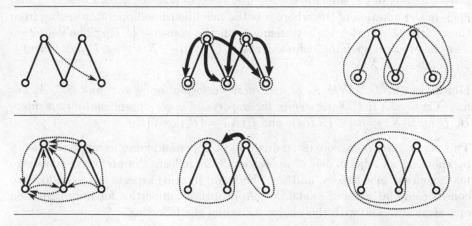

Fig. 4. Relations from Fig. 3 with corresponding ordered partitions and overlapping dilates

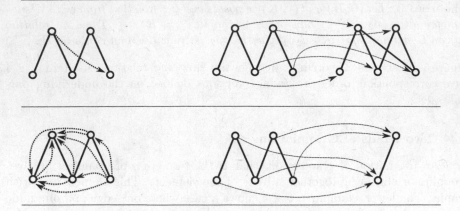

Fig. 5. Relations from Fig. 3 with corresponding quotient structures

5   Approximation Operators

Returning now to the definition of an ordered information system (U, H, A) in Sect. 4.1 we can see that as each subset $B \subseteq A$ provides a partially ordered set of equivalence classes which respects H as in Eq. (1). Thus by Theorem 2 we can construct a relation R_B which is stable with respect to H and is reflexive and transitive. The significance of this is that we can use R_B to define upper and lower approximation operators using opening and closing. Several different pairs of operators can be considered. The pair $\underline{B}$ and $\overline{B}$ suggested below does not generalize any of the three dual pairs discussed in [18] but generalizes apr'_n and $\overline{apr}'_{\check{n}}$ in the notation of [18] where n is the neighbourhood operator arising from the relation R_B and $\check{n}$ is that arising from the converse of R_B. The definition uses the **erosion** operator, defined for $X \subseteq U$ by $R_B \ominus X = \{u \in U : \forall v \, ((u, v) \in R_B \text{ implies } v \in X)\}$.

Definition 8. *Let (U, H, A) be an ordered information system and $B \subseteq A$. For any X where $XH \subseteq X$ we define the upper and lower approximation operators $\overline{B}, \underline{B}$ by $\overline{B}(X) = R_B \ominus (XR_B)$, and $\underline{B}(X) = (R_B \ominus X)R_B$.*

This departs from the more usual dualities of upper and lower in rough set theory by choosing an adjoint pair of operators. This reflects the preferred duality in mathematical morphology, and also the trend in some aspects of modal logic to consider adjoint pairs of modalities. Appropriate properties for approximation operators still hold with this choice including the following.

1. $\overline{B}\overline{B}(X) = \overline{B}(X)$, and $\underline{B}\underline{B}(X) = \underline{B}(X)$
2. $\underline{B}(X) \subseteq X \subseteq \overline{B}$
3. $\underline{B}(X \cap Y) \subseteq \underline{B}(X) \cap \underline{B}(Y)$ and $\underline{B}(X) \cup \underline{B}(Y) \subseteq \underline{B}(X \cup Y)$
4. $\overline{B}(X \cap Y) \subseteq \overline{B}(X) \cap \overline{B}(Y)$ and $\overline{B}(X) \cup \overline{B}(Y) \subseteq \overline{B}(X \cup Y)$
5. $X \subseteq Y$ implies $\overline{B}(X) \subseteq \overline{B}(Y)$ and $\underline{B}(X) \subseteq \underline{B}(Y)$

Some properties here are weaker than in the situation where the universe is unordered and where indiscernability is symmetric as well as reflexive and transitive. For example, we do not have $\overline{B}(X) \cup \overline{B}(Y) = \overline{B}(X \cup Y)$, nor do we have $\underline{B}(\overline{B}(X)) = \overline{B}(X)$. These weakenings are already well known in the case of binary relations on a set (rather than a poset as here) in [18].

6   Conclusions and Further Work

This paper has used a correspondence between partially ordered partitions and certain reflexive transitive relations to find a simple way of defining approximation operators on ordered information systems. Further work is continuing on connections between these approximation operators and the bi-intuitionistic modal logic using relations on hypergraphs that was introduced in [17]. This is likely to provide a way of generalizing information logics [4] to the partially ordered setting. The notion of granulation used in this paper is not the only way parts of graphs can constitute granules. Further work will examine how ideas such as the tree-decompositions described in [5, p. 337] are connected with the reflexive and transitive relations studied here. In [6] Fan considers information systems where instead of functional attributes defined on the universe there are relations. Combining this approach with the present paper suggests it would be interesting to explore relational information systems in the more general case of a partially ordered universe.

References

1. Berge, C.: Hypergraphs: Combinatorics of Finite Sets, vol. 45. North-Holland Mathematical Library, North-Holland (1989)
2. Bloch, I., Heijmans, H., Ronse, C.: Mathematical morphology. In: Aiello, M., Pratt-Hartmann, I., van Benthem, J. (eds.) Handbook of Spatial Logics, vol. 14, pp. 857–944. Springer, Heidelberg (2007)
3. Brown, R., Morris, I., Shrimpton, J., Wensley, C.D.: Graphs of morphisms of graphs. Elect. J. Comb. **15** (#A1). www.combinatorics.org/ojs/ (2008)
4. Demri, S.P., Orłowska, E.S.: Incomplete Information: Structure, Inference, Complexity. Springer, Heidelberg (2002)
5. Diestel, R.: Graph Theory. Graduate Texts in Mathematics, vol. 173, 4th edn. Springer, Heidelberg (2010)
6. Fan, T.-F.: Rough set analysis of relational structures. Inf. Sci. **221**, 230–244 (2013)
7. Lin, T.Y.: Granular computing on binary relations I, II. In: Polkowski, L., Skowron, A. (eds.) Rough Sets in Knowledge Discovery, pp. 107–140. Physica, Heidelberg (1998)
8. Pawlak, Z.: Rough sets. Int. J. Comput. Inf. Sci. **11**, 341–356 (1982)
9. Polkowski, L.: Rough Sets. Mathematical Foundations. Advances in Soft Computing. Physica, Heidelberg (2002)
10. Serra, J.: Image Analysis and Mathematical Morphology. Academic Press, London (1982)
11. Skowron, A., Stepaniuk, J.: Tolerance approximation spaces. Fundamenta Informaticae **27**, 245–253 (1996)

12. Slowinski, R., Vanderpooten, D.: A generalized definition of rough approximations based on similarity. IEEE Trans. Knowl. Data Eng. **12**, 331–336 (2000)
13. Stell, J.G.: Relations in mathematical morphology with applications to graphs and rough sets. In: Winter, S., Duckham, M., Kulik, L., Kuipers, B. (eds.) COSIT 2007. LNCS, vol. 4736, pp. 438–454. Springer, Heidelberg (2007)
14. Stell, J.G.: Relational granularity for hypergraphs. In: Szczuka, M., Kryszkiewicz, M., Ramanna, S., Jensen, R., Hu, Q. (eds.) RSCTC 2010. LNCS, vol. 6086, pp. 267–276. Springer, Heidelberg (2010)
15. Stell, J.G.: Relations on hypergraphs. In: Kahl, W., Griffin, T.G. (eds.) RAMICS 2012. LNCS, vol. 7560, pp. 326–341. Springer, Heidelberg (2012)
16. Stell, J.G.: Symmetric Heyting relation algebras with applications to hypergraphs. J. Logical Algebraic Methods Program. **84**, 440–455 (2015)
17. Stell, J.G., Schmidt, R.A., Rydeheard, D.: A bi-intuitionistic modal logic: foundations and automation. J. Logical Algebraic Methods Program. **85**, 500–519 (2016)
18. Yao, Y.Y.: Relational interpretations of neighborhood operators and rough set approximation operators. Inf. Sci. **111**, 239–259 (1998)

Rough Sets, Non-Determinism and Incompleteness

A Rough Perspective on Information in Extensive Form Games

Georg Peters[1,2(✉)]

[1] Department of Computer Science and Mathematics,
Munich University of Applied Sciences, Lothstrasse 34, Munich, Germany
[2] Australian Catholic University, Sydney, Australia
georg.peters@cs.hm.edu

Abstract. In game theory imperfect and incomplete information have
been intensively addressed. In extensive form games a player faces imperfect information when it cannot identify the decision node it is presently
located at. The player is only aware of an information set consisting of
more than one node. A player faces incomplete information when it is not
aware of, e.g., preferences or payoffs of its opponents. Rough set theory
is a prime method addressing missing and contradicting information in
decision tables where a set of variables induces a decision. In particular,
rough set theory provides a means by which records with identical variable values lead to different, contradicting decisions. To indicate such
situations, these records are assigned to the boundaries of all possible
decisions. Obviously, both situations, games with imperfect or incomplete
information and rough decision tables are similar with respect to their
characteristics and challenges regarding a lack of information. Hence, a
discussion of their relationship could be mutually beneficial. Therefore,
the objective of our paper is to provide a rough set perspective on extensive form games with imperfect and incomplete information.

Keywords: Rough set theory · Game theory · Extensive form games ·
Imperfect information · Incomplete information

1   Introduction

Game theory is widely applied in a diverse range of areas. Its objective is to optimize payoffs in situations with two or more players. Important fields of application of game theory are in economics and social science where it has been used to
investigate and understand human behavior [8]. However, in the past decades, it
suitability for economic analysis has been questioned. It has been observed that
important preassumptions of game theory do not match with human behavior
(basically assuming a homo economicus). Therefore, a new field, behavioral economics [1], has emerged where experiments are performed to understand human
behavior.

Game theory is also applied to a wide range of 'technical' fields, i.e., where
it is irrelevant how humans behave. These fields include engineering and computer science [4] and many others. Its 'technical' applications as an optimization

© Springer International Publishing AG 2016
V. Flores et al. (Eds.): IJCRS 2016, LNAI 9920, pp. 145–154, 2016.
DOI: 10.1007/978-3-319-47160-0_13

method also include rough sets. In game theoretic rough sets [6,7], game theory is applied to reduce the boundaries by moving selected objects from there to positive and negative regions. So, in game theoretic rough sets, rough sets are not integrated into game theory in a sense of rough game theory but applied to optimize rough set approximations. In contrast to this, Xu and Yao [9], for example, integrated rough sets into game theory by developing a rough payoff matrix derived from rough variables.

Since practically any real life situation is characterized by a lack of information intensive attention has been given to games dealing with imperfect and incomplete information. Often the terms (im)perfect and (in)complete are not precisely defined and used interchangeably addressing any lack of information in a systems. For example, at EconPort [3] perfect information is defined as follows: "By perfect information we mean that anything that may impact a buyer or seller's decision making process is known and understood." Hence, imperfect information is given when information is incorrect, incomplete or missing.

However, in game theory the terms imperfect information and incomplete information have different meanings. Imperfect information can be observed in extensive form games, i.e., games in tree forms, when a player does not know its present position in the tree at all times. Such situations occur when it is not aware of all previous decisions taken by the other players. In contrast to this, incomplete information refers to games where one player has only limited information about the preferences, payoffs etc. of the other players. So, imperfect information is associated with past actions of a player's opponents, while incomplete information is linked to future actions of a player's opponents.

Imperfect and incomplete information have been extensively addressed in game theory and have led to several refinements of the equilibria in games (see, e.g., Bonanno [2] for a good introduction). The relationship between imperfect and incomplete information has also raised great attention. When some assumptions are made about the preferences of the players and about probabilities, a game with incomplete information can be transformed into a game with imperfect information (Harsanyi transformation [5]).

Rough set theory addresses missing and contradicting information in decision tables. For example, two objects are indiscernible, i.e., they have identical attribute values. Sometimes these objects lead to different decisions. Reasons may include that a crucial attribute is missing or that the data recorded are inconsistent. To indicate this, such objects are assigned to the boundaries of all possible decisions. When objects with identical attribute values lead to identical decisions, they are regarded as sure objects and are assigned to the lower approximation of the respective decision.

So, regarding the emphasis to deal with information, game theory and rough sets seem to be rather similar. However, little attention has been directed to the relationship of imperfect and incomplete information in game theory and rough sets. Therefore, the objective of the paper is to discuss the relationship of imperfect and incomplete information in extensive form games and rough sets. We limit our presentation on a rough set perspective on games and only address some very key concepts of game theory. So basically, we provide a 'rough' rough perspective on extensive form games in our paper.

The remainder of the paper is organized as follows. In Sect. 2, we discuss imperfect information. In the next section, we deal with incomplete information. In Sect. 4, we merge imperfect and incomplete information into one rough decision table and develop a rough payoff matrix. The paper concludes with a summary in Sect. 5.

2   Imperfect Information

2.1   Imperfect Information in Extensive Form Games

In extensive form games imperfect information is defined when a player does not always know at which decision node it is located. For example see Fig. 1 that shows an extensive form game with two players. Player A, indicated by coarse dotted lines, has two decision nodes (A_1 and A_2) while for Player B, indicated by solid lines, there are three decision nodes (B_1, B_2 and B_3). The results are depicted as circled C_i with $C_i = (c_{Ai}, c_{Bi})$ the payoffs for Players A and B, respectively.

Player A is always aware of its position in the game. It has perfect information about the game, i.e., each node forms a separate information set: $I_{A1} = \{A_1\}$ and $I_{A2} = \{A_2\}$. In contrast to this, Player B can identify decision node B_3 but it cannot distinguish between the decision nodes B_1 and B_2, i.e., it has no information about the decision taken by Player A at node A_1. Therefore, B_1 and B_2 belong to one information set ($I_{B1} = \{B_1, B_2\}$). In Fig. 1 the information set I_{B1} is indicated by a vertical fine dotted line between B_1 and B_2. Node B_3 forms another information set ($I_{B2} = \{B_3\}$).

A strategy is a predefined set of actions that determines how a player will decide at any information set of a game, i.e., it is a complete guide to action. Player A has perfect information, i.e., each decision node forms an information set. Hence, each of its strategies comprises of two predefined actions (for node A_1 with three possible decisions (up: ↑ towards B_1, right: → towards B_2 or down: ↓ towards B_3) and for A_2 with two possible decisions (up: ↑ towards C_1 or down: ↓ towards C_2)). Therefore, Player A has a total of $6 = 3 \cdot 2$ strategies. In contrast to this, Player B faces imperfect information with two information sets only for three nodes. At each information set it can go up (↑) or down (↓) which leads to $4 = 2 \cdot 2$ strategies. Moving up at I_{B1} leads to A_2 or C_4, moving down to C_3 or C_5 depending on the node it is.

Obviously, we obtain different games if a player has perfect information or only imperfect information. In the case of perfect information, we get a game as depicted in Table 1. The left matrix in Table 1 shows the full redundant matrix while the right matrix shows the minimum matrix. The arrows indicate the decisions the players take at a node. E.g., strategy $b_2 = \uparrow\uparrow\downarrow$ for Player B means that it moves up at B_1 and B_2, and it moves down when it is at B_3. In the minimum matrix, a star ∗ indicates that any decision taken at a particular node leads to the same result.

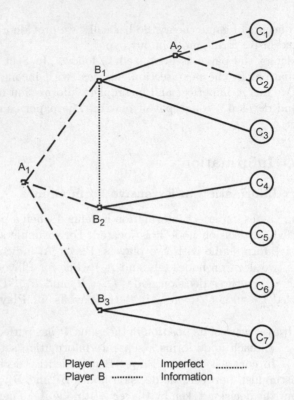

Fig. 1. An extensive form game with imperfect information

In the case of imperfect information, Player B still has three nodes but it cannot distinguish between B_1 and B_2. Hence, it can only decide on the two information sets $I_{B1} = \{B_1, B_2\}$ and $I_{B2} = \{B_3\}$. The corresponding game is shown in Table 2 (again the full redundant on the left and the minimum matrix on the right).

2.2 Imperfect Information and Rough Sets

To obtain a rough perspective on imperfect information in extensive form games we present the original game (Table 2) in a modified way. In Table 3, the 'rough strategies' of Player B are depicted. The decision of Player A at node A_2 depends of the payoff it gets. For simplicity, we assume that Player A always prefers C_1 over C_2: $c_{A1} \succ c_{A2}$. In Sect. 3, we address the preferences of Player A in more detail in the context of incomplete information.

Let us first discuss the strategies b_5 and b_6 of Player B. Since it is aware that it is at node B_3, it can distinguish between the strategies and select the strategy that optimizes its payoff. For $c_{B6} \succ c_{B7}$ it would choose going up ($\uparrow$) and for $c_{B6} \prec c_{B7}$ it would go down ($\downarrow$) while it would be indifferent for $c_{B6} \sim c_{B7}$ ($*$).

Hence, in rough set terms we suggest to assign the strategies b_5 and b_6 to lower approximations.

In contrast to the above, due to imperfect information, it cannot distinguish whether it is at node B_1 or at node B_2. These nodes are indiscernible for the player. If it decides to move up ($\uparrow$) and happens to be at node B_1 the payoffs C_1 will be obtained. If it happens to be at node B_2 it is heading towards C_4. When it decides to move down ($\downarrow$) it ends up at C_3 if it happens to be at node B_1 and at C_5 if it is at node B_2. Hence, in rough set terms we would assign the strategies b_1, b_2, b_3 and b_4 to boundaries.

In rough sets the boundaries are of particular interest. Therefore, we take a look at them and propose to distinguish three different kinds of boundaries in extensive form games with imperfect information:

- Irrelevant Boundaries. Assuming that $c_{A6}, c_{A7} \succ c_{A1}, c_{A2}, c_{A3}, c_{A4}, c_{A5}$, Player A will select to go down ($\downarrow$) at node A_1 which leads to node B_3 of Player B. Node B_3 is identifiable for Player B. Hence, Player B is not challenged by any imperfect information since it will never be at $I_{B1} = \{B_1, B_2\}$. There are still boundaries in the game but they are irrelevant (dominated).
- Weak Boundaries. For $c_{A6}, c_{A7} \prec c_{A1}, c_{A2}, c_{A3}, c_{A4}, c_{A5}$ Player A will go up ($\uparrow$) or to the right ($\rightarrow$) at node A_1. For our discussion, it is irrelevant where it actually goes since both pathes lead to the same information set $\{B_1, B_2\}$ for Player B. Now, Player B face imperfect information. However, if $c_{B1}, c_{B2}, c_{B4} \succ c_{B3}, c_{B5}$ then it decides to move up ($\uparrow$) independently whether it is at B_1 of B_2. Although it does not know how much it gets, it, at least, knows that moving up is the optimal action. To indicate this partial knowledge we call the boundary weak.
- Strong Boundaries. Like before, we assume $c_{A6}, c_{A7} \prec c_{A1}, c_{A2}, c_{A3}, c_{A4}, c_{A5}$, i.e., Player A will go up ($\uparrow$) or to the right ($\rightarrow$). In the case of strong boundaries, Player B does not know what it will get but it also does not know its optimal action at the information set $\{B_2, B_3\}$. E.g., for $c_{B1}, c_{B2}, c_{B5} \succ c_{B3}, c_{B4}$ Player B should move up ($\uparrow$) if it is at node B_2 but should move down ($\downarrow$) if

Table 1. Game with perfect information

		Player A						Player A			
		a'_1 ↑↑	a'_2 ↑↓	a'_3 →↑	a'_4 →↓	a'_5 ↓↑	a'_6 ↓↓	a_1 ↑↑	a_2 ↑↓	a_3 → *	a_4 ↓ *
	b_1 ↑↑↑	C_1	C_2	C_4	C_4	C_6	C_6	C_1	C_2	C_4	C_6
	b_2 ↑↑↓	C_1	C_2	C_4	C_4	C_7	C_7	C_1	C_2	C_4	C_7
	b_3 ↑↓↑	C_1	C_2	C_5	C_5	C_6	C_6	C_1	C_2	C_5	C_6
Player	b_4 ↑↓↓	C_1	C_2	C_5	C_5	C_7	C_7 ⇒	C_1	C_2	C_5	C_7
B	b_5 ↓↑↑	C_3	C_3	C_4	C_4	C_6	C_6	C_3	C_3	C_4	C_6
	b_6 ↓↑↓	C_3	C_3	C_4	C_4	C_7	C_7	C_3	C_3	C_4	C_7
	b_7 ↓↓↑	C_3	C_3	C_5	C_5	C_6	C_6	C_3	C_3	C_5	C_6
	b_8 ↓↓↓	C_3	C_3	C_5	C_5	C_7	C_7	C_3	C_3	C_5	C_7

Table 2. Game with imperfect information

		Player A						⇒		Player A			
		a'_1	a'_2	a'_3	a'_4	a'_5	a'_6			a_1	a_2	a_3	a_4
		$\uparrow\uparrow$	$\uparrow\downarrow$	$\rightarrow\uparrow$	$\rightarrow\downarrow$	$\downarrow\uparrow$	$\downarrow\downarrow$			$\uparrow\uparrow$	$\uparrow\downarrow$	$\rightarrow *$	$\downarrow *$
	b_1 $\uparrow\uparrow$	C_1	C_2	C_4	C_4	C_6	C_6		$\uparrow\uparrow$	C_1	C_2	C_4	C_6
Player	b_2 $\uparrow\downarrow$	C_1	C_2	C_4	C_4	C_7	C_7		$\uparrow\downarrow$	C_1	C_2	C_4	C_7
B	b_3 $\downarrow\uparrow$	C_3	C_3	C_5	C_5	C_6	C_6		$\downarrow\uparrow$	C_3	C_3	C_5	C_6
	b_4 $\downarrow\downarrow$	C_3	C_3	C_5	C_5	C_7	C_7		$\downarrow\downarrow$	C_3	C_3	C_5	C_7

it is at B_3. To indicate this absence of any knowledge we call the boundary strong (dominating).

3 Incomplete Information

3.1 Incomplete Information in Extensive Form Games

To provide an example for a rough interpretation of incomplete information we assume that Player A moves up ($\uparrow$) at node A_1. This is the case for $c_{A1}, c_{A2}, c_{A3} \succ c_{A4}, c_{A5}, c_{A6}, c_{A7}$. We further assume that $c_{A1} \succ c_{A2}$, i.e., Player A would move up when it is at node A_2. For Player B we assume the following order of its payoffs: $c_{B2} \succ c_{B3} \succ c_{B1}$. A possible corresponding sub-game is shown in Fig. 2.

In the case of complete information, Player B knows the possible payoffs of Player A (as in the left sub-figure of Fig. 2). It just needs to evaluate its possible moves at node B_1 and the responses of Player A. For its two possible actions we get:

- Player B moves up ($\uparrow$) $\Rightarrow$ Player A moves also up ($\uparrow$) since $(c_{A1} = 10) \succ (c_{A2} = 5)$. Player B obtains $c_{B1} = 8$.
- Player B moves down ($\downarrow$) and obtains $c_{B3} = 9$.

Obviously, in the case of complete information, Player B would decide to move down ($\downarrow$).

Table 3. Imperfect information: rough strategies of Player B

Rough strategy	Player B		Player A	Payoffs
	Action	at node	Action at A_1	
b_1	$\uparrow$	B_1	$\uparrow$	C_1 or C_2
b_2	$\uparrow$	B_2	$\rightarrow$	C_4
b_3	$\downarrow$	B_1	$\uparrow$	C_3
b_4	$\downarrow$	B_2	$\rightarrow$	C_5
b_5	$\uparrow$	B_3	$\downarrow$	C_6
b_6	$\downarrow$	B_3	$\downarrow$	C_7

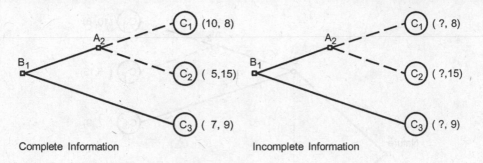

Fig. 2. A sub-game to illustrate incomplete information from the perspective of Player B

Now, we assume that Player B faces incomplete information, i.e., it does not know the possible payoffs of Player A (as in the right sub-figure of Fig. 2). Therefore, it does not know if Player A will move up or down if it is at node A_2.

Under certain assumptions games with incomplete information can be interpreted as games with imperfect information. Harsanyi proposed a method how to transform a game with incomplete information into a game with imperfect information. Basically, Player B assumes two or more possible sets of payoffs that Player A might get. Each of these sets of payoffs lead to different sub-games. To select which of these sub-games is played, a new player, 'nature', is introduced. Nature has the first move and selects the sub-game (see Fig. 3).

3.2 Incomplete Information and Rough Sets

To discuss incomplete information in extensive form games we refrain from the possible transformation, the so called Harsanyi transformation [5], of a game with incomplete information into a game with imperfect information. A rough interpretation would go beyond the scope of our paper.

As already discussed in the previous section, Player B has no knowledge about the payoffs of Player A. Let us assume that Player B decides to move up at node B_1. Then it does not know the next step of Player A; it does not know if Player A will move up or down at node A_2. When Player B moves down it directly reaches payoffs C_3 without any further decision node of Player A. This leads to strategies for Player B as depicted in Table 4. The irrelevant strategy of Player A when Player B moves down is indicated by a star $(*)$.

4 A Rough Payoff Matrix

In the previous sections we discussed rough strategies of Player B in the context of imperfect information (Table 3) and incomplete information (Table 4) separately. In this section, we merge these previously obtained tables into one and get Table 5. Player B has two sets of strategies in boundaries, i.e., due to imperfect and/or incomplete information; there is more than one possible payoff for

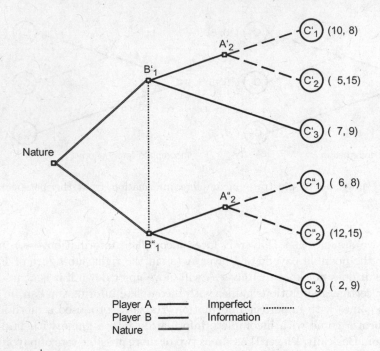

Fig. 3. Transforming a game with incomplete information in a game with imperfect information

some of its actions. The first set of strategies is moving up ($\uparrow$) at information set $I_{B1} = \{B_1, B_2\}$. Depending on the previous and future actions of Player A the payoffs C_1, C_2 or C_4 are obtained. The second set of strategies for Player B is moving down ($\downarrow$). Depending on the decision of Player A the payoffs C_3 or C_5 are obtained. The remaining two strategies b_6 and b_7 lead to C_6 and C_7, respectively. Since Player B is aware of the payoffs of its actions they belong to lower approximations.

Let us summarize the definitions for lower approximations and boundaries:

– Boundary. We define a boundary when a strategy of a player can lead to more than one payoff.

Table 4. Incomplete information: rough strategies of Player B

Rough strategy	Player B		Player A	Payoffs
	Action	at node	Action at A_2	
b_1	$\uparrow$	B_1	$\uparrow$	C_1
b_2	$\uparrow$	B_1	$\rightarrow$	C_2
b_3	$\downarrow$	B_1	$*$	C_3

Table 5. Imperfect and incomplete information: rough strategies of Player B

Rough strategy	Player B		Player A	Payoffs
	Action	at node	Action at A_1, A_2	
b_1	↑	B_1	↑↑	C_1
b_2	↑	B_1	↑↓	C_2
b_3	↑	B_2	→ *	C_4
b_4	↓	B_1	↑ *	C_3
b_5	↓	B_2	→ *	C_5
b_6	↑	B_3	↓ *	C_6
b_7	↓	B_3	↓ *	C_7

– Lower Approximation. Any strategy that is not a member of a boundary belongs to a lower approximation, i.e., a strategy of a player leads to one and only one possible payoff.

In the case of our example all strategies of Player A belong to lower approximations. For Player B we get four regions R (with a hat ($\widehat{R}$) indicating a boundary and an underline ($\underline{R}$) a lower approximation):

$$b_1, b_2, b_3 \in \widehat{R}_1 = C_1 \vee C_2 \vee C_4 \qquad b_6 \in \underline{R}_3 = C_6$$
$$b_4, b_5 \quad \in \widehat{R}_2 = C_3 \vee C_5 \qquad b_7 \in \underline{R}_4 = C_7$$

Note, that Player B just knows that it can move up or down but does not see its particular strategy when it faces imperfect information. Therefore, we finally can derive from Table 2 a rough payoff matrix as shown in Table 6.

Table 6. Rough payoff matrix

Player A

		a_1	a_2	a_3	a_4
		↑↑	↑↓	→ *	↓ *
	b'_1 ↑↑	$(\underline{c}_{A1}, \widehat{c}_{B1})$	$(\underline{c}_{A2}, \widehat{c}_{B2})$	$(\underline{c}_{A4}, \widehat{c}_{B4})$	$(\underline{c}_{A6}, \underline{c}_{B6})$
Player	b'_2 ↑↓	$(\underline{c}_{A1}, \widehat{c}_{B1})$	$(\underline{c}_{A2}, \widehat{c}_{B2})$	$(\underline{c}_{A4}, \widehat{c}_{B4})$	$(\underline{c}_{A7}, \underline{c}_{B7})$
B	b'_3 ↓↑	$(\underline{c}_{A3}, \widehat{c}_{B3})$	$(\underline{c}_{A3}, \widehat{c}_{B3})$	$(\underline{c}_{A5}, \widehat{c}_{B5})$	$(\underline{c}_{A6}, \underline{c}_{B6})$
	b'_4 ↓↑	$(\underline{c}_{A3}, \widehat{c}_{B3})$	$(\underline{c}_{A3}, \widehat{c}_{B3})$	$(\underline{c}_{A5}, \widehat{c}_{B5})$	$(\underline{c}_{A7}, \underline{c}_{B7})$

The rough payoff matrix discloses structures of the corresponding extensive form game regarding imperfect and incomplete information faced by Player B. The payoffs $\widehat{c}_{B1}, \widehat{c}_{B2}$ and $\widehat{c}_{B4}$ form the boundary region $\widehat{R}_1$ and the payoffs $\widehat{c}_{B2}$ and $\widehat{c}_{B4}$ the boundary region $\widehat{R}_2$, while the payoffs $\underline{c}_{B6}$ and $\underline{c}_{B7}$ belong to separate lower approximations: $\underline{R}_3$ and $\underline{R}_4$, respectively.

The payoffs can be characterized with respect to their degree of 'roughness' ρ, the percentage of boundary payoffs of the players. For $C_1 = (\underline{c}_{A1}, \widehat{c}_{B1})$ we would get $\rho(C_1) = 1/2 = 0.5$ indicating that one out of two payoffs belong to boundaries; similarly, e.g., $\rho(C_7) = 0/2 = 0.0$. It also makes sense to distinguish payoffs derived from imperfect and incomplete information. Some implications of the selection of the (rough) equilibria have already been discussed in the previous sections. They can be determined straightforwardly from Table 6. Therefore, we refrain from a detailed discussion here.

5   Conclusion

In this paper, the relationship between imperfect information and incomplete information in game theory and rough sets is discussed. While in both areas great attention has been given how to deal with information a discussion of their relationship is still missing. We showed how imperfect information and incomplete information in game theory can be interpreted in rough set terms. The information sets comprising of two or more nodes in extensive form games are similar to boundaries in rough sets and, therefore, can be interpreted from a rough set perspective. We limited our examples to illustrative and simple cases to motivate for further research in this area. A more detailed discussion on the relationship of imperfect and incomplete information in classic game theory and rough sets could be mutually beneficial to both fields. It possibly leads to applications beyond game theoretic rough sets; in particular, it would be interesting to investigate the potentials of an integrated 'rough game theory'.

References

1. Angner, E.: A Course in Behavioral Economics. Palgrave Macmillan, New York (2012)
2. Bonanno, G.: Game Theory. University of California, Davis (2015). http://faculty. econ.ucdavis.edu/faculty/bonanno. Accessed 15 June 2016
3. EconPort: Imperfect information. Experimental Economics Center, Andrew Young School of Policy Studies, Georgia State University (2006). http://www.econport. org/content/handbook/Imperfect-Information.html. Accessed 15 June 2016
4. Hardesty, L.: Gaming the system. In: MIT News Magazine. MIT, July/August 2013
5. Harsanyi, J.C.: Games with incomplete information played by "Bayesian" players, I-III. Manage. Sci. **14**, 159–183 (Part I), 320–334 (Part II), 486–502 (Part III) (1967/1968)
6. Herbert, J.P., Yao, J.T.: Game-theoretic risk analysis in decision-theoretic rough sets. In: Wang, G., Li, T., Grzymala-Busse, J.W., Miao, D., Skowron, A., Yao, Y. (eds.) RSKT 2008. LNCS (LNAI), vol. 5009, pp. 132–139. Springer, Heidelberg (2008)
7. Herbert, J.P., Yao, J.T.: Game-theoretic rough sets. Fundamenta Informaticae **108**(3–4), 267–286 (2011)
8. Kreps, D.M.: Game Theory and Economic Modelling. Oxford University Press, Oxford, New York (1990)
9. Xu, J., Yao, L.: A class of two-person zero-sum matrix games with rough payoffs. Intl. J. Math. Math. Sci. **2010**, 1–22 (2010)

Towards Coordination Game Formulation in Game-Theoretic Rough Sets

Yan Zhang and Jing Tao Yao[⊠]

Department of Computer Science, University of Regina,
Regina, SK S4S 0A2, Canada
{zhang83y,jtyao}@cs.uregina.ca

Abstract. Game-theoretic rough set model (GTRS) is a recent advancement in determining three rough set regions by formulating games between multiple measures. GTRS has been focusing on researching competitive games in which the involved game players have opposite interests or incentives. There are different types of games that can be adopted in GTRS, such as coordination games, cooperation games, as well as competition games. Coordination games are a class of games in which the involved players have harmonious interests and enforce coordinative behaviors to achieve an efficient outcome. In this paper, we formulate coordination games between measures to determine rough set regions. Especially, we analyze the measures for evaluating equivalence classes. We determine rough regions to which every equivalence class should belong to by formulating coordination games and finding equilibrium for each equivalence class. The motivation and process of formulating coordination games are discussed in detail, and a demonstrative example shows the feasibility of the proposed approach.

Keywords: Game-theoretic rough sets · Coordination games · Rough sets · Measures

1 Introduction

Game-theoretic rough set model (GTRS) is a recent advancement in determining three rough set regions. GTRS has been employing game mechanisms to reach agreements between two or more measures with different interests or incentives [12]. These measures, which can be used to evaluate rough set regions from different point of views, are set as game players. GTRS formulates games between these measures to determine the suitable partition of three rough set regions [1,12]. The essential idea of GTRS is to implement games to obtain rough region thresholds in the rough set context. The aim is to improve the rough sets based decision making by finding a compromise among the involved measures.

In the existing formulations, GTRS is applied in probabilistic rough sets to determine and interpret the probabilistic thresholds that define three rough set regions. It implements competitive games amongst criteria which have the opposite interests when determining balanced probabilistic thresholds.

© Springer International Publishing AG 2016
V. Flores et al. (Eds.): IJCRS 2016, LNAI 9920, pp. 155–165, 2016.
DOI: 10.1007/978-3-319-47160-0_14

Herbert and Yao proposed two probabilistic thresholds (α, β) compete against each other to directly reduce the boundary regions [6]. They also formulated competition games between two classification approximation measures, i.e., accuracy and precision, to improve the classification ability of rough set model [6]. Azam and Yao applied GTRS to formulate competition games between two measures for evaluating positive regions, i.e., confidence and coverage, to solve multiple criteria decision making problems in rough sets [1]. Azam and Yao optimized the probabilistic thresholds with GTRS model by considering the competition between two properties of rough set model, accuracy and generality [2]. Yao and Azam proposed a competition game between immediate and deferred decision regions to improve the overall uncertainty level of the rough set classification [13]. Zhang and Yao used GTRS to find solutions to Gini objective functions by formulating competitive games between impurities of decision regions [18]. These studies not only provide a good beginning for GTRS research, but also build up a solid foundation for future research in GTRS.

There are many types of games involved in game theory, such as coordination games, cooperative games, non-cooperative games, sequential games. Most of GTRS research are focusing on competitive games, e.g. the games between accuracy and generality, the games between immediate and deferred decision regions, the games between thresholds α and β. We investigate coordination games in this paper. In coordination games, the involved measures are set as game players and they have the same interests when evaluating equivalence classes. These measures coordinate together to obtain an equilibrium of the formulated games, which represent a consensus on whether accepting or rejecting equivalence classes as a target concept. The coordination games are formulated between the measures that are used to evaluate equivalence classes, i.e., confidence, coverage, support. The three rough set regions are determined by checking whether to accept or reject each equivalence class.

2   Background Knowledge

In this section, we briefly introduce the background concepts about rough set regions, equivalence classes, as well as measures that can be used to evaluate equivalence classes.

2.1   Rough Sets

Rough set theory is a mathematical approach to deal with inconsistent and uncertain data [5,10]. Suppose the universe of objects U is a finite nonempty set. Let $E \subseteq U \times U$ be an equivalence relation on U, where E is reflexive, symmetric, and transitive [10]. For an element $x \in U$, the equivalence class containing x is given by $[x] = \{y \in U | xEy\}$. The family sets of all equivalence classes defines a partition of the universe U and is denoted by $U/E = \{[x] | x \in U\}$, that is the intersection of any two elements is an empty set and the union of all elements are the universe U [10]. Given a target concept $C \subseteq U$, Pawlak rough sets divide the universe U into three disjoint regions, i.e., positive, negative and boundary regions of C. All equivalence classes are classify to three regions according to if equivalence classes are included in C. Along with the development of rough sets, the extensions and generalizations of rough sets have continued to evolve [3,9].

The probabilistic rough set approach introduces a pair of thresholds (α, β) to extend the Pawlak rough set model [14,17]. For an undefinable target concept $C \subseteq U$, probabilistic rough sets utilize conditional probability and thresholds (α, β) to define three rough set regions of C, i.e., positive, negative and boundary regions [15,17]:

$$POS_{(\alpha,\beta)}(C) = \bigcup\{[x] \mid [x] \in U/E, Pr(C|[x]) \geq \alpha\},$$
$$NEG_{(\alpha,\beta)}(C) = \bigcup\{[x] \mid [x] \in U/E, Pr(C|[x]) \leq \beta\},$$
$$BND_{(\alpha,\beta)}(C) = \bigcup\{[x] \mid [x] \in U/E, \beta < Pr(C|[x]) < \alpha\}. \qquad (1)$$

These three rough set regions are pair-wise disjoint and their union is the universe of objects U according to Eq. (1). They form a tripartition of the universe U. Intuitively speaking, given an equivalence class $[x]$, if the probability of the concept C given $[x]$ is greater than or equal to α, i.e., $Pr(C|[x]) \geq \alpha$, we consider that all objects in $[x]$ belong to the concept C, that is, accept $[x]$ as C. If the probability of the concept C given $[x]$ is less than or equal to β, i.e., $Pr(C|[x]) \leq \beta$, we consider that all objects in $[x]$ do not belong to the concept C, that is, reject $[x]$ as C or accept $[x]$ as C^c. If the probability of the concept C given $[x]$ is between α and β , i.e., $\beta < Pr(C|[x]) < \alpha$, we are not sure whether to accept or to reject $[x]$ as C and make non-commitment decisions.

2.2 Measures for Evaluating Equivalence Classes

We are able to use some measures to evaluate the degree of an equivalence class belonging to a concept C. An equivalence class belonging to C can be described as accepting the equivalence class as the concept C. Similarly, an equivalence class not belonging to C can be described as rejecting the equivalence class as the concept C, or accepting the equivalence class as the concept C^c. Here we discuss three measures, i.e., confidence [16] or certainty [11], coverage [16], and support [8]. The formulas for calculating these measures for accepting an equivalence class $[x]_i$ as a concept C are listed as follows:

$$confidence([x]_i \subseteq C) = \frac{|[x]_i \cap C|}{|[x]_i|} = Pr(C|[x]_i). \qquad (2)$$

$$coverage([x]_i \subseteq C) = \frac{|[x]_i \cap C|}{|C|} = \frac{Pr(C|[x]_i) \times Pr([x]_i)}{\sum_1^n Pr(C|[x]_j) \times Pr([x]_j)}. \qquad (3)$$

$$support([x]_i \subseteq C) = \frac{|[x]_i \cap C|}{|U|} = Pr(C|[x]_i) \times Pr([x]_i). \qquad (4)$$

The formulas for calculating these measures for rejecting an equivalence class $[x]_i$ as a concept C are listed as follows:

$$confidence([x]_i \subseteq C^c) = \frac{|[x]_i \cap C^c|}{|[x]_i|} = Pr(C^c|[x]_i). \tag{5}$$

$$coverage([x]_i \subseteq C^c) = \frac{|[x]_i \cap C^c|}{|C^c|} = \frac{Pr(C^c|[x]_i) \times Pr([x]_i)}{\sum_1^n Pr(C^c|[x]_j) \times Pr([x]_j)}. \tag{6}$$

$$support([x]_i \subseteq C^c) = \frac{|[x]_i \cap C^c|}{|U|} = Pr(C^c|[x]_i) \times Pr([x]_i). \tag{7}$$

3   Coordination Game Formulation

3.1   Framework of Game-Theoretic Rough Sets

The objective of Game-theoretic rough set model is to determine the suitable partition of three rough set regions by formulating games between two measures. The obtained suitable partition of regions may represent a balance between two involved measures. In Game-theoretic rough sets, there are three elements when formulating a game, that is $G = \{O, S, u\}$ [6]:

- a set of players $O = \{o_1, o_2, ..., o_n\}$,
- sets of strategies or actions for players $S = \{S_1, S_2, ..., S_n\}$ where S_i is a set of possible strategies or actions for player o_i ,
- a set of payoff functions resulting from players performing strategies or actions $u = \{u_1, u_2, ..., u_n\}$.

A strategy profile s is a particular play of a game, in which player i performs the strategy or action s_i, that is, $s = \{s_1, s_2, ..., s_n\}$. The payoff of player i under the strategy profile s is denoted as $u_i(s) = u_i(s_1, s_2, ..., s_n)$. The payoff of each player depends on the strategies or actions performed by all involved players.

The game solution of Nash equilibrium is typically used to determine possible game outcomes in GTRS. Considering a strategy profile $(s_1, s_2, ..., s_n)$, let $s_{-i} = (s_1, s_2, ..., s_{i-1}, s_{i+1}, ..., s_n)$ be the same strategy profile without player o_i's strategy. We may write $(s_1, s_2, ..., s_n) = (s_i, s_{-i})$. The strategy profile $(s_1, s_2, ..., s_n)$ is a Nash equilibrium, if for all players o_i, s_i is the best response to s_{-i}, that is [7],

$$\forall i, \forall s_i' \in S_i, \quad u_i(s_i, s_{-i}) \geqslant u_i(s_i', s_{-i}), \quad where \ (s_i' \neq s_i) \tag{8}$$

Equation (8) may be interpreted as a strategy profile such that no player would like to change his strategy, provided he has the knowledge of other players' strategies.

3.2   Coordination Game Formulations

In game theory, coordination games are a class of games in which the involved players cooperate to choose the same strategies in order to achieve an outcome [4]. The main difference between coordination games and competition games is that players in coordination games have harmonious interests and may enforce coordinative behaviors. The players in competitive games have opposite interests and they intend to compete to maximize their own interests. When we evaluate equivalence classes, the measures of confidence, coverage and support have the same interests, which is the motivation to formulate coordination games between those measures. Now we investigate the process of formulating coordination games which includes game players, possible strategies, payoff functions, and decision conditions.

Game Players. Equivalence classes can be evaluated by using measures defined in Eqs. (2)−(7). Different measures may obtain different values when evaluating an equivalence class. The problem we are facing is how to determine whether an equivalence class belongs to the concept C or the positive region of C when considering two measures simultaneously. The measures have different values for one equivalence class, but they all decrease with the decrease of conditional probability $Pr(C|[x]_i)$. In other words, given two equivalence classes $[x]_i$ and $[x]_j$, if $Pr(C|[x]_i) > Pr(C|[x]_j)$, then

$$confidence([x]_i \subseteq C) > confidence([x]_j \subseteq C),$$
$$coverage([x]_i \subseteq C) > coverage([x]_j \subseteq C),$$
$$support([x]_i \subseteq C) > support([x]_j \subseteq C). \tag{9}$$

These measures have harmonious interests when evaluating equivalence classes. In this case, we are able to formulate coordination games between any two of above measures, that is,

$$O = \{o_1, o_2\}, \text{ and } o_1 \in \{con, cov, sup\}, o_2 \in \{con, cov, sup\}, o_1 \neq o_2 \tag{10}$$

here con denotes the measure confidence, cov for coverage, and sup for support. For example, the player o_1 represents the measure confidence and the player o_2 represents the measure coverage, $O = \{con, cov\}$.

Strategies. Each player has two actions or strategies for an equivalence class. One is to accept this equivalence class as the concept C, and the other is to reject this equivalence class as the concept C, that is,

$$S_1 = S_2 = \{s_1, s_2\} = \{acc, rej\}, \tag{11}$$

here acc denotes accept action and rej denotes reject action.

Payoff Functions. Each player has two strategies, thus there are four strategy profiles in a coordination game. The payoff functions of two players are generally definitions of the measures they are representing. Their payoff functions under four strategy profiles are determined as follows,

- when $(s_1, s_1) = (acc, acc)$, two players both accept the equivalence class as the concept C, the payoffs of the players are determined by the corresponding measures of accepting $[x]$ as the concept C, as defined in Eqs. (2)−(4).
- when $(s_2, s_2) = (rej, rej)$, two players both reject the equivalence class as the concept C, or they both accept the equivalence class as the concept C^c, the payoffs of the players are determined by the corresponding measures of rejecting $[x]$ as the concept C or accepting $[x]$ as the concept C^c, as defined in Eqs. (5)−(7).
- when $(s_1, s_2) = (acc, rej)$ or $(s_2, s_1) = (rej, acc)$, the players choose different strategies, the difference makes them doubtful about their own actions, so they adjust their utilities by consulting their partners. When one player chooses to accept and the other chooses to reject, the payoffs are the difference between the measures of accepting $[x]$ and rejecting $[x]$.

Table 1 represents the proposed coordinate game using a payoff table.

Table 1. Payoff table of a coordination game

		o_2	
		accept	*reject*
o_1	*accept*	$u_1(acc, acc), u_2(acc, acc)$	$u_1(acc, rej), u_2(acc, rej)$
	reject	$u_1(rej, acc), u_2(rej, acc)$	$u_1(rej, rej), u_2(rej, rej)$

Decision Conditions. A coordination game is formulated for each equivalence class and then we are able to find equilibria for games. The strategy profile (s_i, s_j) is a Nash equilibrium, if for players o_1 and o_2, s_i and s_j are the best responses to each other. This is expressed as [7],

$$\text{For player } o_1 : \forall s_i' \in S_1, u_{c_1}(s_i, s_j) \geqslant u_{c_1}(s_i', s_j), \text{with } (s_i' \neq s_i),$$
$$\text{For player } o_2 : \forall s_j' \in S_2, u_{c_2}(s_i, s_j) \geqslant u_{c_2}(s_i, s_j'), \text{with } (s_j' \neq s_j). \quad (12)$$

For different game equilibria we may make different decisions about whether accepting an equivalence class $[x]$ as the concept C,

- when $(s_1, s_1) = (acc, acc)$ is the equilibrium, both players agree to accept this equivalence class $[x]$ as the concept C, so we decide to accept this equivalence class as the concept C, or we classify it in the positive region of C;

– when $(s_2, s_2) = (rej, rej)$ is the equilibrium, both players agree to reject this equivalence class $[x]$ as the concept C, so we decide to reject this equivalence class as the concept C, or we classify it in the negative region of C;
– for other situations, two players are not able to reach a consensus about whether accepting or rejecting this equivalence class $[x]$ as the concept C, so we do not make decisions on the equivalence class, and we classify it in the boundary region of C;

4   An Example

In this section, we present an example to demonstrate how to formulate coordination games for equivalence classes to obtain suitable rough set regions. Table 2 summarizes probabilistic data about a concept C. There are 16 equivalence classes denoted by $[x]_i (i = 1, 2, ..., 16)$, which are listed in a decreasing order of the conditional probabilities $Pr(C|[x]_i)$ for convenient computations;

Table 2. Summary of the experimental data

	$[x]_1$	$[x]_2$	$[x]_3$	$[x]_4$	$[x]_5$	$[x]_6$	$[x]_7$	$[x]_8$
$Pr([x]_i)$	0.084	0.072	0.069	0.066	0.063	0.059	0.041	0.042
$Pr(C|[x]_i)$	1	0.978	0.95	0.91	0.89	0.81	0.72	0.61
	$[x]_9$	$[x]_{10}$	$[x]_{11}$	$[x]_{12}$	$[x]_{13}$	$[x]_{14}$	$[x]_{15}$	$[x]_{16}$
$Pr([x]_i)$	0.049	0.049	0.057	0.061	0.063	0.068	0.071	0.086
$Pr(C|[x]_i)$	0.42	0.38	0.32	0.29	0.2	0.176	0.1	0

The Fig. 1 shows the values of three measures for evaluating accepting $[x]$ as the concept C.

The measures confidence, coverage and support are all in the decreasing order with the increase of conditional probability $Pr(C|[x]_i)$.

Game players: We use confidence and coverage as examples to show the formulation of a coordination game. We set the measures confidence and coverage as two game players, i.e., $O = \{con, cov\}$, con represents the measure confidence and con for the measure coverage.

Strategies: The strategy sets for two players are:

$$S_{con} = S_{cov} = \{s_1, s_2\} = \{acc, rej\}.$$

Payoff functions: For each possible strategy profile, the payoff functions of two players are defined as follows:

– when $(s_1, s_1) = (acc, acc)$, the payoffs of the players are determined by the confidence and coverage of accepting $[x]$ as C defined in Eqs. (2) and (3),

$$u_{con}(acc, acc) = \frac{|[x] \cap C|}{|[x]|}, u_{cov}(acc, acc) = \frac{|[x] \cap C|}{|C|}. \tag{13}$$

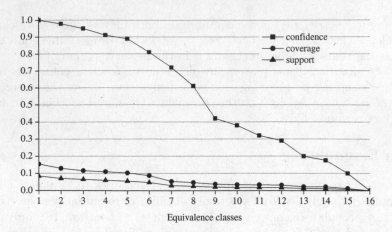

Fig. 1. The values of measures for evaluating all equivalence classes

– when $(s_2, s_2) = (rej, rej)$, the payoffs of the players are determined by the confidence and coverage of rejecting $[x]$ as C or accepting $[x]$ as C^c defined in Eqs. (5) and (6),

$$u_{con}(rej, rej) = \frac{|[x] \cap C^c|}{|[x]|}, u_{cov}(rej, rej) = \frac{|[x] \cap C^c|}{|C^c|}. \tag{14}$$

– when $(s_1, s_2) = (acc, rej)$, the payoffs are determined by the difference of the confidence of accepting $[x]$ as C and the coverage of rejecting $[x]$ as C,

$$u_{con}(acc, rej) = \frac{|[x] \cap C|}{|[x]|} - \frac{|[x] \cap C^c|}{|C^c|},$$

$$u_{cov}(acc, rej) = \frac{|[x] \cap C^c|}{|C^c|} - \frac{|[x] \cap C|}{|[x]|}. \tag{15}$$

– when $(s_2, s_1) = (rej, acc)$, the payoffs are determined by the difference of the confidence of rejecting $[x]$ as C and the coverage of accepting $[x]$ as C,

$$u_{con}(rej, acc) = \frac{|[x] \cap C^c|}{|[x]|} - \frac{|[x] \cap C|}{|C|},$$

$$u_{cov}(rej, acc) = \frac{|[x] \cap C|}{|C|} - \frac{|[x] \cap C^c|}{|[x]|}. \tag{16}$$

Following the above described process, we formulate a coordination game for each equivalence class, and then find the equilibrium for each game based on Eq. (12).

We use the equivalence class $[x]_3$ in Table 2 as an example to show the process of formulation of a game. The confidence of accepting $[x]_3$ as the concept C and the rejecting $[x]_3$ as the concept C are:

$$confidence([x]_3 \subseteq C) = \frac{|[x]_3 \cap C|}{|[x]_3|} = Pr(C|[x]_3) = 0.9500,$$

$$confidence([x]_3 \subseteq C^c) = \frac{|[x]_3 \cap C^c|}{|[x]_3|} = 1 - Pr(C|[x]_3) = 0.0500.$$

The coverage of accepting $[x]_3$ as the concept C and rejecting $[x]_3$ as the concept C are:

$$coverage([x]_3 \subseteq C) = \frac{|[x]_3 \cap C|}{|C|} = \frac{Pr(C|[x]_3) \times Pr([x]_3)}{\sum_1^{16} Pr(C|[x]_i) \times Pr([x]_i)} = 0.1155,$$

$$coverage([x]_3 \subseteq C^c) = \frac{|[x]_3 \cap C^c|}{|C^c|} = \frac{(1 - Pr(C|[x]_3)) \times Pr([x]_3)}{\sum_1^{16}(1 - Pr(C|[x]_i)) \times Pr([x]_i)} = 0.0076.$$

Table 3. Payoff table for $[x]_3$

		cov	
		accept	reject
con	accept	**< 0.9500, 0.1155 >**	< 0.9424, −0.9424 >
	reject	< −0.0655, 0.0655 >	< 0.0500, 0.0076 >

The payoff table for the equivalence class $[x]_3$ is shown in Table 3. The equilibrium of this payoff table is $(s_1, s_1) = (acc, acc)$, so both players accept the equivalence class $[x]_3$ as the concept C. We classify $[x]_3$ in the positive region of C. After formulating 16 coordination games for 16 equivalence classes in Table 2, we get the following partition of rough set regions,

$$POS(C) = \bigcup \{[x]_1, [x]_2, [x]_3, [x]_4, [x]_5, [x]_6, [x]_7, [x]_8\},$$

$$NEG(C) = \bigcup \{[x]_9, [x]_{10}, [x]_{11}, [x]_{12}, [x]_{13}, [x]_{14}, [x]_{15}, [x]_{16}\},$$

$$BND(C) = \emptyset.$$

The above result means the measures confidence and coverage both agree to accept the equivalence classes $[x]_1, [x]_2, ..., [x]_8$ as the concept C, and reject the equivalence classes $[x]_9, [x]_{10}, ..., [x]_{16}$ as the concept C.

5   Conclusion

This paper investigates the coordination game formulations in game-theoretic rough sets. We analyze three measures, i.e., confidence, coverage and support, which can be used to evaluate equivalence classes. These measures have harmonious interests when evaluating equivalence classes, that is, given an equivalence class, a high value of one measure indicates high values of other measures. In this case, coordination games are able to be formulated and the involved measures are set as game players. Each player has two possible strategies, accept and reject. These measures intend to choose the same strategies when deciding whether to accept or reject an equivalence class since they have the harmonious interests. The consensus of the players are the Nash equilibrium of the coordination games. We formulate a coordination game for each equivalence class and the corresponding equilibrium determines which regions each equivalence class belongs to. Then we obtain the three rough set regions comprised by equivalence classes. Formulating coordination games in GTRS provides a feasible and effective approach to obtain three rough set regions by checking each equivalence class. Coordination games and competition games have different features and should be applied in different situations, we will examine suitability of applications to each type games in the future research.

Acknowledgements. This work is partially supported by a Discovery Grant from NSERC Canada, the University of Regina Gerhard Herzberg Fellowship.

References

1. Azam, N., Yao, J.T.: Multiple criteria decision analysis with game-theoretic rough sets. In: Li, T., Nguyen, H.S., Wang, G., Grzymala-Busse, J., Janicki, R., Hassanien, A.E., Yu, H. (eds.) RSKT 2012. LNCS, vol. 7414, pp. 399–408. Springer, Heidelberg (2012)
2. Azam, N., Yao, J.T.: Analyzing uncertainties of probabilistic rough set regions with game-theoretic rough sets. Int. J. Approximate Reason. **55**(1), 142–155 (2014)
3. Ciucci, D.: Attribute dynamics in rough sets. In: International Symposium on Methodologies for Intelligent Systems, pp. 43–51 (2011)
4. Cooper, R.: Coordination Games. Cambridge University Press, New York (1999)
5. Grzymala-Busse, J.W.: Knowledge acquisition under uncertaintya rough set approach. J. intell. Robot. Syst. **1**(1), 3–16 (1988)
6. Herbert, J.P., Yao, J.T.: Game-theoretic rough sets. Fundamenta Informaticae **108**(3–4), 267–286 (2011)
7. Leyton-Brown, K., Shoham, Y.: Essentials of Game Theory: A Concise Multidisciplinary Introduction. Morgan & Claypool Publishers, San Rafael (2008)
8. Liang, J.Y., Li, D.Y.: Uncertainty and Knowledge Acquisition in Information Systems. Science Press, Beijing (2005)
9. Lingras, P., Peters, G.: Applying rough set concepts to clustering. In: Rough Sets: Selected Methods and Applications in Management and Engineering, pp. 23–37. Springer (2012)
10. Pawlak, Z.: Rough Sets: Theoretical Aspects of Reasoning About Data. Kluwer Academic Publishers, Boston (1991)
11. Qian, Y.H., Liang, J.Y., Li, D.Y., Zhang, H.Y., Dang, C.Y.: Measures for evaluating the decision performance of a decision table in rough set theory. Inf. Sci. **178**(1), 181–202 (2008)
12. Yao, J.T., Herbert, J.P.: A game-theoretic perspective on rough set analysis. J. Chongqing Univ. Posts Telecommun. **20**(3), 291–298 (2008)

13. Yao, J.T., Azam, N.: Web-based medical decision support systems for three-way medical decision making with game-theoretic rough sets. IEEE Trans. Fuzzy Syst. **23**(1), 3–15 (2015)
14. Yao, J.T., Ciucci, D., Zhang, Y.: Generalized rough sets. In: Springer Handbook of Computational Intelligence. Springer, Berlin (2015)
15. Yao, Y.Y.: Rough sets and three-way decisions. In: Ciucci, D., Mitra, S., Wu, W.-Z. (eds.) RSKT 2015. LNCS, vol. 9436, pp. 62–73. Springer, Heidelberg (2015). doi:10.1007/978-3-319-25754-9_6
16. Yao, Y.Y., Zhao, Y.: Attribute reduction in decision-theoretic rough set models. Inf. Sci. **178**(17), 3356–3373 (2008)
17. Yao, Y.Y.: Probabilistic rough set approximations. Int. J. Approximate Reason. **49**(2), 255–271 (2008)
18. Zhang, Y., Yao, J.T.: Determining three-way decision regions by combining gini objective functions and GTRS. In: Yao, Y., Hu, Q., Yu, H., Grzymala-Busse, J.W. (eds.) RSFDGrC 2015. LNCS, vol. 9437, pp. 414–425. Springer, Heidelberg (2015). doi:10.1007/978-3-319-25783-9_37

Probabilistic Estimation for Generalized Rough Modus Ponens and Rough Modus Tollens

Ning Yao[1,2], Duoqian Miao[1,2(✉)], Zhifei Zhang[1,2], and Guangming Lang[1,2]

[1] Department of Computer Science and Technology,
Tongji University, Shanghai 201804, China
dqmiao@tongji.edu.cn
[2] The Key Laboratory of Embedded System and Service Computing,
Ministry of Education, Tongji University, Shanghai 201804, China
xinhuperfect@163.com

Abstract. We review concepts and principles of Modus Ponens and Modus Tollens in the areas of rough set theory and probabilistic inference. Based on the upper and the lower approximation of a set as well as the existing probabilistic results, we establish a generalized version of rough Modus Ponens and rough Modus Tollens with a new fact different from the premise (or the conclusion) of "if ... then ..." rule, and address the problem of computing the conditional probability of the conclusion given the new fact (or of the premise given the new fact) from the probability of the new fact and the certainty factor of the rule. The solutions come down to the corresponding interval for the conditional probabilities, which are more appropriate than the exact values in the environment full of uncertainty due to errors and inconsistency existed in measurement, judgement, management, etc., plus illustration analysis.

Keywords: Rough modus ponens · Rough modus tollens · The lower approximation · Conditional probability · Rough sets

1 Introduction

At the center of human intelligence and reasoning lies common sense, gained from experience of life or common knowledge. Knowledge is often acquired from data such as observations and measurements in the form of numbers, words, or images, usually represented in an organized manner with a level of granularity, and pervaded by imprecision or vagueness. However, data, collected for use, are generally disorganized and contain useless details. Therefore, how to obtain the available knowledge or information from data is a central point in data analysis whose goal is finding patterns or regularity hidden in the data. The utilization of statistics was only realizable in the early period of data analysis, then followed by fuzzy sets, rough sets, neural networks, genetic algorithms, cluster analysis and other analysis tools.

Typically encoded as the rule of "if ... then ...", hidden patterns or regularity in data can enable us to make decisions, do prediction and management

© Springer International Publishing AG 2016
V. Flores et al. (Eds.): IJCRS 2016, LNAI 9920, pp. 166–176, 2016.
DOI: 10.1007/978-3-319-47160-0_15

activities, or other reasoning activities in everyday life, and the highly influential comes down to Modus Ponens inference rule (Modus Ponens, for short), as the basis of classical deductive reasoning and an universal rule of inference valid in any logical system. Modus Ponens has the form that, given that a formula or a new fact ϕ is true and the rule "if ϕ then ψ" is also true, then the formula or the conclusion ψ would also be true. To estimate the truth value of ψ relates closely to the formalization of the conditional "if ... then ... ", and in turn, to formalize "if ... then ... " has led to the keen competition between material implication and conditional probability. For the material implication of Modus Ponens, the representative work is Compositional Rule of Inference as an approximate extension of Modus Ponens [17], proposed by Zadeh based on fuzzy sets, where fuzzy relations derived from fuzzy implication operators are employed to compute the truth value of the conclusion ψ and the new fact is allowed to different from the premise of the rule. However, considering lack of the ability to treat the exceptions or counterfactuals for material implication and its inherent paradoxes, for example, the false of the premise does infer the true of material implication only if the conclusion is true, the probabilistic interpretation of "if ... then ... " is more plausible in the reasoning process [5, 13].

In addition, due to the uncertainty in the represented data or the knowledge, inconsistency in information systems, and the limited number of available knowledge obtained for use, directly characterizing the truth values of formulas is not feasible because of many difficulties in the construction of the truth function and can often be influenced by subjective factors such as assumption intervention. The idea of replacing truth values with probabilities was first proposed by Łukasiewicz, who advocates multivalued logic as probability logic and assigns each of indefinite proposition $\phi(x)$ the ratio $\pi(\phi(x))$ of the number of all values of the variable x satisfying $\phi(x)$ to the number of all possible values of x as the truth value of $\phi(x)$. And later, Pawlak, the founder of rough sets with the aim of finding the dependencies or cause-effect relations in data, pursued this idea and introduced Rough Modus Ponens, a generalized version of Modus Ponens in the context of rough set theory where the new fact is of the same form as the premise of the rule [10]. Fuzzy-Rough version of Modus Ponens [4] presented the characterization of the conclusion through gradual decision rules extracted from decision table based on fuzzy rough set theory, plus the fuzzy-rough version for modus tollens (i.e., given that a formula $\neg\psi$ is true and the rule "if ϕ then ψ" is also true, then the formula $\neg\phi$ would also be true), without using any fuzzy logical connectives. Although this approach is successful in the treatment of the difference between the new fact and the premise, it still involves the selection of fuzzy membership function influenced by subjective factors. Probabilistic counterpart of Modus Ponens yields the best possible bounds for the probability of the conclution ψ and even for the update of the bounds on new-found uncertain evidence as well as the bounds for modus tollens [12, 15].

Reasoning based on rough set theory obeys data collected and the inferences stem from the data. Empowered by these motivations and analysis, the central goal of this paper is to investigate the generalized version of Rough Modus

Ponens permitting the new fact different from the premise, and try to allow the solution to make the plausible responds to the new evidence even when the evidence is contradictory or irrelevant to the premise of the rule, as well as the study for Rough Modus Tollens. Section 2 exhibits some definitions about rough set theory as well as some results on probable Modus Ponens and Modus Tollens. In Sect. 3 the generalized rough Modus Ponens and rough Modus Tollens are developed through the consideration for the relations between the new fact and the premise based on the concept of lower approximation of the set, together with some illustration studies depicted. The detailed comments on the present approach are explored in Sect. 4, along with a brief sketch of further research.

2 Basic Concepts on Rough Modus Ponens and Rough Modus Tollens

Subject to measurability requirements, one is led to consider upper and lower approximations defined over any set as follows:

Definition 1 ([8]). *Given an information system $S = (U, A)$ with U a non-empty finite set called the Universe and A a nonempty finite set called the set of Attributes, and let $X \subseteq U$, $B \subseteq A$. The upper approximation $\overline{B}(X)$ and the lower approximation $\underline{B}(X)$ of any set X in terms of attributes B can be defined respectively by $\overline{B}(X) = \bigcup_{x \in U} \{x \in U : [x]_B \cap X \neq \emptyset\}$, $\underline{B}(X) = \bigcup_{x \in U} \{x \in U : [x]_B \subseteq X\}$, where $[x]_B$ (i.e., the set of $\{y \in U : y\ I(B)\ x\}$) denotes the equivalence class of x with respect to the indiscernibility relation $I(B)$ (i.e., $\{(x, y) \in U^2 | a(x) = a(y)$ for every $a \in B\}$, $a(x)$ denotes the value of attribute a for element x), which means the object y and x are indiscernible in terms of attributes in B.*

Definition 2 ([9,11]). *Given a decision table $S = (U, C, D)$ with the attributes A of the system classified into disjoint sets of condition attributes C and decision attributes D, and let $\phi \to \psi$ be a decision rule with ϕ and ψ as logical formulas representing conditions and decisions, respectively. Define the certainty factor $\mu(\phi, \psi)$ of the rule as a number, namely, $\mu(\phi, \psi) = \pi(\psi|\phi) = \frac{\pi(\phi \wedge \psi)}{\pi(\phi)} = \frac{card(\|\phi\| \cap \|\psi\|)}{card(\|\phi\|)}$, where $\|\phi\|$ denote the set of all objects satisfying ϕ in S, $card(\cdot)$ denotes the cardinality or the number of elements in a given set, and $\pi(\cdot)$ represents the corresponding probability (the purpose of using this notation as probability is only to accord with the ones in the rough set literature), $\pi(\phi) := \frac{card(\|\phi\|)}{card(U)}$ and $card(\|\phi\|) \neq 0$.*

The rough modus ponens [10] may be formed from

$$
\begin{array}{ll}
\text{if} \quad \phi \to \psi \text{ is true with probability} & \pi(\psi|\phi) \\
\text{and } \phi \qquad \text{is true with probability} & \pi(\phi) \\
\hline
\text{then} \qquad \psi \text{ is true at least with probability} & \pi(\psi)
\end{array}
$$

where $\pi(\psi) = \pi(\neg\phi \wedge \psi) + \pi(\phi) \cdot \pi(\psi|\phi)$. This formula can be taken as a generalization (e.g., for $\pi(\phi \rightarrow \psi) \neq 0$) of Łukasiewicz' axiom 3 (i.e., if $\pi(\phi \rightarrow \psi) = 1$, then $\pi(\psi) = \pi(\neg\phi \wedge \psi) + \pi(\phi)$).

From a probabilized point of view, as to modus ponens, one has the best possible bounds [12,15] for $\pi(\psi)$, namely, $\pi(\phi)\pi(\psi|\phi) \leq \pi(\psi) \leq \pi(\phi)\pi(\psi|\phi) + 1 - \pi(\phi)$ with $0 < \pi(\phi) \leq 1$ and $0 \leq \pi(\psi|\phi) \leq 1$; as to modus tollens, given $\neg\psi$ and the rule $\phi \rightarrow \psi$ to infer $\neg\phi$, the solution with the best possible bounds is (see the theorem on p. 751 of [12]: 'ϕ' and 'ψ' for 'H' and 'E'; $\neg$ for over-bars; '$\pi(\psi|\phi)$' and '$\pi(\neg\psi)$' for 'a' and 'b')

$$if \ \ 0 < \pi(\neg\psi), \pi(\psi|\phi) < 1, \ then$$
$$\max\left\{ \frac{1-\pi(\psi|\phi)-\pi(\neg\psi)}{1-\pi(\psi|\phi)}, \frac{\pi(\psi|\phi)+\pi(\neg\psi)-1}{\pi(\psi|\phi)} \right\} \leq \pi(\neg\phi) < 1$$
$$if \ \ 0 < \pi(\neg\psi) \leq 1, \pi(\psi|\phi) = 0, \ then \ 1 - \pi(\neg\psi) \leq \pi(\neg\phi) < 1$$
$$if \ \ 0 \leq \pi(\neg\psi) < 1, \pi(\psi|\phi) = 1, \ then \ \pi(\neg\psi) \leq \pi(\neg\phi) < 1.$$

Moreover, concerning the update of the probability for ψ on new-found possibly uncertain evidence, the solution has been obtained as follows [15]:

*Let time $-t$ be, for a person, just before time t probabilistically speaking. Assume that this person is not certain of $\neg\phi$ at t, that is, $\pi_t(\phi) > 0$ and $\pi_t(\phi) \neq \pi_{-t}(\phi)$. Then this person does update his probability for ψ subject to the bounds $\pi_t(\phi)\pi_{-t}(\psi|\phi)$ and $\pi_t(\phi)\pi_{-t}(\psi|\phi) + 1 - \pi_t(\phi)$, **if and only if**, the rigidity-condition for ψ on ϕ, i.e., $\pi_t(\psi|\phi) = \pi_{-t}(\psi|\phi)$, is satisfied.*

Analogous to the update of the probability for ψ, updating $\neg\phi$ on new-found possibly uncertain evidence $\neg\psi$ has been solved to yield [15].

*Let time $-t$ be, for a person, just before time t probabilistically speaking. Assume that $\pi_t(\neg\psi) \neq \pi_{-t}(\neg\psi)$. Then this person does update his probability for $\neg\phi$ on the evidence $\neg\psi$, **if and only if**, the rigidity-condition $\pi_t(\psi|\phi) = \pi_{-t}(\psi|\phi)$ is satisfied. If this condition is satisfied, there is*

$$if \ \ 0 < \pi_t(\neg\psi), \pi_t(\psi|\phi) < 1, \ then$$
$$\max\left\{ \frac{1-\pi_t(\psi|\phi)-\pi_t(\neg\psi)}{1-\pi_t(\psi|\phi)}, \frac{\pi_t(\psi|\phi)+\pi_t(\neg\psi)-1}{\pi_t(\psi|\phi)} \right\} \leq \pi_t(\neg\phi) < 1$$
$$if \ \ 0 < \pi_t(\neg\psi) \leq 1, \pi_t(\psi|\phi) = 0, \ then \ 1 - \pi_t(\neg\psi) \leq \pi_t(\neg\phi) < 1$$
$$if \ \ 0 \leq \pi_t(\neg\psi) < 1, \pi_t(\psi|\phi) = 1, \ then \ \pi_t(\neg\psi) \leq \pi_t(\neg\phi) < 1.$$

3 Generalized Versions of Rough Modus Ponens and Rough Modus Tollens

In this section we continue Pawlak's work [3,10,11] and it is convenient to begin with the case where $\phi^\diamond$ takes a different form of ϕ but the same rule "if ϕ then ψ" as the case of Modus Ponens, associating this rule with a conditional probability $\pi(\psi|\phi) = \frac{\pi(\phi \wedge \psi)}{\pi(\phi)}$ and likewise the formulas ϕ and ψ with their respective unconditional probabilities $\pi(\phi)$ and $\pi(\psi)$.

Consider that, in practice, the new observation is rarely identical to the sample data but to some extent is of particular relevance to the observed sample (i.e., they describe different states of the same attributes or different attributes of different attributes). Here we denote the new fact or observation by $\phi^\diamond$ and mainly deal with the case when $\phi^\diamond$ is not the same as the sample ϕ. The above Rough Modus Ponens may be regarded as the special case of the proposed generalized version:

$$
\begin{array}{ll}
\text{if} \quad \phi \to \psi \text{ is true with probability} & \pi(\psi|\phi) \\
\text{and } \phi^\diamond \quad \text{ is true with probability} & \pi(\phi^\diamond) \\
\hline
\text{then} \qquad ?\psi \text{ is true with probability} & ?\pi_t(\psi) \; ?\pi(\psi|\phi^\diamond)
\end{array}
$$

where the notation $\pi_t(\psi)$ means the probability of ψ when the new fact $\phi^\diamond$ is observed, which is identical to the meaning of the conditional probability of ψ given $\phi^\diamond$. The subscript t is used only to distinguish the probability of ψ when the new fact $\phi^\diamond$ is observed from the prior probability of ψ as well as the posterior probability of ψ when the evidence ϕ is observed.

Lemma 1. *Let $0 < \pi(\psi|\phi) \le 1$ and $0 < \pi(\phi^\diamond) \le 1$. Then the probability of ψ with $\phi^\diamond$ known satisfies*

$$
\pi(\phi^\diamond)\pi(\psi|\phi) \le \pi_t(\psi) \le \pi(\phi^\diamond)\pi(\psi|\phi) + 1 - \pi(\phi^\diamond).
$$

The relation between the new fact $\phi^\diamond$ and the occurrence of ψ is connected closely to the relation of the new fact $\phi^\diamond$ and the premise ϕ. Our solution lies in the detailed description of the relation between $\phi^\diamond$ and ϕ, more specifically, the lower approximation of $\phi^\diamond$ and the lower approximation of ϕ.

Lemma 2. *For $\underline{\phi^\diamond} \subseteq \underline{\phi}$, the probability $\pi(\psi|\phi^\diamond)$ satisfies*

$$
\frac{\pi(\psi|\phi)\pi(\phi)}{\pi(\phi^\diamond)} = \frac{\pi(\phi \wedge \psi)}{\pi(\phi^\diamond)} \le \pi(\psi|\phi^\diamond) = \frac{\pi(\phi^\diamond \wedge \psi)}{\pi(\phi^\diamond)} \le \frac{\pi(\phi^\diamond)}{\pi(\phi^\diamond)} = 1.
$$

Proof. This result follows immediately from the fact that the frequency of the occurrence of one event is usually greater and equal to the frequency of the simultaneous occurrence of this event together with other events. Let $\|\phi^\diamond\|$ and $\|\phi\|$ represent the sets of all objects satisfying respectively $\phi^\diamond$ and ϕ in S, there exist $card(\|\phi^\diamond\|) \ge card(\|\phi\|)$ and $card(\|\phi^\diamond \wedge \psi\|) \ge card(\|\phi \wedge \psi\|)$, plus $\pi(\phi^\diamond \wedge \psi) = \frac{card(\|\phi^\diamond \wedge \psi\|)}{card(U)}$, $\pi(\phi \wedge \psi) = \frac{card(\|\phi \wedge \psi\|)}{card(U)}$ and $\pi(\phi^\diamond \wedge \psi) \le \pi(\phi^\diamond)$.   □

Theorem 1. *If $\underline{\phi^\diamond} \subseteq \underline{\phi}$, the probability of ψ given $\phi^\diamond$ can be solved by*

$$
\max\left\{ \frac{\pi(\psi|\phi)\pi(\phi)}{\pi(\phi^\diamond)}, \pi(\phi^\diamond)\pi(\psi|\phi) \right\} \le \pi(\psi|\phi^\diamond) \le \pi(\phi^\diamond)\pi(\psi|\phi) + 1 - \pi(\phi^\diamond) .
$$

Example 1. From   Table 1   (see   [10]),   we   have   the   rule   "if $\phi = (Headache, yes)$ *and* $(Muscle - pain, no)$ *and* $(Temperature, high)$, *then* $\psi = (Flu, yes)$" with the probability $\pi(\psi|\phi) = \frac{1}{2}$ and $\pi(\phi) = \frac{1}{3}$, and a new fact

Table 1. Characterization of flu

Patient	Headache	Muscle-pain	Temperature	Flu
p1	no	yes	high	yes
p2	yes	no	high	yes
p3	yes	yes	very high	yes
p4	no	yes	normal	no
p5	yes	no	high	no
p6	no	yes	very high	yes

$\phi^\circ = (Headache, yes)$ with $\pi(\phi^\circ) = \frac{1}{2}$. Because $\underline{\phi^\circ} = \{(Headache, yes)\} \subseteq \underline{\phi}$, the probability of $\psi = (Flu, yes)$ given $\phi^\circ = (Headache, yes)$ lies within the interval $[\frac{1}{3}, \frac{3}{4}]$ by Theorem 1 (if possible, based on rough set theory from the table one has $\pi(\psi|\phi^\circ) = \frac{2}{3}$).

When $\underline{\phi^\circ} \subseteq U - \underline{\phi}$ but $\underline{\phi^\circ} \cap \underline{\phi} \neq \emptyset$, $\pi(\psi|\phi^\circ)$ depends on whether or not the element of $\underline{\phi^\circ} - \underline{\phi^\circ} \cap \underline{\phi}$ is the description of the same attribute with the different states from the one of $\underline{\phi} - \underline{\phi^\circ} \cap \underline{\phi}$.

Theorem 2. *If the elements of $\underline{\phi^\circ} - \underline{\phi^\circ} \cap \underline{\phi}$ and $\underline{\phi} - \underline{\phi^\circ} \cap \underline{\phi}$ depict different states of the same attribute, then the probability $\pi(\psi|\phi^\circ)$ can be determined by the interval $[\pi(\phi^\circ)\pi(\psi|\phi), \pi(\phi^\circ)\pi(\psi|\phi) + 1 - \pi(\phi^\circ)]$, more specifically, for $0 < \pi(\psi|\phi) \leq 1$, $\pi(\psi|\phi^\circ)$ can be located inside or outside this interval, which corresponds to the degrees of beliefs for the attribute in the language of $\underline{\phi^\circ} - \underline{\phi^\circ} \cap \underline{\phi}$ and $\underline{\phi} - \underline{\phi^\circ} \cap \underline{\phi}$. By contrast, if the elements of $\underline{\phi^\circ} - \underline{\phi^\circ} \cap \underline{\phi}$ and $\underline{\phi} - \underline{\phi^\circ} \cap \underline{\phi}$ depict different states of different attributes, then the probability of ψ given ψ°, namely, $\pi(\psi|\phi^\circ)$ is generally situated in $[\pi(\phi^\circ)\pi(\psi|\phi), \pi(\phi^\circ)\pi(\psi|\phi) + 1 - \pi(\phi^\circ)]$.*

Example 2. From Table 1, one have the rule "if $\phi = (Headache, no)$ then $\psi = (Flu, yes)$" with $\pi(\psi|\phi) = \frac{2}{3}$ as well as a new fact $\phi^\circ = (Headache, no)$ and $(Temperature, normal)$ with $\pi(\phi^\circ) = \frac{1}{6}$. Additionally we have known that the probability of 'if Temperature is normal then Flu is yes' is 0 and the probability of 'if Temperature is very high then Flu is yes' is 1. According to Theorem 2, one can determine that the value $\pi(\psi|\phi^\circ)$ is outside $[\frac{1}{9}, \frac{17}{18}]$ and specifically in $[0, \frac{1}{9}]$. (From Table 1 one has $\pi(\psi|\phi^\circ) = 0 \in [0, \frac{1}{9}]$, which means that the obtained result acts in accordance with our common sense.)

When $\underline{\phi^\circ} \cap \underline{\phi} = \emptyset$, similarly $\pi(\psi|\phi^\circ)$ is associated with the fact whether ϕ° depicts the same attributes as ϕ does or as the elements of ϕ do. In more details, if they do, then the range of $\pi(\psi|\phi^\circ)$ closely relates to the degrees of beliefs for these attributes of ϕ° and ϕ, specified in the following result.

Theorem 3. *Let $\underline{\phi^\circ} \cap \underline{\phi} = \emptyset$. If the elements of ϕ° and ϕ describe different states of the same attribute, then for $\pi(\psi|\phi) = 1$, $\pi(\psi|\phi^\circ)$ might smaller than or equal to 1 and the specific value will be inside or outside $[\pi(\phi^\circ)\pi(\psi|\phi), 1]$ with a trend of moving from right to left on the horizontal axis according to the degrees of beliefs* *for the attribute; for $0 < \pi(\psi|\phi) < 1$, $\pi(\psi|\phi^\circ)$ might be 0 or 1 and the specific value also might be inside or outside the interval $[\pi(\phi^\circ)\pi(\psi|\phi), \pi(\phi^\circ)\pi(\psi|\phi) + 1 - \pi(\phi^\circ)]$ according to the degrees. If they depict different states of different attributes, $\pi(\psi|\phi^\circ)$ generally lies in $[\pi(\phi^\circ)\pi(\psi|\phi), \pi(\phi^\circ)\pi(\psi|\phi) + 1 - \pi(\phi^\circ)]$.*

Example 3. From Table 1, one can get the rule "if $\phi = (Headache, no)$ and $(Muscle - pain, yes)$ and $(Temperature, high)$, then $\psi = (Flu, yes)$" with the probability $\pi(\psi|\phi) = 1$ and $\pi(\phi) = \frac{1}{6}$, and the new fact $\phi^\circ = (Temperature, very\ high)$ with $\pi(\phi^\circ) = \frac{1}{3}$. By Theorem 3, $\pi(\psi|\phi^\circ)$ falls into $[\frac{1}{3}, 1]$ and from the table one might get $\pi(\psi|\phi^\circ) = 1 \in [\frac{1}{3}, 1]$.

Obviously notice that the case of $\pi(\psi|\phi) = 1$ is the special case of the above result. Clearly in this case there is $\pi(\phi^\circ) \leq \pi_t(\psi) \leq 1$. In particular, let $C(\psi)$ denote the set of all conditions of ψ in the data table about the domain of interest, and $C_*(\psi)$ denote the lower approximation of $C(\psi)$ defined by [11]
$$C_*(\psi) = \bigcup_{\phi^\circ \in C(\psi), \pi(\psi|\phi^\circ)=1} \|\phi^\circ\| = \left\| \bigvee_{\phi^\circ \in C(\psi), \pi(\psi|\phi^\circ)=1} \phi^\circ \right\|.$$ When $\phi^\circ \in C_*(\psi)$, we have $\pi(\psi|\phi^\circ) = 1$ and furthermore, if $\phi^\circ \notin C_*(\psi)$, we need to study the relation between ϕ° and ϕ. If $\phi^\circ \subseteq \phi$, then $\max\left\{\pi(\phi^\circ), \frac{\pi(\phi)}{\pi(\phi^\circ)}\right\} \leq \pi(\psi|\phi^\circ) = \frac{\pi(\phi^\circ \wedge \psi)}{\pi(\phi^\circ)} \leq 1$, otherwise the probability of ψ given ϕ° will be from the inside or the outside of the interval $[\pi(\phi^\circ), 1]$.

Analogous to the discussion of the rough modus ponens, consider the Rough Modus Tollens, formed from [3, 11]

if $\phi \to \psi$ is true with probability $\pi(\phi|\psi)$

and ψ is true with probability $\pi(\psi)$

_____

then ϕ is true with probability $\pi(\phi)$

where $\pi(\phi) = \pi(\phi \wedge \neg\psi) + \pi(\psi)\pi(\phi|\psi)$.
From the conditional probability point of view, there is

if $0 < \pi(\psi|\phi), \pi(\psi) < 1$,   *then* $0 < \pi(\phi) \leq \min\left\{\frac{1-\pi(\psi)}{1-\pi(\psi|\phi)}, \frac{\pi(\psi)}{\pi(\psi|\phi)}\right\}$

if $\pi(\psi|\phi) = 0, 0 \leq \pi(\psi) < 1$, *then* $0 < \pi(\phi) \leq 1 - \pi(\psi)$

if $\pi(\psi|\phi) = 1, 0 < \pi(\psi) \leq 1$, *then* $0 < \pi(\phi) \leq \pi(\psi)$

To put it in another way, one has

$$\pi(\psi)\pi(\phi|\psi) \leq \pi(\phi) \leq \pi(\psi)\pi(\phi|\psi) + 1 - \pi(\psi) \ with \ \pi(\phi|\psi) > 0.$$

The following attention in the remaining part of this section will be given to the case when the fact ψ° is not always the same as the conclusion ψ of the rule $\phi \to \psi$ but can be regarded as the characterization of ψ with different beliefs such as 'if it rained then it was cold' and 'it is very cold', defined by

if $\phi \to \psi$ is true with probability $\pi(\phi|\psi)$

and ψ° is true with probability $\pi_t(\psi^\circ)$

_____

then $?\phi$ is true with probability $?\pi_t(\phi)$ $?$ $\pi_t(\phi|\psi^\circ)$

It is worth mentioning that, ψ° takes the different form from the one of ψ, $\pi_t(\psi^\circ)$ denotes the prior probability of ψ° which is different from the probability of ψ, $\pi_t(\phi|\psi^\circ)$ represents the conditional probability of ϕ given ψ° and is identical to the posterior probability $\pi_t(\phi)$ of ϕ when the new fact ψ is observed. The subscript t means the derived conclusions is inferred under the condition that a new fact ψ° occurs.

Theorem 4. *The estimation for the probability of ϕ given ψ° can be formed in*

$$\left[\pi_t(\psi^\circ)\pi(\phi|\psi), \pi_t(\psi^\circ)\pi(\phi|\psi) + 1 - \pi_t(\psi^\circ)\right] \bigcap \left(0, \min\left\{\tfrac{1-\pi_t(\psi^\circ)}{1-\pi(\psi|\phi)}, \tfrac{\pi_t(\psi^\circ)}{\pi(\psi|\phi)}\right\}\right]$$
$$\bigcap \left[\pi(\phi|\psi)\pi(\psi^\circ|\phi), 1\right]$$

Proof. Given the fact ψ° and the rule $\phi \to \psi$, the calculation of $\pi_t(\phi)$ follows from the intersection of

$$\pi_t(\psi^\circ)\pi(\phi|\psi) \leq \pi_t(\phi) \leq \pi_t(\psi^\circ)\pi(\phi|\psi) + 1 - \pi_t(\psi^\circ)$$

and $0 < \pi_t(\phi) \leq \min\left\{\tfrac{1-\pi_t(\psi^\circ)}{1-\pi(\psi|\phi)}, \tfrac{\pi_t(\psi^\circ)}{\pi(\psi|\phi)}\right\}$, where $\pi(\phi|\psi)$ and $\pi(\psi|\phi)$ can be estimated by the definition of certainty factor of the rule, that is, $\pi(\phi|\psi) = \frac{card(\|\phi \wedge \psi\|)}{card(\|\psi\|)}$ and $\pi(\psi|\phi) = \frac{card(\|\phi \wedge \psi\|)}{card(\|\phi\|)}$, here we postulate that the sizes of data tables or information systems in the domain of interest do not change. In addition, as for $\pi_t(\phi|\psi^\circ)$, we shall get $\pi(\phi|\psi)\pi(\psi^\circ|\phi) \leq \pi_t(\phi|\psi^\circ) \leq 1$, which follows from $\frac{1}{\pi_t(\phi)} \leq \frac{1}{\pi_t(\psi^\circ)\pi(\phi|\psi)}$ and $\frac{\pi(\phi|\psi)\pi_t(\phi \wedge \psi^\circ)}{\pi_t(\phi)} \leq \frac{\pi_t(\phi \wedge \psi^\circ)}{\pi_t(\psi^\circ)}$ with $\pi(\phi|\psi) > 0$. By means of the results of $\pi_t(\phi|\psi^\circ)$ and $\pi_t(\phi)$, the estimation can be obtained.   □

Example 4. From Table 1, given the new fact of $\psi^\circ = (Flu, no)$ and the rule "if $\phi = (Headache, yes)$ and $(Muscle - pain, no)$ and $(Temperature, high)$, then $\psi = (Flu, yes)$" with the probability $\pi(\psi|\phi) = \frac{1}{2}$, one has $\frac{1}{8} \leq \pi_t(\phi) \leq \frac{8}{12}$, which follows from $\pi_t(\psi^\circ) = \frac{card(\|(Flu,no)\|)}{card(U)} = \frac{1}{3}$, $\pi(\phi|\psi) = \frac{1}{4}$ and the intersection of $[\frac{1}{12}, \frac{9}{12}]$ and $(0, \frac{2}{3}]$ and $[\frac{1}{8}, 1]$ according to Theorem 4. If possible, the probability $\pi_t(\phi|\psi^\circ) = \frac{1}{2}$ in the light of rough set theory.

Example 5. Given the new fact $\psi^\circ = (Nationality, Swede)$ and the rule "if $\phi = (Height, medium)$ and $(Hair, dark)$, then $\psi = (Nationality, German)$" with the probability $\pi(\psi|\phi) = \frac{90}{135} = 0.67$ (from the characterization of nationalities in [11]), then it can happen that $0.08 \leq \pi_t(\phi) \leq 0.63$, which follows from $\pi_t(\psi^\circ) = \frac{card(\|(Nationality,Swede)\|)}{card(U)} = \frac{405}{900} = 0.45$, $\pi(\phi|\psi) = \frac{90}{495} = 0.18$ and the intersection of $[0.08, 0.63] \bigcap (0, 0.67] \bigcap [0.06, 1]$ (if possible, $\pi_t(\phi|\psi^\circ) = \frac{45}{405} = 0.11$).

Example 6. If we have known that a new fact $\psi^\circ = (Fly, yes)$ and the rule "if $\phi = (Bird, yes)$ and $(Gregarious, yes)$, then $\psi = (Fly, no)$" with the probability $\pi(\psi|\phi) = \frac{2}{7}$ (from the characterization of birds in [6]), the probability of $\phi = (Bird, yes)$ and $(Gregarious, yes)$ under the condition of (Fly,yes) can be solved by $\pi_t(\psi^\circ) = \frac{8}{20}$, $\pi(\phi|\psi) = \frac{2}{12}$, $\pi(\psi^\circ|\phi) = 1 - \frac{2}{7} = \frac{5}{7}$ and $[\frac{1}{15}, \frac{10}{15}] \bigcap (0, \frac{21}{25}] \bigcap [\frac{5}{42}, 1]$, thereby $\frac{5}{42} \leq \pi_t(\phi|\psi^\circ) \leq \frac{10}{15}$ (if possible, $\pi_t(\phi|\psi^\circ) = \frac{5}{8}$).

Due to space limitation, the descriptions of decision tables in Examples 5 and 6 have been omitted, and case analysis of other data tables can be taken as exercises on top of the illustrations displayed in this paper. Also it is noted that, based on rough set theory, if the data table or information system is available to "if ϕ then ψ" and $\psi^\diamond$, then $\pi_t(\phi|\psi^\diamond) = \frac{card(\|\phi \wedge \psi^\diamond\|)}{card(\|\psi\|)}$ with $\pi(\phi|\psi) > 0$. Moreover, if $\pi_t(\phi|\psi^\diamond) \neq 0$ and $\pi(\phi|\psi) \neq 0$, it means this information system is inconsistent.

4   Conclusion

In this paper, we started with the relationship or dependency between the new fact and the premise (or if clause) of the rules crystallized by human wisdom, and then presented the solutions for every different relations in the cases of rough Modus Ponens generalized by new-found possibly evidence related to the premise, finally turning to the case of rough Modus Tollens.

In light of the difficulty of gaining the exact value of $\pi_t(\phi|\psi^\diamond)$ or $\pi(\psi|\phi^\diamond)$, we got the interval for the possible values on the basis of the available data source, which is relatively believable compared with the subjective judgement of the fuzzy membership functions except that the reasoner is one of the experts or authorities in the domain of interest, but the expert might make false decisions or inconsistent opinions. Of course the hypothesis ensuring the validity of being believable is that the data source gathered is sound and representative so as to preserve the accuracy of the probability estimated in the process of reasoning. As can be seen from the results of examples in Sect. 3, sometimes we can obtain the exact value of the probability $\pi_t(\phi|\psi^\diamond)$ or $\pi(\psi|\phi^\diamond)$ through computing the corresponding certainty factors, but this is not always the case, for instance, the probability of $\phi^\diamond = (Height, short)\ and\ (Hair, dark)$ as well as the conditional probability $\pi(\psi|\phi^\diamond)$ from data table in [11] where there is no simultaneous occurrence for $(Height, short)$ and $(Hair, dark)$. The root cause of this problem lies in the incompleteness of the data source, which is an inevitable factor even in a big data environment.

Another comment in need is that by comparison with the assessment of the rough probability [7] or the measurement of the observability for the new fact in the event involved in the premise, the direct comparing between the elements of the new fact and the premise is clearer and sharper, although the rough probability has the advantage of the uncertainty measure for an event. Besides, nonmonotonic reasoning is of the center tasks of uncertainty reasoning and human reasoning has been proved to be nonmonotonic [5]. Hence the proposed solution in this paper can be viewed as an initial alternative of solving the nonmonotonic reasoning based on Modus Ponens and Modus Tollens inference patterns from the viewpoint of rough set theory. The causal effects [2] among the data collected (or the events considered) perform a crucial role in human thinking. The direct or indirect causal relationships among the data or events closely affect the treatment of the contrary facts or the irrelevant facts in human reasoning. Moreover, probabilistic rough set models such as variable precision rough set model and Bayesian rough set model [14,18], together with game-theoretic rough sets, have showed great strength in analyzing uncertainties [1,16]. Further research will be put on the relations of causal effects, the approximate characterization of sets [19] and probabilistic rough set approach.

Acknowledgments. The authors would like to thank the reviewers for their comments that help improve the manuscript. This research was supported in parts by the National Natural Science Foundation of China (No. 61273304, 61202170, 61573255, 61573259), the Specialized Research Fund for the Doctoral Program of Higher Education of China (No. 20130072130004), and the program of Further Accelerating the Development of Chinese Medicine Three Year Action of Shanghai (2014–2016) (No. ZY3-CCCX-3-6002).

References

1. Azam, N., Yao, J.T.: Analyzing uncertainties of probabilistic rough set regions with game-theoretic rough sets. Int. J. Approximate Reason. **55**(1), 142–155 (2014)
2. Bareinboim, E., Pearl, J.: Causal inference from big data: theoretical foundations and the data-fusion problem. UCLA Cognitive Systems Laboratory, Technical Report (R-450) (2015). Proceedings of the National Academy of Sciences (2016)
3. Greco, S., Pawlak, Z., Słowiński, R.: Generalized decision algorithms, rough inference rules, and flow graphs. In: Alpigini, J.J., Peters, J.F., Skowron, A., Zhong, N. (eds.) RSCTC 2002. LNCS (LNAI), vol. 2475, pp. 93–104. Springer, Heidelberg (2002)
4. Inuiguchi, M., Greco, S., Słowiński, R.: Fuzzy-rough modus ponens and modus tollens as a basis for approximate reasoning. In: Tsumoto, S., Słowiński, R., Komorowski, J., Grzymała-Busse, J.W. (eds.) RSCTC 2004. LNCS (LNAI), vol. 3066, pp. 84–94. Springer, Heidelberg (2004)
5. Johnson-Laird, P.N., Khemlani, S.S., Goodwin, G.P.: Logic, probability, and human reasoning. Trends Cogn. Sci. **19**(4), 201–214 (2015)
6. Jue, W., Yiyu, Y., Feiyue, W.: 'Rule plus Exception' learning based on reduct. Chin. J. Comput. (11), 1778–1789 (2005)
7. Pawlak, Z.: Rough probability. Bull. Polish Acad. Sci. Tech. **33**(9–10), 499–504 (1985)
8. Pawlak, Z.: Rough Sets: Theoretical Aspects of Reasoning about Data. Kluwer, Dordrecht (1991)
9. Pawlak, Z., Skowron, A.: Rough membership functions: a tool for reasoning with uncertainty. In: Algebraic Methods in Logic and in Computer Science, Banach Center Publications, vol. 28, pp. 135–150 (1993)
10. Pawlak, Z.: Rough Modus Ponens. In: Traitement d'information et gestion d'incertitudes dans les systèmesà base de connaissances. Conférence Internationale, pp. 1162–1166 (1998)
11. Pawlak, Z.: Rough sets, decision algorithms and Bayes' theorem. Eur. J. Oper. Res. **136**(1), 181–189 (2002)
12. Wagner, C.G.: Modus tollens probabilized. British J. Philos. Sci. **55**(4), 747–753 (2004)
13. Pfeifer, N., Kleiter, G.D.: Inference in conditional probability logic. Kybernetika **42**(4), 391–404 (2006)
14. Ślęzak, D., Ziarko, W.: The investigation of the bayesian rough set model. Int. J. Approximate Reason. **40**(40), 81–91 (2005)

15. Sobel, J.H.: Modus Ponens and Modus Tollens for conditional probabilities, and updating on uncertain evidence. Theor. Decis. **66**(2), 103–148 (2009)
16. Yao, Y.Y.: The superiority of three-way decisions in probabilistic rough sets models. Inf. Sci. **181**(6), 1080–1096 (2011)
17. Zadeh, L.A.: Outline of a new approach to the analysis of complex systems and decision processes. IEEE Trans. Syst. Man Cybern. **3**, 28–44 (1973)
18. Zhang, H.Y., Zhou, J., Miao, D.Q., Gao, C.: Bayesian rough set model: a further investigation. Int. J. Approximate Reason. **53**(4), 541–557 (2012)
19. Zhang, X.Y., Miao, D.Q., Liu, C.H., Le, M.L.: Constructive methods of rough approximation operators and multigranulation rough sets. Knowl. Based Syst. **91**, 114–125 (2016)

Definability in Incomplete Information Tables

Mengjun Hu[(✉)] and Yiyu Yao[(✉)]

Department of Computer Science, University of Regina,
Regina, SK S4S 0A2, Canada
{hu258,yyao}@cs.uregina.ca

Abstract. This paper investigates the issues related to definability in an incomplete information table by using interval sets. We review the existing results pertaining to definability in a complete information table. We generalize the satisfiability of formulas in a description language in a complete table to a pair of strong and weak satisfiability of formulas in an incomplete table, which leads to an interval-set based interpretation of formulas. While we have definable sets in a complete table, we have definable interval sets in an incomplete table. The results are useful for studying concept analysis and approximations with incomplete tables.

1   Introduction

The theory of rough sets offers a simple and effective approach to concept analysis [7,8,14]. We consider three fundamental issues in rough-set based concept analysis, namely, the representation, definability and approximations of concepts.

The first issue is a formal representation of concepts. Following the school of Port-Royal Logic [1,2], one can represent a concept by a pair of intension and extension [9]. The intension consists of properties that apply to all instances of the concept and the extension includes all instances belonging to the concept. The intension-extension view requires a context within which one can establish connections between the intension and the extension. In rough set analysis (RSA), the context is given in the form of an information table in which each row represents an object, each column represents an attribute, and each cell contains the value of an object on an attribute. With respect to an information table, we use formulas in a description language to describe the intension of concepts and a set of objects to represent the extension. Through the notion of the satisfiability of formulas by objects, one can easily connect the intension and the extension of a concept. More specifically, the set of objects satisfying a formula is the extension of the concept with the formula as its intension.

The second issue is the definability of sets or concepts with respect to the description language. Given a formula, one may easily find the set of objects

M. Hu—Thanked FGSR at the University of Regina, Saskatchewan Innovation Scholarships and Mr. John Spencer Gordon Middleton for the support of this work.

Y. Yao—This work is partially supported by a Discovery Grant from NSERC, Canada.

V. Flores et al. (Eds.): IJCRS 2016, LNAI 9920, pp. 177–186, 2016.
DOI: 10.1007/978-3-319-47160-0_16

satisfying the formula. On the other hand, for an arbitrary set of objects, there may not exist a formula that defines the set of objects, that is, the formula may not be satisfied by exactly all objects in the set. A set that can be described by a formula is a definable set; otherwise, it is an undefinable set, that is, we cannot use a formula to describe precisely the set of objects. Consequently, all sets of objects are divided into two families, namely, the family of definable sets and the family of undefinable sets.

Due to the undefinability of some sets of objects, the third issue is the approximation of undefinable sets. In rough set theory, we approximate an undefinable set by a pair of definable sets called the lower and upper approximations [7,8,14]. The family of all definable sets is a sub-Boolean algebra of the power set of the universe. With the Boolean algebra of definable sets, the lower approximation of an undefinable set is the greatest definable set contained by the set and the upper approximation is the least definable set containing the set. Through the two approximations, one can make an approximate statement about an undefinable set and the corresponding concept.

In this paper, we examine the first two issues in a context given by an incomplete information table. In an incomplete information table, each cell contains a set of values instead of an individual value. Although an object can only take one value, due to a lack of information, we only know a set of possible values, and do not know which one in the set is the actual value. For the first issue of the representation of concepts, formulas in a description language are still used to describe the intension. However, we can no longer determine the exact set of objects that satisfy a formula. We can get its lower and upper bounds by adopting Lipski's framework [5] of interpreting an incomplete table as a family of complete tables. These two bounds form an interval set [10] that we use to describe the extension of a concept. Accordingly, a concept is represented by a pair of a formula and an interval set in an incomplete table. For the second issue of definability, we generalize the notion of definable sets in a complete table into a notion of definable interval sets in an incomplete table. An interval set is definable if it can be described by a formula. For the third issue of approximation, there are several different approaches. For example, Grzymala-Busse considers a characteristic relation and presents three types of approximations with incomplete data, namely, singleton, subset and concept approximations [3,4]. Following the studies in the present paper, one may generalize the pair of approximations by using the family of definable interval sets. Based on the results about the first two issues reported, the third issue will be discussed in a future paper.

In the rest of this paper, we review the related results about interval sets [10] and definability in complete information tables [6,11,13]. We generalize the notions in complete tables to incomplete tables and investigate their properties.

2   An Overview of Interval Sets

An interval set is represented by a pair of sets, namely, its lower and upper bounds [10]. The definition of an interval set is formally given as follows.

Definition 1. *Suppose U is a finite nonempty set. An interval set is a family of sets defined by $\mathcal{A} = [A_l, A_u] = \{A \in 2^U \mid A_l \subseteq A \subseteq A_u\}$, where 2^U is the power set of U, and $A_l \subseteq A_u$. The sets A_l and A_u are the lower and upper bounds of the interval set, respectively.*

By definition, an interval set is a subset of 2^U. It includes all subsets of U lying between the two bounds, that is, those sets that are supersets of the lower bound and subsets of the upper bound. As a subset of the power set lattice 2^U, an interval set is also a lattice with the minimum element A_l, the maximum element A_u, and set-theoretic operations. The family of all interval sets with respect to 2^U is $\mathcal{I}(2^U) = \{[A_l, A_u] \mid A_l \subseteq A_u \text{ and } A_l, A_u \in 2^U\}$. If the lower and upper bounds are the same set, that is, $A_l = A_u = A$, the interval set $[A_l, A_u]$ contains only the set A. Such interval sets are called degenerate interval sets that are equivalent to ordinary sets.

Since an interval set is a set of subsets of U, the set union, intersection and complement operators may be directly applied to interval sets. Alternatively, standard set-theoretic operators may be generalized into interval-set operators, namely, interval-set union, intersection and complement.

Definition 2. *For two interval sets $\mathcal{A}$ and $\mathcal{B}$ on U, the interval-set union $\sqcup$, interval-set intersection $\sqcap$ and interval-set complement $\neg$ are defined in terms of standard set union $\cup$, intersection $\cap$ and complement ^{c} as follows:*

$$\mathcal{A} \sqcup \mathcal{B} = \{A \cup B \mid A \in \mathcal{A} \text{ and } B \in \mathcal{B}\},$$
$$\mathcal{A} \sqcap \mathcal{B} = \{A \cap B \mid A \in \mathcal{A} \text{ and } B \in \mathcal{B}\},$$
$$\neg \mathcal{A} = \{A^c \mid A \in \mathcal{A}\}. \tag{1}$$

The interval-set operators are defined in terms of corresponding set-theoretic operators on the families of sets in the interval sets. That is, the interval-set union and intersection of two interval sets contain all possible set union and intersection between their set members, respectively. The interval-set complement of an interval set contains set complement of all its set members. The interval-set operators may be computed by applying corresponding set-theoretic operators to the lower and upper bounds of the interval sets as given by the next theorem [10].

Theorem 1. *For two interval sets $\mathcal{A} = [A_l, A_u]$ and $\mathcal{B} = [B_l, B_u]$, the interval-set union $\sqcup$, intersection $\sqcap$ and complement $\neg$ can be computed as:*

$$\mathcal{A} \sqcup \mathcal{B} = [A_l \cup B_l, A_u \cup B_u],$$
$$\mathcal{A} \sqcap \mathcal{B} = [A_l \cap B_l, A_u \cap B_u],$$
$$\neg \mathcal{A} = [A_u^c, A_l^c]. \tag{2}$$

The family of interval sets $\mathcal{I}(2^U)$ is closed under the interval-set union, intersection and complement.

An interval set may be used to represent the extension of a partially known concept [10]. Due to a lack of information, it is sometimes difficult to determine whether an object is an instance or a non-instance of a concept. In this case, the extension of the concept is partially known and the concept is called a partially known concept. Suppose a set A_p contains all the objects that are known to be instances of a concept and another set A_n contains all the objects that are known to be non-instances of the concept. The set A_n^c, that is, the complement set of A_n, contains all the objects that may be instances of the concept. It is easy to verify that $A_p \subseteq A_n^c$. The extension of the concept may be represented by the interval set $[A_p, A_n^c]$. Any set belonging to this interval set may be the actual extension of the concept. As the extension of a partially known concept, the interval set $[A_p, A_n^c]$ divides all objects into three regions:

(1) The set of objects A_p that are known to be instances of the concept;
(2) The set of objects A_n that are known to be non-instances of the concept;
(3) The set of objects $A_n^c - A_p$ that are not known to be instances or non-instances of the concept.

These three regions are pair-wise disjoint and some of them may be empty. This point of view links interval sets with three-way decisions [12] with A_p as the positive region, A_n as the negative region, and $A_n^c - A_p$ as the boundary region.

3   Definable Sets in a Complete Table

Complete information tables, usually referred to as information tables, contain all available data with respect to specific applications in a tabular form [7,8].

Definition 3. *A complete information table is defined as a tuple:*

$$T = (U, AT, \{V_a \mid a \in AT\}, \{I_a : U \rightarrow V_a \mid a \in AT\}), \tag{3}$$

where U is a finite nonempty set of objects called the universe, AT is a finite nonempty set of attributes, V_a is the domain of an attribute a, and I_a is an information function that maps each object to one value on an attribute a.

By definition, in a complete information table, each object takes one and only one value from the domain of each attribute. In this sense, the information about all objects is complete.

To describe the properties of objects, a description language is introduced. In this paper, we consider a sublanguage of the description language proposed by Marek and Pawlak [6]. That is, we only use a special class of atomic formulas and logic conjunction and disjunction to construct logic formulas.

Definition 4. *The formulas in a description language DL are defined by:*

(1) *Atomic formulas: $(a = v) \in DL$, where $a \in AT$ and $v \in V_a$;*
(2) *If $p, q \in DL$, then $p \wedge q, p \vee q \in DL$.*

In some cases, it may be useful to consider a sublanguage of DL. By considering conjunctive formulas that contain only the logic conjunction, Yao and Hu introduce conjunctively definable sets and structured rough set approximations that give deep insights into the semantics of approximations [15].

To interpret the semantics of formulas in DL, the notion of satisfiability is introduced [6, 14].

Definition 5. *In a complete table* $(U, AT, \{V_a \mid a \in AT\}, \{I_a \mid a \in AT\})$, *the satisfiability of a formula by an object x, denoted as $\models$, is defined by:*

$$
\begin{align}
&(1) \quad x \models (a = v), \text{ iff } I_a(x) = v, \\
&(2) \quad x \models (p \wedge q), \text{ iff } (x \models p) \wedge (x \models q), \\
&(3) \quad x \models (p \vee q), \text{ iff } (x \models p) \vee (x \models q), \tag{4}
\end{align}
$$

where $a \in AT$, $v \in V_a$ and $p, q \in DL$.

According to the satisfiability, a formula can be interpreted in terms of a set of objects.

Definition 6. *Given a formula $p \in DL$, its meaning set is defined as $m(p) = \{x \in U \mid x \models p\}$.*

That is, the meaning set $m(p) \subseteq U$ of a formula p is the set of objects satisfying p. In a complete table, we may use set intersection and union to interpret logic conjunction and disjunction in a formula through meaning sets.

Theorem 2. *The meaning sets of formulas in DL may be computed by:*

$$
\begin{align}
&(1) \quad m(a = v) = \{x \in U \mid I_a(x) = v\}, \\
&(2) \quad m(p \wedge q) = m(p) \cap m(q), \\
&(3) \quad m(p \vee q) = m(p) \cup m(q). \tag{5}
\end{align}
$$

One can easily compute a meaning set of a formula. A concept is represented by the pair $(p, m(p))$ where $p \in DL$. The meaning set of a formula describes the meaning of the formula. The formula in turn defines its meaning set. Such a set is a definable set.

Definition 7. *A set of objects $X \subseteq U$ is a definable set if there exists a formula $p \in DL$ such that $X = m(p)$.*

Let $\mathrm{DEF}(U) = \{X \subseteq U \mid X \text{ is definable}\}$ denote the family of all definable sets on U. One may easily verify the following properties of $\mathrm{DEF}(U)$.

Theorem 3. *The family of definable sets satisfies the following properties:*

$$
\begin{align}
&(p1) \quad \emptyset \in \mathrm{DEF}(U), \text{ and } U \in \mathrm{DEF}(U), \\
&(p2) \quad \text{If } X, Y \in \mathrm{DEF}(U), \text{ then } X \cup Y, X \cap Y, X^c \in \mathrm{DEF}(U). \tag{6}
\end{align}
$$

By property (p2), the family of definable sets is closed under set union, intersection and complement. From all properties in Theorem 3, one can conclude that the family of definable sets forms a Boolean algebra.

4   Definable Interval Sets in an Incomplete Table

In this section, we adopt Lipski's framework for interpreting an incomplete information table. We generalize the notions of satisfiability and the meaning sets of formulas for incomplete tables.

4.1   An Incomplete Table as a Family of Complete Tables

In a complete table, we know the exact value of an object on an attribute. In many situations, we may not have such complete information in sense that we only partially know the value of an object. Following Lipski [5], we assume that the available information only allows us to give a set of possible values of an object. Although an object must take exactly one value from the set, we are not able to identify the actual value. A formal way to represent such incomplete information is through the introduction of set-based or incomplete information tables.

Definition 8. *An incomplete information table is defined as:*

$$\widetilde{T} = (U, AT, \{V_a \mid a \in AT\}, \{\widetilde{I}_a : U \to 2^{V_a} - \emptyset \mid a \in AT\}), \tag{7}$$

where U is a finite nonempty set of objects called the universe, AT is a finite nonempty set of attributes that apply to all the objects, V_a is the domain of an attribute a, and $\widetilde{I}_a$ is an information function that maps each object in U to a nonempty subset of V_a.

In an incomplete table, the set $\widetilde{I}_a(x) \subseteq V_a$ denotes the set of possible values of x on the attribute a. That is, based on the available information, we can state that (a) x must take one value from $\widetilde{I}_a(x)$, and (b) x cannot take any value from $V_a - \widetilde{I}_a(x)$. However, the available information is insufficient for us to tell exactly which value in $\widetilde{I}_a(x)$ is the actual value of x. The set-valued mapping $\widetilde{I}_a(x)$ expresses our incomplete knowledge about x.

We assume that all attributes are applicable for all objects. In other words, an object must take a nonempty subset of V_a on any attribute a. Based on characteristics of the set of values $\widetilde{I}_a(x)$, we can classify three types of knowingness regarding the value of x on a:

(i) When $\widetilde{I}_a(x)$ is a singleton set, that is, $\widetilde{I}_a(x) = \{v\}$ where $v \in V_a$, we know that x takes the value v on attribute a. This is the case of complete information, in which the value of x on a is known.

(ii) When $\widetilde{I}_a(x)$ is a proper subset of the domain V_a, that is, $\emptyset \neq \widetilde{I}_a(x) = F \subsetneq V_a$, we know that the actual value of x must be in F, or equivalently, cannot be in its complement $F^c = U - F$. In this case, we have partial information about the value of x on a. We say that the value of x is partially known.

(iii) When $\widetilde{I}_a(x)$ is the entire domain, that is, $\widetilde{I}_a(x) = V_a$, we do not have any information about the value of x on a. We say that the value of x is unknown.

A complete table can be viewed as a special case of an incomplete table. One may easily construct an incomplete table from a complete table by setting $\widetilde{I}_a(x) = \{I_a(x)\}$ for all $x \in U$ and $a \in AT$ where I_a is the information function of the complete table. For the reverse direction, Lipski [5] introduces a framework that interprets an incomplete table as a family of complete tables.

Definition 9. *A complete information table T is called a completion of an incomplete information table $\widetilde{T}$ if T satisfies the following condition:*

$$\forall x \in U, a \in AT, I_a^T(x) \in \widetilde{I}_a(x), \tag{8}$$

where $I_a^T(x)$ and $\widetilde{I}_a(x)$ denote the values of x on attribute a in a complete table T and an incomplete table $\widetilde{T}$, respectively.

The family of all completions of $\widetilde{T}$ is denoted by $\mathrm{COMP}(\widetilde{T})$. One may construct the incomplete table from the family $\mathrm{COMP}(\widetilde{T})$ by setting $\widetilde{I}_a(x) = \{I_a^T(x) \mid T \in \mathrm{COMP}(\widetilde{T})\}$, where $I_a^T(x)$ denotes the value of x on attribute a in a completion T. In this way, an incomplete table is equivalently represented by the family of all its completions. Since every value in $\widetilde{I}_a(x)$ may be the actual one, every completion of $\widetilde{T}$ is possibly the actual table. However, only one of them is the actual table with only one actual value for each object on each attribute. An advantage of Lipski's representation is that we can study an incomplete table through its equivalent family of complete tables. By simply lifting a method for processing a complete table to a family of complete tables, we obtain a method for processing an incomplete table.

4.2 Interpretation of the Description Language

By interpreting an incomplete table as a family of complete tables, we study the satisfiability in an incomplete table though the satisfiability in a family of complete tables. This leads to two senses of satisfiability, namely, the strong satisfiability and the weak satisfiability.

Definition 10. *A pair of strong and weak satisfiability of formulas by objects in an incomplete table $\widetilde{T}$ is defined by: for $x \in U, p \in DL$,*

$$(1) \quad (x, \widetilde{T}) \models_* p, \text{ iff } \forall T \in \mathrm{COMP}(\widetilde{T})((x, T) \models p),$$

$$(2) \quad (x, \widetilde{T}) \models^* p, \text{ iff } \exists T \in \mathrm{COMP}(\widetilde{T})((x, T) \models p), \tag{9}$$

where $\models_*$ and $\models^*$ denote the strong and weak satisfiability, respectively, and $(x, T) \models p$ denotes that the object x satisfies p in the complete table T.

In the definition, we use a pair of an object and an information table to explicitly specify the table in which the satisfiability is considered. We will omit the information table if there is no confusion. Intuitively, the strong satisfiability states that an object satisfies a formula in every completion of an incomplete table. In other words, the object definitely satisfies the formula. The weak satisfiability means that an object satisfies a formula in at least one completion, that is, the object possibly satisfies the formula. According to the two kinds of satisfiability, we derive two sets of objects to interpret a formula, which gives two bounds of the actual meaning set of the formula in the actual table.

Definition 11. *In an incomplete table, a formula $p \in DL$ is interpreted by the following pair of sets:*

$$m_*(p) = \{x \in U \mid x \models_* p\}, \text{ and } m^*(p) = \{x \in U \mid x \models^* p\}. \tag{10}$$

By definition, the set $m_*(p)$ consists of objects definitely satisfying p and the set $m^*(p)$ contains objects possibly satisfying p. These two sets can be equivalently computed by the meaning sets in the family $\mathrm{COMP}(\widetilde{T})$ as given in the following theorem.

Theorem 4. *For a formula $p \in DL$, the two sets $m_*(p)$ and $m^*(p)$ can be computed as:*

$$m_*(p) = \{x \in U \mid \forall T \in \mathrm{COMP}(\widetilde{T}), x \in m^T(p)\} = \bigcap_{T \in \mathrm{COMP}(\widetilde{T})} m^T(p),$$

$$m^*(p) = \{x \in U \mid \exists T \in \mathrm{COMP}(\widetilde{T}), x \in m^T(p)\} = \bigcup_{T \in \mathrm{COMP}(\widetilde{T})} m^T(p), \tag{11}$$

where $m^T(p)$ is the meaning set of p in the completion T.

It follows that $m_*(p) \subseteq m^*(p)$. An interval set $[m_*(p), m^*(p)]$ may be accordingly constructed to interpret the formula. The corresponding concept is represented by the pair $(p, [m_*(p), m^*(p)])$. One may verify the following theorem about the interval set $[m_*(p), m^*(p)]$.

Theorem 5. *For every set $X \in [m_*(p), m^*(p)]$, there exists a completion $T \in \mathrm{COMP}(\widetilde{T})$ such that $m^T(p) = X$, where $m^T(p)$ is the meaning set of p in the completion T.*

Unlike the case of a complete table, we can no longer use set intersection and union to truthfully characterize logic conjunction and disjunction respectively.

Theorem 6. *The sets $m_*(\cdot)$ and $m^*(\cdot)$ satisfy the following properties:*

$$\begin{aligned}
&\text{(l1)} \quad m_*(a = v) = \{x \in U \mid \widetilde{I}_a(x) = \{v\}\}, \\
&\text{(u1)} \quad m^*(a = v) = \{x \in U \mid v \in \widetilde{I}_a(x)\}; \\
&\text{(l2)} \quad m_*(p \wedge q) = m_*(p) \cap m_*(q), \\
&\text{(u2)} \quad m^*(p \wedge q) \subseteq m^*(p) \cap m^*(q); \\
&\text{(l3)} \quad m_*(p \vee q) \supseteq m_*(p) \cup m_*(q), \\
&\text{(u3)} \quad m^*(p \vee q) = m^*(p) \cup m^*(q). \tag{12}
\end{aligned}$$

That is, we can compute the lower bound for conjunction and upper bound for disjunction. However, we cannot compute the lower bound for disjunction, nor the upper bound for conjunction.

4.3   Definable Interval Sets

From a formula $p \in DL$, one may derive an interval set $[m_*(p), m^*(p)]$. This interval set expresses the meaning of p in an incomplete table, and p defines the interval set. Such an interval set is definable.

Definition 12. *An interval set $\mathcal{A}$ on the universe U is a definable interval set if there exists a formula $p \in DL$ such that $\mathcal{A} = [m_*(p), m^*(p)]$.*

Let $\mathrm{DEFI}(U) = \{\mathcal{A} \in \mathcal{I}(2^U) \mid \mathcal{A} \text{ is definable}\}$ denote the family of all definable interval sets. The family $\mathrm{DEFI}(U)$ satisfies the properties given in the following theorem.

Theorem 7. *The family $\mathrm{DEFI}(U)$ satisfies the following properties:*

$$(\mathrm{p}'1) \qquad [\emptyset, \emptyset] \in \mathrm{DEFI}(U), \text{ and } [U, U] \in \mathrm{DEFI}(U),$$
$$(\mathrm{p}'2) \qquad \text{If } \mathcal{X} \in \mathrm{DEFI}(U), \text{ then } \neg\mathcal{X} \in \mathrm{DEFI}(U). \tag{13}$$

For property $(\mathrm{p}'1)$, the interval set $[\emptyset, \emptyset]$ may be defined by any formula containing a contradiction, such as a formula in the form of $(a = a_1) \wedge (a = a_2)$ where $a \in AT, a_1, a_2 \in V_a$ and $a_1 \neq a_2$. The interval set $[U, U]$ may be defined by any formula that is always satisfied, such as a formula in the form of $(a = a_1) \vee (a = a_2) \vee \cdots \vee (a = a_n)$ where $a \in AT$ and $V_a = \{a_1, a_2, \cdots, a_n\}$. By property $(\mathrm{p}'2)$, the family of definable interval sets is closed under interval-set complement. However, it is not closed under interval-set union and intersection. For two definable interval sets $\mathcal{X} = [m_*(p), m^*(p)]$ and $\mathcal{Y} = [m_*(q), m^*(q)]$,

$$\mathcal{X} \sqcup \mathcal{Y} = [m_*(p) \cup m_*(q), m^*(p) \cup m^*(q)],$$
$$\mathcal{X} \sqcap \mathcal{Y} = [m_*(p) \cap m_*(q), m^*(p) \cap m^*(q)]. \tag{14}$$

From Theorem 6, we know that $m^*(p) \cup m^*(q) = m^*(p \vee q)$ and $m_*(p) \cap m_*(q) = m_*(p \wedge q)$. However, we can only verify a set inclusion relation between $m_*(p) \cup m_*(q)$ and $m_*(p \vee q)$ as well as $m^*(p) \cap m^*(q)$ and $m^*(p \wedge q)$. By Lipski [5], the equalities will hold instead of these two set inclusion relations if no attribute appears in both p and q. However, this condition brings strong restrictions to the language.

5   Conclusion and Future Work

Interval sets provide a tool for formulating partially known concepts and investigating definability in incomplete information tables. In this paper, we follow the ideas from studies of definability in complete tables and generalize related notions in incomplete tables. We use an interval set to interpret a formula in an incomplete table. The two bounds of the interval set come from a pair of strong and weak satisfiability of formulas by objects. An interval set that is describable by a formula is definable. The notion of definable interval sets is a generalization of the notion of definable sets in complete tables. The family of all definable interval sets is closed under interval-set complement, but not under interval-set union and intersection. This may lead to difficulties in studying the structure of the family of definable interval sets. One possible solution is to introduce some reasonable restrictions on the formulas in the description language. By using the family of definable interval sets, one may continue with a study of defining the lower and upper approximations in incomplete tables.

References

1. Arnauld, A., Nicole, P.: Logic or the Art of Thinking. Cambridge University Press, Cambridge (1996). Buroker, J.V. (Trans.)
2. Buroker, J.: Port Royal Logic. Stanford Encyclopedia of Philosophy. http://plato.stanford.edu/entries/port-royal-logic/. Accessed 16 May 2016
3. Grzymala-Busse, J.W., Clark, P.G., Kuehnhausen, M.: Generalized probabilistic approximations of incomplete data. Int. J. Approximate Reasoning **55**, 180–196 (2014)
4. Grzymala-Busse, J.W.: A rough set approach to incomplete data. In: Ciucci, D., Wang, G.Y., Mitra, S., Wu, W.Z. (eds.) RSKT 2015. LNCS, vol. 9436, pp. 3–14. Springer, Heidelberg (2015). doi:10.1007/978-3-319-25754-9_1
5. Lipski, W.: On semantics issues connected with incomplete information table. ACM Trans. Database Syst. **4**, 262–296 (1979)
6. Marek, W., Pawlak, Z.: Information storage and retrieval systems: mathematical foundations. Theoret. Comput. Sci. **1**, 331–354 (1976)
7. Pawlak, Z.: Rough sets. Int. J. Comput. Inf. Sci. **11**, 341–356 (1982)
8. Pawlak, Z.: Rough Sets: Theoretical Aspects of Reasoning about Data. Kluwer Academic Publishers, Boston (1991)
9. van Mechelen, I., Hampton, J., Michalski, R.S., Theuns, P. (eds.): Categories and Concepts, Theoretical Views and Inductive Data Analysis. Academic Press, New York (1993)
10. Yao, Y.Y.: Interval-set algebra for qualitative knowledge representation. In: Proceedings of the Fifth International Conference on Computing and Information, pp. 370–374 (1993)
11. Yao, Y.Y.: A note on definability and approximations. In: Peters, J.F., Skowron, A., Marek, V.W., Orłowska, E., Słowiński, R., Ziarko, W.P. (eds.) Transactions on Rough Sets VII. LNCS, vol. 4400, pp. 274–282. Springer, Heidelberg (2007)
12. Yao, Y.Y.: An outline of a theory of three-way decisions. In: Yao, J.T., Yang, Y., Słowiński, R., Greco, S., Li, H., Mitra, S., Polkowski, L. (eds.) RSCTC 2012. LNCS, vol. 7413, pp. 1–17. Springer, Heidelberg (2012)
13. Yao, Y.Y.: The two sides of the theory of rough sets. Knowl. Based Syst. **80**, 67–77 (2015)
14. Yao, Y.Y.: Rough set approximations: a concept analysis point of view. In: Ishibuch, H. (ed.) Computational Intelligence - Volume I. Encyclopedia of Life Support Systems (EOLSS), pp. 282–296 (2015)
15. Yao, Y.Y., Hu, M.J.: A definition of structured rough set approximations. In: Kryszkiewicz, M., Cornelis, C., Ciucci, D., Medina-Moreno, J., Motoda, H., Raś, Z.W. (eds.) RSEISP 2014. LNCS, vol. 8537, pp. 111–122. Springer, Heidelberg (2014)

Rough Sets by Indiscernibility Relations in Data Sets Containing Possibilistic Information

Michinori Nakata[1(⊠)] and Hiroshi Sakai[2]

[1] Faculty of Management and Information Science, Josai International University,
1 Gumyo, Togane, Chiba 283-8555, Japan
nakatam@ieee.org
[2] Faculty of Engineering, Department of Mathematics and Computer Aided Sciences,
Kyushu Institute of Technology, Tobata, Kitakyushu 804-8550, Japan
sakai@mns.kyutech.ac.jp

Abstract. Under data sets containing possibilistic information, rough sets are described by directly using indiscernibility relations. First, we give rough sets based on indiscernibility relations under complete information. Second, we address rough sets by applying possible world semantics to data sets with possibilistic information. The rough sets are used as a correctness criterion of approaches extended to deal with possibilistic information. Third, we extend the approach based on indiscernibility relations to handle data sets with possibilistic information. Rough sets in this extension creates the same results as ones obtained under possible world semantics. This gives justification to our extension.

Keywords: Rough sets · Lower and upper approximations · Indiscernibility relations · Possibilistic information · Possible world semantics

1 Introduction

Natural languages that we daily use contain lots of fuzzy terms, as was pointed out by Zadeh [17–19]. A fuzzy term is expressed by a normal possibility distribution [20]. For example, "around 40" is expressed by the possibility distribution $\{(36, 0.4), (37, 0.6), (38, 0.8), \quad (39, 1), (40, 1), (41, 1), (42, 0.8), (43, 0.6), (44, 0.4)\}_p$ in the sentence "Tom's age is around 40." Such a piece of information, which is called possibilistic information, frequently appears in various situations of our everyday life. Therefore, analysis of data that is created from our daily life requires dealing with possibilistic information.

Rough sets whose components are lower and upper approximations were proposed by Pawlak [12–15]. Data analysis based on the rough sets is well-known as an effective method of data mining. The rough sets is usually used under complete information. However, information from our daily life is possibilistic rather than complete.

Using rough sets in data sets containing possibilistic information requires some extensions of the traditional rough sets that deal with only complete

© Springer International Publishing AG 2016
V. Flores et al. (Eds.): IJCRS 2016, LNAI 9920, pp. 187–196, 2016.
DOI: 10.1007/978-3-319-47160-0_17

information. The concept of possible indiscernibility between objects is used by Słowiński and Stefanowski [16]. Lower and upper approximations of a set of objects with possibilistic information are expressed in terms of using possible equivalence classes by Nakata and Sakai [7]. Couso and Dubois express lower and upper approximations to a set of objects in terms of introducing the degree of possibility that objects are characterized with the same values [3]. The approach in [7] addresses lower and upper approximations from the viewpoint of only possibility. Nakata and Sakai describe these approximations in terms of using possible equivalence classes from the viewpoint of not only possibility but also certainty [8]. However, methods using possible equivalence classes are not applicable in the case where we cannot obtain equivalence classes when information is complete. Therefore, Nakata and Sakai formulate the lower and upper approximations by directly using indiscernibility relations without using possible equivalence classes from the viewpoint of both possibility and certainty [10].

Any justification of these approaches is not described at all. Nakata and Sakai use a correctness criterion to justify their extension of lower and upper approximations in information tables containing unknown types of missing values [9]. The correctness criterion is that lower and upper approximations obtained from some extensions give the same results as ones from the approach based on possible world semantics. This type of criterion is usually used in the field of databases with information that is not complete [1,2,5,21]. An approach based on possible world semantics, whose origin is Lipski's work in databases with incomplete information [6], is proposed in dealing with possibilistic information [11]. Therefore, in this paper we justify our extension by using the correctness criterion.

The paper is organized as follows. In Sect. 2, rough sets is briefly addressed under complete information by directly using indiscernibility relations. In Sect. 3, we first show rough sets based on possible world semantics in information tables with possibilistic information, as was done by Lipski in databases with incomplete information. And then we give an extension of formulae proposed by Dubois and Prade to deal with possibilistic information. Subsequently, we show that the extension gives the same results as ones based on possible world semantics. In Sect. 4, conclusions are addressed.

2   Rough Sets in Data Sets Containing Complete Information

A data set is represented as an information table. The information table consists of universe U, set AT of attributes such that attribute $a_i : U \rightarrow D(a_i)$ for every $a_i \in AT$, and set $\{D(a_i) \mid a_i \in AT\}$ of values where $D(a_i)$ is the domain of attribute a_i.

Binary relation R_{a_i} expressing indiscernibility of objects for attribute a_i, which is called the indiscernibility relation of a_i, is:

$$R_{a_i} = \{(o, o') \in U \times U \mid a_i(o) = a_i(o')\}, \tag{1}$$

where $a_i(o)$ is the value of attribute a_i that object o has. From condition $a_i(o) = a_i(o')$ of indiscernibility, this relation is reflexive, symmetric, and transitive.[1] Characteristic function $\chi_{R_{a_i}}$ of R_{a_i} is defined by $\chi_{R_{a_i}}(o, o') = 1$[2] if $(o, o') \in R_{a_i}$, $\chi_{R_{a_i}}(o, o') = 0$ if $(o, o') \notin R_{a_i}$.

Degrees $\chi_{\underline{apr}_{a_i}(\mathcal{O})}(o)$ and $\chi_{\overline{apr}_{a_i}(\mathcal{O})}(o)$ to which object o belongs to a pair of approximations on a_i of set $\mathcal{O}$ of objects, lower approximation $\underline{apr}_{a_i}(\mathcal{O})$ and upper approximation $\overline{apr}_{a_i}(\mathcal{O})$, are:

$$\chi_{\underline{apr}_{a_i}(\mathcal{O})}(o) = \min_{o' \in U} \max(1 - \chi_{R_{a_i}}(o, o'), \chi_{\mathcal{O}}(o')), \tag{2}$$

$$\chi_{\overline{apr}_{a_i}(\mathcal{O})}(o) = \max_{o' \in U} \min(\chi_{R_{a_i}}(o, o'), \chi_{\mathcal{O}}(o')). \tag{3}$$

Lower and upper approximations directly using indiscernibility relations are:

$$\underline{apr}_{a_i}(\mathcal{O}) = \{o \mid \chi_{\underline{apr}_{a_i}(\mathcal{O})}(o) = 1\}, \tag{4}$$

$$\overline{apr}_{a_i}(\mathcal{O}) = \{o \mid \chi_{\overline{apr}_{a_i}(\mathcal{O})}(o) = 1\}. \tag{5}$$

When objects are characterized by values of attributes, a set of objects being approximated is covered by indiscernible classes obtained from the values of attributes being equal. Under this consideration, degrees $\chi_{\underline{apr}_{a_i}(\mathcal{O}/a_j)}(o)$ and $\chi_{\overline{apr}_{a_i}(\mathcal{O}/a_j)}(o)$ to which object o belongs to a pair of approximations on a_i of set $\mathcal{O}$ of objects that are characterized by values of a_j, lower approximation $\underline{apr}_{a_i}(\mathcal{O}/a_j)$ and upper approximation $\overline{apr}_{a_i}(\mathcal{O}/a_j)$, are:

$$\chi_{\underline{apr}_{a_i}(\mathcal{O}/a_j)}(o) = \max_{o'' \in \mathcal{O}} \min_{o' \in U} \max(1 - \chi_{R_{a_i}}(o, o'), \min(\chi_{R_{a_j}}(o', o''), \chi_{\mathcal{O}}(o'))), \tag{6}$$

$$\chi_{\overline{apr}_{a_i}(\mathcal{O}/a_j)}(o) = \max_{o'' \in \mathcal{O}} \max_{o' \in U} \min(\chi_{R_{a_i}}(o, o'), \chi_{R_{a_j}}(o', o''), \chi_{\mathcal{O}}(o')). \tag{7}$$

Lower and upper approximations directly using indiscernibility relations are:

$$\underline{apr}_{a_i}(\mathcal{O}/a_j) = \{o \mid \chi_{\underline{apr}_{a_i}(\mathcal{O}/a_j)}(o) = 1\}, \tag{8}$$

$$\overline{apr}_{a_i}(\mathcal{O}/a_j) = \{o \mid \chi_{\overline{apr}_{a_i}(\mathcal{O}/a_j)}(o) = 1\}. \tag{9}$$

3   Rough Sets in Data Sets Containing Possibilistic Information

Let a data set containing possibilistic information be obtained as an information table. In the information table $a_i : U \rightarrow \pi_{a_i}$ for every $a_i \in AT$ where π_{a_i} is the set of all normal possibility distributions over domain $D(a_i)$ of attribute a_i. Value $a_i(o)$ of attribute a_i for object o is denoted by normal possibility distribution $\{(v, \pi_{a_i(o)}(v)) \mid v \in D(a_i) \wedge \pi_{a_i(o)}(v) > 0 \wedge \max_{v \in D(a_i)} \pi_{a_i(o)}(v) = 1\}_p$, where $\pi_{a_i(o)}(v)$ is the possible degree that $a_i(o)$ may be v in domain $D(a_i)$ of attribute a_i.

---

[1] It is possible to use another condition. For example, $a_i(o) \approx a_i(o')$ that means $a_i(o)$ and $a_i(o')$ are similar. In this case, R_{a_i} is reflexive and symmetric, but not transitive.

[2] $\chi_{R_{a_i}}(o, o')$ is an abbreviation of $\chi_{R_{a_i}}((o, o'))$.

3.1   Rough Sets Based on Possible World Semantics

We obtain a set of possible information tables on A, not the whole set of attributes, from the original information table. A possible information table is obtained by replacing the value of each attribute by a possible value for the attribute value. A possible value of $a_i(o)$ is an element in support set $S(a_i(o))(= \{v \mid \pi_{a_i(o)}(v) > 0\})$ of $a_i(o)$. Each possible information table is accompanied with a possible degree that it may be the actual one. For information table T we obtain possibility distribution π_A^T that consists of pairs of a possible information table on A and its possible degree to which it may be the actual one:

$$\pi_A^T = \{(t, \pi(t)) \mid \pi(t) = \min_{o \in U, a_i \in A} \pi_{a_i(o)}(a_i(o)^t)\}_p, \tag{10}$$

where $a_i(o)^t \in S(a_i(o))$ is the value of a_i that o has in possible information table t.

From every possible information table we obtain a pair of approximations, lower and upper approximations, by using formulae shown in Sect. 2.

Example 1. Let information table T be obtained as follows:

INFORMATION TABLE T

U	a_1	a_2
1	$\{(a,1)\}_p$	$\{(w,1),(z,0.6)\}_p$
2	$\{(a,1),(b,0.8)\}_p$	$\{(w,0.4),(x,1)\}_p$
3	$\{(b,1)\}_p$	$\{(x,1)\}_p$

In the information table, $U = \{o_1, o_2, o_3\}$, where the domains of attributes a_1 and a_2 are $\{a, b, c, d\}$ and $\{w, x, y, z\}$, respectively. We have 2 possible information tables on a_1.

2 POSSIBLE INFORMATION TABLES

t_1

U	a_1	a_2
1	a	$\{(w,1),(z,0.6)\}_p$
2	a	$\{(w,0.4),(x,1)\}_p$
3	b	$\{(x,1)\}_p$

t_2

U	a_1	a_2
1	a	$\{(w,1),(z,0.6)\}_p$
2	b	$\{(w,0.4),(x,1)\}_p$
3	b	$\{(x,1)\}_p$

Possibility distribution $\pi_{a_1}^T$, whose elements are pairs of a possible information table on a_1 and its possible degree to which it may be the actual one, is:

$$\pi_{a_1}^T = \{(t_1, 1), (t_2, 0.8)\}_p.$$

By using formulae shown in the previous section, lower and upper approximations of set $\mathcal{O}(= \{o_2, o_3\})$ of objects on a_1 in possible data sets t_1 and t_2 are as follows:

$$\underline{apr}_{a_1}(\mathcal{O})^{t_1} = \{o_3\}, \overline{apr}_{a_1}(\mathcal{O})^{t_1} = \{o_1, o_2, o_3\}.$$

$$\underline{apr}_{a_1}(\mathcal{O})^{t_2} = \{o_2, o_3\}, \overline{apr}_{a_1}(\mathcal{O})^{t_2} = \{o_2, o_3\}.$$

The information that characterizes objects is not complete in an information table with possibilistic information. In such a case, we cannot obtain the actual information from the information table, as was shown by Lipski [6] in databases containing incomplete information. This is also true for applying rough sets to information tables. Thus, we cannot know the actual membership degrees to which an object actually belongs to lower and upper approximations. We can obtain only certain and possible membership degrees, which are lower and upper bounds of the actual one.

Certain membership degrees $C\mu_{\underline{apr}^\circ_{a_i}(\mathcal{O})}(o)$ and $C\mu_{\overline{apr}^\circ_{a_i}(\mathcal{O})}(o)$, to which object o certainly belongs to lower approximation $\underline{apr}^\circ_{a_i}(\mathcal{O})$ and upper one $\overline{apr}^\circ_{a_i}(\mathcal{O})$, respectively, are:

$$C\mu_{\underline{apr}^\circ_{a_i}(\mathcal{O})}(o) = 1 - \max_t\{\pi(t) \mid o \notin \underline{apr}_{a_i}(\mathcal{O})^t\}, \tag{11}$$

$$C\mu_{\overline{apr}^\circ_{a_i}(\mathcal{O})}(o) = 1 - \max_t\{\pi(t) \mid o \notin \overline{apr}_{a_i}(\mathcal{O})^t\}, \tag{12}$$

where $\underline{apr}_{a_i}(\mathcal{O})^t$ and $\overline{apr}_{a_i}(\mathcal{O})^t$ are the lower and upper approximations in possible information table t. Possible membership degrees $P\mu_{\underline{apr}^\circ_{a_i}(\mathcal{O})}(o)$ and $P\mu_{\overline{apr}^\circ_{a_i}(\mathcal{O})}(o)$, to which object o possibly belongs to lower approximation $\underline{apr}^\circ_{a_i}(\mathcal{O})$ and upper one $\overline{apr}^\circ_{a_i}(\mathcal{O})$, respectively, are:

$$P\mu_{\underline{apr}^\circ_{a_i}(\mathcal{O})}(o) = \max_t\{\pi(t) \mid o \in \underline{apr}_{a_i}(\mathcal{O})^t\}, \tag{13}$$

$$P\mu_{\overline{apr}^\circ_{a_i}(\mathcal{O})}(o) = \max_t\{\pi(t) \mid o \in \overline{apr}_{a_i}(\mathcal{O})^t\}. \tag{14}$$

These four approximations have the following properties:
(1) If $C\mu_{\underline{apr}^\circ_{a_i}(\mathcal{O})}(o) > 0$, then $P\mu_{\underline{apr}^\circ_{a_i}(\mathcal{O})}(o) = 1$,
if $C\mu_{\overline{apr}^\circ_{a_i}(\mathcal{O})}(o) > 0$, then $P\mu_{\overline{apr}^\circ_{a_i}(\mathcal{O})}(o) = 1$,
if $P\mu_{\underline{apr}^\circ_{a_i}(\mathcal{O})}(o) < 1$, then $C\mu_{\underline{apr}^\circ_{a_i}(\mathcal{O})}(o) = 0$, and
if $P\mu_{\overline{apr}^\circ_{a_i}(\mathcal{O})}(o) < 1$, then $C\mu_{\overline{apr}^\circ_{a_i}(\mathcal{O})}(o) = 0$.
(2) $\forall o \in U$ $C\mu_{\underline{apr}^\circ_{a_i}(\mathcal{O})}(o) \leq P\mu_{\underline{apr}^\circ_{a_i}(\mathcal{O})}(o) \leq \mu_{\mathcal{O}}(o) \leq C\mu_{\overline{apr}^\circ_{a_i}(\mathcal{O})}(o) \leq P\mu_{\overline{apr}^\circ_{a_i}(\mathcal{O})}(o)$, where $\mu_{\mathcal{O}}(o) = 1$ if $o \in \mathcal{O}$ and $\mu_{\mathcal{O}}(o) = 0$ if $o \notin \mathcal{O}$.

Using these membership degrees, lower and upper approximations are formulated as follows:

$$\underline{apr}^\circ_{a_i}(\mathcal{O}) = \{(o, [C\mu_{\underline{apr}^\circ_{a_i}(\mathcal{O})}(o), P\mu_{\underline{apr}^\circ_{a_i}(\mathcal{O})}(o)]) \mid P\mu_{\underline{apr}^\circ_{a_i}(\mathcal{O})}(o) > 0\}, \tag{15}$$

$$\overline{apr}^\circ_{a_i}(\mathcal{O}) = \{(o, [C\mu_{\overline{apr}^\circ_{a_i}(\mathcal{O})}(o), P\mu_{\overline{apr}^\circ_{a_i}(\mathcal{O})}(o)]) \mid P\mu_{\overline{apr}^\circ_{a_i}(\mathcal{O})}(o) > 0\}. \tag{16}$$

These expressions show that membership degrees to which each object belongs to lower and upper approximations are expressed by interval values.

Example 2. Let us go back to the information table of Example 1. Let a set $\mathcal{O}$ of objects be $\{o_2, o_3\}$. Using (11)–(14), for example, the membership degrees of object o_1 are:

$$C\mu_{\underline{apr}^\circ_{a_1}}(\mathcal{O})(o_1) = 0, P\mu_{\underline{apr}^\circ_{a_1}}(\mathcal{O})(o_1) = 0,$$

$$C\mu_{\overline{apr}^\circ_{a_1}}(\mathcal{O})(o_1) = 0.2, P\mu_{\overline{apr}^\circ_{a_1}}(\mathcal{O})(o_1) = 1.$$

From using formulae (15) and (16),

$$\underline{apr}^\circ_{a_1}(\mathcal{O}) = \{(o_2, [0, 0.8]), (o_3, [1, 1])\},$$
$$\overline{apr}^\circ_{a_1}(\mathcal{O}) = \{(o_1, [0.2, 1]), (o_2, [1, 1]), (o_3, [1, 1])\}.$$

When set $\mathcal{O}$ consists of objects that are specified by attribute a_j, the family of indiscernible classes on a_j are obtained from $\mathcal{O}$ in each possible information table. Four membership degrees are:

$$C\mu_{\underline{apr}^\circ_{a_i}}(\mathcal{O}/a_j)(o) = 1 - \max_t\{\pi(t) \mid o \notin \underline{apr}_{a_i}(\mathcal{O}/a_j)^t\}, \tag{17}$$

$$C\mu_{\overline{apr}^\circ_{a_i}}(\mathcal{O}/a_j)(o) = 1 - \max_t\{\pi(t) \mid o \notin \overline{apr}_{a_i}(\mathcal{O}/a_j)^t\}, \tag{18}$$

$$P\mu_{\underline{apr}^\circ_{a_i}}(\mathcal{O}/a_j)(o) = \max_t\{\pi(t) \mid o \in \underline{apr}_{a_i}(\mathcal{O}/a_j)^t\}, \tag{19}$$

$$P\mu_{\overline{apr}^\circ_{a_i}}(\mathcal{O}/a_j)(o) = \max_t\{\pi(t) \mid o \in \overline{apr}_{a_i}(\mathcal{O}/a_j)^t\}. \tag{20}$$

Four membership degrees also have the following properties:
(1) If $C\mu_{\underline{apr}^\circ_{a_i}}(\mathcal{O}/a_j)(o) > 0$, then $P\mu_{\underline{apr}^\circ_{a_i}}(\mathcal{O}/a_j)(o) = 1$,
if $C\mu_{\overline{apr}^\circ_{a_i}}(\mathcal{O}/a_j)(o) > 0$, then $P\mu_{\overline{apr}^\circ_{a_i}}(\mathcal{O}/a_j)(o) = 1$,
if $P\mu_{\underline{apr}^\circ_{a_i}}(\mathcal{O}/a_j)(o) < 1$, then $C\mu_{\underline{apr}^\circ_{a_i}}(\mathcal{O}/a_j)(o) = 0$, and
if $P\mu_{\overline{apr}^\circ_{a_i}}(\mathcal{O}/a_j)(o) < 1$, then $C\mu_{\overline{apr}^\circ_{a_i}}(\mathcal{O}/a_j)(o) = 0$.
(2) $\forall o \in U$ $C\mu_{\underline{apr}^\circ_{a_i}}(\mathcal{O}/a_j)(o) \le P\mu_{\underline{apr}^\circ_{a_i}}(\mathcal{O}/a_j)(o) \le \mu_{\mathcal{O}}(o) \le C\mu_{\overline{apr}^\circ_{a_i}}(\mathcal{O}/a_j)(o) \le P\mu_{\overline{apr}^\circ_{a_i}}(\mathcal{O}/a_j)(o)$.

Using these membership degrees, lower and upper approximations are:

$$\underline{apr}^\circ_{a_i}(\mathcal{O}/a_j) = \{(o, [C\mu_{\underline{apr}^\circ_{a_i}}(\mathcal{O}/a_j)(o), P\mu_{\underline{apr}^\circ_{a_i}}(\mathcal{O}/a_j)(o)])$$
$$\mid P\mu_{\underline{apr}^\circ_{a_i}}(\mathcal{O}/a_j)(o) > 0\}, \tag{21}$$

$$\overline{apr}^\circ_{a_i}(\mathcal{O}/a_j) = \{(o, [C\mu_{\overline{apr}^\circ_{a_i}}(\mathcal{O}/a_j)(o), P\mu_{\overline{apr}^\circ_{a_i}}(\mathcal{O})(o)/a_j])$$
$$\mid P\mu_{\overline{apr}^\circ_{a_i}}(\mathcal{O}/a_j)(o) > 0\}. \tag{22}$$

Example 3. Let us go back to the information table in Example 1. Let $\mathcal{O}$ be $\{o_2, o_3\}$ that is characterized by values of attribute a_2. Using (17)–(22),

$$\underline{apr}^\circ_{a_1}(\mathcal{O}/a_2) = \{(o_2, [0, 0.8]), (o_3, [0.6, 1])\},$$
$$\overline{apr}^\circ_{a_1}(\mathcal{O}/a_2) = \{(o_1, [0.2, 1]), (o_2, [1, 1]), (o_3, [1, 1])\}.$$

3.2 Rough Sets Based on Indiscernibility Relations

When values that describe objects are expressed by possibility distributions, indiscernibility relations are expressed by using indiscernibility degrees.

The indiscernibility degree of two objects is an interval value, which comes from an extension of cases in Dubois and Prade [4]. Indiscernibility degree $\mu_{R_{a_i}}(o_k, o_l)$ of two objects o_k and o_l for attribute a_i is expressed by $[C\mu_{R_{a_i}}(o_k, o_l), P\mu_{R_{a_i}}(o_k, o_l)]$ whose lower and upper bounds mean certain and possible degrees, respectively. They are calculated by:

$$\mu_{R_{a_i}}(o_k, o_l) = [C\mu_{R_{a_i}}(o_k, o_l), P\mu_{R_{a_i}}(o_k, o_l)], \tag{23}$$

$$C\mu_{R_{a_i}}(o_k, o_l) = \begin{cases} 1 & \text{if } k = l, \\ 1 - \max_{u \neq v} \min(\pi_{a_i(o_k)}(u), \pi_{a_i(o_l)}(v)) & \text{otherwise.} \end{cases} \tag{24}$$

$$P\mu_{R_{a_i}}(o_k, o_l) = \begin{cases} 1 & \text{if } k = l, \\ \max_v \min(\pi_{a_i(o_k)}(v), \pi_{a_i(o_l)}(v)), & \text{otherwise,} \end{cases} \tag{25}$$

These degrees are reflexive and symmetric, but not max-min transitive.

Example 4. Applying formulae (23)–(25) to the information table of Example 1, the indiscernibility relation on a_1 is:

$$\mu_{R_{a_1}}(o_k, o_l) = \begin{pmatrix} [1, 1] & [0.2, 1] & [0, 0] \\ [0.2, 1] & [1, 1] & [0, 0.8] \\ [0, 0] & [0, 0.8] & [1, 1] \end{pmatrix}.$$

Let $\mathcal{O}$ be a set of objects. Certain membership degrees $C\mu_{\underline{apr}^{\bullet}_{a_i}(\mathcal{O})}(o)$ and $C\mu_{\overline{apr}^{\bullet}_{a_i}(\mathcal{O})}(o)$, to which object o certainly belongs to lower approximation $\underline{apr}^{\bullet}_{a_i}(\mathcal{O})$ and upper one $\overline{apr}^{\bullet}_{a_i}(\mathcal{O})$, respectively, are:

$$C\mu_{\underline{apr}^{\bullet}_{a_i}(\mathcal{O})}(o) = \min_{o' \in U} \max(1 - P\mu_{R_{a_i}}(o, o'), \mu_{\mathcal{O}}(o')), \tag{26}$$

$$C\mu_{\overline{apr}^{\bullet}_{a_i}(\mathcal{O})}(o) = \max_{o' \in U} \min(C\mu_{R_{a_i}}(o, o'), \mu_{\mathcal{O}}(o')). \tag{27}$$

Possible membership degrees $P\mu_{\underline{apr}^{\bullet}_{a_i}(\mathcal{O})}(o)$ and $P\mu_{\overline{apr}^{\bullet}_{a_i}(\mathcal{O})}(o)$, to which object o possibly belongs to lower approximation $\underline{apr}^{\bullet}_{a_i}(\mathcal{O})$ and upper one $\overline{apr}^{\bullet}_{a_i}(\mathcal{O})$, respectively, are:

$$P\mu_{\underline{apr}^{\bullet}_{a_i}(\mathcal{O})}(o) = \min_{o' \in U} \max(1 - C\mu_{R_{a_i}}(o, o'), \mu_{\mathcal{O}}(o')), \tag{28}$$

$$P\mu_{\overline{apr}^{\bullet}_{a_i}(\mathcal{O})}(o) = \max_{o' \in U} \min(P\mu_{R_{a_i}}(o, o'), \mu_{\mathcal{O}}(o')). \tag{29}$$

Four membership degrees have the following properties:
(1) If $C\mu_{\underline{apr}^{\bullet}_{a_i}(\mathcal{O})}(o) > 0$, then $P\mu_{\underline{apr}^{\bullet}_{a_i}(\mathcal{O})}(o) = 1$,
if $C\mu_{\overline{apr}^{\bullet}_{a_i}(\mathcal{O})}(o) > 0$, then $P\mu_{\overline{apr}^{\bullet}_{a_i}(\mathcal{O})}(o) = 1$,
if $P\mu_{\underline{apr}^{\bullet}_{a_i}(\mathcal{O})}(o) < 1$, then $C\mu_{\underline{apr}^{\bullet}_{a_i}(\mathcal{O})}(o) = 0$, and
if $P\mu_{\overline{apr}^{\bullet}_{a_i}(\mathcal{O})}(o) < 1$, then $C\mu_{\overline{apr}^{\bullet}_{a_i}(\mathcal{O})}(o) = 0$.
(2) $\forall o \in U$ $C\mu_{\underline{apr}^{\bullet}_{a_i}(\mathcal{O})}(o) \leq P\mu_{\underline{apr}^{\bullet}_{a_i}(\mathcal{O})}(o) \leq \mu_{\mathcal{O}}(o) \leq C\mu_{\overline{apr}^{\bullet}_{a_i}(\mathcal{O})}(o) \leq P\mu_{\overline{apr}^{\bullet}_{a_i}(\mathcal{O})}(o)$.

Using four membership degrees, lower and upper approximations are

$$\underline{apr}^{\bullet}_{a_i}(\mathcal{O}) = \{(o, [C\mu_{\underline{apr}^{\bullet}_{a_i}(\mathcal{O})}(o), P\mu_{\underline{apr}^{\bullet}_{a_i}(\mathcal{O})}(o)]) \mid P\mu_{\underline{apr}^{\bullet}_{a_i}(\mathcal{O})}(o) > 0\}, \quad (30)$$

$$\overline{apr}^{\bullet}_{a_i}(\mathcal{O}) = \{(o, [C\mu_{\overline{apr}^{\bullet}_{a_i}(\mathcal{O})}(o), P\mu_{\overline{apr}^{\bullet}_{a_i}(\mathcal{O})}(o)]) \mid P\mu_{\overline{apr}^{\bullet}_{a_i}(\mathcal{O})}(o) > 0\}. \quad (31)$$

Proposition 1. The lower and upper approximations expressed by (30) and (31) give the same results as ones expressed by (15) and (16); namely,

$$\underline{apr}^{\bullet}_{a_i}(\mathcal{O}) = \underline{apr}^{\circ}_{a_i}(\mathcal{O}) \text{ and } \overline{apr}^{\bullet}_{a_i}(\mathcal{O}) = \overline{apr}^{\circ}_{a_i}(\mathcal{O}).$$

Subsequently, we show membership degrees in the case where both objects used to approximate and objects approximated are characterized by attributes with possibilistic information. Certain membership degrees $C\mu_{\underline{apr}^{\bullet}_{a_i}(\mathcal{O}/a_j)}(o)$ and $C\mu_{\overline{apr}^{\bullet}_{a_i}(\mathcal{O}/a_j)}(o)$, to which object o certainly belongs to lower approximation $\underline{apr}^{\bullet}_{a_i}(\mathcal{O}/a_j)$ and upper one $\overline{apr}^{\bullet}_{a_i}(\mathcal{O}/a_j)$, respectively, are:

$$C\mu_{\underline{apr}^{\bullet}_{a_i}(\mathcal{O}/a_j)}(o) = \max_{o'' \in \mathcal{O}} \min_{o' \in U} \max(1 - P\mu_{R_{a_i}}(o, o'),$$
$$\min(C\mu_{R_{a_j}}(o', o''), \mu_{\mathcal{O}}(o'))), \quad (32)$$

$$C\mu_{\overline{apr}^{\bullet}_{a_i}(\mathcal{O}/a_j)}(o) = \max_{o'' \in \mathcal{O}} \max_{o' \in U} \min(C\mu_{R_{a_i}}(o, o'), C\mu_{a_j}(o', o''), \mu_{\mathcal{O}}(o')). \quad (33)$$

Possible membership degrees $P\mu_{\underline{apr}^{\bullet}_{a_i}(\mathcal{O}/a_j)}(o)$ and $P\mu_{\overline{apr}^{\bullet}_{a_i}(\mathcal{O}/a_j)}(o)$, to which object o possibly belongs to lower approximation $\underline{apr}^{\bullet}_{a_i}(\mathcal{O}/a_j)$ and upper one $\overline{apr}^{\bullet}_{a_i}(\mathcal{O}/a_j)$, respectively, are:

$$P\mu_{\underline{apr}^{\bullet}_{a_i}(\mathcal{O}/a_j)}(o) = \max_{o'' \in \mathcal{O}} \min_{o' \in U} \max(1 - C\mu_{R_{a_i}}(o, o'),$$
$$\min(P\mu_{R_{a_j}}(o', o''), \mu_{\mathcal{O}}(o'))), \quad (34)$$

$$P\mu_{\overline{apr}^{\bullet}_{a_i}(\mathcal{O}/a_j)}(o) = \max_{o'' \in \mathcal{O}} \max_{o' \in U} \min(P\mu_{R_{a_i}}(o, o'), P\mu_{a_j}(o', o''), \mu_{\mathcal{O}}(o')). \quad (35)$$

Four membership degrees also have the following properties:
(1) If $C\mu_{\underline{apr}^{\bullet}_{a_i}(\mathcal{O}/a_j)}(o) > 0$, then $P\mu_{\underline{apr}^{\bullet}_{a_i}(\mathcal{O}/a_j)}(o) = 1$,
if $C\mu_{\overline{apr}^{\bullet}_{a_i}(\mathcal{O}/a_j)}(o) > 0$, then $P\mu_{\overline{apr}^{\bullet}_{a_i}(\mathcal{O}/a_j)}(o) = 1$,
if $P\mu_{\underline{apr}^{\bullet}_{a_i}(\mathcal{O}/a_j)}(o) < 1$, then $C\mu_{\underline{apr}^{\bullet}_{a_i}(\mathcal{O}/a_j)}(o) = 0$, and
if $P\mu_{\overline{apr}^{\bullet}_{a_i}(\mathcal{O}/a_j)}(o) < 1$, then $C\mu_{\overline{apr}^{\bullet}_{a_i}(\mathcal{O}/a_j)}(o) = 0$.
(2) $\forall o \in U$ $C\mu_{\underline{apr}^{\bullet}_{a_i}(\mathcal{O}/a_j)}(o) \le P\mu_{\underline{apr}^{\bullet}_{a_i}(\mathcal{O}/a_j)}(o) \le \mu_{\mathcal{O}}(o) \le C\mu_{\overline{apr}^{\bullet}_{a_i}(\mathcal{O}/a_j)}(o) \le P\mu_{\overline{apr}^{\bullet}_{a_i}(\mathcal{O}/a_j)}(o)$.

Using four membership degrees, lower and upper approximations are:

$$\underline{apr}^{\bullet}_{a_i}(\mathcal{O}/a_j) = \{(o, [C\mu_{\underline{apr}^{\bullet}_{a_i}(\mathcal{O}/a_j)}(o), P\mu_{\underline{apr}^{\bullet}_{a_i}(\mathcal{O}/a_j)}(o)])$$
$$\mid P\mu_{\underline{apr}^{\bullet}_{a_i}(\mathcal{O}/a_j)}(o) > 0\}, \quad (36)$$

$$\overline{apr}^{\bullet}_{a_i}(\mathcal{O}/a_j) = \{(o, [C\mu_{\overline{apr}^{\bullet}_{a_i}(\mathcal{O}/a_j)}(o), P\mu_{\overline{apr}^{\bullet}_{a_i}(\mathcal{O})}(o)/a_j])$$
$$\mid P\mu_{\overline{apr}^{\bullet}_{a_i}(\mathcal{O}/a_j)}(o) > 0\}. \quad (37)$$

Proposition 2. The lower and upper approximations expressed by (36) and (37) give the same results as ones expressed by (21) and (22); namely,

$$\underline{apr}^{\bullet}_{a_i}(\mathcal{O}/a_j) = \underline{apr}^{\diamond}_{a_i}(\mathcal{O}/a_j) \text{ and } \overline{apr}^{\bullet}_{a_i}(\mathcal{O}/a_j) = \overline{apr}^{\diamond}_{a_i}(\mathcal{O}/a_j).$$

Propositions 1 and 2 justify our extension.

4 Conclusions

We have described an extended version of rough sets in order to deal with data sets where values are expressed by possibility distributions. The extended version is based on directly using indiscernibility relations.

First, we have given rough sets directly using indiscernibility relations under complete information. Second, we have described rough sets under possibilistic information on the basis of possible world semantics, as was done by Lipski in databases with incomplete information, in order to use the rough sets as a correctness criterion of approaches extended under possibilistic information. The set of possible information tables is obtained with possible degrees from the original information table. The lower and upper bounds of membership degrees to which an object belongs to lower and upper approximations are obtained by using the possible degrees that the possible information tables have. As a result, the membership degrees to which an object belongs to lower and upper approximations are expressed by an interval value. Third, we have extended the approach based on indiscernibility relations, which was proposed by Dubois and Prade. Lower and upper approximations obtained from the extension give the same results as ones from possible world semantics. This justifies our extension.

References

1. Abiteboul, S., Hull, R., Vianu, V.: Foundations of Databases. Addison-Wesley Publishing Company, Boston (1995)
2. Bosc, P., Duval, L., Pivert, O.: An initial approach to the evaluation of possibilistic queries addressed to possibilistic databases. Fuzzy Sets Syst. **140**, 151–166 (2003)
3. Couso, I., Dubois, F.: Rough sets, coverings and incomplete information. Fundamenta Informaticae **108**(3–4), 223–347 (2011)
4. Dubois, D., Prade, H.: Rough fuzzy sets and fuzzy rough sets. Int. J. Gen. Syst. **17**, 191–209 (1990)
5. Imielinski, T., Lipski, W.: Incomplete information in relational databases. J. ACM **31**, 761–791 (1984)
6. Lipski, W.: On semantics issues connected with incomplete information databases. ACM Trans. Database Syst. **4**, 262–296 (1979)
7. Nakata, M., Sakai, H.: Lower and upper approximations in data tables containing possibilistic information. In: Peters, J.F., Skowron, A., Marek, V.W., Orłowska, E., Słowiński, R., Ziarko, W. (eds.) Transactions on Rough Sets VII. LNCS, vol. 4400, pp. 170–189. Springer, Heidelberg (2007). doi:10.1007/978-3-540-71663-1_11

8. Nakata, M., Sakai, H.: Rule induction based on rough sets from information tables containing possibilistic information. In: Proceedings of the 2013 Joint IFSA World Congress and NAFIPS Annual Meeting (IFSA/NAFIPS), pp. 91–96. IEEE Press (2013)
9. Nakata, M., Sakai, H.: Twofold rough approximations under incomplete information. Int. J. Gen. Syst. **42**, 546–571 (2013)
10. Nakata, M., Sakai, H.: An approach based on rough sets to possibilistic information. In: Laurent, A., Strauss, O., Bouchon-Meunier, B., Yager, R.R. (eds.) Information Processing and Management of Uncertainty in Knowledge-Based Systems. CCIS, vol. 444, pp. 61–70. Springer, Cham (2014)
11. Nakata, M., Sakai, H.: Rule induction based on rough sets from possibilistic information under Lipski's approach. In: Proceedings of the 2014 IEEE International Conference on Granular Computing (GrC), pp. 218–223. IEEE Computer Society (2014)
12. Pawlak, Z.: Rough sets. Int. J. Comput. Inf. Sci. **11**, 341–356 (1982)
13. Pawlak, Z., Skowron, A.: Rudiments of rough sets. Inf. Sci. **177**(2007), 3–27 (2007)
14. Pawlak, Z., Skowron, A.: Rough sets: some extensions. Inf. Sci. **177**(2007), 28–40 (2007)
15. Pawlak, Z., Skowron, A.: Rough sets and Boolean reasoning. Inf. Sci. **177**, 41–73 (2007)
16. Słowiński, R., Stefanowski, J.: Rough classification in incomplete information systems. Math. Comput. Modell. **12**, 1347–1357 (1989)
17. Zadeh, L.A.: The concept of a linguistic variable and its application to approximate reasoning I. Inf. Sci. **8**, 199–249 (1975)
18. Zadeh, L.A.: The concept of a linguistic variable and its application to approximate reasoning II. Inf. Sci. **8**, 301–357 (1975)
19. Zadeh, L.A.: The concept of a linguistic variable and its application to approximate reasoning III. Inf. Sci. **9**, 43–80 (1975)
20. Zadeh, L.A.: Fuzzy sets as a basis for a theory of possibility. Fuzzy Sets Syst. **1**, 3–28 (1978)
21. Zimányi, E., Pirotte, A.: Imperfect information in relational databases. In: Motro, A., Smets, P. (eds.) Uncertainty Management in Information Systems: From Needs to Solutions, pp. 35-87. Kluwer Academic Publishers (1997)

Matrix-Based Rough Set Approach for Dynamic Probabilistic Set-Valued Information Systems

Yanyong Huang[1], Tianrui Li[1(⊠)], Chuan Luo[2], and Shi-jinn Horng[1]

[1] School of Information Science and Technology, Southwest Jiaotong University,
Chengdu 611756, China
yyhswjtu@163.com, trli@swjtu.edu.cn, horngsj@yahoo.com.tw
[2] College of Computer Science, Sichuan University, Chengdu 610056, China
cluo@scu.edu.cn

Abstract. Set-valued information systems (SvIS), in which the attribute values are set-valued, are important types of data representation with uncertain and missing information. However, all previous investigations in rough set community do not consider the attribute values with probability distribution in SvIS, which may be impractical in many real applications. This paper introduces probabilistic set-valued information systems (PSvIS) and presents an extended variable precision rough sets (VPRS) approach based on λ-tolerance relation for PSvIS. Furthermore, due to the dynamic variation of attributes in PSvIS, viz., the addition and deletion of attributes, we present a matrix characterization of the proposed VPRS model and discuss some related properties. Then incremental approaches for maintaining rough approximations based on matrix operations are presented, which can effectively accelerate the updating of rough approximations in dynamic PSvIS.

Keywords: Information systems · Rough sets · Incremental learning · Matrix

1 Introduction

Rough sets is an efficient mathematical tool for discovering knowledge from the information systems characterized by imprecise, uncertain and vague information [1]. It has been widely applied in different kinds of domains including data mining, machine learning and decision making [2–4].

In order to characterize the multi-values of attributes or fill the missing data by existing information, single-valued information systems are extended to Set-valued Information Systems (SvIS) by replacing the single value with a set value. For example, a language-ability test information system characterized the language ability of candidates through utilizing the set of languages rather than a single language [5]. Qian et al. extended SvIS to Set-valued Ordered Information Systems (SvOIS) by considering the attributes with preference-ordered domains [6]. For instance, the better language capacity of each individual indicate

© Springer International Publishing AG 2016
V. Flores et al. (Eds.): IJCRS 2016, LNAI 9920, pp. 197–206, 2016.
DOI: 10.1007/978-3-319-47160-0_18

the more set values in terms of the conjunctive semantic meaning in aforementioned example. Although SvIS or SvOIS have been successfully explored in a variety of real applications, it can not directly deal with the set-value of objects with probability distribution, which always exist in real-life situations. The information systems with this kind of data are suggested as Probabilistic Set-valued Information Systems (PSvIS) in our study. To our best knowledge, there is no research focuses on discovering knowledge from PSvIS. Hence, the purpose of the paper is to present an extended rough set model for PSvIS. We first present the λ-tolerance relation based on Bhattacharyya distance for concept approximations in PSvIS. Moreover, considering the datasets may exist noisy information in real-applications, variable precision rough sets (VPRS), presented by Ziarko in 1993, can efficiently dealing the scenario by introducing inclusion degree for controlling the degree of misclassification [7]. In this paper, we extend VPRS model by introducing λ-tolerance relation for efficiently mining knowledge from PSvIS.

Another important issue driving this research is that information systems evolve over time. In dynamic information systems, new data may be added or discarded data may be excluded, which will result in the dynamical change of knowledge. Incremental learning is an efficient updating knowledge method by utilizing the accumulated knowledge over time. Recently, many incremental learning algorithms have been developed to deal with the evolving data in rough set theory. Li et al. presented an incremental updating approximations in terms of the characteristic relation under the dynamic attribute generalization [8]. Liang et al. investigated a dynamic attribute reduction approach based on information entropy when a clump of new objects are appended to an information system [9]. Yang et al. discussed a dynamic maintenance multi-granulation approximations considering the addition of granular structures [10]. Luo et al. presented two different updating approximations strategies when the variation of attribute values in SvIS [11]. Since matrix operation has advantages of intuitional representing and simple computation, it has played a key role in rough set-based data analysis. Zhang et al. presented four cut matrices for incremental computing approximations in SvIS [12]. Wang et al. presented two Boolean characteristic matrices for representing covering approximations [13]. Huang et al. presented two matrix operators for representing rough fuzzy approximations and developed a dynamic matrix-based method for computing approximations under the addition of objects and attributes simultaneously [14]. In this paper, we present a matrix-based representation of rough approximations in PSvIS and some incremental mechanisms based on matrix for calculating approximations with the dynamic change of attributes.

The rest of the paper is organized as follows: Sect. 2 reviews the basic concepts of VPRS, and presents the definition of PSvIS and an extension of VPRS model in terms of λ-tolerance relation. Section 3 presents the matrix-based representations of approximations and discusses related properties. Section 4 investigates some incremental mechanisms for computing rough approximations when adding and removing attributes. Section 5 concludes the research work of this paper and our future research.

2   Extended Variable Precision Rough Sets in PSvIS

In this section, we firstly review several basic concepts and preliminaries of VPRS. Then we introduce the concept of PSvIS and extended VPRS model based on λ-tolerance relation for constructing concept approximations in PSvIS.

Definition 1. *[7] Let $S = \{U, AT = A \bigcup D, V, f\}$ be an information system, where U is a non-empty finite set of objects, called the universe; AT is a non-empty finite set of attributes including condition attributes A and decision attributes D; V is the domain of attributes AT; f is an information function from $U \times AT$ to V such that $f : U \times AT \to V$ is a single-valued mapping. Let β denote the proportion of correct classification and $\beta \in (0.5, 1]$. $\forall X \subseteq U$ and $B \subseteq A$, the lower and upper approximations in VPRS are defined as follows.*

$$\underline{R_B}^{\beta}(X) = \{x | P(X | [x]_B) \geq \beta\} \tag{1}$$

$$\overline{R_B}^{\beta}(X) = \{x | P(X | [x]_B) > 1 - \beta\} \tag{2}$$

where $P(X | [x]_B) = \frac{|X \cap [x]_B|}{|[x]_B|}$ and $[x]_B = \{y | (x, y) \in R_B\}$ is the equivalence class determined by the equivalence relation $R_B = \{(x, y) \in U \times U | f(x, b) = f(y, b), \forall b \in B\}$.

Definition 2. *A PSvIS is a sextuple $(U, AT = A \bigcup D, V = V_A \bigcup V_D, f, \sigma, P)$, where $U = \{x_i | i \in \{1, 2, \cdots, n\}\}$ is a non-empty finite set of objects, called the universe. A is a non-empty finite set of condition attributes. D denotes the decision attributes and $A \bigcap D = \emptyset$. $V = V_A \bigcup V_D$ is the domain of attributes set AT, where V_A denotes the domain of condition attribute values, V_D denotes the domain of decision attribute values. $f : U \times A \to 2^{V_A}$ is a set-valued mapping and $f : U \times D \to V_D$ is a single-valued mapping. σ is sigma field of Borel sets in V_A, and P is the probability distribution defined on σ.*

Example 1. Table 1 shows a PSvIS $S = (U, AT = A \bigcup D, V = V_A \bigcup V_D, f, \sigma, P)$ about the election information, where $U = \{x_i | i \in \{1, 2, \cdots, 14\}\}$ denotes fourteen different districts, $A = \{$Economic construction, Social construction, Cultural construction$\}$ indicates three different measure indexes of the candidate about governing capability, D is a decision attribute, $V_A = \{$Dissatisfaction, Neutrality Satisfaction$\} = \{-1, 0, 1\}$ and $V_D = \{$Yes, No$\} = \{$Y, N$\}$. In Table 1, $f(x_1, a_1) = \frac{\{-1, 0, 1\}}{(0.23, 0.45, 0.32)}$ denotes the probability distribution of the set value $\{-1, 0, 1\}$. Other notations are similar.

According to the traditional tolerance relation $T_a = \{(x, y) | f(x, a) \bigcap f(y, a) \neq \emptyset, a \in A\}$ [5], the objects x_1 and x_7 are in the same equivalence class in terms of conditional attribute a_1. But the distance of probability distributions between

Table 1. A probabilistic set-valued information system

U	a_1	a_2	a_3	D	U	a_1	a_2	a_3	D
x_1	{-1,0,1} (0.23,0.45,0.32)	{-1,0,1} (0.10,0.40,0.50)	{0,1} (0.40,0.60)	Y	x_8	{-1,0,1} (0.81,0.14,0.05)	{-1,0,1} (0.03,0.77,0.20)	{-1,0,1} (0.82,0.11,0.07)	Y
x_2	{-1,0,1} (0.20,0.43,0.37)	{-1,0,1} (0.12,0.38,0.5)	{0,1} (0.43,0.57)	Y	x_9	{0,1} (0.32,0.68)	{0,1} (0.33,0.67)	{-1,0,1} (0.44,0.32,0.24)	N
x_3	{-1,0,1} (0.25,0.42,0.33)	{-1,0,1} (0.13,0.39,0.48)	{0,1} (0.44,0.56)	N	x_{10}	{0,1} (0.34,0.66)	{0,1} (0.34,0.66)	{-1,0,1} (0.43,0.33,0.24)	N
x_4	{-1,0,1} (0.24,0.44,0.32)	{-1,0,1} (0.12,0.41,0.47)	{-1,0,1} (0.38,0.52,0.10)	Y	x_{11}	{-1,0,1} (0.82,0.12,0.06)	{-1,0,1} (0.02,0.78,0.20)	{-1,0} (0.90,0.10)	N
x_5	{-1,0,1} (0.22,0.41,0.37)	{-1,0,1} (0.11,0.42,0.47)	{-1,0,1} (0.41,0.53,0.06)	N	x_{12}	{0,1} (0.34,0.66)	{0,1} (0.35,0.65)	{1} (1)	Y
x_6	{-1,0,1} (0.24,0.42,0.34)	{-1,0,1} (0.10,0.44,0.46)	{-1,0,1} (0.40,0.52,0.08)	Y	x_{13}	{-1,0} (0.80,0.20)	{-1,0,1} (0.24,0.29,0.47)	{-1,0,1} (0.47,0.52,0.01)	Y
x_7	{-1,0,1} (0.82,0.12,0.06)	{-1,0,1} (0.02,0.76,0.22)	{-1,0,1} (0.81,0.10,0.09)	Y	x_{14}	{1} (1)	{-1,0,1} (0.30,0.54,0.16)	{0,1} (0.30,0.70)	N

the set-value of x_1 and x_7 is big enough to distinguish. Hence the classical tolerance relation in SvIS can not be used in PSvIS directly. To more reasonably characterize the relation of objects in PSvIS, we present the λ-tolerance relation based on Bhattacharyya distance for PSvIS.

Definition 3. *Let $S = (U, AT = A \bigcup D, V = V_A \bigcup V_D, f, \sigma, P)$ be a PSvIS and the threshold $\lambda \geq 0$. The $\lambda-$tolerance relation BD_a^λ with respect to the attribute $a \in A$ can be defined as follows.*

$$BD_a^\lambda = \{(x,y) \in U \times U | BD_a(x,y) \leq \lambda\} \tag{3}$$

where $BD_a(x,y) = -\ln \left(\sum_{k=1}^{K} \sqrt{p(x_k)p(y_k)} \right)$ is the Bhattacharyya distance which measures the similarity of two discrete probability distributions [15], and $p(x_k)$ and $p(y_k)$ denote the probability distributions of x and y under the attribute a, respectively. Then $\forall B \subseteq A$, the $\lambda-$tolerance relation BD_B^λ is defined by

$$BD_B^\lambda = \{(x,y) \in U \times U | BD_b(x,y) \leq \lambda, \forall b \in B\} = \bigcap_{b \in B} BD_b^\lambda \tag{4}$$

Property 1. λ-tolerance relation is reflexive and symmetric, but not transitive.

Property 2. Let $B_1 \subseteq B_2 \subseteq A$, then we have $BD_{B_2}^\lambda \subseteq BD_{B_1}^\lambda$.

Property 3. For $\lambda_1 \leq \lambda_2$, We have $BD_B^{\lambda_1} \subseteq BD_B^{\lambda_2}$.

Note that PSvIS degenerate to disjunctive SvIS when the probability distribution of objects are $p(x_k) = 1$ and $p(x_i) = 0(i \neq k)(i, k \in \{1, 2, \ldots, K\})$. Furthermore, the λ-tolerance relation will become the traditional tolerance relation in SvIS when $\lambda = 0$.

As we know, classical rough sets is not robust for dealing with PSvIS including some noisy information. However, VPRS could efficiently handle this scenario. In what follows, we extend VPRS through λ-tolerance relation for efficiently characterizing lower and upper approximations in PSvIS.

Definition 4. *Given a PSvIS* $S = (U, AT = A \bigcup D, V = V_A \bigcup V_D, f, \sigma, P)$, $\forall X \subseteq U$ *and* $B \subseteq A$, *then* β *lower and upper approximations with regard to the* λ*-tolerance relation* BD_B^λ *are defined as follows, respectively.*

$$\underline{R_B}^{(\beta,\lambda)}(X) = \{x | P(X | [x]_{BD_B^\lambda}) \geq \beta\} \tag{5}$$

$$\overline{R_B}^{(\beta,\lambda)}(X) = \{x | P(X | [x]_{BD_B^\lambda}) > 1 - \beta\} \tag{6}$$

where $[x]_{BD_B^\lambda} = \{y | (x, y) \in BD_B^\lambda\}$, $\beta \in (0.5, 1]$ *and* $\lambda \geq 0$.

Then the universe U can be partitioned into three regions in terms of lower and upper approximations as follows.

The positive region: $POS_B^{(\beta,\lambda)}(X) = \underline{R_B}^{(\beta,\lambda)}(X)$

The negative region: $NEG_B^{(\beta,\lambda)}(X) = U - \overline{R_B}^{(\beta,\lambda)}(X)$

The boundary region: $BND_B^{(\beta,\lambda)}(X) = \overline{R_B}^{(\beta,\lambda)}(X) - \underline{R_B}^{(\beta,\lambda)}(X) \tag{7}$

Example 2. (Continuation of Example 1) Let $\beta = 0.6$, $\lambda = 0.55$, $B = \{a_1, a_2\}$ and $X = \{x_1, x_2, x_4, x_7, x_8, x_{12}, x_{13}\}$. Then we can compute the lower and upper approximations in terms of the λ-tolerance relation BD_B^λ as follows:

$$\begin{cases} \underline{R_B}^{(\beta,\lambda)}(X) = \{x_7, x_8, x_{11}, x_{13}\}; \\ \overline{R_B}^{(\beta,\lambda)}(X) = \{x_1, x_2, x_3, x_4, x_5, x_6, x_7, x_8, x_{11}, x_{13}\}. \end{cases}$$

3   Matrix-Based Representation of Approximations in PSvIS

In this section, we present a matrix-based representation for depicting the lower and upper approximations in PSvIS intuitively. Then we propose an effective method for calculating approximations by several matrix operators.

Definition 5. *Let* $S = (U, AT = A \bigcup D, V = V_A \bigcup V_D, f, \sigma, P)$ *be a PSvIS, where* $U = \{x_1, x_2, \cdots, x_n\}$. *Let* BD_B^λ *be a* λ-*tolerance relation on* U, *where* $B \subseteq A$. *Then the* λ-*tolerance relation matrix* $M^{BD_B^\lambda} = (m_{ij})_{n \times n}$ *with regard to* BD_B^λ *is defined as follows:*

$$m_{ij} = \begin{cases} 1, & (x_i, x_j) \in BD_B^\lambda \\ 0, & otherwise \end{cases} \tag{8}$$

Property 4. The λ-tolerance relation matrix $M^{BD_B^\lambda}$ is symmetric, and $m_{ii} = 1 (i = 1, \ldots, n)$.

Definition 6. $\forall X \subseteq U$, *the characteristic function* $G(X)$ *with respect to* X *in the PSvIS is defined as:*

$$G(X) = \left(g_1, g_2, \ldots, g_n \right)^T, where \; g_i = \begin{cases} 1, x_i \in X \\ 0, x_i \notin X \end{cases} \tag{9}$$

where "T" denotes the transpose operation.

Definition 7. *Let* $Y = (y_1, y_2, \cdots, y_n)^T$ *be a column vector. The piecewise function* $l_\beta(Y)$ *in terms of* β *is defined as follows:*

$$l_\beta(Y) = \begin{pmatrix} l_\beta(y_1) \\ l_\beta(y_2) \\ \cdots \\ l_\beta(y_n) \end{pmatrix}, where\ l_\beta(y_i) = \begin{cases} 1, & y_i \geq \beta \\ 0, & 1 - \beta < y_i < \beta \\ -1, & y_i \leq 1 - \beta \end{cases} \quad (10)$$

where $\beta \in (0.5, 1]$.

Property 5. *Let* $Q_1 \triangleq M^{BD_B^\lambda} \times G(X)$ *and* $Q_2 \triangleq M^{BD_B^\lambda} \times I$, *where "×" represents matrix multiplication and* $I = (1, 1, \cdots, 1)^T$. *Then we have* $Q_1(i) = |[x_i]_{BD_B^\lambda} \bigcap X|$ *and* $Q_2(i) = |[x_i]_{BD_B^\lambda}|$, *where* $Q_1(i)$ *and* $Q_2(i)$ *denotes the* i*th element of* Q_1 *and* Q_2, *respectively.*

Theorem 1. *Given a PSvIS* $S = (U, AT = A \bigcup D, V = V_A \bigcup V_D, f, \sigma, P)$, $U = \{x_1, x_2, \cdots, x_n\}$. $\forall X \subseteq U$, *let* $Q_3 \triangleq Q_1/.Q_2$, *where "/." denotes matrix dot divide. Then the positive, negative and boundary regions with respect to* $B \subseteq A$ *can be obtained from* $l_\beta(Q_3)$ *as follows:*

$$POS_B^{(\beta,\lambda)}(X) = \{x_i | l_\beta(Q_3(i)) = 1\} \quad (11)$$

$$NEG_B^{(\beta,\lambda)}(X) = \{x_i | l_\beta(Q_3(i)) = -1\} \quad (12)$$

$$BND_B^{(\beta,\lambda)}(X) = \{x_i | l_\beta(Q_3(i)) = 0\} \quad (13)$$

Then the lower approximation $\underline{R_B}^{(\beta,\lambda)}(X) = POS_B^{(\beta,\lambda)}(X)$ and the upper approximation $\overline{R_B}^{(\beta,\lambda)}(X) = U - NEG_B^{(\beta,\lambda)}(X)$.

Example 3. (Continuation of Example 2) Let the parameters λ and β are the same with Example 2. We can compute the λ-tolerance relation matrix $M^{BD_B^\lambda}$ according to Definition 5 and the characteristic function $G(X)$ according to Definition 6. Then $Q_1 = (3, 3, 3, 3, 3, 3, 2, 2, 1, 1, 2, 1, 1, 0)^T$, $Q_2 = (6, 6, 6, 6, 6, 6, 3, 3, 3, 3, 3, 3, 1, 1)^T$ and $Q_3 = Q_1/.Q_2 = (\frac{1}{2}, \frac{1}{2}, \frac{1}{2}, \frac{1}{2}, \frac{1}{2}, \frac{1}{2}, \frac{2}{3}, \frac{2}{3}, \frac{1}{3}, \frac{1}{3}, \frac{2}{3}, \frac{1}{3}, 1, 0)^T$ according to Property 5. Finally we have $l_\beta(Q_3) = (0, 0, 0, 0, 0, 0, 1, 1, -1, -1, 1, -1, 1, -1)^T$, $POS_B^{(\beta,\lambda)}(X) = \{x_7, x_8, x_{11}, x_{13}\}$, $NEG_B^{(\beta,\lambda)}(X) = \{x_9, x_{10}, x_{12}, x_{14}\}$ and $BND_B^{(\beta,\lambda)}(X) = \{x_1, x_2, x_3, x_4, x_5, x_6\}$ according to Theorem 1. Furthermore we have the lower approximation $\underline{R_B}^{(\beta,\lambda)}(X) = \{x_7, x_8, x_{11}, x_{13}\}$ and upper approximation $\overline{R_B}^{(\beta,\lambda)}(X) = \{x_1, x_2, x_3, x_4, x_5, x_6, x_7, x_8, x_{11}, x_{13}\}$.

4   Incremental Updating Approximations Based on Matrix in PSvIS

In dynamic information systems, attribute information will be changed due to new attributes become available or outdated attributes are excluded. In this section, we present incremental approaches based on matrix for computing rough approximations under the addition and deletion of attributes in dynamic PSvIS.

4.1 Dynamic Updating Approximations with the Addition of Attributes

In this section, considering new attributes are added in a PSvIS from time t to $t+1$, the incremental mechanisms based on region relation matrix for dynamically maintaining approximations are proposed in PSvIS.

Let $S = (U, AT = A^t \bigcup D, V = V_{A^t} \bigcup V_D, f, \sigma, P)$ be a PSvIS at time t, where $U = \{x_1, x_2, \cdots, x_n\}$. At time $t+1$, the new attribute set ΔA is added to the condition attribute set A^t, i.e., the PSvIS S is changed as $S^{t+1} = (U, AT^{t+1} = A^{t+1} \bigcup D, V^{t+1} = V_A^{t+1} \bigcup V_D, f^{t+1}, \sigma^{t+1}, P^{t+1})$, where $A^{t+1} = A^t \bigcup \Delta A$. $\forall X \subseteq U$. Let $POS_{A^t}(X) = \{x_i | [x_i]_{BD^\lambda_{A^t}} \subseteq X\}$, $NEG_{A^t}(X) = \{x_i | [x_i]_{BD^\lambda_{A^t}} \bigcap X = \phi\}$ and $BND_{A^t}(X) = \{x_i | [x_i]_{BD^\lambda_{A^t}} \bigcap X \neq \phi$ and $[x_i]_{BD^\lambda_{A^t}} \nsubseteq X\}$.

Definition 8. *Let $M^{BD^\lambda_{A^t}}$ denote the λ-tolerance relation matrix with respect to A^t in the PSvIS S at time t. The region relation matrix $M_R = (m_{ij}^R)$ is defined as follows, where "R" denotes "POS" or "NEG" or "BND".*

$$m_{ij}^R = \begin{cases} 1, & (x_i, x_j) \in BD^\lambda_{A^t}, x_i \in R_{A^t}(X), x_j \in U \\ 0, & otherwise \end{cases} \tag{14}$$

Theorem 2. *Given a PSvIS $S = (U, AT = A^t \bigcup D, V = V_{A^t} \bigcup V_D, f, \sigma, P)$, $U = \{x_1, x_2, \cdots, x_n\}$. Let $M_{R'} = (m_{ij}^{R'})$ denote the region relation matrix when the attribute set ΔA is added to A^t at time $t+1$, where "R" indicates "POS" or "NEG" or "BND", respectively. Then it can be updated by the following mechanism.*

(1) If $m_{ij}^R = 0$, then $m_{ij}^{R'} = m_{ij}^R$;

(2) If $m_{ij}^R = 1$ and $x_i \in [x_j]_{BD^\lambda_{\Delta A}}$ then $m_{ij}^{R'} = m_{ij}^R$;

(3) If $m_{ij}^R = 1$ and $x_i \notin [x_j]_{BD^\lambda_{\Delta A}}$ then $m_{ij}^{R'} = 0$.

Theorem 3. *Let $Q_1' = M_{BND'} \times G(X)$, $Q_2' = M_{BND'} \times I$, $Q_3' = Q_1'/.Q_2'$ and $l_\beta' = l_\beta(Q_3')$. Then $\forall X \subseteq U$, the positive region $POS^{(\beta,\lambda)}_{A^{t+1}}(X)$ and negative region $NEG^{(\beta,\lambda)}_{A^{t+1}}(X)$ at time $t+1$ are updated as follows.*

(1) $POS^{(\beta,\lambda)}_{A^{t+1}}(X) = POS_{A^t}(X) \bigcup \{x_i | l_\beta'(Q_3'(i)) = 1\}$;

(2) $NEG^{(\beta,\lambda)}_{A^{t+1}}(X) = NEG_{A^t}(X) \bigcup \{x_i | l_\beta'(Q_3'(i)) = -1\}$.

Then we have the lower approximation $\underline{R_{A^{t+1}}}^{(\beta,\lambda)}(X) = POS^{(\beta,\lambda)}_{A^{t+1}}(X)$ and upper approximations $\overline{R_{A^{t+1}}}^{(\beta,\lambda)}(X) = U - NEG^{(\beta,\lambda)}_{A^{t+1}}(X)$.

Example 4. (Continuation of Example,3) Let $A^t = B = \{a_1, a_2\}$ at time t and $\Delta A = \{a_3\}$ is appended to A^t at time $t+1$. Firstly, we compute $POS_{A^t}(X) = \{x_{13}\}$, $NEG_{A^t}(X) = \{x_{14}\}$ and $BND_{A^t}(X) = \{x_1, x_2, x_3, x_4, x_5, x_6, x_7, x_8, x_9, x_{10}, x_{11}, x_{12}\}$ based on Pawlak rough sets. Then according to Definition 8 and Theorem 2, we have

$$
M_{BND} = \begin{pmatrix}
1&1&1&1&1&1&0&0&0&0&0&0&0&0 \\
1&1&1&1&1&1&0&0&0&0&0&0&0&0 \\
1&1&1&1&1&1&0&0&0&0&0&0&0&0 \\
1&1&1&1&1&1&0&0&0&0&0&0&0&0 \\
1&1&1&1&1&1&0&0&0&0&0&0&0&0 \\
1&1&1&1&1&1&0&0&0&0&0&0&0&0 \\
0&0&0&0&0&0&1&1&0&0&1&0&0&0 \\
0&0&0&0&0&0&1&1&0&0&1&0&0&0 \\
0&0&0&0&0&0&0&0&1&1&0&1&0&0 \\
0&0&0&0&0&0&0&1&1&0&1&0&0 \\
0&0&0&0&0&0&1&1&0&0&1&0&0&0 \\
0&0&0&0&0&0&0&1&1&0&1&0&0
\end{pmatrix}
\quad
M_{BND'} = \begin{pmatrix}
1&1&1&0&0&0&0&0&0&0&0&0&0&0 \\
1&1&1&0&0&0&0&0&0&0&0&0&0&0 \\
1&1&1&0&0&0&0&0&0&0&0&0&0&0 \\
0&0&0&1&1&1&0&0&0&0&0&0&0&0 \\
0&0&0&1&1&1&0&0&0&0&0&0&0&0 \\
0&0&0&1&1&1&0&0&0&0&0&0&0&0 \\
0&0&0&0&0&0&1&1&0&0&0&0&0&0 \\
0&0&0&0&0&0&1&1&0&0&0&0&0&0 \\
0&0&0&0&0&0&0&0&1&1&0&0&0&0 \\
0&0&0&0&0&0&0&0&1&1&0&0&0&0 \\
0&0&0&0&0&0&0&0&0&0&1&0&0&0 \\
0&0&0&0&0&0&0&0&0&0&0&1&0&0
\end{pmatrix}
$$

Then according to Theorem 3, we have $Q'_1 = (2,2,2,1,1,1,2,2,0,0,0,1)^T$, $Q'_2 = (3,3,3,3,3,3,2,2,2,2,1,1)^T$ and $Q'_3 = (\frac{2}{3},\frac{2}{3},\frac{2}{3},\frac{1}{3},\frac{1}{3},\frac{1}{3},1,1,0,0,0,1)^T$. Finally, $POS^{(\beta,\lambda)}_{A^{t+1}}(X) = \{x_{13}\}\bigcup\{x_1,x_2,x_3,x_7,x_8,x_{12}\} = \{x_1,x_2,x_3,x_7,x_8, x_{12},x_{13}\}$ and $NEG^{(\beta,\lambda)}_{A^{t+1}}(X) = \{x_{14}\}\bigcup\{x_4,x_5,x_6,x_9,x_{10},x_{11}\} = \{x_4,x_5, x_6,x_9,x_{10},x_{11},x_{14}\}$. Furthermore, the lower and upper approximations can be updated by Eq. (7).

Obviously, the proposed method of incremental updating approximations can reduce the computing overhead by utilizing the accumulated information of rough approximations.

4.2 Dynamic Updating Approximations with the Deletion of Attributes

In this section, we present an incremental approach for updating approximations based on region relation matrices and accumulated approximations' information when redundant attributes are deleted from a PSvIS.

Let $S = (U, AT = A^t\bigcup D, V = V_A\bigcup V_D, f, \sigma, P)$ be a PSvIS at time t, where $A^t = A\bigcup \Delta A$. At time $t + 1$, the attribute set ΔA is deleted from the conditional attribute set A^t, i.e., the PSvIS S is changed as $S^{t+1} = (U, AT^{t+1} = A^{t+1}\bigcup D, V^{t+1} = V_A^{t+1}\bigcup V_D, f^{t+1}, \sigma^{t+1}, P^{t+1})$, where $A^{t+1} = A$.

Theorem 4. Let $M_R = (m^R_{ij})$ and $M_{R'} = (m^{R'}_{ij})$ denote the region relation matrix at time t and $t + 1$, respectively. When the attribute set ΔA is deleted from A^t, the region relation matrix $M_{R'} = (m^{R'}_{ij})$ is updated as follows:

(1) If $m^R_{ij} = 1$, then $m^{R'}_{ij} = m^R_{ij}$;

(2) If $m^R_{ij} = 0$ and $x_i \notin [x_j]_{BD^\lambda_A}$ then $m^{R'}_{ij} = m^R_{ij}$;

(3) If $m^R_{ij} = 0$ and $x_i \in [x_j]_{BD^\lambda_A}$ then $m^{R'}_{ij} = 1$.

When removing the attribute set ΔA from A^t, the positive and negative regions of Pawlak rough sets will decrease, but the boundary region will increase. To dynamically updating the positive and negative regions of extended VPRS in PSvIS, let $POS_{A^t}(BND) \triangleq \{x_i | P(X|[x_i]_{BD^\lambda_{A^{t+1}}}) \geq \beta, x_i \in BND_{A^t}(X)\}$ and $NEG_{A^t}(BND) \triangleq \{x_i | P(X|[x_i]_{BD^\lambda_{A^{t+1}}}) \leq 1 - \beta, x_i \in BND_{A^t}(X)\}$. Then for any $X \subseteq U$, let $Q'_{1_R} = M_{R'} \times G(X)$, $Q'_{2_R} = M_{R'} \times I$, $Q'_{3_R} = Q'_{1_R}/.Q'_{2_R}$ and

$l_\beta^R = l_\beta(Q'_{3_R})$, where "R" denotes "POS" and "NEG", respectively. Finally, the incremental mechanisms are listed as follows:

Theorem 5. $\forall X \subseteq U$, the positive region $POS_{A^{t+1}}^{(\beta,\lambda)}(X)$ and the negative region $NEG_{A^{t+1}}^{(\beta,\lambda)}(X)$ at time $t+1$ are updated as follows.

(1) $POS_{A^{t+1}}^{(\beta,\lambda)}(X) = POS_{A^t}(BND) \bigcup \{x_i | l_\beta^{POS}(i) = 1\} \bigcup \{x_i | l_\beta^{NEG}(i) = 1\};$

(2) $NEG_{A^{t+1}}^{(\beta,\lambda)}(X) = NEG_{A^t}(BND) \bigcup \{x_i | l_\beta^{NEG}(i) = -1\} \bigcup \{x_i | l_\beta^{POS}(i) = -1\}.$

Example 5. (Continuation of Example 1) Let $\Delta A = \{a_3\}$ be deleted from $A^t = \{a_1, a_2, a_3\}$ at time $t+1$, then we have $A^{t+1} = \{a_1, a_2\}$. And set $\beta = 0.6$ and $\lambda = 0.55$. Then $POS_{A^t}(X) = \{x_7, x_8, x_{12}, x_{13}\}$, $NEG_{A^t}(X) = \{x_9, x_{10}, x_{11}, x_{14}\}$ and $BND_{A^t}(X) = \{x_1, x_2, x_3, x_4, x_5, x_6\}$. Firstly, we compute $POS_{A^t}(BND) = \emptyset$ and $NEG_{A^t}(BND) = \emptyset$. Then $M_{POS'}$ and $M_{NEG'}$ can be obtained as follows according to Theorem 4.

$$M_{POS} = \begin{pmatrix} 0\,0\,0\,0\,0\,0\,1\,1\,0\,0\,0\,0\,0\,0 \\ 0\,0\,0\,0\,0\,0\,1\,1\,0\,0\,0\,0\,0\,0 \\ 0\,0\,0\,0\,0\,0\,0\,0\,0\,0\,0\,1\,0\,0 \\ 0\,0\,0\,0\,0\,0\,0\,0\,0\,0\,0\,0\,1\,0 \end{pmatrix} \quad M_{NEG'} = \begin{pmatrix} 0\,0\,0\,0\,0\,0\,0\,0\,1\,1\,0\,1\,0\,0 \\ 0\,0\,0\,0\,0\,0\,0\,0\,1\,1\,0\,1\,0\,0 \\ 0\,0\,0\,0\,0\,0\,1\,1\,0\,0\,1\,0\,0\,0 \\ 0\,0\,0\,0\,0\,0\,0\,0\,0\,0\,0\,0\,0\,1 \end{pmatrix}$$

Based on Definition 7, we can compute $l_\beta^{POS} = (1, 1, -1, 1)^T$ and $l_\beta^{NEG} = (-1, -1, 1, -1)^T$. Finally, we have $POS_{A^{t+1}}^{(\beta,\lambda)}(X) = \{x_7, x_8, , x_{11}, x_{13}\}$ and $NEG_{A^{t+1}}^{(\beta,\lambda)}(X) = \{x_9, x_{10}, , x_{12}, x_{14}\}$ according to Theorem 5. Furthermore, the lower and upper approximations can be obtained by Eq. (7).

It is evident that the incremental strategies for computing approximations can reduce the computational cost by partly updating the region relation matrices rather than updating the whole relation matrice when attributes are removed.

5 Conclusions

Previous studies on SvIS assume that the attribute values are set-valued, which ignore the set-values with probability distribution. In this paper, we extended SvIS to PSvIS by introducing probability distribution for more precisely characterizing the relationship of objects. Then we proposed an extended VPRS model based on λ-tolerance relation in PSvIS and investigated the matrix representations of approximations through matrix operators and piecewise function. Furthermore, since the traditional static method could not efficiently tackle the attributes evolution, some incremental mechanisms for maintaining approximations are presented based on the accumulated region relation matrices and approximations information. Future work will extend our method to attribute reduction and rule induction in dynamic PSvIS.

Acknowledgements. This work is supported by the National Science Foundation of China (Nos. 61573292, 61572406).

References

1. Pawlak, Z.: Rough sets. Int. J. Comput. Inf. Sci. **11**, 341–356 (1982)
2. Maji, P.: A rough hypercuboid approach for feature selection in approximation spaces. IEEE Trans. Knowl. Data Eng. **26**, 16–29 (2014)
3. Kotlowski, W., Slowinski, R.: On nonparametric ordinal classification with monotonicity constraints. IEEE Trans. Knowl. Data Eng. **25**, 2576–2589 (2013)
4. Yao, Y.: Rough sets and three-way decisions. In: Ciucci, D., Wang, G., Mitra, S., Wu, W.-Z. (eds.) RSKT 2015. LNCS (LNAI), vol. 9436, pp. 62–73. Springer, Heidelberg (2015). doi:10.1007/978-3-319-25754-9_6
5. Guan, Y.Y., Wang, H.K.: Set-valued information systems. Inf. Sci. **176**, 2507–2525 (2006)
6. Qian, Y.H., Dang, C.Y., Liang, J.Y., Tang, D.W.: Set-valued ordered information systems. Inf. Sci. **179**, 2809–2832 (2009)
7. Ziarko, W.: Variable precision rough set model. J. Comput. Syst. Sci. **46**, 39–59 (1993)
8. Li, T.R., Ruan, D., Geert, W., Song, J., Xu, Y.: A rough sets based characteristic relation approach for dynamic attribute generalization in data mining. Knowl. Based Syst. **20**, 485–494 (2007)
9. Liang, J.Y., Wang, F., Dang, C.Y., Qian, Y.H.: A group incremental approach to feature selection applying rough set technique. IEEE Trans. Knowl. Data Eng. **26**, 294–308 (2014)
10. Yang, X.B., Qi, Y., Yu, H.L., Song, X.N., Yang, J.Y.: Updating multigranulation rough approximations with increasing of granular structures. Knowl. Based Syst. **64**, 59–69 (2014)
11. Luo, C., Li, T.R., Chen, H.M., Lu, L.X.: Fast algorithms for computing rough approximations in set-valued decision systems while updating criteria values. Inf. Sci. **299**, 221–242 (2015)
12. Zhang, J.B., Li, T.R., Ruan, D., Liu, D.: Rough sets based matrix approaches with dynamic attribute variation in set-valued information systems. Int. J. Approximate Reasoning **53**, 620–635 (2012)
13. Wang, S.P., Zhu, W., Zhu, Q.X., Min, F.: Characteristic matrix of covering and its application to boolean matrix decomposition. Inf. Sci. **263**, 186–197 (2014)
14. Huang, Y., Li, T., Horng, S.: Dynamic maintenance of rough fuzzy approximations with the variation of objects and attributes. In: Yao, Y., Hu, Q., Yu, H., Grzymala-Busse, J.W. (eds.) RSFDGrC 2015. LNCS (LNAI), vol. 9437, pp. 173–184. Springer, Heidelberg (2015). doi:10.1007/978-3-319-25783-9_16
15. Kailath, T.: The divergence and bhattacharyya distance measures in signal selection. IEEE Trans. Commun. Technol. **15**, 52–60 (1967)

Rough Sets and Three-way Decisions

Variance Based Determination of Three-Way Decisions Using Probabilistic Rough Sets

Nouman Azam[1]([✉]) and Jing Tao Yao[2]

[1] National University of Computer and Emerging Sciences, Peshawar, Pakistan
nouman.azam@nu.edu.pk
[2] Department of Computer Science, University of Regina, Regina S4S 0A2, Canada
jtyao@cs.uregina.ca

Abstract. The probabilistic rough sets is an important generalization of rough sets where a pair of thresholds is used to form new rough regions. The pair of thresholds controls different quality related criteria such as classification accuracy, precision, uncertainty, costs and risks of rough sets based three-way decision making. In this article, we introduce variance based criteria for determining the thresholds including within region variance, between region variance and ratio of the two variances. In particular, we examine the variance or spread in conditional probabilities of equivalence classes contained in different probabilistic regions. We also show that the determination of thresholds may be considered based on optimization of the proposed criteria.

1 Introduction

In rough set theory, the representation of an undefinable set with three regions has led to the introduction of the theory of three-way decisions [15,16]. A fundamental notion in the theory adopted from rough sets is the division of the universal set into three pair-wise disjoint regions [16]. The theory however goes beyond rough sets by introducing its own notions and ideas. There is a growing interest in the theory from both theoretical and application aspects [6,7,9,11,19].

The probabilistic rough sets is a well studied and useful generalization of rough sets for constructing three-way decisions [13]. It divides the universe into three regions based on a pair of thresholds (α, β). One of the key issues in the application of probabilistic rough sets is the determination of these thresholds [14]. This issue has been addressed by employing different notions, measures and approaches. Some notable attempts in this regards are decision-theoretic rough sets [18], variable precision rough sets [20], game-theoretic rough sets [3, 12], information-theoretic rough sets [2], Bayesain rough sets [10], optimization viewpoint [4] and multilevel approach [5]. In this article, we propose variance based criteria for determining the thresholds.

In contrast to other uncertainty measures, such as entropy or gini index where the focus is on analysis of the probabilities of outcomes, the variance highlights the spread among the outcomes or observations. We consider two types of variances, i.e., within region variance and between region variance. The within

© Springer International Publishing AG 2016
V. Flores et al. (Eds.): IJCRS 2016, LNAI 9920, pp. 209–218, 2016.
DOI: 10.1007/978-3-319-47160-0_19

region variance is defined as a the weighted sum of variances of the three regions. The variance of a certain region is defined in terms of conditional probabilities of equivalence classes included in that region. In particular, it measures the level of difference between individual conditional probabilities and the mean conditional probability of equivalence classes in the region. A high variance for a region suggest a higher variation in conditional probabilities and low variance suggest a lesser variation in conditional probabilities in that region. In general, changing the thresholds to decrease the variance of a certain region may lead to an increase in another. The thresholds may therefore be selected based on optimizing or minimizing the overall within region variance. The between region variance is defined as sum of the weighted differences of the region conditional probability means from the overall conditional probability mean. Again changing the thresholds to increase the difference for one region may lead a decrease in another. It is reasonable to select thresholds that will maximize the overall between region variance. Finally, the ratio of the two criteria may also be used in optimizing the thresholds. The proposed criteria may be incorporated with some machine learning method to obtain effective thresholds.

2   Background

2.1   Three-Way Decisions

The theory of three-way decisions emerged from the need to explain and interpret the three regions in probabilistic rough sets. Recent developments suggest that rough sets is only one way for constructing and inducing three-way decisions. A general theory of three-way decisions was introduced with the aim to extend three-way decisions beyond rough sets and initiate an independent study for investigating three-way decision making and its different aspects [15,16].

The essential idea of three-way decision is to divide the universe into three pair-wise disjoint regions, such as the positive, negative and boundary regions. This division is referred to as trisecting and the resultant three regions is referred to as tripartition. Effective strategies are designed for processing the three regions in order to obtain three-way decisions [16]. Generally speaking, the division of the universe of objects is based on an evaluation function and a pair of thresholds [17].

The evaluation function evaluates each object in the universe and assigns to it an evaluation value. The three regions are created by considering the evaluation of objects whose evaluation values are greater than or equal to an upper threshold, the objects whose evaluation values are lesser than or equal to some lower threshold, and the objects whose evaluation values are between the two thresholds. The three regions based on this division are referred to as positive, negative, boundary or high, low, medium or right, left, middle. How to interpret and determine the evaluation functions and thresholds are fundamental issues in three-way decisions. Other issues of three-way decisions are generation of predictive rules from the three regions for making decisions on new objects, descriptive rules for describing the three regions and design of strategies and actions corresponding to the three regions [15, 16]. By considering different interpretations and realizations of these issues, we may have different three-way decision making models and approaches. We focus on three-way decisions with probabilistic rough sets.

2.2   Probabilistic Rough Sets

The general form of probabilistic rough sets were introduced in the context of decision-theoretic rough sets [18]. According to the general form, probabilistic lower and upper approximations for a concept C are defined based on thresholds (α, β) as [13, 14],

$$\underline{apr}_{(\alpha,\beta)}(C) = \{x \in U \mid P(C|[x]) \geq \alpha\}, \tag{1}$$

$$\overline{apr}_{(\alpha,\beta)}(C) = \{x \in U \mid P(C|[x]) > \beta\}, \tag{2}$$

where U is the universal set of nonempty and finite set of objects and $P(C|[x])$ is the conditional probability of a concept C with an equivalence class $[x]$ provided that an object $x \in [x]$. The conditional probability quantifies the evaluation of an object x to be in C. The three rough set regions are defined based on the lower and upper approximations as,

$$\text{POS}_{(\alpha,\beta)}(C) = \{x \in U | P(C|[x]) \geq \alpha\}, \tag{3}$$

$$\text{NEG}_{(\alpha,\beta)}(C) = \{x \in U | P(C|[x]) \leq \beta\}, \tag{4}$$

$$\text{BND}_{(\alpha,\beta)}(C) = \{x \in U | \beta < P(C|[x]) < \alpha\}. \tag{5}$$

where $\text{POS}_{(\alpha,\beta)}(C), \text{NEG}_{(\alpha,\beta)}(C)$ and $\text{BND}_{(\alpha,\beta)}(C)$ are referred to as positive, negative and boundary regions, respectively and are defined and controlled on thresholds (α, β). The probabilistic rough sets can be explained based on the notions of three-way decisions. The conditional probability $P(C|[x])$ serve as an evaluation function which returns an evaluation value for each object. The division or trisection of the universe is based on conditional probability and thresholds (α, β). The three regions i.e., positive, negative and boundary are processed to induce decisions of acceptance, rejection and deferment, respectively.

3   Variance Based Three-Way Approaches

Statistical measures such as mean, median, percentile and standard deviation were recently being studied to provide an interpretation of three-way decisions [17]. We look at the same statistical measures but from the viewpoint of optimization.

3.1 Basic Formulation

Consider thresholds (α, β) that lead to the three regions based on conditional probability $P(C|[x])$ in the probabilistic rough set framework. The mean conditional probability of the three regions are computed as,

$$\mu_{\mathrm{POS}_{(\alpha,\beta)}}(C) = \sum_{\forall [x] \in \mathrm{POS}_{(\alpha,\beta)}(C)} \frac{P(C|[x]) \times P([x])}{P(\mathrm{POS}_{(\alpha,\beta)}(C))}, \tag{6}$$

$$\mu_{\mathrm{NEG}_{(\alpha,\beta)}}(C) = \sum_{\forall [x] \in \mathrm{NEG}_{(\alpha,\beta)}(C)} \frac{P(C|[x]) \times P([x])}{P(\mathrm{NEG}_{(\alpha,\beta)}(C))}, \tag{7}$$

$$\mu_{\mathrm{BND}_{(\alpha,\beta)}}(C) = \sum_{\forall [x] \in \mathrm{BND}_{(\alpha,\beta)}(C)} \frac{P(C|[x]) \times P([x])}{P(\mathrm{BND}_{(\alpha,\beta)}(C))}. \tag{8}$$

The mean conditional probability of a region is an average conditional probability value for equivalence classes included in that region. For instance, $\mu_{\mathrm{POS}_{(\alpha,\beta)}}(C) = 0.8$ will mean that for all the equivalence classes in the positive region based on Eq. (3), the mean conditional probability is 0.8. The probability of a certain region, say positive region, is computed as,

$$P(\mathrm{POS}_{(\alpha,\beta)}(C)) = \frac{|\mathrm{POS}_{(\alpha,\beta)}(C)|}{|U|}. \tag{9}$$

the probabilities for the other regions are similarly computed. The probability of an equivalence class $[x]$ is $P([x]) = \frac{|[x]|}{|U|}$. The mean conditional probability of all equivalence classes is computed as,

$$\mu = \sum_{\forall [x] \in U} P(C|[x]) \times P([x]). \tag{10}$$

The variance of a certain region will reflect the spread of conditional probabilities of equivalence classes within that region. The variances of the three regions are given by,

$$\sigma^2_{\mathrm{POS}_{(\alpha,\beta)}}(C) = \sum_{\forall [x] \in \mathrm{POS}_{(\alpha,\beta)}(C)} \frac{(P(C|[x] - \mu_{\mathrm{POS}_{(\alpha,\beta)}}(C))^2 \times P([x])}{P(\mathrm{POS}_{(\alpha,\beta)}(C))} \tag{11}$$

$$\sigma^2_{\mathrm{NEG}_{(\alpha,\beta)}}(C) = \sum_{\forall [x] \in \mathrm{NEG}_{(\alpha,\beta)}(C)} \frac{(P(C|[x] - \mu_{\mathrm{NEG}_{(\alpha,\beta)}}(C))^2 \times P([x])}{P(\mathrm{NEG}_{(\alpha,\beta)}(C))} \tag{12}$$

$$\sigma^2_{\mathrm{BND}_{(\alpha,\beta)}}(C) = \sum_{\forall [x] \in \mathrm{BND}_{(\alpha,\beta)}(C)} \frac{(P(C|[x] - \mu_{\mathrm{BND}_{(\alpha,\beta)}}(C))^2 \times P([x])}{P(\mathrm{BND}_{(\alpha,\beta)}(C))} \tag{13}$$

Ideally, we would like to represent each region using lesser variation in conditional probabilities. In the next section, we explain how the region means and variances can be used to define criteria for optimizing thresholds.

3.2   Variance Based Optimization Approaches

According to Deng and Yao [2], optimization is an important approach for determining thresholds. In particular, the determination of thresholds may be formulated as an optimization of different properties of the three region such as classification accuracy, precision, uncertainty, costs and risks [2]. Consider $Q(\alpha, \beta)$ as some measure representing a certain aspect of the quality of the thresholds of the three regions. In some cases, the measure $Q(\alpha, \beta)$ may consist of $Q_{\text{POS}}(\alpha, \beta), Q_{\text{NEG}}(\alpha, \beta)$ and $Q_{\text{BND}}(\alpha, \beta)$ of the measure $Q(\alpha, \beta)$ representing the quality of the positive, negative, and boundary regions, respectively. The determination of thresholds (α, β) may be realized as the minimization or maximization of the overall quality of the three regions, which is defined as [2],

$$Q(\alpha, \beta) = w_1 Q_{\text{POS}}(\alpha, \beta) + w_2 Q_{\text{NEG}}(\alpha, \beta) + w_3 Q_{\text{BND}}(\alpha, \beta) \tag{14}$$

where w_1, w_2 and w_3 are weights associated with different regions. More formally, the determination of thresholds is approached as the following optimization problem [2].

$$\arg \min_{(\alpha, \beta)} Q(\alpha, \beta) \quad \text{or} \quad \arg \max_{(\alpha, \beta)} Q(\alpha, \beta) \tag{15}$$

We consider variance based criteria for determining the thresholds. The first criterion we consider is the within region variance. It is based on the weighted sum of the region based variances (Eqs. (11)–(13)). The region variances of the three regions are given by,

$$\sigma^2_{W_{\text{POS}}}(\alpha, \beta) = \sigma^2_{\text{POS}_{(\alpha, \beta)}(C)}, \tag{16}$$

$$\sigma^2_{W_{\text{NEG}}}(\alpha, \beta) = \sigma^2_{\text{NEG}_{(\alpha, \beta)}(C)}, \tag{17}$$

$$\sigma^2_{W_{\text{BND}}}(\alpha, \beta) = \sigma^2_{\text{BND}_{(\alpha, \beta)}(C)}. \tag{18}$$

where the additional notations are being used for the sake of being consistent with Eq. (15). The overall within region variance is the sum of the weighted region variances, i.e.,

$$\sigma^2_W(\alpha, \beta) = P(\text{POS}_{(\alpha, \beta)}(C)) \times \sigma^2_{W_{\text{POS}}}(\alpha, \beta) + P(\text{NEG}_{(\alpha, \beta)}(C)) \times \sigma^2_{W_{\text{NEG}}}(\alpha, \beta)$$
$$+ P(\text{BND}_{(\alpha, \beta)}(C)) \times \sigma^2_{W_{\text{BND}}}(\alpha, \beta), \tag{19}$$

where the probabilities of the three regions are used as weights w_1, w_2 and w_3 in Eq. (15). Equation (19) measures the overall spread of the conditional probabilities within each region. Minimizing this will lead to compact regions containing equivalence classes having lesser variation in conditional probability values. This leads to following optimization criterion, i.e.,

$$\arg \min_{(\alpha, \beta)} \sigma^2_B(\alpha, \beta). \tag{20}$$

The second criterion we consider is the between region variance. It is based on the differences of region means (Eqs. (6)–(8)) from the overall μ (Eq. (10)). The difference of the region means are computed as,

$$\sigma^2_{B_{POS}}(\alpha, \beta) = (\mu_{POS_{(\alpha,\beta)}(C)} - \mu)^2,$$
$$\sigma^2_{B_{NEG}}(\alpha, \beta) = (\mu_{NEG_{(\alpha,\beta)}(C)} - \mu)^2,$$
$$\sigma^2_{B_{BND}}(\alpha, \beta) = (\mu_{BND_{(\alpha,\beta)}(C)} - \mu)^2. \tag{21}$$

The overall between region mean is the sum of the weighted differences, i.e.,

$$\sigma^2_B(\alpha, \beta) = P(POS_{(\alpha,\beta)}(C)) \times \sigma^2_{B_{POS}}(\alpha, \beta) + P(NEG_{(\alpha,\beta)}(C)) \times \sigma^2_{B_{NEG}}(\alpha, \beta)$$
$$+ P(BND_{(\alpha,\beta)}(C)) \times \sigma^2_{B_{BND}}(\alpha, \beta). \tag{22}$$

Again the probabilities of the three regions are used as weights w_1, w_2 and w_3 in Eq. (15). Equation (22) reflects the overall spread of the region means with respect to the mean of entire population (or global mean). Maximizing this will result in well separated and distinguishable regions in conditional probability. This leads to the following optimization criterion,

$$\arg \max_{(\alpha,\beta)} \sigma^2_B(\alpha, \beta). \tag{23}$$

Finally, we may combine the within region variance and between region variance to approach the threshold determination,

$$\arg \max_{(\alpha,\beta)} \frac{\sigma^2_B(\alpha, \beta)}{\sigma^2_W(\alpha, \beta)}, \tag{24}$$

Equation (24) is a known discriminant criterion that is frequently used in discriminant analysis. The underlying conjecture in these approaches is that well thresholded regions would be separated in conditional probabilities. In other words, the thresholds providing the best separation of regions in conditional probabilities would be the best thresholds. Finally, one may employ a learning algorithm that will search the space of possible thresholds to obtain a pair of optimal thresholds based on any criteria in Eqs. (20) and (23) or (24).

3.3   Analyzing Variances in the Probabilistic Rough Sets

An important observation from the previous section is that the within region and between region variance depend on the mean and variance of conditional probabilities (of equivalence classes include in) the three regions. Since the regions are determined by thresholds (α, β), they also effect the mean and variance of the regions. Considering the typical condition $0 \leq \beta < 0.5 \leq \alpha \leq 1$, the division between the positive-boundary is controlled by threshold α and the division between negative-boundary is controlled by β. The mean and variance of the positive region therefore depend on the threshold α and the mean and variance of the negative region depend on the threshold β. The mean and variance of the boundary depend on both the thresholds. In order to look at how the thresholds effect the within region and between region variances, we consider two extreme settings of thresholds. The first setting is given by $(\alpha, \beta) = (1, 0)$ (also known as the Pawlak model) which typically has large boundary size. The second setting is given by $(\alpha, \beta) = (0.5, 0.5)$ (also known as two-way decision model) which has minimum boundary size [1].

In the first threshold setting, all the equivalence classes in the Pawlak positive region have the same conditional probability of 1. This leads to a zero variance for the positive region. It may be confirmed based on Eq. (11) by considering $\mu_{POS_{(1,0)}(C)} = 1$ and $P(C|[x]) = 1$. In the same way, all the equivalence classes in the Pawlak negative region have the same conditional probability of 0 which also leads to a zero variance for the negative region. Although the Pawlak positive and negative regions have minimum region variances, we may not have overall minimum within region variance due to large boundary. In the second case of threshold setting, where as opposed to Pawlak model we have minimum size of the boundary region thereby leading in minimum variance for boundary. The overall variance however may not be necessarily minimum for the model. A probabilistic model may be defined based on thresholds aiming for minimizing the overall within region variance.

We now examine the between region variance for the two extreme settings of thresholds. For the first threshold setting, i.e., $(\alpha, \beta) = (1, 0)$, the mean conditional probability of the Pawlak positive region according to Eq. (6) is 1 and the mean conditional probability of the Pawlak negative region according to Eq. (7) is 0. The Pawlak positive and negative regions may provide maximum between region variance since the values of the means are at extreme ends of 1 and 0, respectively. However the overall between region variance may not be necessarily optimal due large boundary. For the second threshold setting, again we may not have very effective between region separation due to large positive and negative regions. The thresholds may be configured based on maximizing the overall between region variance. One may also simultaneously consider the optimization of the two criteria as suggested in Eq. (24).

Table 1. Probabilistic information of a concept C

	$[x]_1$	$[x]_2$	$[x]_3$	$[x]_4$	$[x]_5$	$[x]_6$	$[x]_7$
$P([x]_i)$	0.02	0.06	0.08	0.12	0.15	0.06	0.07
$Pr(C|[x]_i)$	0.0	0.0	0.1	0.15	0.2	0.3	0.5
	$[x]_8$	$[x]_9$	$[x]_{10}$	$[x]_{11}$	$[x]_{12}$	$[x]_{13}$	
$P([x]_i)$	0.02	0.1	0.08	0.09	0.06	0.09	
$Pr(C|[x]_i)$	0.7	0.8	0.85	0.9	1.0	1.0	

4 An Example

We illustrate the main ideas of the proposed variance based approaches using an example from [2]. The example is based on Table 1 which contains summarized probabilistic information about a concept C with respect to 13 equivalence classes. The equivalence classes are arranged based on their increasing conditional probability. The possible values for the thresholds (α, β) are represented by conditional probabilities. Considering the typical condition $0 \leq \beta < 0.5 \leq \alpha \leq 1$, we have the following possible domains for the thresholds,

$$D_\alpha = \{0.5, 0.7, 0.8, 0.85, 0.9, 1.0\}, \quad D_\beta = \{0.0, 0.1, 0.15, 0.2, 0.3\}. \quad (25)$$

For the pair of thresholds $(\alpha, \beta) = (1, 0)$, the positive, negative and boundary regions according to Eqs. (3)–(5) are given by $\text{POS}_{(\alpha,\beta)}(C) = \bigcup\{[x]_{12}, [x]_{13}\}$, $\text{NEG}_{(\alpha,\beta)}(C) = \bigcup\{[x]_1, [x]_2\}$ and $\text{BND}_{(\alpha,\beta)}(C) = \bigcup\{[x]_3, \ldots, [x]_{11}\}$. The probability of the three regions are computed as, $\text{POS}_{(\alpha,\beta)}(C) = \sum_{i=12}^{13}[x]_i = 0.06 + 0.09 = 0.15$, $\text{NEG}_{(\alpha,\beta)}(C) = \sum_{i=1}^{2}[x]_i, = 0.02 + 0.06 = 0.08$, and $\text{BND}_{(\alpha,\beta)}(C) = \sum_{i=3}^{11}[x]_i = 0.08 + 0.12 + 0.15 + 0.06 + 0.07 + 0.02 + 0.1 + 0.08 + 0.09 = 0.77$.

The mean conditional probability of the three regions according to Eqs. (6)–(8), are computed as,

$$\mu_{\text{POS}_{(\alpha,\beta)}}(C) = \frac{1 \times 0.06 + 1 \times 0.09}{0.15} = 1.0, \quad (26)$$

$$\mu_{\text{NEG}_{(\alpha,\beta)}}(C) = \frac{0.0 \times 0.02 + 0.0 \times 0.06}{0.08} = 0.0, \quad (27)$$

$$\mu_{\text{BND}_{(\alpha,\beta)}}(C) = \frac{0.1 \times 0.08 + \ldots + 0.9 \times 0.09}{0.77} = 0.457, \quad (28)$$

and the overall mean according to Eq. (10) is given by,

$$\mu = 0.0 \times 0.6 + \ldots + 0.09 \times 0.9 = 0.5020. \quad (29)$$

Table 2. Between region variances based on different thresholds

		α					
		0.5	**0.7**	0.8	0.85	0.9	1.0
β	0.0	0.1205	0.1238	0.1203	0.0961	0.0783	0.0589
	0.1	0.1208	0.1249	0.1219	0.1009	0.0860	0.0699
	0.15	0.1206	0.1260	0.1238	0.1081	0.0978	0.0869
	0.2	0.1199	**0.1279**	0.1272	0.1211	0.1177	0.1139
	0.3	0.1184	0.1277	0.1279	0.1261	0.1252	0.1238

The variances of the three regions according to Eqs. (11)–(13) are,

$$\sigma^2_{\text{POS}_{(\alpha,\beta)}(C)} = \frac{(1-1)^2 \times 0.06 + (1-1)^2 \times 0.09}{0.15} = 0.0 \tag{30}$$

$$\sigma^2_{\text{NEG}_{(\alpha,\beta)}(C)} = \frac{(0-0)^2 \times 0.06 + (0-0)^2 \times 0.02}{0.08} = 0.0 \tag{31}$$

$$\sigma^2_{\text{BND}_{(\alpha,\beta)}(C)} = \frac{(0.1-0.457)^2 \times 0.08 + \ldots + (0.9-0.457)^2 \times 0.09}{0.77} = 0.098 \tag{32}$$

Based on the region means, overall mean and region variances, the between region variance, within region variance are determined as,

$$\sigma^2_B(\alpha, \beta) = 0.15 \times (1.0 - 0.5020)^2 + 0.08 \times (0.0 - 0.5020)^2$$
$$+ \ 0.77 \times (0.45 - 0.5020)^2 = 0.0589. \tag{33}$$

$$\sigma^2_W(\alpha, \beta) = 0.15 \times 0.0 + 0.08 \times 0.0 + 0.77 \times 0.0987 = 0.076. \tag{34}$$

By using the same procedure, we may compute the between and within region variances for all possible pair of thresholds given in Eq. (25). Table 2 shows all the thresholds and the corresponding between region variances. The maximum between region variance is shown by the cell with bold fonts. In this example, the maximum value is against the thresholds $(\alpha, \beta) = (0.7, 0.2)$. One may repeat the same for the other two approaches to obtain thresholds.

Please note that due to extensive computations, the listing of all possible thresholds may not be possible in real applications. We may combine the proposed approaches with some learning methods that will search the space of possible threshold values by utilizing the between region and within region variances as search heuristics. Recently, the gradient descent or genetic algorithms are being used for this purpose [2,8].

5 Conclusion

The three regions and the implied three-way decisions in probabilistic rough sets are controlled by a pair of thresholds. We proposed variance based criteria for determining these thresholds. More specifically, we examined the variance or spread in conditional probabilities of equivalence classes contained in different regions. Three criteria are being introduced including within region variance, between region variance and the ratio of the two. We show that the determination of thresholds can be approached by optimizing these criteria. The relationship with Pawlak model is also discussed. An example is included for demonstrating the determination of thresholds based on the proposed criteria.

As part of future work, the proposed criteria may be included with searching algorithms to learn optimal thresholds. Moreover, the comparison with existing criteria may also provide useful insights.

Acknowledgment. This work was partially supported by a Discovery Grant from NSERC Canada and a Startup Research Grant Program from HEC Pakistan.

References

1. Azam, N., Yao, J.T.: Analyzing uncertainties of probabilistic rough set regions with game-theoretic rough sets. Int. J. Approximate Reasoning **55**(1), 142–155 (2014)
2. Deng, X.F., Yao, Y.Y.: A multifaceted analysis of probabilistic three-way decisions. Fundamenta Informaticae **132**, 291–313 (2014)
3. Herbert, J.P., Yao, J.T.: Game-theoretic rough sets. Fundamenta Informaticae **108**(3–4), 267–286 (2011)
4. Jia, X.Y., Tang, Z.M., Liao, W.L., Shang, L.: On an optimization representation of decision-theoretic rough set model. Int. J. Approximate Reasoning **55**(1), 156–166 (2014)
5. Li, H.X., Zhou, X.Z.: Risk decision making based on decision-theoretic rough set: a three-way view decision model. Int. J. Comput. Intell. Syst. **4**(1), 1–11 (2011)
6. Liu, D., Yao, Y.Y., Li, T.R.: Three-way investment decisions with decision-theoretic rough sets. Int. J. Comput. Intell. Syst. **4**(1), 66–74 (2011)
7. Liu, D., Li, T.R., Liang, D.: Three-way government decision analysis with decision-theoretic rough sets. Int. J. Uncertainty Fuzziness Knowl. Based Syst. **20**(Supp. 1), 119–132 (2012)
8. Majeed, B., Azam, N., Yao, J.T.: Thresholds determination for probabilistic rough sets with genetic algorithms. In: Miao, D., Pedrycz, W., Slezak, D., Peters, G., Hu, Q., Wang, R. (eds.) RSKT 2014. LNCS, vol. 8818, pp. 693–704. Springer, Heidelberg (2014)
9. Nauman, M., Azam, N., Yao, J.T.: A three-way decision making approach to malware analysis. In: Ciucci, D., Mitra, S., Wu, W.-Z. (eds.) RSKT 2015. LNCS, vol. 9436, pp. 286–298. Springer, Heidelberg (2015). doi:10.1007/978-3-319-25754-9_26
10. Slezak, D., Ziarko, W.: The investigation of the Bayesian rough set model. Int. J. Approximate Reason. **40**(1–2), 81–91 (2005)
11. Yao, J.T., Azam, N.: Three-way decision making in Web-based medical decision support systems with game-theoretic rough sets. IEEE Trans. Fuzzy Syst. **23**(1), 3–15 (2014)
12. Yao, J.T., Herbert, J.P.: A game-theoretic perspective on rough set analysis. J. Chongqing Univ. Posts Telecommun. (Nat. Sci. Ed.) **20**(3), 291–298 (2008)
13. Yao, Y.Y.: Probabilistic approaches to rough sets. Expert Syst. **20**(5), 287–297 (2003)
14. Yao, Y.Y.: Two semantic issues in a probabilistic rough set model. Fundamenta Informaticae **108**(3–4), 249–265 (2011)
15. Yao, Y.Y.: An outline of a theory of three-way decisions. In: Yao, J.T., Yang, Y., Słowiński, R., Greco, S., Li, H., Mitra, S., Polkowski, L. (eds.) RSCTC 2012. LNCS, vol. 7413, pp. 1–17. Springer, Heidelberg (2012)
16. Yao, Y.Y.: Rough sets and three-way decisions. In: Ciucci, D., Mitra, S., Wu, W.-Z. (eds.) RSKT 2015. LNCS, vol. 9436, pp. 62–73. Springer, Heidelberg (2015). doi:10.1007/978-3-319-25754-9_6
17. Yao, Y.Y., Gao, C.: Statistical interpretations of three-way decisions. In: Ciucci, D., Mitra, S., Wu, W.-Z. (eds.) RSKT 2015. LNCS, vol. 9436, pp. 309–320. Springer, Heidelberg (2015). doi:10.1007/978-3-319-25754-9_28
18. Yao, Y.Y., Wong, S.K.M., Lingrass, P.: A decision-theoretic rough set model. Methodologies Intell. Syst. **35**, 17–24 (1990)
19. Zhou, B., Yao, Y.Y., Luo, J.G.: Cost-sensitive three-way email spam filtering. J. Intell. Inf. Syst. **42**(1), 19–45 (2014)
20. Ziarko, W.: Variable precision rough set model. J. Comput. Syst. Sci. **46**(1), 39–59 (1993)

Utilizing DTRS for Imbalanced Text Classification

Bing Zhou[1(✉)], Yiyu Yao[2], and Qingzhong Liu[1]

[1] Department of Computer Science, Sam Houston State University,
Huntsville, TX 77341, USA
zhou@shsu.edu
[2] Department of Computer Science, University of Regina,
Regina, SK S4S 0A2, Canada

Abstract. Imbalanced data classification is one of the challenging problems in data mining and machine learning research. The traditional classification algorithms are often biased towards the majority class when learning from imbalanced data. Much work have been proposed to address this problem, including data re-sampling, algorithm modification, and cost-sensitive learning. However, most of them focus on one of these techniques. This paper proposes to utilize both algorithm modification and cost-sensitive learning based on decision-theoretic rough set (DTRS) model. In particular, we use naive Bayes classifier as the base classifier and modify it for imbalanced learning. For cost-sensitive learning, we adopt the systematic method from DTRS to derive required thresholds that have the minimum decision cost. Our experimental results on three well-known text classification databases show that unified DTRS provides similar performance on balanced class distribution, outperforms naive Bayes classifier on imbalanced datasets, and is competitive with other imbalanced learning classifier.

Keywords: Imbalance data · Rough sets · Cost-sensitive · Text classification

1 Introduction

Imbalanced data learning is one of the challenging problems in data classification [18], which refers to one class is under-represented relative to others. It is common in many real world applications such as fraud detection, medical diagnosis, and text classification. The positive class (e.g., fraud) is usually the one that has the highest interest from a learning point of view and it also implies a great cost when it is not well classified. The traditional classification algorithms are often biased towards the negative class (e.g., non-fraud) and therefore there is a higher misclassification rate for the minority class instances [12]. Most classifiers in supervised machine learning are designed to maximize the accuracy of their models. Thus, when learning from imbalanced data, they are usually overwhelmed by the majority class examples. This is the main problem that degrades the performance of such classifiers.

V. Flores et al. (Eds.): IJCRS 2016, LNAI 9920, pp. 219–228, 2016.
DOI: 10.1007/978-3-319-47160-0_20

Over the years, many solutions have been proposed to deal with this problem, they can be categorized into three groups [9]:

1. **Data sampling:** the training samples are modified to produce a more balanced class distribution;
2. **Algorithmic modification:** the traditional classification algorithms are modified to be more attuned to class imbalance issues;
3. **Cost-sensitive learning:** considering different misclassification cost for each class, for example, higher costs occur for misclassifying samples of the positive class with respect to the negative class, and the goal is to find the class with minimum cost.

At the data level, a typical solution is to re-balance the class distribution by re-sampling the training data, including over-sampling the minority class or under-sampling the majority class. Traditional classification algorithms can then be used from the re-sampled data. However, data re-sampling brings extra learning cost for pre-processing data, it may lead to over-fitting because of the extra samples added into the training data, and it also may risk losing information when discarding useful samples [15]. At the algorithm level, typical solutions try to adopt existing classifier learning algorithms to bias towards the small class. For example, a larger weight is assigned to the minority class to balance the data distribution. Some classic learning algorithms, such as decision tree [13] and Support Vector Machine (SVM) [14], have been improved for class imbalance learning. Compared to re-sampling training data, sample weighting can usually be used to achieve better performance.

Decision-theoretic rough set (DTRS) model [20] is a probabilistic generalization of Pawlak's rough set model [11]. It has attracted attentions of many researchers to contribute to its development and applications. Different from other probabilistic rough set models, the main advantage of DTRS in data analysis is that it gives a semantic interpretation of the thresholds in the definition of three probabilistic regions. This semantic interpretation is tightly related to the cost, risk, or benefits occurred during the classification process. In other words, DTRS provides a cost-sensitive approach (i.e., a priori knowledge about training samples) to balance the class distribution. For example, in fraud detection, the non-fraud samples are usually a lot more than fraud samples. Since misclassifying a fraud sample into non-fraud implies higher cost, we can balance the class distribution by finding the class with minimum cost.

Although much work about the imbalance learning problem have been proposed, most of them focus on either re-sampling, modifying the originally classification algorithms, or cost-sensitive learning. In this paper, we propose to utilize both algorithm modification and cost-sensitive learning based on DTRS. In particular, we use naive Bayes classifier as the base classifier and modify it for imbalanced learning. For cost-sensitive learning, we adopt the systematic method from DTRS to derive required thresholds that have the minimum cost. Our experimental results on three well-known text classification databases show that the unified DTRS provides similar performance on balanced datasets, outperforms naive Bayes classiifier on imbalanced data, and is competitive with other imbalanced learning method.

2   Rough Set Theory

In Pawlak's rough set model [11], information about a finite set of objects are represented in an information table with a finite set of attributes. Consider an equivalence relation R on universe U. The equivalence classes induced by the partition U/R are the basic blocks to construct Pawlak's rough set approximations. For a subset $C \subseteq U$, the lower and upper approximations of C with respect to U/R are defined by:

$$\underline{apr}(C) = \{x \in U | [x] \subseteq C\}$$
$$= \bigcup \{[x] \in U/R | [x] \subseteq C\};$$
$$\overline{apr}(C) = \{x \in U | [x] \cap C \neq 0\}$$
$$= \bigcup \{[x] \in U/R | [x] \cap C \neq \emptyset\}. \tag{1}$$

Based on the rough set approximations of C, one can divide the universe U into three pair-wise disjoint regions: the positive region POS(C) is the union of all the equivalence classes that is included in C; the negative region NEG(C) is the union of all equivalence classes that have an empty intersection with C; and the boundary region BND(C) is the difference between the upper and lower approximations.

The definitions of positive, negative and boundary regions in rough sets lead to a three-way classification [19]. In decision-theoretic rough set mode, although the true class is only binary (i.e., C or C^c), we make a three-way decision based on each sensor output. A pair of thresholds (α, β) with $0 \leq \beta < \alpha \leq 1$ is used to distinguish different value ranges of $f(a)$, where $f(a)$ is a discriminant function indicating the confidence level of certain class. The pair of thresholds produces three classification regions, called the positive, boundary, and negative regions as follows:

$$\text{POS}_{(\alpha,\beta)}(C) = \{a | f(a) \geq \alpha\},$$
$$\text{BND}_{(\alpha,\beta)}(C) = \{a | \beta < f(a) < \alpha\},$$
$$\text{NEG}_{(\alpha,\beta)}(C) = \{a | f(a) \leq \beta\}, \tag{2}$$

We determine a sample as positive if $f(a)$ is greater than or equals to α. We determine a sample as negative if $f(a)$ is less than or equals to β. We do not make an immediate decision if $f(a)$ is between α and β, instead, we make a decision of deferment.

In the next two sections, we are going to show how to utilize DTRS for imbalanced learning. Section 3 shows how to realize $f(a)$ for imbalanced data using naive Bayes classifier as the base classifier. Section 4 shows how to calculate the required thresholds α and β based on minimum cost.

3 Modifying Naive Bayes Classifier for Imbalanced Learning

Many popular classification algorithms have been modified to better suit imbalanced learning. The two conventional decision tree algorithms, C4.5 [13] and CART [1], their corresponding splitting criteria information gain and the Gini measure, are considered to be sensitive to data skew. Many authors suggested improved splitting criteria of decision tree induction [2,4,6] for imbalanced data distribution. Modifications have also been made to Support Vector Machine (SVM) algorithm by re-aligning the boundary for the imbalanced data [14,17].

3.1 Naive Bayes Classifier

A naive Bayes classifier is a probabilistic classifier based on Bayesian decision theory with naive independence assumptions [5,7]. It provides an effective method to estimate the likelihood by representing an object as a feature vector and assuming that the features are probabilistically independent. It remains a popular (baseline) method for text classification due to its simplicity which leads to less computational complexity. It is for this reason we choose to use naive Bayes classier as the base classifier for our method.

In naive Bayes classifier, the conditional probability $Pr(C|[x])$ indicating the probability that a document belongs to class C given that the document is described by $[x]$:

$$Pr(C|[x]) = \frac{Pr(C)Pr(v_1, v_2, ...v_n|C)}{Pr(v_1, v_2, ...v_n)}, \tag{3}$$

where $Pr(C)$ is the prior probability of C, $\{v_1, v_2, ...v_n\}$ is a set of keywords that appear in a document, $Pr(v_1, v_2, ...v_n|C)$ is the likelihood of a document given a class C, and $Pr(v_1, v_2, ...v_n)$ is the evidence. Since the denominator of Eq. (3) does not depend on C and the values of the keywords v_i(e.g., word frequency) are given, the denominator can be considered as a constant.

We need to estimate joint probabilities $Pr(v_1, v_2, ..., v_n|C)$. In practice, it is difficult to analyze the interactions between the components of $[x]$, especially when the number n is large. A common solution to this problem is to calculate the likelihood based on the naive conditional independence assumption [7]. That is, we assume each component v_i of $[x]$ to be conditionally independent of every other component v_j for $j \neq i$. Although this assumption may seem overly simplistic, many empirical studies showed its effectiveness for classification problems [3,8,21]. Formally, the probabilistic independence assumptions are given by:

$$Pr([x]|C) = Pr(v_1, v_2, ..., v_n|C) = \prod_{i=1}^{i=n} Pr(v_i|C). \tag{4}$$

By inserting them into Eq. (3), we get:

$$Pr(C|[x]) = \frac{Pr(C)\prod_i Pr(v_i|C)}{Pr(v_1, v_2, ...v_n)}, \tag{5}$$

3.2 Modifying Naive Bayes Classifier for Imbalanced Text Classification

When choosing naive Bayes classifier as the based classifier for imbalanced learning, the discriminate function $f(a)$ in Eq. (2) can be interpreted as the conditional probability $Pr(C|[x]):f(a) = Pr(C|[x])$.

Based on Eq. (2), we have

$$Pr(C|[x]) = \frac{Pr(C)\prod_i Pr(v_i|C)}{Pr(v_1, v_2, ...v_n)} \geq \alpha. \tag{6}$$

Replacing $Pr(v_1, v_2, ...v_n)$ with a constant z, we get:

$$Pr(C|[x]) = Pr(C)\prod_i Pr(v_i|C) \geq \alpha \cdot z. \tag{7}$$

Computing the logarithm of both side, we get:

$$\log Pr(C) + \sum_i \log Pr(v_i|C) \geq \log \alpha + \log z. \tag{8}$$

For text classification problem, we need to consider the frequency of word occurrence. Let n_i denote the number of times word v_i occurs in document d, we get:

$$\log Pr(C) + \sum_i n_i \log Pr(v_i|C) \geq \log \alpha + \log z, \tag{9}$$

where $Pr(C)$ is estimated by the proportion of training documents pertaining to class C and $Pr(v_i|C)$ is estimated as:

$$Pr(v_i|C) = \frac{N_{iC} + 1}{N_C + k}, \tag{10}$$

where N_{iC} is the frequency count of word v_i occurs in the documents in class C, N_C is the total number of word occurrences in class C, and k is the total number of distinct words in all training documents. The additional one in the numerator is called Laplace correction, which initializes each word count to one instead of zero. It requires the addition of k in the denominator to obtain a probability distribution that sums to one. Plugging Eq. (10) to Eq. (9), we get:

$$\log Pr(C) + \sum_i n_i \log \frac{N_{iC} + 1}{N_C + k} \geq \log \alpha + \log z, \tag{11}$$

For imbalanced class distribution, we normalize the word frequency (weight) in each class so that the total size of each class is the same. To do this, we replace N_{iC} (i.e., word frequency of v_i in all training documents of class C) with $\frac{N_{iC}}{N_C}$ and plug into Eq. (11), we get:

$$\log Pr(C) + \sum_i n_i \log \frac{\frac{N_{iC}}{N_C} + 1}{1 + k} \geq \log \alpha + \log z, \qquad (12)$$

where N_c (i.e., the total number of word occurrence in all training documents of class C) in Eq. (11) is replace by 1 after normalization. This way, we achieve a balanced class distribution by normalizing the word count of different class. Similarly, we can calculate the probability of the other two regions/decisions.

4   Computing Thresholds for Cost-Sensitive Learning

Most classifiers assume that the misclassification costs (false negative and false positive cost) are the same. In most real-world applications, this assumption is not true. For example, in medical diagnosis, misclassifying a cancer is much more serious than the false alarm since the patients could lose their life because of a late diagnosis and treatment.

The basic idea of cost-sensitive learning is to introduce misclassification costs into the learning algorithms. For example, Ting [16] introduced C4.5_CS decision tree as a cost-sensitive modification for imbalanced data classification, an inverse class probability weight is assigned to each sample for class imbalance learning. Margineantu [10] proposed C4.5-avg, a version of the C4.5 algorithm [13] modified to accept weighted training examples, each example was weighted in proportion to the average value of the column of the cost matrix. In their approach, the decision tree induction is adapted to minimize the misclassification costs. On the other hand, other approaches based on genetic algorithms can incorporate misclassification costs in the fitness function.

4.1   Cost-Sensitive Learning Framework

With respect to a sample x, there are two classes C and C^c indicating that x is in C (i.e., positive) or not in C (i.e., negative). Two classification decisions are given by $\mathcal{A} = \{a_P, a_N\}$, where a_P and a_N represent x belong to C and x not belong to C, respectively. The cost is given by a 2×2 matrix as shown in Table 1. In the matrix, λ_{PP} and λ_{NP} denote the costs incurred for making decisions a_P and a_N when x belongs to C, and λ_{PN} and λ_{NN} denote the costs incurred for making these decisions when x does not belong to C.

The general cost-sensitive learning framework are described as follows:

1. Define different misclassification costs in cost matrix. There are four types of costs in binary cost matrix as shown in Table 1.

Table 1. Binary cost matrix

	$C\,(P)$(positive)	$C^c\,(N)$(negative)
a_P(accept)	$\lambda_{PP} = \lambda(a_P\|C)$	$\lambda_{PN} = \lambda(a_P\|C^c)$
a_N(reject)	$\lambda_{NP} = \lambda(a_N\|C)$	$\lambda_{NN} = \lambda(a_N\|C^c)$

2. According to the minimum risk decision rules in Bayesian decision theory, we accept a sample as positive if the expected risk is smaller than accept as negative, that is, $R(a_P\|[x]) \leq R(a_N\|[x])$, where

$$R(a_P\|[x]) = \lambda_{PP} Pr(C\|[x]) + \lambda_{PN} Pr(C^c\|[x]),$$
$$R(a_N\|[x]) = \lambda_{NP} Pr(C\|[x]) + \lambda_{NN} Pr(C^c\|[x]). \tag{13}$$

3. Minimizing the overall cost based on Bayes conditional risk. For a given $[x]$, a decision rule is a function $\tau([x])$ that specifies which action to take. The overall risk $\mathbf{R}$ is the expected cost associated with a given decision rule. The overall risk is defined by $\mathbf{R} = \sum_{[x]} R(\tau([x])\|[x])Pr([x])$. If $\tau([x])$ is chosen so that $R(\tau([x])\|[x])$ is as small as possible for every $[x]$, the overall risk $\mathbf{R}$ is minimized. Thus, the optimal Bayesian decision procedure can be formally stated as: for every $[x]$, compute the conditional risk $R(a_i\|[x])$ and select the action for which the conditional risk is minimum.

4.2   Computing Thresholds Based on DTRS

The cost-sensitive learning is utilized by deriving the required threshold values (i.e., α and β in Eq. (2)) based on the systematic method from DTRS model [20]. In DTRS model, although the true class is only binary (i.e., C and C^c), we make a three-way decision. The three-way cost matrix is shown in Table 2. To derive the three classification regions in Eq. (2), the set of decisions is changed to $\mathcal{A} = \{a_P, a_B, a_N\}$, where a_P, a_B, and a_N represent the three decisions in classifying x, namely, deciding $x \in \mathrm{POS}(C)$, deciding $x \in \mathrm{BND}(C)$, and deciding $x \in \mathrm{NEG}(C)$, respectively. The expected costs associated with making different decisions for samples with description $[x]$ can be expressed as:

$$R(a_P\|[x]) = \lambda_{PP} Pr(C\|[x]) + \lambda_{PN} Pr(C^c\|[x]),$$
$$R(a_B\|[x]) = \lambda_{BP} Pr(C\|[x]) + \lambda_{BN} Pr(C^c\|[x]),$$
$$R(a_N\|[x]) = \lambda_{NP} Pr(C\|[x]) + \lambda_{NN} Pr(C^c\|[x]). \tag{14}$$

The Bayesian decision theory suggests the following minimum-risk decision rules:

(P) If $R(a_P\|[x]) \leq R(a_B\|[x])$ and $R(a_P\|[x]) \leq R(a_N\|[x])$, decide $x \in \mathrm{POS}(C)$;

(B) If $R(a_B\|[x]) \leq R(a_P\|[x])$ and $R(a_B\|[x]) \leq R(a_N\|[x])$, decide $x \in \mathrm{BND}(C)$;

(N) If $R(a_N\|[x]) \leq R(a_P\|[x])$ and $R(a_N\|[x]) \leq R(a_B\|[x])$, decide $x \in \mathrm{NEG}(C)$.

Based on DTRS [20], we can express three thresholds using different costs:

$$\alpha = \frac{(\lambda_{PN} - \lambda_{BN})}{(\lambda_{PN} - \lambda_{BN}) + (\lambda_{BP} - \lambda_{PP})},$$

$$\beta = \frac{(\lambda_{BN} - \lambda_{NN})}{(\lambda_{BN} - \lambda_{NN}) + (\lambda_{NP} - \lambda_{BP})},$$

$$\gamma = \frac{(\lambda_{PN} - \lambda_{NN})}{(\lambda_{PN} - \lambda_{NN}) + (\lambda_{NP} - \lambda_{PP})}. \tag{15}$$

After tie-breaking, the parameter γ is no longer needed. The threshold α and β can be systematically calculated from Eq. (15) based on minimum decision rules.

Table 2. Three-way cost matrix

	$C(P)$: positive	$C^c(N)$: negative
a_P: positive	$\lambda_{PP} = \lambda(a_P\|C)$	$\lambda_{PN} = \lambda(a_P\|C^c)$
a_B: deferment	$\lambda_{BP} = \lambda(a_B\|C)$	$\lambda_{BN} = \lambda(a_B\|C^c)$
a_N: negative	$\lambda_{NP} = \lambda(a_N\|C)$	$\lambda_{NN} = \lambda(a_N\|C^c)$

5   Experiments

In this section, we present experiments comparing the unified DTRS approach (UDTRS) with original naive Bayes classifier (NB) and the cost-sensitive learning method, MetaCost, from Weka. Our experiments were based on three well-known text classification database: Industry Sector, 20 Newsgroups, and WebKB. We selected 3680 web pages and 30 classes from Industry Sector, the largest class has over 100 web pages and the smallest has less than 30 web pages. The distribution

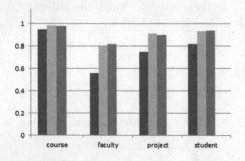

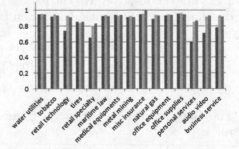

Fig. 1. AUC for NB (dark grey), UDTRS (light grey), and MetaCost (medium grey) on WebKB

Fig. 2. AUC for NB (dark grey), UDTRS (light grey), and MetaCost (medium grey) on Industry Sector

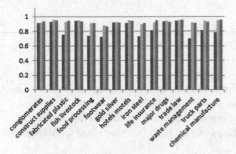
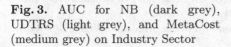
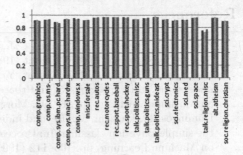

Fig. 3. AUC for NB (dark grey), UDTRS (light grey), and MetaCost (medium grey) on Industry Sector

Fig. 4. AUC for NB (dark grey), UDTRS (light grey), and MetaCost (medium grey) on 20 Newsgroups

of documents per class for 20 Newsgroups is even at around 1000 documents per class. The WebKB contains 4200 web pages collected from university computer science departments and unevenly distributed into 4 classes. For each dataset, the same data pre-processing steps are applied to extract keywords and remove stop words. All results are averages from ten trials with a 30 % test data.

The comparison results are reported in terms of the area under the ROC curve(AUC), which is typically used to evaluate classification performance with unevenly distributed data. Figures 1, 2, 3 and 4 give classification performance on three text datasets. As we can see, UDTRS significantly improves the performance of NB on imbalanced class distribution(3 classes of WebKB, and 14 classes of Industry Sector), provides similar performance of NB on evenly distributed datasets (20 Newsgroups, 1 class of WebKB, and 16 classes of Industry Sector), and is competitive with MetaCost classifier on imbalanced datasets.

6 Conclusions

In this paper, we propose to utilize DTRS for imbalanced learning. Different to other works in the literature, instead of focusing only on data re-sampling, algorithm modification, or cost-sensitive learning, our approach consists a method which utilize both algorithm modification and cost-sensitive learning under a unified framework. At the algorithm level, naive Bayes classifier is chosen as the base classifier because its simplicity and fast speed for text classification. Concretely, we normalize the word frequency count of different class to achieve a balanced class distribution. At the cost-sensitive learning level, we adopt the systematic method from DTRS to derived the required threshold values based on minimum decision cost. That is, both sides of Eq. (2) are justified for imbalanced learning. The main contribution of this paper is to unify two imbalanced learning techniques under a unified framework to archive a better classification performance.

References

1. Breiman, L., Friedman, J., Stone, C.J., Olshen, R.: Classification and Regression Trees. Chapman and Hall, Belmont (1984)
2. Dietterich, T., Kearns, M., Mansour, Y.: Applying the weak learning framework to understand and improve C4.5. In: Proceedings of the 13th International Conference on Machine Learning, pp. 96–104. Morgan Kaufmann (1996)
3. Domingos, P., Pazzani, M.: Beyond independence: conditions for the optimality of the simple Bayesian classifier. In: Proceedings of the 13th International Conference on Machine Learning, pp. 105–112 (1996)
4. Drummond, C., Holte, R.: Exploiting the cost (in)sensitivity of decision tree splitting criteria. In: ICML, pp. 239–246 (2000)
5. Duda, R.O., Hart, P.E.: Pattern Classication and Scene Analysis. Wiley, New York (1973)
6. Flach, P.A.: The geometry of ROC space: understanding machine learning metrics through ROC isometrics. In: ICML, pp. 194–201 (2003)
7. Good, I.J.: The Estimation of Probabilities: An Essay on Modern Bayesian Methods. MIT Press, Cambridge (1965)
8. Langley, P., Wayne, I., Thompson, K.: An analysis of Bayesian classifiers. In: Proceedings of the 10th National Conference on Artificial Intelligence, pp. 223–228 (1992)
9. Lpez, V., Fernndez, A., Garca, S., Palade, V., Herrera, F.: An insight into classification with imbalanced data: empirical results and current trends on using data intrinsic characteristics. Inf. Sci. 250, 113–141 (2013)
10. Margineantu, D.D. When does imbalanced data require cost-sensitive learning? AAAI Technical report WS-00-05 (2000)
11. Pawlak, Z.: Rough Sets, Theoretical Aspects of Reasoning about Data. Kluwer Academic Publishers, Dordrecht (1991)
12. Probost, F. Machine learning from imbalanced data sets 101. Invited Paper for the AAAI 2000 Workshop on Imbalanced Data Sets (2000)
13. Quinlan, J.R.: C4.5 Programs for Machine Learning. Morgan Kaufman, San Mateo (1993)
14. Raskutti, B.: Extreme re-balancing for SVM's: a case study. In: ICML-KDD 2003 Workshop: Learning from Imbalanced Data Sets (2003)
15. Sun, Y., Kamel, M.S., Wong, A.K.C., Wang, Y.: Cost-sensitive boosting for classification of imbalanced data. Pattern Recogn. 40(12), 3358–3378 (2007)
16. Ting, K.M.: An instance-weighting method to induce cost-sensitive trees. IEEE Trans. Knowl. Data Eng. 14(3), 659–665 (2002)
17. Wu, G.: Class-boundary alignment for imbalanced dataset learning. In: ICML-KDD 2003 Workshop: Learning from Imbalanced Data Sets (2003)
18. Yang, Q., Wu, X.D.: 10 challenging problems in data mining research. Int. J. Inf. Technol. Decis. Making 05, 597 (2006)
19. Yao, Y.Y.: Three-way decisions and cognitive computing. Cogn. Comput. 8, 543–554 (2016). doi:10.1007/s12559-016-9397-5
20. Yao, Y.Y., Wong, S.K.M., Lingras, P.: A decision-theoretic rough set model. In: Ras, Z.W., Zemankova, M., Emrich, M.L. (eds.) Methodologies for Intelligent Systems, vol. 5, pp. 17–24. North-Holland, New York (1990)
21. Yao, Y.Y., Zhou, B.: Two Bayesian approaches to rough sets. Eur. J. Oper. Res. 251, 904–917 (2016)

A Three-Way Decision Clustering Approach for High Dimensional Data

Hong Yu[✉] and Haibo Zhang

Chongqing Key Laboratory of Computational Intelligence,
Chongqing University of Posts and Telecommunications, Chongqing 400065, China
yuhong@cqupt.edu.cn

Abstract. In this paper, we propose a three-way decision clustering approach for high-dimensional data. First, we propose a three-way K-medoids clustering algorithm, which produces clusters represented by three regions. Objects in the positive region of a cluster certainly belong to the cluster, objects in the negative region of a cluster definitively do not belong to the cluster, and objects in the boundary region of a cluster may belong to multiple clusters. Then, we propose the novel three-way decision clustering approach using random projection method. The basic idea is to apply the three-way K-medoids several times, increasing the dimensionality of the data after each iteration of three-way K-medoids. Because the center of the project result is used to be the initial center of the next projection, the time of computing is greatly reduced. Experimental results show that the proposed clustering algorithm is suitable for high-dimensional data and has a higher accuracy and does not sacrifice the computing time.

Keywords: Cluster · Three-way decisions · K-medoids · Random projection · High-dimensional data

1 Introduction

Cluster analysis is the task of grouping a set of objects into clusters based on some measure of inherent similarity or distance, and it is considered as an instance of unsupervised learning. It is commonly used in machine learning, pattern recognition, image analysis, information retrieval, bioinformatics, data compression, and computer graphics [12,17]. The sparsity and the problem of the curse of dimensionality of high dimensional data, make the most of traditional clustering algorithms lose their action in high dimensional data. Therefore, the study of clustering high dimensional data has become a key issue in the current.

Random projection [9] is a technique used to reduce the dimensionality of a set of objects which lie in Euclidean space. It is a simple and computationally efficient way to reduce the dimensionality of data by trading a controlled amount of error for faster processing times and smaller model sizes. The dimensions and distribution of random projection matrices are controlled so as to

© Springer International Publishing AG 2016
V. Flores et al. (Eds.): IJCRS 2016, LNAI 9920, pp. 229–239, 2016.
DOI: 10.1007/978-3-319-47160-0_21

approximately preserve the pairwise distances between any two samples of the dataset. For example, Cardoso and Wichert [2] proposed iterative random projections K-means for high dimensional data clustering, which applies K-means several times, increasing the dimensionality of the data after each convergence of K-means. Murtagh and Contreras [14] proposed random projection towards the Baire Metric for high dimensional clustering.

Generally speaking, there exist a lot of uncertain information in high dimensional data due to diversity in features. For example, people's interests is variable, and some feature values are missing due to the difficulty on obtaining. To solve this problem, many scholars have proposed uncertainty clustering methods by combining some uncertainty information processing technologies [1,5–7,16]. Based on the theory of three-way decisions [18,19], Yu et al. [20,21] introduced a framework of three-way cluster analysis. The three-way representation of a cluster is represented by an interval set instead of a single set, and the representation as a triple of positive, boundary and negative regions brings more insight into interpretation of clusters. That is, objects in the positive region certainly belong to the cluster, objects in the negative region definitively do not belong to the cluster, and objects in the boundary region may belong to multiple clusters. Their preliminary results provide us with a tool for studying the problem of clustering high dimensional data.

In this paper, we address the problem of clustering high dimensional data based on the theory of three-way decisions. We first introduce some basic concepts about three-way clustering and random projection in Sect. 2. In Sect. 3, we first propose a new three-way K-medoids clustering algorithm which produces a three-way result of clustering, we design an objective function which evaluate the quality of the result of clustering, and we devise a three-way clustering algorithm for high dimensional data. In Sect. 4, experimental results show that the proposed clustering algorithm is suitable for high dimensional data and has a higher accuracy and does not sacrifice the computing time. Finally, in Sect. 5, we summarize the present study.

2   Basic Concepts

2.1   Three-Way Clustering

In our previous work, we had introduced a framework of three-way cluster analysis [20,21]. Let $U = \{\mathbf{x}_1, \cdots, \mathbf{x}_n, \cdots, \mathbf{x}_N\}$ be a finite set. $\mathbf{x}_n$ is an object which has M attributes, namely, $\mathbf{x}_n = (x_n^1, \cdots, x_n^m, \cdots, x_n^M)$. x_n^m denotes the value of the m-th attribute of the object $\mathbf{x}_n$, where $n \in \{1, \cdots, N\}$, and $m \in \{1, \cdots, M\}$. A cluster is representation by an interval set, namely, $C^k = [\underline{C^k}, \overline{C^k}]$. $\underline{C^k}$ represents the lower bound of C^k, $\overline{C^k}$ represents the upper bound of C^k, and $\underline{C^k} \subseteq \overline{C^k}$. If the clustering result has K clusters, the result scheme of three-way clustering using interval set is $\mathbf{C} = \{[\underline{C^1}, \overline{C^1}], \cdots, [\underline{C^k}, \overline{C^k}], \cdots, [\underline{C^K}, \overline{C^K}]\}$.

Therefore, the sets $\underline{C^k}$, $\overline{C^k} - \underline{C^k}$ and $U - \overline{C^k}$ formed by certain decision rules constitute the three regions of the cluster C^k as the positive region, boundary region and negative region, respectively. The three-way decisions are given as:

$$\text{POS}(C^k) = \underline{C^k},$$
$$\text{BND}(C^k) = \overline{C^k} - \underline{C^k},$$
$$\text{NEG}(C^k) = U - \overline{C^k}. \tag{1}$$

The objects in $\text{POS}(C^k)$ definitely belong to the cluster C^k, objects in $\text{NEG}(C^k)$ definitely do not belong to the cluster C^k, and objects in the region $\text{BND}(C^k)$ might or might not belong to the cluster.

In this paper, we obtain the three regions of the cluster by comparing the distance between an object and the cluster through a pair of threshold values α and β. The distance is calculated by the following Euclidean distance formula:

$$||\mathbf{x} - \mathbf{y}|| = d(\mathbf{x}, \mathbf{y}) = \sqrt{\sum_{i=1}^{M} (x^i - y^i)^2}. \tag{2}$$

Obviously, x, y is more similar while $d(x, y)$ is smaller.

Then, the three-way decisions rules are given as follows.

RulePOS: if $d(\mathbf{x}_n, center(C^k)) \leq \alpha$, $\mathbf{x}_n \in \text{POS}(C^k)$;
RulePOB: if $\beta \geq d(\mathbf{x}_n, center(C^k)) > \alpha$, $\mathbf{x}_n \in \text{BND}(C^k)$;
RuleBOB: if $d(\mathbf{x}_n, center(C^k)) > \beta$, $\mathbf{x}_n \in \text{NEG}(C^k)$.

2.2 Random Projection

Random projection is a simple and powerful dimensionality reduction tool, and its core idea is given in the Johnson-Lindenstrauss lemma [4,9].

Lemma 1 Johnson-Lindenstrauss Lemma (J-L lemma). For any $0 < \varepsilon < 1$ and any integer n such that $d \geq O(\ln(N/(\varepsilon)^2))$, then for any set U of N points with M dimensions there is a map $f : M \to d$ such that for all $\mathbf{u}, \mathbf{v} \in U$,

$$(1 - \varepsilon)||\mathbf{u} - \mathbf{v}||^2 \leq ||f(\mathbf{u}) - f(\mathbf{v})||^2 \leq (1 + \varepsilon)||\mathbf{u} - \mathbf{v}||^2. \tag{3}$$

The original universe U can be denoted as a matrix $\mathbf{X}_{N \times M}$, and its projection on a d-dimensional random subspace ($d \ll M$) is denoted as:

$$\mathbf{X}_{N \times d}^{RP} = \mathbf{X}_{N \times M} \mathbf{R}_{M \times d}, \tag{4}$$

Here, $\mathbf{R}_{M \times d}$ is random $M \times d$ matrix and $\mathbf{X}_{N \times d}^{RP}$ is the projection of $\mathbf{X}$ in d-dimensional subspace. Random projection is computationally very simple: forming the random matrix $\mathbf{R}$ and projecting the $N \times M$ matrix $\mathbf{X}$ into d dimensions is of order $O(dMn)$, and if the matrix $\mathbf{X}$ is sparse with about c nonzero entries per row, the complexity is of order $O(cMN)$ [15].

The effect to which pair-wise distances between points before and after projection are preserved depends upon the projection vectors $r_i \in \mathbf{R}_{M \times d}$. The essential property of the projection matrix $\mathbf{R}_{M \times d}$ in J-L lemma is that its row vectors $r_i \in R$ are required to be orthogonal to each other, because this will have the optimal dimensionality reduction effect. Gram Schmidt orthogonalization process is a technique that is usually applied to transforms the random vector into orthogonal ones, but achieving orthogonality is a very computationally expensive process. On the other hand, Indyk and Motwani [8] noted that the condition of orthogonality can be dropped while using random projections. They computed a random projection matrix $\mathbf{R}_{M \times d}$ whose entries are independent random variables with the standard Gaussian distribution $N(0, 1)$, and they proved that the projection $\mathbf{X}_{N \times d}^{RP}$ of a unit vector $\mathbf{x} \in \mathbf{R}_{M \times d}$ has the chi-square distribution with k-degrees of freedom, and tail estimates for this distribution can be used to prove that the pair-wise distance between any two points is not distorted by a factor more than $(1 \pm \varepsilon)^{17}$. Projection on standard Gaussian distributed random vectors is a distance preserving mapping with less computation costs [11]. According to the properties of normal distribution, the linear projection of a Gaussian remains Gaussian. Hence, a mixture of high dimensional Gaussians onto a single vector will be producing a mixture of univariate normally distributed variables.

3   Clustering Approach for High Dimensional Data

In this section, we first propose a new three-way K-medoids clustering which produces a three-way result of clustering. Then, we propose a three-way decision clustering approach for sparse high dimensional data.

3.1   Three-Way K-medoids Clustering Algorithm

There is a bunch of approaches to deal with uncertainty clustering for high dimensional data. However, these approaches can not show us which objects belong to multi-clusters and which objects are the common ones between clusters. The three-way clustering method can solve the problem very well. Compared with other clustering algorithms, the three-way clustering methods assign objects into the positive region of the cluster and the boundary region of the cluster. The objects in the positive region only belong to the cluster and the objects in the boundary region may belong to more than one cluster. The method gives an intuitionistic view on clustering result scheme.

We first improve the traditional K-medoids algorithm to obtain the three-way clustering result, shorted by TWD-K-medoids, which is described in Algorithm 1. In order to determine the threshold values used in the three-way decisons rules in Subsect. 2.1, we define a pair of threshold values α and β for each cluster C^i, ($1 \le i \le K$). That is, $\alpha[i] = d_{\min}[i] + a * d_{ave}[i]$, and $\beta[i] = d_{\min}[i] + b * d_{ave}[i]$. $d_{\min}[i]$ and $d_{ave}[i]$ are the minimal one, the average one in all the euclidean distances which are between the cluster center $Center[i]$ and one object, respectively. a, b are parameters and $a, b \in (0, 1)$.

Algorithm 1. Three-way K-medoids Clustering algorithm

---

1 **Input:** The matrix $\mathbf{X}_{N \times M}$, N objects and M attributes, the number of clsuters K, the intitial cluster centers $Center_0[K]$, and parameters a, b.

2 **Output:** The clustering result $\mathbf{C} = \{[\underline{C^1}, \overline{C^1}], \cdots, [\underline{C^k}, \overline{C^k}], \cdots, [\underline{C^K}, \overline{C^K}]\}$, the centers of clusters $Center[K]$.

3 **if** $Center_0[K] = \emptyset$ **then**

4     $Random(Center[K])$;

5 **if** $Center_0[K] \neq \emptyset$ **then**

6     $Center[K] = Center_0[K]$;

7 $Prevariance = 0, Variance = 0, iterate = 15, Change = true$;

8 **while** $iterate > 0$ & $Change$ **do**

9   **for** $i=1$ to K **do**

10   Compute $d_{min}[i], d_{ave}[i]$;

11        $\alpha[i] = d_{\min}[i] + a * d_{ave}[i]$;

12        $\beta[i] = d_{\min}[i] + b * d_{ave}[i]$;

13   **for** $i=1$ to N **do**

14   **for** $j=1$ to K **do**

15   Find out the minimal $d(\mathbf{x}_i, Center[j])$;

16           $MinTempDis = d(\mathbf{x}_i, Center[j])$;

17   ClusterNumber=j;

18   **if** $MinTempDis \leq \alpha[ClusterNumber]$ **then**

19           $\mathbf{x}_i \in POS(C^ClusterNumber)$;

20   **if** $MinTempDis > \alpha[ClusterNumber]$ & $MinTempDis \leq \beta[ClusterNumber]$ **then**

21           $\mathbf{x}_i \in BND(C^ClusterNumber)$.

22   **for** $i=1$ to K **do**

23   **for** $every\ objects\ \boldsymbol{x}\ in\ BND(C^i)$ **do**

24   **for** $j=1$ to K **do**

25   **if** $d(\boldsymbol{x}, Center[j]) > \alpha[j]$ & $d(\boldsymbol{x}, Center[j]) \leq \beta[j]$ **then**

26                 $\mathbf{x} \in BND(C^j)$.

27     $iterate - -$;

28   **for** $i=1$ to K **do**

29   **for** $every\ objects\ \boldsymbol{x}\ in\ BND(C^i)\ and\ POS(C^i)$ **do**

30   Compute the euclidean distance $d(\mathbf{x}, Center[i])$;

31           $variance+ = d(\mathbf{x}, Center[i])$;

32   **if** $Prevariance == Variance$ **then**

33        $Change = false$;

34   Prevariance=Variance;

35   **for** $i=1$ to K **do**

36   //Update $Center[K]$;

37   Find out the minimal sum of euclidean distances between one object $\mathbf{x}$ and other all objects in C^i;

38   Put the data object $\mathbf{x}$ in $Center[i]$;

39 Output the result $\mathbf{C} = \{[\underline{C^1}, \overline{C^1}], \cdots, [\underline{C^k}, \overline{C^k}], \cdots, [\underline{C^K}, \overline{C^K}]\}$ and the cluster centers $Center[K]$.

3.2   Three-Way Decision Algorithm Based on Random Projection

As we know, the errors between the clustering result of the projection and the clustering result of the original data are different under different dimensional subspace. The lower the projection dimension is, the bigger the error is. The higher the projection dimension is, the more computing time the algorithm costs.

In this paper, we propose a dynamic random projection approach to make a tradeoff between decreasing errors and reducing computing time. That is, the original data is projected to different dimension for ascending ordering and a three-way clustering result is produced by Algorithm 1 in each dimension subspace. The results of adjacent iterations are compared and the better one is kept. The processing is run until the stop conditions are satisfied. The stop conditions conclude two cases. One is the objective function is small enough, another is the iteration time is maximal. The objective function is defined by considering the tradeoff between decreasing errors and reducing computing time.

We give the related definitions, in order to describe the new Algorithm 2. Let the two projection clustering results be $\mathbf{C}_{N \times d_i}$, $\mathbf{C}_{N \times d_j}$, the number of clusters be K. Of course, we need to establish a correspondence between clusters in the two results. Anyway, it is easy to deal due to lots of existing methods. Then, we define the difference function between clustering results as follows.

Definition 1. *The difference function between clustering results is:*

$$
DCR\left(\mathbf{C}_{N \times d_i}, \mathbf{C}_{N \times d_j}\right) = 1 - \frac{1}{K}\left(\sum_{k=1}^{K} \frac{\left|\mathrm{POS}\left(\mathrm{C}_{N \times d_i}^k\right) \cap \mathrm{POS}\left(\mathrm{C}_{N \times d_j}^k\right)\right|}{\min\left\{\left|\mathrm{POS}\left(\mathrm{C}_{N \times d_i}^k\right)\right|, \left|\mathrm{POS}\left(\mathrm{C}_{N \times d_j}^k\right)\right|\right\}} \right.
$$
$$
\left. + \sum_{k=1}^{K} \frac{\left|\mathrm{BND}\left(\mathrm{C}_{N \times d_i}^1\right) \cap \mathrm{BND}\left(\mathrm{C}_{N \times d_j}^k\right)\right|}{\min\left\{\left|\mathrm{BND}\left(\mathrm{C}_{N \times d_i}^k\right)\right|, \left|\mathrm{BND}\left(\mathrm{C}_{N \times d_j}^k\right)\right|\right\}}\right),
$$

(5)

where $\mathrm{DCR}(\mathbf{C}_{N \times d_i}, \mathbf{C}_{N \times d_j}) \in [0, 1]$.

In order to save computing time, the algorithm is ran on a high level instead of running on the projection dimension just one by one. That is, if the dimension is d in the current iteration, the dimension will be $d + Sl$ in the next iteration not be $d + 1$. Usually, $d \ll M$ and the step length Sl is more than 1. Let d_s be the number of attributes in the sth iteration. Then, we have the following definition.

Definition 2. *The computation cost function $COC(d_s)$ is defined as:*

$$
COC(d_s) = COC(d_{s-1}) + \frac{DCR(\mathbf{C}_{N \times d_s}, \mathbf{C}_{N \times d_{s-1}}) * Sl}{M},
$$

(6)

where $COC(d_1) = \frac{d_1}{M}$, and $COC(d) \in [0, 1]$.

Obviously, the objective function OF is a two-dimensional utility function on the two criteria, DCR and COC, namely $OF = (DCR, COC)$. When the utilities of the two dimensions are maximum, the merge utility is maximum, namely

$OF(1,1) = 1$; when the utilities of the two dimensions are minimum, the merge utility is minimum, namely $OF(0,0) = 0$; when the utility of DCR is maximum, the merge utility is maximum whatever COC is, namely $OF(1, COC) = 1$; when the utility of COC is maximum, the merge utility is maximum whatever DCR is, namely $OF(DCR, 1) = 1$. In other words, the two dimensions are equally important to determine the objective function. Then we have the computational formula by using the rule of replacement [13],

$$OF = DCR + COC - DCR \times COC. \tag{7}$$

Algorithm 2 describes the proposed three-way clustering for high dimensional data based on random projection, shorted by TWD-HD. In the algorithm, we save the cluster centers $Center[K]$ at the current iteration as the initial cluster centers $Center_0[K]$ of the next iteration. It is reasonable because the projection still retains some of the structure of the original data even if it has errors. Thus, compared with to update the initial cluster centers randomly in every projection and clustering, the algorithm will reduce lots of computing time in Line 7. On the other hand, the higher projected dimension is, the better clustering result is. It is more inclined to the final good enough result the method of using the centers to be the next initial cluster centers than choosing the next initial cluster centers randomly. Thus, there is less iterations both in the TWD-K-medoids algorithm and in the TWD-HD algorithm.

Algorithm 2. Three-way decision algorithm based on random projection

---

1 **Input:** $N \times M$ matrix $\mathbf{X}_{N \times M}$, the initial projection dimension d_1, cluster number K;
2 **Output:** Clustering result $\mathbf{C} = \{[\underline{C^1}, \overline{C^1}], \cdots, [\underline{C^k}, \overline{C^k}], \cdots, [\underline{C^K}, \overline{C^K}]\}$
3 **Initialization:** $\mathbf{C}_0 = \emptyset$, $Center_0[K] = \emptyset$;
4 **while** $OF < \lambda$ **do**
5 $\quad$ Generate a Gaussian random matrix $\mathbf{R}_{M \times d_1}$;
6 $\quad$ Compute projection matrix $\mathbf{X}_{N \times d_1}^{RP} = \mathbf{X}_{N \times M} \times \mathbf{R}_{M \times d_1}$;
7 $\quad$ Clustering $\mathbf{X}_{N \times d_1}^{RP}$ by Algorithm 1, obtain $\mathbf{C}_{N \times d_1}$ and $Center[K]$;
8 $\quad$ Compare($\mathbf{C}_0, \mathbf{C}_{N \times d_1}$), and only remain the better one to $\mathbf{C}_0$;
9 $\quad$ $Center_0[K] = Center[K]$;
10 $\quad$ Compute OF;
11 $\quad$ Increase d_1;
12 output the final result $\mathbf{C} = \{[\underline{C^1}, \overline{C^1}], \cdots, [\underline{C^k}, \overline{C^k}], \cdots, [\underline{C^K}, \overline{C^K}]\}$.

---

4   Experimental Results

To validate the performance of the proposed algorithm, we have carried out a number of experiments on three data sets with three compared algorithms such as IRPK-means [2], FSC [6] and K-medoids clustering algorithm [10]. The detail information of data sets is shown in Table 1. The algorithms are programed in C++ with Microsoft Visual Studio 2010 and the experiments are tested on a PC computer with Windows 7 OS, 2.67 GHz CPU and 4 GB Memory.

The two synthetic data sets are Gaussian and Ellipse, which is generated by the generation available tool [3] on the website [23]. The real data set ORL_Faces is from the reference [2], which consists of 10 faces of 40 people, a total of 400 samples, and each sample contains 10304 characteristics.

Three traditional evaluation indicators are used here such as Accuracy, NMI and Time. Let $\mathbf{C} = \{C^1, \cdots, C^k, \cdots, C^K\}$ be the clustering result, and the number of cluster is K. $\mathbf{P} = \{C^1, \cdots, C^l, \cdots, C^L\}$ is the classification of the original data, and the number of cluster is L. Adjust the cluster labels of results $\mathbf{C}$ and $\mathbf{P}$, then we can compute the accuracy as follows:

$$\text{Accuracy} = \sum_{k=1}^{K} \frac{n_l^k}{N} \tag{8}$$

where N is the number of data objects, and n_l^k is the number of common data in the k-th cluster of $\mathbf{C}$ and the corresponding l-th cluster of $\mathbf{P}$.

The normalized mutual information NMI [22] is calculated by the following formula:

$$\text{NMI} = \frac{\sum_{k=1}^{K} \sum_{l=1}^{L} n_l^k \log\left(\frac{n n_l^k}{n^k n^l}\right)}{\sqrt{\left(\sum_{k=1}^{K} n^k \log\left(\frac{n^k}{n}\right)\right)\left(\sum_{l=1}^{L} n^l \log\left(\frac{n^l}{n}\right)\right)}}, \tag{9}$$

where n^k is the number of objects in the k-th cluster of $\mathbf{C}$ and n^l is the number of objects in the l-th cluster of $\mathbf{P}$.

Table 1. The information of data sets

Datasets	Size of datasets	Number of dimension	Number of clusters
Gaussian	100	3000	2
Ellipse	500	5000	10
ORL_Faces	400	10304	40

Each clustering algorithm runs on the every data set 10 times, the average values and variances of Accuracy, NMI and CPU time are recorded in Table 2, Table 3 and Table 4, respectively. The parameters of the proposed algorithm are set as follows. a is 0.7, 0.7, 0.75, b is 0.85, 0.75, 0.8, and λ is 0.5, 0.6, 0.8, for data sets Gaussian, Ellipse and ORL_Faces respectively.

Observe the experimental results, we see that the proposed algorithm is best in the performance of Accuracy. The FSC algorithm is best in the performance

Table 2. The results of the comparison experiments on Accuracy

Data sets	Algorithms			
	TWD-HD	IRPK-means [2]	FSC [6]	K-medoids [10]
Gaussian	**0.968** ± 0.041	0.888 ± 0.067	0.917 ± 0.007	0.860 ± 0.000
Ellipse	**0.811** ± 0.039	0.748 ± 0.047	0.804 ± 0.014	0.732 ± 0.060
ORL_Faces	**0.592** ± 0.011	0.547 ± 0.027	0.592 ± 0.04	0.586 ± 0.041

Table 3. The results of the comparison experiments on NMI

Data sets	Algorithms			
	TWD-HD	IRPK-means [2]	FSC [6]	K-medoids [10]
Gaussian	0.826 ± 0.149	0.629 ± 0.209	**0.862** ± 0.023	0.531 ± 0.000
Ellipse	0.807 ± 0.016	0.797 ± 0.034	**0.857** ± 0.004	0.787 ± 0.322
ORL_Faces	0.757 ± 0.018	0.746 ± 0.013	**0.777** ± 0.013	0.762 ± 0.011

Table 4. The results of the comparison experiments on CPU Time (s)

Data sets	Algorithms			
	TWD-HD	IRPK-means [2]	FSC [6]	K-medoids [10]
Gaussian	4.600 ± 0.699	**3.700** ± 0.675	97.632 ± 0.644	124.400 ± 27.900
Ellipse	53.700 ± 6.816	**34.400** ± 3.239	3242.400 ± 13.789	5771.000 ± 545.028
ORL_Faces	42.000 ± 2.789	**28.200** ± 4.686	29899.200 ± 3588.643	5389.400 ± 985.922

of NMI, and the TWD-HD is almost the second one. The TWD-HD is the second one just a little bit less than IRPK-means. In a word, the proposed method has a higher accuracy and does not sacrifice the computing time.

5   Conclusions

In this paper, we address the problem of clustering high-dimensional data. We utilize the three-way clustering approach to deal with the uncertainty in high-dimensional data, in which a cluster is represented by three regions instead of two regions. That is, objects in the positive region certainly belong to the cluster, objects in the negative region definitively do not belong to the cluster, and objects in the boundary region may belong to multiple clusters. Next, we first propose a three-way K-medoids clustering algorithm, TWD-K-medoids, in which a cluster is represented by three regions. Then, we propose the novel three-way decision clustering approach using random projection method, TWD-HD. The basic idea of the TWD-HD is to apply the TWD-K-medoids several times, increasing the dimensionality of the data after each iteration of three-way K-medoids. The comparison experimental results show that the proposed approach is suitable for high-dimensional data, and it has a higher accuracy and does not sacrifice the computing time.

Acknowledgments. This work was supported in part by the National Natural Science Foundation of China under Grant Nos. 61379114 & 61533020.

References

1. Aggarwal, C.C.: On high himensional projected clustering of uncertain data streams. In: Proceedings of the 25th International Conference on Data Engineering, pp. 1152–1154 (2009)
2. Cardoso, A., Wichert, A.: Iterative random projections for high-dimensional data clustering. Pattern Recogn. Lett. **33**(13), 1749–1755 (2012)
3. Choi, Y.K., Park, C.H., Kweon, I.S.: Accelerated k-means clustering using binary random projection. In: 12th Asian Conference on Computer Vision, pp. 257–272 (2014)
4. Dasgupta, S., Gupta, A.: An elementary proof of a theorem of Johnson and Lindenstrauss. Random Struct. Algorithms **22**(1), 60–65 (2003)
5. Deng, Z.H., Choi, K.S., Jiang, Y.Z., Wang, J., Wang, S.T.: A survey on soft subspace clustering. Inf. Sci. **348**, 84–106 (2016)
6. Gan, G., Wu, J., Yang, Z.-J.: A fuzzy subspace algorithm for clustering high dimensional data. In: Li, X., Zaïane, O.R., Li, Z. (eds.) ADMA 2006. LNCS (LNAI), vol. 4093, pp. 271–278. Springer, Heidelberg (2006)
7. Gunnemann, S., Kremer, H., Seidl, T.: Subspace clustering for uncertain data. In: Proceedings of the SIAM International Conference on Data Mining, SDM 2010, PP. 385–396 (2010)
8. Indyk, P., Motwani, R.: Approximate nearest neighbors: towards removing the curse of dimensionality. In: 30th Annual ACM Symposium on Theory of Computing, pp. 604–613. ACM Press (1998)
9. Johnson, W.B., Lindenstrauss, J.: Extensions of Lipschitz mappings into a Hilbert space. Contemp. Math. **26**, 189C–206 (1984)
10. Kaufman, L., Rousseeuw, P.J.: Finding Groups in Data: An Introduction to Cluster Analysis. Wiley, New York (1990)
11. Kaur, H., Khanna, P.: Gaussian random projection based non-invertible cancelable biometric templates. Procedia Comput. Sci. **54**, 661–670 (2015)
12. Kriegel, H.P., Kroger, P., Zimek, A.: Clustering high-dimensional data: a survey on subspace clustering, pattern-based clustering and correlation clustering. ACM Trans. Knowl. Disc. Data **3**(1), 337–348 (2009)
13. Liu, K., Guo, Y., Pan, Y.: Information system evaluation based on the multidimensional utility mergence method. Inf. Stud. Theor. Appl. **35**(3), 103–108 (2012). (In Chinese)
14. Murtagh, F., Contreras, P.: Random projection towards the baire metric for high dimensional clustering. In: Gammerman, A., Vovk, V., Papadopoulos, H. (eds.) SLDS 2015. LNCS, vol. 9047, pp. 424–431. Springer, Heidelberg (2015). doi:10.1007/978-3-319-17091-6_37
15. Papadimitriou, C.H., Raghavan, P., Tamaki, H., Vempala, S.: Latent semantic indexing: a probabilistic analysis. In: 7th ACM Symposium on Principles of Database Systems, pp. 159–168. ACM Press (1998)

16. Zhang, X., Gao, L., Yu, H.: Constraint based subspace clustering for high dimensional uncertain data. In: Khan, L., Bailey, J., Washio, T., Huang, J.Z., Wang, R. (eds.) PAKDD 2016. LNCS, vol. 9652, pp. 271–282. Springer, Heidelberg (2016). doi:10.1007/978-3-319-31750-2_22

17. Xu, D.K., Tian, Y.J.: A Comprehensive survey of clustering algorithms. Ann. Data Sci. 2(2), 165–193 (2015)

18. Yao, Y.: An outline of a theory of three-way decisions. In: Yao, J.T., Yang, Y., Słowiński, R., Greco, S., Li, H., Mitra, S., Polkowski, L. (eds.) RSCTC 2012. LNCS, vol. 7413, pp. 1–17. Springer, Heidelberg (2012)

19. Yao, Y.Y.: Three-way decisions and cognitive computing. Cogn. Comput. 8, 543–554 (2016)

20. Yu, H., Liu, Z.G., Wang, G.Y.: An automatic method to determine the number of clusters using decision-theoretic rough set. Int. J. Approximate Reason. 55(1), 101–115 (2014)

21. Yu, H., Zhang, C., Wang, G.Y.: A tree-based incremental overlapping clustering method using the three-way decision theory. Knowl. Based Syst. 91, 189–203 (2016)

22. Wang, Y.T., Chen, L.H., Mei, J.P.: Incremental fuzzy clustering with multiple medoids for large data. IEEE Trans. Fuzzy Syst. 22(6), 1557–1568 (2014). IEEE Press

23. http://personalpages.manchester.ac.uk/mbs/Julia.Handl/generators.html

Three-Way Decisions Based Multi-label Learning Algorithm with Label Dependency

Feng Li[1,2], Duoqian Miao[1,2(✉)], and Wei Zhang[1,2]

[1] Department of Computer Science and Technology,
Tongji University, Shanghai 201804, China
tjleefeng@hotmail.com, dqmiao@tongji.edu.cn
[2] Key Laboratory of Embedded Systems and Service Computing,
Ministry of Education, Tongji University, Shanghai 201804, China

Abstract. A great number of algorithms have been proposed for multi-label learning, and these algorithms usually divide the labels with an optimal threshold according to their relevances to an unseen instance. However, it may easily cause misclassification to directly determine whether an unseen instance has the label with relevance close to the threshold. The label with relevance close to the threshold has a high uncertainty. Three-way decisions theory is an efficient method to solve the uncertainty problem. Therefore, based on three-way decisions theory, a multi-label learning algorithm with label dependency is proposed in this paper. Label dependency is an inherent property in multi-label data. The labels with high uncertainty are further handled with a label dependency model, which is represented by the logistic regression in this paper. The experimental results show that this algorithm performs better.

Keywords: Multi-label learning · Label dependency · Three-way decisions · Logistic regression

1 Introduction

Multi-label learning is a challenging problem in machine learning field, because multi-label instances have several possible labels simultaneously and labels have correlations with each other in multi-label data. Given a predefined label space L, the task of multi-label learning algorithm is to predict a set of relevant class labels Y for an unseen instance through analyzing training instances with known label sets, where $Y \subset L$ and $|Y| \geq 1$ [1–3]. Multi-label objects exist widely in various real-world domains. For example, in the image domain [4], a picture may express multiple semantic classes simultaneously, such as *sea*, *beach* and *sky*. In the text domain [5], a document possibly belongs to several topics, such as *society*, *sport* and *politics*. In the biology domain [6], a gene could have a set of functions, such as *transcription* and *metabolism*. In the video domain [7], a movie may be labeled with several genres, such as *horror*, *cartoon* and *family*.

Multi-label classification (MLC) and *multi-label ranking* (MLR) are two major tasks in multi-label learning [1]. MLC predicts binary values for an unseen

© Springer International Publishing AG 2016
V. Flores et al. (Eds.): IJCRS 2016, LNAI 9920, pp. 240–249, 2016.
DOI: 10.1007/978-3-319-47160-0_22

instance instructing relevant or irrelevant to labels, while MLR yields an order of labels according to their relevances to an unseen instance. The outputs of them, especially MLR, greatly depend on the label relevance. There are several ways to measure the label relevance, such as vote, possibility and membership degree. Here, the possibility is used for investigation in this paper, and others have the similar disciplines. Most of the multi-label algorithms firstly predict relevances that an unseen instance has the labels, then find a threshold t to get a bipartition of the labels into relevant or irrelevant. The instance more possibly has the label with greater relevance. On the contrary, those with smaller relevance are more likely to be not associated to the instance. Therefore, it is very certain that the instance has the labels with very great relevance and does not have the labels with very few relevance. However, it is hard to judge whether the instance has the label with a relevance around the threshold, which is full of uncertainty, usually resulting in misclassification.

Three-way decisions theory is an efficient method to solve the uncertainty problem, which is proposed by Yao [8]. The method can improve the algorithm performance, and simplify the complex problem. It divides the problem into three regions, and different decisions are taken for different regions. Normally, the problem in the uncertain region will be further handled to make the right judgement. According to the relevance, the labels can be grouped into three regions in multi-label learning. The region with great relevance is the positive region, the region with few relevance is the negative region, and the region between them is called the boundary region. The labels in the positive region are assigned to the instance, while those in the negative region are not. We are not sure about labels in the boundary region, needing a further learning.

In multi-label data, there usually exists dependency among labels. For example, an action movie is more likely to be an adventure movie than be a romance at the same time. Label dependency is a hot topic, and there are a great number of algorithms about how to explore the label dependency in multi-label learning [10–12]. Hence, the labels in boundary region can be further predicted with the help of labels in the positive and the negative regions by using the label dependency. We propose a multi-label learning algorithm with label dependency based on three-way decisions theory to improve the algorithm performance. A logistic regression model is constructed to represent the label dependency. We experiment the proposed algorithm on multi-label data sets, and the results show that the proposed algorithm can achieve a better performance.

The rest of this paper is organized as follows: Sect. 2 briefly reviews the related work of multi-label learning. In Sect. 3, some basic concepts of three-way decisions theory and multi-label learning are introduced. In Sect. 4, we learn a model to revise the uncertain labels with label dependency. Section 5 displays the experimental results. We conclude the paper in Sect. 6.

2 Related Work

In recent years, multi-label learning has attracted significant attentions from various domains, and been a hot topic in machine learning field. A lot of multi-label learning algorithms have been proposed. These proposed multi-label learning algorithms can be divided into two groups: *problem transportation method* (PTM) and *algorithm adaption method* (AAM) [1]. PTM is independent on algorithm, and transforms the multi-label data into numerous single-label data, such as Binary Relevance (BR) [13], Pairwise Binary (PW) [14], and Label Powerset (LP) [15]. AAM on the other hand extends some specific traditional machine learning algorithms to handle the multi-label data directly, such as decision tree [16], support vector machine [17], neural networks [18] and rough sets [12].

Furthermore, based on rough sets, Yu [12] proposed a multi-label learning with exploiting label correlation, called MLRS-LC. To exploit the label dependency, Zhang [10] proposed a multi-label learning by exploiting label dependency, which uses a Bayesian network structure to efficiently encode the conditional dependencies of the labels as well as the feature set. Kang [11] correlated label propagation with application to multi-label learning, which explicitly models interactions between labels in an efficient manner. In a word, the label dependency should be taken into consideration.

3   Preliminaries

3.1   Three-Way Decisions

Three-way decisions theory is a proper semantic explanation of probabilistic rough sets and decision-theoretic rough sets [8,9]. The main idea is to divide the whole into three regions, and different regions are treated with different ways. Let $Pr(X|[x])$ denote the conditional probability that x belongs to X.

$$Pr(X|[x]) = \frac{|[x] \cap X|}{|[x]|} \tag{1}$$

$[x]$ is the equivalence class of x, and $|\cdot|$ stands for the cardinality. Then, the three regions of three-way decisions can be represented by probabilistic rough sets [19] as follow:

$$
\begin{aligned}
POS(X) &= \{x | Pr(X|[x]) \geq \alpha\}; \\
NEG(X) &= \{x | Pr(X|[x]) \leq 1 - \alpha\}; \\
BND(X) &= \{x | 1 - \alpha < Pr(X|[x]) < \alpha\}.
\end{aligned}
\tag{2}
$$

where $POS(X)$ denotes the positive region of X, $NEG(X)$ denotes the negative region, and $BND(X)$ is the boundary region. α is a threshold and $\alpha \in [0.5, 1]$. When $\alpha = 0.5$, the three-way decisions become the two-way decisions.

The three-way decisions theory is a generalized and efficient model for decisions and information processing, not limited for rough sets. There widely exist three-way phenomena in the real-world.

3.2   Multi-label Learning

Formally, let $F \subset R^b$ represent the input feature space, and $L = \{l_1, l_2, ..., l_q\}$ denote the label space with q possible labels. Given a multi-label training data

$T = \{(X_1, Y_1), (X_2, Y_2), ..., (X_n, Y_n)\}$, X_i is a b-dimensional input feature vector, and $Y_i = \{y_i^1, y_i^2, ..., y_i^q\}$ is the binary label vector of X_i, where y_i^j equals to 1 if X_i has label l_j, and equals to -1, otherwise. The task of multi-label learning is to derive a multi-label classification function $h : F \to \{0,1\}^q$, through the training data T. For an unseen instance X, the multi-label classification function can predict its relevant label vector $Y' = \{y^{1'}, y^{2'}, ..., y^{q'}\}$.

However, most of multi-label learning algorithms do not directly predict whether the instance X has the label l_j, but firstly give a relevance $h^j(X)$ between X and l_j, which is usually a possibility, then, divide the labels with an optimal threshold t as follows:

$$y^{j'} = \begin{cases} 1, & h^j(X) \geq t \\ -1, & h^j(X) < t \end{cases} \tag{3}$$

The relevance $h^j(X)$ is a certainty degree that the label l_j belongs to X. So it is very certain that X belongs to the labels with relevances significantly greater than the threshold t. It is almost impossible that the labels with relevances much less than t are assigned to X, namely, these labels are very certain to not relate to X. It is full of uncertainties that the labels with relevances around t. The closer to t the relevance gets, the more uncertain the label is. Therefore, the three-way decisions theory is used to solve the problem in multi-label learning.

4   The Proposed Algorithm

Label dependency is important information contained in multi-label data, and a hot topic in multi-label learning. Therefore, it is a practicable way to correct the labels with high uncertainties through label dependency which can be represent by a dependency model of a label on the other $q - 1$ labels. Here, the logistic regression is used to construct the dependency model of label l_j on the others.

$$g^j(X) = \frac{1}{1 + e^{-u_j}} \tag{4}$$

where the equation is a sigmod function, and

$$u_j = \theta_{j1} * y^1 + ... + \theta_{jj-1} * y^{j-1} + \theta_{jj+1} * y^{j+1} + ... + \theta_{jq} * y^q + \theta_{jj} \tag{5}$$

$\theta_{ji(i \neq j)}$ is the weight of l_i to l_j which informs the dependency between l_i to l_j, and θ_{jj} is a constant for l_j. Equation (5) can be rewritten as:

$$u_j = \theta_j * y_j^T \tag{6}$$

In the equation, $\theta_j = \{\theta_{j1}, \theta_{j2}, ..., \theta_{jq}\}$ is the weight vector, and $y_j = \{y^1, ..., y^{j-1}, 1, y^{j+1}, ..., y^q\}$ is the input vector where the input for constant is set to 1. Then,

$$g^j(X) = \frac{1}{1 + e^{-\theta_j * y_j^T}} \tag{7}$$

The weight vector θ_j is trained with the label information in training data set.

Given a test instance X, and its label relevance $h(X)$ predicted by a multi-label learning algorithm, the label space L can be grouped into three regions for X according to $h(X)$, namely, the positive region $POS(X)$ in which the labels are assigned to X, the negative region $NEG(X)$ in which the labels are not related to X, and the boundary region $BND(X)$ where the labels are uncertain and need to be further predicted. The three regions can be defined as:

$$\text{if } h^j(X) \geq t + \beta, \text{ then } l_j \in POS(X);$$
$$\text{if } h^j(X) \leq t - \beta, \text{ then } l_j \in NEG(X); \tag{8}$$
$$\text{if } t - \beta < h^j(X) < t + \beta, \text{ then } l_j \in BND(X).$$

where t is the optimal threshold in the original multi-label learning algorithm, and $\beta \in [0, \min(t, 1 - t)]$ determines the width of the boundary region, i.e. the uncertainty region. A three-value label vector $Z = \{z^1, z^2, ..., z^q\}$ can be gotten for X as follows:

$$z^j = \begin{cases} 1, & l_j \in POS(X) \\ 0, & l_j \in BND(X) \\ -1, & l_j \in NEG(X) \end{cases} \tag{9}$$

The labels in $POS(X)$ $[NEG(X)]$ have very high certainty degrees belonging [not belonging] to X, and do not need to be further processed and changed. Suppose $t \geq (1 - t)$, then $\beta \in [0, 1 - t]$. Z is used as an input vector of Eq. (7) to obtain a correction term φ_j for label $l_j \in BND(X)$

$$\varphi_j = \frac{1}{1 + e^{-(\theta_j * Z^T + \theta_{jj})}} \tag{10}$$

For $l_j \in BND(X)$, $z^j = 0$, the constant θ_{jj} is added. In the input vector Z, the values of the labels in $BND(X)$ are 0, means they have no influence on φ_j, because of their high certainties. For $l_j \in BND(X)$, φ_j is added to the original label relevance $h^j(X)$. Therefore the label relevance after correcting $f^j(X)$ is computed as follows:

$$f^j(X) = \begin{cases} (t + \beta) * h^j(X) + (1 - t - \beta) * \varphi_j, & \text{if } l_j \in BND(X) \\ h^j(X), & \text{otherwise} \end{cases} \tag{11}$$

The formula considers the label relevance predicted from the features by the original multi-label learning algorithm and the label relevance from label dependency simultaneously. $(1 - t - \beta)$ is the weight of the correction term, and determines influence of the correction term. The boundary region becomes lager with the increase of β, leading to the rising of number of uncertain labels and decreasing of the reliability of the correction term. Therefore, it can be seen that the weight of the correction term decreases as the β increases. When $\beta = 0$, there is no uncertain label needing to be corrected, so the label relevance keeps the same. When $\beta = 1 - t$, the certain labels is the least, so the weight of the correction term equals to 0, and no change is on the label relevance.

The label l_j can be predicted whether be associated to X or not by using label relevance $f^j(X)$ after correcting as:

$$y^{j'} = \begin{cases} 1, & f^j(X) \geq t \\ -1, & f^j(X) \leq t \end{cases} \tag{12}$$

Algorithm 1. The multi-label learning algorithm with label dependency based on three-way decisions theory

---

Input: Original label relevance $h(X)$; Parameter of the width of boundary region β;
　Optimal threshold t; Label data $W = \{(Y_1), (Y_2), ..., (Y_n)\}$;
Output: Predicted label vector Y'
　for $l_j \in L$ **do**
　　//Initialize variables;
　　$z^j \leftarrow 0$;
　　$\varphi_j \leftarrow 0$;
　　$f^j(X) \leftarrow 0$;
　　$y^{j'} \leftarrow 0$;
　　Compute the value z^j with $h^j(X)$ according to equations (8) and (9);
　end for
　for $l_j \in L$ **do**
　　Construct the logistic regression model with W to get the weight vector θ_j;
　　Count the correction term φ_j according to equation (10);
　　Calculate $f^j(X)$ according to equation (11);
　　Determine $y^{j'}$ according to equation (12)
　end for
　Output the predicted label vector $Y' = \{y^{1'}, y^{2'}, ..., y^{q'}\}$;

---

5 Experimental Results

5.1 Data Sets

We experiment on three real-world multi-label data sets covering different domains from the Mulan Libary [20]. The statistical information is summarized in Table 1. As shown in Table 1, *Medical* [21] data set has 978 instances, of

Table 1. Multi-label data sets in the experiments

Name	Instance	Feature	Label	Cardinality
Medical	978	1449	45	1.245
Enron	1702	1001	53	3.378
CAL500	502	68	174	26.044

which each instance is a radiology text report consisting of the medical history and symptom and is associated with a subset of 45 ICD-9-CM labels. There are 1702 instances in *Enron* [22] data set, and these instances are e-mails of the Enron company and labeled with 53 possible tags. *CAL500* [23] data set contains 502 popular musical tracks, and 174 labels such as style, emotion and instrument.

5.2   Evaluation Criteria

Five example based multi-label learning evaluation criteria are considered, *Hamming loss, Precision, Recall, F1-measure, Accuracy* [1]. The larger the latter four evaluation criteria are, the better the algorithm performs, while *Hamming loss* in contrast. Given a testing multi-label data set $D = \{(X_1, Y_1), (X_2, Y_2), ..., (X_m, Y_m)\}$, the five evaluation criteria are defined as follows:

Hamming loss evaluates how many labels belonging to the instance is not associated, or not belonging to the instance is associated. $\langle \pi \rangle$ equals to 1 if π holds and 0 otherwise. $Hamloss = \frac{1}{mq} \sum_{i=1}^{m} \sum_{j=1}^{q} \langle y'_{ij} \neq y_{ij} \rangle$.

Precision evaluates how many labels actually belong to the instance in the predicted label set. $Precision = \frac{1}{m} \sum_{i=1}^{m} \frac{|Y_i \cap Y'_i|}{|Y'_i|}$.

Recall computes the number of labels that the are correctly predicted in the ground-truth label set. $Recall = \frac{1}{m} \sum_{i=1}^{m} \frac{|Y_i \cap Y'_i|}{|Y_i|}$.

F1-measure is the harmonic mean between *precision* and *recall*, common to information retrieval. $F1 = \frac{1}{m} \sum_{i=1}^{m} \frac{|Y_i \cap Y'_i|}{|Y_i| + |Y'_i|}$.

Accuracy measures the average degree of similarity between the predicted and the ground-truth label sets of all testing instances. $Accuracy = \frac{1}{m} \sum_{i=1}^{m} \frac{|Y_i \cap Y'_i|}{|Y_i \cup Y'_i|}$.

5.3   Results and Discussion

The *ten-fold cross-validations* evaluation is used to evaluate algorithms in the experiment. ML-KNN is a popular multi-label algorithm and chosen to produce the original label relevance. As recommended in [24], the number of neighbors is 10, and the threshold t is set to be 0.5. The β arranges from 0 to 0.5 with a step of 0.05. In the following tables, the symbol '↓' represents that the smaller the evaluation criterion value is, the better the performance is, while the symbol '↑' in contrast. Furthermore, the best result is marked in boldface on each evaluation criterion by considering the mean value.

When β is set to be 0, there are no labels changed on all data sets. Therefore, the algorithm results with β equal to 0 are the same as the original results achieved by the ML-KNN. Tables 2, 3 and 4 show that when β is 0.5, the algorithm performance is the same as the original performance, too, the reason of which has been discussed in Sect. 4. In more detail, the performance is improved

Table 2. Experimental results (mean ± std. deviation) on the *Medical* data set.

β	Hamming loss↓	Precision↑	Recall↑	F1↑	Accuracy↑
0	0.0155 ± 0.0070	0.6232 ± 0.2132	0.5888 ± 0.1727	0.2970 ± 0.0929	0.0163 ± 0.0049
0.05	0.0154 ± 0.0074	0.6428 ± 0.1966	0.6012 ± 0.1481	0.3048 ± 0.0823	0.0166 ± 0.0048
0.10	0.0154 ± 0.0076	0.6536 ± 0.2019	0.6064 ± 0.1633	0.3086 ± 0.0872	0.0167 ± 0.0052
0.15	**0.0153 ± 0.0081**	**0.6665 ± 0.2049**	**0.6153 ± 0.1807**	**0.3139 ± 0.0930**	**0.0169 ± 0.0056**
0.20	0.0157 ± 0.0078	0.6624 ± 0.1811	0.6134 ± 0.1591	0.3125 ± 0.0813	0.0168 ± 0.0047
0.25	0.0157 ± 0.0085	0.6598 ± 0.1781	0.6082 ± 0.1870	0.3106 ± 0.0885	0.0166 ± 0.0057
0.30	0.0161 ± 0.0087	0.6412 ± 0.2011	0.5921 ± 0.2113	0.3021 ± 0.1011	0.0161 ± 0.0063
0.35	0.0174 ± 0.0083	0.5649 ± 0.2163	0.5210 ± 0.2476	0.2656 ± 0.1149	0.0142 ± 0.0071
0.40	0.0189 ± 0.0052	0.4665 ± 0.1622	0.4299 ± 0.1886	0.2190 ± 0.0851	0.0119 ± 0.0052
0.45	0.0201 ± 0.0060	0.3804 ± 0.2210	0.3503 ± 0.2039	0.1781 ± 0.1033	0.0099 ± 0.0060
0.50	0.0155 ± 0.0070	0.6232 ± 0.2132	0.5888 ± 0.1727	0.2970 ± 0.0929	0.0163 ± 0.0049

Table 3. Experimental results (mean ± std. deviation) on the *Enron* data set.

β	Hamming loss↓	Precision↑	Recall↑	F1↑	Accuracy↑
0	0.0539 ± 0.0074	0.5644 ± 0.0915	0.3443 ± 0.0731	0.1997 ± 0.0354	0.0239 ± 0.0038
0.05	0.0540 ± 0.0067	0.5715 ± 0.1037	0.3549 ± 0.1116	0.2047 ± 0.0499	0.0243 ± 0.0054
0.10	0.0546 ± 0.0063	0.5746 ± 0.1025	0.3627 ± 0.1466	0.2083 ± 0.0638	**0.0244 ± 0.0083**
0.15	0.0548 ± 0.0058	0.5685 ± 0.1056	**0.3669 ± 0.1825**	**0.2091 ± 0.0774**	0.0241 ± 0.0106
0.20	0.0553 ± 0.0069	0.5477 ± 0.1560	0.3411 ± 0.2450	0.1967 ± 0.1104	0.0232 ± 0.0136
0.25	0.0543 ± 0.0086	0.5631 ± 0.1078	0.3441 ± 0.1930	0.1998 ± 0.0850	0.0232 ± 0.0113
0.30	0.0539 ± 0.0080	0.5705 ± 0.0996	0.3426 ± 0.1699	0.2002 ± 0.0759	0.0230 ± 0.0096
0.35	0.0539 ± 0.0071	0.5765 ± 0.0865	0.3518 ± 0.1120	0.2046 ± 0.0485	0.0238 ± 0.0065
0.40	**0.0536 ± 0.0072**	**0.5780 ± 0.0851**	0.3442 ± 0.1106	0.2015 ± 0.0491	0.0236 ± 0.0055
0.45	0.0538 ± 0.0069	0.5549 ± 0.1034	0.3345 ± 0.0917	0.1952 ± 0.0430	0.0238 ± 0.0046
0.50	0.0539 ± 0.0074	0.5644 ± 0.0915	0.3443 ± 0.0731	0.1997 ± 0.0354	0.0239 ± 0.0038

to reach the best, then decreases gradually. That is because when β is too small, not many labels are corrected, while the certain labels are not enough to produce a reliable correction term, if β is too large. Thus, it is a proper β that could produce a balance between the number of the certain labels and the number of the uncertain to obtain the best performance.

As shown in Table 2, the proposed algorithm obtains the best performance on all evaluation criteria on *Medical* data set, when β is equal to 0.15. The proposed algorithm improves the performance of *Medical*, especially on *precision*, *recall* and *F1*. On the *Enron* data set, the proposed algorithm performs best on *hamming loss* and *precision* when β is 0.4, while it achieves the best results on *recall* and *F1* when β is 0.15 and *accuracy* when β is 0.1. On the *CAL500* data set, the proposed algorithm performs best when β is 0.3 on all evaluation criteria except for *precision*, which is the best when β is 0.05. All the best results of the proposed algorithm are better than the original ones without correcting, and it promotes the original performance.

Table 4. Experimental results (mean ± std. deviation) on the *CAL500* data set.

β	Hamming loss↓	Precision↑	Recall↑	F1↑	Accuracy↑
0	0.1399 ± 0.0201	0.5927 ± 0.0732	0.2247 ± 0.0370	0.1604 ± 0.0219	0.0394 ± 0.0038
0.05	0.1390 ± 0.0212	**0.6119 ± 0.0702**	0.2117 ± 0.0456	0.1547 ± 0.0277	0.0369 ± 0.0068
0.10	0.1394 ± 0.0240	0.6077 ± 0.0849	0.2059 ± 0.0906	0.1511 ± 0.0562	0.0359 ± 0.0142
0.15	0.1400 ± 0.0215	0.5992 ± 0.0779	0.2078 ± 0.0894	0.1513 ± 0.0559	0.0364 ± 0.0149
0.20	0.1411 ± 0.0211	0.5844 ± 0.0844	0.2187 ± 0.0633	0.1561 ± 0.0384	0.0385 ± 0.0090
0.25	0.1409 ± 0.0202	0.5831 ± 0.0771	0.2292 ± 0.0544	0.1615 ± 0.0319	0.0404 ± 0.0076
0.30	**0.1389 ± 0.0211**	0.5962 ± 0.0696	**0.2362 ± 0.0502**	**0.1666 ± 0.0290**	**0.0416 ± 0.0047**
0.35	0.1394 ± 0.0210	0.5990 ± 0.0781	0.2253 ± 0.0469	0.1612 ± 0.0283	0.0397 ± 0.0051
0.40	0.1390 ± 0.0211	0.6077 ± 0.0901	0.2122 ± 0.0519	0.1549 ± 0.0322	0.0372 ± 0.0070
0.45	0.1392 ± 0.0181	0.6038 ± 0.0722	0.2162 ± 0.0278	0.1568 ± 0.0183	0.0380 ± 0.0045
0.50	0.1399 ± 0.0201	0.5927 ± 0.0723	0.2247 ± 0.0370	0.1604 ± 0.0219	0.0394 ± 0.0038

6   Conclusions

By using a logistic regression model of label dependency, this paper proposed a multi-label learning algorithm based on three-way decisions theory to further handle those labels with high uncertainty. The experimental results show that it is helpful to correct the labels near the threshold through the proposed algorithm. How to theoretically choose the best β is not researched, hence in the next step, we will propose a theory analysis to choose an optimal width of the boundary region. Furthermore, two variables instead of a single variable β are taken into consideration to determine the width of boundary region, which is more generalized and not restricted by the threshold.

Acknowledgments. The work is partially supported by the National Natural Science Foundation of China (Nos. 61273304, 61573259), the Specialized Research Fund for the Doctoral Program of Higher Education of China (No. 20130072130004), and the program of Further Accelerating the Development of Chinese Medicine Three Year Action of Shanghai (2014–2016) (No. ZY3-CCCX-3-6002).

References

1. Tsoumakas, G., Katakis, I., Vlahavas, I.: Mining multi-label data. In: Maimon, O., Rokach, L. (eds.) Data Mining and Knowledge Discovery Handbook, pp. 667–685. Springer, New York (2010)
2. Tsoumakas, G., Katakis, I.: Multi label classification: an overview. Int. J. Data Warehouse. Min. **3**(3), 1–13 (2007)
3. Zhang, M.L., Zhou, Z.H.: A review on multi-label learning algorithms. IEEE Trans. Knowl. Data Eng. **26**(8), 1819–1837 (2014)
4. Yu, Y., Pedrycz, W., Miao, D.Q.: Neighborhood rough sets based multi-label classification for automatic image annotation. Int. J. Approximate Reason. **54**(9), 1373–1387 (2013)
5. Schapire, R.E., Singer, Y.: BoosTexter: a boosting-based system for text categorization. Mach. Learn. **39**(2), 135–168 (2000)

6. Pavlidis, P., Weston, J., Cai, J., Grundy, W.N.: Combining microarray expression data and phylogenetic profiles to learn functional categories using support vector machines. In: Proceedings of the Fifth Annual International Conference on Computational Biology, Montreal, Canada, pp. 242–248 (2001)

7. Snoek, C.G.M., Worring, M., Van Gemert, J.C., et al.: The challenge problem for automated detection of 101 semantic concepts in multimedia. In: Proceedings of the 14th Annual ACM International Conference on Multimedia, pp. 421–430 (2006)

8. Yao, Y.: An outline of a theory of three-way decisions. In: Yao, J.T., Yang, Y., Słowiński, R., Greco, S., Li, H., Mitra, S., Polkowski, L. (eds.) RSCTC 2012. LNCS, vol. 7413, pp. 1–17. Springer, Heidelberg (2012). doi:10.1007/978-3-642-32115-3_1

9. Pawlak, Z.: Rough sets. Int. J. Parallel Prog. **11**(5), 341–356 (1982)

10. Zhang, M.L., Zhang, K.: Multi-label learning by exploiting label dependency. In Proceedings of the 16th ACM SIGKDD International Conference on Knowledge Discovery and Data Mining, pp. 999–1008. ACM, New York (2010)

11. Kang, F., Jin, R., Sukthankar, R.: Correlated label propagation with application to multi-label learning. In: Proceedings of the 2006 IEEE Computer Society Conference on Computer Vision and Pattern Recognition, pp. 1719–1726 (2006)

12. Yu, Y., Predrycz, W., Miao, D.Q.: Multi-label classification by exploiting label correlations. Expert Syst. Appl. **41**(6), 2989–3004 (2014)

13. Boutell, M.R., Luo, J., Shen, X., et al.: Learning multi-label scene classification. Pattern Recogn. **37**(9), 1757–1771 (2004)

14. Hllermeier, E., Frnkranz, J., Cheng, W., et al.: Label ranking by learning pairwise preferences. Artif. Intell. **172**(16), 1897–1916 (2008)

15. Tsoumakas, G., Vlahavas, I.P.: Random k-labelsets: an ensemble method for multilabel classification. In: Kok, J.N., Koronacki, J., Lopez de Mantaras, R., Matwin, S., Mladenič, D., Skowron, A. (eds.) ECML 2007. LNCS (LNAI), vol. 4701, pp. 406–417. Springer, Heidelberg (2007). doi:10.1007/978-3-540-74958-5_38

16. Clare, A., King, R.D.: Knowledge discovery in multi-label phenotype data. In: Raedt, L., Siebes, A. (eds.) PKDD 2001. LNCS (LNAI), vol. 2168, pp. 42–53. Springer, Heidelberg (2001). doi:10.1007/3-540-44794-6_4

17. Elisseeff, A., Weston, J.: A kernel method for multi-labelled classification. In: Advances in Neural Information Processing Systems, vol. 14, pp. 681–687 (2001)

18. Zhang, M.L., Zhou, Z.H.: Multilabel neural networks with applications to functional genomics and text categorization. IEEE Trans. Knowl. Data Eng. **18**(10), 1338–1351 (2006)

19. Yao, Y.Y.: Three-way decisions with probabilistic rough sets. Inf. Sci. **180**(3), 341–353 (2010)

20. Tsoumakas, G., Spyromitros-Xiousfis, E., Vilcek, I.V.J.: Mulan: a Java library for multi-label learning. J. Mach. Learn. Res. **12**(7), 2411–2414 (2011)

21. Pestian, J., Brew, C., Matykiewicz, P., et al.: A shared task involving multi-label classification of clinical free text. In: Proceedings of the Workshop on BioNLp 2007, pp. 97–104. Association for Computational Linguistics, Stroudsburg (2007)

22. UC Berkeley Enron Email Analysis Project. http://bailando.sims.berkeley.edu/enron_email.html

23. Turnbull, D., Barrington, L., Torres, D., et al.: Semantic annotation and retrieval of music and sound effects. IEEE Trans. Audio Speech Lang. Process. **16**(2), 467–476 (2008)

24. Zhang, M.L., Zhou, Z.H.: ML-kNN: a lazy learning approach to multi-label learning. Pattern Recogn. **40**(7), 2038–2048 (2007)

A Decision-Theoretic Rough Set Approach to Multi-class Cost-Sensitive Classification

Guojian Deng and Xiuyi Jia[✉]

School of Computer Science and Engineering,
Nanjing University of Science and Technology, Nanjing 210094, China
dengguojian_njust@163.com, jiaxy@njust.edu.cn

Abstract. As a kind of probabilistic rough set model, decision-theoretic rough set is usually used to deal with binary classification problems. This paper provides a new formulation of multi-class decision-theoretic rough set by combining decision-theoretic rough set model with classical cost-sensitive learning. Upper approximation, lower approximation, positive region, negative region and boundary region can be derived from the $n \times n$ cost matrix of classical multi-class situation. The probability thresholds for three-way decisions making are defined. A cost-sensitive classification algorithm based on multi-class decision-theoretic rough set model is presented. The experimental results on several UCI data sets indicate that the proposed algorithm can get a better performance on classification accuracy and total cost.

Keywords: Decision-theoretic rough set model · Cost-sensitive learning · Multi-class classification problems · Three-way decisions

1 Introduction

Rough set theory is a mathematical tool which is proposed by Pawlak to deal with the imprecise and uncertain problems [1]. In Pawlak rough set model, the lower and upper approximations are determined by the algebraic relation between sets [5]. The lower and upper approximations can divide the universe into three pair-wise disjoint regions: positive, boundary and negative regions.

Unfortunately, since the definition of lower and upper approximations in Pawlak rough set is a qualitative description and strict, makes Pawlak rough set lack of fault tolerance in classification problems [2]. Therefore, Pawlak rough set tends to produce more misclassification information when it deals with the uncertain data in the real world, as well as different misclassification may lead to different classification costs. Fortunately, Yao et al. [4] proposed decision-theoretic rough set model (**DTRS**) to overcome above problems. The qualitative relation of set inclusion in Pawlak rough set can be replaced by a quantitative probability inclusion relation in DTRS. In the process of solving classification problems, DTRS introduces the cost-sensitive analysis mechanism and the decision is made on the basis of Bayesian decision theory, which makes DTRS have fault tolerance and cost sensitivity.

V. Flores et al. (Eds.): IJCRS 2016, LNAI 9920, pp. 250–260, 2016.
DOI: 10.1007/978-3-319-47160-0_23

Cost matrix plays an important role in DTRS. In general, cost matrix is given by experts, and the two probability thresholds of dividing positive, negative and boundary regions can be calculated based on Bayesian decision theory [3]. If the cost matrix cannot be obtained from experts, some self-learning methods were proposed [8]. After getting probability thresholds, we can obtain positive, negative and boundary rules, corresponding to classify an object into positive, negative and boundary regions, respectively. The positive region means making a decision of acceptance, the negative region means making a decision of rejection and the boundary region means making a deferred decision [2]. Then we can get the cost that is composed of three types of costs: costs of the positive, boundary and negative rules of the whole table [9].

As a kind of decision making method, DTRS is usually used to address two-class classification problems, but decision makers may encounter some multi-class problems in reals applications. In recent years, the research of multi-class decision-theoretic rough set model has made some progress, Zhou [7] proposed a $3n \times n$ (suppose we have n values in the decision class) cost matrix to deduce a multi-class decision-theoretic rough set model. Liu et al. [6] presented a $n \times 6$ cost matrix and a two stages method with Bayesian decision procedure to solve the multiple-category classification problems based on DTRS. Lingras et al. [12] proposed a rough multi-category decision theoretic framework with $2^n - 1$ cost functions. However, the cost matrix of their studies are different from the cost matrix in classical cost-sensitive learning.

Cost-sensitive learning is an important research direction in machine learning. When the different types of misclassification costs are unequal, the objective of classification is to minimize the total classification cost instead of minimizing the classification error rate based on a $n \times n$ cost matrix. Much attention has been paid to the study of cost-sensitive learning and its classification approaches in recent years [10,13].

Motivated by above analysis, this paper combines DTRS with classical cost-sensitive learning, proposes a multi-class decision-theoretic rough set model which imports a $n \times n$ cost matrix as the input, and present a multi-class decision-theoretic rough set cost-sensitive classification algorithm. Finally, the effectiveness of the proposed model and algorithm is verified by the comparison experiments on several UCI data sets.

2 Decision-Theoretic Rough Set Model

In the DTRS model, the set of states $\Omega = \{X, X^C\}$ indicates that an object is in a decision class X and not in X, respectively. With respect to the three regions of rough set, let $A = \{a_P, a_B, a_N\}$ be a set of actions, where a_P, a_B and a_N denote three actions in classifying an object x into positive, boundary and negative regions, namely, deciding $x \in POS(X)$, $x \in BND(X)$ and $x \in NEG(X)$, respectively. When an object x belongs to X, let λ_{PP}, λ_{BP} and λ_{NP} denote the cost of taking actions a_P, a_B and a_N, respectively. When an object x does not belong to X, let λ_{PN}, λ_{BN} and λ_{NN} denote the cost of taking actions a_P, a_B

and a_N, respectively. Let $p(X|x)$ be the conditional probability of an object x being in class X.

Suppose the cost functions satisfy the condition: $\lambda_{PP} \leq \lambda_{BP} < \lambda_{NP}$ and $\lambda_{NN} \leq \lambda_{BN} < \lambda_{PN}$. We can get following decision rules based on the Bayesian minimum cost decision theory:

(P) If $p(X|x) \geq \alpha$ and $p(X|x) \geq \gamma$, decide $x \in POS(X)$,

(B) If $p(X|x) \geq \beta$ and $p(X|x) \leq \alpha$, decide $x \in BND(X)$,

(N) If $p(X|x) \leq \beta$ and $p(X|x) \leq \gamma$, decide $x \in NEG(X)$.

Where the parameters α, γ and β are defined as:

$$
\begin{aligned}
\alpha &= \frac{\lambda_{PN} - \lambda_{BN}}{(\lambda_{PN} - \lambda_{BN}) + (\lambda_{BP} - \lambda_{PP})}, \\
\gamma &= \frac{\lambda_{PN} - \lambda_{NN}}{(\lambda_{PN} - \lambda_{NN}) + (\lambda_{NP} - \lambda_{PP})}, \\
\beta &= \frac{\lambda_{BN} - \lambda_{NN}}{(\lambda_{BN} - \lambda_{NN}) + (\lambda_{NP} - \lambda_{BP})}.
\end{aligned}
\tag{1}
$$

When $(\lambda_{PN} - \lambda_{BN}) \cdot (\lambda_{NP} - \lambda_{BP}) > (\lambda_{BP} - \lambda_{PP}) \cdot (\lambda_{BN} - \lambda_{NN})$, we have $0 \leq \beta < \gamma < \alpha \leq 1$. After tie-breaking, we obtain:

(P1) If $p(X|x) \geq \alpha$, decide $x \in POS(X)$,

(B1) If $\beta < p(X|x) < \alpha$, decide $x \in BND(X)$,

(N1) If $p(X|x) \leq \beta$, decide $x \in NEG(X)$.

3   Multi-class Decision-Theoretic Rough Set and Cost-Sensitive Classification

In this section, we will propose a multi-class decision-theoretic rough set model by a $n \times n$ cost matrix that is the same with cost-sensitive learning in machine learning. The method for determining the values of $\lambda_{PP}, \lambda_{BP}, \lambda_{NP}, \lambda_{NN}, \lambda_{BN}$ and λ_{PN} of each class is given in the semantic perspective. The expressions of upper and lower approximations, positive, boundary and negative regions are defined on the basis of the probability thresholds derived from Bayesian decision theory, and provide a multi-class decision-theoretic rough set approach to multi-class cost-sensitive classification.

Definition 1. Let $\Omega = \{C_1, C_2, ..., C_n\}$ be a finite decision set of n classes and $A = \{a_1, a_2, ..., a_n\}$ be a finite set of n possible actions that classify an object into corresponding class. The cost matrix of multi-class classification problems can be denoted as Table 1. Where $\lambda_{ij} = \lambda(a_i|C_j)$ denotes the cost of classifying

Table 1. The cost matrix of multi-class classification problems.

	C_1	C_2	...	C_j	...	C_n
a_1	$\lambda_{11} = \lambda(a_1\|C_1)$	$\lambda_{12} = \lambda(a_1\|C_2)$	...	$\lambda_{1j} = \lambda(a_1\|C_j)$	...	$\lambda_{1n} = \lambda(a_1\|C_n)$
a_2	$\lambda_{21} = \lambda(a_2\|C_1)$	$\lambda_{22} = \lambda(a_2\|C_2)$	...	$\lambda_{2j} = \lambda(a_2\|C_j)$	...	$\lambda_{2n} = \lambda(a_2\|C_n)$
...	...	...	...	...	...	...
a_i	$\lambda_{i1} = \lambda(a_i\|C_1)$	$\lambda_{i2} = \lambda(a_i\|C_2)$	...	$\lambda_{ij} = \lambda(a_i\|C_j)$	...	$\lambda_{in} = \lambda(a_i\|C_n)$
...	...	...	...	...	...	...
a_n	$\lambda_{n1} = \lambda(a_n\|C_1)$	$\lambda_{n2} = \lambda(a_n\|C_2)$	...	$\lambda_{nj} = \lambda(a_n\|C_j)$	...	$\lambda_{nn} = \lambda(a_n\|C_n)$

an object x belonging to C_j into C_i. $\lambda_{ii} = 0$ indicates zero cost for correct classification, $\lambda_{ij} > 0$ denotes the cost of misclassification when $i \neq j$.

In decision-theoretic rough set model, we must identify the values of λ_{PP}, λ_{BP}, λ_{NP}, λ_{NN}, λ_{BN} and λ_{PN} in calculating the costs of classifying an object x into positive, negative and boundary regions. However, the cost functions appeared in Table 1 represent the misclassification cost between two classes only, it is not easy to transfer these $n * n$ values into 6 values directly. Thus, we need to reconsider the multi-class cost-sensitive learning problems in both the semantic and the calculation perspectives. Let λ_{PP}^i, λ_{BP}^i and λ_{NP}^i denote the cost incurred for taking actions of classifying an object into positive, boundary and negative regions, respectively, when the object belongs to C_i, let λ_{PN}^i, λ_{BN}^i and λ_{NN}^i denote the cost incurred for taking the same actions when the object does not belong to C_i (the object belongs to C_j actually, $i \neq j$). Similar to two-class decision-theoretic rough set, these parameters should satisfy the condition (denoted by c_1): $\lambda_{PP}^i \leq \lambda_{BP}^i < \lambda_{NP}^i$ and $\lambda_{NN}^i \leq \lambda_{BN}^i < \lambda_{PN}^i$.

In the followings, we will give the explanation of these cost functions in the semantic perspective and show how to compute them from the $n \times n$ cost matrix in the calculation perspective.

Consider every value in the i-th row of the cost matrix in Table 1, λ_{ij} $(i \neq j)$ denotes the cost of classisying an object x into C_i when x does not belong to C_i, namely, x is assigned into the positive region of C_i, satisfy the semantic of λ_{PN}, therefore, let $\lambda_{PN}^i = \sum_{j=1}^n \lambda_{ij} \cdot p(C_j|x)$. Due to the constraint $\lambda_{BN}^i < \lambda_{PN}^i$, λ_{BN}^i can be calculated as: $\lambda_{BN}^i = \sum_{j=1}^n \theta_{BN}^{ij} \cdot \lambda_{ij} \cdot p(C_j|x)$, where $\theta_{BN}^{ij} \in (0,1)$. Consider every value in the i-th column of the cost matrix in Table 1, λ_{ji} $(i \neq j)$ denotes the cost of classifying an object x into C_j when x belongs to C_i, C_j can be seen an opposite class of C_i, satisfy the semantic of λ_{NP}. Therefore, let $\lambda_{NP}^i = \sum_{j=1}^n \lambda_{ji} \cdot p(C_j|x)$, because of $\lambda_{BP}^i < \lambda_{NP}^i$, λ_{BP}^i can be expressed as: $\lambda_{BP}^i = \sum_{j=1}^n \theta_{BP}^{ji} \cdot \lambda_{ji} \cdot p(C_j|x)$, where $\theta_{BP}^{ji} \in (0,1)$. The coefficients θ_{BN}^{ij} and θ_{BP}^{ji} can be called adjustment factors, they can be determined by experts or adjusted through experiments. In general, the cost of correct classification is zero, namely, $\lambda_{PP}^i = \lambda_{NN}^i = 0$. In summary, the parameters of multi-class decision-theoretic rough set model can be denoted as:

$$\lambda_{PP}^i = \lambda_{NN}^i = 0,$$
$$\lambda_{PN}^i = \sum\nolimits_{j=1}^{n} \lambda_{ij} \cdot p(C_j|x),$$
$$\lambda_{BN}^i = \sum\nolimits_{j=1}^{n} \theta_{BN}^{ij} \cdot \lambda_{ij} \cdot p(C_j|x), \qquad i \neq j \qquad (2)$$
$$\lambda_{NP}^i = \sum\nolimits_{j=1}^{n} \lambda_{ji} \cdot p(C_j|x),$$
$$\lambda_{BP}^i = \sum\nolimits_{j=1}^{n} \theta_{BP}^{ji} \cdot \lambda_{ji} \cdot p(C_j|x).$$

$p(C_j|x)$ denotes the probability distribution of an object x in each class.

Since $\sum_{i=1}^{n} p(C_i|x) = 1$, the expected cost of classifying an object x into the positive, boundary and negative regions can be expressed as following:

$$\Re_{P_i} = p(C_i|x) \cdot \lambda_{PP}^i + (1 - p(C_i|x)) \cdot \lambda_{PN}^i,$$
$$\Re_{B_i} = p(C_i|x) \cdot \lambda_{BP}^i + (1 - p(C_i|x)) \cdot \lambda_{BN}^i, \qquad (3)$$
$$\Re_{N_i} = p(C_i|x) \cdot \lambda_{NP}^i + (1 - p(C_i|x)) \cdot \lambda_{NN}^i.$$

We can get decision rules based on the Bayesian decision theory:

(P) If $p(C_i|x) \geq \alpha_i$ and $p(C_i|x) \geq \gamma_i$, decide $x \in POS(C_i)$,

(B) If $p(C_i|x) \geq \beta_i$ and $p(C_i|x) \leq \alpha_i$, decide $x \in BND(C_i)$,

(N) If $p(C_i|x) \leq \beta_i$ and $p(C_i|x) \leq \gamma_i$, decide $x \in NEG(C_i)$.

Where the parameters α_i, β_i and γ_i are defined as:

$$\alpha_i = \frac{\sum_{j=1}^{n} (1 - \theta_{BN}^{ij}) \cdot \lambda_{ij} \cdot p(C_j|x)}{\sum_{j=1}^{n} \left[(1 - \theta_{BN}^{ij}) \cdot \lambda_{ij} + \theta_{BP}^{ji} \cdot \lambda_{ji}\right] \cdot p(C_j|x)},$$

$$\beta_i = \frac{\sum_{j=1}^{n} \theta_{BN}^{ij} \cdot \lambda_{ij} \cdot p(C_j|x)}{\sum_{j=1}^{n} \left[\theta_{BN}^{ij} \cdot \lambda_{ij} + (1 - \theta_{BP}^{ji}) \cdot \lambda_{ji}\right] \cdot p(C_j|x)}, \qquad i \neq j \qquad (4)$$

$$\gamma_i = \frac{\sum_{j=1}^{n} \lambda_{ij} \cdot p(C_j|x)}{\sum_{j=1}^{n} (\lambda_{ij} + \lambda_{ji}) \cdot p(C_j|x)}.$$

When parameters satisfy the condition (denoted by c_2): $(\lambda_{NP}^i - \lambda_{BP}^i) \cdot (\lambda_{PN}^i - \lambda_{BN}^i) > (\lambda_{BP}^i - \lambda_{PP}^i) \cdot (\lambda_{BN}^i - \lambda_{NN}^i)$, we have $\alpha_i > \gamma_i > \beta_i$. Aftering tie-breaking, we obtain:

(P1) If $p(C_i|x) \geq \alpha_i$, decide $x \in POS(C_i)$,

(B1) If $\beta_i < p(C_i|x) < \alpha_i$, decide $x \in BND(C_i)$,

(N1) If $p(C_i|x) \leq \beta_i$, decide $x \in NEG(C_i)$.

The upper and lower approximations based on α_i and β_i can be denoted as:

$$\overline{apr}_{(\alpha_i, \beta_i)}(C_i) = \{x \in U | p(C_i|x) > \beta_i\},$$
$$\underline{apr}_{(\alpha_i, \beta_i)}(C_i) = \{x \in U | p(C_i|x) \geq \alpha_i\}. \qquad (5)$$

The probabilistic positive, boundary and negative regions are defined by

$$POS_{(\alpha_i,\beta_i)}(C_i) = \{x \in U | p(C_i|x) \geq \alpha_i\},$$
$$BND_{(\alpha_i,\beta_i)}(C_i) = \{x \in U | \beta_i < p(C_i|x) < \alpha_i\}, \quad (6)$$
$$NEG_{(\alpha_i,\beta_i)}(C_i) = \{x \in U | p(C_i|x) \leq \beta_i\}.$$

Let π_D be a partition of the universe U, defined by the decision attribute D. The three regions of the partition π_D can be defined as following if there are no duplicate objects in the three regions of each class:

$$POS_{(\alpha_i,\beta_i)}(\pi_D) = \bigcup_{1 \leq i \leq n} POS_{(\alpha_i,\beta_i)}(C_i),$$
$$BND_{(\alpha_i,\beta_i)}(\pi_D) = \bigcup_{1 \leq i \leq n} BND_{(\alpha_i,\beta_i)}(C_i), \quad (7)$$
$$NEG_{(\alpha_i,\beta_i)}(\pi_D) = U - POS_{(\alpha_i,\beta_i)}(\pi_D) \bigcup BND_{(\alpha_i,\beta_i)}(\pi_D).$$

In the following, we will present a multi-class cost-sensitive classification algorithm on the basis of above multi-class decision-theoretic rough set model.

In Algorithm 1, we calculate the probability $p(C_j|x_i)$ of x_i belongs to C_j by the base classifier in Weka [11], then we use the probability $p(C_j|x_i)$ and cost matrix to calculate the six cost functions and the probability thresholds of each class, where α_j^i and β_j^i denote the probability thresholds of each object x_i in each class C_j. We give priority to classifying an object into the positive region of one class, that is to say, if an object x_i is classified into $POS(C_j)$, meanwhile, x_i is classified into $BND(C_k)$, we will finally assign x_i into $POS(C_j)$ rather than $BND(C_k)$. In the loop statements of our algorithm, an object x_i may belong to $BND(C_j)$, besides, x_i may belong to $BND(C_k)$, we classified x_i into $BND(C_j)$ rather than $BND(C_k)$ if $j < k$, it is just our handled method in the classification process. We can obtain the three regions and total classification cost of the test set X by above algorithm, the constraints in judgement statements can ensure that there are no duplicate objects in the three regions of each class, therefore, we can obtain $POS(X)$, $BND(X)$ and $NEG(X)$ based on the *formula* (7), the total classification cost can be defined as the sum of costs of classifying objects into positive, boundary and negative regions. The time complexity of Algorithm 1 is $O(m * n)$.

4   Experiments

In this section, we will illustrate the effectiveness of multi-class decision-theoretic rough set cost-sensitive classification algorithm (denoted by **Mcrsca**) by some comparison experiments. Mcrsca is compared with a cost-blind machine learning method that is C4.5 decision tree and a cost-sensitive machine learning method that is an improved *Rescaling* method (we choose instance weighting method, denoted by **New_IW**) [10] to solve multi-class classification problems.

Information of data sets are summarized in Table 2. For each data set, the cost matrix are randomly generated, adjustment factors θ_{BN}^{ij} and θ_{BP}^{ji} are also

Algorithm 1. Multi-class decision-theoretic rough set cost-sensitive classification algorithm

---

Input: a train set S contains n classes; a test set X contains m objects, $x_i \in X$; a cost matrix $M = (\lambda_{ij})_{n \times n}$; adjustment factors θ_{BN}^{ij} and θ_{BP}^{ji}.

Output: $POS(X)$, $BND(X)$ and $NEG(X)$, total classification cost $cost_{total}$.

1: train a classifier C based on the train set S;
2: $cost_{total} = cost_{pos} = cost_{bnd} = cost_{neg} = 0$;
3: $POS(X) = BND(X) = NEG(X) = \varnothing$;
4: **for** $(i = 1$ to $m)$ **do**
5:   **for** $(j = 1$ to $n)$ **do**
6:     get the true class label of x_i, suppose it is C_t;
7:     calculate the probability $p(C_j|x_i)$ of x_i belongs to C_j based on C;
8:     calculate the cost functions λ_{BP}^j and λ_{BN}^j of C_j based on *formula* (2);
9:     calculate the probability thresholds α_j^i and β_j^i on the basis of *formula* (4) ;
10:    **if** $(p(C_j|x_i) \geq \alpha_j^i$ and $POS(X)$ does not contain $x_i)$ **then**
11:      decide $x_i \in POS(C_j)$;
12:      $POS(X) = POS(X) \bigcup POS(C_j)$, $cost_{pos} = cost_{pos} + \lambda_{jt}$;
13:      **if** $(BND(X)$ contains $x_i)$ **then**
14:      $$BND(X) = BND(X) - \{x_i\}, cost_{bnd} = cost_{bnd} - \begin{cases} \lambda_{BP}^j, & t = j \\ \lambda_{BN}^j, & t \neq j \end{cases};$$
15:      **end if**
16:    **end if**
17:    **if** $(\beta_j^i < p(C_j|x_i) < \alpha_j^i$ and $BND(X)$ does not contain $x_i)$ **then**
18:      decide $x_i \in BND(C_j)$;
19:      $$BND(X) = BND(X) \bigcup BND(C_j), cost_{bnd} = cost_{bnd} + \begin{cases} \lambda_{BP}^j, & t = j \\ \lambda_{BN}^j, & t \neq j \end{cases};$$
20:    **end if**
21:   **end for**
22: **end for**
23: $NEG(X) = X - POS(X) \bigcup BND(X)$;
24: **for** (each x_i in $NEG(X)$) **do**
25:   **for** $(j = 1$ to $n)$ **do**
26:     get the true class label of x_i, suppose it is C_t;
27:     calculate the probability $p(C_j|x_i)$ of x_i belongs to C_j based on C;
28:     calculate the cost functions λ_{NP}^j and λ_{NN}^j of C_j based on *formula* (2);
29:     calculate the probability thresholds α_j^i and β_j^i on the basis of *formula* (4);
30:    **if** $(p(C_j|x_i \leq \beta_j^i)$ and $NEG(C_j)$ does not contain $x_i)$ **then**
31:      decide $x_i \in NEG(C_j)$; $cost_{neg} = cost_{neg} + \begin{cases} \lambda_{NP}^j, & t = j \\ \lambda_{NN}^j, & t \neq j \end{cases};$
32:    **end if**
33:   **end for**
34: **end for**
35: $cost_{total} = cost_{pos} + cost_{bnd} + cost_{neg}$;
36: **return** $POS(X)$, $BND(X)$ and $NEG(X)$, total classification cost $cost_{total}$;

Table 2. Brief description of UCI data sets (A: # attributes, C: # classes).

Data sets	Size	A	C
Balance-scale	625	4	3
Waveform-5000	5000	40	3
Connect-4	67557	42	3
Splice	3190	60	3
cmc	1473	9	3
Car	1728	6	4
Vehicle	846	18	4
Segment	2310	19	7
Vowel	990	13	11
Letter	20000	16	26

generated randomly with respect to constraint conditions c_1 and c_2. Ensuring that each of comparison experiments have a same cost matrix and same adjustment factors, experiments are carried out on ten data sets. Ten times 10-fold cross validation are employed and the following evaluation criteria are recorded:

(1) Accuracy (E_{Acc}) represents the proportion of accepted objects correctly identified by the classifier, and it is defined as: $E_{Acc} = n_{PP}/|POS(X)|$.
(2) Deferment rate (E_{Def}) means the proportion of deferred objects identified by the classifier, and it is defined as: $E_{Def} = |X - POS(X)|/|X|$.
(3) Total cost (E_{Cost}) means the total cost of the classification, and it is defined as: $E_{Cost} = cost_{total}$.

Where X is a test set, $POS(X)$ is the positive region of universe X, $|*|$ denotes the number of elements in a set, n_{PP} denotes the number of objects for correct classification in $POS(X)$, $cost_{total}$ has been defined in Algorithm 1.

In our experiments, cost-sensitive classification method is applied by using C4.5 as the base classifier. All approaches are implemented based on J48 in Weka [11] with default settings. The results of comparisons are summarized in Tables 3 and 4:

In all tables, all of values are the mean of results in ten times 10-fold cross validation, and the best performance of each row is boldfaced.

From the Table 3, we can draw three conclusions. First, the target of cost-sensitive learning methods is to minimize the total classification cost, therefore, New_IW and Mcrsca based on the cost-sensitive learning method own lower total cost than cost-blind method C4.5. Second, essentially, both C4.5 and New_IW are two-way decisions methods, they do not have the concept of boundary region and deferred decision, thus, all of the deferment rate E_{Def} of them are zero in Table 3, but Mcrsca is a three-way decisons method, E_{Def} in Mcrsca denotes the proportion of objects that have not be made an immediate decision, the lower of its value the better. Third, as two kinds of cost-sensitive learning methods,

Table 3. Comparison of overall evaluation criteria for C4.5, New_IW and Mcrsca.

Data sets	E_{Acc} (Mean)			E_{Def} (Mean)			E_{Cost} (Mean)		
	C4.5	New_IW	Mcrsca	C4.5	New_IW	Mcrsca	C4.5	New_IW	Mcrsca
Balance-scale	0.779	0.731	**0.806**	0	0	0.227	47.724	35.348	**30.827**
Waveform-5000	0.751	0.757	**0.764**	0	0	0.059	325.152	274.407	**268.068**
Connect-4	0.809	0.799	**0.831**	0	0	0.086	3545.514	3308.69	**2969.836**
Saplice	0.942	0.945	**0.949**	0	0	0.064	32.229	28.859	**27.583**
cmc	0.515	0.482	**0.536**	0	0	0.221	167.709	156.024	**118.822**
Car	0.924	0.934	**0.971**	0	0	0.099	53.776	42.906	**22.683**
Vehicle	0.708	0.688	**0.745**	0	0	0.111	37.408	34.415	**30.017**
Segment	0.97	0.958	**0.974**	0	0	0.018	29.369	38.700	**26.364**
Vowel	0.803	0.810	**0.832**	0	0	0.123	136.074	118.548	**99.908**
Letter	0.880	0.861	**0.915**	0	0	0.088	1955.217	1797.350	**1267.090**
Average	0.808	0.797	**0.832**	0	0	0.109	633.017	583.525	**486.121**

Table 4. Comparison of evaluation criteria for Mcrsca on different adjustment factors.

Data sets	E_{Acc} (Mean)		E_{Def} (Mean)		E_{Cost} (Mean)	
	(0.3,0.5)	(0.2,0.4)	(0.3,0.5)	(0.2,0.4)	(0.3,0.5)	(0.2,0.4)
Balance-scale	0.790	**0.827**	**0.111**	0.238	71.100	**57.877**
Waveform-5000	0.754	**0.759**	**0.013**	0.035	617.686	**592.891**
Connect-4	0.830	**0.846**	**0.146**	0.317	5106.312	**3513.598**
Splice	0.943	**0.946**	**0.006**	0.020	330.988	**251.831**
cmc	0.555	**0.572**	**0.289**	0.474	94.676	**90.128**
Car	0.945	**0.953**	**0.039**	0.058	38.186	**30.128**
Vehicle	0.752	**0.805**	**0.115**	0.267	127.310	**93.293**
Segment	0.969	**0.971**	**0.001**	0.005	42.402	**40.315**
Vowel	0.811	**0.823**	**0.037**	0.079	95.180	**86.747**
Letter	0.891	**0.901**	**0.022**	0.045	1210.254	**1086.404**
Average	0.824	**0.840**	**0.078**	0.154	773.409	**584.321**

Mcrsca almost always owns higher accuracy, deferment rate and lower total cost than New_IW, this conclusion can be explained that Mcrsca is a three-way decisions method, the three-way decisions theory put the blurred objects into boundary region rather than misclassified into positive or negative regions. Therefore, the classifiers based on Mcrsca are more accurate than the classifiers based on New_IW. The higher accuracy and deferment rate for Mcrsca means it has lower error rate than New_IW, all these facts result in lower total classification cost for Mcrsca. The experimental results indicate that Mcrsca can get a better performance on classification accuracy and total cost.

In order to compare overall evaluation criteria for Mcrsca on different adjustment factors, we record the results by using different adjustment factors (0.3,0.5) and (0.2,0.4) in Table 4. As can be seen, after reducing the adjustment factors, we can get higher accuracy and deferment rate, lower total cost. The results suggest that the decision makers can increase the deferment rate by reducing the adjustment factors to a certain extent to ensure make decision with higher accuracy and lower total cost when the current information is insufficient to make a right decision.

5   Conclusion

The cost-sensitive analysis mechanism is introduced into decision-theoretic rough set to solve the classification problems when the different types of misclassification costs are unequal. This paper propose a multi-class decision-theoretic rough set model by combining decision-theoretic rough set with cost-sensitive learning. We determine several parameters in multi-class decision-theoretic rough set based on the cost matrix in semantic and calculation perspectives, derive the decision rules from the Bayesian decision procedure, and present a multi-class decision-theoretic rough set cost-sensitive classification algorithm. Experimental results indicate that proposed model and algorithm can get a better performance on classification accuracy and total cost.

In the furture, we will further study multi-class decision-theoretic rough set model from the cost-sensitive learning view and apply the model to some real applications, such as text categorization, emotion analysis and so on.

Acknowledgments. This work is supported by the National Natural Science Foundation of China under Grant Nos. 61403200 and Natural Science Foundation of Jiangsu Province under Grant No. BK20140800.

References

1. Pawlak, Z.: Rough sets. Int. J. Comput. Inf. Sci. **11**, 341–356 (1982)
2. Yao, Y.Y.: Three-way decisions with probabilistic rough sets. Inf. Sci. **180**, 341–353 (2010)
3. Yao, Y.: Decision-theoretic rough set models. In: Yao, J.T., Lingras, P., Wu, W.-Z., Szczuka, M.S., Cercone, N.J., Ślęzak, D. (eds.) RSKT 2007. LNCS (LNAI), vol. 4481, pp. 1–12. Springer, Heidelberg (2007)
4. Yao, Y.Y., Wong, S.K.M., Lingras, P.: A decision-theoretic rough set model. In: Proceedings of ISMIS 1990, vol. 5, pp. 17–24 (1990)
5. Li, H.X., Zhou, X., Huang, B., Zhao, J.: Decision-theoretic rough set and cost-sensitive classification. J. Front. Comput. Sci. Technol. **7**(2), 126–135 (2013) (in Chinese)
6. Liu, D., Li, T.R., Li, H.X.: A mutliple-category classification approach with decision-theoretic rough sets. Fundamenta Informaticae **115**(2–3), 173–188 (2012)
7. Zhou, B.: Multi-class decision-theoretic rough sets. Int. J. Approx. Reason. **55**(1), 211–224 (2014)
8. Jia, X.Y., Tang, Z.M., Liao, W.H., Shang, L.: On an optimization representation of decision-theoretic rough set model. Int. J. Approx. Reason. **55**(1), 156–166 (2014)
9. Jia, X.Y., Liao, W.H., Tang, Z.M., Shang, L.: Minimum cost attribute reduction in decision-theoretic rough set models. Inf. Sci. Int. J. **219**(1), 151–167 (2013)

10. Zhou, Z.H., Liu, X.Y.: On multi-class cost-sensitive learning. Comput. Intell. **26**(3), 232–257 (2010)
11. Hall, M., Frank, E., Holmes, G., et al.: The WEKA data mining software: an update. ACM SIGKDD Explor. Newsl. **11**(1), 10–18 (2008)
12. Lingras, P., Chen, M., Miao, D.: Rough multi-category decision theoretic framework. In: Wang, G., Li, T., Grzymala-Busse, J.W., Miao, D., Skowron, A., Yao, Y. (eds.) RSKT 2008. LNCS (LNAI), vol. 5009, pp. 676–683. Springer, Heidelberg (2008)
13. Zhou, Z.H., Liu, X.Y.: Training cost-sensitive neural networks with methods addressing the class imbalance problem. IEEE Trans. Knowl. Data Eng. **18**(1), 63–77 (2006)

Research on Cost-Sensitive Method for Boundary Region in Three-Way Decision Model

Yanping Zhang[1,2], Gang Wang[1,2], Jie Chen[1,2](✉), Liandi Fang[1,2],
Shu Zhao[1,2], Ling Zhang[1,2], and Xiangyang Wang[3]

[1] Key Laboratory of Intelligent Computing and Signal Processing
of Ministry of Education, Hefei 230601, Anhui Province, People's Republic of China
chenjie200398@163.com
[2] School of Computer Science and Technology, Anhui University,
Hefei 230601, Anhui Province, People's Republic of China
[3] Anhui Electrical Engineering Professional Technique College,
Hefei 230051, Anhui Province, People's Republic of China
http://ailab.ahu.edu.cn

Abstract. The three-way decision theory (3WD) is constructed based on the notions of the acceptance, rejection or non-commitment, which can be directly generated by the three regions: positive region (POS), negative region (NEG) and boundary region (BND). At present, how to process the boundary region has become a hot topic in the field of three-way decision theory. Although several methods have been proposed to address this problem, most of them don't take cost-sensitive classification into consideration. In this paper, we adopt a cost-sensitive method to deal with the boundary region. Under the principle of reducing loss of classification, we adjust the border distance which is between sample of boundary region and the cover through introducing a cost-sensitive distance coefficient η. The coefficient η can be automatically calculated according to the distribution characteristics of samples. Compared with other models, experimental results show that our model can obtain high correct classification rate. What's more, our model can reduce loss of classification by improving the recall rate of high cost sample when dealing with the boundary region.

Keywords: The three-way decision · Constructive covering algorithm · Processing boundary region · Cost-sensitive classification

1 Introduction

Yao put forward the three-way decision theory [1,2] in the study of rough sets and decision-theoretic rough sets. It extends two-way decision theory by incorporating an additional choice: boundary decision. In recent years, researches on three-way decision theory are mainly focused on the three-way decision theory based on rough sets. The most representative one is Decision Theoretic Rough

© Springer International Publishing AG 2016
V. Flores et al. (Eds.): IJCRS 2016, LNAI 9920, pp. 261–271, 2016.
DOI: 10.1007/978-3-319-47160-0_24

Set model(DTRS) [3], which is proposed by Yao et al. in 1990. From then on, DTRS has been introduced into the incomplete systems [4,5] and the multi-agent systems [6]. It has made great achievements in the investment decision-making [7], text classification [8], information filtering [9], email spam filtering [10,11], social judgment theory [12,13] and et al. However, there are still two problems that need to be solved in DTRS. One is how to compute the thresholds α, β that generally rely on the experience of experts, which is subjective and empirical. The other is that the boundary region is not properly dealt with.

Constructive Covering Algorithm (CCA) [14] was put forward by Ling Zhang and Bo Zhang. On the basis of CCA, Zhang and Xing proposed the three-way decision model based on CCA [15]. The model can automatically generate the three regions without any given parameters. It also provides three methods to deal with boundary region, but non of them is cost-sensitive. Then, Zhang and Zou proposed the cost-sensitive three-way decisions model based on CCA(CCTDM) [16]. The new model combines loss function with three-way decisions model based on CCA. By changing the radius of the cover according to the loss function, it gets the purpose of reducing loss of classification and computing the thresholds α, β. This method essentially makes a part of sample of boundary region divided into covers, there is no further discussion on the rest samples.

In this paper, we put forward a cost-sensitive method to deal with boundary region. Compared with other methods, our method fully takes cost-sensitive classification into account. More specifically, we firstly adopt the maximum radius principle to form covers, which is named MinCA in literature [17]. Then we introduce a cost-sensitive distance coefficient η, which can be automatically calculated according to the distribution characteristics of samples. According to η, we adjust the border distance between samples of boundary region and the covers to process the samples with lower loss. Finally, all samples of boundary region will be divided into positive region and negative region. The rest of the paper is organized as follows: In Sect. 2, we briefly review the traditional three-way decision models. In Sect. 3, we introduce the cost-sensitive three-way decision model for processing boundary region in detail. In Sect. 4, we analyze the experimental results. We draw our conclusion in Sect. 5.

2 The Traditional Three-Way Decision Models

2.1 Decision-Theoretic Rough Set Model

Yao introduced Bayesian decision procedure into rough set theory (RST) and proposed a decision-theoretic rough set model (DTRS). According to the principle of minimum conditional risk, all the samples are divided into positive region, negative region and boundary region by a pair of thresholds (α, β) [3].

Let $\Omega = \{C, C^c\}$ denotes a set of two states, indicating that an object belongs to C or C^c. With respect to the three regions, the set of actions is given by $A = \{a_P, a_B, a_N\}$, where a_P, a_N, a_B represent the three actions in classifying an object, deciding POS(C), deciding NEG(C), and deciding BND(C), respectively. The λ_{PP}, λ_{BP} and λ_{NP} denote the losses incurred for taking action a_P, a_B and

a_N respectively, when an object belongs to C. λ_{PN}, λ_{BN} and λ_{NN} denote the losses for taking the same actions when an object belongs to C^c. The thresholds (α, β) can be calculated by following formula.

$$\alpha = \frac{\lambda_{PN} - \lambda_{BN}}{(\lambda_{PN} - \lambda_{BN}) + (\lambda_{BP} - \lambda_{PP})} \qquad \beta = \frac{\lambda_{PN} - \lambda_{NN}}{(\lambda_{BN} - \lambda_{NN}) + (\lambda_{NP} - \lambda_{BP})} \quad (1)$$

After deducing from the Bayesian decision procedure, fundamental result of DTRS is that the positive, boundary and negative regions are defined by the thresholds (α, β) which is shown as follows.

$$POS_{(\alpha,\beta)}(C) = \{x \epsilon U | P(C|[x]) \geq \alpha\}$$
$$BND_{(\alpha,\beta)}(C) = \{x \epsilon U | \beta < P(C|[x]) < \alpha\} \qquad (2)$$
$$NEG_{(\alpha,\beta)}(C) = \{x \epsilon U | P(C|[x]) \leq \beta\}$$

Where the equivalence class $[x]$ of x is viewed as description of x and $P(C|[x])$ denotes the conditional probability of the classification, U is the universe.

2.2 The Three-Way Decision Model Based on CCA

The three-way decision model based on CCA was proposed by Zhang and Xing [15], which produces three regions automatically according to the samples and does not need any given parameters.

Assume a training samples set $X = \{(x_1, y_1), (x_2, y_2), ..., (x_p, y_p)\}$. X is a set in n-dimensional Euclidean space, containing p samples. $x_i = (x_i^1, x_i^2, ..., x_i^n)$ represents n-dimensional feature attribute of the i-th sample. y_i is the decision attribute, i.e., category. According to Geometrical Representation of M-P Neural Model, cover is a spherical space in accord with neuron. The specific formation process of the covers has been introduced in Ref. [15]. CCA Finally obtained a set of covers $C = \{C_1^1, C_1^2, ..., C_1^{n_1}, C_2^1, C_2^2, ..., C_2^{n_2}, ..., C_m^1, ..., C_m^{n_m}\}$, where C_i^j represents the jth cover of the ith category. Each category has a cover at least. We assume that θ_i^j and w_i^j are the radius and center of C_i^j. Usually there are three methods to form covers. The maximum radius regards the max distance between the center and the similar points as the radius while the minimum radius adopts the min distance between the center and the dissimilar. The compromised radius takes the average value of the two. As are shown in Fig. 1 and formula (3).

$$d_1(k) = \max_{y_i=y_k} \{dist(x_i, x_k) | dist(x_i, x_k) < d_2(k)\}, i \in \{1, 2, ..., p\}$$
$$d_2(k) = \min_{y_i \neq y_k} dist(x_i, x_k), i \in \{1, 2, ..., p\} \qquad (3)$$

According to the formula (3), the three methods to calculate the radius (θ_i^j) are as follows.

1. Maximum radius: $\theta_i^j = d_1(k)$
2. Minimum radius: $\theta_i^j = d_2(k)$
3. Compromised radius: $\theta_i^j = [d_1(k) + d_2(k)]/2$.

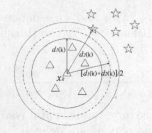

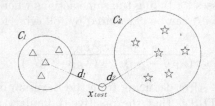

Fig. 1. Cover's radius (θ_i^j) **Fig. 2.** Nearest to the Boundary Principle

In this paper, we adopt the maximum radius in the formation of covers and CCTDM adopts the compromised radius. Compared with the compromised radius, the covers generated by the maximum radius contain fewer samples. But there is no any dissimilar points in the covers [20], it's beneficial to reduce loss of classification.

Besides, the model provides three methods to deal with boundary region, the most representative one is the Nearest to the Boundary Principle (NBP). Assuming that w_t^k denotes the decision attribute of x_i. As is shown in formula (4).

$$w_t^k = \arg\min_{w_i^j} dist(x_i, w_i^j) - \theta_i^j \qquad (4)$$

Where $x_i \in X, C_t^k \in C, dist(x_i, w_i^j) - \theta_i^j$ is the border distance between x_i and C_i^j.

In Fig. 2, $x_{text} \in BND$. d_1 denotes the border distance between x_{text} and C_1 and d_2 denotes the border distance between x_{text} and C_2. It is obvious that $d_1 > d_2$. According to the NBP, x_{text} will be divided into C_2.

2.3 Cost-Sensitive Three-Way Decision Model Based on CCA

The Cost-sensitive Three-way Decision Model based on CCA (CCTDM)[16] combined loss functions with three-way decisions model based on CCA.

Assume only two categories C_1 and C_2. According to CCA, CCTDM obtains a set of cover $C = \{C_1^1, C_1^2, ..., C_1^{n_1}, C_2^1, C_2^2, ..., C_2^{n_2}\}$. The radius of those covers are $\theta = \{\theta_1^1, \theta_1^2, ..., \theta 1^{n_1}, \theta_2^1, \theta_2^2, ..., \theta_2^{n_2}\}$. Samples in $C_1^i(i = (1, 2, ..., n_1))$ belong to positive region and $C_2^j(j = (1, 2, ..., n_2))$ belong to negative region. For covers of C_1, CCTDM regards the k-nearest distance between the center and the dissimilar point as radius ($k = 0, 1, 2, 3, 4...$). For covers of C_2, CCTDM regards the t-nearest distance between the center and the dissimilar point as radius ($t = 0, 1, 2, 3, 4...$). When the radius increase, the number of cover decrease. After increase the radius, the cover $C = \{C_1^1, C_1^2, ..., C_1^{m_1}, C_2^1, C_2^2, ..., C_2^{m_2}\}$ and the radius $\theta = \{\theta_{1k}^1, \theta_{1k}^2, ..., \theta_{1k}^{m_1}, \theta_{2t}^1, \theta_{2t}^2, ..., \theta_{2t}^{m_2}\}$. The average value of radius increase can be computed according to following formula.

$$\Delta\theta_1 = \frac{\theta_1^1 + \theta_1^2 + ... + \theta_1^{n_1}}{n_1} - \frac{\theta_{1k}^1 + \theta_{1k}^2 + ... + \theta_{1k}^{m_1}}{m_1} \tag{5}$$

$$\Delta\theta_2 = \frac{\theta_2^1 + \theta_2^2 + ... + \theta_2^{n_2}}{n_2} - \frac{\theta_{2t}^1 + \theta_{2t}^2 + ... + \theta_{2t}^{m_2}}{m_2} \tag{6}$$

$R(C_1)$ is the ratio of C_1's radius increase and $R(C_2)$ is the ratio of C_2's radius increase. They can be expressed by following formula.

$$R(C_1) = \frac{\Delta\theta_1}{\frac{\theta_1^1 + \theta_1^2 + ... + \theta_1^{n_1}}{n_1}} \qquad R(C_2) = \frac{\Delta\theta_2}{\frac{\theta_2^1 + \theta_2^2 + ... + \theta_2^{n_2}}{n_2}} \tag{7}$$

Then the thresholds (α, β) can be computed according to the following formula:

$$\alpha = 1 - R(C_1) \qquad \beta = 0 + R(C_2) \tag{8}$$

CCTDM reduces loss of classification by increasing the covers' radius, which will make some samples of BND divided into positive region or negative region. For example, if $\lambda_{NP} > \lambda_{PN}$, CCTDM increases the radius of covers in positive region to make some samples of BND divided into positive region. Meanwhile, the correct classification rate of samples in positive region increases.

Essentially, this method only reduces the number of sample of BND and the remaining samples still need to be dealt with. Compared with CCTDM, our model can thoroughly deal with boundary region. In addition, the cost-sensitive processing methods of two models are different.

3   The Cost-Sensitive Three-Way Decision Model for Processing Boundary Region

Cost-sensitive classification refers to that the losses are different for different actions in the same classification task. Although several methods have been proposed for processing boundary region, most of them don't take cost-sensitive classification into consideration. On the basis of the three-way decision model based on CCA, a Cost-sensitive three-way decision Model for Processing Boundary region (CPBM) is proposed in this section. CPBM can not only deal with boundary region, but also takes fully the cost-sensitive classification into account. In order to reduce loss of classification, we introduced an important cost-sensitive distance coefficient η. It can be automatically calculated by the distribution characteristics of sample. The calculation procedure is as follows.

Definition 1. Cost-sensitive distance coefficient $\eta(\eta > 1)$

Among the Euclidean distances between all samples of BND and all covers of POS, we assume that d_{1max} denotes the maximum value and d_{1min} denotes the minimum value. Among the Euclidean distances between all samples of BND and all covers of NEG, we assume that d_{2max} denotes the maximum value and

d_{2min} denotes the minimum value. $\Delta d_1 = d_{1max} - d_{1min}$, $\Delta d_2 = d_{2max} - d_{2min}$. Then η can be calculated by following formula.

$$\eta = \begin{cases} \Delta d_1/\Delta d_2, \Delta d_1 > \Delta d_2 \\ \Delta d_2/\Delta d_1, \Delta d_1 < \Delta d_2 \end{cases} \tag{9}$$

In this way, η computed by the distribution of sample can prevent the samples of BND from being led to one region(POS or NEG) when the value of η is illogical. If $\Delta d_1 = \Delta d_2$, namely $\eta = 1$, CPBM will become the Nearest to the Boundary Principle (NBP).

The detail of CPBM is presented as below.

Algorithm 1. Cost-sensitive three-way decision Model for Processing Boundary region(CPBM)

Input: A set of objects $X = \{x_1, x_2, ..., x_n\}$, a set of attributes $A = \{A_1, A_2, ..., A_m\}$
  and a set of classes $\Omega = \{C, C^c\}$.
Output: two regions POS and NEG.
1. Train sample set X with attribute set A based on the maximum radius principle.
Generate cover set $C = \{C_1^1, C_1^2, ..., C_1^{n_1}, C_2^1, C_2^2, ..., C_2^{n_2}\}$ and the boundary region BND.
2. Calculate the value of η according to the above method.
while *BND is not empty* **do**
    Randomly select an object x_d from BND and t^* is the decision attribute of x_d;
    foreach *cover C_i^j in C* **do**
        Let w_i^j and θ_i^j be the center and radius of C_i^j
        respectively.$(i = (1,2), j = (1, 2, ..., n_i))$;
        Let d_i^j be the border distance between x_d and C_i^j after adjusted;
        if $(\lambda_{NP} > \lambda_{PN}$ *and* $i = 2)$ *or* $(\lambda_{NP} < \lambda_{PN}$ *and* $i = 1)$ **then**

$$d_i^j = \eta * [dist(x_d, w_i^j) - \theta_i^j]$$

        else

$$d_i^j = dist(x_d, w_i^j) - \theta_i^j$$

        end
    end
    After all covers are traversed, obtain a set of border distance
    $D = \{d_1^1, d_1^2, ..., d_1^{n_1}, d_2^1, ..., d_2^{n_2}\}$, define $d_t^k \epsilon D$, then

$$t^* = \arg\min_t d_t^k$$

    if $t^* == 1$ **then**
        | $POS = POS \cup x_d$, $BND = BND - x_d$;
    else
        | $NEG = NEG \cup x_d$, $BND = BND - x_d$;
    end
end
Return POS and NEG

In Algorithm 1, the three regions and η are automatically produced according to the samples. The change on border distance depends on the size of λ_{NP} and λ_{PN}. When $\lambda_{NP} > \lambda_{PN}$, in order to decrease loss, we just increase the border distance between samples of BND and covers of NEG. When $\lambda_{NP} < \lambda_{PN}$, in order to decrease loss, we just increase the border distance between samples of BND and covers of POS. When $\lambda_{NP} = \lambda_{PN}$, the algorithm is equivalent to NBP.

4   Experiment

Our experiments were performed on six data sets from UCI Machine Learning Repository (http://archive.ics.uci.edu/ml/datasets.html). The verification method used in all experiments is 10-fold cross-validation. Table 1 shows the details of the data sets. Except for Car, the number of classes of other data sets is two. So for Car, we regard the samples that belong to *good* and *vgood* as one class and samples that belong to *acc* and *unacc* as the other class. The number of classes of Car is two after preprocessed. All the samples used in experiment have complete attribute values. We did three groups of comparative experiments to evaluate our model's performance. The two comparative models are DTRS and CCTDM.

Table 1. Data sets information from UCI

Data set	Number of sample	Attributes	Classes
Chess	3196	36	2
Wdbc	198	34	2
Wpdc	569	32	2
Spambase	4601	58	2
Car	1728	6	2
Ads	3279	1558	2

The size relation of the values of two loss functions is man-made. For users, if an emergency legitimate email assigned to the spam, it may bring huge loss to them. But if a spam marked as legitimate email, which would just spend the user a little time to check the mail. So for spambase, we think $\lambda_{NP} > \lambda_{PN}$. Since the loss functions have subjective character, which of the two is biggest doesn't influence the anticipate experiment result. To facilitate the discussions, for chess,wdbc and wpdc, we assume they satisfy $\lambda_{NP} < \lambda_{PN}$. That is to say, the loss of classifying spam email as legitimate is bigger than classifying legitimate email as spam. In this case, we need to increase the border distance between samples of BND and covers of POS to prevent samples from being misclassified into POS. For spambase, car and ads, we assume they meet $\lambda_{NP} > \lambda_{PN}$. Similarly, we need to increase the border distance between samples of BND and covers of NEG to prevent samples from being misclassified into NEG. Assume

Table 2. Loss functions and calculation results of η

Data Set	Loss function	η	Δd_1	Δd_2	NSB
Chess	$\lambda_{NP} < \lambda_{PN}$	1.23490	0.94239	1.16376	395
Wdbc	$\lambda_{NP} < \lambda_{PN}$	1.42679	1.02325	0.71717	21
Wpdc	$\lambda_{NP} < \lambda_{PN}$	1.68489	1.05884	0.62843	56
Spambase	$\lambda_{NP} > \lambda_{PN}$	1.34582	1.40938	1.04723	383
Car	$\lambda_{NP} > \lambda_{PN}$	1.56999	1.12251	0.71498	42
Ads	$\lambda_{NP} > \lambda_{PN}$	1.51101	1.05786	0.91451	154

NSB denotes the number of sample of BND. Table 2 shows the loss functions and the calculation results of η.

All the values of η are calculated according to the distribution characteristics of sample. In the experiment, we also found that the experimental error will be large if the value of η is more than 2.0.

After the boundary region being processed, all the samples in BND will be divided into POS and NEG. Among these samples, we assume that N_{NN} denotes the number of spam emails classified as spam. N_{PN} denotes the number of spam emails classified as legitimate. N_{PP} denotes the number of legitimate emails classified as legitimate. N_{NP} denotes the number of legitimate emails classified as spam. $recall(l)$ denotes the recall rate of legitimate email and $recall(s)$ denotes the recall rate of spam email. $recall(l)$ and $recall(s)$ are expressed as follows.

$$recall(l) = \frac{N_{PP}}{N_{PP} + N_{NP}} \qquad recall(s) = \frac{N_{NN}}{N_{NN} + N_{PN}} \qquad (10)$$

Figure 3 shows recall rate of high cost sample when CPBM and NBP dealt with the same boundary region. As a result of the different values of loss functions, the loss of classification of chess,wdbc and wpdc mainly depend on N_{PN} while spambase,car and ads mainly depend on N_{NP}. So in order to evaluate CPBM's performance more accurately, we adopt $recall(s)$ for chess,wdbc and wpdc and $recall(l)$ for spambase,car and ads in the experiment.

From Fig. 3, we can see that the results of CPBM are all higher than NBP. For example, compared with NBP, the $recall(s)$ of CPBM rises by 21 % (77 %–56 % = 21 %) on chess and the $recall(l)$ of CPBM rises by 15 % (85 %–70 % = 15 %) on ads. For the same boundary region, N_{NN} increases with the decrease of N_{PN} and N_{PP} increases with the decrease of N_{NP}. So for chess,wdbc and wpdc, fewer samples are wrongly classified into POS when we increase the border distance between the samples and the covers of POS. Then, N_{PN} decreases and N_{NN} increases, which is why $recall(s)$ of CPBM is higher. For spambase,car and ads, fewer samples are wrongly classified into NEG when we increase the border distance between the samples and the covers of NEG. Then, N_{NP} decrease and N_{PP} increase, which is why $recall(l)$ of CPBM is higher. When similar in other classification results, the loss of classification of CPBM is smaller.

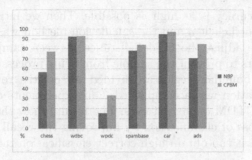

Fig. 3. Recall rate of high cost sample

We assume that the correct classification rate of sample (CCR) is the ratio of the number of correct classification of sample (NCC) and the number of all samples(NAS) in POS, NEG and BND. The formula is as follow.

$$CCR = NCC/NAS \qquad (11)$$

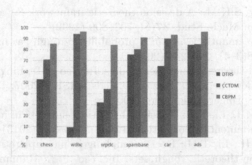

Fig. 4. Correct classification rate of sample

Figure 4 shows the CCR of DTRS, CCTD and CPBM on six datasets. Through it, we can see that the CCR of CPBM is higher than other two models. That is because the number of correct classification of sample increases when the boundary region is dealt with.

5 Conclusion

Processing the boundary region is usually associated with cost-sensitive classification, but traditional methods don't take it into consideration. In this paper, we combined cost-sensitive classification with CCA and put forward the cost-sensitive three-way decision model for processing boundary region. First of all, the model adopts the maximum radius principle to generate covers, making sure that the covers' accuracy is as high as possible. Then we introduced the cost-sensitive distance coefficient η and the calculation method was given in Sect. 3. According to η, we adjust the border distance between samples of boundary region and the covers to prevent the sample from being processed with high loss. Until all samples of boundary region are divided into positive region and negative region, the classification was accomplished. Our experiments compared CPBM with DTRS and CCTDM on six datasets. The results show that our model can effectively reduce loss of classification by improving the recall rate of high cost sample. Meanwhile, it can obtain high correct classification rate.

Acknowledgments. This work is supported by the National Natural Science Foundation of China (No. 61175046, No. 61402006), supported by Provincial Natural Science Research Program of Higher Education Institutions of Anhui Province (No. KJ2013A016), and supported by Open Funding Project of Co-Innovation Center for Information Supply & Assurance Technology of Anhui University (No. ADXXBZ201410), and supported by the Provincial Natural Science Foundation of Anhui Province (No. 1508085MF113).

References

1. Yao, Y.Y., Wong, S.K.M.: A decision theoretic framework for approximating concepts. Int. J. Man Mach. Stud. **37**(6), 793–809 (1992)
2. Yao, Y.Y.: Two semantic issues in a probabilistic rough set model. Fundamenta Informaticae **108**(3–4), 249–265 (2011)
3. Yao, Y.: Decision-theoretic rough set models. In: Yao, J.T., Lingras, P., Wu, W.-Z., Szczuka, M.S., Cercone, N.J., Ślęzak, D. (eds.) RSKT 2007. LNCS (LNAI), vol. 4481, pp. 1–12. Springer, Heidelberg (2007)
4. Yang, X.P., Lu, Z.J., Li, T.J.: Decision-theoretic rough sets in incomplete information system. Fundamenta Informaticae **126**(4), 353–375 (2013)
5. Yang, X., Song, H., Li, T.-J.: Decision making in incomplete information system based on decision-theoretic rough sets. In: Yao, J.T., Ramanna, S., Wang, G., Suraj, Z. (eds.) RSKT 2011. LNCS, vol. 6954, pp. 495–503. Springer, Heidelberg (2011)
6. Yang, X., Yao, J.: Modelling multi-agent three-way decisions with decision-theoretic rough sets. Fundamenta Informaticae **115**(2), 157–171 (2012)
7. Liu, D., Yao, Y.Y., Li, T.R.: Three-way investment decisions with decision-theoretic rough sets. Int. J. Comput. Intell. Syst. **4**(1), 66–74 (2011)
8. Li, W.: Hierarchical rough decision theoretic framework for text classification. In: 2010 9th IEEE International Conference on Cognitive Informatics (ICCI). IEEE (2010)
9. Li, Y.F., Zhang, C.Q.: An information filtering model on the web and its application in job agent. Knowl. Based Syst. **13**(5), 285–296 (2000)
10. Jia, X.Y.: Three-way decisions solution to filter spam email: an empirical study. In: International Conference on Rough Sets and Current Trends in Computing. Springer, Berlin (2012)
11. Zhou, B., Yao, Y., Luo, J.: A Three-Way Decision Approach to Email Spam Filtering. In: Farzindar, A., Kešelj, V. (eds.) AI 2010. LNCS (LNAI), vol. 6085, pp. 28–39. Springer, Heidelberg (2010). doi:10.1007/978-3-642-13059-5_6

12. Li, H., Zhou, X., Zhao, J., Huang, B.: Cost-sensitive classification based on decision-theoretic rough set model. In: Li, T., Nguyen, H.S., Wang, G., Grzymala-Busse, J., Janicki, R., Hassanien, A.E., Yu, H. (eds.) RSKT 2012. LNCS (LNAI), vol. 7414, pp. 379–388. Springer, Heidelberg (2012). doi:10.1007/978-3-642-31900-6_47
13. Jia, X.Y., Lin, S., Chen, J.J.: Attribute reduction based on minimum decision cost. Jisuanji Kexue yu Tansuo **5**(2), 155–160 (2011)
14. Zhang, L., Zhang, B.: A geometrical representation of McCulloch-Pitts neural model and its applications. IEEE Trans. Neural Netw. **10**(4), 925–929 (1999)
15. Zhang, Y., Xing, H., Zou, H., Zhao, S., Wang, X.: A three-way decisions model based on constructive covering algorithm. In: Lingras, P., Wolski, M., Cornelis, C., Mitra, S., Wasilewski, P. (eds.) RSKT 2013. LNCS, vol. 8171, pp. 346–353. Springer, Heidelberg (2013)
16. Zhang, Y., Zou, H., Chen, X., Wang, X., Tang, X., Zhao, S.: Cost-sensitive three-way decisions model based on CCA. In: Cornelis, C., Kryszkiewicz, M., Ślęzak, D., Ruiz, E.M., Bello, R., Shang, L. (eds.) RSCTC 2014. LNCS, vol. 8536, pp. 172–180. Springer, Heidelberg (2014)
17. Chen, J., Zhao, S., Zhang, Y.: A multi-view decision model based on CCA. In: Ciucci, D., et al. (eds.) RSKT 2015. LNCS, vol. 9436, pp. 266–274. Springer, Heidelberg (2015). doi:10.1007/978-3-319-25754-9_24

Determining Thresholds in Three-Way Decisions with Chi-Square Statistic

Cong Gao$^{(\boxtimes)}$ and Yiyu Yao

Department of Computer Science, University of Regina,
Regina, SK S4S 0A2, Canada
{gao266,yyao}@cs.uregina.ca

Abstract. In an evaluation function based three-way decisions model, a pair of thresholds divides a universal set into three regions called a trisection or tri-partition of the universe: a region consists of objects whose values are at or above one threshold, a region of objects whose values are at or below the other threshold, and a region of objects whose values are between the two thresholds. An optimization based method for determining the pair of thresholds is to minimize or maximize an objective function that quantifies the quality, cost, or benefit of a trisection. In this paper, we use the chi-square statistic to interpret and establish an objective function in the context of classification. The maximization of the chi-square statistic searches for a strong correlation between the trisection and the classification.

Keywords: Three-way decisions · Chi-square statistic · Probabilistic rough sets · Maximally selected chi-square statistics

1 Introduction

Three-way decisions can be formulated as a two step, trisecting-and-acting framework [21,24]. The trisecting step divides a universal set of objects into three pair-wise disjoint regions. The acting step adopts strategies to process objects in different regions. Three-way decisions are widely used in many applications [1,2,5,9,10,12–17,25,28–30,32].

An evaluation function based three-way decisions model uses an evaluation function $e(\cdot)$ to map all objects into a totally ordered set $(\mathbb{V}, \succeq)$. According to a pair of thresholds $(\alpha, \beta) \in \mathbb{V} \times \mathbb{V}$ with $\alpha \succ \beta$ (i.e., $\alpha \succeq \beta$ and $\neg(\beta \succeq \alpha)$), we divide objects into three regions: a region consists of objects whose values are at or above one threshold, a region of objects whose values are at or below the other threshold, and a region of objects whose values are between the two thresholds. To determine an optimal pair of thresholds, one method is to construct a meaningful objective function measuring the quality of trisections; the required pair of thresholds maximizes or minimizes the objective function. Examples of quality measures of a trisection are cost [26], Gini index [31], and information entropy [6].

© Springer International Publishing AG 2016
V. Flores et al. (Eds.): IJCRS 2016, LNAI 9920, pp. 272–281, 2016.
DOI: 10.1007/978-3-319-47160-0_25

Consider a classification problem in which all objects in U are classified into one of the two categories $\{C, \bar{C}\}$, where C is a set of objects belonging to the given class and $\bar{C}$ is the set of objects not belonging to the given class. A fundamental task is to construct rules or a description function to achieve such a classification. Two-way classification models are typically used for such a task. However, these models may not produce a desirable results with acceptable classification errors. In three-way classification [23], a trisection $\pi_{(\alpha,\beta)}(C) = (\mathrm{POS}_{(\alpha,\beta)}(C), \mathrm{BND}_{(\alpha,\beta)}(C), \mathrm{NEG}_{(\alpha,\beta)}(C))$ as an approximation of $\{C, \bar{C}\}$ is obtained by a pair of thresholds (α, β) on an evaluation function. Different choices of thresholds lead to different three-way approximations. A good approximation shows a strong association or correlation of $\pi_{(\alpha,\beta)}(C)$ and $\{C, \bar{C}\}$. In other words, $\pi_{(\alpha,\beta)}(C)$ and $\{C, \bar{C}\}$ are correlated or dependent. The chi-square statistic is a measure of correlation and can be used as an objective function for measuring the goodness of a trisection $\pi_{(\alpha,\beta)}(C)$. The maximization of the chi-square statistic suggests the strongest correlation between a trisection and $\{C, \bar{C}\}$. Therefore, the optimal pair of thresholds can be determined by maximizing chi-square statistic.

The rest of this paper is organized as follows. Section 2 reviews a trisecting-and-acting framework of three-way decisions. Section 3 examines the chi-square statistic as an objective function and briefly discusses how to search for an optimal pair of thresholds. Section 4 demonstrates the proposed methods by using an example.

2   Basic Concepts of Three-Way Decisions

This section reviews a general model of three-way decisions and an evaluation based model. The problem of determining the required thresholds is discussed.

2.1   A Trisecting-and-Acting Framework of Three-Way Decisions

The trisecting-and-acting framework of three-way decisions [23,24] includes two steps: trisecting and acting. The trisecting step divides a finite nonempty set U into three pair-wise disjoint regions. The acting step constructs and applies strategies to process objects in different regions produced by trisecting.

Depending on different applications, the three regions can be named specifically. For example, in Pawlak rough sets [19] and probabilistic rough sets [22], the three regions are called positive region, negative region, and boundary region, respectively, denoted by POS, NEG, and BND. In general, they may be named as left region, middle region, and right region, respectively [23]. In this paper, we adopt notations from probabilistic rough sets.

Consider probabilistic rough sets as an example. Given a universal set of objects U and a set $C \subseteq U$, one considers an equivalence relation on U. Let $[x]$ denote the equivalence class containing x and let $Pr(C|[x])$ denote the conditional probability that an object is in C given that the object is in $[x]$. The trisecting step uses a pair of thresholds (α, β) on the conditional probability $Pr(C|[x])$

to divide U into $\text{POS}_{(\alpha,\beta)}(C)$, $\text{BND}_{(\alpha,\beta)}(C)$, and $\text{NEG}_{(\alpha,\beta)}(C)$. The acting step deals with the three types of actions for the three regions: accept objects in $\text{POS}_{(\alpha,\beta)}(C)$, reject objects in $\text{NEG}_{(\alpha,\beta)}(C)$, and make a non-commitment decision to objects in $\text{BND}_{(\alpha,\beta)}(C)$. Take election as another example. By surveying voters about their intended voting decisions, the trisecting step divides surveyed voters into three regions: those who support a candidate, those who oppose the candidate, and those who are undecided or unwilling to tell their decisions. The acting step may aim at retaining supporters and transforming those who are undecided or oppose the candidate into supporters.

2.2   An Evaluation Function Based Model

An evaluation function based three-way decision model can be derived based on evaluating objects in the universe [21]. Let $e(\cdot) : U \longrightarrow \mathbb{V}$ denote an evaluation function that maps each object in the universe to an evaluation status value (ESV) from a totally ordered set $(\mathbb{V}, \succeq)$. Given a pair of thresholds $(\alpha, \beta) \in \mathbb{V} \times \mathbb{V}$ with $\alpha \succ \beta$ (i.e., $\alpha \succeq \beta \wedge \neg(\beta \succeq \alpha)$), we divide U into three regions as follows:

$$\text{POS}_{(\alpha,\beta)}(C) = \{x \in U \mid e(x) \succeq \alpha\},$$
$$\text{BND}_{(\alpha,\beta)}(C) = \{x \in U \mid \beta \prec e(x) \prec \alpha\},$$
$$\text{NEG}_{(\alpha,\beta)}(C) = \{x \in U \mid e(x) \preceq \beta\}, \tag{1}$$

where the $e(x) \preceq \beta$ means $\neg(e(x) \succ \beta)$. Consider again probabilistic rough sets. The probability of C given an equivalence class of object $[x] \subseteq U$ is used as an evaluation function, i.e., $e(x) = Pr(C|[x])$, where C is a subset of U. The conditional probability $Pr(C|[x])$ is the ESV of object x and all ESVs are real numbers between 0 and 1. The relation $\succeq$ is the "greater than or equal" relation $\geq$. Under the assumption $0 \leq \beta < 0.5 \leq \alpha \leq 1$, one easily obtains probabilistic three regions by Eq. (1).

Other evaluation functions can be used in different applications, such as the stanford-binet test [20] that maps a person to an IQ value, word frequency that maps each letter to the number of its appearance in a test.

2.3   The Problem of Determining the Thresholds

The determination of thresholds can be implemented by establishing an objective function and minimizing or maximizing the objective function. An objective function is used to measure the quality or goodness of trisection and can be formulated by a linear combination of qualities of three regions [6]:

$$Q(\pi_{(\alpha,\beta)}(C)) = w_1 Q_P(\alpha, \beta) + w_2 Q_B(\alpha, \beta) + w_3 Q_N(\alpha, \beta), \tag{2}$$

where $Q_P(\alpha, \beta)$, $Q_B(\alpha, \beta)$ and $Q_N(\alpha, \beta)$ are qualities of the positive, boundary, and negative regions, respectively, and w_1, w_2, and w_3 are weights associated to different regions, representing their relative importance.

Examples of objective function include cost [26], information entropy [6], and Gini index of different regions [31]. According to the interpretation of a measure of quality, we may minimize it or maximize it to obtain the optimal pair of thresholds. In decision-theoretic rough sets model [26], the objective function is a measure of cost and we want to minimize it. That is, we minimize the objective function to determine the thresholds:

$$(\alpha^*, \beta^*) = \arg \min_{(\alpha, \beta)} Q(\pi_{(\alpha, \beta)}(C)), \tag{3}$$

where (α^*, β^*) is the optimal pair of thresholds.

3   A Framework of Chi-Square Statistic Based Interpretation of Three-Way Decisions

In this section, we give the contingency table of three-way decisions and argue that chi-square statistic may be used as an objective function to determine the pair of thresholds (α, β).

3.1   Contingency Table of Three-Way Decisions

Given a class C, all objects in U are classified into one of the two categories $\{C, \bar{C}\}$, where C is the set of objects belonging to the given class, $\bar{C}$ is the set of objects not belonging to the class, and $C \cup \bar{C} = U$. Suppose we use an evaluation function to determine the probability or possibility that an object is an instance of the class. By using a pair of thresholds (α, β) on the evaluation function, we trisect U into three pair-wise disjoint regions as an approximation of $\{C, \bar{C}\}$, namely, $\text{POS}_{(\alpha, \beta)}(C)$, $\text{BND}_{(\alpha, \beta)}(C)$, $\text{NEG}_{(\alpha, \beta)}(C)$, respectively. The connection of the actual classification $\{C, \bar{C}\}$ and a three-way approximation $\pi_{(\alpha, \beta)}(C) = (\text{POS}_{(\alpha, \beta)}(C), \text{BND}_{(\alpha, \beta)}(C), \text{NEG}_{(\alpha, \beta)}(C))$ of $\{C, \bar{C}\}$ can be represented by a contingency table [7] as shown in Table 1. The two factors, i.e., the class C and the pair of thresholds (α, β), form the rows and columns, respectively, are called two variables of the contingency table. A contingency table has two directions, i.e., row and column; it is also called a cross-classification table.

Table 1. A contingency table of three-way decision.

	$\text{POS}_{(\alpha, \beta)}(C)$	$\text{BND}_{(\alpha, \beta)}(C)$	$\text{NEG}_{(\alpha, \beta)}(C)$	Total
C	n_{CP}	n_{CB}	n_{CN}	$n_{C \cdot}$
$\bar{C}$	$n_{\bar{C}P}$	$n_{\bar{C}B}$	$n_{\bar{C}N}$	$n_{\bar{C} \cdot}$
Total	$n_{\cdot P}$	$n_{\cdot B}$	$n_{\cdot N}$	n

The numbers in the table such as n_{CP} and $n_{\bar{C}N}$ represent the numbers of objects in the corresponding category of a class and a region. Numbers with subscripts having a dot such as $n_{\bar{C}}.$ and $n._N$ are called marginal totals, denoting the numbers of objects in the corresponding row or column. The number n is the grand total. It is the number of all objects in the table, i.e., $n = |U|$, where $|\cdot|$ is cardinality of a set. In probabilistic rough sets, numbers in first column of Table 1 are $n_{CP} = |C \cap \text{POS}_{(\alpha,\beta)}(C)|$, $n_{\bar{C}P} = |\bar{C} \cap \text{POS}_{(\alpha,\beta)}(C)|$, and $n._P = |\text{POS}_{(\alpha,\beta)}(C)|$, respectively. Additionally, we can estimate probabilities, such as $Pr(\text{POS}_{(\alpha,\beta)}(C)) = n._P/n$, $Pr(C|\text{POS}_{(\alpha,\beta)}(C)) = n_{CP}/n._P$, and $Pr(\bar{C}|\text{POS}_{(\alpha,\beta)}(C)) = n_{\bar{C}P}/n._P$.

3.2   Chi-Square Statistic as an Objective Function

The chi-square statistic, also referred to as χ^2 statistic, plays an important role in testing independence of two variables. Given a contingency table, the χ^2 statistic is computed by:

$$\chi^2 = \sum \frac{(observed - expected)^2}{expected}, \tag{4}$$

where the "observed" is the actual observed number in a contingency table cell and the "expected" is the corresponding expected number under the independence assumption. For example, consider the cell $(C, \text{POS}_{(\alpha,\beta)}(C))$, the observed number of objects is n_{CP} and the expected number of objects is computed by assuming independence of $\{C, \bar{C}\}$ and $\pi_{(\alpha,\beta)}(C)$. With the marginal numbers $n._P$ and $n_{C.}$, the expected number is computed by:

$$Pr(C) * Pr(\text{POS}_{(\alpha,\beta)}(C)) * |U| = \left(\frac{n._P}{n} \frac{n_{C.}}{n}\right)n = \frac{n_{C.}.n._P}{n}. \tag{5}$$

The value $(n_{CP} - n_{C.}.n._P/n)^2$ measures the divergence of the observed number n_{CP} from the expected number $n_{C.}.n._P/n$ under the independent assumption. If the observed value is close or equal to the expected number, then $(n_{CP} - n_{C.}.n._P/n)^2$ is close or equal to 0 and $(n_{CP} - n_{C.}.n._P/n)^2/(n_{C.}.n._P/n)$ is close or equal to 0 as well. This suggests that the actual number is highly probable due to chance and there is a lack of dependence of C and $\text{POS}_{(\alpha,\beta)}(C)$. By summing up all cells, the chi-square statistics can be used to measure the independence/dependence of $\{C, \bar{C}\}$ and $\pi_{(\alpha,\beta)}(C)$. A higher value of chi-statistic suggests a stronger dependency. Therefore, the chi-square statistic may be used as a measure of the goodness of a three-way approximation $\pi_{(\alpha,\beta)}(C)$.

We can demonstrate the appropriateness of chi-square statistics as an objective function by relating it to the general formulation of objective function as given by Eq. (2). Each region occupies a column with two cells in the contingency table. We may quantify the quality of each region as a sum of two cell's divergencies of observed numbers from their expected numbers as follows:

$$Q_P(\alpha, \beta) = \frac{(n_{CP} - n_{C.}.n._P/n)^2}{n_{C.}.n._P/n} + \frac{(n_{\bar{C}P} - n_{\bar{C}.}.n._P/n)^2}{n_{\bar{C}.}.n._P/n},$$

$$Q_B(\alpha, \beta) = \frac{(n_{CB} - n_{C.}.n._B/n)^2}{n_{C.}.n._B/n} + \frac{(n_{\bar{C}B} - n_{\bar{C}.}.n._B/n)^2}{n_{\bar{C}.}.n._B/n},$$

$$Q_N(\alpha, \beta) = \frac{(n_{CN} - n_C . n_{.N}/n)^2}{n_C . n_{.N}/n} + \frac{(n_{\bar{C}N} - n_{\bar{C}} . n_{.N}/n)^2}{n_{\bar{C}} . n_{.N}/n}. \tag{6}$$

By summing up the three quantities with $w_1 = w_2 = w_3 = 1$, we have:

$$Q(\pi_{(\alpha, \beta)}(C)) = Q_P(\alpha, \beta) + Q_B(\alpha, \beta) + Q_N(\alpha, \beta)$$
$$= \chi^2_{(\alpha, \beta)}. \tag{7}$$

That is, the χ^2 statistic of contingency table of three-way decisions may be viewed as a special case of a measure of the quality of a three-way approximation $\pi_{(\alpha, \beta)}(C)$ as defined by Eq. (2).

If the χ^2 statistic is statistically significant, that means $\{C, \bar{C}\}$ and $\pi_{(\alpha, \beta)}(C)$ are correlated or dependent; otherwise, they are independent. A larger χ^2 statistic indicates a stronger correlation. Each pair of thresholds $(\alpha, \beta) \in R \times R$ induces a trisection of U. We want to find a pair of thresholds that provides the strongest correlation. In other words, we search for a pair of thresholds by maximizing the χ^2 statistic:

$$(\alpha^*, \beta^*) = \arg\max_{(\alpha, \beta)} \chi^2_{(\alpha, \beta)} \tag{8}$$

where (α^*, β^*) is the optimal pair of thresholds. As pointed out by Miller and Siegmund [18], "if the chi-square value is statistically significant, then it can be judged that a predictor variable has been found". In the context of three-way decisions, a good pair of (α, β) is obtained.

3.3 Maximizing Chi-Square Statistic to Find Thresholds

Based on the framework shown in Eqs. (6) and (7), we take a look at every component of the objective function. When $0 \le \beta < 0.5 \le \alpha \le 1$, $Q_P(\alpha, \beta)$ is only related to the threshold α and $Q_N(\alpha, \beta)$ is only related to the threshold β. When α changes from 0.5 to 1, $n_{.P}$, n_{CP}, and $n_{\bar{C}P}$ become smaller. However, $Q_P(\alpha, \beta)$ may either increase or decrease, that is, $Q_P(\alpha, \beta)$ is non-monotonic with respect to α. Similarly, $Q_N(\alpha, \beta)$ and $\chi^2_{(\alpha, \beta)}$ are non-monotonic as well. Thus, a pair of thresholds (α, β) that maximizes the statistic cannot be obtained in a simple analytical expression like in a decision-theoretic rough sets model [26] (i.e., in decision-theoretic rough sets, once the cost matrix is given, the pair of thresholds can be computed directly by equations). Fortunately, given a finite universe, the number of possible values for α and β are limited. The exhaustive search method may work well in these cases.

Many studies [3, 4, 11, 18, 27] discussed the computation of maximally selected chi-square statistic. Boulesteix [3] analyzed maximally selected chi-square statistics in the case of one binary response and nominal predictor. Miller and Siegmund [18], and Boulesteix and Strobl [4] discussed the situation that a predictor variable is generated by two cut-points that have some relationships between each other and combined the two columns of contingency table together. Hothorn and Zeileis [11] explained the general maximally selected statistics and proposed an efficient algorithm that can be applied to compute the maximally selected χ^2 statistic for a 2 by 2 contingency table.

4   An Illustrative Example

We use an example from [6] to demonstrate the main idea of the proposed method. Suppose that we have a partition of a universal set with 15 equivalence classes $X_1, X_2, \cdots, X_{15}$. Table 2 gives the conditional probability of a class C given an equivalence class X_i, that is $Pr(C|X_i)$. To derive three-way decisions to approximate C, we use a pair of thresholds (α, β) with $0 \leq \beta < 0.5 \leq \alpha \leq 1$. The three regions are given by:

$$\text{POS}_{(\alpha,\beta)}(C) = \bigcup\{X_i \mid Pr(C|X_i) \geq \alpha\},$$

$$\text{BND}_{(\alpha,\beta)}(C) = \bigcup\{X_i \mid \beta < Pr(C|X_i) < \alpha\},$$

$$\text{NEG}_{(\alpha,\beta)}(C) = \bigcup\{X_i \mid Pr(C|X_i) \leq \beta\}. \tag{9}$$

According to Table 2, the sets of possible values of α and β for consideration are $D_\alpha = \{0.5, 0.6, 0.8, 0.9, 1.0\}$ and $D_\beta = \{0.0, 0.1, 0.2, 0.4\}$, respectively.

Table 2. Probabilistic information of a class C [6].

	X_1	X_2	X_3	X_4	X_5	X_6	X_7	X_8
$Pr(X_i)$	0.0177	0.1285	0.0137	0.1352	0.0580	0.0069	0.0498	0.1070
$Pr(C|X_i)$	1.0	1.0	1.0	1.0	0.9	0.8	0.8	0.6
	X_9	X_{10}	X_{11}	X_{12}	X_{13}	X_{14}	X_{15}	
$Pr(X_i)$	0.1155	0.0792	0.0998	0.1299	0.0080	0.0441	0.0067	
$Pr(C|X_i)$	0.5	0.4	0.4	0.2	0.1	0.0	0.0	

Given a sample size n, a pair of thresholds (α, β) produces a contingency table and the corresponding chi-square statistic. For example, n_{CP} can be computed as the closest integer of the following expression:

$$\left(\sum_{X_i \in \text{POS}_{(\alpha,\beta)}(C)} Pr(C \mid X_i)Pr(X_i) \right) n.$$

The numbers in other cells can be similarly computed. Table 3 shows the contingency table for $(\alpha, \beta) = (0.6, 0.4)$ and $n = 1000$. All computed numbers in Table 3 are modified to their nearest integers, since some computed numbers are not integers when n is set to some numbers. The χ^2 statistic of Table 3 is 351.18. By computing contingency tables and the corresponding χ^2 statistics for all possible combinations of α and β, we obtain Table 4. Accordingly, $(\alpha = 0.8, \beta = 0.2)$ is selected as the optimal pair of thresholds due to its maximal χ^2 statistic. This means $(\alpha = 0.8, \beta = 0.2)$ provides the strongest correlation between the class C and approximation of $\pi_{(\alpha,\beta)}(C)$. The Gini index method [31] also chooses $(\alpha = 0.8, \beta = 0.2)$, the game theory method [1] chooses $(\alpha = 0.5, \beta = 0)$ (using initial search point $(\alpha = 1, \beta = 0.5)$), while the information entropy method [6] chooses $(\alpha = 0.9, \beta = 0.2)$ that provides the second largest χ^2 statistic.

Table 3. The contingency table for $(\alpha, \beta) = (0.6, 0.4)$ and $n = 1000$.

	$\text{POS}_{(0.6,0.4)}(C)$	$\text{BND}_{(0.6,0.4)}(C)$	$\text{NEG}_{(0.6,0.4)}(C)$	Total
C	457	58	98	613
$\bar{C}$	60	58	269	387
Total	517	116	367	1000

Table 4. χ^2 statistics for all combinations of (α, β).

	$\beta = 0.0$	$\beta = 0.1$	$\beta = 0.2$	$\beta = 0.4$
$\alpha = 1.0$	311.24	316.04	368.05	373.31
$\alpha = 0.9$	355.18	358.97	397.12	389.58
$\alpha = \mathbf{0.8}$	381.39	384.36	**411.35**	394.72
$\alpha = 0.6$	356.15	358.20	374.02	351.18
$\alpha = 0.5$	310.29	311.53	318.29	292.50

5   Conclusion

Chi-square statistic is widely used in independence test of a contingency table. In the context of three-way decisions, it measures the correlation between the real classification $\{C, \bar{C}\}$ and a three-way approximation $\pi_{(\alpha,\beta)}(C)$. Therefore, chi-square statistic can be used as an objective function to quantify the goodness of a three-way approximation. According to the meaning of chi-square statistic, the largest chi-square statistic suggests a high probability of correlation between $\{C, \bar{C}\}$ and $\pi_{(\alpha,\beta)}(C)$. An optimal pair of (α, β) is determined by maximizing the statistic.

We use a simple example to show the main idea. As future research, several additional topics may be discussed, such as analyzing thresholds in 2×2 contingency table by combining two columns together, developing a heuristic algorithm to find the thresholds, considering Fisher's exact test [8] if magnitudes of some cells are less than 5, using likelihood ratio statistic or phi coefficient instead of chi-square statistic, and using log-linear model to determine the pair of thresholds.

The maximally selected χ^2 statistic method for three-way decisions can be easily extended to other applications. For example, the area around the decision hyper plane in any classifier has higher impurity, i.e., it includes instances from different classes that are difficult to distinguish. The suggested method in this paper provides an option to abstract a boundary region between two hyper planes located on both sides of and parallel with the decision hyper plane for further analysis. These two hyper planes are determined by a pair of distances from the decision hyper plane and can be found by maximizing χ^2 statistic.

Acknowledgements. This work is partially supported by a Discovery Grant from NSERC, Canada, Saskatchewan Innovation and Opportunity Graduate Scholarship, and Sampson J. Goodfellow Scholarship.

References

1. Azam, N., Yao, J.T.: Multiple criteria decision analysis with game-theoretic rough sets. In: Li, T., Nguyen, H.S., Wang, G., Grzymala-Busse, J., Janicki, R., Hassanien, A.E., Yu, H. (eds.) RSKT 2012. LNCS, vol. 7414, pp. 399–408. Springer, Heidelberg (2012). doi:10.1007/978-3-642-31900-6_49
2. Baram, Y.: Partial classification: the benefit of deferred decision. IEEE Trans. Pattern Anal. Mach. Intell. **20**(8), 769–776 (1998)
3. Boulesteix, A.L.: Maximally selected chi-square statistics and binary splits of nominal variables. Biometrical J. **48**(5), 838–848 (2006)
4. Boulesteix, A.L., Strobl, C.: Maximally selected chi-square statistics and non-monotonic associations: an exact approach based on two cutpoints. Comput. Stat. Data Anal. **51**(12), 6295–6306 (2007)
5. Ciucci, D.: Orthopairs and granular computing. Granular Comput. (2016). doi:10.1007/s41066-015-0013-y
6. Deng, X.F., Yao, Y.Y.: A multifaceted analysis of probabilistic three-way decisions. Fundamenta Informaticae **132**(3), 291–313 (2014)
7. Everitt, B.S.: The Analysis of Contingency Tables. Wiley, New York (1977)
8. Fisher, R.A.: On the interpretation of χ^2 from contingency tables, and the calculation of P. J. Roy. Stat. Soc. **85**(1), 87–94 (1922)
9. Gao, C., Yao, Y.: An addition strategy for reduct construction. In: Miao, D., Pedrycz, W., Slezak, D., Peters, G., Hu, Q., Wang, R. (eds.) RSKT 2014. LNCS, vol. 8818, pp. 535–546. Springer, Heidelberg (2014). doi:10.1007/978-3-319-11740-9_49
10. Herbert, J.P., Yao, J.T.: Game-theoretic rough sets. Fundamenta Informaticae **108**(3–4), 267–286 (2011)
11. Hothorn, T., Zeileis, A.: Generalized maximally selected statistics. Biometrics **64**, 1263–1269 (2008)
12. Hu, B.Q.: Three-way decisions space and three-way decisions. Inf. Sci. **281**, 21–52 (2014)
13. Li, H.X., Zhang, L.B., Huang, B., Zhou, X.Z.: Sequential three-way decision and granulation for cost-sensitive face recognition. Knowl. Based Syst. **91**, 241–251 (2016)
14. Li, M.Z., Wang, G.Y.: Approximate concept construction with three-way decisions and attribute reduction in incomplete contexts. Knowl. Based Syst. **91**, 165–178 (2016)
15. Li, W.W., Huang, Z.Q., Li, Q.: Three-way decisions based software defect prediction. Knowl. Based Syst. **91**, 263–274 (2016)
16. Liang, D.C., Liu, D.: Deriving three-way decisions from intuitionistic fuzzy decision theoretic rough sets. Inf. Sci. **300**, 28–48 (2015)
17. Liu, D., Liang, D.C., Wang, C.C.: A novel three-way decision model based on incomplete information system. Knowl. Based Syst. **91**, 32–45 (2016)
18. Miller, R., Siegmund, D.: Maximally selected chi square statistics. Biometrics **38**(4), 1011–1016 (1982)
19. Pawlak, Z.: Rough Sets: Theoretical Aspects of Reasoning About Data. Kluwer Academic Publishers, Dordrecht (1991)

20. Sattler, J.M.: Assessment of Children's Intelligence. W.B. Saunders Company, Philadelphia (1975)
21. Yao, Y.Y.: An outline of a theory of three-way decisions. In: Yao, J.T., Yang, Y., Słowiński, R., Greco, S., Li, H., Mitra, S., Polkowski, L. (eds.) RSCTC 2012. LNCS, vol. 7413, pp. 1–17. Springer, Heidelberg (2012). doi:10.1007/978-3-642-32115-3_1
22. Yao, Y.Y.: Probabilistic rough set approximations. Int. J. Approximate Reasoning **49**(2), 255–271 (2008)
23. Yao, Y.Y.: Rough sets and three-way decisions. In: Ciucci, D., Mitra, S., WU, W.-Z. (eds.) RSKT 2015. LNCS, vol. 9436, pp. 62–73. Springer, Heidelberg (2015). doi:10.1007/978-3-319-25754-9_6
24. Yao, Y.Y.: Three-way decisions and cognitive computing. Cognitive Comput. (2016). doi:10.1007/s12559-016-9397-5
25. Yao, Y.Y., Gao, C.: Statistical interpretations of three-way decisions. In: Ciucci, D., Mitra, S., Wu, W.-Z. (eds.) RSKT 2015. LNCS, vol. 9436, pp. 309–320. Springer, Heidelberg (2015). doi:10.1007/978-3-319-25754-9_28
26. Yao, Y.Y., Wong, S.K.M.: A decision theoretic framework for approximating concepts. Int. J. Man Mach. Stud. **37**(6), 793–809 (1992)
27. Yenigun, C.D., Szekely, G.J., Rizzo, M.L.: A test of independence in two-way contingency tables based on maximal correlation. Commun. Stat. Theory Methods **40**(12), 2225–2242 (2011)
28. Yu, H., Su, T., Zeng, X.: A three-way decisions clustering algorithm for incomplete data. In: Miao, D., Pedrycz, W., Slezak, D., Peters, G., Hu, Q., Wang, R. (eds.) RSKT 2014. LNCS, vol. 8818, pp. 765–776. Springer, Heidelberg (2014). doi:10.1007/978-3-319-11740-9_70
29. Yu, H., Zhang, C., Wang, G.Y.: A tree-based incremental overlapping clustering method using the three-way decision theory. Knowl. Based Syst. **91**, 189–203 (2016)
30. Zhang, H.R., Min, F., Shi, B.: Regression-based three-way recommendation. Inf. Sci. (2016). doi:10.1016/j.ins.2016.03.019
31. Zhang, Y.: Optimizing Gini coefficient of probabilistic rough set regions using game-theoretic rough sets. In: 26th Canadian Conference of Electrical and Computer Engineering (CCECE), pp. 1–4 (2013)
32. Zhou, B., Yao, Y.Y., Luo, J.G.: Cost-sensitive three-way email spam filtering. J. Intell. Inf. Syst. **42**(1), 19–45 (2014)

Optimistic Decision-Theoretic Rough Sets in Multi-covering Space

Caihui Liu[1(✉)] and Meizhi Wang[2]

[1] Department of Mathematics and Computer Science, Gannan Normal University,
Ganzhou 341000, Jiangxi Province, People's Republic of China
liu_caihui@163.com
[2] Department of Physical Education, Gannan Normal University,
Ganzhou 341000, China
dei2002@163.com

Abstract. This paper discusses optimistic multigranulation decision-theoretic rough sets in multi-covering space. First, by using the strategy "seeking commonality while preserving difference", we propose the notion of optimistic multigranulation decision-theoretic rough sets on the basis of Bayesian decision procedure. Then, we investigate some important properties of the model. Finally, we investigate the relationships between the proposed model and other related rough set models.

Keywords: Multigranulation · Decision-theoretic rough sets · Optimistic · Multi-covering space

1 Introduction

Based on the Bayesian decision procedure, Yao and Wong [1] presented the notion of decision-theoretic rough sets (DTRS), which provides reasonable semantic interpretation for decision-making process and gives an effective approach for selecting the threshold parameters. Since the DTRS was proposed, it has attracted a substantial level of detail. Herbert and Yao [2] explored the game-theoretic rough set model by combining game theory with decision making. Liu et al. [3] proposed a multiple-category classification approach with decision-theoretic rough sets, which can effectively reduce misclassification rate. Yu et al. [4] studied an automatic method of clustering analysis with the decision-theoretic rough set theory. Li et al. [5] studied an axiomatic characterization of decision-theoretic rough sets. Jia et al. [6] proposed an optimization representation of decision-theoretic rough set model and developed a heuristic approach and a particle swarm optimization approach for searching an attribute reduction with a minimum cost. Based on the DTRS, Yao [7,8] presented a new decision making method, where a universe is divided into three pairwise disjoint regions, positive, negative and boundary regions by using an evaluation function and a pair of thresholds. Three-way decisions have been applied to many domains, such as email filtering [9], cost-sensitive face recognition [10], recommender system design [11], and so on.

ⓒ Springer International Publishing AG 2016
V. Flores et al. (Eds.): IJCRS 2016, LNAI 9920, pp. 282–293, 2016.
DOI: 10.1007/978-3-319-47160-0_26

From the viewpoint of Granular Computing [12], most existing extensions and generalizations of Pawlak rough set model [13] are based on extracting knowledge from a single relation on the universe, which may be impractical in the circumstance where a target concept need to be described concurrently through multi binary relations. To resolve such issues, Qian and Liang [14] proposed multigranulation rough sets (MGRS) that uses multiple equivalence relations on the universe. These relations can be chosen according to a user's requirements or targets of problem solving. Since the introduction of MGRS, the theoretical framework has been largely enriched, and many extended multigranulation rough set models and relative properties and applications have also been proposed and studied [15–21]. The study on decision-theoretic rough set in a multigranulation environment is a new and interesting topic. Qian et al. [22] developed the multigranulation decision-theoretic rough set and proved that many existing multigranulation rough set models can be derived from the multigranulation decision-theoretic rough set framework. However, Qian's model [22] has its own limitations: (1) All granular structures in the model are based on equivalence relations, hence the model is not suitable for coverings or neighborhoods based environments. (2) The model evaluates the multigranulation approximations in a quantitative way, so it is not suitable for the situations where general binary relations are considered. To tackle the problems above, in this paper we propose optimistic multigranulation decision-theoretic rough set model in a multi-covering approximation space, which may help to build a more reasonable and suitable decision environment in real problem.

The remainder of the paper is organized as follows. Section 2 reviews some basic notions and notations. Section 3 proposes the optimistic multigranulation decision-theoretic rough set model and discusses the interrelationships with the other generalized rough sets. Section 4 concludes the paper.

2   Preliminaries

This section will review notions and notations used in the paper.

2.1   Covering-Based Rough Sets

We review some fundamental concepts about covering-based rough sets in this subsection.

Definition 1 [23]. Let U be a universe of discourse and C a family of nonempty subsets of U. If $\cup C = U$, then C is called a covering of U. The ordered pair $\langle U, C \rangle$ is called a covering approximation space.

Definition 2 [23]. Let $\langle U, C \rangle$ be a covering approximation space, $x \in U$, then $md_C(x) = \{ K \in C_x | \forall S \in C_x (S \subseteq K \Rightarrow K = S) \}$ is called the minimal description of x, where $C_x = \{ K \in C | x \in K \}$.

2.2 Qian's MGRS

In this subsection, we will briefly outline the definition of optimistic multi-granulation rough sets.

Definition 3.   Let $K = (U, \mathbf{R})$ be a knowledge base, where $\mathbf{R}$ is a family of equivalence relations on the universe U. Let $A_1, A_2, ..., A_m \in \mathbf{R}$, where m is a natural number. For any $X \subseteq U$, its optimistic lower and upper approximations with respect to $A_1, A_2..., A_m$ are defined as follows.

$$\underline{\sum_{i=1}^{m} A_i(X)} = \{x \in U \mid [x]_{A_1} \subseteq X \; or \; [x]_{A_2} \subseteq X \; or \; \cdots \; or \; [x]_{A_m} \subseteq X\}$$

$$\overline{\sum_{i=1}^{m} A_i(X)} = \sim \underline{\sum_{i=1}^{m} A_i(\sim X)}$$

where $\sim X$ denotes the complement set of X. $(\underline{\sum_{i=1}^{m} A_i(X)}, \overline{\sum_{i=1}^{m} A_i(X)})$ is called the optimistic multi-granulation rough sets of X. Here, the word "optimistic" means that only a single granular structure is needed to satisfy the inclusion condition between an equivalence class and a target concept when multiple independent granular structures are available in the problem.

2.3 Decision-Theoretic Rough Sets

In [8], Yao proposed the theory of three-way decisions. Compared with two-way decisions, three-way decisions exhibit a third option, that is, non-commitment in addition to acceptance and rejection. The theory of three-way decisions can be described as follows.

Within the frame of three-way decisions, the set of states is given by $\Omega = \{X, \neg X\}$ (where $\neg X$ denotes the complement of X), the set of actions is given by $A = \{a_P, a_B, a_N\}$, where a_P, a_B and a_N represent the three actions in classifying an object x, namely, deciding $x \in POS(X)$, deciding x should be further investigated $x \in BND(X)$, and deciding $x \in NEG(X)$. $\lambda_{PP}, \lambda_{BP}$ and λ_{NP} denote the loss incurred for taking actions of a_P, a_B and a_N, respectively, when an object belongs to X. Similarly, $\lambda_{PN}, \lambda_{BN}$ and λ_{NN} denote the loss incurred for taking the correspondence actions when the object belongs to $\neg X$. By Bayesian decision procedure, for an object x, the expected loss $R(a_\bullet|[x])$ associated with taking the individual actions can be expressed as

$$R(a_P|[x]) = \lambda_{PP}P(X|[x]) + \lambda_{PN}P(\neg X|[x]),$$
$$R(a_N|[x]) = \lambda_{NP}P(X|[x]) + \lambda_{NN}P(\neg X|[x]),$$
$$R(a_B|[x]) = \lambda_{BP}P(X|[x]) + \lambda_{BN}P(\neg X|[x]).$$

Then the Bayesian decision procedure suggests the following three minimum-risk decision rules.

(P1) If $R(a_P|[x]) \leq R(a_B|[x])$ and $R(a_P|[x]) \leq R(a_N|[x])$, decide $x \in POS(X)$,
(N1) If $R(a_N|[x]) \leq R(a_P|[x])$ and $R(a_N|[x]) \leq R(a_B|[x])$, decide $x \in NEG(X)$,
(B1) If $R(a_B|[x]) \leq R(a_P|[x])$ and $R(a_B|[x]) \leq R(a_N|[x])$, decide $x \in BND(X)$.

By considering $0 \leq \lambda_{PP} \leq \lambda_{BP} < \lambda_{NP}$ and $0 \leq \lambda_{NN} \leq \lambda_{BN} < \lambda_{PN}$, (P1)–(B1) can be expressed concisely as:

(P2) If $P(X|[x]) \geq \alpha$ and $P(X|[x]) \geq \gamma$, decide $x \in POS(X)$,
(N2) If $P(X|[x]) \leq \gamma$ and $P(X|[x]) \leq \beta$, decide $x \in NEG(X)$,
(B2) If $P(X|[x]) \leq \alpha$ and $P(X|[x]) \geq \beta$, decide $x \in BND(X)$,

where:

$$\alpha = \frac{\lambda_{PN} - \lambda_{BN}}{(\lambda_{PN} - \lambda_{BN}) + (\lambda_{BP} - \lambda_{PP})},$$

$$\beta = \frac{\lambda_{BN} - \lambda_{NN}}{(\lambda_{BN} - \lambda_{NN}) + (\lambda_{NP} - \lambda_{BP})},$$

$$\gamma = \frac{\lambda_{PN} - \lambda_{NN}}{(\lambda_{PN} - \lambda_{NN}) + (\lambda_{NP} - \lambda_{PP})}.$$

If $0 \leq \beta < \gamma < \alpha \leq 1$, (P2)–(B2) can be rewritten as follows:

(P3) If $P(X|[x]) \geq \alpha$, decide $x \in POS(X)$,
(N3) If $P(X|[x]) \leq \beta$, decide $x \in NEG(X)$,
(B3) If $\beta < P(X|[x]) < \alpha$, decide $x \in BND(X)$.

Based on the decision rules above, we obtain lower and upper approximations of the decision-theoretic rough sets as follows.

$$\underline{PR}(X) = \{x \in U \mid P(X|[x]) \geq \alpha\} \text{ and } \overline{PR}(X) = \{x \in U \mid P(X|[x]) > \beta\}.$$

3   Optimistic Multigranulation Decision-Theoretic Rough Sets in Multi-covering Space

We define the pair $\langle U, \mathbf{C} \rangle$ as a multi-covering approximation space in this paper, where U is a universe of discourse and $\mathbf{C}$ is a family of coverings on the universe U.

Let $\langle U, \mathbf{C} \rangle$ be a multi-covering approximation space, $C_1, C_2 \in \mathbf{C}$ are two granular structures on U. $\Omega_i = \{X, \neg X\}$ is state set for i-th granular structure ($i = 1, 2$), indicating that an element is in X or not in X. $A = \{a_P, a_B, a_N\}$ denotes the set of actions, where a_P means deciding $x \in POS(X)$, a_B means deciding $x \in BND(X)$ and a_N deciding $x \in NEG(X)$. $\lambda_{PP}^i, \lambda_{BP}^i$ and λ_{NP}^i denote the loss, or cost, for a_P, a_B and a_N, respectively, when an object x belongs to X under i-th granular structure. Analogously, $\lambda_{PN}^i, \lambda_{BN}^i$ and λ_{NN}^i denote the loss, or cost, for taking the corresponding actions when x belongs to $\neg X$. According to the Bayesian decision proceduce, for the i-th granular structure, the expected loss of taking actions a_P, a_B and a_N for x by employing the minimal descriptions of x can be defined as follows.

$$R(a_P|\cup md_{C_i}(x)) = \lambda^i_{PP}P(X|\cup md_{C_i}(x)) + \lambda^i_{PN}P(\neg X|\cup md_{C_i}(x)),$$
$$R(a_B|\cup md_{C_i}(x)) = \lambda^i_{BP}P(X|\cup md_{C_i}(x)) + \lambda^i_{BN}P(\neg X|\cup md_{C_i}(x)),$$
$$R(a_N|\cup md_{C_i}(x)) = \lambda^i_{NP}P(X|\cup md_{C_i}(x)) + \lambda^i_{NN}P(\neg X|\cup md_{C_i}(x)).$$

By the strategy "seeking commonality while preserving difference" and suppose all $\lambda^i_{\bullet\bullet}$ are equal, the expected overall loss of taking actions a_P, a_B and a_N for the object x can be computed as follows.

$$R(a_P|(\cup md_{C_1}(x), \cup md_{C_2}(x))) = \lambda_{PP} \bigwedge_{i=1}^{2} P(X|\cup md_{C_i}(x))$$
$$+ \lambda_{PN} \bigwedge_{i=1}^{2} P(\neg X|\cup md_{C_i}(x)),$$

$$R(a_B|(\cup md_{C_1}(x), \cup md_{C_2}(x))) = \lambda_{BP} \bigwedge_{i=1}^{2} P(X|\cup md_{C_i}(x))$$
$$+ \lambda_{BN} \bigwedge_{i=1}^{2} P(\neg X|\cup md_{C_i}(x)),$$

$$R(a_N|(\cup md_{C_1}(x), \cup md_{C_2}(x))) = \lambda_{NP} \bigwedge_{i=1}^{2} P(X|\cup md_{C_i}(x))$$
$$+ \lambda_{NN} \bigwedge_{i=1}^{2} P(\neg X|\cup md_{C_i}(x)).$$

where "$\bigwedge$" denotes the operation "minimum".

If $\bigwedge_{i=1}^{2} P(X|\cup md_{C_i}(x)) \neq 0$ or $\bigwedge_{i=1}^{2} P(\neg X|\cup md_{C_i}(x)) \neq 0$, then we obtain the following three minimum-risk decision rules.

(OP1) If $R(a_P|(\cup md_{C_1}(x), \cup md_{C_2}(x))) \leq R(a_B|(\cap md_{C_1}(x), \cup md_{C_2}(x)))$ and $R(a_P|(\cup md_{C_1}(x), \cup md_{C_2}(x))) \leq R(a_N|(\cup md_{C_1}(x), \cup md_{C_2}(x)))$, decide $x \in POS_{C_1+C_2}{}^O(X)$;

(ON1) If $R(a_N|(\cup md_{C_1}(x), \cup md_{C_2}(x))) \leq R(a_P|(\cup md_{C_1}(x), \cup md_{C_2}(x)))$ and $R(a_N|(\cup md_{C_1}(x), \cup md_{C_2}(x))) \leq R(a_B|(\cup md_{C_1}(x), \cup md_{C_2}(x)))$, decide $x \in NEG_{C_1+C_2}{}^O(X)$;

(OB1) If $R(a_B|(\cup md_{C_1}(x), \cup md_{C_2}(x))) \leq R(a_P|(\cup md_{C_1}(x), \cup md_{C_2}(x)))$ and $R(a_B|(\cup md_{C_1}(x), \cup md_{C_2}(x))) \leq R(a_N|(\cup md_{C_1}(x), \cup md_{C_2}(x)))$, decide $x \in BND_{C_1+C_2}{}^O(X)$.

If the loss function satisfies $0 \leq \lambda_{PP} \leq \lambda_{BP} < \lambda_{NP}$ and $0 \leq \lambda_{NN} \leq \lambda_{BN} < \lambda_{PN}$, and noting that $P(X|\cup md_{C_i}(x)) + P(\neg X|\cup md_{C_i}(x)) = 1$, we have:

(1) For rule (OP1):

$$R(a_P|(\cup md_{C_1}(x), \cup md_{C_2}(x))) \leq R(a_B|(\cup md_{C_1}(x), \cup md_{C_2}(x)))$$

$$\Longleftrightarrow \frac{\bigwedge_{i=1}^2 P(X| \cup md_{C_i}(x))}{1 + \bigwedge_{i=1}^2 P(X| \cup md_{C_i}(x)) - \bigvee_{i=1}^2 P(X| \cup md_{C_i}(x))}$$

$$\geq \frac{\lambda_{PN} - \lambda_{BN}}{(\lambda_{PN} - \lambda_{BN}) + (\lambda_{BP} - \lambda_{PP})} \text{ and}$$

$$R(a_P|(\cup md_{C_1}(x), \cup md_{C_2}(x))) \leq R(a_N|(\cup md_{C_1}(x), \cup md_{C_2}(x)))$$

$$\Longleftrightarrow \frac{\bigwedge_{i=1}^2 P(X| \cup md_{C_i}(x))}{1 + \bigwedge_{i=1}^2 P(X| \cup md_{C_i}(x)) - \bigvee_{i=1}^2 P(X| \cup md_{C_i}(x))}$$

$$\geq \frac{\lambda_{PN} - \lambda_{NN}}{(\lambda_{PN} - \lambda_{NN}) + (\lambda_{NP} - \lambda_{PP})}.$$

where "$\bigvee$" denotes the operation "maximum".

(2) For rule (ON1):

$$R(a_N|(\cup md_{C_1}(x), \cup md_{C_2}(x))) \leq R(a_P|(\cup md_{C_1}(x), \cup md_{C_2}(x)))$$

$$\Longleftrightarrow \frac{\bigwedge_{i=1}^2 P(X| \cup md_{C_i}(x))}{1 + \bigwedge_{i=1}^2 P(X| \cup md_{C_i}(x)) - \bigvee_{i=1}^2 P(X| \cup md_{C_i}(x))}$$

$$\leq \frac{\lambda_{PN} - \lambda_{NN}}{(\lambda_{PN} - \lambda_{NN}) + (\lambda_{NP} - \lambda_{PP})} \text{ and}$$

$$R(a_N|(\cup md_{C_1}(x), \cup md_{C_2}(x))) \leq R(a_B|(\cup md_{C_1}(x), \cup md_{C_2}(x)))$$

$$\Longleftrightarrow \frac{\bigwedge_{i=1}^2 P(X| \cup md_{C_i}(x))}{1 + \bigwedge_{i=1}^2 P(X| \cup md_{C_i}(x)) - \bigvee_{i=1}^2 P(X| \cup md_{C_i}(x))}$$

$$\leq \frac{\lambda_{BN} - \lambda_{NN}}{(\lambda_{BN} - \lambda_{NN}) + (\lambda_{NP} - \lambda_{BP})}.$$

(3) For rule (OB1):

$$R(a_B|(\cup md_{C_1}(x), \cup md_{C_m}(x))) \leq R(a_P|(\cup md_{C_1}(x), \cup md_{C_2}(x)))$$

$$\Longleftrightarrow \frac{\bigwedge_{i=1}^2 P(X| \cup md_{C_i}(x)}{1 + \bigwedge_{i=1}^2 P(X| \cup md_{C_i}(x)) - \bigvee_{i=1}^2 P(X| \cup md_{C_i}(x))}$$

$$\leq \frac{\lambda_{PN} - \lambda_{BN}}{(\lambda_{PN} - \lambda_{BN}) + (\lambda_{BP} - \lambda_{PP})} \text{ and}$$

$$R(a_B|(\cup md_{C_1}(x), \cup md_{C_2}(x))) \leq R(a_N|(\cup md_{C_1}(x), \cup md_{C_2}(x)))$$

$$\Longleftrightarrow \frac{\bigwedge_{i=1}^2 P(X| \cup md_{C_i}(x))}{1 + \bigwedge_{i=1}^2 P(X| \cup md_{C_i}(x)) - \bigvee_{i=1}^2 P(X| \cup md_{C_i}(x))}$$

$$\geq \frac{\lambda_{BN} - \lambda_{NN}}{(\lambda_{BN} - \lambda_{NN}) + (\lambda_{NP} - \lambda_{BP})}.$$

Therefore, the rules (OP1)–(OB1) can be rewritten as:

(OP2) If $\dfrac{\bigwedge_{i=1}^{2} P(X|\cup md_{C_i}(x))}{1+\bigwedge_{i=1}^{2} P(X|\cup md_{C_i}(x))-\bigvee_{i=1}^{2} P(X|\cup md_{C_i}(x))} \geq \alpha$ and

$$\frac{\bigwedge_{i=1}^{2} P(X|\cup md_{C_i}(x))}{1+\bigwedge_{i=1}^{2} P(X|\cup md_{C_i}(x)) - \bigvee_{i=1}^{2} P(X|\cup md_{C_i}(x))} \geq \gamma$$

decide $x \in POS_{C_1+C_2}{}^{O}(X)$;

(ON2) If $\dfrac{\bigwedge_{i=1}^{2} P(X|\cup md_{C_i}(x))}{1+\bigwedge_{i=1}^{2} P(X|\cup md_{C_i}(x))-\bigvee_{i=1}^{2} P(X|\cup md_{C_i}(x))} \leq \gamma$ and

$$\frac{\bigwedge_{i=1}^{2} P(X|\cup md_{C_i}(x))}{1+\bigwedge_{i=1}^{2} P(X|\cup md_{C_i}(x)) - \bigvee_{i=1}^{2} P(X|\cup md_{C_i}(x))} \leq \beta$$

decide $x \in NEG_{C_1+C_2}{}^{O}(X)$;

(OB2) If $\dfrac{\bigwedge_{i=1}^{2} P(X|\cup md_{C_i}(x))}{1+\bigwedge_{i=1}^{2} P(X|\cup md_{C_i}(x))-\bigvee_{i=1}^{2} P(X|\cup md_{C_i}(x))} \leq \alpha$ and

$$\frac{\bigwedge_{i=1}^{2} P(X|\cup md_{C_i}(x))}{1+\bigwedge_{i=1}^{2} P(X|\cup md_{C_i}(x)) - \bigvee_{i=1}^{2} P(X|\cup md_{C_i}(x))} \geq \beta$$

decide $x \in BND_{C_1+C_2}{}^{O}(X)$.

Where

$$\alpha = \frac{\lambda_{PN} - \lambda_{BN}}{(\lambda_{PN} - \lambda_{BN}) + (\lambda_{BP} - \lambda_{PP})},$$

$$\beta = \frac{\lambda_{BN} - \lambda_{NN}}{(\lambda_{BN} - \lambda_{NN}) + (\lambda_{NP} - \lambda_{BP})},$$

$$\gamma = \frac{\lambda_{PN} - \lambda_{NN}}{(\lambda_{PN} - \lambda_{NN}) + (\lambda_{NP} - \lambda_{PP})}.$$

If $(\lambda_{PN} - \lambda_{BN})(\lambda_{NP} - \lambda_{BP}) > (\lambda_{BN} - \lambda_{NN})(\lambda_{BP} - \lambda_{PP})$, we have $0 \leq \beta < \gamma < \alpha \leq 1$. Then (OP2)–(OB2) can be rewritten as:

(OP3) If $\dfrac{\bigwedge_{i=1}^{2} P(X|\cup md_{C_i}(x))}{1 + \bigwedge_{i=1}^{2} P(X|\cup md_{C_i}(x)) - \bigvee_{i=1}^{2} P(X|\cup md_{C_i}(x))} \geq \alpha$,

decide $x \in POS_{C_1+C_2}{}^{O}(X)$;

(ON3) If $\dfrac{\bigwedge_{i=1}^{2} P(X|\cup md_{C_i}(x))}{1 + \bigwedge_{i=1}^{2} P(X|\cup md_{C_i}(x)) - \bigvee_{i=1}^{2} P(X|\cup md_{C_i}(x))} \leq \beta$,

decide $x \in NEG_{C_1+C_2}{}^{O}(X)$;

(OB3) If $\beta < \dfrac{\bigwedge_{i=1}^{2} P(X|\cup md_{C_i}(x))}{1 + \bigwedge_{i=1}^{2} P(X|\cup md_{C_i}(x)) - \bigvee_{i=1}^{2} P(X|\cup md_{C_i}(x))} < \alpha$,

decide $x \in BND_{C_1+C_2}{}^{O}(X)$.

Based on (OP3–OB3), we can get the definitions of the optimistic multigranulation positive, negative, and boundary regions of X as follows.

Definition 4. Let $\langle U, \mathbf{C} \rangle$ be a covering approximation space, $C_1, C_2 \in \mathbf{C}$, and $P : 2^U \longrightarrow [0, 1]$ is a probability function defined on the power set 2^U. For any $X \subseteq U$, the positive, negative, and boundary regions of X of covering-based optimistic multigranulation decision-theoretic rough set are defined as:

$$POS_{C_1+C_2}{}^O(X)$$
$$= \{x \in U \mid \frac{\bigwedge_{i=1}^2 P(X \mid \cup md_{C_i}(x))}{1 + \bigwedge_{i=1}^2 P(X \mid \cup md_{C_i}(x)) - \bigvee_{i=1}^2 P(X \mid \cup md_{C_i}(x))} \geq \alpha\}$$

$$NEG_{C_1+C_2}{}^O(X)$$
$$= \{x \in U \mid \frac{\bigwedge_{i=1}^2 P(X \mid \cup md_{C_i}(x))}{1 + \bigwedge_{i=1}^2 P(X \mid \cup md_{C_i}(x)) - \bigvee_{i=1}^2 P(X \mid \cup md_{C_i}(x))} \leq \beta\}$$

$$BND_{C_1+C_2}{}^O(X) =$$
$$\{x \in U \mid \beta < \frac{\bigwedge_{i=1}^2 P(X \mid \cup md_{C_i}(x))}{1 + \bigwedge_{i=1}^2 P(X \mid \cup md_{C_i}(x)) - \bigvee_{i=1}^2 P(X \mid \cup md_{C_i}(x))} < \alpha\}.$$

Moreover, the lower and upper approximations of X of optimistic multigranulation decision-theoretic rough sets can be defined as follows.

Definition 5. Let $\langle U, \mathbf{C} \rangle$ be a covering approximation space, $C_1, C_2 \in \mathbf{C}$, and $P : 2^U \longrightarrow [0, 1]$ is a probability function defined on the power set 2^U. For any $X \subseteq U$, the lower and upper approximations of X are defined as follows.

$$\underline{C_1 + C_2}^{O,\alpha}(X)$$
$$= \{x \in U \mid \frac{\bigwedge_{i=1}^2 P(X \mid \cup md_{C_i}(x))}{1 + \bigwedge_{i=1}^2 P(X \mid \cup md_{C_i}(x)) - \bigvee_{i=1}^2 P(X \mid \cup md_{C_i}(x))} \geq \alpha\}$$

$$\overline{C_1 + C_2}^{O,\beta}(X)$$
$$= \{x \in U \mid \frac{\bigwedge_{i=1}^2 P(X \mid \cup md_{C_i}(x))}{1 + \bigwedge_{i=1}^2 P(X \mid \cup md_{C_i}(x)) - \bigvee_{i=1}^2 P(X \mid \cup md_{C_i}(x))} > \beta\}$$

The pair $(\underline{C_1 + C_2}^{O,\alpha}(X), \overline{C_1 + C_2}^{O,\beta}(X))$ is called an optimistic multigranulation decision-theoretic rough set.

By the definition of optimistic multigranulation decision-theoretic lower and upper approximations, we have the following properties.

Proposition 1. Let $\langle U, \mathbf{C} \rangle$ be a covering approximation space, $C_1, C_2 \in \mathbf{C}$, and $P : 2^U \longrightarrow [0, 1]$ is a probability function defined on the power set 2^U. For any $0 \leq \beta < \alpha \leq 1$, and $X, Y \subseteq U$, we have

(1)

$$\underline{C_1 + C_2}^{O,\alpha}(\emptyset) = \overline{C_1 + C_2}^{O,\beta}(\emptyset) = \emptyset,$$
$$\underline{C_1 + C_2}^{O,\alpha}(U) = \overline{C_1 + C_2}^{O,\beta}(U) = U;$$

(2)

$$\underline{C_1 + C_2}^{O,\alpha}(X) \subseteq X \subseteq \overline{C_1 + C_2}^{O,\beta}(X);$$

(3) If $X \subseteq Y$, we have $\underline{C_1 + C_2}^{O,\alpha}(X) \subseteq \underline{C_1 + C_2}^{O,\alpha}(Y)$ and $\overline{C_1 + C_2}^{O,\beta}(X) \subseteq \overline{C_1 + C_2}^{O,\beta}(Y)$;

(4) If $\alpha > 0.5$, we have $\underline{C_1 + C_2}^{O,\alpha}(X) = \neg\overline{C_1 + C_2}^{O,1-\alpha}(\neg X)$;
    If $\beta < 0.5$, we have $\overline{C_1 + C_2}^{O,\beta}(X) = \neg\underline{C_1 + C_2}^{O,1-\beta}(\neg X)$.

Proof. We only offer the proofs of (4) here, others can be easily proved according to Definition 5.

For given $\alpha > 0.5$,

$$\neg\overline{C_1 + C_2}^{O,1-\alpha}(\neg X)$$

$$= \neg\{x \in U \mid \frac{\bigwedge_{i=1}^{2} P(\neg X \mid \cup md_{C_i}(x))}{1 + \bigwedge_{i=1}^{2} P(\neg X \mid \cup md_{C_i}(x)) - \bigvee_{i=1}^{2} P(\neg X \mid \cup md_{C_i}(x))} > 1 - \alpha\}$$

$$= \{x \in U \mid \frac{1 - \bigvee_{i=1}^{2} P(X \mid \cup md_{C_i}(x))}{1 - \bigvee_{i=1}^{2} P(X \mid \cup md_{C_i}(x)) + \bigwedge_{i=1}^{2} P(X \mid \cup md_{C_i}(x))} \leq 1 - \alpha\}$$

$$= \{x \in U \mid \frac{\bigwedge_{i=1}^{2} P(X \mid \cup md_{C_i}(x))}{1 + \bigwedge_{i=1}^{2} P(X \mid \cup md_{C_i}(x)) - \bigvee_{i=1}^{2} P(X \mid \cup md_{C_i}(x))} \geq \alpha\}$$

$$= \underline{C_1 + C_2}^{O,\alpha}(X).$$

Other part of (4) can be proved in a similar way. $\square$

Theorem 1. Let $\langle U, \mathbf{C} \rangle$ be a covering approximation space, $C_1, C_2 \in \mathbf{C}$, for any $0 \leq \beta < \alpha \leq 1$, and $X \subseteq U$, we have

(1)

$$\underline{C_1 + C_2}^{O,\alpha}(\underline{C_1 + C_2}^{O,\alpha}(X)) \subseteq \overline{C_1 + C_2}^{O,\beta}(\underline{C_1 + C_2}^{O,\alpha}(X));$$

(2)

$$\overline{C_1 + C_2}^{O,\beta}(\overline{C_1 + C_2}^{O,\beta}(X)) \supseteq \underline{C_1 + C_2}^{O,\alpha}(\overline{C_1 + C_2}^{O,\beta}(X)).$$

Theorem 2. Let $\langle U, \mathbf{C} \rangle$ be a covering approximation space, $C_1, C_2 \in \mathbf{C}$, for any $0 \leq \beta_2 \leq \beta_1 \leq \alpha_1 \leq \alpha_2 \leq 1$, and $X \subseteq U$, we have

$$\underline{C_1 + C_2}^{O,\alpha_2}(X) \subseteq \underline{C_1 + C_2}^{O,\alpha_1}(X) \subseteq \overline{C_1 + C_2}^{O,\beta_1}(X) \subseteq \overline{C_1 + C_2}^{O,\beta_2}(X)$$

Theorem 3. Let $\langle U, \mathbf{C} \rangle$ be a covering approximation space, $C_1, C_2 \in \mathbf{C}$, for any $0 \leq \beta < \alpha \leq 1$, and $X \subseteq U$, we have

(1) If $\alpha = 1$, $\underline{C_1 + C_2}^{O,\alpha}(X) = \underline{O}_{C_1 + C_2}(X)$
(2) If $\beta = 0$, $\overline{C_1 + C_2}^{O,\alpha}(X) = \overline{O}_{C_1 + C_2}(X)$

Where $\underline{O}_{C_1 + C_2}(X)$ and $\overline{O}_{C_1 + C_2}(X)$ are defined by Liu et al. in [24].

Proof. If $\alpha = 1$, noting that $P(X| \cup md_{C_i}(x)) = \frac{|X \cup (\cup md_{C_i}(x))|}{|\cup md_{C_i}(x)|}$, then

$$\underline{C_1 + C_2}^{O,1}(X) = \{x \in U \mid \frac{\bigwedge_{i=1}^2 P(X| \cup md_{C_i}(x))}{1 + \bigwedge_{i=1}^2 P(X| \cup md_{C_i}(x)) - \bigvee_{i=1}^2 P(X| \cup md_{C_i}(x))} \geq 1\}$$

$$= \{x \in U \mid \bigvee_{i=1}^2 P(X| \cup md_{C_i}(x)) \geq 1\}$$

$$= \{x \in U \mid P(X| \cup md_{C_1}(x)) = 1 \text{ or } P(X| \cup md_{C_2}(x)) = 1\}$$

$$= \{x \in U \mid \cup md_{C_1}(x) \subseteq X \text{ or } \cup md_{C_2}(x) \subseteq X\}$$

$$= \underline{O_{C_1+C_2}}(X);$$

If $\beta = 0$, we have that

$$\overline{C_1 + C_2}^{O,0}(X) = \{x \in U \mid \frac{\bigwedge_{i=1}^2 P(X| \cup md_{C_i}(x))}{1 + \bigwedge_{i=1}^2 P(X| \cup md_{C_i}(x)) - \bigvee_{i=1}^2 P(X| \cap md_{C_i}(x))} > 0\}$$

$$= \{x \in U \mid \bigwedge_{i=1}^2 P(X| \cap md_{C_i}(x)) > 0\}$$

$$= \{x \in U \mid P(X| \cup md_{C_1}(x)) > 0 \text{ and } P(X| \cup md_{C_2}(x)) > 0\}$$

$$= \{x \in U \mid \cup md_{C_1}(x) \cup X \neq \emptyset \text{ and } \cap md_{C_2}(x) \cap X \neq \emptyset\}$$

$$= \overline{O_{C_1+C_2}}(X).$$

$\square$

Theorem 3 implies that, in this case, the optimistic multigranulation decision-theoretic rough set model will degenerate to the covering-based multigranulation rough set model in [24].

Remark 1. If **C** is a set of partitions of U, the optimistic multigranulation decision-theoretic rough set model in multi-covering space will degenerate to the optimistic multigranulation decision-theoretic rough set model in [22].

4 Conclusion

This paper studies multigranulation decision-theoretic rough sets in the multi-covering space. Under the assumptions that all $\lambda_{\bullet\bullet}^i$ are equal, we proposed the optimistic multigranulation decision-theoretic rough sets based on the Bayesian decision procedure. The relationships between covering-based multigranulation decision-theoretic rough sets and other rough sets are disclosed.

Acknowledgements. This work was supported by the China National Natural Science Foundation of Youth Science Foundation under Grant No.: 61305052, 61403329, the Key Technology Research and Development Program of Education Bureau of Jiangxi Province of China under Grant No.: GJJ14660, the Key Technology Research and Development Program of Jiangxi Province of China under Grant No.: 20142BBF60010, 20151BBF60071.

References

1. Yao, Y.Y., Wong, S.K.M.: A decision theoretic framework for approximating concepts. Int. J. Man Mach. Stud. **37**, 793–809 (1992)
2. Herbert, J.P., Yao, J.T.: Game-theoretic rough sets. Fundamenta Informaticae. **108**(3–4), 267–286 (2011)
3. Liu, D., Li, T.R., Li, H.X.: A multiple-category classification approach with decision-theoretic rough sets. Fundamenta Informaticae. **115**(2–3), 173–188 (2012)
4. Yu, H., Liu, Z.G., Wang, G.Y.: An automatic method to determine the number of clusters using decision-theoretic rough set. Int. J. Approximate Reasoning **55**(1), 101–115 (2014)
5. Li, T.J., Yang, X.P.: An axiomatic characterization of probabilistic rough sets. Int. J. Approximate Reasoning **55**(1), 130–141 (2014)
6. Jia, X.Y., Tang, Z.M., Liao, W.H., Shang, L.: On an optimization representation of decision-theoretic rough set model. Int. J. Approximate Reasoning **55**(1), 156–166 (2014)
7. Yao, Y.Y.: Three-way decisions with probabilistic rough sets. Inf. Sci. **180**, 341–353 (2010)
8. Yao, Y.: An outline of a theory of three-way decisions. In: Yao, J.T., Yang, Y., Słowiński, R., Greco, S., Li, H., Mitra, S., Polkowski, L. (eds.) RSCTC 2012. LNCS, vol. 7413, pp. 1–17. Springer, Heidelberg (2012)
9. Zhou, B., Yao, Y., Luo, J.: A three-way decision approach to email spam filtering. In: Farzindar, A., Kešelj, V. (eds.) AI 2010. LNCS (LNAI), vol. 6085, pp. 28–39. Springer, Heidelberg (2010). doi:10.1007/978-3-642-13059-5_6
10. Li, H., Zhang, L., Huang, B., Zhou, X.: Sequential three-way decision and granulation for cost-sensitive face recognition. Knowl.-Based Syst. **91**, 241–251 (2016)
11. Zhang, H.R., Min, F.: Three-way recommender systems based on random forests. Knowl.-Based Syst. **91**, 275–286 (2016)
12. Pedrycz, W.: Granular Computing: Analysis and Design of Intelligent Systems. CRC Press, Boca Raton (2013)
13. Pawlak, Z.: Rough sets. Int. J. Comput. Inform. Sci. **11**, 341–356 (1982)
14. Qian, Y.H., Liang, J.Y., Yao, Y.Y., Dang, C.Y.: MGRS: A multi-granulation rough set. Inf. Sci. **180**, 949–970 (2010)
15. Qian, Y.H., Liang, J.Y., Dang, C.Y.: Incomplete multigranulation rough set. IEEE Trans. Syst. Man Cybern. Part A **20**, 420–430 (2010)
16. Xu, W.H., Sun, W.X., Zhang, X.Y., Zhang, W.X.: Multiple granulation rough set approach to ordered information systems. Int. J. Gen. Syst. **41**(5), 475–501 (2012)
17. Yang, X.B., Qi, Y., Song, X.N., Yang, J.Y.: Test cost sensitive multigranulation rough set: model and minimal cost selection. Inf. Sci. **250**, 184–199 (2013)
18. Huang, B., Guo, C.X., Zhuang, Y.L., Li, H.X., Zhou, X.Z.: Intuitionistic fuzzy multigranulation rough sets. Inf. Sci. **277**, 299–320 (2014)
19. She, Y.H., He, X.L.: On the structure of the multigranulation rough set model. Knowl. Based Syst. **36**, 81–92 (2012)
20. Yao, Y., She, Y.: Rough set models in multigranulation spaces. Inf. Sci. **327**, 40–56 (2016)
21. Zhang, X.H., Miao, D.Q., Liu, C.H., Le, M.L.: Constructive methods of rough approximation operators and multigranulation rough sets. Knowl. Based Syst. **91**, 114–125 (2016)
22. Qian, Y.H., Zhang, H., Sang, Y.L., Liang, J.L.: Multigranulation decision-theoretic rough sets. Int. J. Approximate Reasoning **55**(1), 225–237 (2014)

23. Zakowski, W.: Approximations in the space (U, Π). Demonstratio Math. **16**, 761–769 (1983)
24. Liu, C.H., Cai, K.C.: Multi-granulation covering rough sets based on the union of minimal descriptions of elements. In: CAAI Transactions on Intelligent Systems, Accepted. (In Chinese)

Fuzziness and Similarity in Knowledge Representation

Interpretations of Lower Approximations in Inclusion Degrees

Wei-Zhi Wu[1,2]([✉]), Chao-Jun Chen[1,2], and Xia Wang[1,2]

[1] School of Mathematics, Physics and Information Science,
Zhejiang Ocean University, Zhoushan, Zhejiang 316022, China
wuwz@zjou.edu.cn, chenchaojun718@163.com, bblylm@126.com
[2] Key Laboratory of Oceanographic Big Data Mining and Application of Zhejiang
Province, Zhejiang Ocean University, Zhoushan, Zhejiang 316022, China

Abstract. The nature of uncertainty inference is to give evaluations on
inclusion relationships by means of various measures. In this paper we
introduce the concept of inclusion degrees into rough set theory. It is
shown that the lower approximations of the rough set theory in both the
crisp and the fuzzy environments can be represented as inclusion degrees.

Keywords: Approximation operators · Fuzzy implicators · Fuzzy rough
sets · Inclusion degrees · Rough sets

1 Introduction

Rough set theory, introduced by Pawlak [7], is a powerful tool for reasoning about
data. The basic notions of rough set theory are lower and upper approximations
constructed by an approximation space. When the rough set approach is used
to extract decision rules from a given information table, two types of decision
rules may be unravelled. Based on the lower approximation of a decision class,
certain information can be discovered and certain rules can be derived whereas,
by using the upper approximation of a decision class, uncertain or partially
certain information is discovered and possible rules may be induced.

In order to analyze data effectively, many qualitative measures, such as accuracy measure of rough set, accuracy of approximation of classification, measure
of dependency of attributes, measure of importance of attributes, and accuracy
and coverage of decision rule, were defined in rough set data analysis. It is well-
known that the nature of uncertainty inference is to give evaluations on inclusion
relationships by means of various measures. Based on this observation, Polkowski
and Skowron [8,9] proposed a new paradigm for approximate reasoning called
rough mereology which includes a formal treatment of the hierarchy of relations
of being a part in a degree. On the other hand, on the basis of abstracting the
existing methods of uncertainty inferences, Zhang et al. [17,18] developed the
theory of inclusion degrees for approximate reasoning. The degree of inclusion
is in fact a particular case of inclusion in a degree (rough inclusion) basic for
rough mereology. Xu et al. [15] introduced the concept of inclusion degree into

© Springer International Publishing AG 2016
V. Flores et al. (Eds.): IJCRS 2016, LNAI 9920, pp. 297–306, 2016.
DOI: 10.1007/978-3-319-47160-0_27

rough set data analysis and showed that many measures in Pawlak's information tables on rough set data analysis can be reduced to inclusion degrees.

In this paper, we provide a further study on the interpretation of rough approximations in inclusion degrees. We will show that the lower approximations of the rough set theory in both the crisp and the fuzzy environments can be represented by inclusion degrees.

2   Interpretations of Crisp Rough Set Approximations

Throughout this paper, U will be a nonempty set called the universe of discourse. The class of all subsets of U (resp. all fuzzy sets of U) will be denoted by $\mathcal{P}(U)$ (resp. $\mathcal{F}(U)$). For any $A \in \mathcal{F}(U)$, $\sim A$ will be used to denote the fuzzy complement of A in U, i.e. $(\sim A)(x) = 1 - A(x)$ for all $x \in U$.

2.1   Inclusion Degrees on Crisp Sets

Definition 1 [18]. *Let U be a universe of discourse. If for any $A, B \in \mathcal{P}(U)$, there is a real number $D(A, B)$ with the following properties:*

(D1) $0 \le D(A, B) \le 1$.
(D2) $A \subseteq B$ implies $D(A, B) = 1$.
(D3) For any $A, B, C \in \mathcal{P}(U)$, $A \subseteq B \subseteq C$ implies

$$D(C, A) \le D(B, A). \tag{1}$$

Then D is called an inclusion degree on $\mathcal{P}(U)$. Furthermore, if for any $A, B, C \in \mathcal{P}(U)$,
(D4) $A \subseteq B$ implies

$$D(C, A) \le D(C, B), \tag{2}$$

then D is referred to as a strongly inclusion degree on $\mathcal{P}(U)$.

Example 1. Let U be a universe of discourse, for $A, B \in \mathcal{P}(U)$, define

$$D_1(A, B) = \begin{cases} 1, \text{ if } A \subseteq B, \\ 0, \text{ otherwise.} \end{cases} \tag{3}$$

Then, it can be verified that D_1 is a strongly inclusion degree on $\mathcal{P}(U)$.

Example 2. Let U be a finite universe of discourse, for $A, B \in \mathcal{P}(U)$, define

$$D_2(A, B) = \begin{cases} \frac{|A \cap B|}{|A|}, \text{ if } A \ne \emptyset, \\ 1, \quad \text{ otherwise.} \end{cases} \tag{4}$$

where $|A|$ is the cardinality of the set A, then, it can be verified that D_2 is a strongly including degree on $\mathcal{P}(U)$.

2.2   Approximations of Sets in Generalized Approximation Spaces

Definition 2 [13,16]. *Let U and W be two nonempty universes of discourse. A subset $R \in \mathcal{P}(U \times W)$ is referred to as a binary relation from U to W, and the triple (U, W, R) is called a generalized approximation space. For any set $A \in \mathcal{P}(W)$, the lower and upper approximations of A w.r.t. (U, W, R), denoted as $\underline{R}(A)$ and $\overline{R}(A)$, are, respectively, defined by*

$$\underline{R}(A) = \{x \in U : R_s(x) \subseteq A\}, \quad \overline{R}(A) = \{x \in U : R_s(x) \cap A \neq \emptyset\}, \quad (5)$$

where $R_s(x) = \{y \in W : (x, y) \in R\}$ is the successor neighborhood of x in R. The pair $(\underline{R}(A), \overline{R}(A))$ is referred to as a generalized rough set, and $\underline{R}$ and $\overline{R} : \mathcal{P}(W) \to \mathcal{P}(U)$ are called the lower and upper generalized approximation operators, respectively.

From the definitions of approximation operators, the following theorem can be easily derived [13,16]:

Theorem 1. *For a given approximation space (U, W, R), the lower and upper approximation operators defined by Eq. (5) satisfy the following properties: for all $A, B, A_i \in \mathcal{P}(W), i \in J, J$ is an index set,*

(LD) $\underline{R}(A) = \sim \overline{R}(\sim A)$,   (UD) $\overline{R}(A) = \sim \underline{R}(\sim A)$;

(L1) $\underline{R}(W) = U$,   (U1) $\overline{R}(\emptyset) = \emptyset$;

(L2) $\underline{R}(\bigcap_{i \in J} A_i) = \bigcap_{i \in J} \underline{R}(A_i)$,   (U2) $\overline{R}(\bigcup_{i \in J} A_i) = \bigcup_{i \in J} \overline{R}(A_i)$;

(L3) $A \subseteq B \implies \underline{R}(A) \subseteq \underline{R}(B)$, (U3) $A \subseteq B \implies \overline{R}(A) \subseteq \overline{R}(B)$;

(L4) $\underline{R}(\bigcup_{i \in J} A_i) \supseteq \bigcup_{i \in J} \underline{R}(A_i)$,   (U4) $\overline{R}(\bigcap_{i \in J} A_i) \subseteq \bigcap_{i \in J} \overline{R}(A_i)$.

Properties (LD) and (UD) show that $\underline{R}$ and $\overline{R}$ are dual approximation operators.

Theorem 2. *Let (U, W, R) be a generalized approximation space, $A \in \mathcal{P}(W)$, and $x \in U$. Then*

$$x \in \underline{R}(A) \iff D_1(R_s(x), A) = 1. \quad (6)$$

Remark 1. Theorem 2 shows that an object $x \in U$ is in the lower approximation of a set $A \in \mathcal{P}(W)$ w.r.t. an approximation space (U, W, R) iff the inclusion degree of which the successor neighborhood of x in R, $R_s(x)$, is included in A is 1. In other words, the lower approximation can be represented by the inclusion degree D_1 defined in Example 1, i.e.

$$\underline{R}(A) = \{x \in U : D_1(R_s(x), A) = 1\}, A \in \mathcal{P}(W). \quad (7)$$

Theorem 3. *If (U, W, R) is a finite generalized approximation space, $A \in \mathcal{P}(W)$, and $x \in U$. Then*

$$x \in \underline{R}(A) \iff D_2(R_s(x), A) = 1. \quad (8)$$

Remark 2. Theorem 3 states that when U and W are finite sets, the lower approximation can be represented by the inclusion degree D_2 defined in Example 2, i.e.

$$\underline{R}(A) = \{x \in U : D_2(R_s(x), A) = 1\}, A \in \mathcal{P}(W). \quad (9)$$

3   Interpretations of Fuzzy Rough Approximations

3.1   Fuzzy Logical Operators

Definition 3 [3]. *A triangular norm (t-norm for short) is a binary operation T on the unit interval $[0,1]$, i.e., a function $T : [0,1]^2 \to [0,1]$, such that for all $x, y, z \in [0,1]$ the following four axioms are satisfied:*

(T1) $T(x,y) = T(y,x)$.   *(commutativity)*
(T2) $T(x, T(y,z)) = T(T(x,y), z)$.   *(associativity)*
(T3) $T(x,y) \leq T(x,z)$ *whenever* $y \leq z$.   *(monotonicity)*
(T4) $T(x,1) = x$.   *(boundary condition)*

The most popular continuous *t*-norms are:

- The standard min operator $T_{\mathrm{M}}(\alpha, \beta) = \min\{\alpha, \beta\}$ (the largest *t*-norm [3]),
- The algebraic product $T_{\mathrm{P}}(\alpha, \beta) = \alpha * \beta$,
- The bold intersection (also called the Łukasiewicz *t*-norm) $T_{\mathrm{L}}(\alpha, \beta) = \max\{0, \alpha + \beta - 1\}$.

Definition 4 [3]. *A triangular conorm (t-conorm for short) is a binary operation S on the unit interval $[0,1]$, i.e., a function $S : [0,1]^2 \to [0,1]$, which, for all $x, y, z \in (0,1]$, satisfies (T1)–(T3) and*
      (S4) $S(x,0) = x$.   *(boundary condition)*

Three well-known continuous *t*-conorms are:

- The standard max operator $S_{\mathrm{M}}(\alpha, \beta) = \max\{\alpha, \beta\}$ (the smallest *t*-conorm),
- The probabilistic sum $S_{\mathrm{P}}(\alpha, \beta) = \alpha + \beta - \alpha * \beta$,
- The bounded sum $S_{\mathrm{L}}(\alpha, \beta) = \min\{1, \alpha + \beta\}$.

Definition 5. *A function $\mathcal{I} : [0,1]^2 \to [0,1]$ is referred to as an implicator (fuzzy implication operator) if it satisfies $\mathcal{I}(1,0) = 0$ and $\mathcal{I}(1,1) = \mathcal{I}(0,1) = \mathcal{I}(0,0) = 1$. An implicator $\mathcal{I}$ is called left monotonic (resp. right monotonic) iff for every $\alpha \in I, \mathcal{I}(\cdot, \alpha)$ is decreasing (resp. $\mathcal{I}(\alpha, \cdot)$ is increasing). If $\mathcal{I}$ is both left monotonic and right monotonic, then it is called hybrid monotonic. $\mathcal{I}$ is semicontinuous if*

$$\mathcal{I}\left(\bigvee_j a_j, \bigwedge_k b_k\right) = \bigwedge_{j,k} \mathcal{I}(a_j, b_k) \tag{10}$$

for all index family $\{a_j : j \in J\}$ and $\{b_k : k \in K\}$ of real numbers in $[0,1]$.

Remark 3. It is easy to verify that $\mathcal{I}(\alpha, 1) = 1$ for all $\alpha \in [0,1]$ when $\mathcal{I}$ is a left monotonic implicator, and if $\mathcal{I}$ is right monotonic then $\mathcal{I}(0, \alpha) = 1$ for all $\alpha \in [0,1]$.

An implicator $\mathcal{I}$ is said to be a *border implicator* (or it satisfies the neutrality principle [2]) if $\mathcal{I}(1, x) = x$ for all $x \in [0,1]$.

An implicator $\mathcal{I}$ is said to be *a CP implicator* (CP stands for confinement principle [2]) if it satisfies for all $\alpha, \beta \in [0,1]$

$$\alpha \leq \beta \Longleftrightarrow \mathcal{I}(\alpha, \beta) = 1. \tag{11}$$

Definition 6. *A unary operation* $\mathcal{N} : [0,1] \to [0,1]$ *is called a negator if it is a decreasing mapping satisfying* $\mathcal{N}(0) = 1$ *and* $\mathcal{N}(1) = 0$. *The negator* $\mathcal{N}_s(\alpha) = 1 - \alpha$ *is usually referred to as the standard negator. A negator* $\mathcal{N}$ *is called involutive iff* $\mathcal{N}(\mathcal{N}(\alpha)) = \alpha$ *for all* $\alpha \in [0,1]$.

It is well-known that every involutive negator is continuous [4]. For a left monotonic implicator $\mathcal{I}$, the function $\mathcal{N}_{\mathcal{I}}$, defined by $\mathcal{N}_{\mathcal{I}}(x) = \mathcal{I}(x,0), x \in [0,1]$, is a negator, called a negator induced by $\mathcal{I}$. For example, the Łukasiewicz implicator $\mathcal{I}_{\mathrm{L}}(x,y) = \min\{1, 1 - x + y\}$ induces the standard negator $\mathcal{N}_s$.

Several classes of implicators have been studied in the literature. We recall here the definitions of two main classes of operators [4].

Let T, S and $\mathcal{N}$ be a t-norm, a t-conorm and a negator, respectively. An implicator $\mathcal{I}$ is called

- an *S-implicator* based on S and $\mathcal{N}$ iff

$$\mathcal{I}(x,y) = S(\mathcal{N}(x),y) \text{ for all } x,y \in [0,1]. \tag{12}$$

- an *R-implicator* (residual implicator) based on a left-continuous t-norm T iff for every $x,y \in [0,1]$,

$$\mathcal{I}(x,y) =: \theta_T(x,y) = \sup\{\lambda \in [0,1] : T(x,\lambda) \leq y\}. \tag{13}$$

Three most popular S-implicators are:

- the Łukasiewicz implicator $\mathcal{I}_{\mathrm{L}}(x,y) = \min\{1, 1 - x + y\}$, based on S_{L} and $\mathcal{N}_s$,
- the Kleene-Dienes implicator $\mathcal{I}_{\mathrm{KD}}(x,y) = \max\{1 - x, y\}$, based on S_{M} and $\mathcal{N}_s$,
- the Kleene-Dienes-Łukasiewicz implicator $\mathcal{I}_*(x,y) = 1 - x + x * y$, based on S_{p} and $\mathcal{N}_s$.

The most popular R-implicators are:

- the Łukasiewicz implicator $\mathcal{I}_{\mathrm{L}}$, based on T_{L},
- the Gödel implicator $\mathcal{I}_{\mathrm{G}}(x,y) = 1$ for $x \leq y$ and $\mathcal{I}_{\mathrm{G}}(x,y) = y$ elsewhere, based on T_{M},
- the Gaines implicator $\mathcal{I}_{\Delta}(x,y) = 1$ for $x \leq y$ and $\mathcal{I}_{\Delta}(x,y) = y$ elsewhere, based on T_{p}.

Proposition 1 [1,10]. *Every S-implicator and every R-implicator is a hybrid monotonic, border implicator. And every R-implicator is a CP implicator.*

Given a negator $\mathcal{N}$ and a border implicator $\mathcal{I}$, one can define an $\mathcal{N}$-dual operator of $\mathcal{I}$, $\theta_{\mathcal{I},\mathcal{N}} : [0,1]^2 \to [0,1]$, as follows:

$$\theta_{\mathcal{I},\mathcal{N}}(x,y) = \mathcal{N}(\mathcal{I}(\mathcal{N}(x),\mathcal{N}(y))), \quad x,y \in [0,1]. \tag{14}$$

It can be verified that $\theta_{\mathcal{I},\mathcal{N}}$ satisfies following properties [12]:

(1) $\theta_{\mathcal{I},\mathcal{N}}(1,0) = \theta_{\mathcal{I},\mathcal{N}}(1,1) = \theta_{\mathcal{I},\mathcal{N}}(0,0) = 0$.

(2) $\theta_{\mathcal{I},\mathcal{N}}(0,1) = 1$.

(3) If $\mathcal{N}$ is involutive, then $\theta_{\mathcal{I},\mathcal{N}}(0,x) = x$ for all $x \in [0,1]$.

(4) $\theta_{\mathcal{I},\mathcal{N}}$ is left monotonic (resp. right monotonic) whenever $\mathcal{I}$ is left monotonic (resp. right monotonic).

(5) If $\mathcal{I}$ is left monotonic, then $\theta_{\mathcal{I},\mathcal{N}}(x,0) = 0$ for all $x \in [0,1]$; and if $\mathcal{I}$ is right monotonic, then $\theta_{\mathcal{I},\mathcal{N}}(1,x) = 0$ for all $x \in [0,1]$.

(6) If $\mathcal{I}$ is a CP implicator, then $y \leq x$ iff $\theta_{\mathcal{I},\mathcal{N}}(x,y) = 0$.

If $\mathcal{I}$ is an S-implicator, and T the t-norm dual to S w.r.t. $\mathcal{N}$, then it can be verified that

$$\theta_{\mathcal{I},\mathcal{N}}(x,y) = T(\mathcal{N}(x),y). \tag{15}$$

If $\mathcal{I}$ is an R-implicator, and S the t-conorm dual to S w.r.t. $\mathcal{N}$, then it can be checked that

$$\theta_{\mathcal{I},\mathcal{N}}(x,y) = \inf\{\lambda \in [0,1] : S(x,\lambda) \geq y\}. \tag{16}$$

Proposition 2. *Given a negator $\mathcal{N}$ and an implicator $\mathcal{I}$, for $A,B \in \mathcal{F}(U)$, define*

$$D_{\mathcal{I}}(A,B) = \bigwedge_{x \in U} \mathcal{I}(A(x), B(x)). \tag{17}$$

Then $D_{\mathcal{I}}$ satisfies following properties:

(1) *If $\mathcal{I}$ is left monotonic, then, for any $A,B,C \in \mathcal{F}(U)$, $A \subseteq B$ implies $D_{\mathcal{I}}(A,C) \geq D_{\mathcal{I}}(B,C)$.*

(2) *If $\mathcal{I}$ right monotonic, then, for any $A,B,C \in \mathcal{F}(U)$, $A \subseteq B$ implies $D_{\mathcal{I}}(C,A) \leq D_{\mathcal{I}}(C,B)$.*

(3) *If $\mathcal{I}$ is a CP implicator, then, for any $A,B \in \mathcal{F}(U)$,*

$$A \subseteq B \Longleftrightarrow D_{\mathcal{I}}(A,B) = 1. \tag{18}$$

(4) *If $\mathcal{I}$ is a border implicator, then $D_{\mathcal{I}}(U,A) = \bigwedge_{x \in U} A(x)$ for all $A \in \mathcal{F}(U)$.*

(5) $D_{\mathcal{I}}(U,\emptyset) = 0$.

(6) $D_{\mathcal{I}}(U,U) = D_{\mathcal{I}}(\emptyset,U) = D_{\mathcal{I}}(\emptyset,\emptyset) = 1$.

(7) *If $\mathcal{I}$ is hybrid monotonic, then $D_{\mathcal{I}}(A,U) = D_{\mathcal{I}}(\emptyset,A) = 1$ for all $A \in \mathcal{F}(U)$.*

(8) *For any $A, A_j \in \mathcal{F}(U)$, $j \in J$, where J is an index set, then*

$$D_{\mathcal{I}}\left(A, \bigcap_{j \in J} A_j\right) = \bigwedge_{j \in J} D_{\mathcal{I}}(A, B_j). \tag{19}$$

3.2   Inclusion Degrees on Fuzzy Sets

Definition 7 [18]. *Let U be a universe of discourse. If for any $A,B \in \mathcal{F}(U)$, there is a real number $D(A,B)$ with the following conditions:*

(FD1) $0 \leq D(A,B) \leq 1$.

(FD2) $A \subseteq B$ implies $D(A,B) = 1$.

(FD3) *For any $A, B, C \in \mathcal{F}(U)$, $A \subseteq B \subseteq C$ implies*

$$D(C, A) \le D(B, A). \tag{20}$$

Then D is called an inclusion degree on $\mathcal{F}(U)$. Furthermore, if for any $A, B, C \in \mathcal{F}(U)$,
(FD4) $A \subseteq B$ *implies*

$$D(C, A) \le D(C, B), \tag{21}$$

then D is referred to as a strongly inclusion degree on $\mathcal{F}(U)$.

Definition 8 [18]. *Let U be a universe of discourse. If for any $A, B \in \mathcal{F}(U)$, there is a real number $D(A, B)$, which satisfies conditions (FD1), (FD3) and (FD2)':*
(FD2)' *For any $A, B \in \mathcal{F}(U) \cap \mathcal{P}(U)$, that is, A and B are crisp subsets of U, $A \subseteq B$ implies $D(A, B) = 1$.*
Then D is called a weakly inclusion degree on $\mathcal{F}(U)$.

By Proposition 2, we can conclude following Theorem 4.

Theorem 4. *Let $\mathcal{I}$ be an implicator, if the binary operation $D_{\mathcal{I}} : \mathcal{F}(U) \times \mathcal{F}(U) \to [0, 1]$ is defined as Eq. (17), then*

(1) $D_{\mathcal{I}}$ *is a weakly inclusion degree on $\mathcal{F}(U)$ whenever $\mathcal{I}$ is left monotonic.*
(2) $D_{\mathcal{I}}$ *is an inclusion degree on $\mathcal{F}(U)$ whenever $\mathcal{I}$ is a left monotonic and CP implicator.*
(3) $D_{\mathcal{I}}$ *is a strongly inclusion degree on $\mathcal{F}(U)$ whenever $\mathcal{I}$ is hybrid monotonic and CP implicator.*

Since every R-implicator is a hybrid monotonic, border, and CP implicator, by Theorem 4, we can conclude following Corollary 1.

Corollary 1. *If $\mathcal{I} = \theta_T$ is the R-implicator determined by a t-norm T defined as Eq. (13), denote*

$$D_{\theta_T}(A, B) = \bigwedge_{x \in U} \theta_T\big(A(x), B(x)\big), \ A, B \in \mathcal{F}(U). \tag{22}$$

Then D_{θ_T} is a strongly inclusion degree on $\mathcal{F}(U)$.

Since every S-implicator is a hybrid monotonic and border implicator, by Theorem 4, we can obtain following result.

Corollary 2. *If $\mathcal{I}$ is the S-implicator determined by a t-conorm S and a negator $\mathcal{N}$, denote*

$$D_{S,\mathcal{N}}(A, B) = \bigwedge_{x \in U} S\big(\mathcal{N}(A(x)), B(x)\big), \ A, B \in \mathcal{F}(U). \tag{23}$$

Then $D_{S,\mathcal{N}}$ is a weakly inclusion degree on $\mathcal{F}(U)$.

3.3 Approximations of Fuzzy Sets in Fuzzy Approximation Spaces

Throughout this section, $\mathcal{I}$ will be a border implicator and $\mathcal{N}$ an involutive negator on $[0, 1]$.

Definition 9. *Let U and W be two non-empty universes of discourse and R a fuzzy relation from U to W, then the triple (U, W, R) is called a generalized fuzzy approximation space. For any $A \in \mathcal{F}(W)$, the lower and upper $\mathcal{I}$-fuzzy rough approximations of A w.r.t. (U, W, R), denoted as $\underline{R}_{\mathcal{I}}(A)$ and $\overline{R}_{\mathcal{I}}(A)$ respectively, are fuzzy sets of U whose membership functions are defined respectively by*

$$
\begin{aligned}
\underline{R}_{\mathcal{I}}(A)(x) &= \bigwedge_{y \in W} \mathcal{I}\big(R(x, y), A(y)\big), \quad x \in U. \\
\overline{R}_{\mathcal{I}}(A)(x) &= \bigvee_{y \in W} \theta_{\mathcal{I}, \mathcal{N}}\big(\mathcal{N}(R(x, y)), A(y)\big), \quad x \in U.
\end{aligned}
\tag{24}
$$

The operators $\underline{R}_{\mathcal{I}}$ and $\overline{R}_{\mathcal{I}}$ from $\mathcal{F}(W)$ to $\mathcal{F}(U)$ are referred to as lower and upper $\mathcal{I}$-fuzzy rough approximation operators of (U, W, R) respectively, and the pair $(\underline{R}_{\mathcal{I}}(A), \overline{R}_{\mathcal{I}}(A))$ is called the $\mathcal{I}$-fuzzy rough set of A w.r.t. (U, W, R).

It has been proved that $\underline{R}_{\mathcal{I}}$ and $\overline{R}_{\mathcal{I}}$ are dual with each other [12], i.e.

$$
\begin{aligned}
&\text{(DFL)} \ \underline{R}_{\mathcal{I}}(A) = \sim_{\mathcal{N}} \overline{R}_{\mathcal{I}}(\sim_{\mathcal{N}} A), \forall A \in \mathcal{F}(W). \\
&\text{(DFU)} \ \overline{R}_{\mathcal{I}}(A) = \sim_{\mathcal{N}} \underline{R}_{\mathcal{I}}(\sim_{\mathcal{N}} A), \forall A \in \mathcal{F}(W).
\end{aligned}
\tag{25}
$$

Remark 4.(1) When $\mathcal{N} = \mathcal{N}_s$, $\mathcal{I}$ is an R-implicator determined by a t-norm T, and S the t-conorm dual to T, then it can be verified that the lower and upper approximation operators in Definition 9 degenerate to the dual fuzzy rough approximation operators defined by Mi and Zhang in [6], i.e.

$$
\begin{aligned}
\underline{R}_{\mathcal{I}}(A)(x) &= \underline{R}_{\theta}(A)(x) = \bigwedge_{y \in W} \theta_T\big(R(x, y), A(y)\big), \ A \in \mathcal{F}(W), x \in U, \\
\overline{R}_{\mathcal{I}}(A)(x) &= \overline{R}_{\sigma}(A)(x) = \bigvee_{y \in W} \sigma_S\big(1 - R(x, y), A(y)\big), \ A \in \mathcal{F}(W), x \in U,
\end{aligned}
\tag{26}
$$

where

$$
\begin{aligned}
\theta_T(a, b) &= \sup\{c \in [0, 1] : T(a, c) \leq b\}, \ a, b \in [0, 1]. \\
\sigma_S(a, b) &= \inf\{c \in [0, 1] : S(a, c) \geq b\}, \ a, b \in [0, 1].
\end{aligned}
\tag{27}
$$

(2) When $\mathcal{N} = \mathcal{N}_s$, T is a t-norm, S the t-conorm dual to T, and $\mathcal{I}$ the S-implicator determined by the t-conorm S, i.e. $\mathcal{I}(a, b) = S(1 - a, b)$, then it can be verified that the lower and upper approximation operators in Definition 9 degenerate to the dual fuzzy rough approximation operators defined by Mi et al. [5] and Wu [11], i.e.

$$
\begin{aligned}
\underline{R}_{\mathcal{I}}(A)(x) &= \bigwedge_{y \in W} S\big(1 - R(x, y), A(y)\big), A \in \mathcal{F}(W), \ x \in U, \\
\overline{R}_{\mathcal{I}}(A)(x) &= \bigvee_{y \in W} T\big(R(x, y), A(y)\big), \ A \in \mathcal{F}(W), x \in U,
\end{aligned}
\tag{28}
$$

More specifically, when U and W are finite sets, if $T = \min$ and $S = \max$, then the lower and upper approximation operators in Definition 9 are no other than the dual fuzzy rough approximation operators defined by Wu and Zhang [14], i.e.

$$\underline{R_{\mathcal{I}}}(A)(x) = \bigwedge_{y \in W} \big((1 - R(x,y)) \vee A(y) \big), \ A \in \mathcal{F}(W), x \in U,$$
$$\overline{R_{\mathcal{I}}}(A)(x) = \bigvee_{y \in W} R(x,y) \wedge A(y), \ A \in \mathcal{F}(W), x \in U. \tag{29}$$

Theorem 5. *Let (U, W, R) be a fuzzy approximation space, for $x \in U$, we define a fuzzy set $R(x)$ on W as follows:*

$$R(x)(y) = R(x,y), \ y \in W. \tag{30}$$

For $A \in \mathcal{F}(W)$, if $\underline{R_{\mathcal{I}}}(A)$ is the lower $\mathcal{I}$-fuzzy rough approximation of A w.r.t. (U, W, R), then

$$\underline{R_{\mathcal{I}}}(A)(x) = D_{\mathcal{I}}\big(R(x), A \big), \ x \in U. \tag{31}$$

Remark 5.(1) If $\mathcal{I}$ is left monotonic, then, by Theorems 4 and 5, it can be seen that the membership $\underline{R_{\mathcal{I}}}(A)(x)$ of an object $x \in U$ in the lower $\mathcal{I}$-fuzzy rough approximation $\underline{R_{\mathcal{I}}}(A)$ can be interpreted as the weakly inclusion degree $D_{\mathcal{I}}(R(x), A)$ of which the fuzzy set $R(x)$ is included in A. If $\mathcal{I}$ is a left monotonic and CP implicator, then the membership $\underline{R_{\mathcal{I}}}(A)(x)$ of the object $x \in U$ in the lower fuzzy approximation $\underline{R_{\mathcal{I}}}(A)$ can be described as the inclusion degree of which the fuzzy set $R(x)$ is included in A. If $\mathcal{I}$ is hybrid monotonic and CP implicator, then the membership $\underline{R_{\mathcal{I}}}(A)(x)$ of the object $x \in U$ in the lower fuzzy approximation $\underline{R_{\mathcal{I}}}(A)$ can be represented as the strongly inclusion degree of which the fuzzy set $R(x)$ is included in A.
(2) If θ_T is the R-implicator based on a left-continuous t-norm T, then, by Corollary 1 and Theorem 5, it can be observed that the membership $\underline{R_{\theta}}(A)(x)$ of an object $x \in U$ in the lower fuzzy approximation $\underline{R_{\theta}}(A)$ can be interpreted as the strongly inclusion degree $D_{\theta_T}(R(x), A)$ of which the fuzzy set $R(x)$ is included in A.
(3) If $\mathcal{I}$ is the S-implicator based on a t-conorm S and a negator $\mathcal{N}$, then, by Corollary 2 and Theorem 5, the membership $\underline{R_S}(A)(x)$ of an object $x \in U$ in the lower fuzzy approximation $\underline{R_S}(A)$ can be interpreted as the weakly inclusion degree $D_{S,N}(R(x), A)$ of which the fuzzy set $R(x)$ is included in A.

4   Conclusion

Rough set data analysis is one of the main application techniques arising from rough set theory. Based on the lower approximation of decision classes in a decision table, certain information can be discovered and certain rules can be derived. In this paper, the concept of inclusion degrees has been used to interpret the lower approximations of the rough set theory in both the crisp and the fuzzy environments. These results will be very helpful for people to understand more semantic meaning of rough set data analysis.

Acknowledgments. This work was supported by grants from the National Natural Science Foundation of China (Nos. 61573321, 61272021, 61602415, and 41631179) and the Open Foundation from Marine Sciences in the Most Important Subjects of Zhejiang (No. 20160102).

References

1. Abdel-Hamid, A.A., Morsi, N.N.: On the relationship of extended necessity measures to implication operators on the unit interval. Inf. Sci. **82**, 129–145 (1995)
2. Cornelis, C., Deschrijver, G., Kerre, E.E.: Implication in intuitionistic fuzzy and interval-valued fuzzy set theory: construction, classification, application. Int. J. Approx. Reason. **35**, 55–95 (2004)
3. Klement, E.P., Mesiar, R., Pap, E.: Triangular Norms. Trends in Logic, vol. 8. Kluwer Academic Publishers, Dordrecht (2000)
4. Klir, G.J., Yuan, B.: Fuzzy Logic: Theory and Applications. Prentice-Hall, Englewood Cliffs (1995)
5. Mi, J.-S., Leung, Y., Zhao, H.-Y., et al.: Generalized fuzzy rough sets determined by a triangular norm. Inf. Sci. **178**, 3203–3213 (2008)
6. Mi, J.-S., Zhang, W.-X.: An axiomatic characterization of a fuzzy generalization of rough sets. Inf. Sci. **160**, 235–249 (2004)
7. Pawlak, Z.: Rough sets. Int. J. Comput. Inf. Sci. **11**, 341–356 (1982)
8. Polkowski, L., Skowron, A.: Rough mereology. In: Raś, Z.W., Zemankova, M. (eds.) ISMIS 1994. LNCS, vol. 869, pp. 85–94. Springer, Heidelberg (1994). doi:10.1007/3-540-58495-1_9
9. Polkowski, L., Skowron, A.: Rough mereology: a new paradigm for approximate reasoning. Int. J. Approx. Reason. **15**, 333–365 (1996)
10. Radzikowska, A.M., Kerre, E.E.: A comparative study of fuzzy rough sets. Fuzzy Sets Syst. **126**, 137–155 (2002)
11. Wu, W.-Z.: On some mathematical structures of T-fuzzy rough set algebras in infinite universes of discourse. Fundamenta Informaticae **108**, 337–369 (2011)
12. Wu, W.-Z., Leung, Y., Shao, M.-W.: Generalized fuzzy rough approximation operators determined by fuzzy implicators. Int. J. Approx. Reason. **54**, 1388–1409 (2013)
13. Wu, W.-Z., Mi, J.-S.: Some mathematical structures of generalized rough sets in infinite universes of discourse. In: Peters, J.F., Skowron, A., Chan, C.-C., Grzymala-Busse, J.W., Ziarko, W.P. (eds.) Transactions on Rough Sets XIII. LNCS, vol. 6499, pp. 175–206. Springer, Heidelberg (2011)
14. Wu, W.-Z., Zhang, W.-X.: Constructive and axiomatic approaches of fuzzy approximation operators. Inf. Sci. **159**, 233–254 (2004)
15. Xu, Z.B., Liang, J.Y., Dang, C.Y., et al.: Inclusion degree: a perspective on measures for rough set data analysis. Inf. Sci. **141**, 229–238 (2002)
16. Yao, Y.Y.: Generalized rough set model. In: Polkowski, L., Skowron, A. (eds.) Rough Sets in Knowledge Discovery 1. Methodology and Applications, pp. 286–318. Physica-Verlag, Heidelberg (1998)
17. Zhang, W.X., Leung, Y.: Theory of including degrees and its applications to uncertainty inferences. In: Soft Computing in Intelligent Systems and Information Processing, pp. 496–501. IEEE, New York (1996)
18. Zhang, W.X., Liang, Y., Xu, P.: Uncertainty Reasoning Based on Inclusion Degree. Tsinghua Univerity Press, Bejing (2007)

Multigranulation Rough Sets in Hesitant Fuzzy Linguistic Information Systems

Chao Zhang, De-Yu Li[(✉)], and Yan-Hui Zhai

Key Laboratory of Computational Intelligence and Chinese Information Processing
of Ministry of Education, School of Computer and Information Technology,
Shanxi University, Taiyuan 030006, Shanxi, China
lidysxu@163.com

Abstract. Based on lower and upper approximations induced by multiple binary relations, multigranulation rough set theory has become one of the most promising research topics in the domain of rough set theory. Through combining multigranulation rough sets with hesitant fuzzy linguistic term sets, this article introduces a hybrid model of multigranulation rough sets, named a hesitant fuzzy linguistic (HFL) multigranulation rough set. In the framework of granular computing, we first give basic definitions of optimistic and pessimistic hesitant fuzzy linguistic multigranulation rough sets. Then, we explore some important properties about hesitant fuzzy linguistic multigranulation rough sets. Lastly, uncertainty measures for the hesitant fuzzy linguistic multigranulation approximation space are addressed.

Keywords: Granular computing · Hesitant fuzzy linguistic term sets · Multigranulation rough sets · Uncertainty measures

1 Introduction

Rough set theory [1], due to Zdzislaw Pawlak, is known as a widely used mathematical tool to cope with various uncertainties in numerous real-life applications related to data mining, knowledge discovery, information processing, machine learning and so on. It can be represented by pairs of sets that give lower and upper approximations of original sets. In Pawlak's rough set, the approximations are defined in terms of an equivalence relation. Since the equivalence relation in Pawlak's rough set is too restrictive to be utilized in many kinds of applications, various extended forms of classical rough sets have been developed during the past few years.

Hesitant fuzzy linguistic term sets, initially developed in [2], act as a significant extended form of hesitant fuzzy sets [3] in the qualitative environment. The motivation of introducing the hesitant fuzzy set is that in various practical decision-making procedures, the task of providing the membership degree of an object belonging to a certain set is relatively difficult for many experts. And hesitant fuzzy sets (HFS) are more likely to be utilized in the quantitative environment. However, when we are confronted with situations which are

© Springer International Publishing AG 2016
V. Flores et al. (Eds.): IJCRS 2016, LNAI 9920, pp. 307–317, 2016.
DOI: 10.1007/978-3-319-47160-0_28

ill-defined to be addressed through utilizing quantitative expressions, it may be suitable to evaluate membership degrees of the alternatives by using qualitative expressions. The method of fuzzy linguistic approach is generally regarded as a reasonable model to deal with the difficulties. Therefore, by taking advantages of the hesitant fuzzy set and fuzzy linguistic approach, the hesitant fuzzy linguistic term set was presented to deal with complicated problems in the qualitative environment. Ever since the establishment of hesitant fuzzy linguistic term set theory, many scholars have studied the new model from different points of view and obtained an increasing number of achievements [4–6]. Since the theories of hesitant fuzzy sets and rough sets both aim to settle uncertainties in information systems, studies of the fusion of these two theories is being regarded as a meaningful direction to the rough set framework [7,8].

To introduce the notion of granular computing [9] in practical situations, it is useful to view a problem through multiple binary relations according to the objectives of the problem solving. Under this circumstance, we can enlarge the application domains of classical rough sets by computing approximation operators through multiple granules induced by multiple binary relations. Thus, multigranulation rough set theory (MGRS) was first introduced by Qian's research group [10]. And two types of multigranulation rough sets were developed (optimistic MGRS and pessimistic MGRS) [11]. Later, several viewpoints of multigranulation rough sets have been put forward during these years [12–15]. Although the theories of hesitant fuzzy rough sets and multigranulation rough sets act as significant extended forms of classical rough sets, and considering that there are few researches on the fusion of multigranulation rough sets and hesitant fuzzy linguistic term sets. Therefore, we intend to develop a hesitant fuzzy linguistic multigranulation rough set model from aspects of basic definitions, some useful properties and uncertainty measures, respectively.

The presentation of the article is organized below: in the next section, we review some preliminaries with respect to the hesitant fuzzy linguistic term set and multigranulation rough set. In Sect. 3 we introduce the concept of hesitant fuzzy linguistic multigranulation rough sets and some related properties are explored. Section 4 presents uncertainty measures on hesitant fuzzy linguistic multigranulation rough sets. Finally, we summarize this study and point out future study directions.

2   Preliminaries

In this section, we start by introducing the fuzzy linguistic approach. Suppose that $S = \{s_0, s_1, \ldots, s_g\}$ is a linguistic term set that owns some special features such as finite and totally ordered. Generally, for any linguistic term set, the linguistic terms must satisfy the following additional characteristics:

(1) The set is ordered: If $i \leq j$, then $s_i \leq s_j$;
(2) We have $\min(s_i, s_j) = s_i$ holds when $s_i \leq s_j$; otherwise, $\max(s_i, s_j) = s_i$ holds when $s_i \geq s_j$;

(3) There is a negation operator: $neg(s_i) = s_{g-i}$, where $g + 1$ denotes the total number of linguistic terms in S.

Since Rodriguez et al. [2] does not give specific mathematical form of the hesitant fuzzy linguistic term set (HFLTS), Liao et al. [6] presented a new definition for HFLTS mathematically that is much easier to be understood.

Definition 1. [6] Suppose that U is a universe of discourse and we have a linguistic term set denoted as $S = \{s_0, s_1, \ldots, s_g\}$. Then, a hesitant fuzzy linguistic term set $\mathcal{A}$ on U is defined in terms of a function $h_{\mathcal{A}}(x)$ that when applied to U returns a subset of S, we express the HFLTS which is shown as follows:

$$\mathcal{A} = \{\langle x, h_{\mathcal{A}}(x)\rangle \,|\, x \in U\},$$

where $h_{\mathcal{A}}(x)$ is a set of some different ordered finite values in S, representing the possible membership degrees of the element $x \in U$ to the set $\mathcal{A}$. Usually, we denote $h_{\mathcal{A}}(x)$ a hesitant fuzzy linguistic element (HFLE).

Based on the above description, we denote the set which includes all HFLTSs on U as $HFL(U)$ in this paper. Moreover, we have $\forall \mathcal{A} \in HFL(U)$.

(1) For all $x \in U$, we call $\mathcal{A}$ an empty HFLTS if and only if $h_{\mathcal{A}}(x) = \{s_0\}$. Under that circumstance, the empty HFLTS is represented by $\emptyset$ in this article.
(2) For all $x \in U$, we call $\mathcal{A}$ a full HFLTS if and only if $h_{\mathcal{A}}(x) = \{s_g\}$. Under that circumstance, the full HFLTS is represented by U in this article.

Similar to the hesitant fuzzy set, Wei et al. [5] established some novel operations on HFLTSs.

Definition 2. [5] Suppose that we have a linguistic term set denoted as $S = \{s_0, s_1, \ldots, s_g\}$, and U denotes the universe of discourse, $\forall \mathcal{A}, \mathcal{B} \in HFL(U)$,

(1) We represent the negation of $\mathcal{A}$ as $\mathcal{A}^c$:

$$h_{\mathcal{A}^c}(x) = \sim h_{\mathcal{A}}(x) = \{s_{g-i} \,|\, i \in Ind(h_{\mathcal{A}}(x))\};$$

where $Ind(s_i)$ denotes the index i of a linguistic term s_i in S. Similarly, $Ind(h_{\mathcal{A}}(x))$ denotes the set of indexes of the linguistic terms in an HFLE $h_{\mathcal{A}}(x)$.
(2) We represent the max-union of $\mathcal{A}$ and $\mathcal{B}$ as $\mathcal{A} \cup \mathcal{B}$:

$$h_{\mathcal{A} \cup \mathcal{B}}(x) = h_{\mathcal{A}}(x) \vee h_{\mathcal{B}}(x) = \{\max\{s_i, s_j\} \,|\, s_i \in h_{\mathcal{A}}(x), s_j \in h_{\mathcal{B}}(x)\};$$

(3) We represent the min-intersection of $\mathcal{A}$ and $\mathcal{B}$ as $\mathcal{A} \cap \mathcal{B}$:

$$h_{\mathcal{A} \cap \mathcal{B}}(x) = h_{\mathcal{A}}(x) \wedge h_{\mathcal{B}}(x) = \{\min\{s_i, s_j\} \,|\, s_i \in h_{\mathcal{A}}(x), s_j \in h_{\mathcal{B}}(x)\}.$$

In above definition, the operations $^c, \cup, \cap$ are defined on HFLTSs, while operations $\sim, \vee, \wedge$ are defined on the corresponding HFLEs. In the following part, we present the notion of HFL subsets for comparing two HFLTSs. At first, we denote the kth largest value in $h_{\mathcal{A}}(x)$ is expressed as $h_{\mathcal{A}}^{\sigma(k)}(x)$, the kth largest value in $h_{\mathcal{B}}(x)$ is expressed as $h_{\mathcal{B}}^{\sigma(k)}(x)$.

Definition 3. Suppose that U is a universe of discourse, $\forall \mathcal{A}, \mathcal{B} \in HFL(U)$, if $h_{\mathcal{A}}(x) \preceq h_{\mathcal{B}}(x)$ holds for each $x \in U$ such that $h_{\mathcal{A}}(x) \preceq h_{\mathcal{B}}(x) \Leftrightarrow h_{\mathcal{A}}^{\sigma(k)}(x) \leq h_{\mathcal{B}}^{\sigma(k)}(x)$, then we call $\mathcal{A}$ an HFL subset of $\mathcal{B}$. Finally, it is expressed as $\mathcal{A} \subseteq \mathcal{B}$.

Multigranulation rough sets were initially founded by Qian's research group [10], and it has become a new and hot research issue in the rough set domain. Next, we introduce basic definitions of optimistic and pessimistic MGRSs.

Definition 4. [10] Suppose that U is a universe of discourse and $R_1, R_2, ..., R_m$ are m crisp binary relations. For any $X \subseteq U$, the optimistic lower approximation and upper approximation of X are expressed by $\sum\limits_{i=1}^{m} R_i{}^O(X)$ and $\overline{\sum\limits_{i=1}^{m} R_i{}^O}(X)$, where:

$$\sum\limits_{i=1}^{m} R_i{}^O(X) = \{x \in U : [x]_{R_1} \subseteq X \vee ... \vee [x]_{R_m} \subseteq X\}, \overline{\sum\limits_{i=1}^{m} R_i{}^O}(X) = \sim \sum\limits_{i=1}^{m} R_i{}^O(\sim X).$$

We denote $[x]_{R_i}\,(1 \leq i \leq m)$ the equivalence class of X with respect to the equivalence relation $R_i\,(1 \leq i \leq m)$, and $\sim X$ denotes the complement of set X. Based on that, $\langle \sum\limits_{i=1}^{m} R_i{}^O(X), \overline{\sum\limits_{i=1}^{m} R_i{}^O}(X) \rangle$ is the classical optimistic multi-granulation rough set. Similarly, the pessimistic lower approximation and upper approximation of X are expressed by $\sum\limits_{i=1}^{m} R_i{}^P(X)$ and $\overline{\sum\limits_{i=1}^{m} R_i{}^P}(X)$, where:

$$\sum\limits_{i=1}^{m} R_i{}^P(X) = \{x \in U : [x]_{R_1} \subseteq X \wedge ... \wedge [x]_{R_m} \subseteq X\}, \overline{\sum\limits_{i=1}^{m} R_i{}^P}(X) = \sim \sum\limits_{i=1}^{m} R_i{}^P(\sim X).$$

Then, we call $\langle \sum\limits_{i=1}^{m} R_i{}^P(X), \overline{\sum\limits_{i=1}^{m} R_i{}^P}(X) \rangle$ the classical pessimistic multigranulation rough set.

3   Hesitant Fuzzy Linguistic Multigranulation Rough Sets

In this section, based on the constructive approach, we extend the hesitant fuzzy linguistic relation into the multigranulation rough set background. Both the definitions and some basic properties of optimistic and pessimistic hesitant fuzzy linguistic (HFL) multigranulation rough sets will be presented. At first, we present the definition of hesitant fuzzy linguistic relations.

Definition 5. Suppose that U is a universe of discourse and we have a linguistic term set denoted as $S = \{s_0, s_1, ..., s_g\}$. Then, a hesitant fuzzy linguistic (HFL) relation over U is an HFL subset of $U \times U$. Then, we represent the HFL relation $\mathcal{R}$ which is shown below:

$$\mathcal{R} = \{\langle (x, y), h_{\mathcal{R}}(x, y) \rangle | (x, y) \in U \times U\},$$

where $h_\mathcal{R}(x, y)$ is a set of some different ordered finite values in S, representing the possible membership degrees of the relationship between x and y to the relation $\mathcal{R}$. Moreover, the family of all HFL relations over $U \times U$ is denoted as $HFLR(U \times U)$.

Definition 6. Suppose that U is a universe of discourse, and S is a linguistic term set. $\mathcal{R}_i (i = 1, \ldots, m)$ is an HFL relation on U, and $(U, \mathcal{R}_i)$ is an HFL approximation space. For any $\mathcal{A} \in HFL(U)$, the optimistic and pessimistic HFL multigranulation rough lower approximation and upper approximation of $\mathcal{A}$ are expressed in the following:

$$\underline{\sum_{i=1}^{m} \mathcal{R}_i}^{O}(A)(x) = \overset{m}{\underset{i=1}{\vee}} \wedge_{y \in U} \{h_{\mathcal{R}_i^c}(x, y) \vee h_\mathcal{A}(y)\},$$

$$\overline{\sum_{i=1}^{m} \mathcal{R}_i}^{O}(\mathcal{A})(x) = \overset{m}{\underset{i=1}{\wedge}} \vee_{y \in U} \{h_{\mathcal{R}_i}(x, y) \wedge h_\mathcal{A}(y)\},$$

$$\underline{\sum_{i=1}^{m} \mathcal{R}_i}^{P}(\mathcal{A})(x) = \overset{m}{\underset{i=1}{\wedge}} \wedge_{y \in U} \{h_{\mathcal{R}_i^c}(x, y) \vee h_\mathcal{A}(y)\},$$

$$\overline{\sum_{i=1}^{m} \mathcal{R}_i}^{P}(\mathcal{A})(x) = \overset{m}{\underset{i=1}{\vee}} \vee_{y \in U} \{h_{\mathcal{R}_i}(x, y) \wedge h_\mathcal{A}(y)\}.$$

We call the pair $[\underline{\sum_{i=1}^{m} \mathcal{R}_i}^{O}(\mathcal{A}), \overline{\sum_{i=1}^{m} \mathcal{R}_i}^{O}(\mathcal{A})]$ an optimistic HFL multigranulation rough set of $\mathcal{A}$, and the pair $[\underline{\sum_{i=1}^{m} \mathcal{R}_i}^{P}(\mathcal{A}), \overline{\sum_{i=1}^{m} \mathcal{R}_i}^{P}(\mathcal{A})]$ a pessimistic HFL multigranulation rough set of $\mathcal{A}$.

Next, we investigate some significant properties of optimistic HFL multigranulation rough sets, the pessimistic version of HFL multigranulation rough sets is obtained in an identical fashion.

Theorem 1. Suppose that U is a universe of discourse, S is a linguistic term set. $\mathcal{R}_i (i = 1, \ldots, m)$ is an HFL relation on U, and $(U, \mathcal{R}_i)$ is an HFL approximation space. For any $\mathcal{A} \in HFL(U)$, the following properties are true:

(1) $\underline{\sum_{i=1}^{m} \mathcal{R}_i}^{O}(\mathcal{A}) \subseteq \mathcal{A} \subseteq \overline{\sum_{i=1}^{m} \mathcal{R}_i}^{O}(\mathcal{A})$;

(2) $\underline{\sum_{i=1}^{m} \mathcal{R}_i}^{O}(\emptyset) = \overline{\sum_{i=1}^{m} \mathcal{R}_i}^{O}(\emptyset) = \emptyset$, $\underline{\sum_{i=1}^{m} \mathcal{R}_i}^{O}(U) = \overline{\sum_{i=1}^{m} \mathcal{R}_i}^{O}(U) = U$;

(3) $\underline{\sum_{i=1}^{m} \mathcal{R}_i}^{O}(\mathcal{A}) = \overset{m}{\underset{i=1}{\bigcup}} \underline{\mathcal{R}_i}(\mathcal{A})$, $\overline{\sum_{i=1}^{m} \mathcal{R}_i}^{O}(\mathcal{A}) = \overset{m}{\underset{i=1}{\bigcap}} \overline{\mathcal{R}_i}(\mathcal{A})$;

(4) $\underline{\sum_{i=1}^{m} \mathcal{R}_i}^{O}(\underline{\sum_{i=1}^{m} \mathcal{R}_i}^{O}(\mathcal{A})) = \underline{\sum_{i=1}^{m} \mathcal{R}_i}^{O}(\mathcal{A})$, $\overline{\sum_{i=1}^{m} \mathcal{R}_i}^{O}(\overline{\sum_{i=1}^{m} \mathcal{R}_i}^{O}(\mathcal{A})) = \overline{\sum_{i=1}^{m} \mathcal{R}_i}^{O}(\mathcal{A})$;

(5) $\sum_{i=1}^{m} \mathcal{R}_i{}^O(\sim\mathcal{A}) = \sim\overline{\sum_{i=1}^{m}\mathcal{R}_i{}^O}(\mathcal{A}),\ \overline{\sum_{i=1}^{m}\mathcal{R}_i{}^O}(\sim\mathcal{A}) = \sim\sum_{i=1}^{m}\mathcal{R}_i{}^O(\mathcal{A});$

(6) $\mathcal{A} \subseteq \mathcal{A}' \Rightarrow \sum_{i=1}^{m}\mathcal{R}_i{}^O(\mathcal{A}) \subseteq \sum_{i=1}^{m}\mathcal{R}_i{}^O(\mathcal{A}'),\ \mathcal{A} \subseteq \mathcal{A}' \Rightarrow \overline{\sum_{i=1}^{m}\mathcal{R}_i{}^O}(\mathcal{A}) \subseteq$

$\overline{\sum_{i=1}^{m}\mathcal{R}_i{}^O}(\mathcal{A}').$

Proof. They can be directly derived from Definitions 3 and 6.

In above theorem, (1) indicates the optimistic hesitant fuzzy linguistic multi-granulation lower and upper approximations satisfy the contraction and extension respectively; (2) represents the optimistic hesitant fuzzy linguistic multi-granulation rough set satisfies the normality and conormality; (3) shows the relationships between hesitant fuzzy linguistic multigranulation and hesitant fuzzy linguistic single granulation rough sets; (4) and (5) show the idempotency and complement of hesitant fuzzy linguistic multigranulation rough set respectively; while (6) shows the monotone of optimistic hesitant fuzzy linguistic multigranulation rough approximation in terms of the variety of hesitant fuzzy linguistic target.

Theorem 2. Suppose that U is a universe of discourse, S is a linguistic term set. $\mathcal{R}_i\,(i = 1,\ldots,m)$ is an HFL relation on U. And we let $\mathcal{R}_i, \mathcal{R}_i' \in HFLR\,(U \times U)$ be two HFL relations over $U \times U$. If $\mathcal{R}_i \subseteq \mathcal{R}_i'$, for any $\mathcal{A} \in HFL\,(U)$, the following properties are true:

(1) $\underline{\sum_{i=1}^{m}\mathcal{R}_i'{}^O}(\mathcal{A}) \subseteq \underline{\sum_{i=1}^{m}\mathcal{R}_i{}^O}(\mathcal{A})$, for all $\mathcal{A} \in HFL\,(U)$;

(2) $\overline{\sum_{i=1}^{m}\mathcal{R}_i'{}^O}(\mathcal{A}) \supseteq \overline{\sum_{i=1}^{m}\mathcal{R}_i{}^O}(\mathcal{A})$, for all $\mathcal{A} \in HFL\,(U)$.

Proof. They can be directly derived from Definitions 3 and 6.

In above theorem, the lower and upper approximations in optimistic hesitant fuzzy linguistic multigranulation rough sets are monotonic in terms of the monotonic forms of the multiple binary hesitant fuzzy linguistic relations.

Theorem 3. Suppose that U is a universe of discourse, S is a linguistic term set. $\mathcal{R}_i\,(i = 1,\ldots,m)$ is an HFL relation on U, and $(U, \mathcal{R}_i)$ is an HFL approximation space. For any $\mathcal{A}_j\,(j = 1,\ldots,n) \in HFL\,(U)$, the following properties are true:

(1) $\underline{\sum_{i=1}^{m}\mathcal{R}_i{}^O}(\bigcap_{j=1}^{n}\mathcal{A}_j) = \bigcup_{i=1}^{m}(\bigcap_{j=1}^{n}\underline{\mathcal{R}_i}(\mathcal{A}_j)),\ \overline{\sum_{i=1}^{m}\mathcal{R}_i{}^O}(\bigcup_{j=1}^{n}\mathcal{A}_j) = \bigcap_{i=1}^{m}(\bigcup_{j=1}^{n}\overline{\mathcal{R}_i}(\mathcal{A}_j));$

(2) $\underline{\sum_{i=1}^{m}\mathcal{R}_i{}^O}(\bigcap_{j=1}^{n}\mathcal{A}_j) = \bigcap_{j=1}^{n}(\underline{\sum_{i=1}^{m}\mathcal{R}_i{}^O}(\mathcal{A}_j)),\qquad \overline{\sum_{i=1}^{m}\mathcal{R}_i{}^O}(\bigcup_{j=1}^{n}\mathcal{A}_j) =$

$\bigcup_{j=1}^{n}(\overline{\sum_{i=1}^{m}\mathcal{R}_i{}^O}(\mathcal{A}_j));$

(3) $\sum\limits_{i=1}^{m} \mathcal{R}_i{}^O(\bigcup\limits_{j=1}^{n} \mathcal{A}_j) \supseteq \bigcup\limits_{j=1}^{n}(\overline{\sum\limits_{i=1}^{m} \mathcal{R}_i{}^O}(\mathcal{A}_j)), \quad \overline{\sum\limits_{i=1}^{m} \mathcal{R}_i{}^O}(\bigcap\limits_{j=1}^{n} \mathcal{A}_j) \subseteq$

$\bigcap\limits_{i=1}^{n}(\overline{\sum\limits_{i=1}^{m} \mathcal{R}_i{}^O}(\mathcal{A}_j)).$

Proof. They can be directly derived from Definition 6 and Theorem 1.

The above theorem indicates the relationship between the optimistic hesitant fuzzy linguistic rough approximations of a single set with the optimistic hesitant fuzzy linguistic rough approximations of multi-sets under the hesitant fuzzy linguistic multigranulation environment.

Theorem 4. Suppose that U is a universe of discourse, S is a linguistic term set. $\mathcal{R}_i$ $(i = 1, \ldots, m)$ is an HFL relation on U, and $(U, \mathcal{R}_i)$ is an HFL approximation space. For any $\mathcal{A} \in HFL(U)$, we have the following properties:

(1) $\sum\limits_{i=1}^{m} \mathcal{R}_i{}^P(\mathcal{A}) \subseteq \sum\limits_{i=1}^{m} \mathcal{R}_i{}^O(\mathcal{A}), \overline{\sum\limits_{i=1}^{m} \mathcal{R}_i{}^P}(\mathcal{A}) \supseteq \overline{\sum\limits_{i=1}^{m} \mathcal{R}_i{}^O}(\mathcal{A});$

(2) $\underline{\mathcal{R}_i}(\mathcal{A}) \subseteq \sum\limits_{i=1}^{m} \mathcal{R}_i{}^O(\mathcal{A}), \underline{\mathcal{R}_i}(\mathcal{A}) \supseteq \sum\limits_{i=1}^{m} \mathcal{R}_i{}^P(\mathcal{A});$

(3) $\overline{\mathcal{R}_i}(\mathcal{A}) \supseteq \overline{\sum\limits_{i=1}^{m} \mathcal{R}_i{}^O}(\mathcal{A}), \overline{\mathcal{R}_i}(\mathcal{A}) \subseteq \overline{\sum\limits_{i=1}^{m} \mathcal{R}_i{}^P}(\mathcal{A}).$

Proof. They can be directly derived from Definition 6 and Theorem 1.

From above theorem, it is noted that the hesitant fuzzy linguistic lower approximation includes the pessimistic hesitant fuzzy linguistic multigranulation lower approximation, while the optimistic hesitant fuzzy linguistic multigranulation upper approximation includes the hesitant fuzzy linguistic lower approximation. Parallel, the pessimistic hesitant fuzzy linguistic multigranulation upper approximation includes the hesitant fuzzy linguistic upper approximation, the hesitant fuzzy linguistic upper approximation includes the optimistic hesitant fuzzy linguistic multigranulation upper approximation.

4 Uncertainty Measures of HFL Multigranulation Rough Sets

According to the proposed hesitant fuzzy linguistic multigranulation rough sets, we mainly discuss some roughness measures of hesitant fuzzy linguistic multigranulation rough sets in the corresponding hesitant fuzzy linguistic multigranulation approximation space. And we only discuss the optimistic version of hesitant fuzzy linguistic MGRSs. Similarly, the pessimistic version of hesitant fuzzy linguistic MGRSs could be handled according to the optimistic version. Prior to the introduction of roughness measures of hesitant fuzzy linguistic MGRSs, we present the notion of level sets to optimistic hesitant fuzzy linguistic multigranulation rough lower and upper approximations.

Definition 7. Suppose that U is a universe of discourse, S is a linguistic term set. $\mathcal{R}_i\,(i=1,\ldots,m)$ is an HFL relation on U, and $(U,\mathcal{R}_i)$ is an HFL approximation space. For any $\mathcal{A} \in HFL\,(U)$ and $s_0 < s_\beta \leq s_\alpha \leq s_g$, we denote $\underline{\sum\limits_{i=1}^{m} \mathcal{R}_i{}^O(\mathcal{A})}_{s_\alpha} = \{h_{\underline{\sum\limits_{i=1}^{m} \mathcal{R}_i{}^O(\mathcal{A})}}(x) \succeq \{s_\alpha\}\} = \{h^{\sigma(k)}_{\underline{\sum\limits_{i=1}^{m} \mathcal{R}_i{}^O(\mathcal{A})}}(x) \geq s_\alpha\}$ the $s_\alpha - level$ set of the optimistic HFL multigranulation rough lower approximations of $\mathcal{A}$. And we call $\overline{\sum\limits_{i=1}^{m} \mathcal{R}_i{}^O(\mathcal{A})}_{s_\beta} = \{h_{\overline{\sum\limits_{i=1}^{m} \mathcal{R}_i{}^O(\mathcal{A})}}(x) \succeq \{s_\beta\}\} = \{h^{\sigma(k)}_{\overline{\sum\limits_{i=1}^{m} \mathcal{R}_i{}^O(\mathcal{A})}}(x) \geq s_\beta\}$ the $s_\beta - level$ set of the optimistic HFL multigranulation rough upper approximations of $\mathcal{A}$.

Theorem 5. Suppose that U is a universe of discourse, S is a linguistic term set. $\mathcal{R}_i\,(i=1,\ldots,m)$ is an HFL relation on U, and $(U,\mathcal{R}_i)$ is an HFL approximation space. Additionally, we have $s_0 < s_\beta \leq s_\alpha \leq s_g$. For any $\mathcal{A}, \mathcal{A}' \in HFL\,(U)$, then:

(1) $\underline{\sum\limits_{i=1}^{m} \mathcal{R}_i{}^O(\mathcal{A})}_{s_\alpha} \subseteq \overline{\sum\limits_{i=1}^{m} \mathcal{R}_i{}^O(\mathcal{A})}_{s_\beta}$;

(2) $\underline{\sum\limits_{i=1}^{m} \mathcal{R}_i{}^O(\mathcal{A} \cap \mathcal{A}')}_{s_\alpha} = \underline{\sum\limits_{i=1}^{m} \mathcal{R}_i{}^O(\mathcal{A})}_{s_\alpha} \cap \underline{\sum\limits_{i=1}^{m} \mathcal{R}_i{}^O(\mathcal{A}')}_{s_\alpha}$,

$\overline{\sum\limits_{i=1}^{m} \mathcal{R}_i{}^O(\mathcal{A} \cup \mathcal{A}')}_{s_\beta} = \overline{\sum\limits_{i=1}^{m} \mathcal{R}_i{}^O(\mathcal{A})}_{s_\beta} \cup \overline{\sum\limits_{i=1}^{m} \mathcal{R}_i{}^O(\mathcal{A}')}_{s_\beta}$;

(3) $\underline{\sum\limits_{i=1}^{m} \mathcal{R}_i{}^O(\mathcal{A} \cup \mathcal{A}')}_{s_\alpha} \supseteq \underline{\sum\limits_{i=1}^{m} \mathcal{R}_i{}^O(\mathcal{A})}_{s_\alpha} \cup \underline{\sum\limits_{i=1}^{m} \mathcal{R}_i{}^O(\mathcal{A}')}_{s_\alpha}$,

$\overline{\sum\limits_{i=1}^{m} \mathcal{R}_i{}^O(\mathcal{A} \cap \mathcal{A}')}_{s_\beta} \subseteq \overline{\sum\limits_{i=1}^{m} \mathcal{R}_i{}^O(\mathcal{A})}_{s_\beta} \cap \overline{\sum\limits_{i=1}^{m} \mathcal{R}_i{}^O(\mathcal{A}')}_{s_\beta}$;

(4) $\mathcal{A} \subseteq \mathcal{A}' \Rightarrow \underline{\sum\limits_{i=1}^{m} \mathcal{R}_i{}^O(\mathcal{A})}_{s_\alpha} \subseteq \underline{\sum\limits_{i=1}^{m} \mathcal{R}_i{}^O(\mathcal{A}')}_{s_\alpha}$,

$\mathcal{A} \subseteq \mathcal{A}' \Rightarrow \overline{\sum\limits_{i=1}^{m} \mathcal{R}_i{}^O(\mathcal{A})}_{s_\beta} \subseteq \overline{\sum\limits_{i=1}^{m} \mathcal{R}_i{}^O(\mathcal{A}')}_{s_\beta}$.

Proof. It is not difficult to obtain the results according to Definition 7 and Theorem 1.

Definition 8. Suppose that U is a universe of discourse, S is a linguistic term set. $\mathcal{R}_i\,(i=1,\ldots,m)$ is an HFL relation on U, and $(U,\mathcal{R}_i)$ is an HFL approximation space. Additionally, we have $s_0 < s_\beta \leq s_\alpha \leq s_g$. For any $\mathcal{A} \in HFL\,(U)$, we define the roughness measure of $\mathcal{A}$ in the following:

$$\rho_{\mathcal{A}}^{s_\alpha, s_\beta} = 1 - \frac{|\underline{\sum\limits_{i=1}^{m} \mathcal{R}_i{}^O(\mathcal{A})}_{s_\alpha}|}{|\overline{\sum\limits_{i=1}^{m} \mathcal{R}_i{}^O(\mathcal{A})}_{s_\beta}|}.$$

In above definition, $|X|$ represents the cardinality of X, and we call $\eta_{\mathcal{A}}^{s_\alpha,s_\beta} = \dfrac{|\sum_{i=1}^{m} \mathcal{R}_i{}^O(\mathcal{A})_{s_\alpha}|}{|\sum_{i=1}^{m} \mathcal{R}_i{}^O(\mathcal{A})_{s_\beta}|}$ as the approximate precision of $\mathcal{A}$, and we have $\rho_{\mathcal{A}}^{s_\alpha,s_\beta}, \eta_{\mathcal{A}}^{s_\alpha,s_\beta} \in [0,1]$.

Theorem 6. Suppose that U is a universe of discourse, S is a linguistic term set. $\mathcal{R}_i\,(i = 1, \ldots, m)$ is an HFL relation on U, and $(U, \mathcal{R}_i)$ is an HFL approximation space. For any $\mathcal{A}, \mathcal{A}' \in HFL(U)$, $\mathcal{A} \subseteq \mathcal{A}'$ and $s_0 < s_\beta \leq s_\alpha \leq s_g$, then:

(1) If $\sum_{i=1}^{m} \overline{\mathcal{R}_i{}^O(\mathcal{A})}_{s_\alpha} = \sum_{i=1}^{m} \overline{\mathcal{R}_i{}^O(\mathcal{A}')}_{s_\alpha}$, then $\rho_{\mathcal{A}}^{s_\alpha,s_\beta} \leq \rho_{\mathcal{A}'}^{s_\alpha,s_\beta}$;

(2) If $\sum_{i=1}^{m} \mathcal{R}_i{}^O(\mathcal{A})_{s_\beta} = \sum_{i=1}^{m} \mathcal{R}_i{}^O(\mathcal{A}')_{s_\beta}$, then $\rho_{\mathcal{A}}^{s_\alpha,s_\beta} \geq \rho_{\mathcal{A}'}^{s_\alpha,s_\beta}$.

Proof. It is not difficult to obtain the results according to Definition 8 and proof (3) of Theorem 5.

Theorem 7. Suppose that U is a universe of discourse, S is a linguistic term set. $\mathcal{R}_i\,(i = 1, \ldots, m)$ is an HFL relation on U, and $(U, \mathcal{R}_i)$ is an HFL approximation space. If $\mathcal{R}_i \subseteq \mathcal{R}'_i$, for any $\mathcal{A} \in HFL(U)$ and $s_0 < s_\beta \leq s_\alpha \leq s_g$, then:

(1) $\sum_{i=1}^{m} \mathcal{R}_i{}^O(\mathcal{A})_{s_\alpha} \supseteq \sum_{i=1}^{m} \mathcal{R}'_i{}^O(\mathcal{A})_{s_\alpha}$, $\sum_{i=1}^{m} \overline{\mathcal{R}_i{}^O(\mathcal{A})}_{s_\beta} \subseteq \sum_{i=1}^{m} \overline{\mathcal{R}'_i{}^O(\mathcal{A})}_{s_\beta}$;

(2) $\overline{\rho_{\mathcal{R}_i}^{s_\alpha,s_\beta}}(\mathcal{A}) \leq \rho_{\mathcal{R}'_i}^{s_\alpha,s_\beta}(\mathcal{A})$.

Proof. It is not difficult to obtain the results according to Definitions 7, 8 and Theorem 2.

Theorem 8. Suppose that U is a universe of discourse, S is a linguistic term set. $\mathcal{R}_i\,(i = 1, \ldots, m)$ is an HFL relation on U, and $(U, \mathcal{R}_i)$ is an HFL approximation space. For any $\mathcal{A} \in HFL(U)$ and $s_0 < s_{\beta_1} \leq s_{\beta_2} \leq s_{\alpha_1} \leq s_{\alpha_2} \leq s_g$, then:

(1) $\sum_{i=1}^{m} \mathcal{R}_i{}^O(\mathcal{A})_{s_{\alpha_1}} \supseteq \sum_{i=1}^{m} \mathcal{R}_i{}^O(\mathcal{A})_{s_{\alpha_2}}$, $\sum_{i=1}^{m} \overline{\mathcal{R}_i{}^O(\mathcal{A})}_{s_{\beta_2}} \subseteq \sum_{i=1}^{m} \overline{\mathcal{R}_i{}^O(\mathcal{A})}_{s_{\beta_1}}$;

(2) $\overline{\rho_{\mathcal{A}}^{s_{\alpha_1},s_{\beta_2}}} \leq \rho_{\mathcal{A}}^{s_{\alpha_2},s_{\beta_1}}$.

Proof. It is not difficult to obtain the results according to Definitions 7 and 8.

5   Conclusions and Future Perspectives

The focal point of interest in the present article is to establish a comprehensive framework for the research of multigranulation rough sets in hesitant fuzzy linguistic information systems, and the newly proposed rough set model seems to be of great significance to analyse hesitant fuzzy linguistic data and complex problem solving in artificial intelligence and cognitive science fields. It is necessary to investigate hesitant fuzzy linguistic multigranulation rough sets model. In this paper, using the idea of granular computing, we have first given the definitions and some useful properties of optimistic and pessimistic hesitant fuzzy linguistic multigranulation rough sets. Then, we have further presented the relationship between these two types of hesitant fuzzy linguistic multigranulation rough sets. Finally, some uncertainty measures for hesitant fuzzy linguistic multigranulation rough sets are discussed. This study develops a general framework of hesitant fuzzy linguistic multigranulation rough sets that enriches the multigranulation rough sets theory.

In decision-making field, by adopting decision-making strategies: seeking common ground while reserving differences and seeking common ground while eliminating differences, multigranulation rough sets could be regarded as an ideal information fusion model when facing large-group and inconsistent decision-making situations by providing optimistic and pessimistic risk decision-making strategies. Thus, hesitant fuzzy linguistic multigranulation rough sets can open a door for the hesitant fuzzy linguistic data analysis and hesitant fuzzy linguistic information fusion, these impacts will play a significant place in solving various complex decision-making problems from hesitant and qualitative views. Thus, our future study will focus on practical applications related to the newly proposed multigranulation rough set model.

Acknowledgments. The work was supported from the National Natural Science Foundation of China (No. 61272095, 61303107, 61432011, 61573231, U1435212) and the Shanxi Science and Technology Infrastructure (No. 2015091001-0102).

References

1. Pawlak, Z.: Rough sets. Int. J. Comput. Inf. Sci. **11**(5), 341–356 (1982)
2. Rodriguez, R.M., Martinez, L., Herrera, F.: Hesitant fuzzy linguistic term sets for decision making. IEEE. Trans. Fuzzy Syst. **20**(1), 109–119 (2012)
3. Torra, V.: Hesitant fuzzy sets. Int. J. Intell. Syst. **25**(6), 529–539 (2010)
4. Rodriguez, R.M., Martinez, L., Herrera, F.: A group decision making model dealing with comparative linguistic expressions based on hesitant fuzzy linguistic term sets. Inf. Sci. **241**(12), 28–42 (2014)
5. Wei, C.P., Zhao, N., Tang, X.J.: Operators and comparisons of hesitant fuzzy linguistic term sets. IEEE. Trans. Fuzzy Syst. **22**(3), 575–585 (2014)
6. Liao, H.C., Xu, Z.S., Zeng, X.J., Merigo, J.M.: Qualitative decision making with correlation coefficients of hesitant fuzzy linguistic term sets. Knowl.-Based Syst. **76**, 127–138 (2015)
7. Yang, X.B., Song, X.N., Qi, Y.S., Yang, J.Y.: Constructive and axiomatic approaches to hesitant fuzzy rough set. Soft. Comput. **18**(6), 1–11 (2014)
8. Liang, D.C., Liu, D.: A novel risk decision-making based on decision-theoretic rough sets under hesitant fuzzy information. IEEE. Trans. Fuzzy Syst. **23**(2), 237–247 (2015)
9. Zadeh, L.A.: Toward a theory of fuzzy information granulation and its centrality in human reasoning and fuzzy logic. Fuzzy Sets Syst. **90**(2), 111–127 (1997)
10. Qian, Y.H., Liang, J.Y., Yao, Y.Y., Dang, C.Y.: MGRS: a multi-granulation rough set. Inf. Sci. **180**(6), 949–970 (2010)

11. Qian, Y.H., Li, S.Y., Liang, J.Y., Shi, Z.Z., Wang, F.: Pessimistic rough set based decisions: a multigranulation fusion strategy. Inf. Sci. **264**(6), 196–210 (2014)

12. Liang, J.Y., Wang, F., Dang, C.Y., Qian, Y.H.: An efficient rough feature selection algorithm with a multi-granulation view. Int. J. Approx. Reason. **53**(6), 912–926 (2012)

13. Yang, X.B., Zhang, Y.Q., Yang, J.Y.: Local and global measurements of MGRS rules. Int. J. Comput. Int. Syst. **5**(6), 1010–1024 (2012)

14. Zhang, C., Li, D.Y., Yan, Y.: A dual hesitant fuzzy multigranulation rough set over two-universe model for medical diagnoses. Comput. Math. Method Med. **2015**(5), 1–12 (2015)

15. Zhang, C., Li, D.Y., Ren, R.: Pythagorean fuzzy multigranulation rough set over two universes and its applications in merger and acquisition. Int. J. Intell. Syst. **31**(9), 921–943 (2016)

Multi-granularity Similarity Measure of Cloud Concept

Jie Yang, Guoyin Wang[(✉)], and Xukun Li

Chongqing Key Laboratory of Computational Intelligence,
Chongqing University of Posts and Telecommunications,
Chongqing 400065, People's Republic of China
wanggy@ieee.org

Abstract. Cloud model achieves bidirectional transformation between qualitative concepts and quantitative values using the forward and backward cloud transformation algorithms. In a cognition process, the similarity measure of cloud concepts is a crucial issue. Traditional similarity measures of cloud concept based on single granularity fail to measure the similarity of multi-granularity concepts. Based on a combination of Earth Movers Distance (EMD) and Kullback-Leibler Divergence (KLD), a multi-granularity similarity measure - EMDCM based on Adaptive Gaussian Cloud Transformation (AGCT) is proposed. Wherein, AGCT realizes multiple granularity concept generation and uncertain extraction between cloud models automatically. EMD is used to measure the similarity between different concepts. Experiments have been done to evaluate this method and the results show its performance and validity.

Keywords: Cloud model · Multi-granularity · AGCT · EMD · KLD

1 Introduction

It is known to all that difference and similarity simultaneously exist in human cognition process. For a person, the cognition of the same concept changes with the increase of his knowledge and experience. For different people, their cognition for the same concept are also different due to the influence of congenital factors. The cloud model proposed by Li realizes the uncertain transformation between qualitative concept and quantitative values [1, 2]. It can be further used to realize the bidirectional cognition from concept intention to extension [3]. If a concept is characterized by the cloud model, we call this concept as could concept.

Since cloud concept combined vagueness with randomness, Cloud similarity measurement (CSM) enjoys more advantages. For instance, in data mining, quantitative data can achieve conceptual transformation by cloud model. In system performance evaluation, the results are more convincing when CSM is applied. In the detection and control of Internet Public Opinion (IPO), the use of CSM could be more accurate to judge whether the concept has matched public opinion. In collaborative filtering recommendation system, using CMS according to the user's preferences can improve the accuracy of recommendation. Therefore, CSM in the cognitive process is an issue worthy of study.

© Springer International Publishing AG 2016
V. Flores et al. (Eds.): IJCRS 2016, LNAI 9920, pp. 318–330, 2016.
DOI: 10.1007/978-3-319-47160-0_29

Currently, there is no consensus on how to measure the similarity of cloud concepts. However, an excellent similarity measure algorithm for cloud concept requires not only strong stability and high efficiency, but the ability to highlight the differences among different types of clouds and ensure greater distinction, under the premise of guaranteeing correct similarity conclusion. Besides, a similarity measure of cloud model with good performance should be universal.

Previous studies [4–8] tried to solve the similarity measure, however, most of them focused on single cloud model, On the other hand, when two concepts is characterized by several cloud models, multiple granularity concepts. Then previous studies failed. Therefore, we address the task of multi-granularity similarity measurement of cloud concept.

The remainder is organized as follows. The following section introduces related works about CSM. Section 4 gives an introduction to granular computing (GrC) and cloud model and its relevant multi-granularity mechanism - Adapt Gaussian Cloud Transformation (AGCT). Section 4 proposes EMDCM after illustrating the feasibility of EMD, and then proves the universality of the algorithm. Section 5 demonstrates accuracy and robustness of EMDCM through experiments. The paper is briefly summarized in Sect. 6.

2   Related Works

Previous studies main focused on the similarity measure of single cloud model. Zhang [4] proposed an Interactive Drops Certainty (IDC) algorithm to measure the similarity of two cloud concepts, and the results are unstable and the accuracy relies on the number of cloud drops. Reference [5] uses the Euclidean distance to measure the similarities between cloud concepts, ignores the weights of the three numerical features of cloud model. Likeness comparing method based on Cloud Model (LICM) measures the similarities by calculating the cosine of the intersection angle [6]. Although LICM performs well in the collaborative filtering system, the cosine difference is not obvious when En and He are much smaller than expected, which reduces the uncertainty. Expectation based Cloud Model (ECM) and Max boundary based Cloud Model (MCM) overcome the problems of high time complexity [7]. However, ECM ignores the function of He when doing the calculation. MCM takes He into consideration, but it fails to describe the similarity of cloud concepts when He takes very large value. Based on Kullback-Leibler Divergence (KLD) in information theory, Reference [8] combines with the Maximum Boundary Curve of Cloud Model to describe the similarity of cloud model. Cloud Measure based on Kullback-Leibler Divergence (KLDCM) perform well in reflecting the similarity between cloud concepts. However, when a data set or an evaluation set is composed of multiple qualitative concepts, KLDCM fails to settle this condition of multiple different granularities by a single cloud model. Table 1 gives the comparison of above similarity measures.

For the problems of similarity measures mentioned above, especially they could not meet the demand of multiple granularity concepts. We propose a multi-granularity similarity measure in this paper.

Table 1. A list of similarity measures of cloud concept

	CS	LICM	ECM	MCM	KLDCM
Efficiency	Low	High	High	High	High
Discrimination	Medium	Low	Medium	High	High
Stability	Low	Low	High	High	High
Universality	High	Low	Medium	Medium	High

3   Multi-granularity Cloud Model

3.1   Granular Computing

Granularity is originally a concept of physics, referring to the mean metric of the substantial particle size. It aims to figure the amount of information in different granularities [9]. Gradual granulation method of perceptions counts in marvelous capabilities that human intrinsically possess [10]. Figure 1 shows the entire granular structure, Layer$_k$ represents the finest layer, and each dot represents the finest data. As the objects of processing, granules are any subsets, objects, clusters, and elements of a universe as they are drawn together by distinguishability,

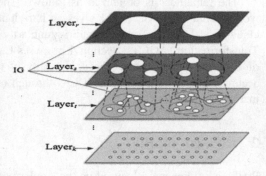

Fig. 1. Information granule, granularity layer, and granularity structure

similarity, or functionality [11]. GrC is a label of the family of any theories, methodologies, techniques and tools, which is an information processing theory for using "granules" effectively to build an efficient computational model for dealing with the problems [12]. Especially when problems are with uncertainty, it can solve them approximately [13]. The granular structure is usually used for representing and interpreting a problem. Granulation is the heart of any knowledge representation system [14]. It aims to achieve the right granule from the raw data. The first step of granulation is to select a specific model and then conduct granulation according to the corresponding granularity expression. The granulation models mainly include: fuzzy set [15], rough set [16], quotient space [17], cloud model [18], etc. These four granular computing models describe the human ability to solve the problem from different granularities. They come with their own methodologies, relative Granularity structure, comprehensive design framework and a large body of knowledge supporting analysis, design, and processing of constructs developed therein. Table 2 indicates the common features of the four granular computing models.

Theory of Fuzzy information granulation (TFIG) based on the fuzzy set is an important granulation model, which is inspired by human granulation and information processing and is based on mathematics. The point of departure in TFIG is the concept of a generalized constraint. Granule is characterized by the generalized constraint that is

Table 2. Common features of four granulation models

Granulation model	Granule	Granularity structure
Fuzzy set	Fuzzy information granule	If-then rule
Rough set	Equivalence class	Hierarchical rough set
Quotient space	Quotient set	Quotient structure
Cloud model	Cloud generated by characteristic parameters	Multi-granularity cloud generated by cloud transformation

used to define it. The principal types of granules include possibilistic, veristic and probabilistic. The principal modes of generalization in fuzzy information granulation theory can be mainly generalized to fuzzification; granulation and fuzzy granulation [19].

The main idea of rough set is to build a division in the universe of discourse according to the equivalence relation and get indistinguishable equivalence classes, namely granules, thus forming an approximate space composed by granules with different granularities [20].

Quotient space theory was first proposed by Chinese scholars Ling Zhang and Bo Zhang in literature [17], which is a model for solving the problem from different perspectives and shifting the focus of thinking onto different abstract level by the idea of granularity.

The extension of qualitative concept is often uncertain, ambiguous and dynamic, including randomness and fuzziness. Since the granulation models mentioned above only achieve hard partition of information, the knowledge acquired from raw data is lack of generalization ability.

3.2 Cloud Model

The cloud model [1] theory was proposed by Prof D.Y. Li in 1995, by using forward cloud algorithm and backward cloud algorithm to achieve a bi-directional transformation between qualitative concepts and quantitative values, revealing the randomness and fuzziness of the objective things. The characteristics stated above determines that the cloud model can be used as the basic model of the concept. Wherein, Ex corresponds to the core of the model, and En reflects the degree of discreteness of the data relative to the core, and He can be used as a measure of the maturity of concepts. As shown in Fig. 2, the cloud model can achieve the bi-directional transformation of intension and extension through the transformation between forward cloud and backward cloud.

As shown in Fig. 3, y_1 is the outer envelope curve of the cloud model and y_2 is the inner envelope curve of the cloud model:

$$y_1 = \exp\left\{ -\frac{(x - Ex)^2}{2(En + 3He)^2} \right\} \tag{1}$$

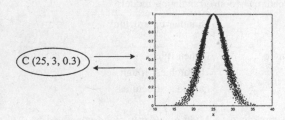

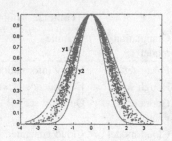

Fig. 2. Bi-directional cognitive transformation

Fig. 3. Inner and outer envelope curves

$$y_2 = \exp\left\{-\frac{(x - Ex)^2}{2(En - 3He)^2}\right\} \qquad (2)$$

When En − 3He > 0, 0 < He < En/3, then 99.74 % certainty of cloud drops fall between the inner and outer envelope curves. From the Ref. [21], we know that even the characteristic parameters of the cloud model change, outer envelope curve remains and contains almost all of the cloud drops. Therefore, the outer envelope curve can be used to depict the distribution characteristics of different cloud models.

3.3 Gaussian Cloud Transformation

Gaussian Mixture Model (GMM) is used to transfer an original data set to a sum of Gaussian distributions [22]. The process of GMM parameters estimated by the EM algorithm does not consider the concept cognition law, because many Gaussian distributions are overlapped. This causes concept confusion when GMM is used to express concepts. Based on the Gaussian hybrid model, Adaptive Gaussian Cloud Transformation (AGCT) uses hyper entropy to solve the problem of the soft partition in the uncertain areas among concepts, and it also

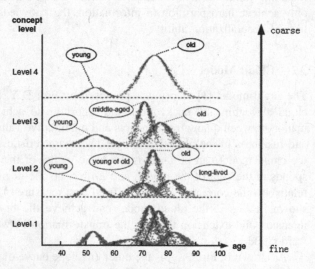

Fig. 4. MGrR generated by AGCT in the experiment of ACAE [23]

uses concepts' ambiguity to measure the effects that overlapping concepts had on the consensus of concepts [23]. AGCT also realizes the generation, selection and optimization of concept's quantity, granularity and hierarchy. As shown in Fig. 4, based on the definition of parameter concept clarity, AGCT generates multi-granular concepts by clustering academicians in Chinese Academy of Engineering with regards to age. It can be seen from the figure that AGCT is able to realize variable granularity's cognitive map from five concepts to two concepts, so it finally generated two clear concepts, that is, the elder and younger academicians. Therefore, AGCT is a clustering process, and also a variable granular computing process, and even can be defined as a process of deep learning. For any given data sets, it is difficult to describe it with only one qualitative concept. Compared with the simple backward cloud algorithm, Gaussian cloud transformation can generate multiple concepts of different granularity, which is more universal.

Algorithm 1 **AGCT** [23]

Input: Data sets $\{x_i \,|\, i = 1, 2, \cdots, N\}$, Concept clarity β

Output: Gaussian clouds $C(Ex_k, En_k, He_k)|k=1,2,\cdots,m$

1. Count the wave number of data frequency distribution, as an initial concepts quantity m

2. Using GMM to transfer X to M Gaussian distributions: $G(\mu_k, \sigma_k)|k=1,...,m$

3. For each $G(\mu_k, \sigma_k)$ compute α_k and parameters of Gaussian cloud:

$Ex_k = \mu_k, En_k = (1+\alpha_k) \times \sigma_k / 2, He_k = (1-\alpha_k) \times \sigma_k / 6, He_k / En_k = (1-\alpha_k)/3(1+\alpha_k)$

Given a strategy as $He / En \le \beta$ input, A-GCT will form several concepts which satisfy β.

4. For each $C(Ex_k, En_k, He_k)$ if $He_k / En_k > \beta$, then m=m-1

5. Loop step 3-4 until form m Gaussian in which $He_k / En_k \le \beta, k=1,...,m$

The time complexity of AGCT is:

$$O \ (m^*N) + O \ ((m-1)^*N) + ... + O \ (M^*N) = O \ (m^2 {}^*N)$$

Wherein, m is the number of peaks according to the frequency distribution of data samples, M is the number of the concept of the final generation.

End Algorithm

3.4   The Earth Movers Distance

The Earth Movers Distance (EMD) [24, 25] proposed by Rubner et al. in 2000, was originally arisen from the transportation problem, which can measure the differences of two probability distributions. The EMD uses the minimum costs of moving, but not the real distances, so as to avoid the quantization (from continuous values to discrete values), that is, to avoid the generation of quantization error. Hence, the EMD is of robustness [25]. Besides, the results of EMD could be more close to human's judgment. As shown in Fig. 5, P and Q are two distributions, P1 and P2 are the signatures of P, and Q1, Q2 and Q3 are the signatures of Q, while w denotes the weight of signature and d denotes the distance of every signature. The target of EMD is minimizing the cost of transformation of P to Q. In other words, if we regard the two distributions as two mountains stacked by two different ways within the region, the EMD is to figure out the minimum costs of moving from one mountain to the other.

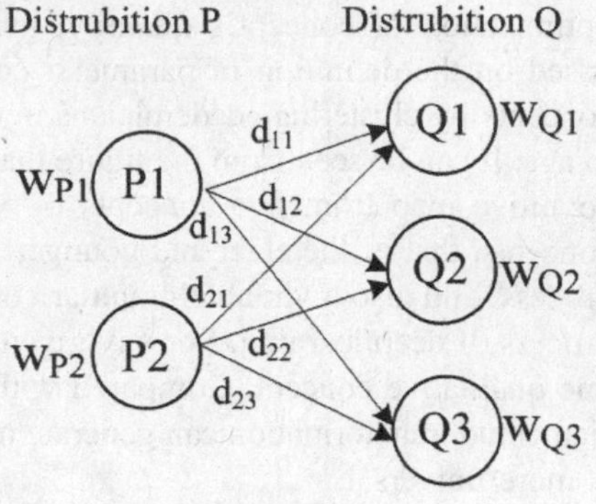

Fig. 5. Schematic diagram of Earth Movers Distance

Nowadays, the EMD is widely used in computer vision, machine learning [26] etc. For the consideration of the importance of different signatures, the EMD minimizes the total separation distance of signatures, which is a many-to-many matching calculation and so that it can calculate partial match. The objective function of EMD is as follows [24]:

$$\arg \min \sum_{i=1}^{n} \sum_{j=1}^{m} c_{ij} f_{ij} \tag{3}$$

s.t.

$$\sum_{j=1}^{m} f_{ij} \leq w_i; 1 \leq i \leq n \tag{4}$$

$$\sum_{i=1}^{n} f_{ij} \leq w_j; 1 \leq i \leq m \tag{5}$$

$$\sum_{i=1}^{n} \sum_{j=1}^{m} f_{ij} = 1 \tag{6}$$

$$f_{ij} \geq 0; 1 \leq i \leq n, 1 \leq j \leq m \tag{7}$$

Define *EMD* as:

$$EMD = \frac{\sum_{i=1}^{n} \sum_{j=1}^{m} c_{ij} f_{ij}}{\sum_{i=1}^{n} \sum_{j=1}^{m} f_{ij}} \tag{8}$$

Formula (4) allows moving "supplies" from P to Q and not vice versa. Formula (5) limits the amount of supplies that can be sent by the clusters in P to their weights. Formula (6) limits the clusters in Q to receive no more supplies than their weights; and formula (7) forces to move the maximum amount of supplies possible. Obviously, EMD meets the four requirements of distance measurement metric:

1. Positivity: $EMD(x_i, x_j) \geq 0$;
2. Symmetry: $EMD(x_i, x_j) = EMD(x_j, x_i)$;
3. Reflexivity: $EMD(x_i, x_j) = 0 \; x_i = x_j$;
4. Triangle inequality: $EMD(x_i, x_j) \leq EMD(x_i, x_k) + EMD(x_j, x_k)$.

4 Multi-granularity Similarity Measure of Cloud Concept

4.1 Multi-granularity Similarity Measure Based on AGCT and EMD

As we know, the cloud drops are generated by three characteristic parameters of cloud model through the forward cloud generator (FCG). So the distributions of cloud drops are unfixed since characteristic parameters would be randomized for two times, leading to great errors with traditional similarity measure. As the EMD is of great robustness, it can measure the similarity of cloud concepts precisely by avoiding the errors mentioned above. Combined with the EMD and KLD, we first proposed the multi-granularity similarity measure based on AGCT, and we name this method as EMDCM.

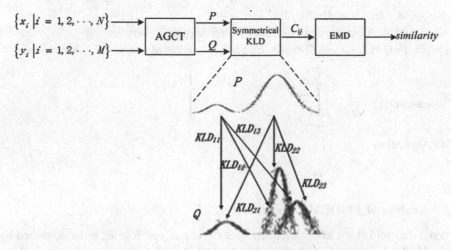

Fig. 6. The algorithm flowchart of EMDCM

The algorithm flowchart of EMDCM is shown in Fig. 6. Firstly, we use AGCT to achieve the transformation from the conceptual intension to the extension of each data set. Secondly, the cost matrix of each concept is calculated by symmetrical KLD. Finally, the similarity of the cloud concept of each data set is computing by EMD combined with cost matrix acquired from step 2. The detailed steps of EMDCM are as follows:

Algorithm 2 EMDCM

Input: two data sets $\{x_i | i = 1, 2, \cdots, N\}$、$\{y_i | i = 1, 2, \cdots, M\}$ and respective concept clarity β_1 and β_2

Output: the similarity of the two data sets

1. Using AGCT algorithm to extract cloud concepts from $\{x_i | i = 1, 2, \cdots, N\}$ and $\{y_i | i = 1, 2, \cdots, M\}$ respectively, noted as $C(Ex_i, En_i, He_i) | i = 1, 2, \cdots, n$ and $C(Ex_j, En_j, He_j) | j = 1, 2, \cdots, m$, the weights of each cloud concept are $w_i | i = 1, 2, \cdots, n$ and $w_j | j = 1, 2, \cdots, m$

2. Putting $C(Ex_i, En_i, He_i) | i = 1, 2, \cdots, n$ and $C(Ex_j, En_j, He_j) | j = 1, 2, \cdots, m$ into the cloud model's outer envelope curve formula:

$$p(x) = \frac{1}{\sqrt{2\pi}(En_i + 3He_i)} \exp\left\{-\frac{(x - Ex_i)^2}{2(En_i + 3He_i)^2}\right\}$$

and compute the outer envelope curve of each cloud concept.

3. Using symmetric KL-distance formula.

$$D_J(P \square Q) = D_{KL}(P \square Q) + D_{KL}(Q \square P) = \frac{1}{2}[(Ex_i - Ex_j)^2 + (\sigma_i^2 + \sigma_j^2)](\frac{1}{\sigma_i^2} + \frac{1}{\sigma_j^2}) - 2$$

Wherein, $\delta_i = En_i + 3He_i$, $\delta_j = En_j + 3He_j$

And, would work out the distance among the concepts as: between $C(Ex_i, En_i, He_i) | i = 1, 2, \cdots, n$ and $C(Ex_j, En_j, He_j) | j = 1, 2, \cdots, m$, noted as $c_{ij} | i = 1, 2, \cdots, n, j = 1, 2, \cdots, m$

4. $similarity(X, Y) = \dfrac{1}{EMD} = \dfrac{\sum\limits_{i=1}^{n}\sum\limits_{j=1}^{m} f_{ij}}{\sum\limits_{i=1}^{n}\sum\limits_{j=1}^{m} c_{ij} f_{ij}}$

End Algorithm

4.2   Analysis of EMDCM's Universality

When a data set is hard to be described by a cloud concept, it tends to be described by multiple cloud concepts. A method used the KLD to describe the similarity among concepts was put forward by the Ref. [27]. Since KLD uses MBCT to transform data sets to single cloud concept, KLD would not be able to achieve the purpose of measuring similarity of data sets described by multiple cloud concepts. Combined with the EMD and KLD, EMDCM can describe the similarity between concepts based on

AGCT, whether the concepts of data set are multiple or not. So EMDCM would be more universality. Especially, we prove that KLDCM is the special case of EMDCM:

Proof:

There is only one cloud model for each comparison objects, so the condition of the EMD formula is:

$$\sum_{i=1}^{n}\sum_{j=1}^{m} f_{ij} = 1; n = m = 1$$

$$w_i = w_j = 1; i = j = 1$$

$$similarity(X,Y) = \frac{1}{EMD} = \frac{1}{c} = \frac{1}{D_J(P\|Q)}$$

Wherein, c is the KLD between concepts.

End Proof

According to the above proof, KLDCM proposed by Ref. [12] is the exceptional case (when the sample concept is described only by one cloud model) of the EMDCM_AGCT. Therefore, compared to the method that merely uses KLD to describe the similarity of two concepts, EMDCM_AGCT is more universal.

As shown in Fig. 7, this experiment is to verify that KLDCM is the special case of the EMDCM, that is to say, EMDCM has universality, two different concepts are measured by KLDCM and EMDCM, and the results are shown as Table 3. From the results of Table 3, we see it clearly that the similarities of the two measures turn out to be the same, thereby proving the universality of the EMDCM.

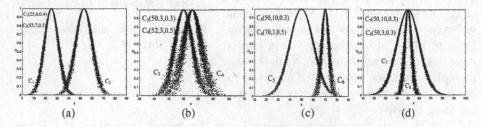

(a)   (b)   (c)   (d)

Fig. 7. The similarity measure of two cloud concepts

Table 3. Results of two measures

Similarity Measures	Similarity (C_1&C_2)	Similarity (C_3&C_4)	Similarity (C_5&C_6)	Similarity (C_7&C_8)
KLDCM	0.0712	0.9604	0.0588	0.2519
EMDCM	0.0712	0.9604	0.0588	0.2519

5 Experiments

In order to validate universality and accuracy of the EMDCM, based on the principle that the concepts come from the same field, we take pictures in the COREL image database as the experimental objects to design two experiments respectively to verify the advantages of EMDCM. The experiments are as follows:

As shown in Fig. 8, by comparing the retrieval results of the first picture in the top left corner by EMDCM, KLD and GMM respectively, the first 16 results by EMDCM turn out to be similar with the target picture, namely, the 16 pictures belong to the same class, thereby further verifying that EMDCM owns greater accuracy. As shown in Fig. 9(a), the experimental objects are two pictures that look very similar, namely the two data sets are close in distributions, so the result of similarity measurement should be very close. As shown in Fig. 9(b), the experimental objects are two pictures that look very different, so the value of similarity measure should be rather small. Our work is transforming the data sets with MBCT and AGCT respectively, representing two different thinking modes of extracting concepts. Since MBCT and other traditional backward cloud transformation (BCT) algorithms could only generate a single concept, and the universality of EMDCM has been proven in Exp. 1, regardless of the final abstracting result is one or multiple cloud concepts, EMDCM could be applied to the similarity measure of two concepts. Therefore, the experimental results are more objective.

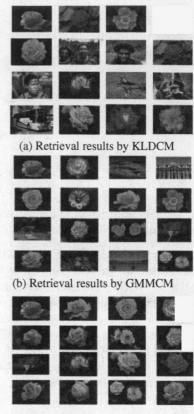

(a) Retrieval results by KLDCM

(b) Retrieval results by GMMCM

(c) Retrieval results by EDMCM

Fig. 8. Accuracy verification experiment of EMDCM

Table 4 clearly shows that the result is larger after being measured by AGCT than that of other backward cloud algorithm, which demonstrates the high accuracy of EMDCM.

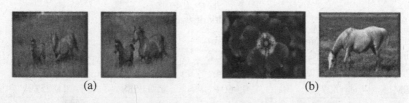

(a)   (b)

Fig. 9. Universality verification experiment of EMDCM

Table 4. Result of two concept extracting methods

Similarity Measures	Extracting methods	Numbers of formed cloud concepts	Similarity Fig. 9(a)	Similarity Fig. 9(b)
KLDCM	MBCT	1	0.7962	0.2058
EMDCM	AGCT	Adaptive extracting	0.8223	0.0899

6 Conclusion

In this paper, we propose a similarity measure of cloud concept, using AGCT to extract the concept adaptively, which simulate the cognitive process of human being to abstract data sets to multi-granularity concepts. It is more universal when compared with the methods that use traditional backward cloud algorithms to accomplish the cognitive process from extension to intension. Besides, it combines KLD with EMD to measure the similarity between concepts. From the results of experiment, it is evident that EMDCM is more universal and precise, and more suitable to describe the similarity between concepts. The following work of this paper is to focus on the improvements of Gaussian cloud transformation algorithm, i.e. trying to design a backward cloud algorithm based on Gaussian cloud transformation, doing simulation experiments for the cognitive drift of concepts transmission that the algorithm acts on, and testing whether the algorithm is stable during the concepts transmission.

Acknowledgements. This work is supported by the National Natural Science Foundation of China under Grant (No. 61572091 and No. 61272060) and Chong Qing postgraduate scientific research and innovation projects (No. CYB16106).

References

1. Li, D.Y., Shi, X.M., Meng, H.J.: Membership cloud and membership cloud generators. J. Comput. Res. Dev. **32**(6), 15–20 (1995)
2. Li, D.Y., Liu, C.Y.: The University of normal cloud model. Chin. Eng. Sci. **6**, 28–34 (2004)
3. Li, D.Y.: Uncertainty in knowledge representation. Chin. Eng. Sci. **10**, 73–79 (2000)
4. Zhang, S.B., Xu, C.X.: Study on the trust evaluation approach based on cloud model. Chin. J. Comput. **36**, 422–431 (2013)
5. Liao, L.F., Li, C., Meng, X.M.: Cloud model collaborative filtering algorithm based on similarity of Euclidean space. Comput. Eng. Sci. **37**, 1977–1982 (2015)
6. Zhang, G.W., Li, D.Y., Li, P., et al.: A collaborative filtering recommendation algorithm based on cloud model. J. Softw. **18**, 2403–2411 (2007)
7. Li, H.L., Guo, C.H., Qiu, W.R.: Similarity measurement between normal cloud models. J. Electron. **39**, 2561–2567 (2011)
8. Xu, C.L., Wang, G.Y.: Excursive measurement and analysis of normal cloud concept. Comput. Sci. **41**, 9–14 (2014)
9. Livi, L., Sadeghian, A.: Granular computing, computational intelligence, and the analysis of non-geometric input spaces. Granular Comput. **1**, 13–30 (2016)
10. Pedrycz, W.: Granular Computing: Analysis and Design of Intelligent Systems. CRC Press, Boca Raton (2013)

11. Yao, J., Vasilakos, A.V., Pedrycz, W.: Granular computing: perspectives and challenges. IEEE Trans. Cybern. **43**, 1977–1989 (2013)

12. Loia, V., Aniello, G.D., Gaeta, A., Orciuoli, F.: Enforcing situation awareness with granular computing: a systematic overview and new perspectives. Granular Comput. **1**, 127–143 (2016)

13. Bazan, J.: Hierarchical classifiers for complex spatio-temporal concepts. In: Peters, J.F., Skowron, A., Rybiński, H. (eds.) Transactions on Rough Sets IX. LNCS, vol. 5390, pp. 474–750. Springer, Heidelberg (2008). doi:10.1007/978-3-540-89876-4_26

14. Dubois, D., Prade, H.: Bridging gaps between several forms of granular computing. Granular Comput. **1**, 1–12 (2006)

15. Zadeh, L.A.: Fuzzy sets. Inf. Control **8**, 338–353 (1965)

16. Pawlak, Z.: Rough sets. Int. J. Comput. Inf. Sci. **11**, 341–356 (1982)

17. Zhang, B., Zhang, L.: Theory and Applications of Problem Solving. Elsevier Science Inc., New York (1992)

18. Zadeh, L.A.: Fuzzy sets and information granulation. In: Advances in Fuzzy Set Theory and Applications. North-Holland Publishing, Amsterdam (1979)

19. Zadeh, L.A.: Toward a theory of fuzzy information granulation and its centrality in human reasoning and fuzzy logic. Fuzzy Sets Syst. **90**, 111–127 (1997)

20. Wang, G.Y., Yao, Y.Y., Yu, H.: A survey on rough set theory and applications. Chin. J. Comput. **32**, 1229–1246 (2009)

21. Li, D.Y., Du, Y.: Uncertain Artificial Intelligence. National Defense Industry University Press, Beijing (2014)

22. Bilmes, J., et al.: A gentle tutorial of the EM algorithm and its application to parameter estimation for Gaussian mixture and hidden Markov models, vol. 4. International Computer Science Institute (1998)

23. Liu, Y.C., Li, D.Y., He, W., Wang, G.Y.: Granular computing based on Gaussian cloud transformation. Fundam. Inf. **127**, 385–398 (2013)

24. Rubner, Y., Tomasi, C., Guibas, L.J.: The earth mover's distance as a metric for image retrieval. Int. J. Comput. Vis. **40**, 99–121 (2000)

25. Ren, Z., Yuan, J., Meng, J., et al.: Robust part-based hand gesture recognition using kinect sensor. IEEE Trans. Multimedia **15**, 1110–1120 (2013)

26. Wang, F., Jiang, Y.G., Ngo, C.W.: Video event detection using motion relativity and visual relatedness. In: Proceedings of the 16th ACM International Conference on Multimedia, pp. 239–248. ACM (2008)

27. Yu, D., Yao, K., Su, H., et al.: KL-divergence regularized deep neural network adaptation for improved large vocabulary speech recognition. In: 2013 IEEE International Conference on Acoustics, Speech and Signal Processing (ICASSP). IEEE, pp. 7893–7897 (2013)

Multi-adjoint Concept Lattices, Preferences and Bousi Prolog

M. Eugenia Cornejo[1(✉)], Jesús Medina[2], Eloísa Ramírez-Poussa[2],
and Clemente Rubio-Manzano[3]

[1] Department of Statistics and O.R, University of Cádiz, Cádiz, Spain
mariaeugenia.cornejo@uca.es
[2] Department of Mathematics, University of Cádiz, Cádiz, Spain
{jesus.medina,eloisa.ramirez}@uca.es
[3] Department of Information Systems, University of the Bio-Bio, Concepción, Chile
clrubio@ubiobio.cl

Abstract. The use of preferences is usual in the natural language and it must be taken into account in the diverse theoretical frameworks focused on the knowledge management in databases. This paper exploits the possibility of considering preferences in a (discrete) fuzzy concept lattice framework.

Keywords: Fuzzy logic programming · Discrete t-norm · Residuated implications

1 Introduction

Preferences are a very important resource for the personalization of knowledge-based systems. Nowadays, personalization is a crucial task in those applications in which a large amount of data are available. Preferences have been used as a means to address this challenge through supporting the expression of user interests, likes and dislikes. In this regard, the number of units of information that a person can store in their working memory is seven plus or minus two [12]. Hence, assuming a finite and short number of preferences seems logical.

On the other hand, Formal Concept Analysis (FCA) is a theory of data analysis that identifies conceptual structures among data sets. Due to the need to allow a certain kind of uncertainty in this framework, different fuzzy extensions of FCA have been introduced [1–3]. The relation between FCA and Rough Set Theory, for example, has been widely study in diverse papers [4,7,16,17], which shows that advances developed in one also provide benefits in the other one.

Recently, the philosophy of the multi-adjoint paradigm has been applied to FCA giving rise to one of the most general fuzzy approaches, multi-adjoint concept lattices [8–10]. Adjoint triples are the basic operators to make computations in the multi-adjoint concept lattice framework and different adjoint triples can be taken into account. This last property provides the possibility of considering different preferences among the sets of objects and attributes, which is really

© Springer International Publishing AG 2016
V. Flores et al. (Eds.): IJCRS 2016, LNAI 9920, pp. 331–341, 2016.
DOI: 10.1007/978-3-319-47160-0_30

interesting as previously was highlighted. Regarding the useful of this property, different preferences need to be fixed associated with a family of adjoint triples. Since the implications in an adjoint triple are the operators involved in the definitions of the concept-forming operators, an ordering among different selected implications will offer different preferences.

A correct interpretation and visualization of discrete fuzzy implications allows designers to get a better understanding of them and properly select a suitable set of family of implications. With this aim, we have developed a tool called 3D Preferences Fuzzy Concept Lattices (3D Preferences FCL, for short) whose objective is to provide designers with graphical tools for interpreting and visualizing families of fuzzy implications when they are employed to establish a formal method for modelling a set of ordered user preferences during a process of knowledge representation. This tool has been incorporated into the Bousi Prolog system which allows us to develop knowledge-based systems in which the users preferences are employed during a process of inference.

Furthermore, in this paper the implementation of FCA theory based on preferences by using a logic programming system is detailed on an example introduced in [10]. We focus on operational aspects by leaving apart a deeper formal study on the relation between both topics. In order to implement FCA theory based on preferences in Bousi Prolog, we must perform two steps: (i) to extend the syntax of the language for defining inference rules; (ii) to enhance the unification mechanism for computing with preferences (see Sect. 4).

2   Multi-adjoint Concept Lattices

First of all, we will introduce the notion of adjoint triple [5,6] which are a generalization of a triangular norm and its residuated implication.

Definition 1. *Let* $(P_1, \leq_1)$, $(P_2, \leq_2)$, $(P_3, \leq_3)$ *be posets and* $\&: P_1 \times P_2 \to P_3$, $\swarrow: P_3 \times P_2 \to P_1$, $\nwarrow: P_3 \times P_1 \to P_2$ *be mappings, then* $(\&, \swarrow, \nwarrow)$ *is an adjoint triple with respect to* P_1, P_2, P_3, *if the equivalence:*

$$x \leq_1 z \swarrow y \quad \text{iff} \quad x \& y \leq_3 z \quad \text{iff} \quad y \leq_2 z \nwarrow x$$

holds, for all $x \in P_1$, $y \in P_2$ *and* $z \in P_3$. *The previous equivalence is known as* adjoint property.

In principle, we can observe that the domain and codomain of the operators of an adjoint triple may have three different sorts, this feature provides more flexibility into the language. In addition, monotonic properties are satisfied but no boundary conditions are required.

Note that, when the conjunctor of an adjoint triple satisfies the commutative property, both residuated implications coincide and, in this case, only the implication $\nwarrow$ will be defined. An interesting example of commutative adjoint triple is shown in the next definition.

Definition 2. *Given $\alpha \in [0,1]$, the operators $\&_{L^\alpha}, \diagdown_{L^\alpha} : [0,1] \times [0,1] \to [0,1]$, defined as*

$$x \,\&_{L^\alpha} y = \sqrt[1+\alpha]{\max(0, x^{1+\alpha} + y^{1+\alpha} - 1)}$$

$$z \diagdown_{L^\alpha} x = \sqrt[1+\alpha]{\min(1, 1 + z^{1+\alpha} - x^{1+\alpha})}$$

for all $x, y, z \in [0,1]$, form the adjoint triple $(\&_{L^\alpha}^*, \diagup_{L^\alpha}^*, \diagdown_{L^\alpha}^*)$. The set of pairs $\{(\&_{L^\alpha}, \diagdown_{L^\alpha})\}_{\alpha \in [0,1]}$ is called Łukasiewicz family.

A discretization of this family can be considered based on granular intervals $[0,1]_G$, which correspond to a particular granularity G [11]. For example, $[0,1]_4$ is the set of values $\{0/4, 1/4, 2/4, 3/4, 4/4\} = \{0, 0.25, 0.5, 0.75, 1\}$.

Example 1. Given $\alpha \in [0,1]$ and the granular intervals $[0,1]_{20}, [0,1]_8, [0,1]_{100}$, a discretization of the Łukasiewicz operator $\&_{L^\alpha}$ is $\&_{L^\alpha}^* : [0,1]_{20} \times [0,1]_8 \to [0,1]_{100}$ defined, for each $x \in [0,1]_{20}$ and $y \in [0,1]_8$ as:

$$x \,\&_{L^\alpha}^* y = \frac{\left\lceil 100 \cdot \sqrt[1+\alpha]{\max(0, x^{1+\alpha} + y^{1+\alpha} - 1)} \right\rceil}{100}$$

where $\lceil \_ \rceil$ is the ceiling function and its residuated implications $\diagup_{L^\alpha}^* : [0,1]_{100} \times [0,1]_8 \to [0,1]_{20}, \diagdown_{L^\alpha}^* : [0,1]_{100} \times [0,1]_{20} \to [0,1]_8$ are defined as: $z \diagup_{L^\alpha}^* y = \frac{\left\lfloor 20 \cdot \sqrt[1+\alpha]{\min(1, 1 + z^{1+\alpha} - x^{1+\alpha})} \right\rfloor}{20}, z \diagdown_{L^\alpha}^* x = \frac{\left\lfloor 8 \cdot \sqrt[1+\alpha]{\min(1, 1 + z^{1+\alpha} - x^{1+\alpha})} \right\rfloor}{8}$, where $\lfloor \_ \rfloor$ is the floor function.

Therefore, the triple $(\&_{L^\alpha}^*, \diagup_{L^\alpha}^*, \diagdown_{L^\alpha}^*)$ is an adjoint triple and the operator $\&_{L^\alpha}^*$ is neither commutative nor associative. Similar adjoint triples can be obtained from other t-norms.

The adjoint implication introduced in Example 1 has been used as a computational operator in the software application presented in Sect. 3. Once we have introduced the definition of the calculus operators, it is important to note that adjoint triples will be used as the underlying structures of the multi-adjoint concept lattice. For that reason, we will require the lattice structure on some of the posets in the definition of adjoint triple.

Definition 3 [10].

(a) A multi-adjoint frame $(L_1, L_2, P, \preceq_1, \preceq_2, \leq, \&_1, \diagup^1, \diagdown_1, \ldots, \&_n, \diagup^n, \diagdown_n)$ is a tuple composed by two complete lattices $(L_1, \preceq_1)$, $(L_2, \preceq_2)$, a poset $(P, \leq)$ and several adjoint triples $(\&_i, \diagup^i, \diagdown_i)$, with respect to L_1, L_2, P and $i \in \{1, \ldots, n\}$. Multi-adjoint frames are denoted as $(L_1, L_2, P, \&_1, \ldots, \&_n)$.

(b) A context is a tuple (A, B, R, σ) such that A and B are non-empty sets (usually interpreted as attributes and objects, respectively), R is a P-fuzzy relation $R: A \times B \to P$ and $\sigma: A \times B \to \{1, \ldots, n\}$ is a mapping which associates any element in $A \times B$ with some particular adjoint triple in the multi-adjoint frame $(L_1, L_2, P, \&_1, \ldots, \&_n)$.

Given a multi-adjoint frame and a context for that frame, the concept-forming operators $^{\uparrow\sigma}: L_2^B \longrightarrow L_1^A$ and $^{\downarrow^\sigma}: L_1^A \longrightarrow L_2^B$ are defined, for all $g \in L_2^B$, $f \in L_1^A$ and $a \in A$, $b \in B$, as

$$g^{\uparrow_\sigma}(a) = \inf\{R(a,b) \swarrow^{\sigma(a,b)} g(b) \mid b \in B\}$$

$$f^{\downarrow^\sigma}(b) = \inf\{R(a,b) \nwarrow_{\sigma(a,b)} f(a) \mid a \in A\}$$

These concept-forming operators form a Galois connection [10], and the notion of concept is defined as usual: a *multi-adjoint concept* is a pair $\langle g, f \rangle$ satisfying that $g \in L_2^B$, $f \in L_1^A$ and that $g^{\uparrow_\sigma} = f$ and $f^{\downarrow^\sigma} = g$; with $(^{\uparrow_\sigma}, ^{\downarrow^\sigma})$ being the Galois connection defined above.

Definition 4. *The* multi-adjoint concept lattice *associated with a multi-adjoint frame* $(L_1, L_2, P, \&_1, \ldots, \&_n)$ *and a context* (A, B, R, σ) *is the set*

$$\mathcal{M} = \{\langle g, f \rangle \mid g \in L_2^B, f \in L_1^A \text{ and } g^{\uparrow_\sigma} = f, f^{\downarrow^\sigma} = g\}$$

in which the ordering is defined by $\langle g_1, f_1 \rangle \preceq \langle g_2, f_2 \rangle$ *if and only if* $g_1 \preceq_2 g_2$ *(equivalently* $f_2 \preceq_1 f_1$*).*

One of the main properties of this concept lattice framework is that different preferences can be considered in the set of attributes and objects. This property is very interesting, since different levels of preference among the set of objects can be considered depending on the importance for the proposed user goal. More details about this feature was given in [10].

For example, considering the granular interval $[0, 1]_4$ and the Łukasiewicz family, since the operators used in the definition of the concept-forming operators are implications, we need an increasing chain of that operators, such as $\nwarrow_{L^{0.0}}$, $\nwarrow_{L^{0.25}}$, $\nwarrow_{L^{0.5}}$, $\nwarrow_{L^{0.75}}$, $\nwarrow_{L^{1.0}}$, which, for all $z, x \in [0, 1]_4$, verify

$$(z \nwarrow_{L^{0.0}} x) \leq (z \nwarrow_{L^{0.25}} x) \leq (z \nwarrow_{L^{0.5}} x) \leq (z \nwarrow_{L^{0.75}} x) \leq (z \nwarrow_{L^{1.0}} x)$$

From this change we can consider that the objects related to $\nwarrow_{L^{1.0}}$ have a "strong preference", to $\nwarrow_{L^{0.75}}$ has "preference", to $\nwarrow_{L^{0.5}}$ has a "normal" preference, etc. The following sections present an implementation of discrete fuzzy implications which will be used in the design of a procedural mechanism for the query answering process in the multi-adjoint concept lattice framework.

3   Implementation of Discrete Fuzzy Implications

3D Preferences FCL is a useful software application written in Java. We have employed the library Jzy3d which is an open source java library that allows us to easily draw 3d scientific data. The implementation of this tool requires the design and implementation of a set of specialized data structures that provide programmer with the necessary functionality for establishing the best set of preferences via fuzzy implications, for example in the multi-adjoint concept lattice framework.

Firstly, we have implemented a data structure called "3Rational" which allows programmer to define and store three rational numbers (x, y, and z) on a discretization of the unit interval. That is, the result y of applying a particular implication I on x and z, $I(z, x) = y$.

Secondly, we have designed a data structure called "Implication" which provides programmer with a computational tool for defining and storing all the values obtained from a particular implication with a particular granularity.

We provide users with two different ways of creating implications. In the first one, the programmer only must indicate the size of the interval on which the implications will be generated and a constant α. At the present time, this type of implications are only Łukasiewicz implications.

Each implication can easily be visualized in the 3D viewer, the values for each implication can be checked on the left panel of the main window (see Fig. 1) and the difference obtained between the different implications can be consulted after the computation.

The input of our application is a discrete interval obtained by means of the discretization of the unit interval with a particular granularity G. For instance, for a granularity $G = 6$, we generate the following discrete interval: $[0, 1]_6 = \{0/6, 1/6, 2/6, 3/6, 4/6, 5/6, 6/6\}$. After that, we use these values to compute the values for each implication. For example, for the family of Łukasiewicz implications with $\alpha = 0.0$, we obtain a list of forty nine 3Rationals: $\{0/6, 0/6, 6/6\}^{0.0}$, $\{0/6, 1/6, 6/6\}^{0.0}$, ..., $\{6/6, 6/6, 6/6\}^{0.0}$. Note that, the table showed on the left-panel in the main window is difficult to interpret. Hence, a 3d Viewer has been implemented in order to make easier the reading. Moreover, in any time the user can fix the z values, and a 2d graph is provided by the system.

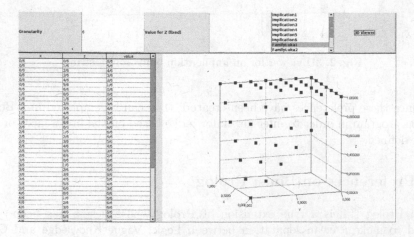

Fig. 1. 3D viewer for Łukasiewicz implication with $\alpha = 0.0$

The second type of implications are those whose values are provided at initialization time as a string. We automatically create the discretized interval generating the values for z and x. An important function in this process is a method called ``StringToImplication'' which transforms a string in a complete data structure implication, that is, a list of 3Rational numbers. For example, the following implication is built from a string "1, 5/6, 4/6, 3/6, 2/6, 1/6, 0, 1, 1, 5/6, 4/6, 3/6, 2/6, 1/6, 1 ,1, 1, 5/6, 4/6, 3/6, 2/6, 1, 1, 1, 1, 5/6, 4/6, 3/6, 1 ,1,1, 1, 1, 5/6, 4/6, 1 ,1, 1, 1, 1, 1, 5/6, 4/6, 1 ,1, 1, 1, 1, 1, 5/6, 1 ,1, 1, 1, 1, 1, 1" and by indicating a particular granularity, in this case $G = 6$.

```
Implication implicacion1=new Implication("1, 5/6, 4/6, 3/6, 2/6, 1/6, 0,
                  1, 1, 5/6, 4/6, 3/6, 2/6, 1/6,
                  1 ,1, 1, 5/6, 4/6, 3/6, 2/6, 1,
                  1, 1, 1, 5/6, 4/6, 3/6, 1 ,1,1,
                  1, 1, 5/6, 4/6, 1 ,1, 1, 1, 1,
                  1, 5/6, 1 ,1, 1, 1, 1, 1, 1",6);
```

The implication is automatically built and it can be visualized on the 3D Preferences FCL application (see Fig. 2).

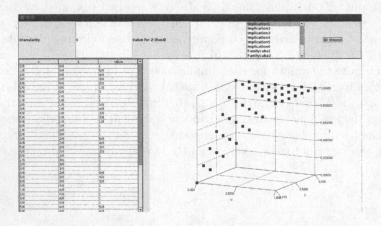

Fig. 2. 3D viewer for an implication built from a string

Once these preferences have been created, they can be employed in a Bousi Prolog program in order to infer new knowledge based on the preferences previously defined.

4   Preferences on Bousi Prolog

Bousi Prolog [14] is a true extension of Prolog whose design has been conceived to make a clean separation between Logic, Vague Knowledge and Control. This is well-suited to making the query answering process more flexible. It can be employed for advanced pattern matching, flexible deductive databases, knowledge-based systems or approximate reasoning. Also, it has been used in a number of real applications: text cataloguing [13]; knowledge discovery [15] and linguistic feedback in computer games.

This section implements the concept-forming operators using Bousi Prolog and so, considering a logic mechanism for query answering process in a multi-adjoint context. The given implementation will also be interesting in the future when attribute implications will be computed in this framework. Notice that the results of soundness on the continuous interval $[0, 1]$ have been proved in [14] and, since the proposal defined in this paper is a specialization of it, those properties are preserved.

In the particular case of user preferences, they can easily be defined in a Bousi Prolog program since it allows programmer to define a fuzzy relation between the objects and their attributes. Note that, preferences are automatically handled by the system, this is a distinctive feature of the declarative programming languages, the designer only needs to focus on the logic of the problem, in our case, the definition of the preferences by using a family of discrete implications, then the system performs the operations in an automatic and transparent way. Concerning the concept-forming operators given in the multi-adjoint framework we need to extend the language in order to support them. Moreover, we must extend the unification mechanism in order to take into account the preferences in the inference. For that, we need to recall the notion of atom. Given a first order language $\mathfrak{F}$, an *atom* is the expression $p(t_1, \ldots, t_n)$, where p is a predicate symbol and $t_1, \ldots, t_n$ are terms, which are formed by constants, variables and function symbols. An *annotated atom* is an atom $p(t_1, \ldots, t_n)$ together with a truth value α in the considered support, which can be a general lattice $(L, \preceq)$, and it is denoted as $p(t_1, \ldots, t_n)[\alpha]$.

Definition 5. *Given two annotated atoms $E_1 \equiv p(t_1, \ldots, t_n)[\alpha]$ and $E_2 \equiv q(s_1, \ldots, s_n)[\beta]$ with $\alpha, \beta \in L$ and a fuzzy implication $\leftarrow$ on L. If $p = q$ and there exists a substitution τ such that $p(\tau(t_1), \ldots, \tau(t_n)) = q(s_1, \ldots, s_n)$, we say that E_2 is $\leftarrow$-unified by E_1, with a truth-value $\beta \leftarrow \alpha$.*

Note that the usual unification procedure considers conjunctors instead of implications, this is why we are writing $\leftarrow$-unify instead of unify.

Now, from this notion of unification, a procedural semantics can be defined for the query answering process in a multi-adjoint context, in which a preference among the set of objects and attributes can be considered. This procedure will be explained next on a particular example about the determination of the suitable journal for a written paper. See more details in [10].

The multi-adjoint frame is $([0, 1]_6, \&^*_{\text{L}0.0}, \&^*_{\text{L}0.25}, \&^*_{\text{L}0.5}, \&^*_{\text{L}0.75}, \&^*_{\text{L}1.0})$, and the considered formal context is formed by the set of attributes $A = \{\text{Impact Factor, Immediacy Index, Cited Half-Life, Best Position}\}$, the set of objects $B = \{\text{AMC, CAMWA, FSS, IEEE-FS, IJGS, IJUFKS}\}$ and the relations showed in Table 1.

The σ mapping is originally defined as $\sigma(a, b) = \&^*_{\text{L}0.5}$, which assigns the same operator to every pair $a \in A$, $b \in B$. Hence, every pair has the same preference, specifically, a "normal" preference. From this context and frame a

Table 1. Fuzzy relation between the objects and the attributes.

R	AMC	CAMWA	FSS	IEEE-FS	IJGS	IJUFKS
Impact Factor	$2/6$	$1/6$	$4/6$	$5/6$	$3/6$	$2/6$
Immediacy Index	$1/6$	0	$2/6$	$1/6$	$1/6$	0
Cited Half-Life	$2/6$	$4/6$	1	$4/6$	$5/6$	$3/6$
Best Position	$4/6$	$3/6$	1	1	$3/6$	$2/6$

first order logic program is obtained. The considered set of truth-values is $[0, 1]_6$ and the set of predicates and constants of the language is given by A and B, respectively.

The relation is transformed into a set of facts of a multi-adjoint logic program. For example, the annotated fact ''impact_factor(amc) [2/6]'' must be read as follows: the journal ''AMC'' has an impact factor with a truth degree of ''2/6''. For instance, the following facts are obtained with respect to predicate ''impact_factor''.

```
%%impact_factor
impact_factor(amc)   [2/6].      impact_factor(camwa) [1/6].
impact_factor(fss)   [4/6].      impact_factor(ieee_fs) [5/6].
impact_factor(ijgs)  [3/6].      impact_factor(ijufks) [2/6].
```

Concerning the determination of the suitable journal an inference rule is added to the program. Let us assume the same definition of "suitable journal" given in [10], that is, a journal with a high impact factor, a medium immediacy index, a relatively big half-life and with not a bad position in the listing of the category. Regarding the linguistic variables: "high", "medium", "relatively big" and "not a bad", which can be related to the following truth-values: 5/6, 3/6, 4/6 and 3/6, respectively, considering the variables "medium" and "not a bad" with a similar meaning. The values 1, 2/6 and 1/6 can correspond to "very high", "bad" and "very bad", respectively.

Now, we introduce in the program the following inference rule:

```
suitable_journal(X) :- impact_factor(X) [5/6],
                       immediacy_index(X) [3/6],
                       cited_half_life(X) [4/6],
                       best_position(X) [3/6].
```

If we ask to the system about the suitable journals by launching the query ''?.-suitable_journal(X)'', the system verifies whether each annotated term in the body $\diagdown_{L\alpha}^{*}$-unifies a fact in the program. For example, since a constant preference is demanded (by the σ mapping) we consider the implication $\diagdown_{L0.5}^{*}$ and we obtain:

– "impact_factor(amc)[2/6]" is $\diagdown_{L0.5}^{*}$-unified by "impact_factor(X)[5/6]", with the truth value $2/6 \diagdown_{L0.5}^{*} 5/6 = 3/6$.

- "immediacy_index(amc)[1/6]" is $\nwarrow_{L^{0.5}}^*$-unified by "immediacy_index(X)[3/6]", with the truth value $1/6 \nwarrow_{L^{0.5}}^* 3/6 = 4/6$.
- " cited_half_life(amc)[2/6]" is $\nwarrow_{L^{0.5}}^*$-unified by " cited_half_life(X)[4/6]", with the truth value $2/6 \nwarrow_{L^{0.5}}^* 4/6 = 4/6$.
- "best_position(amc)[4/6]" is $\nwarrow_{L^{0.5}}^*$-unified by "best_position(X)[3/6]", with the truth value $4/6 \nwarrow_{L^{0.5}}^* 3/6 = 1$.

Finally, we compute the value for `suitable_journal(amc)` as the infimum of the obtained results, that is, $\inf\{3/6, 4/6, 4/6, 1\} = 3/6$. Next, the final value for the journals is introduced:

```
X=amc with=3/6;  X=camwa with=2/6; X=fss with 5/6;
X=ieee_fs with 3/6; X=ijgs with 3/6; X=ijufks with 3/6.
```

Therefore, the most suitable journal is FSS.

Now, we can consider different preferences in the mechanism of inference. For example, if the user prefers the Artificial Intelligence journals (IEEE-FS and IJUFKS), then (s)he must choose the linguistic label preference for the journals in this category and normal for the rest of journals, which correspond to the implications $\nwarrow_{L^{0.75}}^*$ and $\nwarrow_{L^{0.5}}^*$, respectively. Therefore, we need to consider a new mapping σ', such that, $\sigma'(a, b_1) = \&_{L^{0.75}}^*$ and $\sigma'(a, b_2) = \&_{L^{0.5}}^*$, for all $a \in A$, $b_1 \in \{\text{IEEE-FS}, \text{IJUFKS}\}$, $b_2 \in \{\text{AMC}, \text{CAMWA}, \text{FSS}, \text{IJGS}\}$.

We have incorporated a new directive which allows user to establish the implications for modelling user preferences.

```
:-multi_adjoint_frame ([0, 1], 6, [P1,P2,P3,P4,P5]).
:-preferences(normal(P3,[amc,camwa,fss,ijgs]),
              preference(P2,[ieee-fs,ijufks]).
```

where P1, P2, P3, P4, P5 represent the preferences given by $\nwarrow_{L^{1.0}}^*$, $\nwarrow_{L^{0.75}}^*$, $\nwarrow_{L^{0.5}}^*$, $\nwarrow_{L^{0.25}}^*$ and $\nwarrow_{L^{0.0}}^*$, respectively.

Now we take into account the implication preferences and the same rule. In this case, the computation for AMC, CAMWA, FSS and IJGS is the same, since they have the same associated preference ("normal"). Since IEEE-FS and IJUFKS has a greater preference, the truth-values obtained for the queries ``?.-suitable_journal(ieee_fs)'' and ``?.-suitable_journal(ijufks)'' can change. Let us display the value associated with IEEE-FS.

Considering the preference P2 the system proceeds as follows:

1. `impact_factor(ieee_fs)[5/6]` is $\nwarrow_{L^{0.75}}^*$-unified by `impact_factor(X)` `[5/6]`, with truth value $5/6 \nwarrow_{L^{0.75}}^* 5/6 = 1$.
2. `immediacy_index(ieee_fs)[1/6]` is $\nwarrow_{L^{0.75}}^*$-unified by `immediacy_index(X)` `[3/6]`, with $1/6 \nwarrow_{L^{0.75}}^* 3/6 = 5/6$.
3. `cited_half_life(ieee_fs)[1/6]` is $\nwarrow_{L^{0.75}}^*$-unified by `cited_half_life(X)` `[4/6]`, with $1/6 \nwarrow_{L^{0.75}}^* 4/6 = 4/6$.
4. `best_position(ieee_fs)[1]` is $\nwarrow_{L^{0.75}}^*$-unified by `best_position(X)[3/6]`, with $1 \nwarrow_{L^{0.75}}^* 3/6 = 1$.

Hence, the system responses with the substitution represented by $X = ieee\_fs$, and with the truth-value $\inf\{1, 5/6, 4/6, 1\} = 4/6$. Therefore, the obtained values are:

```
X=amc with=3/6;  X=camwa with=2/6; X=fss with 5/6;
X=ieee_fs with 4/6;  X=ijgs with 3/6; X=ijufks with 3/6.
```

Thus, the best journal is FSS, although we prefer a journal in the Artificial Intelligence category (the value for IEEE-FS has been increased to 4/6). In [10] another notion of "suitable journal" was proposed in order to provide an example from which the preferred journal is the best, although, as we have noted in this example, this is not mandatory.

5   Future Work

In the future, an implementation of the concept-forming operators of property-oriented and object-oriented concept lattices will be introduced. In addition, we will study the implications associated with the preferences in order to know the degree in which they are different and provide an efficient degree of preference. Moreover, we will provide a mechanism based on Bousi Prolog in order to built a multi-adjoint concept lattice.

References

1. Antoni, L., Krajci, S., Kridlo, O., Macek, B., Pisková, L.: On heterogeneous formal contexts. Fuzzy Sets Syst. **234**, 22–33 (2014)
2. Bělohlávek, R., Funiokova, T., Vychodil, V.: Galois connections with hedges. In: Proceedings of IFSA World Congress, vol. II, pp. 1250–1255. Springer (2005)
3. Burusco, A., Fuentes-González, R.: Construction of the L-fuzzy concept lattice. Fuzzy Sets Syst. **97**(1), 109–114 (1998)
4. Chen, J., Li, J., Lin, Y., Lin, G., Ma, Z.: Relations of reduction between covering generalized rough sets and concept lattices. Inf. Sci. **304**, 16–27 (2015)
5. Cornejo, M.E., Medina, J., Ramírez-Poussa, E.: A comparative study of adjoint triples. Fuzzy Sets Syst. **211**, 1–14 (2013)
6. Cornejo, M.E., Medina, J., Ramírez-Poussa, E.: Multi-adjoint algebras versus non-commutative residuated structures. Int. J. Approximate Reasoning **66**, 119–138 (2015)
7. Medina, J.: Relating attribute reduction in formal, object-oriented and property-oriented concept lattices. Comput. Math. Appl. **64**(6), 1992–2002 (2012)
8. Medina, J., Ojeda-Aciego, M.: Multi-adjoint t-concept lattices. Inf. Sci. **180**(5), 712–725 (2010)
9. Medina, J., Ojeda-Aciego, M.: On multi-adjoint concept lattices based on heterogeneous conjunctors. Fuzzy Sets Syst. **208**, 95–110 (2012)
10. Medina, J., Ojeda-Aciego, M., Ruiz-Calviño, J.: Formal concept analysis via multi-adjoint concept lattices. Fuzzy Sets Syst. **160**(2), 130–144 (2009)

11. Medina, J., Ojeda-Aciego, M., Valverde, A., Vojtáš, P.: Towards biresiduated multi-adjoint logic programming. In: Conejo, R., Urretavizcaya, M., Pérez-de-la-Cruz, J.-L. (eds.) CAEPIA/TTIA -2003. LNCS (LNAI), vol. 3040, pp. 608–617. Springer, Heidelberg (2004). doi:10.1007/978-3-540-25945-9_60

12. Miller, G.A.: The magical number seven, plus or minus two: some limits on our capacity for processing information. Psychol. Rev. **63**(2), 81–87 (1956)

13. Romero, F.P., Julián-Iranzo, P., Soto, A., Ferreira-Satler, M., Gallardo-Casero, J.: Classifying unlabeled short texts using a fuzzy declarative approach. Lang. Resour. Eval. **47**(1), 151–178 (2013)

14. Rubio-Manzano, C., Julián-Iranzo, P.: A fuzzy linguistic prolog and its applications. J. Intell. Fuzzy Syst. **26**(3), 1503–1516 (2014)

15. Rubio-Manzano, C., Julián-Iranzo, P.: Incorporation of abstraction capability in a logic-based framework by using proximity relations. J. Intell. Fuzzy Syst. **29**(4), 1671–1683 (2015)

16. Yao, Y.: Rough-set concept analysis: Interpreting RS-definable concepts based on ideas from formal concept analysis. Inf. Sci. **346–347**, 442–462 (2016)

17. Zhang, R., Xiong, S., Chen, Z.: Construction method of concept lattice based on improved variable precision rough set. Neurocomputing **188**, 326–338 (2016)

Modified Generalized Weighted Fuzzy Petri Net in Intuitionistic Fuzzy Environment

Sibasis Bandyopadhyay[1], Zbigniew Suraj[2(✉)], and Piotr Grochowalski[2]

[1] Department of Mathematics, Visva Bharati, Santiniketan, India
sibasisbanerjee@rediffmail.com

[2] Chair of Computer Science, University of Rzeszów, Rzeszów, Poland
{zbigniew.suraj,piotrg}@ur.edu.pl

Abstract. In this paper, a modification for the generalized weighted fuzzy Petri net in intuitionistic fuzzy environment has been proposed with the help of inverted fuzzy implication as an output operator in operator binding function. It provides a way to optimize the truth values at the output places. Approximate reasoning algorithms for such Petri net have been proposed. A numerical example is provided to logically establish the proposed theory.

Keywords: Generalized weighted fuzzy production rule · Generalized weighted intuitionistic fuzzy petri net · Knowledge representation · Approximate reasoning · Weighted composite average operator · Inverted fuzzy implication

1 Introduction

Fuzzy Petri net provides an efficient way to represent fuzzy production rule graphically in an inexact environment. For effective knowledge representation in decision support system, IF-THEN rule is very efficient to represent fuzzy production rule. The ground breaking researches in intelligent system have taken place in recent past with substantial contribution on fuzzy production rule in forward and backward reasoning [1–3] with fuzzy Petri net representation. Scarpelli et al. [3] proposed a reasoning algorithm involving formation of a subnet and evaluation process with high level fuzzy Petri net. Suraj [2,4] proposed simple fuzzy Petri net unlike to usual fuzzy Petri nets as it defines input/output operators. In Fryc et al. [5] the fuzzy reasoning process involves matrix representation of extended fuzzy Petri nets. Yuan et al. [6] proposed forward reasoning with some improvement in reasoning efficiency. Liu et al. [8] proposed an approach with intuitionistic fuzzy number in fuzzy production rule which helps in more generalization with the consideration of crisp weights associated with input values. Suraj and Bandyopadhyay [9] proposed more generalization of reasoning process with intuitionistic fuzzy number subjected to intuitionistic fuzzy weights using a dual structured (N, N') fuzzy Petri net. This represents a simple model with $S^*(.)$, $T^*(.)$ operators based on $t-$ norm and $s-$ norm. In [10] a modified generalized fuzzy Petri (mGFP) net has been discussed as an extension of generalized

© Springer International Publishing AG 2016
V. Flores et al. (Eds.): IJCRS 2016, LNAI 9920, pp. 342–351, 2016.
DOI: 10.1007/978-3-319-47160-0_31

fuzzy Petri (GFP) net where basic difference lies in the operator binding function δ. Unlike to GFP, mGFP has transitions with operator binding function δ as triples of operators (In, Out_1, Out_2) such that Out_1 belongs to inverted fuzzy implications [10].

In this paper, the main objective is to describe an extended version of weighted generalized fuzzy Petri net to more modified one with the help of inverted fuzzy implications. Since uncountably many fuzzy implications are available in the literature of fuzzy logic, and it effects the nature of the marking in modified generalized weighted fuzzy Petri nets (MGWIFPN) in intuitionistic fuzzy environment, it is quite challenging to obtain a suitable implication function for the applications concerned. The approach proposed in [11] provides a direction towards obtaining a suitable fuzzy implication function. Since the net model proposed in this paper provides a chance to define both input/output operators as well as transition operators depending on own needs, truth values of all output places connected with a given transition t can be optimized using the method described in [10,12]. The approach proposed in this paper is also concerned with the speed of reasoning process, especially in real-time decision support systems under the paradigm of incomplete, imprecise and/or vague information and also useful for knowledge representation and strong reasoning process in decision support systems.

The structure of the paper is as follows: In Sect. 2 some preliminaries regarding intuitionistic fuzzy set, fuzzy implications, weighted composite average operator and score function are described. Section 3 proposes Modified Generalized Weighted Fuzzy Petri Net in intuitionistic fuzzy environment such that the weights are also intuitionistic fuzzy set and corresponding firing rule is provided as based on modified generalized weighted fuzzy production rules. In Sect. 4 computational algorithms have been worked out in support of the theory process. Section 5 establishes the theory proposed on a reasonably strong ground with an elaborated numerical example concerning train traffic problem. In Sect. 6 a discussion on comparison with the existing literature has been made. Section 7 provides the conclusion.

2   Preliminaries

We first recall formal basic concepts which are quite helpful in discussing modified weighted generalized fuzzy Petri net in intuitionistic fuzzy environment. In fuzzy logic, fuzzy implications play an important role as operators. A fuzzy implication [10,13] is defined as a function $I : [0,1] \times [0,1] \to [0,1]$ satisfying, for all $x, x_1, x_2, y, y_1, y_2, \in [0,1]$, the following conditions: (1) $I(.,y)$ is decreasing, (2) $I(x,.)$ is increasing, (3) $I(0,0) = 1, I(1,1) = 1$, and $I(1,0) = 0$.

We have uncountably many such fuzzy implications in literature. Table 1 shows some basic fuzzy implications and extended list is available in [11].

Here, a structure of lattice is formed ([13], p. 186) by incomparable fuzzy implications generating new fuzzy implication using $min(inf)$ and $max(sup)$ operations. Now, finding correct function [12] among basic fuzzy implications is

Table 1. Some basic fuzzy implications

Name	Year	Formula of basic fuzzy implication
Łukasiewicz	1923	$I_{LK}(x,y) = min(1, 1 - x + y)$
Kleene-Dienes	1938	$I_{KD}(x,y) = max(1 - x, y)$
Goguen	1969	$I_{GG}(x,y) = \begin{cases} 1 & if \ x \leq y \\ \frac{y}{x} & if \ x > y \end{cases}$

Table 2. A list of inverse fuzzy implications and their domains from Table 1

Inverted fuzzy implications	Domain of inverted fuzzy implication
$InvI_{LK}(x,z) = z + x - 1$	$1 - x \leq z < 1, x \in (0,1]$
$InvI_{KD}(x,z) = z$	$1 - x < z \leq 1, x \in (0,1]$
$InvI_{GG}(x,z) = xz$	$0 \leq z < 1, x \in (0,1]$

very challenging. Since this method involves comparing two fuzzy implications, the fuzzy implication with greatest truth value can be easily chosen and hence a new modified fuzzy Petri net model is proposed in this paper in this direction. In Table 2 a list of inverse fuzzy implications and their domains are provided corresponding to the fuzzy implications from Table 1.

An intuitionistic fuzzy set $\widetilde{A}$ is defined [7] as an ordered triplet $\{(x, \mu_{\widetilde{A}}(x), \nu_{\widetilde{A}}(x))| \ x \in X\}$, where X represents the universe of discourse, $\mu_{\widetilde{A}}(x) : X \rightarrow [0,1]$ represents a membership function, $\nu_{\widetilde{A}}(x) : X \rightarrow [0,1]$ denotes a non-membership function, satisfying the condition $0 \leq \mu_{\widetilde{A}}(x) + \nu_{\widetilde{A}}(x) \leq 1$.

An intuitionistic fuzzy number (IFN) $\widetilde{A}^i$ is a convex normalized fuzzy set $\widetilde{A}^i$ defined over the real line $\Re$ for membership function and concave normalized fuzzy set Table 2 over real line $\Re$ for non-membership function satisfying the following conditions:

1. $\exists$ exactly one point x_0 such that for $x_0 \in \Re$, $\mu_{\widetilde{A}^i}(x_0) = 1$ and $\nu_{\widetilde{A}^i}(x_0) = 0$.
   Here x_0 is said to be mean value of $\widetilde{A}^i$.
2. $\mu_{\widetilde{A}^i}(x)$ and $\nu_{\widetilde{A}^i}(x)$ are piecewise continuous.

In this paper, we mainly use Einstein $s-$ norm and $t-$ norm [8,15] for describing operational laws of IFS which are as follows

$$(1) \ \widetilde{a}_1 \oplus \widetilde{a}_2 = \left(\frac{\mu_{\widetilde{a}_1} + \mu_{\widetilde{a}_2}}{1 + \mu_{\widetilde{a}_1}\mu_{\widetilde{a}_2}}, \frac{\nu_{\widetilde{a}_1}\nu_{\widetilde{a}_2}}{1 + (1 - \nu_{\widetilde{a}_1})(1 - \nu_{\widetilde{a}_2})} \right) \tag{1}$$

$$(2) \ \widetilde{a}_1 \otimes \widetilde{a}_2 = \left(\frac{\mu_{\widetilde{a}_1}\mu_{\widetilde{a}_2}}{1 + (1 - \mu_{\widetilde{a}_1})(1 - \mu_{\widetilde{a}_2})}, \frac{\nu_{\widetilde{a}_1} + \nu_{\widetilde{a}_2}}{1 + \nu_{\widetilde{a}_1}\nu_{\widetilde{a}_2}} \right) \tag{2}$$

Here, A weighted composite average operator (WCAO) [8] and Weighted S-score and weighted H-score for an weighted IFN $\widetilde{A} = (a, b)^{(W_a, W_b)}$ [9] are also used in this paper in modelling MGWIFPN.

3   Modified Generalized Weighted Intuitionistic Fuzzy Petri Net

This section discusses main part of this paper. We provide the definition of a modified generalized weighted fuzzy Petri net based on fuzzy implications as described in Sect. 2 which is nothing but a modification to that given in [2].

Definition 1. *A modified generalized weighted intuitionistic fuzzy Petri net (MGWIFPN) is expressed as a tuple* $N = (P, T, S, I, O, \alpha, \beta, \gamma, O_p^*, \delta, M_0)$, *where: (1)* $P = \{p_1, p_2, \cdots, p_n\}$ *is a finite set of places,* $n > 0$; *(2)* $T = \{t_1, t_2, \cdots, t_m\}$ *is a finite set of transitions,* $m > 0$; *(3)* $S = \{s_1, s_2, \cdots, s_n\}$ *is a finite set of statements and P,T,S are pairwise disjoint, i.e.,* $P \cap T = S \cap T = P \cap S = \emptyset$ *and* $card(P) = card(S)$; *(4)* $I : T \to 2^P$ *is the input function; (5)* $O : T \to 2^P$ *is the output function; (6)* $\alpha : P \to S$ *is the statement binding function; (7)* $\beta : T \to [0,1]$ *is the truth degree function; (8)* $\gamma : T \to [0,1]$ *is the threshold function; (9)* O_p^* *is a finite set of* $S_{N'}^*$ *and* $T_{N'}^*$ *operators and fuzzy implications whose domain consists of the elements as weighted IFS of the form* $(a,b)^{(W_a, W_b)}$; *(10)* $\delta : T \to O_p^* \times O_p^* \times O_p^*$ *is the operator binding function, (11)* $M_0 : P \to [0,1]$ *is the initial marking, and* 2^P *denotes a family of all subsets of the set P.*

Here, operator binding function δ associates triples of operators (In, Out_1, Out_2) with the transitions T, the first one In being the input operator and the other two operators are output operators. In initial marking M_0, the input values are weighted IFN of the form $(a, b)^{(W_a, W_b)}$. Similar to [9], here also we propose a parallel weight propagation FPN to obtain the resultant weight after each fire and this weight propagation FPN is finally fused with the original FPN to obtain resulting MGWIFPN.

Definition 2. *A weight propagation intuitionistic fuzzy Petri net (WPIFPN) can be expressed as a tuple* $N' = (P, T, S, I, O, \alpha, \beta, \gamma, \bar{O}_p^*, \bar{\delta}, M_0)$, *where: (1)* $P, T, S, I, O, \alpha, \beta, \gamma, M_0$ *have the same meaning as given in Definition 1; (2)* $\bar{O}_p^*$ *is a finite set of* $S_{N'}^*$ *and* $T_{N'}^*$ *operators and fuzzy implications whose domain consists of the elements as weights associated with each element in initial marking in the the form IFS* (W_a, W_b); *(3)* $\bar{\delta} : T \to \bar{O}_p^* \times \bar{O}_p^* \times \bar{O}_p^*$ *is the operator binding function.*

Here, the operators $S_{N'}^*$ and $T_{N'}^*$ have the same meaning as [9] and given as follow

$$S_{N'}^*(\widetilde{A}, \widetilde{B}) = (S(a, \bar{a}), S(b, \bar{b})) \tag{3}$$

and

$$T_{N'}^*(\widetilde{A}, \widetilde{B}) = (T(a, \bar{a}), T(b, \bar{b})) \tag{4}$$

where $\widetilde{A} = (a, b)$ and $\widetilde{B} = (\bar{a}, \bar{b})$. Based on the theory given so far, we propose the firing rule as follows.

3.1   Firing Rule

We consider that $N = (P, T, S, I, O, \alpha, \beta, \gamma, O_p^*, \delta, M_0)$ is a MGWIFPN and $N' = (P, T, S, I, O, \alpha, \beta, \gamma, \bar{O}_p^*, \bar{\delta}, M_0)$ is a WPIFPN with marking $M : P \to [0, 1]$ and $M_w : P \to [0, 1]$, respectively. Now, analogous to [4] a transition $t \in T$ is said to be enabled for firing at the marking M if following conditions are satisfied

$$In(M(p_{i1}), M(p_{i2}), \cdots, M(p_{in})) \geq \gamma(t) > 0 \tag{5}$$

$$In(M_w(p_{i1}), M_w(p_{i2}), \cdots, M_w(p_{in})) \geq \gamma(t)|_{weight} > 0 \tag{6}$$

where Eq. (5) gives the condition for firing for generalized weighted IFPN and (6) provides the firing rule for weight propagation fuzzy Petri net, $I(t) = \{p_{i1}, p_{i2}, \cdots, p_{in}\}$ represents a set of input places corresponding to a transition $t \in T$ and $\beta(t) \in [0, 1]$. Here, In is represented as input operator and Out_1, Out_2 are considered as output operators for transition t. Basing on these assumptions, we propose two modes of firing as follow.

Mode 1. Let M and M_w are markings of N and N' respectively enabling transition t and M', M'_w are the markings derived from M and M_w respectively by firing transition t. Then for each $t \in T$

$$M'(p) = \begin{cases} 0 \text{ if } p \in I(t), \\ Out_2(Out_1(In(M(p_{i1}), M(p_{i2}), \cdots, M(p_{in})), \beta(t)), M(p)) \text{ if } p \in O(t), \\ M(p) \text{ otherwise.} \end{cases}$$

$$M'_w(p) = \begin{cases} 0 \text{ if } p \in I(t), \\ Out_2(Out_1(In(M_w(p_{i1}), M_w(p_{i2}), \cdots, M_w(p_{in})), \beta(t)), M_w(p)) \\ \text{ if } p \in O(t), \\ M(p) \text{otherwise.} \end{cases}$$

Mode 2. Suppose M and M_w represent the markings of N and N' respectively enabling transition t and M', M'_w represent the markings derived from M and M_w respectively by firing transition t. Then for each $t \in T$

$$M'(p) = \begin{cases} Out_2(Out_1(In(M(p_{i1}), M(p_{i2}), \cdots, M(p_{in})), \beta(t)), M(p)) \text{ if } p \in O(t), \\ M(p) \text{otherwise.} \end{cases}$$

$$M'_w(p) = \begin{cases} Out_2(Out_1(In(M_w(p_{i1}), M_w(p_{i2}), \cdots, M_w(p_{in})), \beta(t)), M_w(p)) \\ \text{ if } p \in O(t), \\ M(p) \text{otherwise.} \end{cases}$$

In the modes as described above, the second $M'_w(p)$ part is related to the firing of weight propagation fuzzy Petri net and it is finally fused to the first $M'(p)$ part. Now, modified generalized weighted fuzzy Petri net is mainly based on fuzzy production rule which is of the IF *premise*(s_i) THEN *conclusion*(s_k) (CF) form, where CF denotes degree of certainty. Now, the generalized weighted FPRs are represented as follow:

1. Type 1: A generalized weighted fuzzy simple rule can be represented as GR :
   IFATHEN$C(CF = t), \lambda, Gw$
2. Type 2: A generalized weighted fuzzy conjunctive (disjunctive) rule for the antecedent can be represented by GR : IFA_1AND(OR)$A_2 \cdots$ AND(OR) A_nTHENC $(CF = t), \lambda_i, Gw_i, i = 1, 2, \cdots, n$
3. Type 3: A generalized weighted fuzzy conjunctive (disjunctive) rule for the consequent is represented by GR : IFATHENC_1AND(OR)$C_2 \cdots$ AND(OR) C_n $(CF = t), \lambda, Gw$

Here, a MGWIFPN represents [8, 9] the model of the system based on production rules. Corresponding notations are provided as below: q_i - certainty factor $\in$ $[0, 1]$, r_i - threshold value $\in [0, 1]$, In - input operator, Out_1, Out_2 - output operators. Here, the modifications are mainly done on the basis of use of fuzzy implication as Out_1 operator. The figures are similar to those given in [9].

4   Computational Algorithms

In this section, two algorithms for construction of a MGWIFPN are proposed that describes an approximate reasoning process based on the production rules given in Sect. 3.1 which resemble those given in [2] but it is more modified and generalized.

Algorithm 1. Constructing a MGWIFPN based on production rules as provided in Sect. 3.1
Input: Finite set R of production rules with parameters.
Output: A MGWIFPN net (N, N'), N' being the weight propagation net corresponding to N.
begin
    $F := \emptyset$;
    for each $r \in R$ **do**
      begin
        if r *is a rule of type* 1 **then** *construct a subnet* (N_r, N'_r)
        else if r *is a rule of type* 2 **then** *construct a subnet* (N_r, N'_r)
        else if r *is a rule of type* 3 **then** *construct a subnet* (N_r, N'_r)
        $F := F \cup (N_r, N'_r)$;
      end;
      Integrate all subnets from a family F on joint places and create
      a result net (N, N');
      return (N, N');
end

Notice: Symbol := denotes the assignment operator and the subnets are similar to those given in [9].

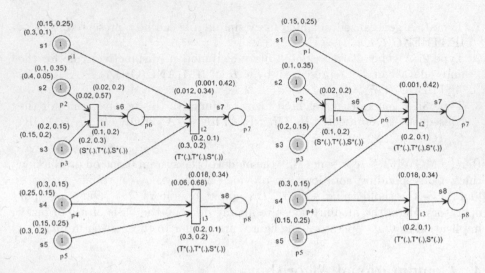

Fig. 1. (a) MGWIFPN, (b) MWPIFPN - before firing

Algorithm 2. Approximate reasoning based on MGWIFPN.

Input: *The initial marking of the starting places with elements of the form* $(a, b)^{(W_a, W_b)}$.

Output: *The final marking of the goal places with elements of the form* $(a, b)^{(W_a, W_b)}$.

begin

    while *it is not the end of simulation* **do**

      begin

        Determine transitions enabled for firing based on firing rule in Sect. 3.1;

        while *There is a transition enabled for firing* **do**

          begin

            Compute a new marking of all places after firing the transition;

            Determine a new transition enabled for firing;

        end;

        Read final marking of goal places;

        Read final marking of all places;

      end;

end

5 Numerical Example

In this section, we discuss a real life train traffic problem as an extension and modification of [2,9] to establish the relevance of the theory proposed in this paper. Here, the situation is represented logically as follows: (1) IF s_2 OR s_3

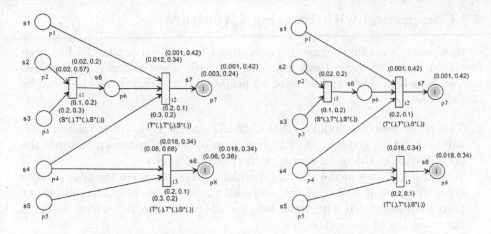

Fig. 2. (a) MGWIFPN, (b) MWPIFPN - after firing (*Mode* 1)

THEN s_6; (2) IF s_1 AND s_4 AND s_6 THEN s_7; (3) IF s_4 AND s_5 THEN s_8, where we may interpret the variables as follow: s_1 - ('Train B is the last train towards the destination today.', quite sure); s_2 - ('The delay in arrival of train A is quite large.', very sure); s_3 - ('There is a need for the track of the train B.', very urgent); s_4 - ('Passengers would like to change for train B.', large number); s_5 - ('There is a delay of train A.', quite short); s_6 - ('Train B departs according to schedule.', almost); s_7 - ('Employing an additional train in the direction of B.', quite sure); s_8 - ('Train B wait for train A.', possibly).

Here, one should observe that the second part in every proposition is the weight related to the first part and corresponding MGWIFPN can be provided as follows. Figures 1(a) and (b) show the MGWIFPN and the MWPIFPN before firing with input value as $(0.3, 0.1)^{(0.15, 0.25)}$, $(0.4, 0.05)^{(0.1, 0.35)}$, $(0.15, 0.2)^{(0.2, 0.15)}$, $(0.25, 0.15)^{(0.3, 0.15)}$, $(0.3, 0.2)^{(0.15, 0.25)}$ at the places $p_1, p_2, p_3, \ p_4, p_5$. Now, based on firing rule as provided in Sect. 3.1 we get the final marking as $(0.003, 0.24)^{(0.001, 0.42)}$, $(0.06, 0.38)^{(0.018, 0.34)}$ at the places p_7 and p_8 with the *mode* 1 as given in Figs. 2(a) and (b). Here, all the three transitions t_1, t_2, t_3 as appeared in the figures, are enabled to fire with $\beta_1, \beta_2, \beta_3$ as $(0.02, 0.57)^{(0.02, 0.2)}$, $(0.012, 0.34)^{(0.001, 0.42)}$, $(0.06, \ 0.68)^{(0.018, 0.34)}$ and $\gamma_1, \gamma_2, \gamma_3$ as $(0.2, 0.3)^{(0.1, 0.2)}$, $(0.3, 0.2)^{(0.2, 0.1)}$, $(0.3, 0.2)^{(0.2, 0.1)}$ based on the operators S^* and T^* and fuzzy implications as defined in Sect. 2 and operators $S_{N'}^*, T_{N'}^*$ as defined in Sect. 3. If sequence of transitions $t_1 t_2$ are chosen then the final value, corresponding to statement s_7 can be obtained as $(0.003, 0.24)^{(0.001, 0.42)}$. Again, if they select transition t_3 only then the final value corresponding to the statement s_8 is obtained as $(0.06, 0.38)^{(0.018, 0.34)}$. This result is based on Goguen implication and the operational laws follow from Einstein $s-$ norm and $t-$ norm. One may try other implications e.g. Łukasiewicz, Kleene-Dienes implications etc. also.

6   Comparison with Existing Literature

In this paper, a modified generalized weighted fuzzy Petri net model has been proposed in intuitionistic fuzzy environment with inverted fuzzy implications having some benefits compared to those proposed in the literature which can be stated as follows:

1. This paper uses fuzzy implications instead $t-$ norm $s-$ norm or related average operator as given in [8,9] and as such opens an approach towards the optimization of the truth degree at the output places.
2. Weighted Petri net model with intuitionistic fuzzy set makes the system more generalized comparing to [2,4,6] since all the markings in input and output places are associated with some weights which are also IFS which concerns the reliability of the system.
3. Since dual structured Petri net model (N, N') has been used in this paper, so it involves lesser computational complexity.

7   Conclusion

In this paper, a well structured knowledge representation and strong reasoning process has been proposed with the help of modified generalized fuzzy Petri net model based on fuzzy production rule. The main feature of this paper is that firstly, all the input and output values are IFS and they are subjected to some weights which are also IFS and secondly, an implication function has been used as Out_1 operator in the operator binding function δ which provides a space towards the optimization of truth values at the output places. One may choose an appropriate fuzzy implication among all other fuzzy implications which provides maximum truth values. There still exists an open space where we can think about a generalized methodology for optimization of truth values in forward and backward reasoning.

Acknowledgments. This work was partially supported by the Center for Innovation and Transfer of Natural Sciences and Engineering Knowledge at the University of Rzeszów. We would like to thank the anonymous referees for critical remarks and useful suggestions to improve the presentation of the paper.

References

1. Suraj, Z., Lasek, A.: Inverted fuzzy implications in backward reasoning. In: Kryszkiewicz, M., Bandyopadhyay, S., Rybinski, H., Pal, S.K. (eds.) PReMI 2015. LNCS, vol. 9124, pp. 354–364. Springer, Heidelberg (2015)
2. Suraj, Z.: A new class of fuzzy Petri nets for knowledge representation and reasoning. Fundam. Inform. **128**(1–2), 193–207 (2013)
3. Scarpelli, H., Gomide, F., Yager, R.R.: A reasoning algorithm for high-level fuzzy Petri net. IEEE Trans. Fuzzy Syst. **4**(3), 282–294 (1996)

4. Suraj, Z.: Knowledge representation and reasoning based on generalised fuzzy Petri nets. In: Proceedings of the 12th International Conference on Intelligent Systems Design and Applications (ISDA), Kochi, India, 27–29 November, pp. 101–106, IEEE Press (2012)
5. Fryc, B., Pancerz, K., Peters, J.F., Suraj, Z.: On fuzzy reasoning using matrix representation of extended fuzzy Petri nets. Fundam. Inform. **60**(1–4), 143–157 (2004)
6. Yuan, J., Shi, H., Liu, C., Shang, W.: Improved basic inference models of fuzzy Petri nets. In: Proceedings of the 7th World Congress on Intelligent Control and Automation, Chongqing, China, 25–27 June (2008)
7. Atanassov, K.: Intuitionistic fuzzy sets. Fuzzy Sets Syst. **20**, 87–96 (1986)
8. Liu, H.C., You, J.X., You, X.Y., Su, Q.: Fuzzy Petri nets using intuitionistic fuzzy sets and ordered weighted averaging operators. IEEE Trans. Cybern. **46**(8), 1839–1850 (2015)
9. Suraj, Z., Bandyopadhyay, S.: Generalized weighted fuzzy Petri net in intuitionistic fuzzy environment. IEEE World Congress on Computational Intelligence, Vancouver, Canada (2016, to appear)
10. Suraj, Z.: Modified generalised fuzzy Petri nets for rule-based systems. In: Yao, Y., Hu, Q., Yu, H., Grzymala-Busse, J.W. (eds.) RSFDGrC 2015. LNCS (LNAI), vol. 9437, pp. 196–206. Springer, Heidelberg (2015). doi:10.1007/978-3-319-25783-9_18
11. Suraj, Z., Lasek, A., Lasek, P.: Inverted fuzzy implications in approximate reasoning. Fundam. Inform. **143**, 151–171 (2015)
12. Suraj, Z., Lasek, A.: Toward optimization of approximate reasoning based on rule knowledgde. In: Proceedings of the 2nd International Conference on Systems and Informatics (ICSAI), pp. 281–285 (2014)
13. Baczyński, M., Jayaram, B.: Fuzzy Implications. Springer, Heidelberg (2008)
14. Yeung, D.S., Tsang, E.C.C.: Weighted fuzzy production rules. Fuzzy Sets Syst. **88**(3), 299–313 (1997)
15. Wang, W., Liu, X.: Intuitionistic fuzzy geometric aggregation operators based on Einstein operations. Int. J. Intell. Syst. **26**(11), 1049–1075 (2011)
16. Fay, A., Schnieder, E.: Fuzzy Petri nets for knowledge modelling in expert systems. In: Cardoso, J., Camargo, H. (eds.) Fuzziness in Petri Nets, pp. 300–318. Physica-Verlag, Berlin (1999)
17. Petri, C.A.: Kommunikation mit Automaten, Schriften des IIM, Nr. 2, Institut für Instrumentelle Mathematik, Bonn. English translation: Technical report RADC-TR-65-377, vol. 1, suppl. 1, Griffiths Air Force Base, New York 1966 (1962)
18. Seikh, M.R., Pal, M., Nayak, P.K.: Application of triangular intuitionistic fuzzy numbers in bi-matrix games. Int. J. Pure Appl. Math. **79**(2), 235–247 (2012)
19. Bandyopadhyay, S., Nayak, P.K., Pal, M.: Solution of matrix game with triangular intuitionistic fuzzy pay-off using score function. Open J. Optim. **2**, 9–15 (2013)
20. Chen, S.M., Tan, J.M.: Handling multicriteria fuzzy decision making problems based on vague set theory. Fuzzy Sets Syst. **67**(2), 163–172 (1994)
21. Hong, D.H., Choi, C.H.: Multicriteria fuzzy decision making problems based on vague set theory. Fuzzy Sets Syst. **144**(1), 103–113 (2000)

Machine Learning and Decision Making

Similarity-Based Classification with Dominance-Based Decision Rules

Marcin Szeląg[1]([✉]), Salvatore Greco[2,3], and Roman Słowiński[1,4]

[1] Institute of Computing Science, Poznań University of Technology,
60-965 Poznań, Poland
{mszelag,rslowinski}@cs.put.poznan.pl
[2] Department of Economics and Business, University of Catania,
Corso Italia, 55, 95129 Catania, Italy
salgreco@unict.it
[3] Portsmouth Business School, University of Portsmouth, Portsmouth, UK
[4] Systems Research Institute, Polish Academy of Sciences, 01-447 Warsaw, Poland

Abstract. We consider a similarity-based classification problem where a new case (object) is classified based on its similarity to some previously classified cases. In this process of case-based reasoning (CBR), we adopt the Dominance-based Rough Set Approach (DRSA), that is able to handle monotonic relationship "the more similar is object y to object x with respect to the considered features, the closer is y to x in terms of the membership to a given decision class X". At the level of marginal similarity concerning single features, we consider this similarity in ordinal terms only. The marginal similarities are aggregated within induced decision rules describing monotonic relationship between comprehensive similarity of objects and their similarities with respect to single features.

Keywords: Classification · Similarity · Case-based reasoning · Dominance-based rough set approach · Decision rules

1 Introduction

People tend to solve new problems using the solutions of *similar* problems encountered in the past. This process if often referred to as *case-based reasoning* (CBR) [9]. As observed by Gilboa and Schmeidler [3], the basic idea of CBR can be found in the following sentence of Hume [8]: "From causes which appear similar we expect similar effects. This is the sum of all our experimental conclusions." We can rephrase this sentence by saying: "The more similar are the causes, the more similar one expects the effects".

We consider classification performed according to the (broadly construed) CBR paradigm, i.e., a *similarity-based classification*. In the similarity-based classification problem, there is given a finite set of training objects (case base), described by a set of *features*, a set of *marginal similarity functions* (one for each feature), and a set of predefined *decision classes*. This information is used to suggest membership of a new (unseen) object to particular decision classes.

© Springer International Publishing AG 2016
V. Flores et al. (Eds.): IJCRS 2016, LNAI 9920, pp. 355–364, 2016.
DOI: 10.1007/978-3-319-47160-0_32

In case-based reasoning, one needs a *similarity model* aggregating marginal similarities into comprehensive similarity. Traditionally, this model has the form of a real-valued aggregation function (e.g., Euclidean norm) or binary relation (e.g., fuzzy relation). In this paper, we present a method based on the Dominance-based Rough Set Approach (DRSA) [4,11,12], using a new similarity model in terms of a set of *if-then decision rules* employing dominance relation in the space created by marginal similarity functions. The first method concerning application of DRSA to CBR was introduced in [5–7], and then extended in [14]. The method presented in this paper, first described in an unpublished PhD thesis [13], concerns revision and improvement of the approach given in [14].

The proposed rule-based similarity model makes it possible to avoid an arbitrary aggregation of marginal similarity functions. In this approach, comprehensive similarity is represented by decision rules *induced* from classification examples. These rules underline the monotonic relationship "the more similar is object y to object x with respect to the considered features, the closer is y to x in terms of the membership to a given decision class X". Violation of this principle causes an inconsistency in the set of objects, which is handled using DRSA. An important characteristic of the proposed approach is that induced rules employ only ordinal properties of marginal similarity functions. Thus, this approach is invariant to ordinally equivalent marginal similarity functions.

We improve over [14] by proposing a way o inducing decision rules, and by introducing a new rule-based classification scheme extending the one given in [1].

This paper is organized as follows. Section 2 describes problem setting. In Sect. 3, we discuss basic notions and assumptions. Section 4 defines considered similarity learning task. In Sect. 5, we introduce two *comprehensive closeness relations*. Section 6 defines rough approximations of the sets of objects being in either kind of comprehensive closeness relation with a reference object x. In Sect. 7, we describe induction of monotonic decision rules from the rough approximations. Section 8 concerns application of induced rules. In Sect. 9, we present an illustrative example. Section 10 concludes the paper.

2   Problem Setting

We consider the following classification problem setting. There is given a finite set of objects U (*case base*) and a finite family of pre-defined decision classes $\mathcal{D}$. An object $y \in U$ (a "case") is described in terms of features $f_1, \ldots, f_n \in F$. For each feature $f_i \in F$, there is given a *marginal similarity function* $\sigma_{f_i} : U \times U \to [0,1]$, such that the value $\sigma_{f_i}(y, x)$ expresses the similarity of object $y \in U$ to object $x \in U$ with respect to (w.r.t.) feature f_i, and for all $x, y \in U$, $\sigma_{f_i}(y, x) = 1 \Leftrightarrow f_i(y) = f_i(x)$. Moreover, for each object $y \in U$ there is given an information concerning *credibility* of its membership to each of the considered classes. To admit graded credibilities, each class $X \in \mathcal{D}$ is modeled as a fuzzy set in U [15], characterized by membership function $\mu_X : U \to [0,1]$. Thus, each object $y \in U$ can belong to different decision classes with different degrees of membership. The above input information is processed to produce a recommendation concerning a new object z, in terms of a degree of membership of z to particular classes.

3   Basic Notions and Assumptions

Pairwise fuzzy information base. Given the problem setting introduced in Sect. 2, a *pairwise fuzzy information base* **B** [5–7] is the 3-tuple

$$\mathbf{B} = <U, F, \Sigma>, \tag{1}$$

where U is a finite set of objects (a case base), F is a finite set of n *features*, and $\Sigma = \{\sigma_{f_1}, \sigma_{f_2}, \ldots, \sigma_{f_n}\}$ is a finite set of n marginal similarity functions.

Marginal similarity functions. Different marginal similarity functions can be used, depending on the value set V_{f_i} of feature $f_i \in F$. For a *numeric feature* f_i, with values on interval or ratio scale, similarity can be defined using a function, e.g., $\sigma_{f_i} = 1 - \frac{|f_i(x) - f_i(y)|}{\max_{v_i \in V_{f_i}} - \min_{v_i \in V_{f_i}}}$. For a *nominal feature* f_i, similarity can be defined using a table, like Table 1. The marginal similarity functions create an n-dimensional *similarity space*.

Table 1. Exemplary definition of similarity for a nominal feature $f_i \in F$

$f_i(x) \setminus f_i(y)$	Low	Medium	High
Low	1.0	0.6	0.3
Medium	0.6	1.0	0.5
High	0.3	0.5	1.0

Problem decomposition. We consider the decision classes belonging to family $\mathcal{D}$ to be mutually independent in the sense of membership function values. Then, we decompose the original multi-class problem π to a set of *single-class subproblems* π_X, where $X \in \mathcal{D}$. Thus, each subproblem concerns a single decision class $X \in \mathcal{D}$ with membership function $\mu_X : U \to [0, 1]$. In each subproblem, let

$$V_{\mu_X} = \{\mu_X(y) : y \in U\}. \tag{2}$$

Reference objects. We assume that for each subproblem π_X, there is given a set of so-called *reference objects* $U_X^R \subseteq U$. These are objects to which objects from set U are going to be compared. The reference objects may be indicated by a user, and thus, the set of reference objects should be relatively small. If such information is not available, one can use clustering to choose a suitable set of reference objects, sample U, or treat all the objects from U as the reference ones.

4   Similarity Learning

The method proposed in this paper is designed for the following learning task. Given: (i) the pairwise fuzzy information base **B**, (ii) the family $\mathcal{D}$ of decision classes, implying subproblems π_X, $X \in \mathcal{D}$, (iii) the membership functions

$\mu_X : U \to [0,1]$, $X \in \mathcal{D}$, and (iv) the sets of reference objects $U_X^R \subseteq U$, $X \in \mathcal{D}$, learn, for each subproblem π_X, set of decision rules

$$R_X = \bigcup_{x \in U_X^R} R_X(x), \qquad (3)$$

where $R_X(x)$ is the set of rules describing membership of an object $y \in U$ to class $X \in \mathcal{D}$ based on similarity of y to reference object $x \in U_X^R$.

5   Comprehensive Closeness of Objects

Given a decision class X being a fuzzy set in U, we define two kinds of binary *comprehensive closeness relations* on U:

$$y \underset{\sim\alpha,\beta}{\succ}^X x \Leftrightarrow \mu_X(x) \in [\alpha,\beta] \text{ and } \mu_X(y) \in [\alpha,\beta], \qquad (4)$$

$$y \underset{\sim\alpha,\beta}{\prec}^X x \Leftrightarrow \mu_X(x) \in [\alpha,\beta] \text{ and } \mu_X(y) \notin (\alpha,\beta), \qquad (5)$$

where $y, x \in U$ and $-\delta \le \alpha \le \beta \le 1+\delta$, where $\delta \in \mathbb{R}_+$ is any fixed positive value (a technical parameter, e.g., 0.01). When $y \underset{\sim\alpha,\beta}{\succ}^X x$, then $\alpha \le \mu_X(y) \le \mu_X(x) \le \beta$ or $\alpha \le \mu_X(x) \le \mu_X(y) \le \beta$, i.e., looking from the perspective of y, $\mu_X(y)$ is on the left side of $\mu_X(x)$ but not farther than α, or $\mu_X(y)$ is on the right side of $\mu_X(x)$ but not farther than β. When $y \underset{\sim\alpha,\beta}{\prec}^X x$, then $\mu_X(y)$ is on the left side of $\mu_X(x)$ but not closer than α, or $\mu_X(y)$ is on the right side of $\mu_X(x)$ but not closer than β. Thus, α and β play roles of limiting levels of membership to X.

The "special" values $-\delta$ and $1+\delta$, where $\delta \in \mathbb{R}_+$, are considered in (5) to allow, respectively, $\mu_X(y) \notin (-\delta,\beta)$ (i.e., $\mu_X(y) \ge \beta$) and $\mu_X(y) \notin (\alpha, 1+\delta)$ (i.e., $\mu_X(y) \le \alpha$). This is crucial, e.g., when X is crisp – one can then consider two meaningful relations $\underset{\sim 0,1+\delta}{\prec}^X$ and $\underset{\sim -\delta,1}{\prec}^X$, composed of pairs $(y,x) \in U \times U$ such that $\mu_X(y) \le 0$ and $\mu_X(y) \ge 1$, respectively.

Let us observe that $\underset{\sim\alpha,\beta}{\succ}^R$ is reflexive, symmetric and transitive and thus it is an equivalence relation. Moreover, $\underset{\sim\alpha,\beta}{\prec}^X$ is only transitive.

Given a class X and a reference object $x \in U_X^R$, we are interested in characterizing, in terms of similarity-based decision rules, the objects $y \in U$ being in:

- $\underset{\sim\alpha,\beta}{\succ}^X$ relation with x, where $\alpha, \beta \in V_{\mu_X}$,
- $\underset{\sim\alpha,\beta}{\prec}^X$ relation with x, where $\alpha, \beta \in V_{\mu_X} \cup \{-\delta\} \cup \{1+\delta\})$, $\alpha < \mu_X(x) < \beta$.

Let $V_{\mu_X}^\delta = V_{\mu_X} \cup \{-\delta\} \cup \{1+\delta\}$, where $\delta \in \mathbb{R}_+$. We define two types of sets:

$$S(\underset{\sim\alpha,\beta}{\succ}^X, x) = \{y \in U : y \underset{\sim\alpha,\beta}{\succ}^X x\}, \quad \text{where } \alpha, \beta \in V_{\mu_X}, \alpha \le \mu_X(x) \le \beta, \qquad (6)$$

$$S(\underset{\sim\alpha,\beta}{\prec}^X, x) = \{y \in U : y \underset{\sim\alpha,\beta}{\prec}^X x\}, \quad \text{where } \alpha, \beta \in V_{\mu_X}^\delta, \alpha < \mu_X(x) < \beta. \qquad (7)$$

The strict constraint $\alpha < \mu_X(x) < \beta$ in (7) prevents from considering not meaningful sets $S(\underset{\sim\alpha,\beta}{\prec}^X, x)$ [13]. From this point of view, it is crucial that when $\mu_X(x) = 0$ (or $\mu_X(x) = 1$), one can take $\alpha = -\delta$ (or $\beta = 1+\delta$, respectively).

The sets of objects defined by (6) and (7) are to be approximated using dominance cones in the similarity space created by functions $\sigma_{f_1}, \dots, \sigma_{f_n}$.

6   Rough Approximation by Dominance Relation

Let us define the dominance relation w.r.t. the similarity to an object $x \in U$, called in short x-*dominance relation*, defined over U, and denoted by D_x. For any $x, y, w \in U$, y is said to x-*dominate* w (denotation yD_xw) if for every $f_i \in F$,

$$\sigma_{f_i}(y, x) \geq \sigma_{f_i}(w, x). \tag{8}$$

Thus, object y is said to x-dominate object w iff for every feature $f_i \in F$, y is at least as similar to x as w is.

Given an object $y \in U$, x-*positive* and x-*negative dominance cones* of y in the similarity space are defined as follows:

$$D_x^+(y) = \{w \in U : wD_xy\}, \tag{9}$$

$$D_x^-(y) = \{w \in U : yD_xw\}. \tag{10}$$

In order to induce meaningful certain and possible decision rules concerning similarity to a reference object $x \in U_X^R$, we structure the objects $y \in U$ by calculation of lower and upper approximations of sets $S(\succsim_{\alpha,\beta}^X, x)$ and $S(\precsim_{\alpha,\beta}^X, x)$.

The *lower approximations* of sets $S(\succsim_{\alpha,\beta}^X, x)$ and $S(\precsim_{\alpha,\beta}^X, x)$ are defined as:

$$\underline{S(\succsim_{\alpha,\beta}^X, x)} = \{y \in U : D_x^+(y) \subseteq S(\succsim_{\alpha,\beta}^X, x)\}, \tag{11}$$

$$\underline{S(\precsim_{\alpha,\beta}^X, x)} = \{y \in U : D_x^-(y) \subseteq S(\precsim_{\alpha,\beta}^X, x)\}, \tag{12}$$

and the *upper approximations* of sets $S(\succsim_{\alpha,\beta}^X, x)$ and $S(\precsim_{\alpha,\beta}^X, x)$ are defined as:

$$\overline{S(\succsim_{\alpha,\beta}^X, x)} = \{y \in U : D_x^-(y) \cap S(\succsim_{\alpha,\beta}^X, x) \neq \emptyset\}, \tag{13}$$

$$\overline{S(\precsim_{\alpha,\beta}^X, x)} = \{y \in U : D_x^+(y) \cap S(\precsim_{\alpha,\beta}^X, x) \neq \emptyset\}. \tag{14}$$

With respect to the three basic properties of set approximations defined for rough sets in [10], it follows from definitions (6), (7), (11), (12), (13), and (14), that lower and upper approximations defined above fulfill properties of *rough inclusion* and *monotonicity of the accuracy of approximation*. Moreover, these approximations enjoy also *complementarity* property, as shown in [14].

Using (11), (12), (13), and (14), one can define the *boundary* of set $S(\succsim_{\alpha,\beta}^X, x)$ (or set $S(\precsim_{\alpha,\beta}^X, x)$), as the difference between its upper and lower approximation. It is also possible to perform further DRSA-like analysis by calculating the quality of approximation, reducts, and the core (see, e.g., [4,11,12]).

7   Induction of Decision Rules

Lower (or upper) approximations of considered sets $S(\succsim_{\alpha,\beta}^X, x)$ and $S(\precsim_{\alpha,\beta}^X, x)$ are the basis for induction of *certain* (or *possible*) *decision rules* belonging to set $R_X(x)$, $x \in U_X^R$. We distinguish two types of rules:

(1) *at least rules*:
     if $\sigma_{f_{i1}}(y,x) \geq h_{i1} \ldots$ and $\sigma_{f_{ip}}(y,x) \geq h_{ip}$, then certainly (or possibly) $y \succsim_{\alpha,\beta}^{X} x$,
(2) *at most rules*:
     if $\sigma_{f_{i1}}(y,x) \leq h_{i1} \ldots$ and $\sigma_{f_{ip}}(y,x) \leq h_{ip}$, then certainly (or possibly) $y \precsim_{\alpha,\beta}^{X} x$,

where $\{f_{i1},\ldots,f_{ip}\} \subseteq F$, $h_{i1},\ldots,h_{ip} \in [0,1]$, and α,β satisfy $0 \leq \alpha \leq \mu_X(x) \leq \beta \leq 1$ in case of at least rules, and $-\delta \leq \alpha < \mu_X(x) < \beta \leq 1 + \delta$ in case of at most rules, $\delta \in \mathbb{R}_+$.

Remark that according to Definitions (4) and (5), the decision part of the rule of type (1) and (2) can be rewritten, respectively, as:

(1) "then certainly (or possibly) $\mu_X(y) \in [\alpha,\beta]$", i.e., the conclusion is that the membership of object y to decision class X is inside the interval $[\alpha,\beta]$,
(2) "then certainly (or possibly) $\mu_X(y) \notin (\alpha,\beta)$", i.e., the conclusion is that the membership of object y to decision class X is outside the interval (α,β).

A certain rule of type (1) is read as: "if similarity of object y to reference object x w.r.t. feature f_{i1} is at least $h_{i1} \ldots$ and similarity of y to x w.r.t. feature f_{ip} is at least h_{ip}, then certainly y belongs to class X with credibility between α and β. A possible rule of type (2) is read as: "if similarity of object y to reference object x w.r.t. feature f_{i1} is at most $h_{i1} \ldots$ and similarity of y to x w.r.t. feature f_{ip} is at most h_{ip}, then possibly y belongs to class X with credibility at most α or at least β.

Decision rules of type (1) and (2) can be induced using the VC-DomLEM algorithm [2]. On one hand, these rules reveal similarity-based patterns present in the training data. On the other hand, set $R_X = \bigcup_{x \in U_X^R} R_X(x)$ of induced certain/possible rules can be applied to *classify* new objects (new cases).

8   Application of Decision Rules

The rules from R_X can be applied to a new object z, described in terms of features $f_1,\ldots,f_n \in F$, to predict its degree of membership to class X. Then, the rules covering z may give an ambiguous classification suggestion (intervals of μ_X instead of a crisp value). In order to resolve this ambiguity, we adapt and revise the rule classification scheme described in [1]. In this way, one can obtain a precise (crisp) value of membership $\mu_X(z)$. Let us consider three situations, assuming that $Cov_z \subseteq R_X$ denotes the set of rules covering object z, $Cond_\rho \subseteq U$ denotes the set of objects covered by rule ρ, $U_X^t = \{y \in U : \mu_X(y) = t\}$, and $|\cdot|$ denotes cardinality of a set.

Situation (i). No rule from R_X covers object z (i.e., $Cov_z = \emptyset$), so there is no reliable suggestion concerning $\mu_X(z)$. If a concrete answer is expected, one can suggest that $\mu_X(z)$ equals to the most frequent value $\mu_X(y)$, where $y \in U$.

Situation (ii). Exactly one rule $\rho \in R_X(x) \subseteq R_X$, $x \in U_X^R$, covers object z (i.e., $|Cov_z| = 1$). Then, we calculate value $Score_X^\rho(t,z)$ for each membership value $t \in V_{\mu_X}$ covered by the decision part of this rule:

$$Score_X^\rho(t,z) = \frac{|Cond_\rho \cap U_X^t|^2}{|Cond_\rho||U_X^t|}. \tag{15}$$

Then, $\mu_X(z)$ is calculated as $\mu_X(z) = \max_t Score_X^\rho(t,z)$. Let us observe that $Score_X^\rho(t,z) \in [0,1]$. It can be interpreted as the degree of certainty of the suggestion that $\mu_X(z)$ equals t.

Situation (iii). Several rules from R_X cover object z (i.e., $|Cov_z| > 1$). Then, we calculate value $Score_X(t,z)$ for each $t \in V_{\mu_X}$ covered by the decision part of any of the covering rules:

$$Score_X(t,z) = Score_X^+(t,z) - Score_X^-(t,z), \tag{16}$$

where $Score_X^+(t,z)$ and $Score_X^-(t,z)$ represent the positive and negative part of $Score_X(t,z)$, respectively. $Score_X^+(t,z)$ takes into account rules $\rho_1,\ldots,\rho_k \in Cov_z$ whose decision part covers t:

$$Score_X^+(t,z) = \frac{|(Cond_{\rho_1} \cap U_X^t) \cup \ldots \cup (Cond_{\rho_k} \cap U_X^t)|^2}{|Cond_{\rho_1} \cup \ldots \cup Cond_{\rho_k}||U_X^t|}. \tag{17}$$

Let us observe that $Score_X^+(t,z) \in [0,1]$. $Score_X^-(t,z)$ takes into account the rules $\rho_{k+1},\ldots,\rho_h \in Cov_z$ whose decision part does not cover t. If there is no such rule, then $Score_X^-(t,z) = 0$. Otherwise:

$$Score_X^-(t,z) = \frac{|(Cond_{\rho_{k+1}} \cap U_X^{\rho_{k+1}}) \cup \ldots \cup (Cond_{\rho_h} \cap U_X^{\rho_h})|^2}{|Cond_{\rho_{k+1}} \cup \ldots \cup Cond_{\rho_h}||U_X^{\rho_{k+1}} \cup \ldots \cup U_X^{\rho_h}|}, \tag{18}$$

where U_X^ρ is subset of U containing objects whose membership to class X is covered by the decision part of rule ρ. Let us observe that $Score_X^-(t,z) \in [0,1]$.

After calculating $Score_X(t,z)$ for all considered values of t, we take $\mu_X(z) = \max_t Score_X(t,z)$. It can be interpreted as a net balance of the arguments in favor and against the suggestion "the membership of object z to class X equals t".

9   Illustrative Example

Let us consider set U composed of five objects described by two features: f_1, with value set $[0,8]$, and f_2, with value set $[0,1]$. Moreover, let us consider decision class X, with membership function μ_X. The five objects are presented in Fig. 1.

We assume that object x is a reference object, and that there are given two marginal similarity functions σ_{f_1}, σ_{f_2} defined as:

$$\sigma_{f_i}(y,x) = 1 - \frac{|f_i(y) - f_i(x)|}{f_i^{max} - f_i^{min}},$$

where $i = 1,2$, and f_i^{max}, f_i^{min} denote max and min value in the value set of f_i.

Functions σ_{f_1} and σ_{f_2} create a 2-dimensional similarity space. Figure 2 shows pairs of objects $(\cdot, x)$ in this space.

First, using (9) and (10), we calculate x-positive and x-negative dominance cones in the similarity space. Two such cones are shown in Fig. 1 and in Fig. 2.

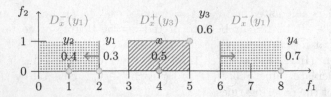

Fig. 1. Set of objects considered in the illustrative example; the number below an object id denotes the value of function μ_X for this object; the hatched area corresponds to dominance cone $D_x^+(y_3)$, and the two dotted areas (one for $f_1(y) \leq 2$, and the other for $f_1(y) \geq 6$) correspond to dominance cone $D_x^-(y_1)$

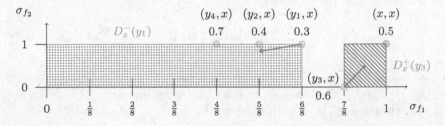

Fig. 2. Pairs of objects $(\cdot, x)$ in the similarity space created by σ_{f_1} and σ_{f_2}; the number below a pair of object ids denotes the value of function μ_X for the object whose id is the first in the pair; the hatched area corresponds to dominance cone $D_x^+(y_3)$, and the dotted area corresponds to dominance cone $D_x^-(y_1)$

Second, we calculate sets of objects $S(\succsim_{\alpha,\beta}^X, x)$ according to (6), for $\alpha \in \{0.3, 0.4, 0.5\}$ and $\beta \in \{0.5, 0.6, 0.7\}$. Moreover, we calculate $S(\precsim_{\alpha,\beta}^X, x)$ according to (7), for $\alpha \in \{-\delta, 0.3, 0.4\}$ and $\beta \in \{0.6, 0.7, 1 + \delta\}$, where $\delta \in \mathbb{R}_+$.

Third, sets $S(\succsim_{\alpha,\beta}^X, x)$ and $S(\precsim_{\alpha,\beta}^X, x)$ are approximated using the x-positive and x-negative dominance cones in the similarity space – see Table 2.

One can observe several inconsistencies w.r.t. the x-dominance relation in the similarity space. e.g., objects $y_2, y_4 \in S(\succsim_{0.4,0.7}^X, x)$ are inconsistent since they are x-dominated by object y_1, and $y_1 \notin S(\succsim_{0.4,0.7}^X, x)$ (because $\mu_X(y_1) = 0.3$).

Table 3 presents minimal decision rules induced by VC-DomLEM algorithm from the non-empty lower and upper approximations shown in Table 2.

Example of application of induced decision rules. Consider a new object z such that $f_1(z) = 5.5$, $f_2(z) = 0.5$, and thus, $\sigma_{f_1}(z, x) = 6.5/8$, $\sigma_{f_2}(z, x) = 1/2$. Object z is covered by rules ρ_2 and ρ_{15}, suggesting $\mu_X(z) \in [0.3, 0.6]$ and $\mu_X(z) \notin (0.4, 0.6)$, respectively. Applying (17), (18), and (16) for each membership degree $t \in \{0.3, 0.4, 0.5, 0.6, 0.7\}$, we get the result shown in Table 4. Consequently, one can conclude that $\mu_X(z)$ is equal to 0.3, 0.4, or 0.6.

Table 2. Approximations of sets $S(\succsim^X_{\alpha,\beta}, x)$, $S(\precsim^X_{\alpha,\beta}, x)$; $\delta \in \mathbb{R}_+$; objects struck through belong to respective set but not to its lower approximation; underlined objects do not belong to respective set but belong to its upper approximation

$S(\succsim^X_{\alpha,\beta}, x)$	$\beta = 0.5$	$\beta = 0.6$	$\beta = 0.7$	$S(\precsim^X_{\alpha,\beta}, x)$	$\beta = 0.6$	$\beta = 0.7$	$\beta = 1 + \delta$
$\alpha = 0.3$	$\{y_1, y_2, x\}$	$\{y_1, y_2, y_3, x\}$	U	$\alpha = -\delta$	$\{y_3, y_4\}$	$\{y_4\}$	$\emptyset$
$\alpha = 0.4$	$\{\cancel{y_2}, x\}$	$\{\cancel{y_2}, y_3, x\}$	$\{\cancel{y_2}, y_3, \cancel{y_4}, x\}$	$\alpha = 0.3$	$\{\cancel{y_1}, y_3, y_4\}$	$\{\cancel{y_1}, y_4\}$	$\{\cancel{y_1}\}$
$\alpha = 0.5$	$\{x\}$	$\{y_3, x\}$	$\{y_3, \cancel{y_4}, x\}$	$\alpha = 0.4$	$\{y_1, y_2, y_3, y_4\}$	$\{y_1, y_2, y_4\}$	$\{\cancel{y_1}, \cancel{y_2}\}$
$S(\succsim^X_{\alpha,\beta}, x)$	$\beta = 0.5$	$\beta = 0.6$	$\beta = 0.7$	$S(\precsim^X_{\alpha,\beta}, x)$	$\beta = 0.6$	$\beta = 0.7$	$\beta = 1 + \delta$
$\alpha = 0.3$	$\{y_1, y_2, x\}$	$\{y_1, y_2, y_3, x\}$	U	$\alpha = -\delta$	$\{y_3, y_4\}$	$\{y_4\}$	$\emptyset$
$\alpha = 0.4$	$\{y_1, y_2, x\}$	$\{y_1, y_2, y_3, x\}$	$\{y_1, y_2, y_3, y_4, x\}$	$\alpha = 0.3$	$\{y_1, y_2, y_3, y_4\}$	$\{y_1, y_2, y_4\}$	$\{y_1, y_2, \underline{y_4}\}$
$\alpha = 0.5$	$\{x\}$	$\{y_3, x\}$	$\{y_1, y_2, y_3, y_4, x\}$	$\alpha = 0.4$	$\{y_1, y_2, y_3, y_4\}$	$\{y_1, y_2, y_4\}$	$\{y_1, y_2, y_4\}$

Table 3. Rules induced for referent x such that $f_1(x) = 4$, $f_2(x) = 0$; 'Supp.' ('¬ Supp.') presents ids of objects supporting a rule (covered by a rule but not supporting it); 8 possible rules, identical (except for "possibly") to respective certain rules, are skipped

Id	Decision rule	Supp.	¬ Supp.
ρ_1	if $\sigma_{f_1}(y, x) \geq \frac{5}{8}$ and $\sigma_{f_2}(y, x) \geq 1$, then certainly $\mu_X(y) \in [0.3, 0.5]$	$\{y_1, y_2, x\}$	
ρ_2	if $\sigma_{f_1}(y, x) \geq \frac{5}{8}$, then certainly $\mu_X(y) \in [0.3, 0.6]$	$\{y_1, y_2, y_3, x\}$	
ρ_3	if $\sigma_{f_1}(y, x) \geq 1$, then certainly $\mu_X(y) \in [0.5, 0.5]$	$\{x\}$	
ρ_4	if $\sigma_{f_1}(y, x) \geq \frac{7}{8}$, then certainly $\mu_X(y) \in [0.5, 0.6]$	$\{y_3, x\}$	
ρ_7	if $\sigma_{f_1}(y, x) \geq \frac{5}{8}$ and $\sigma_{f_2}(y, x) \geq 1$, then possibly $\mu_X(y) \in [0.4, 0.5]$	$\{y_2, x\}$	$\{y_1\}$
ρ_8	if $\sigma_{f_1}(y, x) \geq \frac{5}{8}$, then possibly $\mu_X(y) \in [0.4, 0.6]$	$\{y_2, y_3, x\}$	$\{y_1\}$
ρ_9	if $\sigma_{f_1}(y, x) \geq \frac{4}{8}$, then possibly $\mu_X(y) \in [0.4, 0.7]$	$\{y_2, y_3, y_4, x\}$	$\{y_1\}$
ρ_{12}	if $\sigma_{f_1}(y, x) \geq \frac{4}{8}$, then possibly $\mu_X(y) \in [0.5, 0.7]$	$\{y_3, y_4, x\}$	$\{y_1, y_2\}$
ρ_{13}	if $\sigma_{f_2}(y, x) \leq 0$, then certainly $\mu_X(y) \geq 0.6$	$\{y_3\}$	
ρ_{14}	if $\sigma_{f_1}(y, x) \leq \frac{4}{8}$, then certainly $\mu_X(y) \geq 0.7$	$\{y_4\}$	
ρ_{15}	if $\sigma_{f_1}(y, x) \leq \frac{7}{8}$, then certainly $\mu_X(y) \notin (0.4, 0.6)$	$\{y_1, y_2, y_3, y_4\}$	
ρ_{16}	if $\sigma_{f_1}(y, x) \leq \frac{6}{8}$, then certainly $\mu_X(y) \notin (0.4, 0.7)$	$\{y_1, y_2, y_4\}$	
ρ_{19}	if $\sigma_{f_1}(y, x) \leq \frac{7}{8}$, then possibly $\mu_X(y) \notin (0.3, 0.6)$	$\{y_1, y_3, y_4\}$	$\{y_2\}$
ρ_{20}	if $\sigma_{f_1}(y, x) \leq \frac{6}{8}$, then possibly $\mu_X(y) \notin (0.3, 0.7)$	$\{y_1, y_4\}$	$\{y_2\}$
ρ_{21}	if $\sigma_{f_1}(y, x) \leq \frac{6}{8}$, then possibly $\mu_X(y) \leq 0.3$	$\{y_1\}$	$\{y_2, y_4\}$
ρ_{24}	if $\sigma_{f_1}(y, x) \leq \frac{6}{8}$, then possibly $\mu_X(y) \leq 0.4$	$\{y_1, y_2\}$	$\{y_4\}$

Table 4. Scores of a new object z resulting from application of induced decision rules

$t \in V_{\mu_X}$	0.3	0.4	0.5	0.6	0.7
$Score^+_X(t, z)$	$\frac{1}{5}$	$\frac{1}{5}$	$\frac{1}{4}$	$\frac{1}{5}$	$\frac{1}{4}$
$Score^-_X(t, z)$	0	0	1	0	1
$Score_X(t, z)$	$\frac{1}{5}$	$\frac{1}{5}$	$-\frac{3}{4}$	$\frac{1}{5}$	$-\frac{3}{4}$

10 Conclusions

We presented a method of similarity-based classification using the Dominance-based Rough Set Approach. This method exploits only ordinal properties of marginal similarity functions and membership functions of decision classes. It avoids arbitrary aggregation of marginal similarities into one comprehensive similarity. Instead, it uses a rule-based similarity model employing rules describing monotonic relationship between comprehensive similarity of objects and their similarities with respect to single features. Thus, our case-based reasoning approach is as much "neutral" and "objective" as possible. Moreover, our method provides more insight when determining membership of a new object z to class X – one can see the rules matching z and the objects supporting these rules.

Acknowledgments. The first author wishes to acknowledge financial support from the Faculty of Computing at Poznań University of Technology, grant 09/91/DSMK/0609.

References

1. Błaszczyński, J., Greco, S., Słowiński, R.: Multi-criteria classification - a new scheme for application of dominance-based decision rules. Eur. J. Oper. Res. **181**(3), 1030–1044 (2007)
2. Błaszczyński, J., Słowiński, R., Szeląg, M.: Sequential covering rule induction algorithm for variable consistency rough set approaches. INS **181**, 987–1002 (2011)
3. Gilboa, I., Schmeidler, D.: A Theory of Case-Based Decisions. Cambridge University Press, Cambridge (2001). No. 9780521003117 in Cambridge Books
4. Greco, S., Matarazzo, B., Słowiński, R.: Rough sets theory for multicriteria decision analysis. Eur. J. Oper. Res. **129**(1), 1–47 (2001)
5. Greco, S., Matarazzo, B., Słowiński, R.: Dominance-based rough set approach to case-based reasoning. In: Torra, V., Narukawa, Y., Valls, A., Domingo-Ferrer, J. (eds.) MDAI 2006. LNCS (LNAI), vol. 3885, pp. 7–18. Springer, Heidelberg (2006)
6. Greco, S., Matarazzo, B., Słowiński, R.: Case-based reasoning using gradual rules induced from dominance-based rough approximations. In: Wang, G., Li, T., Grzymala-Busse, J.W., Miao, D., Skowron, A., Yao, Y. (eds.) RSKT 2008. LNCS (LNAI), vol. 5009, pp. 268–275. Springer, Heidelberg (2008)
7. Greco, S., Matarazzo, B., Słowiński, R.: Granular computing for reasoning about ordered data: the dominance-based rough set approach. In: Pedrycz, W., et al. (eds.) Handbook of Granular Computing, chap. 15, pp. 347–373. Wiley (2008)
8. Hume, D.: An Enquiry Concerning Human Understanding. Clarendon Press, Oxford (1748)
9. Kolodner, J.: Case-Based Reasoning. Morgan Kaufmann, San Mateo (1993)
10. Pawlak, Z.: Rough Sets: Theoretical Aspects of Reasoning about Data. Kluwer, Dordrecht (1991)
11. Słowiński, R., Greco, S., Matarazzo, B.: Rough sets in decision making. In: Meyers, R.A. (ed.) Encyclopedia of Complexity and Systems Science, pp. 7753–7786. Springer, New York (2009)
12. Słowiński, R., Greco, S., Matarazzo, B.: Rough set based decision support. In: Burke, E.K., Kendall, G. (eds.) Search Methodologies: Introductory Tutorials in Optimization and Decision Support Techniques, pp. 557–609. Springer, New York (2014)
13. Szeląg, M.: Application of the dominance-based rough set approach to ranking and similarity-based classification problems. Ph.D. thesis, Poznań University Of Technology (2015). http://www.cs.put.poznan.pl/mszelag/Research/MSzPhD.pdf
14. Szeląg, M., Greco, S., Błaszczyński, J., Słowiński, R.: Case-based reasoning using dominance-based decision rules. In: Yao, J.T., Ramanna, S., Wang, G., Suraj, Z. (eds.) RSKT 2011. LNCS, vol. 6954, pp. 404–413. Springer, Heidelberg (2011)
15. Zadeh, L.: Fuzzy sets. Inf. Control **8**(3), 338–353 (1965)

Representative-Based Active Learning
with Max-Min Distance

Fu-Lun Liu, Fan Min[(✉)], Liu-Ying Wen, and Hong-Jie Wang

School of Computer Science, Southwest Petroleum University,
Chengdu 610500, China
minfanphd@163.com

Abstract. Active learning has been a hot topic because labeled data
are useful, however expensive. Many existing approaches are based on
decision trees, Naïve Bayes algorithms, etc. In this paper, we propose
a representative-based active learning algorithm with max-min distance.
Our algorithm has two techniques interacting with each other. One is the
representative-based classification inspired by covering-based neighborhood rough sets. The other is critical instance selection with max-min
distance. Experimental results on six UCI datasets indicate that, with
the same number of labeled instances, our algorithm is comparable with
or better than the ID3, C4.5 and Naïve Bayes algorithms.

Keywords: Active learning · Classifier · Distance · Representative ·
Similarity

1 Introduction

Active learning is an algorithm that can perform better with less training if
it is allowed to choose the data from which it learns [1]. It is a special case of
semi-supervised learning [2] in which a learning algorithm is able to interactively
query the users to obtain the desired outputs at new data points. In addition
to classification, informative active learning is also employed in collaborative
filtering [3] and ranking [4]. The basic idea of active learning is that a number
of labeled data are necessary to build a trustworthy classifier. However, labeled
data are scarce or expensive in many applications. Therefore we should choose
the data to be labeled deliberately. This issue will be called critical instance
selection. Another issue for active learning is classification. Currently, decision
trees such as ID3 [5], C4.5 [6], and Naïve Bayes algorithms [7] are often employed
(see, e.g., [8–10]).

Recently, a representative-based classifier [11] was proposed to take advantages of both lazy and model-based learning. This classifier is inspired by
covering-based neighborhood rough sets [12–14]. In the training stage, representatives are selected, and their neighborhood thresholds are computed. In
the testing stage, the distances between an unlabeled instance and existing
representatives are computed. The closest representatives determine the class

© Springer International Publishing AG 2016
V. Flores et al. (Eds.): IJCRS 2016, LNAI 9920, pp. 365–375, 2016.
DOI: 10.1007/978-3-319-47160-0_33

of the unlabeled instance. With this classifier, representatives and outliers are quite clear. Consequently, it is an ideal foundation for active learning. Though some researches (see, e.g., [15,16]) focus on active learning with representatives, there is a significant difference between theirs and ours. They find the valuable instances as representatives, however we use these existing representatives.

In this paper, we propose a representative-based active learning algorithm. For the classification issue, we adopt the representative-based classifier (RC) [17]. For the critical instance selection issue, we propose the max-min distance and design the selection technique. First, we define the distance between an unlabeled instance and a representative. It is determined by the similarity and the threshold of the representative. A positive distance means the unlabeled instance is out of the neighborhood of the representative. Second, an unlabeled instance is defined as an outlier if all its distances are positive. We record the minimum distance of each outlier. The critical instance is the outlier with the maximum minimum distance. Third, we obtain the class label of critical instance directly, and move it from testing set to training set. This operation will be repeated until the given number of instances are selected. Classification and active learning are interdependent. The classifier determines the critical instance, while the critical instance influences the classifier construction.

Two sets of experiments are undertaken on six UCI datasets [18] including Mushroom, Voting, Zoo, Wine, Tic-tac-toe and Dermatology. One is to compare with the original representative-based classification algorithm. Our algorithm has significantly higher accuracies on most datasets. The other one is to compare with other active learners based on ID3, C4.5, and Naïve Bayes. Our algorithm generally outperforms the counterparts with the same number of labeled instances.

2   Preliminaries

This section reviews some basic knowledge about decision systems. Some concepts such as similarity, neighborhood, and indiscernibility are also discussed.

Decision systems [19] are fundamental for classification through covering-based neighborhood rough sets.

Definition 1. *A decision system is a 3-tuple:*

$$S = (U, C, d), \tag{1}$$

where U is a finite set of instances called the universe, C is the set of conditional attributes, d is the decision attribute.

We only consider symbolic decision system. Moreover, there is no cost information such as misclassification cost [20] and test cost [21].

The semantic interpretation of weak indiscernibility relation [22] is that two instances are considered indistinguishable if they have the same values on at least one attribute in C.

Definition 2. *Let S be a decision system, the weak indiscernibility relation $WIND(C)$ is a relation on $U \times U$ defined as*

$$WIND(C) = \{(x,y) \in U \times U | \exists a \in C, st. a(x) = a(y)\} \qquad (2)$$

Definition 3. *Let S be a decision system, the similarity between $x, y \in U$ with respect to $\varnothing \subset A \subseteq C$ is*

$$sim(x, y, A) = \frac{sam(x, y, A)}{|A|}, \qquad (3)$$

where

$$sam(x, y, A) = |\{a \in A | a(x) = a(y)\}|. \qquad (4)$$

When $A = C$, we also denote $sim(x, y, A)$ as $sim(x, y)$. Note that this concept is the same as the measure for the quantitative indiscernibility relation.

The core of standard rough set theory is the notion of a partition, where equivalence classes cover the universe and are disjoint. If we remove this second requirement, we obtain a covering.

Definition 4. *The neighborhood of $x \in U$ with similarity θ is:*

$$n(x, \theta) = \{y \in U | sim(x, y) \geq \theta\}. \qquad (5)$$

Definition 5. *Let S be a decision system, and $U/\{d\} = X_1, X_2, ..., X_k$. The minimal similarity threshold value for $x \in X_i$ is:*

$$\theta_x^+ = \min\{\theta | n(x, \theta) \subseteq X_i\}. \qquad (6)$$

Now, θ_x^+ is computed by decision system and x itself. Meanwhile, instances in this neighborhood which determined by θ_x^+ should be consistent with x.

Definition 6. *The positive neighborhood of $x \in U$ is:*

$$n^+(x) = n(x, \theta_x^+). \qquad (7)$$

Indeed, the similarity threshold of each instance in a decision system will be computed. After all instances similarity thresholds are computed, their neighborhoods are determined too. These instances are used to represent their neighbors.

3   The Proposed Algorithm

In this section, we firstly define the active learning problem. Our algorithm framework for general active learning is presented secondly. Finally the instance selection with max-min distance approach is discussed in detail.

3.1 Problem Definition

The active learning problem can be defined as follows.

Problem 1. **Active learning**

Input: A decision system $S = (U, C, d)$, the set of labeled instances $U_r \subset U$, the number of additional labels m.

Output: New labeled instance set U_r', a classifier, and the classification accuracy on the unlabeled set.

Optimization objective: Maximize the classification precision.

3.2 Algorithm Framework

The algorithm framework is illustrated in Fig. 1. It is applicable to general active learning algorithms. When we build the original classifier, only instances in U_r are considered since they are already labeled. When this classifier is applied for classification, we could actively select a critical instance and obtain its class label. Therefore we have more labeled instances. Sometimes, the added labeled instance may update the classifier. We will continue to select critical instance until the termination condition is satisfied, and then the final classification accuracy could be calculated. In this paper, the termination condition is decided by a given condition. For instance, when the number of labeled instances $|U_r|$ is equal to the given number m, the termination condition is fulfilled.

3.3 Classifier Building

We revise the classifier proposed in paper [17]. As illustrated by Fig. 2, classifier construction has two subtasks. The similarity between each pair of instances could be computed by Eq. (3). Next, the similarity threshold θ^+ and neighborhood of each instance can be computed by Eqs. (6) and (7). As shown in Fig. 2(a), in order to determine the similarity threshold of x_9, we obtained the similarity relationship between x_9 and other instances. With decreased similarity threshold, some similar instances will become the neighbors of x_9. Until a contradictory instance has become a neighbor of x_9, e.g., x_5 and x_6, threshold decrease will

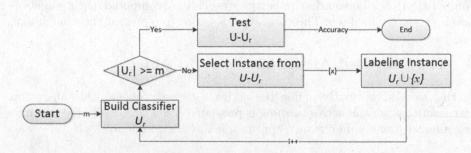

Fig. 1. The active learning framework

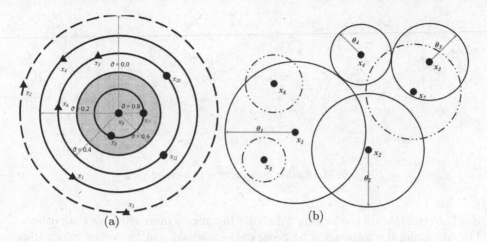

Fig. 2. Classifier building

be stopped. In this way, the neighborhood of each instance in the training set is computed. A neighborhood could contain a lot of instances, an instance may belong to several neighborhoods. Furthermore, a neighborhood may be a subset of a larger neighborhood or the union of some other neighborhoods. Thus, we will remove these redundant neighborhoods by greedy strategy. Figure 2(b) indicates the process of redundant neighborhoods reduction. The core instances of these remain neighborhoods are called representatives.

3.4   Instance Selection with Max-Min Distance

The training set is usually a small part of the universe, hence selected representatives could not represent the whole universe. For classification, these unlabeled instances will be distributed around these representatives. Figure 3 illustrates 3 main distributions. Let R be the set of representatives. The distance between instance x and a representative $r \in R$ is:

$$distance(x, r) = \frac{1}{sim(x, r)} - \frac{1}{\theta_x^+}. \tag{8}$$

Though each unlabeled instance has many different distances from representatives, the minimum one is more important for our algorithm. A positive minimum distance can confirm the unlabeled instance is a outlier, as illustrated in Fig. 3(c). If an unlabeled instance is not an outlier, it can be represented by respective representatives. In Fig. 3(a), x' is represented by x_1, x'' is represented by x_1 and x_2. In Fig. 3(b), x' is represented by x_1, x_2 and x_3 at the same time. A outlier with smaller distance is more likely to be classified correctly. The outlier with max-min distance is far from all representatives, hence it is not similar to any of them.

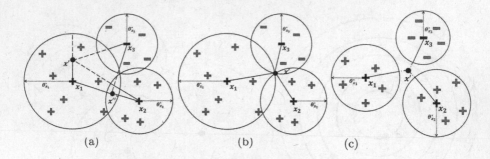

Fig. 3. Main different situations for classifying

The class label of the min-max distance instance is more valuable than others. On one hand, outlier cannot be accurately classified. On the other hand, that outlier may be selected as a representative and change the model of classifier. Therefore we obtain its class label directly.

3.5   Representative-Based Classification

In this section, we discuss how to predict the class label of an unlabeled instance x. We will compute the distance between x and each $r \in R$. The set of representatives that have the minimal distance from x is

$$X = \{r \in R | distance(x, r) = min(x, R)\}, \qquad (9)$$

where

$$min(x, R) = \min\{distance(x, r) | r \in R\}. \qquad (10)$$

Given an unlabeled instance x, representatives with the same minimum distance are called conclusive representatives. If there is only one conclusive representative, this unlabeled instance will be predicted directly. Otherwise we use standard voting with all conclusive representatives. In summarize, the predicted class label of x is:

$$d(x) = \arg \max_{1 \le i \le |vd|} |\{r \in R | d(r) = i\}|. \qquad (11)$$

With these conclusive representatives, the majority class label will assigned to the unlabeled instance.

Moreover, these classification principles can be applied in any situation, no matter the unlabeled instance is an outlier or not. Because the minimum distance is equal to the highest similarity, we consider the higher similarity could increase the probability of correct prediction. When all those unlabeled instances has been classified, we can calculate the classification accuracy.

Table 1. Description of experimental datasets

Datesets	Features	Class	Instance
Mushroom	23	2	8124
Voting	17	2	435
Wine	14	3	178
Zoo	17	8	101
Tic-tac-toe	10	3	958
Dermatoloy	34	4	366

4   Experiments

In this section, we present the experimental results to answer these following questions.

1. How does the performance of our active learning approach (ALRC) compare with original representative-based algorithm (RC)?
2. How does the performance of our active learning approach (ALRC) compare with ID3, C4.5 and Naïve Bayes algorithms?

4.1   Experimental Setup

Experiments are performed on 6 UCI datasets, namely Voting, Zoo, Mushroom, Wine, Tic-tac-toe and Dermatology [18]. Class labels of testing instances are actually invisible for us until all unlabeled instances have been classified. Because Wine is a numeric dataset, so we have to discretize it before using it. The description of these datasets is given in Table 1.

The general experiment is same for each dataset. Each dataset will be divided into two parts randomly according to the division proportion. The first one is training set and the other one is testing set. In order to ensure the scale of the training set is small, the division proportion is set 0.1. Considering the scale of Mushroom is big relatively, so we narrow the divided proportion to 0.01. During the cause of the experiment, we will record the updated accuracy when a critical instance has been selected and learned. However, the updated accuracy is useless for the next critical instance selection and classifier reconstruction. For Mushroom, if the labeled instances accounted for 5 % of universe, we terminate that experiment. We terminate the respect experiments when the labeled instances accounted for 50 % of other datasets. We undertook 10 experiments for each dataset, and computed the average accuracy.

4.2   Results

Table 2 lists the classification accuracies between ALRC and RC of each dataset. When the RC algorithm is employed, the classification accuracies of all datasets

increases as the labeled instances proportion increases. As for the ALRC algorithm, all datasets are still accordance with those above tendencies except Tic-
tac-toe. In this regard, the ALRC algorithm is unsuited for Tic-tac-toe. Moreover, the tendencies of classification accuracy about these two algorithms are
obviously different. Even a number of critical instances are more useful than multiple random instances. For example, 80 critical instances are worth 320 random
instances for classification on Mushroom. When the tendencies of classification
accuracies become smooth, the average accuracies of ALRC are almost higher
than RC. For the first question, we know that ALRC generally outperforms RC.

Table 2. Comparison between active-learning algorithm and original algorithm on
classification accuracies

Labeled	Algorithm	0.1	0.2	0.3	0.4	0.5
Wine	ALRC	89.69 ± 7.08	*96.92 ± 8.81*	98.88 ± 1.28	*99.44 ± 1.31*	*99.56 ± 1.78*
	RC	89.69 ± 7.08	94.97 ± 4.76	97.62 ± 3.12	96.26 ± 3.74	96.67 ± 3.33
Voting	ALRC	89.39 ± 4.95	*93.04 ± 2.95*	*94.67 ± 1.73*	*·96.54 ± 1.48*	*96.91 ± 1.27*
	RC	89.39 ± 4.95	91.23 ± 2.69	91.96 ± 2.81	92.09 ± 2.21	92.32 ± 2.77
Zoo	ALRC	76.15 ± 31.10	*93.83 ± 4.94*	*96.34 ± 3.66*	*98.52 ± 1.80*	*99.02 ± 2.94*
	RC	76.15 ± 31.10	85.93 ± 11.85	90.42 ± 8.73	91.80 ± 6.56	94.71 ± 5.29
Tic-tac-toe	ALRC	72.55 ± 3.72	*80.60 ± 3.94*	79.27 ± 5.20	77.79 ± 9.10	74.23 ± 5.06
	RC	72.55 ± 3.72	77.41 ± 4.26	*79.51 ± 2.46*	*82.64 ± 3.10*	*84.71 ± 4.08*
Dermatology	ALRC	83.76 ± 7.70	*89.15 ± 3.03*	93.84 ± 2.67	96.62 ± 2.48	98.55 ± 0.91
	RC	83.76 ± 7.70	88.37 ± 4.15	90.19 ± 3.76	91.44 ± 2.70	93.12 ± 2.26
Labeled	Algorithm	0.01	0.02	0.03	0.04	0.05
Mushroom	ALRC	95.62 ± 2.92	*99.25 ± 0.71*	*99.36 ± 0.83*	*99.56 ± 0.79*	*99.68 ± 0.41*
	RC	95.62 ± 2.92	97.65 ± 1.93	97.96 ± 2.19	98.85 ± 0.89	99.18 ± 0.57

Figure 4 compares the classification accuracies for different classifiers and different datasets. Through Figs. 4(a), (b) and (c), we can find the classification performances of our algorithm are significantly better than others. Meanwhile, the
tendencies of Mushroom, Voting and Zoo illustrate our algorithm can improve
the classification accuracies obviously. Figures 4(d) and (f) illustrate the classification accuracies of Wine and Dermatology, the performance of Naïve Bayes is
comparable with ours, but ID3 and C4.5 are slightly inferior. However, Fig. 4(e)
indicates ID3 and C4.5 algorithms could obtain higher classification accuracies
than our algorithm and Naïve Bayes on Tic-tac-toe. For Tic-tac-toe, the class
label is related to the values of conditional attributes and their distribution. If
we disorder the conditional attributes sequence of Tic-tac-toe, there will be a
distinct classification result. This suggests ALRC is not suited to some datasets
such as Tic-tac-toe. Overall, our algorithm is comparable with even better than
other algorithms for most datasets. For the second question, ALRC also outperforms algorithms based on ID3, C4.5 and Naïve Bayes.

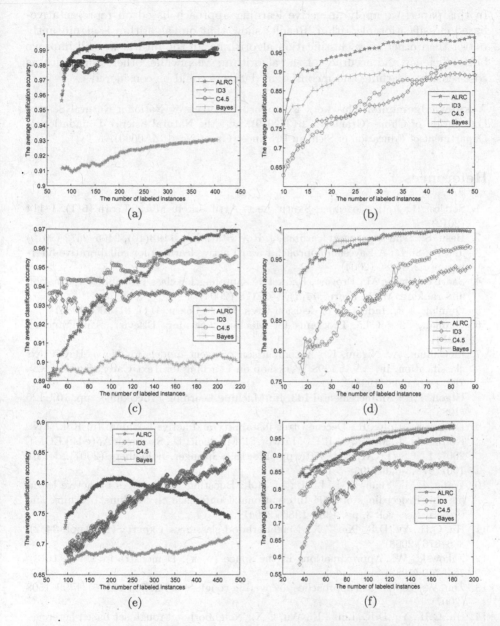

Fig. 4. Comparison of classification accuracy using ALRC, ID3, C4.5 and Bayes with **a** Mushroom, **b** Zoo, **c** Voting, **d** Wine, **e** Tic-tac-toe, **h** Dermatology

5   Conclusions and Future Work

In this paper, we apply an active learning approach based on representative-based classification algorithm. Results show that our algorithm is significantly better than other exist classification algorithms. In the future, we will improve both efficiency and accuracy of the algorithm. Meanwhile, other issues such as cost-sensitive learning and inconsistent datasets could be considered.

Acknowledgements. This work is supported in part by National Natural Science Foundation of China (Grant No. 61379089), and the Natural Science Foundation of Department of Education of Sichuan Province (Grant No. 16ZA0060).

References

1. Settles, B.: Active learning. Synth. Lect. Artif. Intell. Mach. Learn. **6**(1), 1–114 (2012)
2. Basu, S.: Semi-supervised learning. J. Roy. Stat. Soc. **6493**(10), 2465–2472 (2010)
3. Jin, R., Si, L.: A Bayesian approach toward active learning for collaborative filtering, pp. 278–285 (2004)
4. Saartsechansky, M., Provost, F.: Active sampling for class probability estimation and ranking. Mach. Learn. **54**(2), 153–178 (2004)
5. Quinlan, J.R.: Induction of decision trees. Mach. Learn. **1**(1), 81–106 (1986)
6. Quinlan, J.R.: C4.5: Programs for Machine Learning. Elsevier, San Francisco (2014)
7. McCallum, A., Nigam, K.: A comparison of event models for Naive Bayes text classification. In: AAAI 1998 Workshop on Learning for Text Categorization, vol. 752, pp. 41–48 (1998)
8. Utgoff, P.E.: An incremental Id3. In: Machine Learning Proceedings, pp. 107–120 (1988)
9. Dwyer, K., Holte, R.: Decision tree instability and active learning. In: Kok, J.N., Koronacki, J., Mantaras, R.L., Matwin, S., Mládenič, D., Skowron, A. (eds.) ECML 2007. LNCS (LNAI), vol. 4701, pp. 128–139. Springer, Heidelberg (2007). doi:10.1007/978-3-540-74958-5_15
10. Wang, L.M., Yuan, S.M., Li, L., Li, H.J.: Boosting Naïve Bayes by active learning. In: Proceedings of 2004 International Conference on Machine Learning and Cybernetics, vol. 3, pp. 1383–1386 (2004)
11. Hu, Q.H., Yu, D.R., Xie, Z.X.: Neighborhood classifiers. Expert Syst. Appl. **34**(2), 866–876 (2008)
12. Zakowski, W.: Approximations in the space (u, π). Demonstratio Math. **16**(3), 761–769 (1983)
13. Zhu, W.: Topological approaches to covering rough sets. Inf. Sci. **177**(6), 1499–1508 (2007)
14. Hu, Q.H., Yu, D.R., Liu, J.F., Wu, C.X.: Neighborhood rough set based heterogeneous feature subset selection. Inf. Sci. **178**(18), 3577–3594 (2008)
15. Xu, Z., Yu, K., Tresp, V., Xu, X., Wang, J.: Representative sampling for text classification using support vector machines. ECIR **2633**, 393–407 (2013)
16. Huang, S.J., Jin, R., Zhou, Z.H.: Active learning by querying informative and representative examples. IEEE Trans. Pattern Anal. Mach. Intell. **36**(10), 1936–1949 (2014)

17. Zhang, B.W., Min, F., Ciucci, D.: Representative-based classification through covering-based neighborhood rough sets. Appl. Intell. **43**(4), 840–854 (2015)
18. Blake, C., Merz, C.J.: UCI repository of machine learning databases (1998)
19. Ea, S., Peterson, R.: Decision Systems for Inventory Management and Production Planning. Wiely, New York (1985)
20. Zhao, H., Zhu, W.: Optimal cost-sensitive granularization based on rough sets for variable costs. Knowl.-Based Syst. **65**, 72–82 (2014)
21. Min, F., Zhu, W.: Attribute reduction of data with error ranges and test costs. Inf. Sci. **211**, 48–67 (2012)
22. Yao, Y.Y., Zhao, Y.: Conflict analysis based on discernibility and indiscernibility. In: IEEE Symposium on Foundations of Computational Intelligence, pp. 302–307 (2007)

Fuzzy Multi-label Classification of Customer Complaint Logs Under Noisy Environment

Tirthankar Dasgupta[✉], Lipika Dey, and Ishan Verma

Innovation Labs, Tata Consultancy Services, New Delhi, India
{gupta.tirthankar,lipika.dey,ishan.verma}@tcs.com
http://www.tcs.com

Abstract. Analyzing and understanding customer complaints has become and important issue in almost all enterprises. With respect to this, one of the key factors involve is to automatically identify and analyze the different causes of the complaints. A single complaint may belong to multiple complaint domains with fuzzy associations to each of the different domains. Thus, single label or multi-class classification techniques may not be suitable for classification of such complaint logs. In this paper, we have analyzed and classified customer complaints of some of the leading telecom service providers in India. Accordingly, we have adopted a fuzzy multi-label text classification approach along with different language independent statistical features to address the above mentioned issue. Our evaluation shows combining the features of point-wise mutual information and unigram returns the best possible result.

1 Introduction

Telecom companies receives a high volume of electronic customer complaints that are required to be responded quickly. Such voluminous feedbacks are extremely difficult to be handled manually by the individual customer care operatives. Thus, it is important to automatically process such huge amount of complaint logs [2,4,5,9,12,14,15,18]. Automatic analysis of such complaints can help organizations to extract invaluable information about the customers including maintaining customer retention and word-of-mouth recommendations [3,19].

One of the key issues involved in the complaint analysis process is to identify the specific type of problem a customer is encountering. For example, whether a complaint belongs to customer care representative, internet services or account maintenance services, can help the respective customer care operative to direct the customer to the appropriate domain expert inside the company. It also help the organization to analyze the different problem areas in a particular product or service.

It has been observed that most of the customer complaints belong to multiple domain. Naturally, traditional approaches of multi-class classification techniques may not be appropriate for the present problem. Therefore, it becomes important to device a multilabel classification framework for such a task. Moreover, due to the noisiness of the complaints, standard NLP tools like syntactic or

© Springer International Publishing AG 2016
V. Flores et al. (Eds.): IJCRS 2016, LNAI 9920, pp. 376–385, 2016.
DOI: 10.1007/978-3-319-47160-0_34

semantic parsers, Parts-of-Speech(POS) tagger and chunkers cannot be used for identification and extraction of relevant linguistic features.

In this paper, we have focused ourselves towards the analysis and classification of customer complaint logs related to the telecommunication domain. We have analyzed customer complaint logs of some of the leading telecommunication service providers in India and applied fuzzy multilabel classification technique to classify them into their respective domains. The key contributions of this paper are as follows:

– We propose a fuzzy multilabel classification technique to analyze textual customer complaint logs and classify them into their respective domains.
– given the noisy nature of the dataset we seek to identify which features and feature combinations returns maximum classification accuracy
– We also show how trend analysis of complaint logs can be used for monitoring the customer relationship.

Accordingly, we have adopted a supervised machine learning based multi-label text categorization approach to address the aforementioned task. In order to perform supervised classification, we have collected customer complaint logs of some of the leading telecommunication service providers in India. These complaint logs are then manually analyzed and annotated by a group of domain experts. Based on the annotated dataset, we have applied the fuzzy multi-label classification technique that primarily addresses whether a complaint log belongs to a particular domain or not.

The rest of the paper is organized as follows: Sect. 2 briefly presents the state of the art in the customer complaint classification task. Section 3 presents the proposed fuzzy KNN classification framework including the extraction and preprocessing of the data, multi-label expert annotation and feature selection. Section 4 presents the evaluation methodologies and results and finally, in Sect. 5 we conclude the paper.

2   Related Works

Plethora of works have been done in automatic customer feedback type classification for the purpose of categorizing the complaints [6,16]. In most of the cases supervised algorithms are applied to learn the feature sets used to classify the complaint logs [8] and generation of domain models for call centers from noisy transcriptions [11]. Several feature combinations have been used to classify complaints into respective domains [1,13]. Different approaches use n-gram features and minimum support [7] and language model techniques [16]. Most of the work discussed above considers customer complaints/feedbacks to be related to a single domain. However, we have observed that in the telecom domain scenario, almost all the feedbacks belong to multiple domains. This restricts us to use the aforementioned techniques in our present classification task.

3   The Fuzzy KNN Classification Framework

In order to perform the multi-label classification task, we have chose the fuzzy classification framework. Here, the classifier outputs the degrees of membership in each category C_i for an input test data instance t_i. The classification task can be formally defined as:

Definition 1. *For a set of discrete class Y, consider X to be the domain of data instances in the classification task. Let τ be the training dataset represented as $\tau = (x_1, \boldsymbol{Y_1}), (x_2, \boldsymbol{Y_2}), ..., (x_m, \boldsymbol{Y_m})$. Therefore, the fuzzy classification task aims to find a classifier $f : X \rightarrow Y$ where $\boldsymbol{Y}$ is a vector and the ith element y_l (i = 1, 2, . . . ,|Y|) of $\boldsymbol{Y}$ is the degree of membership of a test item in the ith class and can take a value from the range $[0, 1]$.*

Fuzzy K Nearest Neighbor (FKNN) algorithm is a fuzzy adaptation of traditional crisp k nearest neighbor algorithm. For an unknown data instance, the FKNN assigns class memberships values instead of directly assigning the test data into a particular class. Both the crisp and fuzzy methods search for k nearest neighbors in the training set of the test instance to predict the class of the test instance. Due to its large popularity and effectiveness in various classification tasks, we have selected FKNN algorithm as fuzzy classifier in our work. The FKNN algorithm is described using the following notations.

- Test instance (t) is the dataset for which the membership vector has to be predicted
- Training set $(T) = x_1, x_2, ..., x_m$ is the set of n labeled data instances used for training the classifier.
- Training instance membership (μ_{ij}) of the ith training instance in j^{th} label is denoted by μ_{ij}.
- Test instance membership $(\mu_j(t))$ of test instance t in j^{th} label is denoted by $\mu_j(t)$.
- The number of nearest neighbors denoted by k where $1 \leq k \leq m$.
- The set of nearest neighbor of t is denoted by $N(t)$.

It may be noted that the test instance closer to its neighbors is assigned with higher membership than the test instance which is farther from its neighbors. Further, each of its neighbors is weighted with the membership of the considered neighbor. Thus, the membership of a test instance t in a class is a function of inverse of the distances from its nearest neighbors.

3.1   Extraction and Preprocessing of the Dataset

We have obtained the customer complaint data from www.consumercomplaints. in. The website officially maintains customer complaints of several different industrial sectors ranging from airline services to business and finance, car services, electronic goods and telecommunication services. In the present work we have only focused on the analysis of complaints related to telecommunication

---

* Algorithm for Fuzzy K Nearest Neighbor classification technique (FKNN)

1: **Input:** τ, t, k; **Output:** Predicted membership vector for t ((t);
2: $N(t) \leftarrow \phi$; $c \leftarrow 0$;
3: **for** $i \leftarrow 1$ to n **do** /*Compute the distance between x_i and t*/
4:     **if** $(c \le k)$ **then**
5:         $N(t) \leftarrow N(t) \cup x$; $c \leftarrow c + 1$;
6:     **else if** x_i is closer to t than any $x_i \in$ to $N(t)$ **then**
7:         $N(t) \leftarrow N(t) \setminus x$;
8:         $N(t) \leftarrow N(t) \cup x_i$
9:     **end if**
10: **end for**/* Assigning membership to t in all classes*/
11: **for** $j \leftarrow 1$ to $|Y|$ **do**
12:     $\mu_j(t) = \dfrac{\sum_{p=1}^{k} \mu_{pj}(1/||t-x_p||^{\frac{2}{\mathcal{T}-1}})}{\sum_{p=1}^{k}(1/||t-x_p||^{\frac{2}{\mathcal{T}-1}})}$
13: **end for**

---

services. A customized web crawler was developed to download and extract the customer complaint logs. The complaint logs also contains attributes like:

    <*name of the service provider, date, complaint header, description of the complaint, location of the customer*>

The extracted complaints are then passed to the preprocessing unit where some of the noisy elements like, html tags and junk characters which are removed.

3.2   Multi-label Annotation of the Dataset

For annotating the complaint logs, we have followed the telecom taxonomy as defined in [10]. Here, the authors have developed a hierarchical taxonomy of telecom domain based on the analysis of different domain factors. For simplification, we have considered only the higher order five domains that includes:

Network quality (NT): That involves issues related to network coverage, network category, network features, as well as the relationships of various networks.

Customer Care Service(CC): This involves issues related to customer care and other service related problems.

Sim(S): This includes all the issues related to mobile sim and its portability.

Balance and Bill(B): This includes the different charging-related problems and rules about telecommunications services, including payment methods (such as prepaid and postpaid), charging types (such as time-based, volume-based, event-based, and content based), billing rates, as well as account balances.

Internet Quality(IQ): This involves issues related to internet services like basic internet service, value-added service, voice, data, conference, download, and browsing service.

For a given complaint, the annotators are asked to assign the degree of association of that complaint in the problem domain classes. The membership of a

Table 1. Illustration of the annotation process.

Complain	NT	CC	S	B	IQ
my mobile no. is <mobile no>. on <date> at <time> pm a 3g internet service is activated on my phone although my phone does not support 3g service. they deducted 67 rs. from my account. when i called to customer care and explained him about problem although he knows its <service provider name> fault but he can't refund my balance. when i insisted to talk to their senior he told to call me after 2 hours and when i called my called is not transferred to the customer care executive. i hope to get my money back asap.	0.9	0.7	0.1	0.5	0.2

complaint $c \in C$ in a problem domain category $d \in D$ can be represented as: $\mu_d(c) : C \rightarrow [0.1, 0.9]$. Where 0.1 implies no association of the complaint c with the domain d and 0.9 implies high association of c to d. An example annotation is presented in Table 1. The annotation task is formally defined as:

Let $\varepsilon = 1, 2, ..., L$ be the set of discrete telecom domain labels and S be the domain of data instances. Then, the task is to assign a label set $E_i \in \varepsilon$ for the data instance S_i.

Given the above definition, the multi-label nature of a data set can be determined in terms of two measures, *Label Density* and *Label Cardinality*.
The **Label Density (LD)**, or the average density of the labels is computed as:

$$LD = \frac{1}{S} * \sum_{i=1}^{|S|} \frac{|E_i|}{\varepsilon} \tag{1}$$

Label Cardinality (LC), or the average number of labels associated per item:

$$LC = \frac{1}{|S|} * \sum_{i=1}^{|S|} |E_i| \tag{2}$$

Table 2 presents label density and label cardinality values for individual annotators.

3.3 Feature Selection

As discussed earlier, the given dataset is extremely noisy. In general, the following three types of noises are commonly observed: (a) Complaint texts containing mixed languages. For example, complaints contains Hindi words written in rommanized English, (b) Complaint texts containing grammatical errors (c) Complaint texts containing typographical errors and acronyms, (d) Complaint

Table 2. Label density and label cardinality values for annotations provided by the annotators.

Annotator	Label density	Label cardinality
Annotator 1	0.198	1.186
Annotator 2	0.203	1.216

texts containing texting languages. As a result of this, standard NLP tools like syntactic or semantic parsers, POS tagger and chunkers cannot be used for identification and extraction of relevant features. Moreover, a limited number of studies have been done regarding classification of noisy, and unstructured customer complaints data. Thus, the feature set suitable for such classification task has not been explored much.

In this work, we have used the following features to perform the multi-label classification task.

1. **Unigram** considers the all the unique content words in the text as features.
2. **Bigram** features considers the all the unique bigram in the text as features.
3. **TF-IDF:** The term frequency and inverted document frequency of words are popularly used as features in different information retrieval applications.
4. **Latent semantic analysis (LSA)** are used to compute semantic association between two concepts.
5. **Point-wise mutual information (PMI)** between two given word pairs are also used as feature in many information retrieval applications.

4 Evaluation and Results

As mentioned earlier, we intend to perform complaint classification in fuzzy classification framework. The classification model outputs a membership vector for each test instance where the jth entry $\mu_j (0 \leq \mu_j \leq 1)$ is the predicted membership value in the jth class. The evaluation of the fuzzy membership value prediction is performed by measuring distance between the predicted and actual membership vector. In this work we have used four different distance measures as depicted in Table 3 (column 1 and 2). Here, A_i is the actual membership vector and B_i is the predicted membership vector for the ith test instance.

The actual data set is fuzzy annotated, i.e., each sentence have been assigned with membership values in the emotion classes. Thus fuzzy classification algorithms are best suited for analysis of the fuzzy annotated emotion data. We have used FKNN algorithm. All the experiments have been performed with $f = 1.5$ and $K = 5$. Results have been reported based on 5-fold cross validation setting for each experiment. The evaluation of the fuzzy membership value prediction is performed by measuring the distance between the predicted and actual membership vector. In Table 3, we provide comparative performance of different feature models based on the different distance measures. We have observed that throughout all the four different distance measure, the lowest distance between the actual

Table 3. Illustration of the different distance based similarity measures and the relative performance of the different feature combinations in FKNN framework.

Distance	Descriptions	U	B	TF/IDF	LSA	PMI	U+LSA	U+PMI	B+LSA	B+PMI
Euclidean	$\frac{1}{|T|} * \sum_{i=1}^{|T|} ||A_i - B_i||$	0.45	0.48	0.49	0.48	0.38	0.41	**0.21**	0.32	0.31
Cosine	$sim(A,B) = \frac{A.B}{||A||||B||}$	0.38	0.34	0.43	0.47	0.39	0.41	**0.2**	0.43	0.34
Jaccard	$J(A,B)= \frac{|A \cap B|}{|A \cup B|}$	0.3	0.33	0.42	0.41	0.3	0.47	**0.23**	0.43	0.36

Table 4. Comparison of features in FKNN with (a) *example* and (b) *ranking* based evaluation measures.

Feature comb.	HL	P-Acc	S-Acc	P	R	F-1
U	0.166	0.631	0.538	0.753	0.745	0.743
B	0.221	0.537	0.423	0.619	0.558	0.581
TF/IDF	0.137	0.652	0.559	0.528	0.616	0.771
LSA	0.176	0.674	0.664	0.714	0.724	0.727
PMI	0.108	0.754	0.647	0.817	0.827	0.823
U+LSA	0.157	0.703	0.594	0.755	0.758	0.756
U+PMI	**0.102**	**0.789**	**0.675**	**0.856**	**0.854**	**0.833**
B+LSA	0.138	0.673	0.543	0.625	0.641	0.649
B+PMI	0.119	0.734	0.616	0.853	0.828	0.821

(a)

Feature comb.	OE	COV	RL	AVP
U	0.221	0.754	0.127	0.763
B	0.329	0.908	0.276	0.761
TF/IDF	0.176	0.524	0.116	0.831
LSA	0.223	0.745	0.134	0.872
PMI	0.123	0.534	0.034	0.921
U+LSA	0.148	0.656	0.151	0.786
U+PMI	**0.131**	**0.571**	**0.091**	**0.931**
B+LSA	0.143	0.934	0.223	0.623
B+PMI	0.111	0.622	0.162	0.899

(b)

and predicated fuzzy membership vector has been obtained with unigram and PMI feature combination.

Apart from the distance based measures, we have also used the *example* and *ranking* based measures [17] to evaluate the classifier. For this, we have converted the fuzzy membership values into crisp set of binary values. Accordingly, any membership value greater than 0.5 is represented by 1 and rest to be 0. The conversion of fuzzy memberships to crisp help us to assign class labels to each of the complain logs. The example based measure evaluates the extent of similarity between the actual and predicted label sets for a test instance [17]. Here, T be the test data set containing examples $(t_i, Y_i), i = 1, 2, ..., |T|, y_i \subset Y$ and say h be the classifier which assigns $Z_i = h(t_i)$ to the test instance t_i as the predicted label set. The evaluation metrics belonging to this category are:

(a) *The Hamming Loss* (HL),
(b) *Subset Accuracy* (S-Acc),
(c) *Precision* (P),
(d) *Recall* (R),
(e) F-measure, and
(f) *Partial match acuracy* (P-Acc).

Table 4 depicts the results of the example based evaluation measure.

Apart from producing multi-label prediction, a multi-label learning system outputs a real valued function of the form $f : T \times Y \rightarrow R$. For an test instance

(t_i, Y_i), an ideal learning system output larger values for labels in Y_i than those not in Y_i , i.e., $f(t_i, y) > f(t_i, y')$ for any $y \in Y_i$ and $y' \notin Y_i$. The ranking based measures evaluates how good is a ranking function. The ranking based measures can be defined in terms of (a) One Error (OE), (b) coverage(cov), (c) Ranking Loss (RL) and (d) Average Precision (P). The results are reported in Table 4.

Apart from classification of the complaints, it is equally important to extract and analyze the individual complaints of different organization across each complaint class and time interval. For this, the proposed classification technique is used to classify around 8000 new customer complaints of five leading telecom service providers (denoted as $C_1 to C_5$) across the period of six months. The complaint logs are extracted from the same source as discussed in the earlier section. We have applied the fuzzy KNN classification technique to compute membership values of each complaint logs with respect to the five problem classes. Figure 1 plots the distribution of complaints across the five telecom service providers

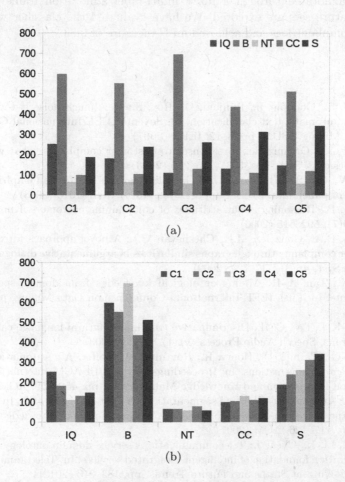

Fig. 1. Distribution of complaints across the five telecom service providers (a) and the distribution of service providers across each complaints class(b)).

(Fig. 1(a)) and the distribution of service providers across each complaints class (Fig. 1(b)).

5   Conclusion

Automatic multi-label classification of customer complaints are becoming critical for online customer service solutions and electronic customer relationship management systems. Most of the existing approaches have treated the problem as a crisp and single label classification task. However, we have observed that most of the customer complaints belong to multiple domains. Thus, it becomes important to device a fuzzy multi-label classification framework for such a task. In this paper, we have used the fuzzy KNN classification technique to classify customer complaint logs into their respective problem domain. The collected dataset is annotated through a fuzzy multi-label annotation framework and different feature sets are explored. We have evaluated the classifier with 2000 customer complaint logs and achieved an F measure of 0.83.

References

1. Agarwal, S., Godbole, S., Punjani, D., Roy, S.: How much noise is too much: a study in automatic text classification. In: Seventh IEEE International Conference on Data Mining, ICDM, pp. 3–12. IEEE (2007)
2. Davidow, M.: Organizational responses to customer complaints: what works and what doesnt. J. Serv. Res. 5(3), 225–250 (2003)
3. Duan, W., Gu, B., Whinston, A.B.: Do online reviews matter? an empirical investigation of panel data. Decis. Support Syst. 45(4), 1007–1016 (2008)
4. Galitsky, B.: Reasoning about attitudes of complaining customers. Knowl.-Based Syst. 19(7), 592–615 (2006)
5. Galitsky, B.A., González, M.P., Chesñevar, C.I.: A novel approach for classifying customer complaints through graphs similarities in argumentative dialogues. Decis. Support Syst. 46(3), 717–729 (2009)
6. Jiang, X., Tan, A.-H.: Mining ontological knowledge from domain-specific text documents. In: Fifth IEEE International Conference on Data Mining, p. 4. IEEE (2005)
7. Kuo, H.-K.J., Lee, C.-H.: Discriminative training of natural language call routers. IEEE Trans. Speech Audio Process. 11(1), 24–35 (2003)
8. Mishne, G., Carmel, D., Hoory, R., Roytman, A., Soffer, A.: Automatic analysis of call-center conversations. In: Proceedings of the 14th ACM International Conference on Information and Knowledge Management, pp. 453–459. ACM (2005)
9. Park, Y.: Automatic call section segmentation for contact-center calls. In: Proceedings of the Sixteenth ACM Conference on Information and Knowledge Management, pp. 117–126. ACM (2007)
10. Qiao, X., Li, X., Chen, J.: Telecommunications service domain ontology: semantic interoperation foundation of intelligent integrated services. In: Telecommunications Networks-Current Status and Future Trends, pp. 183–210 (2012)

11. Roy, S., Subramaniam. L.V.: Automatic generation of domain models for call centers from noisy transcriptions. In: Proceedings of the 21st International Conference on Computational Linguistics and the 44th Annual Meeting of the Association for Computational Linguistics, pp. 737–744. Association for Computational Linguistics (2006)

12. Subramaniam, L.V., Faruquie, T., Ikbal, S., Godbole, S., Mohania, M.K., et al.: Business intelligence from voice of customer. In: IEEE 25th International Conference on Data Engineering, ICDE 2009, pp. 1391–1402. IEEE (2009)

13. Subramaniam, L.V., Roy, S., Faruquie, T.A., Negi, S.: A survey of types of text noise and techniques to handle noisy text. In: Proceedings of The Third Workshop on Analytics for Noisy Unstructured Text Data, pp. 115–122. ACM (2009)

14. Takeuchi, H., Subramaniam, L.V., Nasukawa, T., Roy, S.: Automatic identification of important segments and expressions for mining of business-oriented conversations at contact centers. In: EMNLP-CoNLL, pp. 458–467 (2007)

15. Takeuchi, H., Subramaniam, L.V., Nasukawa, T., Roy, S.: Getting insights from the voices of customers: conversation mining at a contact center. Inf. Sci. 179(11), 1584–1591 (2009)

16. Tang, M., Pellom, B., Hacioglu, K.: Call-type classification and unsupervised training for the call center domain. In: IEEE Workshop on Automatic Speech Recognition and Understanding, ASRU 2003, pp. 204–208. IEEE (2003)

17. Tsoumakas, G., Katakis, I.: Multi-label classification: an overview. Aristotle University of Thessaloniki, Department of Informatics, Greece (2006)

18. Yu, J., Zha, Z.-J., Wang, M., Wang, K., Chua, T.-S.: Domain-assisted product aspect hierarchy generation: towards hierarchical organization of unstructured consumer reviews. In: Proceedings of the Conference on Empirical Methods in Natural Language Processing, pp. 140–150. Association for Computational Linguistics (2011)

19. Yuksel, A., Kilinc, U., Yuksel, F.: Cross-national analysis of hotel customers attitudes toward complaining and their complaining behaviours. Tour. Manag. 27(1), 11–24 (2006)

Some Weighted Ranking Operators with Interval Valued Intuitionistic Fuzzy Information Applied to Outsourced Software Project Risk Assessment

Zhen-hua Zhang[1(✉)], Yong Hu[2], Zhao Chen[1], Shenguo Yuan[1], and Kui-xi Xiao[1]

[1] School of Economics and Trade, Guangdong University of Foreign Studies, Guang-Zhou 510006, China
{zhangzhenhua, chenzhao, yuanshenguo, xiaokuixi}@gdufs.edu.cn

[2] Institute of Big Data and Decision, Jinan University, Guangzhou 510632, China
huyonghenry@163.com

Abstract. Some weighted ranking operators of interval valued intuitionistic fuzzy sets (IVIFS) are presented in this paper. By analyzing the interval of membership degree, the interval of non-membership degree and the interval of hesitant degree, we provide two types of weighted ranking operators with IVIFS information. And we prove some mathematical properties of these ranking operators. Finally, a multiple attribute decision-making example applied to outsourced software project risk assessment is given to demonstrate the application of this multiple attribute decision making method. The simulation results show that two-dimensional operator with IVIFS information is more effective than three-dimensional operator.

Keywords: Interval valued intuitionistic fuzzy sets · Multiple attribute decision making · Ranking operator · Outsourced software project · Risk assessment

1 Introduction

In 1965, Zadeh launched fuzzy sets (FS), which has influenced many researchers and has been applied to many application fields, such as pattern recognition, fuzzy reasoning, decision making, etc. In 1980s, Atanassov [1, 2] introduced membership function, non-membership function and hesitant function, and presented intuitionistic fuzzy sets (IFS) and interval valued intuitionistic fuzzy sets (IVIFS), which generalized the FS theory. In the research field of IFS and IVIFS, Yager [3] discussed its characteristics, more scholars applied it to decision making (Chen and Tan, [4]; Hong and Choi, [5]; Xu and Xia, [6]; Wei et al. [8]; Zhang et al. [9, 10]) and pattern recognition (Zhang, Hu, et al. [11]). Though many scholars studied IFS and IVIFS, few refer-ences related to the study of outsourced software project risk assessment based on IFS and IVIFS were proposed. In this paper, we present some novel operators of IVIFS, and apply them to outsourced software project risk assessment.

V. Flores et al. (Eds.): IJCRS 2016, LNAI 9920, pp. 386–395, 2016.
DOI: 10.1007/978-3-319-47160-0_35

First, we introduce the definition of IVIFS and some conventional operators of IVIFS. And then, we present two types of operators with IVIFS information. Finally, we apply these operators to outsourced software project risk assessment. The simulation results show that the method introduced in this paper is an effective method.

2 Conventional Weighted Operators of IVIFS

Definition 1. An IVIFS A in universe X is given by the following formula [2]:

$$A = \{ <x, M_A(x), N_A(x) > | x \in X\} \tag{1}$$

Where $M_A(x): X \rightarrow [0, 1]$, $N_A(x): X \rightarrow [0, 1]$ with the condition:

$$\forall x \in X, M_A(x) = [u_A^-(x), u_A^+(x)] \subseteq [0,1], N_A(x) = [v_A^-(x), v_A^+(x)] \subseteq [0,1], u_A^+(x) + v_A^+(x) \leq 1.$$

The numbers $M_A(x) \in [0, 1]$, $N_A(x) \in [0, 1]$ denote the interval of membership degree and the interval of non-membership degree of x to A, respectively.

Suppose that $\pi_A^-(x) = 1 - u_A^+(x) - v_A^+(x) \in [0,1], \pi_A^+(x) = 1 - u_A^-(x) - v_A^-(x) \in [0,1]$, and we define the interval of hesitant degree $H_A(x) = [\pi_A^-(x), \pi_A^+(x)]$, and then we get IVIFS.

Definition 2.   A and B are two IVIFSs over X. For each $x \in X$, we obtain:

$$A \subseteq B \text{ iff } M_A(x) \leq M_B(x), N_A(x) \geq N_B(x),$$

$$M_A(x) \leq M_B(x) \Leftrightarrow \mu_A^-(x) \leq \mu_B^-(x), \mu_A^+(x) \leq \mu_B^+(x),$$

$$N_A(x) \geq N_B(x) \Leftrightarrow v_A^-(x) \geq v_B^-(x), v_A^+(x) \geq v_B^+(x).$$

From IVIFS [2], we define a weighted operator of the interval of membership degree and a weighted operator of the interval of non-membership degree, respectively:

$$R_M(A) = \sum_{x \in X} w_A(x)(\mu_A^+(x) + \mu_A^-(x)). \tag{2}$$

$$R_{NM}(A) = \sum_{x \in X} w_A(x)(v_A^+(x) + v_A^-(x)). \tag{3}$$

Based on a dominant ranking function (Chen and Tan, [4]), a weighted operator on dominant ranking function can be expressed as follows:

$$R_{CT}(A) = \sum_{x \in X} w_A(x)((\mu_A^+(x) - v_A^+(x)) + (\mu_A^-(x) - v_A^-(x))). \tag{4}$$

Derived from Hong and Choi [5], the following weighted operator can be achieved:

$$R_{HC}(A) = \sum_{x \in X} w_A(x)((\mu_A^+(x) + v_A^+(x)) + (\mu_A^-(x) + v_A^-(x))). \tag{5}$$

Where $\mu_A^+(x)$ and $\mu_A^-(x)$ are the degree of membership function, and $v_A^+(x)$ and $v_A^-(x)$ the degree of non-membership function.

Using four distance measures, Xu presented four models from formula (6) [6, 7]:

$$R_{Xu}(A) = \frac{m(A^+, A)}{m(A^+, A) + m(A^-, A)}. \tag{6}$$

Where $A^+ = \{ <x, [\max_{A \in \Omega_A}(\mu_A^-(x)), \max_{A \in \Omega_A}(\mu_A^+(x))], [\min_{A \in \Omega_A}(v_A^-(x)), \min_{A \in \Omega_A}(v_A^+(x))], |x \in$
$X\}$, and $A^- = \{ <x, [\min_{A \in \Omega_A}(\mu_A^-(x)), \min_{A \in \Omega_A}(\mu_A^+(x))], [\max_{A \in \Omega_A}(v_A^-(x)), \max_{A \in \Omega_A}(v_A^+(x))], |x \in X\}$.

3   New Weighted Operators of IVIFS and Their Properties

According to Xu's formula (6), we define a basic operator for each variable $x \in X$.

Definition 3. Suppose that T and F are two types of extreme IVIFSs in X, where $T = \{<x, [1, 1], [0, 0] > | x \in X\}$ means $M_T(x) = [1, 1]$ and $N_T(x) = [0, 0]$ and $F = \{<x, [0, 0], [1, 1] > |x \in X\}$ means $M_F(x) = [0, 0]$ and $N_F(x) = [1, 1]$. We note $R(A(x))$ to be an index of IVIFS A for each $x \in X$. And we define:

$$R(A(x)) = \frac{Dis \tan ce(A(x), F(x))}{Dis \tan ce(A(x), F(x)) + Dis \tan ce(A(x), T(x))},$$

$$Dis \tan ce(A(x), F(x)) = \sum_{x \in X} w(x) \cdot d(A(x), F(x)), \tag{7}$$

$$Dis \tan ce(A(x), T(x)) = \sum_{x \in X} w(x) \cdot d(A(x), T(x)).$$

For example, we can use Minkowski distance to define the formula (7). According to (7), we know that distance $(A(x), F(x))$ and distance $(A(x), T(x))$ are both functions on $\mu_A^-(x), \mu_A^+(x), v_A^-(x), v_A^+(x), \pi_A^-(x)$ and $\pi_A^+(x)$. Thus, we define expression (7) as follows. Where we use the power function to define the distance function. And we have:

$$R_3^k(A(x)) = \sum_{x \in X} w(x)(\mu_A^-(x)^k + (1 - v_A^-(x))^k + \pi_A^-(x)^k + \mu_A^+(x)^k + (1 - v_A^+(x))^k + \pi_A^+(x)^k)$$

$$\div \sum_{x \in X} w(x)((\mu_A^-(x)^k + (1 - v_A^-(x))^k + \pi_A^-(x)^k + \mu_A^+(x)^k + (1 - v_A^+(x))^k + \pi_A^+(x)^k) \tag{8}$$

$$+ ((1 - \mu_A^-(x))^k + v_A^-(x)^k + \pi_A^-(x)^k + (1 - \mu_A^+(x))^k + v_A^+(x)^k + \pi_A^+(x)^k).$$

$$R_2^k(A(x)) = \sum_{x \in X} w(x)(\mu_A^-(x)^k + (1 - v_A^-(x))^k + \mu_A^+(x)^k + (1 - v_A^+(x))^k)$$

$$\div \sum_{x \in X} w(x)((\mu_A^-(x)^k + (1 - v_A^-(x))^k + \mu_A^+(x)^k + (1 - v_A^+(x))^k) \qquad (9)$$

$$+ ((1 - \mu_A^-(x))^k + v_A^-(x)^k + (1 - \mu_A^+(x))^k + v_A^+(x)^k).$$

Definition 3. A and B are two IVIFSs over X. For each $x \in X$, we obtain:

$$A \subseteq B \text{ iff } M_A(x) \le M_B(x), N_A(x) \ge N_B(x),$$

$$M_A(x) \le M_B(x) \Leftrightarrow \mu_A^-(x) \le \mu_B^-(x), \mu_A^+(x) \le \mu_B^+(x),$$

$$N_A(x) \ge N_B(x) \Leftrightarrow v_A^-(x) \ge v_B^-(x), v_A^+(x) \ge v_B^+(x).$$

Definition 4. A and B are two IVIFSs over X. R is an operator keeping order if and only if R satisfies: when $A \subseteq B$, we have $R(A) \le R(B)$.

From Definition 3, we infer two Dimensional operators are operators keeping order.

$$A \subseteq B \rightarrow \begin{cases} \mu_A^-(x) \le \mu_B^-(x) \\ \mu_A^+(x) \le \mu_B^+(x) \\ v_A^-(x) \ge v_B^-(x) \\ v_A^+(x) \ge v_B^+(x) \end{cases} \rightarrow \begin{cases} \mu_A^-(x)^k \le \mu_B^-(x)^k, (1 - \mu_A^-(x))^k \ge (1 - \mu_B^-(x))^k \\ \mu_A^+(x)^k \le \mu_B^+(x)^k, (1 - \mu_A^+(x))^k \ge (1 - \mu_B^+(x))^k \\ v_A^-(x)^k \ge v_B^-(x)^k, (1 - v_A^-(x))^k \le (1 - v_B^-(x))^k \\ v_A^+(x)^k \ge v_B^+(x)^k, (1 - v_A^+(x))^k \le (1 - v_B^+(x))^k \end{cases}$$

$$\rightarrow \begin{cases} \frac{\mu_A^-(x)^k + \mu_A^+(x)^k}{\mu_A^-(x)^k + \mu_A^+(x)^k + (1 - \mu_A^-(x))^k + (1 - \mu_A^+(x))^k} \le \frac{\mu_B^-(x)^k + \mu_B^+(x)^k}{\mu_B^-(x)^k + \mu_B^+(x)^k + (1 - \mu_B^-(x))^k + (1 - \mu_B^+(x))^k} \\ \frac{(1 - v_A^-(x))^k + (1 - v_A^+(x))^k}{v_A^-(x)^k + v_A^+(x)^k + (1 - v_A^-(x))^k + (1 - v_A^+(x))^k} \le \frac{(1 - v_B^-(x))^k + (1 - v_B^+(x))^k}{v_B^-(x)^k + v_B^+(x)^k + (1 - v_B^-(x))^k + (1 - v_B^+(x))^k} \\ \frac{\pi_A^-(x)^k + \pi_A^+(x)^k}{\pi_A^-(x)^k + \pi_A^+(x)^k + \pi_A^-(x)^k + \pi_A^+(x)^k} = \frac{1}{2} = \frac{\pi_B^-(x)^k + \pi_B^+(x)^k}{\pi_B^-(x)^k + \pi_B^+(x)^k + \pi_B^-(x)^k + \pi_B^+(x)^k} \end{cases}$$

Theorem 1. R_2^k & $R_{Minkowski2}^k$ are operators keeping order, which means for two IVIFSs A and B, if $A \subseteq B$ then $R_2^k(A(x)) \le R_2^k(B(x)), R_{Minkowski2}^k(A(x)) \le R_{Minkowski2}^k(B(x))$.

4 Application Example

Next we will introduce the methodology on the application of these operators with IVIFS information above to outsourced software project risk assessment. Considering the specialty of the outsourced software project, we use three first-level attributes to make decision according to references [12–26]: project complexity risks, contractor risks, and customer support and collaboration risks.

Example 1. A manager wants to assess the outsourced software project risk in the process of software development. Given A_i, $(i = 1, 2, 3, 4, 5)$ should be sorted. Assume that three attributes C_1(contractor risks), C_2(customer support and collaboration risks), and C_3(project complexity risks) are taken into consideration, the weight vector of the attributes C_j $(j = 1,2,3)$ is $w = (0.5,0.3,0.2)^{\mathrm{T}}$. Suppose that the data and the characteristics of the options $A_i(i = 1,2,3,4,5)$ are shown by IVIFS as follows:

$$A_1 = \{<C_1,[0.6,0.7],[0,0.1]>,\ <C_2,[0.1,0.2],[0.3,0.4]>,\ <C_3,[0.5,0.6],\}[0.2,0.3]>\},$$
$$A_2 = \{<C_1,[0.4,0.5],[0.1,0.2]>,\ <C_2,[0.3,0.4],[0.1,0.2]>,\ <C_3,[0.7,0.8],\}[0,0.1]>\},$$
$$A_3 = \{<C_1,[0.5,0.6],[0.1,0.2]>,\ <C_2,[0.4,0.5],[0.3,0.4]>,\ <C_3,[0.8,0.9],\}[0,0]>\},$$
$$A_4 = \{<C_1,[0.7,0.8],[0,0.1]>,\ <C_2,[0.2,0.3],[0.4,0.5]>,\ <C_3,[0.6,0.7],\}[0.1,0.2]>\},$$
$$A_5 = \{<C_1,[0.6,0.7],[0,0]>,\ <C_2,[0.7,0.8],[0.1,0.2]>,\ <C_3,[0,0.1],\}[0.5,0.6]>\}.$$

From formulas (8), we obtain the results as follows:

$$R^1_{Minkovski3}(A_1) = \frac{0.5 \times (2-0-0.1)+0.3 \times (2-0.3-0.4)+0.2 \times (2-0.2-0.3)}{0.5 \times (4-0.6-0.7-0-0.1)+0.3 \times (4-0.1-0.2-0.3-0.4)+0.2 \times (4-0.5-0.6-0.2-0.3)} \approx 0.61194.$$

And we can also define: $R^k_{Minkovski3}(A_i) = R^k_{M3}(A_i), R^k_{Minkovski2}(A_k) = R^k_{M2}(A_k)$.

For example, we note: $R^1_{M3}(A_1) = 0.61194 \approx 0.6119$. Similarly, we get Table 1.

$R_M(A_3) = R_M(A_4) > R_M(A_5) > R_M(A_1) = R_M(A_2), R_{NM}(A_2) < R_{NM}(A_5) < R_{NM}(A_1) = R_{NM}(A_3) < R_{NM}(A_4)$, thus we get $A_2 \succ A_1, A_5 \succ A_1, A_3 \succ A_1$, and $A_3 \succ A_4$. For example, from the membership degree $R_M(A_5) > R_M(A_1) = R_M(A_2)$ and the non-membership degree $R_{NM}(A_2) < R_{NM}(A_5) < R_{NM}(A_1)$, we obtain $A_5 \succ A_1, A_2 \succ A_1$. Similarly, we have $A_3 \succ A_1, A_3 \succ A_4$. Hence, the optimal decision-making is from set $\{A_2, A_3, A_5\}$.

From Table 1, R_{HC}, R_M, R^2_{M3} and R^2_3 don't satisfy $A_3 \succ A_4$ and $A_2 \succ A_1, R_{NM}, R_{CT}, R^1_{M2}, R^1_{M3}, R^2_{M2}$, and R^2_{M3} satisfy all four conditions on membership degree and non-membership degree. If $\mu^-_{A_k}(x) = \mu^+_{A_k}(x), v^-_{A_k}(x) = v^+_{A_k}(x), \pi^-_{A_k}(x) = \pi^+_{A_k}(x)$, then IVIFS will become IFS, the results will be similar to that from references [8, 9].

Table 1. Evaluation results based on some operators of IVIFS.

Operators	A_1	A_2	A_3	A_4	A_5	Decision-making
R^1_{M2}	0.6500	0.6750	0.7000	0.6950	**0.7025**	$A_5 \succ A_3 \succ A_4 \succ A_2 \succ A_1$
R^1_{M3}	0.6119	0.6259	**0.6613**	0.6585	0.6576	$A_3 \succ A_4 \succ A_5 \succ A_2 \succ A_1$
R^2_{M2}	0.5969	0.5971	**0.6541**	0.6475	0.6276	$A_3 \succ A_4 \succ A_5 \succ A_2 \succ A_1$
R^2_{M3}	0.5287	0.5272	**0.5403**	0.5407	0.5364	$A_4 \succ A_3 \succ A_5 \succ A_1 \succ A_2$
R^2_2	0.6869	0.6871	**0.7814**	0.7813	0.7395	$A_3 \succ A_4 \succ A_5 \succ A_2 \succ A_1$
R^2_3	0.5571	0.5542	**0.5801**	0.5810	0.5723	$A_4 \succ A_3 \succ A_5 \succ A_1 \succ A_2$
R_M	0.480	0.480	**0.580**	**0.580**	0.560	$A_3 = A_4 \succ A_5 \succ A_2 = A_1$
R_{NM}	0.180	**0.130**	0.180	0.190	0.155	$A_2 = A_5 \succ A_3 \succ A_1 \succ A_4$
R_{CT}	0.300	0.350	0.400	0.390	**0.405**	$A_5 \succ A_3 \succ A_4 \succ A_2 \succ A_1$
R_{HC}	0.660	0.610	0.760	**0.770**	0.715	$A_4 \succ A_3 \succ A_5 \succ A_1 \succ A_2$

5 Applications to Outsourced Software Risk Assessment

In the following, we will apply the operators of IVIFS above to outsourced software project risk assessment. We design the risks assessment process framework as follows:

(1) Step 1: Attributes selection.

In our previous research (references [23–26]), we have presented the attribute framework of outsourced software project risk analysis, in which we use Bayesian networks to set up risk assessment model. We use three first-level condition attributes: project complexity risks, contractor risks, and customer support and collaboration risks. And all the second-level condition attributes are shown in Table 2. The decision attribute is Target attribute, including 8 output attributes: Function, Performance, Information Quality, Maintainability, Satisfaction of Customer and User, Company Profits, Completion Degree in Time, Completion Degree in Budget [21, 22]. All the answers are Yes or No. If and only if 8 output values are all Yes then the project is successful.

(2) Step 2: Structural equation modeling.

We use structural equation modeling to select the appropriate condition attributes, and the results are shown as Fig. 1. All the attributes that are not significant will be cancelled, where p-value is 0.05.

(3) Step 3: Weights determined in every step.

By structural equation modeling, we get the effect level between attributes in Fig. 1. Thus we define the weight as follows:

$$w_{C_{ij}}(x) = \frac{w_{ij}w_i}{\sum\limits_{i=1}^{3}\sum\limits_{j=1}^{n_i} w_{ij}w_i}. \tag{10}$$

Where w_i means the effect level of first-level condition attribute influencing target. For example, w_1 means the effect level of Contractor Risks (Development Risks) affecting target, which is 0.604 (in Fig. 1). Similarity, w_{ij} means the effect level of second-level condition attribute in first-level condition attribute. For example, $w_{31} = 0.862$ means the importance of the Estimated Cost in Project Complexity Risks.

(4) Step 4: Fuzzy membership and non-membership degree intervals determined.

According to five-level survey results, we select triangle module to define membership interval, non-membership interval, and hesitation interval.

(5) Step 5: Making final decision.

Semi- supervised method: All results are ranked based on the results from formulas (8 and 9). And we determine a threshold value R_0 according to the success rate of all the outsourced software projects. If the operator value $R(A) > R_0$ then the project is judged as success project, otherwise failure project.

Table 2. Framework of outsourced software project risk analysis

Project complexity risks	References
1 Estimated cost	[13]
2 Lines(KLOC)	[17]
3 Number of team members	[18]
4 Estimated time	[17]
5 Technology complexity	[13]
6 Fun point	[17]
7 Real-time and security	[17]
8 Requirement stability	[12, 15, 19]
9 Number of collaborators	[14, 15]
10 Schedule and budget	[13, 15]
11 Industry experience	[14, 15]
Customer risks (support and collaboration risks)	References
1 Client team collaboration	[14, 15, 18]
2 Top management support	[14, 15, 18]
3 Client department support	[14, 15, 18]
4 Client development experiment	[14, 18]
5 Business environment	[14]
6 Level of IT application	[15]
7 Business process	[14]
Contractor risks	References
1 Project manager	[15]
2 Development team	[14–16]
3 Plan and control	[15, 16]
4 Development and test	[20]
5 Engineering support	[15, 19]

Supervised method: We can also define the threshold value R_0 according to the maximum prediction accuracy. And the assessment standard of supervised method is the same as that of the Semi- supervised method.

6   Experiment Results Analysis

We collect 293 sample data, 260 of them are complete data and 33 incomplete data. In all 260 complete data, 191 of them are success projects and 69 failure projects. And in the experiment, we use 200 to be training sample and the other testing sample. The following Table 3 shows the average accuracy of 10 sampling tests.

It is shown in Table 3 that the prediction effect of two-dimensional operators with membership interval and non-membership interval information are better than that of three-dimensional operators with membership interval, non-membership interval, and hesitant interval information. Considering the formulas, we conclude that it is easier for

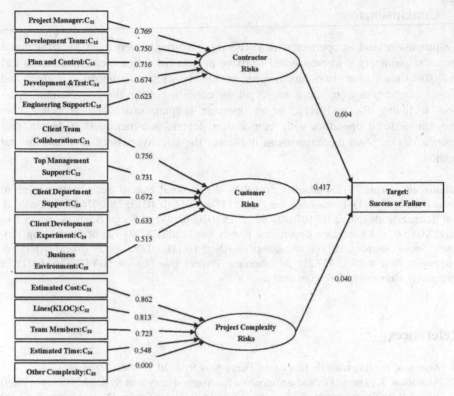

Fig. 1. Effect level of attributes based on structural equation modeling

Table 3. Average accuracy of IVIFS.

Prediction Accuracy	Semi-supervised method	Supervised method
R_{M2}^1	0.8462	0.8577
R_{M3}^1	0.8192	0.8269
R_{M2}^2	0.8231	0.8346
R_{M3}^2	0.7885	0.7962
R_2^2	0.8385	0.8462
R_3^2	0.7923	0.8038

the latter to be influenced than the former by hesitant degree. Thus, its order preservation performance is poor, and further its decision reliability is lower. From Table 3, we also know that the prediction accuracy of supervised model is a little higher than that of semi- supervised model. The predicted results of R_3^2 and R_{M3}^2 are not so good, which verifies the conclusion of the theoretical analysis.

7 Conclusions

We propose a kind of operator with IVIFS information derived from Xu's fractional operator, and apply it to outsourced software project risk assessment. The theoretical analysis shows that two-dimensional operators with membership degree and non-membership degree, which satisfy all the conditions from the conventional operators including R_M and R_{NM}, is an operator keeping order and is better than three-dimensional operators with membership degree, non-membership degree, and hesitant degree. And the experiment illustrates the effectiveness of two-dimensional method.

Acknowledgements. This paper is funded by the National Natural Science Foundation of China (No. 71271061), Natural Science Projects (No. 2014A030313575, 2016A030313688) & Soft Science funds (No. 2015A070704051, 2015A070703019) & Social Sciences project (No GD12XGL14) and Education Department Project (No. 2013KJCX0072) of Guangdong Province, Social Sciences Project of Guangzhou (No. 14G41), Team Fund (No. TD1605) & Innovation Project (No. 15T21) & Education Project (No. 16Z04, GYJYZDA14002) of Guangdong University of Foreign Studies.

References

1. Atanasso, K.: Intuitionistic fuzzy sets. Fuzzy Sets Syst. **20**, 87–96 (1986)
2. Atanassov, K.: Interval valued intuitionistic fuzzy sets. Fuzzy Sets Syst. **31**, 343–349 (1989)
3. Yager, R.R.: Some aspects of intuitionistic fuzzy sets. Fuzzy Optim. Decis. Making **8**, 67–90 (2009)
4. Chen, S.M., Tan, J.M.: Handling multicriteria fuzzy decision-making problems based on vague set theory. Fuzzy Sets Syst. **67**, 163–172 (1994)
5. Hong, D.H., Choi, C.H.: Multicriteria fuzzy decision-making problems based on vague set theory. Fuzzy Sets Syst. **114**, 103–113 (2000)
6. Xu, Z.S.: Some Similarity Measures of Intuitionistic Fuzzy Sets and Their Applications to Multiple Attribute Decision Making[J]. Fuzzy Optim. Decis. Making **6**, 109–121 (2007)
7. Xia, M.M., Xu, Z.S.: Some new similarity measures for intuitionistic fuzzy values and their application in group decision making. J. Syst. Sci. Syst. Eng. **19**, 430–452 (2010)
8. Wei, C.P., Wang, P., Zhang, Y.Z.: Entropy, similarity measure of interval-valued intuitionistic fuzzy sets and their applications. Inf. Sci. **181**, 4273–4286 (2011)
9. Zhang, Z.H., Yang, J.Y., Ye, Y.P., Hu, Y., Zhang, Q.S.: Intuitionistic fuzzy sets with double parameters and its application to dynamic multiple attribute decision making. Inf. Int. Interdisc. J. **15**, 2479–2486 (2012)
10. Zhang, Z.H., Yang, J.Y., Ye, Y.P., Hu, Y., Zhang, Q.S.: A scoring function of intuitionistic fuzzy sets with double parameters and its application to multiple attribute decision making. Inf. Int. Interdisc. J. **15**(11), 4443–4450 (2012)
11. Zhang Z.H., Wang M., Hu Y., Yang J.Y., Ye Y.P., Li Y.F.: A dynamic interval-valued intuitionistic fuzzy sets applied to pattern recognition. Mathematical Problems in Engineering, **2013**(6), 1–16 (2013). Article id 408012
12. Jones, C.: Assessment and Control of Software Risks. Yourdon Press Computing **15**(1), 19–26 (1994)

13. Xu, Z.W., Khoshgoftaar, T.M., Allen, E.B.: Application of fuzzy expert systems in assessing operational risk of software. Inf. Softw. Technol. **45**(7), 373–388 (2003)
14. Xia, W.D., Lee, G.: Complexity of information systems development projects: conceptualization and measurement development. J. Manag. Inf. Syst. **22**(1), 45–84 (2005)
15. Wallace, L., Keil, M.: Software project risks and their effect on outcomes. Commun. ACM **47**(4), 68–73 (2004)
16. Boehm, B.W.: Software risk management: principles and practices. IEEE Softw. **8**(1), 32–41 (1991)
17. Boehm, B.W.: Software Engineering Economics, 1st edn. Prentice-hall, New Jersey (1981)
18. Schmidt, R., Lyytinen, K., Keil, M.: Identifying software project risks: an international delphi study. J. Manag. Inf. Syst. **17**(4), 5–36 (2001)
19. Karolak, D.W.: Software Engineering Risk Management. IEEE Computer Society Press, Los Alamitos (1996)
20. Jiang, J.J., Klein, G.: An exploration of the relationship between software development process maturity and project performance. Inf. Manag. **41**(3), 279–288 (2004)
21. Wallace, L., Keil, M., Rai, A.: Understanding software project risk: a cluster analysis. Inf. Manag. **42**(1), 115–125 (2004)
22. Nidumolu, S.: The effect of coordination and uncertainty on software project performance: residual performance risk as an intervening variable. Inf. Syst. Res. **6**(3), 191–219 (1995)
23. Hu, Y., Mo, X.Z., Zhang, X.Z., Zeng, Y.R., Du, J.F., Xie, K.: Intelligent analysis model for outsourced software project risk using constraint-based bayesian network. J. Softw. **7**(2), 440–449 (2012)
24. Hu, Y., Zhang, X.Z., Nagi, E.W.T., Cai, R.C., Liu, M.: Software project risk analysis using bayesian networks with causality constraints. Decis. Support Syst. **56**(12), 439–449 (2013)
25. Hu, Y., Du, J.F., Zhang, X.Z., Hao, X.L., Nagi, E.W.T., Fan, M., Liu, M.: An Integrative framework for intelligent software project risk planning. Decis. Support Syst. **55**(11), 927–937 (2013)
26. Hu, Y., Feng, B., Mo, X.Z., Zhang, X.Z., Nagi, E.W.T., Fan, M., Liu, M.: Cost-sensitive and ensemble-based prediction model for outsourced software project risk prediction. Decis. Support Syst. **72**(2), 11–23 (2015)

Formal Analysis of HTM Spatial Pooler Performance Under Predefined Operation Conditions

Marcin Pietroń[(✉)], Maciej Wielgosz, and Kazimierz Wiatr

ACK Cyfronet AGH, ul. Nawojki 11, 30-950 Krakow, Poland
{pietron,wielgosz,wiatr}@agh.edu.pl

Abstract. This paper introduces mathematical formalism for Spatial Pooler (SP) of Hierarchical Temporal Memory (HTM) with a spacial consideration for its hardware implementation. Performance of HTM network and its ability to learn and adjust to a problem at hand is governed by a large set of parameters. Most of parameters are codependent which makes creating efficient HTM-based solutions challenging. It requires profound knowledge of the settings and their impact on the performance of system. Consequently, this paper introduced a set of formulas which are to facilitate the design process by enhancing tedious trial-and-error method with a tool for choosing initial parameters which enable quick learning convergence. This is especially important in hardware implementations which are constrained by the limited resources of a platform.

Keywords: Hierarchical temporal memory · Machine learning · Biologically inspired algorithms

1 Introduction

Recent years witnessed huge progress in deep learning architecture driven mostly by abundance of training data and huge performance of parallel GPU processing units [3,6]. This sets a new path in a development of intelligent systems and ultimately puts us on a track to general artificial intelligence solutions. It is worth noting that in addition to well-established Convolutional Neural Networks (CNN) architectures there is a set of biologically inspired solutions such as Hierarchical Temporal Memories [1,2,4]. Those architectures as well as CNNs suffer from lack of well-defined mathematical formulation of rules for their efficient hardware implementation. Large range of heuristics and rules of thumb are used instead. This was not very harmful except for a long training time when most of the algorithm were executed on CPUs without hardware acceleration. However, nowadays most of biologically inspired are ported to hardware for a sake of performance efficiency [10,11]. This in turn requires a profound consideration and analysis of resources consumption to be able to predict both the capacity of the platform and the ultimate performance of the system. Consequently, the authors

© Springer International Publishing AG 2016
V. Flores et al. (Eds.): IJCRS 2016, LNAI 9920, pp. 396–405, 2016.
DOI: 10.1007/978-3-319-47160-0_36

of the papers analyzed HTM design constrains on the mathematical ground and formulated a set of directives for building hardware modules.

The rest of the paper is organized as follows. Section 2 provides the background and related works. Section 3 describes mathematical formalism of Spatial Pooler. Finally, we present our conclusions in Sect. 4.

2   HTM Architecture

Hierarchical Temporal Memory (HTM) replicates the structural and algorithmic properties of the neocortex [8]. It can be regarded as a memory system which is not programmed and it is trained through exposing them to data i.e. text. HTM is organized in the hierarchy which reflects the nature of the world and performs modeling by updating the hierarchy. The structure is hierarchical in both space and time, which is the key in natural language modeling since words and sentences come in sequences which describe cause and effect relationships between the latent objects. HTMs may be considered similar to Bayesian Networks, HMM and Recurrent Neural Networks, but they are different in the way hierarchy, model of neuron and time is organized.

At any moment in time, based on current and past input, an HTM will assign a likelihood that given concepts are present in the examined stream. The HTM's output constitutes a set of probabilities for each of the learned causes. This moment-to-moment distribution of possible concepts (causes) is denoted as a belief. If the HTM covers a certain number of concepts it will have the same number of variables representing those concepts. Typically HTMs learn about many causes and create a structure of them which reflects their relationships.

Even for human beings, discovering causes is considered to be a core of perception and creativity, and people through course of their life learn how to find causes underlying objects in the world. In this sense HTMs mimic human cognitive abilities and with a long enough training, proper design and implementation, they should be able to discover causes humans can find difficult or are unable to detect.

3   Mathematical Formalism

3.1   Spatial Pooler

This section will concentrate on mathematical formalism of Spatial Pooler. The functionality of spatial pooler can be described in a vector and matrix representation, this format of data can improve the efficiency of the operations. In article vectors are defined by lowercase names with an arrow hat (the transpose of the vector will produce a column vector). All matrices will be uppercase. Subscripts on vectors and matrices are presented as right bottom indexes to denote internal elements (e.g. $X_{i,j}$ refers to element in i row and j column). Element wise operations are defined by $\odot$ and $\oplus$ operators. The I(k) function is indicator function that returns 1 if event k given as a parameter is true and 0 otherwise.

The input of this function can be matrix or vector, than the output is matrix or vector, respectively.

The user-defined input parameters are defined in (Table 1). These are parameters that must be defined before the initialization of the algorithm.

Table 1. Input SP parameters

Symbol	Meaning
n	Number of patterns
p	Number of inputs (features) in a pattern
m	Number of columns
q	Number of proximal synapses per column
ϕ_+	Permanence increment amount
ϕ_-	Permanence decrement amount
ϕ_σ	Window of permanence initialization
ρ_s	Proximal synapse activation threshold
ρ_d	Proximal dendrite segment activation threshold
ρ_c	Desired column activity level
s_{duty}	Minimum activity level scaling factor
s_{boost}	Permanence boosting scaling factor
β_0	Maximum boost
τ	Duty cycle period

The terms s, r, i, j and k are integer indices used in article. Theirs values are bounded as follows: $s \in [0, n), r \in [0, p), i \in [0, m), j \in [0, m), k \in [0, q)$.

3.2   Initialization

Competitive learning networks have typically each node fully connected to each input. The other architectures and techniques like self organizing maps, stochastic gradient networks have single input connected to single node. In Spatial Pooler the inputs connecting to a particular column are determined randomly. The density of inputs visible by Spatial Pooler can be computed by using input parameters which defines SP architecture. These rules and dependencies formulas will be described in this section. Let $c \in Z^{1 \times m}$ be the vector of columns indices. The c_i where $i \in [0, m)$ is the column's index at i. Let $I \in \{0, 1\}^{n \times p}$ be the set of input patterns, such that $I_{s,r}$ indicates s index of r pattern.

The initial probabilities of connecting inputs to columns are defined in [7]. After connecting columns to input, the permanences of synapses must be initialized. Permanences were defined to be initialized with a random value close to ρ_s. Permanences should be randomly initialized, with approximately half of the permanences creating connected proximal synapses and the remaining permanences creating potential (unconnected) proximal synapses.

As initial parameters are set the activation process can be described by mathematical formulas. Let $X \in \{0,1\}^{m \times q}$ is the set of inputs for each column, X_i set of inputs for column c_i. Let $ac_i = \sum_k^{q-1} X_{i,k}$ be the random variable of number of active inputs on column i. The average number of active inputs on a column is defined by: $ac = \frac{1}{m} \sum_{i=0}^{m-1} \sum_{k=0}^{q-1} X_{i,k}$. The $P(X_{i,k})$ is defined as the probability of the input connected to column i via proximal synapses. Therefore expected number of active proximal synapses can be computed as follows 1:

$$E[ac_i] = q * P(X_i) \tag{1}$$

Let $ActCon_{i,k} = X_{i,k} \cap I(\phi_{i,k} \geq \rho_s)$ defines the event that proximal synapse k is active and connected to column i. Random variable of number of active and connected synapses for column i is define by $actcon_i$ variable.

The probability that synapse is active and connected: $P(ActCon_i) = P(X_{i,k}) * \frac{1}{2}$. Expected number of active and connected synapses for single column is defined as 2:

$$E[actcon_i] = q * P(ActCon_i) \tag{2}$$

$Bin(k,n,p)$ is the probability mass function of a binomial distribution (k number of successes, n number of trials, p success probability in each trial). Number of columns with more active inputs than threshold 3:

$$acts = \sum_{i=0}^{m-1} I(\sum_{k=0}^{q-1} X_{i,k} > \rho_d) \tag{3}$$

Number of columns with more active and connected proximal synapses than threshold 4:

$$actcol = \sum_{i=0}^{m-1} I(\sum_{k=0}^{q-1} ActCon_{i,k} > \rho_d) \tag{4}$$

Let π_x be the mean of P(x) and π_{ac} the mean of $P(ActCon)$ than by 5 and 6, the summation computes the probability of having less than ρ_d active connected and active proximal synapses. To obtain the desired probability, the complement of that probability is taken.

$$E[acts] = m * [1 - \sum_{t=0}^{\rho_d-1} Bin(t,q,\pi_x)] \tag{5}$$

$$E[actcol] = m * [1 - \sum_{t=0}^{\rho_d-1} Bin(t,q,\pi_{ac})] \tag{6}$$

3.3   Learning

After initialization and during the training process the overlap of each column is calculated by equations defined in 7. Each column is a candidate to be active if his overlap value is greater than specified threshold ρ_c, Then can be activated if its overlap value is greater or equal $k - th$ overlap value in its inhibition radius neighbourhood. Learning phase consists of updating permanence values, inhibition radius and boosting factors updating and duty cycle parameters computing. The permanence values of synapses are updated only when column is activated. Therefore update of synapse can be defined as element wise multiplication of transposed vector of column activations and matrix of values of inputs connected to columns synapses. If inputs are active than permanences are increased by value θ_+ otherwise decreased by θ_-:

$$\delta\phi = r\_actCol^T \odot (\theta_+ X - (\theta_- \neg X)) \tag{7}$$

$$\delta\phi_{i,k} = r\_actCol_i^T \odot (\theta_+ X_{CSI_{i,k}} - (\theta_- \neg X_{CSI_{i,k}})) \tag{8}$$

The permanence values must been in $[0, 1]$ range. The following equation is a rule of updating final permanence values:

$$\phi = clip(\phi \oplus \delta\phi, 0, 1) \tag{9}$$

The clip function clips the permanence values in $[0, 1]$ range:

$$clip(k, l, u) = \begin{cases} u & \text{if } k \geq u \\ l & \text{if } \leq l \\ k & \text{otherwise} \end{cases} \tag{10}$$

Each column in learning phase updates activeDutyCycle parameter - μ_i^a. The set of these parameters is represented by vector μ^a. It is worth to notice that history of activation of the columns activation should be stored in an additional structure to remember and update duty cycle parameter in each cycle - $ActDCHist = \{0,1\}^{m \times history}$, (only set number of steps before should be remember, history parameter is sliding window width). The activeDutyCycle is computed as follows:

$$\mu_i^a = \sum_{k=0}^{\tau} ActDCHist_{i,k} \tag{11}$$

The procedure of *update_active_duty_cycle* in each cycle can be done by organizing above matrix as cycle list. In each cycle the whole single column is updated. Then the index to the column which will be update in next cycle is incremented by one. If the index would be greater than the matrix width it is set to 0. The minimum active duty cycle μ^{min} is computed for boosting purposes by the following equation:

$$\mu^{min} = s_{duty} * max(H_i \odot \mu^a) \forall i \tag{12}$$

The maximal active duty cycle of columns in neighborhood is scaled by s_{duty} factor.

The boost factor computation is base on μ^a, μ^{min} parameters. The boost function should be used when $\mu^a \leq \mu^{min}$. It should be monotonically decreasing due to μ^a:

$$b = \beta(\mu^a, \mu^{min}) \forall i \tag{13}$$

$$\beta(\mu^a, \mu^{min}) = \begin{cases} \beta_0 & \text{for } \mu^{min} = 0 \\ 1 & \text{for } \mu^a > \mu^{min} \\ \text{boost function otherwise} \end{cases} \qquad (14)$$

The next parameter μ^o is *overlapDutyCycle*. It is computed by the same manner like *activeDutyCycle*. Apart from activation indicators the overlap are used. The similar matrix of overlap history is used - $OvlpDCHist$. The permanence boosting definition is based on comparing $\mu^o < \mu^{min}$. If it is true than all input synapses permanence are increased by constant factor:

$$\phi = clip(\phi \oplus s_{pboost} * I(\mu^o < \mu^{min}), 0, 1) \qquad (15)$$

The original inhibition radius presented by Numenta is based on distances between columns and active connected synapses (Eq. 16). Equation 17 presents how inhibition is computed (sum of distances divided by sum of connected synapses). The inhibition radius can be constant during learning phase or can be changed. It depends of SP mode. Both modes what will be described later should converge to the stable value of inhibition radius or to value with minimal variance.

$$D = d(pos(c_i, 0), pos(CSI_{i,k})) \odot ConSyn_i \forall i \forall k \qquad (16)$$

$$inhibition_i = max(1, \lfloor \frac{\sum_i^m \sum_k^q D_{i,k}}{max(1, \sum_i^m \sum_k^q ConSyn_{i,k})} \rfloor) \qquad (17)$$

It should be noticed that the spectrum of inhibition radius in case of hardware implementation can be shifted or decreased in some situations. In GPU when columns are processed by thread blocks, boundary threads compare theirs *overlap* and *activeDutyCycle* in spectrum of reduced inhibition radius to avoid device memory synchronization [9]. During initialization process the mean distance and inhibition is defined as follows:

$$mean\_dist(c_i) = \frac{\frac{end-pos}{input\_size} * q * av_{left} + \frac{pos-start}{input\_size} * q * av_{right}}{q} \qquad (18)$$

$$inhibition = \frac{\sum mean_i}{\frac{1}{2} * q} \qquad (19)$$

The initial probability of column activation based on inhibition radius is defined as:

$$P(r\_ActCol) = \frac{k}{inhibition} * P(ActCol) \qquad (20)$$

The probability of boosting at initial stage can be computed using:

$$P(boost) = \frac{inhibition - k}{inhibition} * P(ActCol) \qquad (21)$$

3.4   Quality of SP

The Spatial Pooler output representation is in sparse format so number of active columns is $k << n$:

$$|\sum I(ActCol == 1)| = k \tag{22}$$

$$max(k) \simeq 4 - 5\,\%(minoverlap == 1 \text{ if sparse input}) \tag{23}$$

As SP pattern is learnt we can estimate what input give the same output and its probability. The probability is equals to product of probabilities that for active columns for given pattern the overlap is greater or equal to *minoverlap*:

$$\prod_{i=0}^{N-k} P(ovlp_i < minovlp) * \prod_{i=0}^{k} P(ovlp_i \geq minovlp) \tag{24}$$

The single probability can be computed by following equation:

$$P(ovlp_i \geq minovlp) = \prod_{k=0}^{q} P(X_{CSI_{i,k} -> (\phi_{i,k} \geq \rho s)} == 1) \tag{25}$$

The number of unique patterns that can be represented on n bits with w bits on is defined as:

$$\binom{n}{w} = \frac{p!}{w!(p-w)!} \tag{26}$$

Then we can define the number of codings that can give similar output as learnt pattern by SP (Eq. 27 is a number of input codings for active columns and 28 is number of input codings for non active columns).

$$\prod_{i=0}^{act} (2^{(q-\sum I(\phi_{i,k} \geq \rho s))} * \binom{\sum I(\phi_{i,k} \geq \rho s)}{minoverlap} \tag{27}$$

$$\prod_{i=0}^{N-act} (2^{(q-\sum I(\phi_{i,k} \geq \rho s))} * \sum_{g=0}^{minoverlap-1} \binom{\sum I(\phi_{i,k} \geq \rho s)}{g-1} \tag{28}$$

The $2^{(q-\sum I(\phi_{i,k} \geq \rho s))}$ is the number of codings on input to synapses that are learnt zero bit pattern ($\phi_{i,k} < \rho s$). The $\binom{\sum I(\phi_{i,k} \geq \rho s)}{minoverlap}$ is number of codings that input has more bits on than minoverlap on synapses learnt for receiving bits with value 1 ($\phi_{i,k} \geq \rho s$).

3.5   Convergence of SP

In this section the convergence of SP learning process will be described. We divided the process of learning SP to two different modes. The first one consists of learning each pattern separately. In this case for each column c_i the final state of SP after learning process should be as follows (for $t \rightarrow \infty$):

- $\sum_0^q X_{i,k} > minovlp$

  $\phi_{i,k} \to 1.0$ for $X_{i,k} == 1$

  $\phi_{i,k} \to 0.0$ for $X_{i,k} == 0$

- $\sum_0^q X_{i,k} < minovlp$, $\phi_{i,k} \to 1.0$

There are three possible starting states at the beginning of learning, the possible transitions from state to other state are as follows (indicated by $\to$):

- not overlap $\to$ $(t \to \infty)$ permanence boosting $\to$ (overlap $\geq$ minoverlap) if $\sum_{k=0}^q X_{i,k} \geq \rho_s$
- overlap $\geq$ minOverlap & no activation $\to$ overlap boosting (activeDutyCycle value) $\to$ activation $\to$ permanence updating
- overlap $\geq$ minOverlap & activation $\to$ permanence updating

It can be noticed that if there are more columns than k parameter with overlap greater or equal than *minoverlap* value in spectrum of constant inhibition radius than columns are in priority queue (priority is activeDutyCycle) in which they are will be activated in cyclic way (because of overlap boosting).

Process of single pattern learning can be run further for next pattern. Before this process learnt columns (columns activated by learnt pattern) should be blocked from permanence boosting (avoid boosting of learnt synapses). The columns activated (learnt) by previous pattern can be activated by new pattern only when overlap between inputs of patterns to this column is greater or equal *minoverlap*. Overlap function is defined as follows:

$$overlap(x, y) = x \times y \tag{29}$$

where: x and y are binary vectors.

In case of SP learning process of multiple patterns simultaneously there can exist some other situation which can speed up or slow down process of convergence. These all situation are mentioned below:

- detraction of 1 on single synapse when multiple patterns activate the same column with opposite input value on single synapse

  $(((r\_ovlp_{i,s} < minoverlap)$ & $(X_{i,k} == 1)) \| (ovlpDC_{i,s} < minActDC_{i,s}))$
  $\&((r\_ovlp_{i,s+1} < minoverlap)$ & $(X_{i,k} == 0))$
- $P(\text{detraction of } 0) = (P(Act = 1) * P(synapse = 1) + P(boost))$
- attraction of 1 on single synapse when multiple patterns activate the same column with the same input value on single synapse

  $(((r\_ovlp_{i,s} < minoverlap)$ & $(X_{i,k} == 1)) \| (ovlpDC_{i,s} < minActDC_{i,s}))$
  $\&((r\_ovlp_{i,s+1} > minoverlap)$ & $(X_{i,k} == 1))$
- attraction of 0 on single synapse when multiple patterns activate the same column with the same input value on single synapse

  $((r\_ovlp_{i,s} < minoverlap)$ & $(X_{i,k} == 0))$ &
  $((r\_ovlp_{i,s+1} > minoverlap)$ & $(X_{i,k} == 0))$

There are three possible situations during learning process (multiple pattern learning with constant inhibition radius):

- permanence boosting of inputs of columns activated by different patterns, harmful effect but if duty cycle big enough (almost bigger than number of patterns), inputs will be learned
- attraction (equations above) – speeding up learning
- detraction (equations above) – slowing down learning

In both presented situations (single and multi pattern learning) constant inhibition radius is used. According to the original Numenta algorithm the fluctuations of inhibition radius should decrease during learning process [7], but there is possibility that inhibition radius never converge to constant value. In our approach the constant inhibition radius allows to show convergence of learning process. This situation can be achieved by stopping the inhibition radius changing after some learning steps or to change algorithm by the one with radius convergence to stable value. It should be noticed that values of inhibition radius and k should guarantee sparse output at the end of learning.

4   Conclusions and Future Work

The presented HTM model is a new architecture in deep learning domain inspired by human brain. Initial results show [5] that it can be at least efficient like other machine and deep learning models. Additionally, our earlier research [9] showed that it can be significantly speed up by hardware accelerators. The presented formalism is one of the first article with full mathematical description of HTM Spatial Pooler. The formal description will help to parameterized the model. According to given encoder and its input distribution characteristic it is possible by formal model to estimate number of learning cycles, probability of patterns attraction, detraction etc. Further work the should concentrate on extending the formalism by accurate proofs of convergence of learning process. Then formal description of temporal pooler should be added.

References

1. Ahmad, S., Hawkins, J.: Properties of sparse distributed representations and their application to hierarchical temporal memory. arXiv preprint arXiv:1503.07469 (2015)
2. Ahmad, S., Hawkins, J.: How do neurons operate on sparse distributed representations? A mathematical theory of sparsity, neurons and active dendrites. arXiv preprint arXiv:1601.00720 (2016)
3. Bengio, Y., Courville, A., Vincent, P.: Representation learning: a review and new perspectives. IEEE Trans. Pattern Anal. Mach. Intell. **35**(8), 1798–1828 (2013)
4. Chen, X., Wang, W., Li, W.: An overview of hierarchical temporal memory: a new neocortex algorithm. In: 2012 Proceedings of International Conference on Modelling, Identification and Control (ICMIC), pp. 1004–1010. IEEE (2012)

5. Cui, Y., Surpur, C., Ahmad, S., Hawkins, J.: Continuous online sequence learning with an unsupervised neural network model. arXiv preprint arXiv:1512.05463 (2015)
6. Liu, Q., Huang, Z., Hu, F.: Accelerating convolution-based detection model on GPU. In: 2015 International Conference on Estimation, Detection and Information Fusion (ICEDIF), pp. 61–66. IEEE (2015)
7. Mnatzaganian, J., Fokoué, E., Kudithipudi, D.: A mathematical formalization of hierarchical temporal memory cortical learning algorithm's spatial pooler. arXiv preprint arXiv:1601.06116 (2016)
8. Mountcastle, V.B.: The columnar organization of the neocortex. Brain **120**(4), 701–722 (1997)
9. Pietron, M., Wielgosz, M., Wiatr, K.: Parallel implementation of spatial pooler in hierarchical temporal memory. In: 8th International Conference on Agents and Artificial Intelligence (ICAART), pp. 346–353 (2016)
10. Thomas, D., Luk, W.: FPGA accelerated simulation of biologically plausible spiking neural networks. In: 17th IEEE Symposium on Field Programmable Custom Computing Machines, FCCM 2009, pp. 45–52. IEEE (2009)
11. Woodbeck, K., Roth, G., Chen, H.: Visual cortex on the GPU: biologically inspired classifier and feature descriptor for rapid recognition. In: IEEE Computer Society Conference on Computer Vision and Pattern Recognition Workshops, CVPRW 2008, pp. 1–8. IEEE (2008)

Performance Comparison to a Classification Problem by the Second Method of Quantification and STRIM

Yuya Kitazaki[1], Tetsuro Saeki[2], and Yuichi Kato[1(✉)]

[1] Shimane University, 1060 Nishikawatsu-cho, Matsue, Shimane 690-8504, Japan
ykato@cis.shimane-u.ac.jp
[2] Yamaguchi University, 2-16-1 Tokiwadai, Ube, Yamaguchi 755-8611, Japan
tsaeki@yamaguchi-u.ac.jp

Abstract. STRIM is used for inducing if-then rules hidden behind a database called the decision table. Meanwhile, the second method of quantification is also often used as a method for summarizing and arranging such a database. This paper first summarizes both methods, next compares their performance in a learning and classification problem by applying them to a simulation dataset, and lastly considers features and clarifies differences of both methods based on the simulation results.

1 Introduction

Rough Sets theory as introduced by Pawlak [1] is used for inducing if-then rules from a database called the decision table and determining the structure of rating and/or knowledge in the database. Such rule induction methods are needed for disease diagnosis systems, discrimination problems, decision problems, and other aspects, and consequently, many effective algorithms for rule induction by rough sets have been reported in the literature [2–7]. We also have presented such a rule induction method named STRIM (Statistical Test Rule Induction Method) [8–12] which extends VPRS [3] from a statistical view and deepens the concept of rules. That is, STRIM shows that a rule is what makes a biased decision and the biased decision is detected by a statistical test, using the sample set of the decision table, and derives the accuracy proposed by VPRS. Rule sets once derived are used for various classification problems, as mentioned above.

On the other hand, the second method of quantification (SMQ) [13] corresponding to the linear discriminant analysis (LDA) with quantified variables in the conventional multivariate statistical analysis is often used for the same classification problems after learning the parameters of SMQ from the same dataset as the decision table. The criterion for learning the parameters is maximizing the correlation ratio, which makes the biased sample scores by each group, which are the output of the linear model.

STRIM and SMQ are used for the same purpose, and have the same feature of learning criterion of the biased decision and/or the biased sample scores, but they are different models from each other. This paper first summarizes STRIM

© Springer International Publishing AG 2016
V. Flores et al. (Eds.): IJCRS 2016, LNAI 9920, pp. 406–415, 2016.
DOI: 10.1007/978-3-319-47160-0_37

and SMQ, then applies them to a simulation dataset, shows their performance comparisons, and clarifies the features and points to keep in mind when applying them. It should be noted that this study is completely different from the literature [14] in which PCA is used for the rule induction.

2 Conventional Rough Sets and STRIM

Rough Sets theory is used for inducting if-then rules hidden in the decision table S. S is conventionally denoted as $S = (U, A = C \cup \{D\}, V, \rho)$. Here, $U = \{u(i)|i = 1, ..., |U| = N\}$ is a sample set, A is an attribute set, $C = \{C(j)|j = 1, ..., |C|\}$ is a condition attribute set, $C(j)$ is a member of C and a condition attribute, and D is a decision attribute. Moreover, V is a set of attribute values denoted by $V = \bigcup_{a \in A} V_a$ and is characterized by the information function $\rho: U \times A \to V$.

Rough Sets theory focuses on the following equivalence relation and equivalence set of indiscernibility: $I_C = \{(u(i), u(j)) \in U^2 | \rho(u(i), a) = \rho(u(j), a), \forall a \in C\}$. I_C is an equivalence relation in U and derives the quotient set $U/I_C = \{[u_i]_C | i = 1, 2, ...\}$. Here, $[u_i]_C = \{u(j) \in U | (u(j), u_i) \in I_C, u_i \in U\}$. $[u_i]_C$ is an equivalence set with the representative element u_i. Let $X = D_d = \{u(i)|(\rho(u(i), D) = d\}$, then X can be approximated like $C_*(X) \subseteq X \subseteq C^*(X)$ by use of $[u_i]_C$. Here,

$$C_*(X) = \{u_i \in U | [u_i]_C \subseteq X\}, \tag{1}$$

$$C^*(X) = \{u_i \in U | [u_i]_C \cap X \neq \emptyset\}, \tag{2}$$

$C_*(X)$ is called the lower approximation of X by C, is surely a set which satisfies $D = d$ and derives if-then rules of $D = d$ with necessity. In the same way, $C^*(X)$ is the upper and derives if-then rules of $D = d$ with possibility.

Ziarko [3] further introduced the variable precision rough set (VPRS) by expanding the concept of the lower and upper approximation by use of an admissible classification error $\varepsilon \in [0, 0.5)$ as follows:

$$\underline{C}_\varepsilon(U(d)) = \{u(i)| \quad accuracy \geq 1 - \varepsilon\}, \tag{3}$$

$$\overline{C}_\varepsilon(U(d)) = \{u(i)| \quad accuracy \geq \varepsilon\}. \tag{4}$$

(3) and (4) coincide with the ordinary lower and upper approximations by $\varepsilon = 0$ respectively. VPRS has been widely used for solving real-world problems [5–7].

On the other hand, STRIM [8–12] considers the decision table to be a sample dataset obtained from an input-output system including a rule box (see Fig. 1), and a hypothesis regarding the decision attribute values (see Table 1). The sample $u(i)$ consists of its condition attribute values of $|C|$-tuple $u^C(i)$, and its decision attribute $u^D(i)$. $u^C(i)$ is the input into the rule box and is transformed into the output $u^D(i)$ using the rules and the hypotheses.

STRIM induces if-then rules by assuming $CP(k) = \bigwedge_j (C(j_k) = v_j (\in V_{C(j_k)}))$ as the condition part of the if-then rule, and derives the set $U(CP(k)) = \{u(i)|u^C(i)$ satisfies $CP(k)$, which is denoted by $u^{C=CP(k)}(i)\}$. Also, $U(m) = \{u(i)|u^{D=m}(i)\}$ $(m = 1, ..., |V_D| = M_D)$ should be derived. The distribution

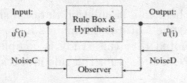

Fig. 1. Rough Sets system contaminated with noise. Rule box contains if-then rules $R(d)$: if Rd then $D = d$ $(d = 1, 2, ...)$.

Table 1. Hypotheses with regard to the decision attribute value.

Hypothesis 1	$u^C(i)$ coincides with $R(d)$, and $u^D(i)$ is uniquely determined as $D = d$ (uniquely determined data)
Hypothesis 2	$u^C(i)$ does not coincide with any $R(d)$, and $u^D(i)$ can only be determined randomly (indifferent data)
Hypothesis 3	$u^C(i)$ coincides with several $R(d)(d = d1, d2, ...)$, and their outputs of $u^C(i)$ conflict with each other. Accordingly, the output of $u^C(i)$ must be randomly determined from the conflicted outputs (conflicted data)

$f = (n_1, n_2, ..., n_{M_D})$ of the decision attribute values of $U(CP(k))$, where $n_m = |U(CP(k)) \cap U(m)|$ $(m = 1, ..., M_D)$ should also be calculated. If the assumed $CP(k)$ does not satisfy the condition $U(Rd) \supseteq U(CP(k))$ (sufficient condition of specified rule Rd) or $U(CP(k)) \supseteq U(Rd)$ (necessary condition), $CP(k)$ generates the indifferent large dataset based on Hypothesis 2 in Table 2, and the distribution f is not biased. Conversely, if $CP(k)$ satisfies either condition, f is biased, since $u^D(i)$ is determined by Hypothesis 1 or 3. Accordingly, whether f is biased or not determines whether the assumed $CP(k)$ is neither a necessary nor sufficient condition. Whether f is biased or not can be determined objectively by a statistical test with a proper test statistic and a standard of the significance level under the following null hypothesis $H0$ and its alternative hypothesis $H1$: $H0$: f is not biased; $H1$: f is biased.

If $H0$ is rejected, then the assumed $CP(k)$ becomes a candidate for the rules in the rule box. After changing $CP(k)$ systematically and extracting all the candidates, the final set of rules in the rule box is obtained by arranging the candidates satisfying the relationship: $CP(ki) \subseteq CP(kj) \subseteq CP(kl) \cdots$ (see [8–11] for more detailed procedures). It should be noted that the sufficient and necessary conditions from the statistical view correspond to the lower and upper approximation respectively and STRIM develops the notion of VPRS into a statistical principle [10].

3   The Second Method of Quantification

The second method of quantification (SMQ) [13] is conventionally applied to the data table, as shown in Table 2, which is basically the same as the decision table

Table 2. An example of the data table for the second method of quantification.

Item variable (explanatory variable) u_j					Group
1	$\cdots$	j	$\cdots$	Q	
$\cdots$	$\cdots$	$\cdots$	$\cdots$	$\cdots$	1
$\cdots$	$\cdots$	$\cdots$	$\cdots$	$\cdots$	$\cdots$
$u_{11}^{(g)}$	$\cdots$	$u_{1j}^{(g)}$	$\cdots$	$u_{1Q}^{(g)}$	g
$\cdots$	$\cdots$		$\cdots$	$\cdots$	
$u_{i1}^{(g)}$	$\cdots$	$u_{ij}^{(g)}$	$\cdots$	$u_{1Q}^{(g)}$	
$\cdots$	$\cdots$		$\cdots$	$\cdots$	
$u_{n_g1}^{(g)}$	$\cdots$	$u_{n_gj}^{(g)}$	$\cdots$	$u_{n_gQ}^{(g)}$	
$\cdots$	$\cdots$	$\cdots$	$\cdots$	$\cdots$	$\cdots$
$\cdots$	$\cdots$	$\cdots$	$\cdots$	$\cdots$	G

S. In Table 2, u_j ($\in \{1, ..., C_j\}$, ($j = 1, ..., Q$)) is called the j-th item's variable which corresponds to $C(j)$ and $u_{ij}^{(g)}$ ($i = 1, ..., n_g$) is the i-th sample of belonging to the g-th group which corresponds to $D = g$. SMQ prepares an intermediate variable called the sample score:

$$y_i^{(g)} = \sum_{j=1}^{Q}\sum_{k=1}^{C_j} a_{jk}x_{ijk}^{(g)}. \tag{5}$$

Here, $x_{ijk}^{(g)}$ is called a dummy variable, $x_{ijk}^{(g)} = 1$ (if $u_{ij}^{(g)} = k$), $= 0$ (if $u_{ij}^{(g)} \neq k$), and a_{jk} is called a category score. SMQ estimates a_{jk} by maximizing the correlation ratio $\eta^2 = a^T Ba/a^T Ta = S_B/S_T$ about a under the constraint: $S_T = S_B + S_W$.

Here, $a = \begin{bmatrix} \vdots \\ a_{jk} \\ \vdots \end{bmatrix} \in R^C$, $X = \begin{bmatrix} \vdots \\ X^{(g)} - 1_g\bar{x}^T \\ \vdots \end{bmatrix}$, $X^{(g)} = \begin{bmatrix} \ddots & \vdots & \ddots \\ \cdots & x_{ijk}^{(g)} & \cdots \\ \ddots & \vdots & \ddots \end{bmatrix} \in R^{n_g \times C}$,

$1_g = \begin{bmatrix} 1 \\ \vdots \\ 1 \end{bmatrix} \in R^{n_g}$, $\bar{x} = \begin{bmatrix} \vdots \\ \bar{x}_{jk} \\ \vdots \end{bmatrix} \in R^C$, $C = \sum_{j=1}^{Q} C_j$, $\bar{x}_{jk} = \frac{1}{n}\sum_{g=1}^{G}\sum_{i=1}^{n_g} x_{ijk}^{(g)}$,

$n = \sum_{g=1}^{G} n_g$, $T = X^T X$, $B = X_B^T X_B$, $X_B = \begin{bmatrix} \vdots \\ \bar{X}^{(g)} - 1_g\bar{x}^T \\ \vdots \end{bmatrix}$,

$$\bar{X}^{(g)} = 1_g \left[\begin{array}{c} \vdots \\ \bar{x}_{jk}^{(g)} \\ \vdots \end{array} \right]^T = 1_g (\bar{x}^{(g)})^T, \ \bar{x}_{jk}^{(g)} = \frac{1}{n_g} \sum_{i=1}^{n_g} x_{ijk}^{(g)}, \ S_W = a^T W a,$$

$$W = X_W^T X_W, \ X_W = \left[\begin{array}{c} \vdots \\ X^{(g)} - 1_g(\bar{x}^{(g)})^T \\ \vdots \end{array} \right].$$

The problem of maximizing η^2 leads to the operation: $\frac{d\eta^2}{da} = 0$ and drives the eigenvalue problem:

$$(T^{-1}B - \eta^2 I)a = 0. \tag{6}$$

That is, a is obtained as the eigenvector of (6) and η^2 is proved to be the eigenvalue. See the literature [13] for a more detailed procedure.

It should be noted that this procedure leads the sample between-group variation S_B of $y_i^{(g)}$ to a maximum relative to the sample total variation S_T of $y_i^{(g)}$, while the sample within-group variation S_W of $y_i^{(g)}$ is led to a minimum, that is, SMQ is a method of making the biased sample scores by each group. The estimated sample scores $\hat{y}_i^{(g)}$ $(i = 1, ..., n_g)$ by estimated $\hat{a}$ and (5) cause each area of group g having the minimized $\hat{S}_W = \hat{a}^T W \hat{a}$ and the maximized $\hat{S}_B = \hat{a}^T B \hat{a}$, which enables an input to predict the belonging group of the input.

Table 3. The correspondence of STRIM and SMQ.

–	Input variables and their values		Output	Input–output relation	Criterion for estimate
SMQ	u_j $(j = 1, ..., Q)$	$u_{ij}^{(g)}$ $(i = 1, ..., n_g)$ $(g = 1, ..., G)$ (derivative variable: $x_{ijk}^{(g)}$)	g (derivative variable: $y_i^{(g)}$)	Linear regression expressions	To maximize S_B making the biased sample scores by each group
STRIM	$C(j)$ $(j = 1, ..., |C|)$	$\rho(u(i), C(j))$ $(-)$	$\rho(u(i), D = d)$ $(-)$	If-then rules	To find $CP(k)$ making the biased f
Correspondence	$Q = |C|$ $u_j = C(j)$	$C_j = |V_{C(j)}|,$ $n_g = |U(g)|,$ $n = \sum_{g=1}^{G} n_g$ $= N$	$g = d,$ $G = |V_D|$	–	To bias f of D or sample scores by g

4   Relationship Between STRIM and SMQ

Table 3 shows the correspondence of STRIM and SMQ introduced in Sects. 2 and 3 by their conventional notation respectively. In the table, "–" denotes that the corresponding relationship doesn't exist. Both methods have basically the same dataset form and the same criterion of making the biased outputs for learning although their models for the input-output conversion are different from each other: STRIM uses if-then rules, SMQ linear regression expressions.

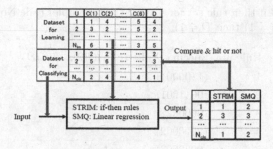

Fig. 2. An image of the simulation experiment.

5 Generate Simulation Dataset by Data Generation Model

Generating a simulation dataset is needed in order to apply it to a learning and classification problem by both methods, and to clarify their features and performance. Although there may be various methods of generating the dataset, it is ideal to adopt the model shown in Fig. 1 since models based on those like (5) seem to artificially generate and/or arrange a dataset not based on the judgements by human beings. Then, the following two cases of if-then rules and parameters $|C| = 6$, $|V_{C(j)}| = M_C = 6$, $|V_D| = M_D = 6$ should be specified in the rule box in Fig. 1:

Case 1: $R(d)$: if Rd then $D = d$, $(d = 1, ..., M_D = 6)$, where $Rd = (C(1) = d) \wedge (C(2) = d) \vee (C(3) = d) \wedge (C(4) = d)$.
Case 2: $R(d)$ is shown in Table 4.

The differences between Case 1 and 2 are that Case 1 has global reduct attributes of $C(5)$ and $C(6)$ while Case 2 does not [12]. Generation of $u^C(i) = (v_{C(1)}(i), v_{C(2)}(i), ..., v_{C(|C|)}(i))$ of $u(i)$ is completed by use of random numbers with a uniform distribution, and then $u^D(i)$ is determined using the rules specified in the rule box and the hypothesis without both noises of NoiseC and NoiseD in Fig. 1 for simplicity. It should be noted that the generated dataset is separated into three types according to the hypotheses in Table 1: the set of uniquely determined data U_{UD}, that of indifferent data U_{ID}, and that of conflicted data U_{CF}.

6 Simulation Experiment and Considerations

6.1 Simulation Experiment and Its Results

The learning and classification experiments for Case 1 and 2 were conducted by use of the dataset of $N = 10000$ generated in Sect. 5. First, the learning processes were executed by the dataset randomly selected by N_{lrn} from the dataset of N, and respectively induced the results: if-then rules for STRIM, $\hat{a}^{(p)}$

Table 4. Specified if-then rule set for Case 2 (For example, Rule No. 1 means that if $(C(1) = 1) \wedge (C(2) = 1)$ then $D = 1$).

Rule no.	Specified rule: $(C(1)C(2)...C(6)D)$
1	(1100001)
2	(0011001)
3	(0000222)
4	(2200002)
5	(0033003)
6	(0000333)
7	(4400004)
8	(0044004)
9	(0000555)
10	(5500005)
11	(0066006)
12	(0000666)

(the estimated p-th eigenvector corresponding to η_p^2 satisfying the relationship: $\eta_1^2 > \cdots > \eta_p^2 > \cdots > \eta_P^2$, $P = \min(G-1, m)$, $m = \sum_{j=1}^{Q} C_j - Q$) for SMQ. Then, the classification problem was studied using the results of the learning processes and the dataset also randomly selected by $N_{cls} = N - N_{lrn}$. See the image of the experiment in Fig. 2. These learning and classification experiments were repeated $N_r = 100$ times (Bootstrap method). Figure 3 shows the classification results by the average of hitting rates of N_r times based on the results of the learning dataset of N_{lrn} for two cases; those by STRIM were calculated by use of only U_{UD} and U_{CF}, that is $U_R = U_{UD} \cup U_{CF}$, while those by SMQ were the hitting rates by all of the datasets of N_{cls}. STRIM could distinguish U_R from U_{ID} by use of the estimated if-then rules and classified them. On other hand, SMQ classified each input of N_{cls} into the group g in order to have the shortest Mahalanobis' generalized distance to the average sample score of group g: $D_{(g)} = \min_g (y - \bar{y}^{(g)})^T \Sigma_g^{-1} (y - \bar{y}^{(g)})$, where $y = [\cdots, y_{(p)}, \cdots]^T \in R^P$, $\bar{y}^{(g)} = [\cdots, \bar{y}_{(p)}^{(g)}, \cdots]^T \in R^P$, $y_{(p)} = (\hat{a}^{(p)})^T x$, $\bar{y}_{(p)} = (\hat{a}^{(p)})^T \bar{x}^{(g)}$, $\Sigma_{(g)} = \sum_{i=1}^{n_g} (y(i)^{(g)} - \bar{y}^{(g)})^T (y(i)^{(g)} - \bar{y}^{(g)})/n_g$, $y(i)^{(g)} = [\cdots, y_{(p)i}^{(g)}, \cdots]^T \in R^P$, $y_{(p)i}^{(g)} = (\hat{a}^{(p)})^T x_i^{(g)}$, $x_i^{(g)} = [\cdots, x_{ijk}^{(g)}, \cdots]^T \in R^C$, and x is the vector of the dummy variable of the input, since it does not have such an ability to distinguish U_{ID}.

From Fig. 3, the followings were found:

(1) STRIM has already shown that it can induce all the if-then rules specified in advance, that is, true rules, with other additional rules, except true ones by use of a dataset of more than around $N_{lrn} = 3000$ [9]. However, the hitting rate $R_{hitting}$ of STRIM was around 83 [%] for Case 1 and 72 [%] for Case 2 due to the influences of U_{CF} and the other additional rules. On the other hand,

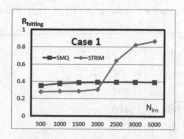 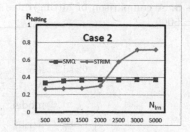

Fig. 3. Comparison of classification results between STRIM and SMQ.

that of SMQ was around 38 [%] due to not having the ability to distinguish U_{ID}.

(2) The $R_{hitting}$ of STRIM fell down to around 28 [%] with decreases of N_{lrn} due to getting difficulties to induce true rules and more additional rules while that of SMQ hardly changed.

From the above experimental knowledge, if the $R_{hitting}$ of STRIM is not high, whether N_{lrn} is enough or not should be doubted. If that of SMQ is not high, whether S contains a large amount of U_{ID} or not should be doubted. However, methods by if-then rules will be generally favorable for learning and classifying problems since they can distinguish U_{ID} from U. Moreover, those by if-then rules will be also favorable for learning problems since their methods are understandable for human beings and easily applicable for classification problems, while methods by linear regression like SMQ need a large number of numerical parameters of $\hat{a}^{(p)}$ ($M_C \cdot |C| \cdot P = 180$ in this specification), which are hard to understand the meanings for human beings.

6.2 Considerations to Experimental Results on Model

A model based on rules specified in advance approximately predicts the $R_{hitting}$ since the rules in Case1 and 2 are relatively simple. For example, $|\Omega|$, $|\Omega_{UD}|$, $|\Omega_{ID}|$ and $|\Omega_{CF}|$ of Case 1 corresponding to $|U|$, $|U_{UD}|$, $|U_{ID}|$ and $|U_{CF}|$, respectively, are given as follows: $|\Omega| = M_C^{|C|}$, $|\Omega_{CF}| = M_C^2 M_D (M_D - 1)$, $|\Omega_{UD}| = |\Omega_R| - |\Omega_{CF}| = 2M_D M_C^4 - M_D^2 M_C - M_C^2 M_D (M_D - 1)$, $|\Omega_{ID}| = |\Omega| - |\Omega_R| = M_C^6 - (2M_D M_C^4 - M_D^2 M_C^2)$, where $|\Omega_R|$ is the cardinality of the members belonging to the specified rules and corresponds to $|U_R|$. The rate of a uniquely determined, indifferent and conflicted sample are given by $P_{UD} = |\Omega_{UD}|/|\Omega| = 0.282$, $P_{ID} = 1 - |\Omega_R|/|\Omega| = 0.695$ and $P_{UD} = |\Omega_{CF}|/|\Omega| = 0.023$ respectively. The hitting rate of STRIM is also approximately given by $P_{hitting}(\Omega_R) = (|\Omega_{UD}| + |\Omega_{CF}|/2)/|\Omega_R| = 0.962$, contributing the half of $|\Omega_{CF}|$ for the hitting. In the same way, that of SMQ is given by $P_{hitting}(\Omega) = (|\Omega_{UD}| + |\Omega_{CF}|/2 + |\Omega_{ID}|/6)/|\Omega| = 0.410$, contributing $|\Omega_{ID}|/6$ for the hitting.

Table 5 shows a comparison of hitting rates on the experiment and model between STRIM and SMQ at $N_{lrn} = 5000$ ($N_{cls} = 5000$) with the results of

Table 5. A comparison of hitting rates on the model between STRIM and SMQ at $N_{lrn} = 5000$.

–		STRIM	SMQ
Case 1	$|U_R|$	0.868	0.862
	$N_{cls} = 5000$	–	0.394
	model	0.962	0.410
Case 2	$|U_R|$	0.720	0.728
	$N_{cls} = 5000$	–	0.375
	model	0.949	0.399

Case 2 studied in the same way. The $R_{hitting}$ of SMQ in the row of $|U_R|$ was calculated by use of the same dataset of STRIM, which suggested the same ability of classification as that of STRIM if SMQ had the ability of distinguishing U_{ID} from U_R. Both $R_{hitting}$ based on the model approximately grasped the experimental $R_{hitting}$ of $|U_R|$, taking into consideration that STRIM also induced the additional rules, except true trues.

7   Conclusion

We conducted a simulation experiment to examine the ability and performance of STRIM and SMQ in the problems of learning and classification after summarizing their methods, since both methods have been used for the same aims with the same dataset form and the same criterion of making the biased outputs for learning. The results showed the following:

(1) SMQ is a kind of linear discriminant analysis (LDA) with quantified variables and has low hitting rates for datasets containing a large amount of indifferent data due to not having any ability to distinguish the indifferent ones from the effective.

(2) The method by use of if-then rules which distinguishes those indifferent has the possibilities to resolve the classification problems at high hitting rates. If the hitting rates are low, it should be examined whether the data size for learning is an appropriate amount or not.

(3) Generally, methods by regression need a large number of numerical parameters for accounting for a dataset as shown in this study, which is hard to understand the meanings for human beings comparing with those by if-then rules.

The above results suggest that STRIM is highly applicable to real-world datasets, and carries advantages for learning and classification problems. In addition, SMQ is contained in Hayashi's first-fourth methods of quantification [13] which are widely applied in Japan and can be used for trial by the free software R [15]. And you can easily confirm the results for SMQ mentioned in (1) and (3).

References

1. Pawlak, Z.: Rough sets. Int. J. Inf. Comput. Sci. **11**(5), 341–356 (1982)
2. Grzymala-Busse, J.W.: LERS – a system for learning from examples based on rough sets. In: Słowiński, R. (eds.) Intelligent Decision Support. Handbook of Applications and Advances of the Rough Sets Theory, pp. 3–18. Kluwer Academic Publishers (1992)
3. Ziarko, W.: Variable precision rough set model. J. Comput. Syst. Sci. **46**, 39–59 (1993)
4. Shan, N., Ziarko, W.: Data-based acquisition and incremental modification of classification rules. Comput. Intell. **11**(2), 357–370 (1995)
5. Xiw, G., Zhang, J., Lai, K.K., Yu, L.: Variable precision rough set group decision-making: an application. Int. J. Approximate Reasoning **49**, 331–343 (2008)
6. Inuiguchi, M., Yoshioka, Y., Kusunoki, Y.: Variable-precision dominance-based rough set approach and attribute reduction. Int. J. Approximate Reasoning **50**, 1199–1214 (2009)
7. Huang, K.Y., Chang, T.-H., Chang, T.-C.: Determination of the threshold β of variable precision rough set by fuzzy algorithms. Int. J. Approximate Reasoning **52**, 1056–1072 (2011)
8. Matsubayashi, T., Kato, Y., Saeki, T.: A new rule induction method from a decision table using a statistical test. In: Li, T., Nguyen, H.S., Wang, G., Grzymala-Busse, J., Janicki, R., Hassanien, A.E., Yu, H. (eds.) RSKT 2012. LNCS, vol. 7414, pp. 81–90. Springer, Heidelberg (2012). doi:10.1007/978-3-642-31900-6_11
9. Kato, Y., Saeki, T., Mizuno, S.: Studies on the necessary data size for rule induction by STRIM. In: Lingras, P., Wolski, M., Cornelis, C., Mitra, S., Wasilewski, P. (eds.) RSKT 2013. LNCS, vol. 8171, pp. 213–220. Springer, Heidelberg (2013). doi:10.1007/978-3-642-41299-8_20
10. Kato, Y., Saeki, T., Mizuno, S.: Considerations on rule induction procedures by STRIM and their relationship to VPRS. In: Kryszkiewicz, M., Cornelis, C., Ciucci, D., Medina-Moreno, J., Motoda, H., Raś, Z.W. (eds.) RSEISP 2014. LNCS, vol. 8537, pp. 198–208. Springer, Heidelberg (2014). doi:10.1007/978-3-319-08729-0_19
11. Kato, Y., Saeki, T., Mizuno, S.: Proposal of a statistical test rule induction method by use of the decision table. Appl. Soft Comput. **28**, 160–166 (2015)
12. Kato, Y., Saeki, T., Mizuno, S.: Proposal for a statistical reduct method for decision tables. In: Ciucci, D., Wang, G., Mitra, S., Wu, W.-Z. (eds.) RSKT 2015. LNCS (LNAI), vol. 9436, pp. 140–152. Springer, Heidelberg (2015). doi:10.1007/978-3-319-25754-9_13
13. Tanaka, Y.: Review of the methods of quantification. Environ. Health Perspect. **32**, 113–123 (1979)
14. Liu, D., Li, T., Hu, P.: A new rough sets decision method based on PCA and ordinal regression. In: Chan, C.-C., Grzymala-Busse, J.W., Ziarko, W.P. (eds.) RSCTC 2008. LNCS (LNAI), vol. 5306, pp. 349–358. Springer, Heidelberg (2008). doi:10.1007/978-3-540-88425-5_36
15. The Comprehensive R Archive Network (CRAN). https://cran.r-project.org/

Outlier Detection and Elimination in Stream Data – An Experimental Approach

Mateusz Kalisch[1], Marcin Michalak[2,3]([✉]), Piotr Przystałka[1],
Marek Sikora[2,3], and Łukasz Wróbel[2,3]

[1] Institute of Fundamentals of Machinery Design, Silesian University of Technology,
ul. Konarskiego 18a, 44-100 Gliwice, Poland
{Mateusz.Kalisch,Piotr.Przystalka}@polsl.pl
[2] Institute of Informatics, Silesian University of Technology,
ul. Akademicka 16, 44-100 Gliwice, Poland
{Marcin.Michalak,Marek.Sikora}@polsl.pl
[3] Institute of Innovative Technologies EMAG,
ul. Leopolda 31, 40-186 Katowice, Poland
Lukasz.Wrobel@ibemag.pl

Abstract. In the paper the issue of outlier detection and substitution (correction) in stream data is raised. The previous research showed that even a small number of outliers in the data influences the prediction model application quality in a significant way. In this paper we try to find a proper complex method of outliers proceeding for stream data. The procedure consists of a method of outlier detection, a statistic used for the outstanding values replacement, a historic horizon for the replacing value calculation. To find the best strategy, a wide grid of experiments were prepared. All experiments were performed on semi–artificial data: data coming from the underground coal mining environment with an artificially introduced dependent variable and randomly introduced outliers. In the paper a new approach for the local outlier correction is presented, that in several cases improved the classification quality.

Keywords: Outlier detection · Data analysis · Classification · Time series

1 Introduction

The missing, incomplete and outlier data issue has been intensively studied by many researchers, and a variety of imputation methods have been developed. The problem of outlier treatment and missing value imputation in data streams has been investigated much more extensively for the last decade. The authors of [7] have been one of the first to notice a lack of research on estimating missing values in a data stream. They showed that the general problem of estimating missing values was well studied in the statistics field, but the derived techniques lacked software implementation and were rarely used in practice.

© Springer International Publishing AG 2016
V. Flores et al. (Eds.): IJCRS 2016, LNAI 9920, pp. 416–426, 2016.
DOI: 10.1007/978-3-319-47160-0_38

For our purpose the outlier is considered an observation that differs from the other significantly. In the context of real environment conditions this kind of observation should be interpreted rather as the result of a measuring, coding or transmission error than as an interesting observation that needs special treatment (as it can occure in medicine). With this interpretation of an outlier, further processing of this element of data becomes analogical to the missing value imputation. This process will be called further an outlier substitution.

The paper is a continuation of our previous works in the area of outlier detection in stream data, that focused on checking the influence of outliers level on the prediction quality [9] and ability of outlier detection in the data [10]. The results of these experiment helped us to provide the data with the statistically important level of outliers and to select the proper and fast algorithm of outlier detection.

In this paper a comparison of well known outlier substitution techniques with the new proposed one is presented. All experiments were performed on data, in which the locations of outliers were known, so it was possible to compare the quality of models derived from data without and with outliers. Additionally, it was also possible to compare the quality of models derived from data with outliers with and without application of outliers detection/substitution procedures. For the purpose of outlier detection typical methods were used.

The paper is organized as follows: it starts from the brief review of outlier detection and substitution methods with the division on approaches for static and stream data. This section is followed by the description of the data used in our experiments. Next part of the paper describes the plan of experiments, an overview of used outlier detection and substitution methods, including our proposition of a local correction of observation recognized as an outlier. For a better clarity of results presentation two reference models (0 and IA) were proposed and are also described. Afterwards, results of experiments are presented. The paper ends with some short discussion on obtained results, including some remarks and tries of results explanation. Also perspectives of future works are marked.

2   Related Works

The issue of outlier detection and substitution in data is very common in many applications of data analysis [6,11]. A review of several kinds of methods for missing data is given in [12]. Three main approaches of outlier analysis can be found in the literature [8]:

- with no prior knowledge about the data, which is similar to unsupervised clustering,
- modelling the known normality and known abnormality,
- modelling the known normality with a very few cases of abnormality.

According their nature, a several types of outliers are distinguished [14]: local (type I), context (type II) and group (type III).

Knowledge discovery from data streams is a very important field, which has recently attracted much attention especially in the context of data stream systems such as STREAM [2], Borealis [1], TelegraphCQ [4], etc. The main issues in this area can be categorized into the following groups [5]: time series data streams, data stream clustering and classification, frequent pattern mining, change detection in data streams, stream cube analysis of multi-dimensional streams, sliding window computations and synopsis construction in data streams, dimensionality reduction and forecasting in data streams, distributed mining of data streams.

A description of used outlier detection methods will be presented in the further part of the paper.

3   Data Description

Our experiments were performed on semi–artificial data sets. The original time series was a five-dimensional data set from over 100 h of underground atmosphere conditions monitoring. Five variables were taken into consideration:

- AN: air flow (in [m/s]),
- MM: methane concentration (in [% CH_4]),
- TP_1, TP_2: air temperature (in [°C]),
- BA: air pressure (in [hPa]).

Original values were aggregated into 60 second intervals (calculated from 20 to 30 real observations). The method of aggregation depended on the variable: a minimum (anemometer), a maximum (methane meter) or an average (thermometers and barometer). To avoid fast changes of values, the aggregated series were also smoothed with the 24 previous values window.

The presented time series was a base of 100 time series that were built as "noised" copies of the pattern, including a dependent variable. Noising of each variable was performed by adding a random value from normal distribution (mean value equal to zero and standard deviation equal to a difference between raw and smoothed values of the considered variable[1]).

The data contained an artificially introduced dependent variable which was implied by a simple decision tree. The tree was created manually and arbitrarily; starting from its structure, through variables in nodes, to the decisions in leaves. A more detailed description of this decision tree can be found in [9] or [10]. What is very important, the introduced dependent variable had no interpretation and meaning.

For each noised time series a procedure of outlier (type I) introduction was applied. A variety of outlier types were used due to their dimensionality (one- or multi–dimensional) and duration (several consecutive observations or several dozen of them). For a selected variable two outlier values were calculated: the

---

[1] In case of exceeding the range of the variable, an appropriate boundary value was used.

lower outlier was $Q1 - 1.5IQR$ while the upper outlier was $Q3 + 1.5IQR$. 50 levels of outliers content in the data were taken into consideration: starting from 1 % up to 50 %. To assure the same level of outliers occurrence it was assumed that time series is divided in a ratio 30:70 and each part contain the same level of outliers. More details of the outliers introduction procedure can be found in [9].

4   Experiments and Results

4.1   Experiments

To avoid the effect of unballanced classes the ballanced accuracy was used as the quality measure for the models. It was planned to performed a threedimensional grid of models of prediction. The grid contained the following dimensions:

- outlier detection method (two algorithms),
- outlier substitution method (two approaches with two measures used for the substituting value calculation),
- history of observations for the outlier substitution method (three cases).

The given set of grid dimensions and the cardinality of their domains lead to the final number of 24 experiments for a single prediction method.

As an additional dimension, A model of prediction could be considered as an additional dimension, because four different classifiers were examined. Their selection is a continuation of a selection made in our previous works [9]. The following classifiers were used: decision trees, naive bayes, kNN (with $k \in \{1, 3, 5\}$), logistic regression.

Six prediction models with the grid of 24 nodes gave a number of 144 experiments in total. However, in our research we also took into consideration 50 levels of outliers content in the test data (starting from 1 % up to 50 %, increasing by 1 %) which finally gave an impressive number of 7,200 performed experiments of prediction.

In addition, two reference models were planned: zero model (0: no outliers in the test data); "no action" model (IA: there are outliers in the test data, but no methods of their detection and further substitution were applied). A more detailed description of these models is presented in the further part of the paper.

The analysis of outlier detection, provided in [10], led to the selection of two algorithms characterized by the best performance results (in the average sense): GAS and LOF. Their descriptions can be found in [3, 13] respectively. It is worth to stress that the first of them was several times faster and not statistically significantly worse.

In our research several methods of outliers substitution were taken into consideration: moving average, moving median and our proposition presented below.

The proposed method of outlier substitution can be called local as it works on each data attribute separately. When an observation is classified as an outlier— information that is given by a GAS or LOF algorithm—each of its dimensions (variables) is examined in terms whether the value extends the interquartile range. If it does, than the value is replaced by a specified value. This approach avoids losing a real (not outlying) values of attributes from the data.

The other two ways of outlier substitution are global: for any observation pointed as an outlier all of its attributes values are replaced in one of the following ways: replaced by an average or a median of the historic observations. The notion of historic observations will be explained in the next subsection.

All mentioned methods of outlier replacement require some assumptions about the amount of a previous observation that will be considered histori-cal. On the basis of the historical data a substituting value will be calculated. According to the conclusions from the previous works [10], a constant window of a previous observation and an incremental window, taking into consideration all observed data, were applied. We decided to check two constant windows of 50 and 100 previous observations. This leads to three variants of this experiment grid dimension.

For each prediction method—decision trees, naive Bayes, kNN and logis-tic regression—reference models of prediction results were developed. The first model—called 0—was represented as a result of the following experiment: train-ing and test data did not contain outliers. This model was applied for all 100 datasets and the average or median prediction quality was taken into considera-tion as the reference value. This value is constant due to the increasing number of outliers in other models, so it is represented on charts as a horizontal line.

Another reference model is a model tested on data that contain outliers but no procedure of their detection and substitution is applied. As a result of that, the quality of this model depends on the percentage of outliers in the test data. So, for a specific prediction method, the model is a series of averaged prediction qualities.

Both reference models are presented in Fig. 1. As it can be expected, the model IA is a series that decreases with the increase of the outlier number in the data.

Experiments were performed on 100 time series with 50 different levels of outliers: from 1 % up to 50 % (0 % level of outliers in the test data are treated a reference model 0). Prediction models were trained on 30 % of the data and then consequently applied for the remaining 70 % of observations. Due to the new

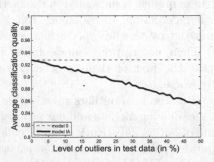

Fig. 1. Visualization of two reference models for decision tree classifier.

observations no re–learning procedure of a classification model was performed. Only in the case of an outlier substitution, new values of means, medians or quartiles were recalculated.

In the first step, it was necessary to tune the values of behavioural para-meters of outlier detection algorithms. In the case of the GAS algorithm, these parameters were arbitrarily chosen taking into account the results of the analysis presented in our previous paper [10] as well as the current trials conducted with the use of the training data (data without outliers). As a result, n-top outliers for each level variant of outliers were selected on the basis of the Euclidean distance measure that was computed using the training data for five nearest neighbors. Based on the current analysis, it was also decided that this value had to be divided by 2.5 in order to select n-top outliers correctly. On the other hand, in the case of the LOF algorithm a similar analysis was carried out taking into con-sideration the training data. The tuning procedure led us to determine the best values of relevant parameters, meaning the lower and upper bounds in specifying k-nearest neighbors were set to 10 and 20 respectively, and the Euclidean metric was used. The threshold value of the outlier factor was experimentally set to 1.1.

In the case when a new observation was qualified as an outlier, the procedure of outlier substitution was applied and the corrected object was classified to one of the classes 0 or 1 (the artificial dependent variable). The corrected object was also taken into consideration as historical data for the further outlier detection and substitution.

For every model, considered the combination of an outlier detection algo-rithm, the outlier substitution procedure and a type of historic data, a chart is presented with the balanced accuracy of prediction. As each model was applied to 100 of time series, the chart presents statistics of these 100 results. On the X axis the increasing level of outlier content in the data is presented. The Y axis is the value of a selected statistic (average or median) of the prediction quality (balanced accuracy) for 100 datasets.

4.2   Results

As it was mentioned at the beginning of this section, for each of six classifiers 24 models of outlier detection and substitution were developed. In Fig. 2 all 24 models are presented against the background of two reference models (0 and IA). The 0 model is a constant line while IA model is represented with the solid one. As a statistic for representing 100 experiments, an average was used. For better clarity of presentation, all of them have the same range of axes X and Y. A similar set of charts—presented in Fig. 3—uses a median instead of an average as an aggregation statistic.

The main goal of this paper was an experimental check of the influence of data improvement (substitution of outlier observations) on the prediction quality. As it can be observed in Fig. 2, none of outlier detection and substitution strategies improved the average prediction quality in any level of outlier number in the data, when decision trees were used as a prediction model. The same effect takes place in the case of the naive Bayes classifier. In the case of logistic regression we can

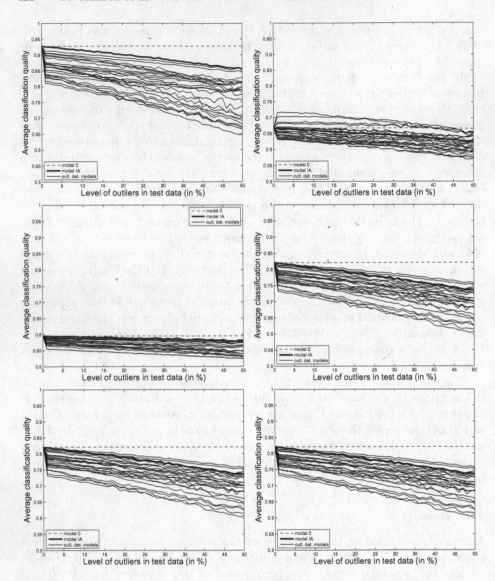

Fig. 2. Visualization of average quality prediction of models on the background of reference models. Upper (from left to right): decision trees, logistic regression, naive Bayes; Lower (from left to right): 1NN, 3NN, 5NN.

observe a set of models which give better results than both reference models. It occurred also that several models of outlier detection and substitution improved an average prediction accuracy for kNN classifiers. The improvement is visible only in reference to a IA reference model, that assumes outliers in the data but none outlier processing procedure.

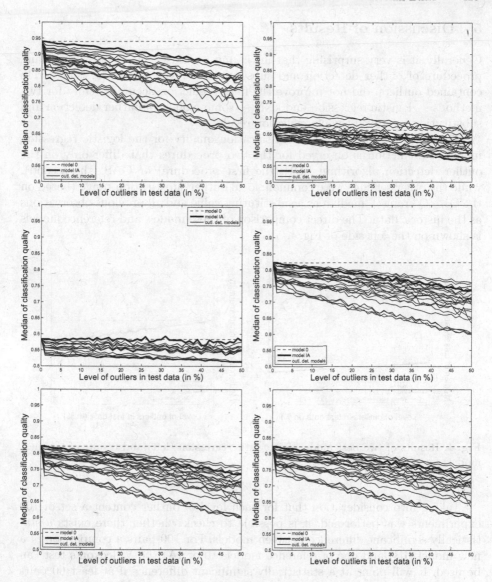

Fig. 3. Visualization of median of quality prediction of models on the background of reference models. Upper (from left to right): decision trees, logistic regression, naive Bayes; Lower (from left to right): 1NN, 3NN, 5NN.

When the medians of classification accuracies are taken into consideration — Fig. 3— corresponding remarks on classification improvement or its lack can be stated.

5 Discussion of Results

Generally, it is very surprising that in the most of cases the application of any procedure of outlier detection and substitution for the test data, that surely contained outliers, did not improve the final prediction accuracy. Only for two methods — logistic regression and kNN — some models of outlier detection and substitution improved the classification results.

The highest improvement of classification quality for the logistic regression model of prediction is observed for the two procedures that differ only on the outlier detection algorithm: LOF (the first procedure) or GAS (the second), while the other settings were common: local outlier substitution range (based on the Q1 – Q3 range), median as a substituting value and all previous observations as the historic data. The direct comparison of these models and reference models is shown on the left side of Fig. 4.

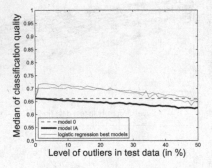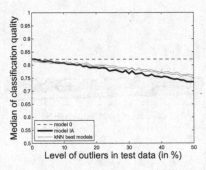

Fig. 4. Results of two best models for logistic regression (left) and results of two best models for kNN classifier (right)

Taking into consideration that for each level of outlier content a set of 100 experiments was performed, it is possible to check whether there exists a statistically significant difference between models. For 100 pairs a comparison of a new model (IB) and one of reference models (0 or IA) the Wilcoxon's test can be used. It will point at a statistically significant difference if at least 60 pairs find a model better.

In the case of the first model (with LOF used in outlier detection) up to the level of 43 % outliers in the test data, the presented approach causes a statistically significant improvement of classification accurracy when compared to the model applied to data without outliers. When compared with the IA model, the presented one is always statistically better.

The second model (with GAS used in outlier detection) is statistically better than the 0 reference model up to 30 % of outliers in the test data, including 32 %, 35 %. It is also always statistically better than the IA model.

Two best models derived from the kNN classifier differ only in the length of the historic data window. They take into consideration 50 or 100 previous observations. The detection of outliers is performed with the LOF algorithm, and the detected outlier is replaced with the median on variables whose values exeed the Q1–Q3 range. The direct comparison of these models with reference models is shown on the right side of Fig. 4.

From the statistical point of view, the wnd50 model is never better than the 0 reference model. It occurs better for the level of outliers equal to 8 % or greater than 10 %. The model with a longer history (wnd100) is statistically worse for the level of outliers smaller than 5 %. When the level of outliers exceeds 17 %, the proposed model classifies objects (statistically) better than the IA model.

But why did only several of over one hundred (144) methods of outliers detection and substitution give such unsatisfactory results? As we observed on the mentioned four winning solutions, all of them replaced outlying values locally. It means, in general, that for multidimensional data this approach should be applied to improve the input data quality. Substitution of all values in all dimensions of the data leads to the worsening of the prediction quality. Probably this is the main reason why the half of models worsened the prediction quality.

Another common feature of the best models is the median of a variable as the value substituting outstanding measurements. It is also worth to be considered how the nature of the data influences the outlier detection and substitution models. The data were taken from real environmental conditions observed in underground coal mining but the dependent variable was introduced artificially and arbitrarily while outliers were introduced randomly.

6   Conclusions and Further Works

In this paper the issue of outliers detection and substitution in stream data was raised. On the basis of semi–artificial data — real data with an artificially introduced dependent variable and randomly introduced outliers— several approaches for outlier detection and substitution were developed. In the total number of 144 approaches only 8 (two for logistic regression and $\{1, 3, 5\}$–NN) gave satisfactory results. It occurred that it was possible to improve the prediction quality with the outlier analysis. On the other hand, however, the most of the models behaved similarly to the reference model IA, which assumed that there exist outliers in the new–coming data but no methods of their identification and correction were applied. This remark is significant for the stream data analysis application, because resignation from the outlier analysis decreases the time of classification significantly. The paper does not exhaust the possible research in this area. From the winning solution it can be observed that the local strategy of outlier correction has the biggest influence on the classification quality improvement. Our future works will focus on developing better local algorithms of outlier correction. The locality is considered a correction of only those object attributes whose values make the object an outlier in the data. The future experiments will take place on much bigger datasets. In this research we used the data that were strongly connected with the project DISESOR (decision support system for mining industry) and from the project performance point of view the set target was reached.

Acknowledgements. This work was partially supported by Polish National Centre for Research and Development (NCBiR) grant PBS2/B9/20/2013 within Applied Research Programmes.

References

1. Abadi, D., Carney, D., Çetintemel, U., et al.: Aurora: a new model and architecture for data stream management. VLDB J. **12**(2), 120–139 (2003)
2. Arvind, A., Brian, B., Shivnath, B., John, C., Keith, I., Rajeev, M., Utkarsh, S., Jennifer, W.: Stream: The stanford data stream management system (2004)
3. Breunig, M., Kriegel, H.P., Ng, R., Sander, J.: LOF: identifying density-based local outliers. In: Proceedings of ACM SIGMOD International Conference on Management of Data, pp. 93–104 (2000)
4. Chandrasekaran, S., Cooper, O., Deshpande, A., et al.: TelegraphCQ: continuous dataflow processing. In: Proceedings of the 2003 ACM SIGMOD International Conference on Management of Data, p. 668 (2003)
5. Gama, J.: Knowledge Discovery from Data Streams. Chapman & Hall/CRC, New York (2010)
6. Gupta, M., Gao, J., Aggarwal, C., Han, J.: Outlier detection for temporal data: a survey. IEEE Trans. Knowl. Data Eng. **26**(9), 2250–2267 (2014)
7. Halatchev, M., Gruenwald, L.: Estimating missing values in related sensor data streams. In: Haritsa, J., Vijayaraman, T. (eds.) COMAD, pp. 83–94 (2005)
8. Hodge, V., Austin, J.: A survey of outlier detection methodologies. Artif. Intell. Rev. **22**(2), 85–126 (2004)
9. Kalisch, M., Michalak, M., Sikora, M., Wróbel, Ł., Przystałka, P.: Influence of outliers introduction on predictive models quality. Commun. Comput. Inf. Sci. **613**, 79–93 (2016)
10. Kalisch, M., Michalak, M., Sikora, M., Wróbel, Ł., Przystałka, P.: Data intensive vs sliding window outlier detection in the stream data — an experimental approach. In: Rutkowski, L., Korytkowski, M., Scherer, R., Tadeusiewicz, R., Zadeh, L.A., Zurada, J.M. (eds.) ICAISC 2016. LNCS (LNAI), vol. 9693, pp. 73–87. Springer, Heidelberg (2016). doi:10.1007/978-3-319-39384-1_7
11. Kuna, H., Garcia-Martinez, R., Villatoro, F.: Outlier detection in audit logs for application systems. Inf. Syst. **44**, 22–33 (2014)
12. Pigott, T.: A review of methods for missing data. Educ. Res. Eval. **7**(4), 353–383 (2001)
13. Ramaswamy, S., Rastogi, R., Shim, K.: Efficient algorithms for mining outliers from large data sets. In: Proceedings of ACM SIGMOD International Conference on Management of Data, pp. 427–438 (2000)
14. Sadik, S., Gruenwald, L.: Research issues in outlier detection for data streams. ACM SIGKDD Explor. Newsl. **1**(15), 33–40 (2013)

Ranking and Clustering

Exploiting Group Pairwise Preference Influences for Recommendations

Kunlei Zhu[1], Jiajin Huang[1(✉)], and Ning Zhong[1,2]

[1] International WIC Institute, Beijing University of Technology, Beijing 100024,
China
s201302176@emails.bjut.edu.cn
[2] Department of Life Science and Informatics, Maebashi Institute of Technology,
Maebashi-City 371-0816, Japan

Abstract. Recommender systems always recommend items to a user based on predicted ratings. However, due to biases of different users, it is not easy to know a user's preference through the predicted ratings. This paper defines a user preference relationship based on the user's ratings to improve the recommendation accuracy. By considering group information, we extend the preference relationship to form four types of correlations including (user, item), (user group, item), (user, item group), and (user group, item group). And then, this paper exploits pair-wise comparisons between two items or two group of items for a singer user or a group of users. The gradient descent algorithm is used to learn latent factors on partial orders to make recommendations. Experimental results show the effectiveness of the proposed method.

Keywords: User's preference · Group information · Choice model

1 Introduction

Recommender systems aim to help users find what they may like. Recommendation algorithms usually assume that the higher rating indicates that the more satisfaction of a user to an item [1]. For example, a collaborative filtering algorithm assumes that similar users like similar items. Therefore, rating vectors for both users and items are utilized to calculate similarities of items or users [2,12]. Based on users' ratings, matrix factorization methods [6] have been proposed for collaborative filtering. The main underlying idea is to learn a latent feature vector to represent each user and item, and predict ratings by the inner product of the user and item latent feature vectors.

Instead of considering one certain type of correlations in a specific situation, recent studies incorporate the group information into recommender systems to

This work was partly supported by National Natural Science Foundation of China (61562009, 61420106005).

V. Flores et al. (Eds.): IJCRS 2016, LNAI 9920, pp. 429–438, 2016.
DOI: 10.1007/978-3-319-47160-0_39

improve recommendation quality [8,13]. In these methods, a set of users, a set of items, clusters of users, and clusters of items form four types of correlations, including (user, item), (user, item group), (user group, item), and (user group, item group) [5,13]. Specially, Wang et al. [13] combined all those four types of correlations into a single unified algorithm.

In the above methods, ratings of users on items are directly utilized for calculation. However, only from ratings, we sometimes cannot tell the actual user preference on this item. As mentioned in [7] , a user gave three stars to an item which he may like, but another user gave three stars to the item which he may not like. In order to solve the problem, Rendle et al. [11] proposed the bayesian personalized ranking (BPR) method based on pairwise comparisons between rated and non-rated items. Pan et al. [9,10] extended the BPR method by considering (user group, item) or (user, item group). In fact, the numeric order of ratings given by a user provides a preference order of items for the user. The user prefers one item with a higher rating to another item with a lower rating. Under such assumption, Li et al. [7] proposed an approach to model users preference by employing utility theory for (user, item). Tran et al. [14] proposed a probabilistic model over ordered partitions by considering (user group, item). Huang et al. [5] proposed a framework of unifying (user, item), (user, item group), (user group, item), and (user group, item group) by using user preference from a qualitative view.

Following the assumption in [7] that a user's rating behavior is indeed a choice process, we propose a model through employing a choice model from the utility theory based on user preferences by considering four types of correlations, including (user, item), (user, item group), (user group, item), and (user group, item group). The group information is added into the model to improve recommendation quality.

2   Formulation of the Model

2.1   Preference Relations

Let us recall the following theorem with respect to a weak order $\succ$ satisfying the totality, antisymmetry, and transitivity.

Theorem 1 [3]. Let X be a nonempty set and $\succ$ a binary relation on X. If and only if $\succ$ is a weak order, there exists a real-valued function $f : X \longrightarrow R$ satisfying $a \succ b \Longleftrightarrow f(a) > f(b)$.

The function f is defined as a strictly monotonic increasing transformation. f is commonly known as a utility function measuring. For an ordinal scale, it is only meaningful to examine the order induced by the utility function rather the actual values of the function.

In the paper, we define a user set $\mathcal{U} = \{u_1, u_2, \cdots, u_m\}$ and an item set $\mathcal{V} = \{p_1, p_2, \cdots, p_n\}$, where u_m indicates the number of users is m and p_n indicates the number of items is n. A user u_i can give a rating r_{ij} to an item p_j.

For user $u \in \mathcal{U}$, we consider a simple partition based on a clustering method. Let E_u be an equivalence relation on $\mathcal{U}$, and E_p an equivalence relation on $\mathcal{V}$. They produce quotient user and item spaces $\mathcal{U}/E_u$ and $\mathcal{V}/E_p$, respectively. Let $[u]$ denote the equivalence class containing the user u, and $[p]$ the equivalence class containing the item p. Let M be the number of user clusters $\mathcal{U}/E_u$, and N the number of item clusters $\mathcal{V}/E_p$. We have:

$$
\begin{aligned}
& p_j \succ_u p_{j'} \Leftrightarrow u \; prefers \; item \; p_j \; to \; p_{j'} \\
& [p_j] \succ_u [p_{j'}] \Leftrightarrow \; user \; u \; prefers \; item \; group \; [p_j] \; to \; [p_{j'}] \\
& p_j \succ_{[u]} p_{j'} \Leftrightarrow \; user \; group \; [u] \; prefers \; item \; p_j \; to \; p_{j'} \\
& [p_j] \succ_{[u]} [p_{j'}] \Leftrightarrow \; user \; group \; [u] \; prefers \; item \; group \; [p_j] \; to \; [p_{j'}]
\end{aligned}
\tag{1}
$$

In fact, they correspond to the ordered user and item spaces $(\mathcal{V}, \succ_u)$ denoted by $\succ_{up}$, $(\mathcal{V}/E_p, \succ_u)$ denoted by $\succ_{u[p]}$, $(\mathcal{V}, \succ_{[u]})$ denoted by $\succ_{[u]p}$, and $(\mathcal{V}/E_p, \succ_{[u]})$ denoted by $\succ_{[u][p]}$, respectively. From Theorem 1, for preference relations satisfying weak orders, we have following real-valued functions.

$$
\begin{aligned}
& p_j \succ_u p_{j'} \Leftrightarrow f_u(p_j) > f_u(p_{j'}) \\
& [p_j] \succ_u [p_{j'}] \Leftrightarrow f_u([p_j]) > f_u([p_{j'}]) \\
& p_j \succ_{[u]} p_{j'} \Leftrightarrow f_{[u]}([p_j]) > f_{[u]}([p_{j'}]) \\
& [p_j] \succ_{[u]} [p_{j'}] \Leftrightarrow f_{[u]}([p_j]) > f_{[u]}([p_{j'}])
\end{aligned}
\tag{2}
$$

2.2 Estimation of Functions

According to [7], the utility function is decomposed into two parts based on the random utility model. The formulation is as follows:

$$
\begin{aligned}
f_u(p) &= v(u,p) + \varepsilon_{up} \\
f_u([p]) &= \frac{1}{\|[p]\|} \sum_{p_j \in [p]} v(u, p_j) + \varepsilon_{u[p]} = v(u, [p]) + \varepsilon_{u[p]} \\
f_{[u]}(p) &= \frac{1}{\|[u]\|} \sum_{u_i \in [u]} v(u_i, p) + \varepsilon_{[u]p} = v([u], p) + \varepsilon_{[u]p} \\
f_{[u]}([p]) &= \frac{1}{\|[u]\|} \frac{1}{\|[p]\|} \sum_{u_i \in [u]} \sum_{p_j \in [p]} v(u_i, p_j) + \varepsilon_{[u][p]} = v([u], [p]) + \varepsilon_{[u][p]}
\end{aligned}
\tag{3}
$$

where the first part is what we observed and the second part is some unobserved factors like emotion, weather or even some occurrent events [7]. In our case, the latent factor based predicted rating can be used to qualify the observed part by borrowing matrix factorization techniques, i.e. $v(c, p) = r_{up}$, $v([u], p)$ is a average rating value of a group of users $[u]$ on an item p, $v(u, [p])$ is a average rating value of a user u on a group of items $[p]$, $v([u], [p])$ is a average rating value of a group of users $[u]$ on a group of items $[p]$.

Following the work in [7], the probability of user preference over alternatives can be defined in terms of the utility of choice. We can extend the above result to the following probability of preference of a single user on a single item, a single user on a group of items, a group of users on a single item, and a group of users on a group of items, respectively.

$$Pr(p_j \succ_u p_{j'}) = \frac{e^{v(u,p_j)}}{\sum_n e^{v(u,p_{j'})}}$$

$$Pr([p_j] \succ_u [p_{j'}]) = \frac{e^{v(u,[p_j])}}{\sum_N e^{v(u,[p_{j'}])}}$$

$$Pr(p_j \succ_{[u]} p_{j'}) = \frac{e^{v([u],p_j)}}{\sum_n e^{v([u],p_{j'})}} \qquad (4)$$

$$Pr([p_j] \succ_{[u]} [p_{j'}]) = \frac{e^{v([u],[p_j])}}{\sum_N e^{v([u],[p_{j'}])}}$$

For the whole observations, we have the the probabilities as follows:

$$Pr(\succ_{up}) = \prod_{u_i \in \mathcal{U}} \prod_{p_j,p_{j'} \in \mathcal{V}} Pr(p_j \succ_{u_i} p_{j'})$$

$$Pr(\succ_{u[p]}) = \prod_{u_i \in \mathcal{U}} \prod_{[p_j],[p_{j'}] \in \mathcal{V}/E_p} Pr([p_j] \succ_{u_i} [p_{j'}])$$

$$Pr(\succ_{[u]p}) = \prod_{[u_i] \in \mathcal{U}/E_u} \prod_{p_j,p_{j'} \in \mathcal{V}} Pr(p_j \succ_{[u_i]} p_{j'}) \qquad (5)$$

$$Pr(\succ_{[u][p]}) = \prod_{[u_i] \in \mathcal{U}/E_u} \prod_{[p_j],[p_{j'}] \in \mathcal{V}/E_p} Pr([p_j] \succ_{[u_i]} [p_{j'}])$$

2.3   Optimization Problem

As the matrix factorization can predict the rating of user u_i on item p_j by the inter product of the user latent factor vector U_i and item latent factor vector V_j, we can use the predicted rating $U_i V_j$ to qualify the observed utility. We assume U, V is in accordance with multivariate Gaussian distribution as shown in [7]:

$$Pr(\Omega|\Theta) = N(\Omega|0, \sigma I) = \lambda e^{-\frac{\sum_l^L \Omega_l}{2\sigma^2}} \qquad (6)$$

where $\Omega = \{U, V\}$, Θ denotes some hyper-parameters, Ω_l is a component of Ω, σ is the standard deviation of Gaussian distribution, U, V as latent matrices.

The preference space is the feature space of users and items. Then we choose the K-means method to cluster users or items to form item group and user group. Considering the four correlations of (user,item), (user, item group), (user group, item), and (user group, item group), we use the matrix factorization technique to get four types of latent factors of users and items, namely U_{global} and V_{global}, U_{pg} and V_{pg}, U_{ug} and V_{ug}, U_{upg} and V_{upg}, respectively. The preferences of users on items can be represented by a product of user and item latent matrices.

As the optimization method comes from the studies of Li et al. [7] on the correlations of (user, item), we start with the extended version on the correlations of (user, item group) and briefly introduce learning methods from other relations.

Learn from the Relation of a Single User and a Group of Items. According to [7], we can get the following equation for the (user, item group) correlation:

$$Pr(U_{pg}, V_{pg}| \succ_{u[p]}) \propto Pr(\succ |U_{pg}, V_{pg})Pr(U_{pg}|\Theta)Pr(V_{pg}|\Theta) \qquad (7)$$

So, we can minimize the negative log form of the posterior to learn latent variable:

$$\Omega = argmin_\Omega - [logPr(\succ_{u[p]} |\Omega) + logPr(\Omega|\Theta)] \qquad (8)$$

where $Pr(\succ_{u[p]} |\Omega)$ is deemed as regularizer to alleviate overfitting problem.

We use a gradient descent algorithm to learn latent factors. In this method, we start with an arbitrary variable Ω_l, and compute the corresponding gradient $\nabla\Omega_l$ as shown in Eq. (9). We update the variable in the direction of steepest descent (i.e., along the negative of the gradient), $\Omega_l \leftarrow \Omega_l - \eta\nabla\Omega_l$, with the step size η. Then we repeat the process.

$$\nabla\Omega_l(\succ_{u[p]}) = -\frac{\partial logPr(\succ_{u[p]} |\Theta)}{\partial\Omega_l} - \frac{\partial logPr(\Omega|\Theta)}{\partial\Omega_l} \qquad (9)$$

Because $\frac{logPr(\Omega|\Theta)}{\partial\Omega_l}$ consists of regularizer, so we can get

$$-\frac{logPr(\Omega|\Theta)}{\partial\Omega_l} = \lambda\Omega_l \qquad (10)$$

Based on $Pr(\succ_{u[p]})$ in Eq. (5), we can present the first part of Eq. (9) as

$$\sum_{u_i\in\mathcal{U}} \sum_{[p_j],[p_{j'}]\in\mathcal{V}/E_p} \frac{\partial log\ Pr([p_j]\succ_{u_i} [p_{j'}])}{\partial\Omega_l}$$
$$= \sum_{u_i\in\mathcal{U}} \sum_{[p_j]\in\mathcal{V}/E_p} \frac{\partial[log(\sum_N e^{v(u_i,[p_{j'}])})-v(u_i,[p_j])]}{\partial\Omega_l} \qquad (11)$$

The descent can be obtained by

$$\nabla U_{pg_i} = \sum_N (\frac{\sum_N(e^{U_{pg_i}\cdot V_{pg_{j'}}} \cdot V_{pg_{j'}})}{\sum_N e^{U_{pg_i}\cdot V_{pg_{j'}}}} - V_{pg_j}) + \lambda U_{pg_i}$$
$$\nabla V_{pg_j} = \sum_m (\frac{e^{U_{pg_i}\cdot V_{pg_j}} \cdot V_{pg_j}}{\sum_N e^{U_{pg_i}\cdot V_{pg_{j'}}}} - U_{pg_i}) + \lambda V_{pg_j} \qquad (12)$$

According to the gradient descent algorithm, we can get $U_{pg_i} \leftarrow U_{pg_i} - \eta\nabla U_{pg_i}$ and $V_{pg_j} \leftarrow V_{pg_j} - \eta\nabla V_{pg_j}$.

Learn from the Relation of a Single User and a Single Item. We regard every item belongs to a same item collection $\mathcal{V}$. Similarly, we use $\succ_{up}$ instead of $\succ_{u[p]}$ in Eq. (11), and U_{global}, V_{global} instead of U_{pg}, V_{pg} in Eq. (12). For $\nabla \Omega_l(\succ_{up})$, we can get:

$$
\begin{aligned}
\nabla U_{global_i} &= \sum_n \left(\frac{\sum_n (e^{U_{global_i} \cdot V_{global_j}} \cdot V_{global_j})}{\sum_n e^{U_{global_i} \cdot V_{global_j}}} - V_{global_j} \right) + \lambda U_{global_i} \\
\nabla V_{global_j} &= \sum_m \left(\frac{e^{U_{global_i} \cdot V_{global_j}} \cdot U_{global_i}}{\sum_n e^{U_{global_i} \cdot V_{global_j}}} - U_{global_i} \right) + \lambda V_{global_j}
\end{aligned}
\tag{13}
$$

According to the gradient descent algorithm, we can get $U_{global_i} \leftarrow U_{global_i} - \eta \nabla U_{global_i}$ and $V_{global_j} \leftarrow V_{global_j} - \eta \nabla V_{global_j}$.

Learn from the Relation of a Group of Users and a Single Item. Similarly, learning from the relation of a user and a group of items, we use $\succ_{[u]p}$ instead of $\succ_{u[p]}$ in Eq. (11) , and U_{ug}, V_{ug} instead of U_{pg}, V_{pg} in Eq. (12). The descent can be obtained by

$$
\begin{aligned}
\nabla U_{ug_i} &= \sum_n \left(\frac{\sum_n (e^{U_{ug_i} \cdot V_{ug_{j'}}} \cdot V_j)}{\sum_n e^{U_{ug_i} \cdot V_{ug_{j'}}}} - V_{ug_j} \right) + \lambda U_{ug_i} \\
\nabla V_{ug_j} &= \sum_M \left(\frac{e^{U_{ug_i} \cdot V_{ug_j}} \cdot U_{ug_i}}{\sum_n e^{U_{ug_i} \cdot V_{ug_{j'}}}} - U_{ug_i} \right) + \lambda V_{ug_j}
\end{aligned}
\tag{14}
$$

According to the gradient descent algorithm, we can get $U_{ug_i} \leftarrow U_{ug_i} - \eta \nabla U_{g_i}$ and $V_{ug_j} \leftarrow V_{ug_j} - \eta \nabla V_{g_j}$.

Learn from the Relation of a Group of Users and a Group of Items. Similarly, learned from the relation of a user and a group of items, we use $\succ_{[u][p]}$ instead of $\succ_{u[p]}$ in Eq. (11) , and U_{global}, V_{global} instead of U_g, V_g in Eq. (12). The descent can be obtained by

$$
\begin{aligned}
\nabla U_{upg_i} &= \sum_N \left(\frac{\sum_N (e^{U_{upg_i} \cdot V_{upg_{j'}}} \cdot V_{upg_{j'}})}{\sum_N e^{U_{upg_{i'}} \cdot V_{upg_{j'}}}} - V_{upg_j} \right) + \lambda U_{upg_i} \\
\nabla V_{upg_j} &= \sum_M \left(\frac{e^{U_{upg_i} \cdot V_{upg_j}} \cdot U_{upg_i}}{\sum_N e^{U_{upg_{i'}} \cdot V_{upg_{j'}}}} - U_{upg_i} \right) + \lambda V_{upg_j}
\end{aligned}
\tag{15}
$$

According to the gradient descent algorithm, we can get $U_{upg_i} \leftarrow U_{upg_i} - \eta \nabla U_{upg_i}$ and $V_{upg_j} \leftarrow V_{upg_j} - \eta \nabla V_{upg_j}$.

2.4   Recommendation

According to the gradient descent algorithm, we can learn U, V. And then, we have $E_{global} = U_{global} \cdot V_{global} + C$, $E_{item} = U_{pg} \cdot V_{pg} + C$, $E_{user} = U_{ug} \cdot V_{ug} + C$, and $E_{userAndItem} = U_{uig} \cdot V_{uig} + C$, where C is a constant.

This paper defines U_{global} and V_{global} which contain information of all users and items as the global matrix, so we can get the utility E_{global}. The utility including information of user group is denoted by E_{user}, the utility including the information of item group is denoted by E_{item}, and the utility including the information of user and item group is denoted by $E_{userAndItem}$.

Because the global matrix contains the information of all users and items, we linearly combine the result of E_{global} with other utility by $E_{item} \leftarrow E_{item} + E_{global}$, $E_{user} \leftarrow E_{user} + E_{global}$, and $E_{userAndItem} \leftarrow E_{userAndItem} + E_{global}$.

Then, we use the hybrid recommendation technology to fuse the results of these utilities. We set α, β, and γ for weighting E_{item}, E_{user}, and $E_{userAndItem}$, respectively, where $\alpha + \beta + \gamma = 1$. So we can get

$$E = \alpha E_{item} + \beta E_{user} + \gamma E_{userAndItem} \tag{16}$$

3 Experiments

3.1 Data Sets

Our experiments was carried out using a public data set Movielens. The data set contains 100,000 anonymous ratings of approximately 1,682 movies made by 943 MovieLens users. We randomly divide the data set into a training set and a test set. For each user we randomly selected only 80 % of rated movies for training and withheld from the result 20 % of the data for testing. We perform a five-fold cross validation in order to avoid any algorithm achieving good results by chance.

3.2 Metrics

We use the recall and precision rate [4] to evaluate the effectiveness of the recommender system. Let $R(u)$ be the set of the recommended items for user u, $T(u)$ be of the set of items which user u rated in the test set, and $|\cdot|$ be the cardinality of a set. We have the recall and precision by $recall = \frac{\sum_{u \in \mathcal{U}} |R(u) \cap T(u)|}{\sum_{u \in \mathcal{U}} |T(u)|}$ and $precision = \frac{\sum_{u \in \mathcal{U}} |R(u) \cap T(u)|}{\sum_{u \in \mathcal{U}} |R(u)|}$.

3.3 Experimental Results

Impact of the Number of Clusters of K-means. In the K-means clustering algorithm, the number of clusters K is indeterminacy. Too small K may make the similarity between users or items small, and too large K may cause that a cluster includes few items or users, which makes data spare. The K-means clustering algorithm is sensitive to the number of clusters, which will affect recommendation quality. Therefore, we vary the number of K-means clusters by varying the number K from 20 to 60. As shown in Fig. 1, when $30 < K < 50$, with the increasing K, the values of precision and recall increase. When K is more than 50, the values of precision and recall decrease. So we have the best result with $K = 50$.

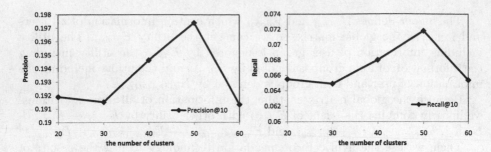

Fig. 1. Impact of the number of clusters

Impact of the Number of Iterations. As discussed in the earlier section, we use gradient descent to get U, V. In the gradient descent method, the number of iterators will affect the convergence rate. In order to confirm the value of the number of iterators, we set the number of iterators as $40, 100, 150, 200$.

In this part, we plot the result of recommending top 10 items with the increment of the iteration step. From Fig. 2, we can see that the best result is obtained in about 100 iteration rounds. For the values of precision and recall, both of them increase at first rapidly as the iterator step increases. At around 100 iterations, they reach the peak and after that both drop, which means that more iterator steps lead to worse results. After 150 iterators, the values of them are constant, which means the proposed approach is able to converge in a limited number of iterators.

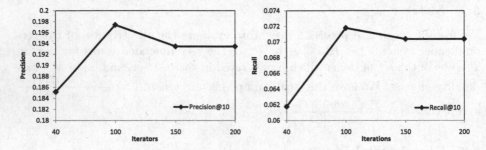

Fig. 2. Impact of the number of iterations

Impact of Parameters. From Eq. (16), we know that the weights of four correlations will affect the finally result. So we set four groups of these parameters to evaluate those impacts. At the time, we recommend top 20 items for a user, and we set the number of clusters as 50. From Table 1, we can see when $\alpha = 0.1$, $\beta = 0.1$, $\gamma = 0.1$, $\lambda = 0.7$, the recommendation result is the best.

Table 1. the result of different $\alpha, \beta, \gamma, \lambda$

Parameters	Precision	Recall
$\alpha = 0.4\ \beta = 0.2\ \gamma = 0.2\ \lambda = 0.2$	0.1935	0.0692
$\alpha = 0.5\ \beta = 0.2\ \gamma = 0.2\ \lambda = 0.1$	0.1906	0.0672
$\alpha = 0.1\ \beta = 0.6\ \gamma = 0.2\ \lambda = 0.1$	0.1924	0.0687
$\alpha = 0.1\ \beta = 0.1\ \gamma = 0.1\ \lambda = 0.7$	0.1974	0.0718
$\alpha = 0.1\ \beta = 0.1\ \gamma = 0.0\ \lambda = 0.8$	0.1963	0.0710

Comparative Results of Different Models. In our experiments, we use the following state-of-the-art methods to evaluate for comparison. NEVM is comparative choice model, which is a simplified version of the study of Li et al. [7] without the influence of items recently used by users. The model that we propose is called EVM. EVM is compared with NEVM, user-based collaborative filtering (UCF) [2], item-based collaborative filtering (ICF) [12] and vector space model (VSM) to verify recommendation quality.

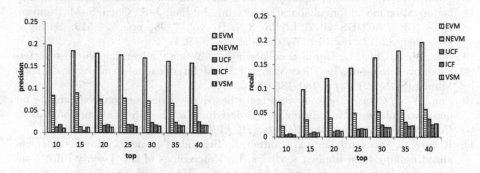

Fig. 3. Precision and recall of different models

The experimental result is shown in Fig. 3. From Fig. 3, we can see that among that baseline methods, our model achieve the best recommendation result in both metrics on the data set. As to precision and recall, when we recommend top 10 items, the improvements compared with NEVM are 11.35 % and 5 %. Through the result of our experiments, we can easily observe that the preference model and four correlations play important roles in personalized recommendations.

4   Conclusion

In this paper, we define user's preference relationships based on rating scores. Considering group information, we extend the preference relationship to form four type of relationships. And then, we use utility theory functions to measure preference relationships. In the future, we will exploit the user group preferences to other models instead of only the choice model proposed in [7].

References

1. Adomavicius, G., Tuzhilin, A.: Toward the next generation of recommender systems: a survey of the state-of-the art and possible extensions. IEEE Trans. Knowl. Data Eng. **17**(6), 734–749 (2005)
2. Breese, J., Hecherman, D., Kadie, C.: Empirical analysis of predictive algorithms for collaborative filtering. In: Proceedings of the 14th Conference on Uncertainty in Artificial Intelligence (UAI 1998), pp. 43–52 (1998)
3. French, S.: Decision theory - an introduction to the mathematics of rationality. J. Am. Stat. Assoc. **29**(2), 212–213 (1986)
4. Herlocker, J., Konstan, J.A., Terveen, L.G., Riedl, J.T.: Evaluating collaborative filtering recommender systems. ACM Trans. Inf. Syst. **22**(1), 5–53 (2004)
5. Huang, J.J., Zhong, N., Yao, Y.Y.: A unified framework of targeted marketing using user preferences. Comput. Intell. **30**(3), 451–472 (2014)
6. Koren, Y., Bell, R.M., Volinsky, C.: Matrix factorization techniques for recommender systems. IEEE Comput. **42**(8), 30–37 (2009)
7. Li, X., Xu, G.D., Chen, E.H., Zhong, Y.: Learning recency based comparative choice towards point-of-interest recommendation. Expert Syst. Appl. **42**(2015), 4274–4283 (2015)
8. Mashhoori, A., Hashemi, S.: Incorporating hierarchical information into the matrix factorization model for collaborative filtering. In: Pan, J.-S., Chen, S.-M., Nguyen, N.T. (eds.) ACIIDS 2012. LNCS (LNAI), vol. 7198, pp. 504–513. Springer, Heidelberg (2012). doi:10.1007/978-3-642-28493-9_53
9. Pan, W.K., Chen, L.: CoFiSet: collaborative filtering via learning pairwise preferences over item-sets. In: Proceedings of SIAM International Conference on Data Mining (SDM 2013), pp. 180–188 (2013)
10. Pan, W.K., Chen, L.: Group preference based Bayesian personalized ranking for one-class collaborative filtering. In: Proceedings of the 23rd International Joint Conference on Artificial Intelligence (IJCAI 2013), pp. 2691–2697 (2013)
11. Rendle, S., Freudenthaler, C., Gantner, Z., Schmidt-Thieme, L.: Bayesian personalized ranking from implicit feedback. In: Proceedings of the Twenty-Fifth Conference on Uncertainty in Artificial Intelligence (UAI 2009), pp. 452–461 (2009)
12. Sarwar, B., Karypis, G., Konstan, J., Riedl, J.: Item-based collaborative filtering recommendation algorithms. In: Proceedings of the 10th International World Wide Web Conference (WWW 2001), pp. 285–295 (2001)
13. Wang, X., Pan, W.K., Xu, C.: Hierarchical group matrix factorization for collaborative recommendation. In: Proceedings of the 23rd ACM International Conference on Information and Knowledge Management (CIKM), pp. 769–778 (2014)
14. Tran, T., Phung, D., Venkatesh, S.: Modelling human preferences for ranking and collaborative filtering: a probabilistic ordered partion approach. Knowl. Inf. Syst. **74**(1), 157–188 (2016)

Discrete Group Search Optimizer
for Community Detection in Social Networks

Moustafa Mahmoud Ahmed[2,3](✉), Mohamed M. Elwakil[1,4],
Aboul Ella Hassanien[1,3], and Ehab Hassanien[1]

[1] Faculty of Computers and Information, Cairo University, Giza, Egypt
{m.elwakil,e.ezat}@fci-cu.edu.eg, aboitcairo@gmail.com
[2] Faculty of Computer and Information, Minia University, Minya, Egypt
moustafa.ali@mu.edu.eg
[3] Scientific Research Group in Egypt (SRGE), Cairo, Egypt
[4] Software Engineering Lab, Innopolis University, Innopolis, Russia
m.elwakil@innopolis.ru
http://www.egyptscience.net

Abstract. Discovering community structure in complex networks has
been intensively investigated in recent years. Community detection can
be treated as an optimization problem in which an objective fitness
function is optimized. Intuitively, the objective fitness function captures
the subgraphs in the network that has densely connected nodes with
sparse connections between subgraphs. In this paper, we propose Discrete Group Search Optimizer (DGSO) which is an efficient optimization algorithm to solve the community detection problem without any
prior knowledge about the number of communities. The proposed DGSO
algorithm adopts the locus-based adjacency representation and several
discrete operators. Experiments in real life networks show the capability of the proposed algorithm to successfully detect the structure hidden
within complex networks compared with other high performance algorithms in the literature.

Keywords: Social network · Community detection · Complex network ·
Unsupervised learning · Group search optimizer

1 Introduction

Discovering communities hidden within the structure of complex networks has
a significant practical importance for many fields such as sociology, physics, and
biology. Community detection in networks can be defined as dividing a network
into a set of internally densely connected groups of nodes, that has sparse connections in-between. Over the last few years, the problem of community detection
has received a lot of attention and many different approaches have been proposed
in different fields of research: computer science, physics, sociology, and others.
Results of a recent survey can be seen in [1].

Recently, He et al. proposed a swarm intelligence optimization algorithm,
called group search optimizer (GSO) [2]. This algorithm mimics the searching

© Springer International Publishing AG 2016
V. Flores et al. (Eds.): IJCRS 2016, LNAI 9920, pp. 439–448, 2016.
DOI: 10.1007/978-3-319-47160-0_40

behavior of animals. Considering the efficiency of GSO algorithm, we propose to extend it into a discrete group search optimizer (DGSO) algorithm for the community detection problem. We employ the optimization mechanism of the basic GSO algorithm with two modifications. First, we avoid the angle evolution strategy. Second, we propose new evolution operations in the producer, scrounger, and ranger phases. Experiments on real life networks show the ability of the DGSO algorithm to correctly detect communities with results comparable to the state-of the-art approaches.

The rest of the paper is organized as follows. In Sect. 2, we define the community detection problem and introduce the objective functions adapted in this paper as well as we describe the basic GSO algorithm. In Sect. 3, we describe our proposed algorithm. In Sect. 4, the results of the method on synthetic and real life networks are presented and discussed. In Sect. 5, we give concluding remarks.

2   Preliminaries

In this section, we will provide a brief background on the community detection problem and optimization problem, and the group search optimizer algorithm.

2.1   The Community Detection Problem

A network can be defined as a graph $G = (V, E)$, in which V is the set of nodes, and E is a set of ties that connect nodes. In the field of social networks, nodes represent persons or actors within the network, and ties represent the relationships or the interaction between those persons. A community structure S in a network is a set of groups of nodes such that each group is densely connected internally and sparsely connected with other groups. So this problem can be defined as dividing network's nodes into k disjoint communities, where the number k is unknown, that best satisfy a given quality measure of communities $F(S)$. Thus, we treated this problem as an optimization problem in which one usually wants to optimize the given quality measure $F(S)$. A single objective optimization problem $(\Omega; F)$ is formulated as in the Eq. 1.

$$min \ f(S), \ s.t \ S \in \Omega \tag{1}$$

Where $F(S)$ is an objective function that needs to be optimized, and $\Omega = \{S_1, S_2, .., S_r\}$ is the set of feasible community structures in a network.

2.2   Group Search Optimizer

The GSO algorithm was proposed by He [2]. This algorithm simulates animal searching (foraging) behavior. The basic variant of the GSO algorithm works by having a population (called a group) of candidate solutions (called members). Each member in the group has its own position, search angle, and search direction. In the GSO algorithm, a group contains three kinds of members:

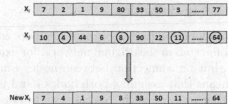

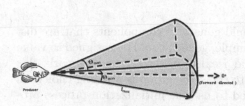

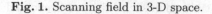

Fig. 1. Scanning field in 3-D space.   **Fig. 2.** The movement from x_i to x_j with Step =4.

producers and scroungers whose behaviors are based on the *producer-scrounger* (PS) model [3], and rangers who perform random transitions in the search space. At each iteration, the producers perform producing strategy to search for the positions containing the best resources. The producer's scanning field of vision is generalized to an n-dimensional space, which is specified by maximum pursuit angle $\theta_{max} \in R^1$ and maximum pursuit distance $l_{max} \in R^1$ as illustrated in a 3D space [4] in Fig. 1. The scroungers perform a following strategy to join resources found by the producers: the remaining members are the rangers that walk randomly in the searching space to stay in new positions. In the *GSO* algorithm, a position of the individual represents a solution of the optimization problem, and the fitness of the position represents the fitness of the solution. The basic *GSO* algorithm is discussed in [4].

3 The Proposed Discrete Group Search Optimizer *DGSO* for the Community Detection Problem

Owing to the continuous nature of the *GSO* algorithm, this algorithm doesn't directly fit for the community detection problem. So it's necessary to develop a suitable mapping which can efficiently convert individuals to solutions. In this paper we propose a discrete version of the *GSO* algorithm for the community detection problem. A detailed description of the proposed algorithm is introduced below.

3.1 Individual Representation

The DGSO algorithm used locus-based adjacency representation proposed in [5] to encode group members, a detailed description of this representation strategy can be found in [6]. To detect community structure, a decoding step is necessary to discover connected components. Each of these components corresponds to community, So the number of these components equals the number of communities in the discovered structure. Thus, there is no need to know in advance the number of communities.

3.2 Initialization

Randomly initializing group members could generates components that are disconnected in the original network, for example, gene g_i could be assigned to value j, but no connection between nodes i and j exists in the original network, this means that assigning both nodes i and j to the same group is a wrong choice. In order to avoid such this case, we proposed to use the initialization process proposed in [7] (safe initialization), which takes in account the effective connections of nodes in the social network. Using safe initialization such this case is avoided by substituting value j with one of the neighbors of i.

3.3 Producer

Group members that obtain the best fitness values are chosen as the producers. A producer tries to guide other group members to the food sources (optima). In nature, animals use vision or other senses, to realize the concentration of food in the environment, to determine the direction of the next movement. In our algorithm, the scanning field of vision is simplified and limited by maximum pursuit distance l_{max}, which is a selected constant number $\in [0, 1]$. In our algorithm, the producer behaves as follows:

1. A producer scans the search space by randomly selecting three points in the scanning field, let x_p is the producer's current state and x_1, x_2, x_3 are the randomly selected states in the x_p's visual, where distance $(x_p, x_i) < l_{max}$ and $i \in \{1, 2, 3\}$.
2. Then, the producer selects the fittest point with the best resource. If this point has a better resource than producer's current position, then it will move a step to this point $Move(x_p, x_i)$. Otherwise it will stay in its current position.

Distance: Since there is no straightforward method to measure distance between two group members (solutions), we adopted the distance measure proposed in [8]. This measure uses Normalized Mutual Information (NMI) [9] to valuate the degree to which two solutions are close to each other as calculated in Eq. 2. NMI is a similarity measure proved to be robust and accurate by Danon et al. [9].

$$dis(x_i, x_j) = 1 - NMI(C(x_i), C(x_j)) \tag{2}$$

where $C(x)$ is the decode functions used to interpret group member state back to a community structure and $NMI(C(x_i), C(x_j))$ calculates the NMI similarity between the two community structures x_i and x_j.

Step: Represents the number of nodes copied from a solution x_p to a solution x_i to move a solution x_i a step in the direction of a solution x_p as illustrated in Fig. 2, where $Step \in [1..n]$.

Movement: We use a crossover operator used previously in genetic algorithms [6] where the mixing ratio is the step size of the move. So in order to move a group member x_i to group member x_j $Move(x_i, x_j)$, the two group member in the crossover operator are considered as the parents of the new offspring (new member state). The new group member state has randomly chosen Steps optimizing variables from x_j and the rest are from x_i as illustrated in Fig. 2.

Recently, Couzin et al. [10] found that, for large groups, only a very small proportion of informed individuals is needed to guide the group to achieve a high accuracy. So, for accuracy and simplicity, there is only one producer in the $DGSO$ algorithm, which means that the best member is the producer and the remaining members in the group are scroungers or rangers.

3.4   Scrounger

After selecting members that will perform producing behavior, the remaining members are distributed into scroungers and rangers, with the probability of P and $(1-P)$, respectively. The scroungers will continue searching for opportunities to join the resources found by the producer. In our algorithm each scrounger x_s randomly selects a producer x_p to move a step towards $Move(x_s, x_p)$.

3.5   Ranger

Rangers are the group members that randomly search in the search space, seeking to find other promising solutions that are yet to be refined. The purpose of this operation is to diverse the search in order to avoid getting trapped in a local optimum. Here each ranger x_r randomly selects a point x_i in the total search space to move a step towards $Move(x_r, x_i)$. If the rangers cannot find a better area after A iterations, a percent RP of the rangers are randomly selected to be mutated with a mutation rate MR, $Mutate(x_r)$. Regardless of whether the movement or the mutation process leads rangers to a better position (fitness value) than the original one, the rangers will do enhance the global search ability.

Mutation: Randomly changing values of a randomly chosen member's genes might causes a useless exploration of the search space. So, as in the initialization step, we propose to randomly select a percentage of the genes and for each selected gene i we randomly change its value to j such that node i and j are neighbors.

Similar to the course of evolution, once the fitness of new member generated by scroungers or rangers is better than the fitness of the producer, the producer will be updated.

3.6   Fitness Function

We decided to use Modularity [11], which is an effective quality function, to quantify and measure how "good" the discovered community structure is. Studies in the literature proved that modularity is effective in many kinds of complex networks [11].

The pseudo code of the DGSO algorithm processes are shown in Algorithm 1.

Data: A Network G =(V, E)

Result: Community membership assignment for each node in the
network G

1 initialization Population size *popsize*, Randomly initialize group
members, Maximum pursuit distance l_{max}, *Step*, Scrounging percent P,
Mutation percent MP, Mutation rate MR, Ranging trials A, Maximum
number of iterations $Max\_Iterations$

2 Calculate the fitness values of initial group members.

3. **while** *(Iteration number $\leq$ Max_Iterations)* **do**

    /* Perform producing.   */

4   Find the producer x_p of the group(the fittest member).

5   The producer randomly sample three points in the scanning field
using (2).

6   The fittest point with the best resource is chosen. If this point has a
better resource than producer's current position, then it will move a
Step to this point. Otherwise it will stay in its current position.

    /* Perform scrounging.   */

7   Randomly select P percent from the rest of the members to perform
scrounging, by moving a *Step* to words the producer.

    /* Perform ranging.   */

8   The remainder members leave their current position to perform
ranging, by randomly selecting a point x_i in the total search space to
move a step towards.

9   If the rangers can not find a better area after A iterations, a percent
RP of the rangers are randomly selected to be mutated with a
mutation rate MR.

10 **end**

Algorithm 1. DGSO Algorithm.

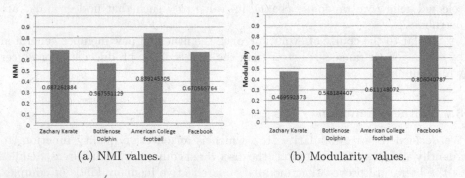

(a) NMI values.   (b) Modularity values.

Fig. 3. Average NMI and Modularity values of the result community structure on each
social network.

4   Experimental Results and Discussion

We tested our algorithm on four real life social networks: The Zachary Karate Club [12], The Bottlenose Dolphin network [13], American College football network [14], and Facebook Dataset [15]. The ground truth communities partitions for these networks are known. To compare the accuracy of the resulting community structures, we used Normalized Mutual Information (NMI) [9] to calculate the similarity between the true community structures and the discovered ones.

For each dataset, we applied the algorithm ten times. In each trial we calculated the NMI and Modularity values of the best solution. Then, we calculated the average NMI and average Modularity over the ten trials. The *DGSO* algorithm was applied with the following parameters values; $L_{max} = 0.8$, $Step = 0.2 * n$, population size *popsize* $= 200$, Scrounging percent $P = 80\%$ of popsize (Ranging percent$= 20\%$ of popsize), the maximum number of iterations $Max\_Iterations = 200$, and the number of ranging trials $A = 5$. Figure 3a and b show the average NMI value and the average Modularity values, respectively, for the community structures detected in each dataset. We can observe that our algorithm achieves high NMI values for all social networks. The Modularity value of the community structure detected by our algorithm is higher than the corresponding Modularity value of the ground truth division of those networks as shown in Fig. 7a. This means that, according to Modularity measure, our algorithm detects more modular community structures than the original ones.

To understand the results produced by the algorithm we visualized the community structure detected on the small size dataset. Figure 4 shows a visualization of the discovered structure for the Zachary network. The original structure of the network is indicated by the black thick line and the structure detected by our algorithm is indicated by nodes'colors. From this figure we can observe in the top level the result is similar to the original division of the network, however in the result structure each group is farther subdivided into two groups.

Figure 5 visualizes the result for the Dolphin network. The original structure of the network is indicated by the black thick line and the detected structure is indicated by nodes'colors. From this figure we can observe in the top level the result is similar to the original division of the network, however in the result structure, the right group is further subdivided into four groups.

Figure 6 visualizes the result obtained for the College football network. The original division of the network is visualized in Fig. 6a; where nodes' labels refer to the groups they assigned to. From Fig. 6a; we can observe that some groups such as 5,10 are sparsely connected internally, however they densely connected with other groups. This problem disappears in the community structure detected by our algorithm. From Fig. 6b; we can observe that our algorithm discovered a community structure with 9 groups which assigns nodes from the smaller groups, such as 5,10 into a larger groups leading to a more modular community structure.

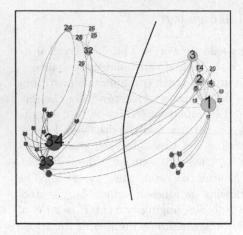

Fig. 4. Visualization of the result for the Zachary network.

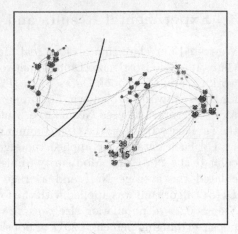

Fig. 5. Visualization of the result for the Dolphin network.

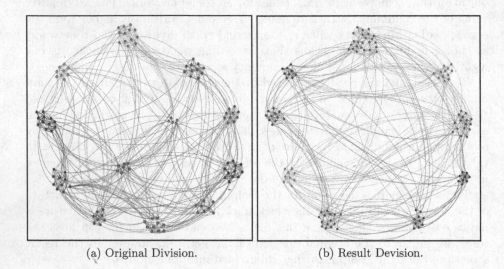

(a) Original Division.   (b) Result Devision.

Fig. 6. Visualizations of the result for the American College football network.

4.1   Comparison Analysis

Here, we can show practical comparison between the results obtained by DGSO algorithm and other seven well-known methods proposed in the literature, which are Infomap [16], Fast greedy [17], Label propagation [18], Maulilevel [19], Walktrap [20], leading Eigenvector [11], and Artificial fish swarm algorithm [8]. We applied each method 10 times on each dataset and the average NMI and the average Modularity of the best community structure is reported. Figure 7 summarizes the NMI and Modularity values for all methods. In terms of Modularity,

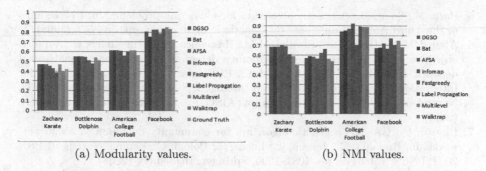

(a) Modularity values.   (b) NMI values.

Fig. 7. NMI and Modularity values for each dataset reported by each method.

DGSO is very competitive with other methods as shown in Fig. 7a. For the small size datasets we can observe that DGSO detects a community structure with a high Modularity value compared to all other methods. Regarding the Facebook datasets, DGSO competes with the seven methods with a very small difference. In terms of NMI, DGSO produces results seems to be bad compared to other methods as shown in Fig. 7b. However we could return this to the different high modular community structures our algorithm produced compared to the ground truth divisions.

5   Conclusion and Future Work

DGSO is an optimization technique that suits the community detection problem. Experiments with real world networks showed the ability of this method to correctly detect community structures based on the quality function used (modularity). DGSO has the advantage that, number of communities is not required to be specified as a prior setting. A comparison with other recently proposed methods shows that DGSO is very competitive with such methods. Enhancing the capabilities of this algorithm to discover communities in multi-dimensional social networks is a necessary task that can be investigated in future work.

References

1. Khatoon, M., Banu, W.A.: A survey on community detection methods in social networks. Int. J. Educ. Manage. Eng. (IJEME) **5**(1), 8 (2015)
2. He, S., Wu, Q., Saunders, J.: A novel group search optimizer inspired by animal behavioural ecology. In: Evolutionary Computation, 2006, CEC 2006, IEEE Congress, pp. 1272–1278. IEEE (2006)
3. Barnard, C.J., Sibly, R.M.: Producers and scroungers: a general model and its application to captive flocks of house sparrows. Anim. Behav. **29**(2), 543–550 (1981)
4. He, S., Wu, Q.H., Saunders, J.: Group search optimizer: an optimization algorithm inspired by animal searching behavior. IEEE Trans. Evol. Comput. **13**(5), 973–990 (2009)

5. Park, Y., Song, M.: A genetic algorithm for clustering problems. In: Proceedings of the Third Annual Conference on Genetic Programming, pp. 568–575 (1998)
6. Ahmed, M.M., Hafez, A.I., Elwakil, M.M., Hassanien, A.E., Hassanien, E.: A multi-objective genetic algorithm for community detection in multidimensional social network. In: Gaber, T., Hassanien, A.E., El-Bendary, N., Dey, N. (eds.) The 1st International Conference on Advanced Intelligent System and Informatics (AISI), November 28–30, 2015, Beni Suef, Egypt. AISC, pp. 129–139. Springer, Heidelberg (2016)
7. Pizzuti, C.: GA-Net: a genetic algorithm for community detection in social networks. In: Rudolph, G., Jansen, T., Lucas, S., Poloni, C., Beume, N. (eds.) PPSN 2008. LNCS, vol. 5199, pp. 1081–1090. Springer, Heidelberg (2008)
8. Hassan, E.A., Hafez, A.I., Hassanien, A.E., Fahmy, A.A.: Community detection algorithm based on artificial fish swarm optimization. In: Filev, D., et al. (eds.) Intelligent Systems'2014. AISC, vol. 323, pp. 509–521. Springer, Heidelberg (2015). doi:10.1007/978-3-319-11310-4_44
9. Danon, L., Diaz-Guilera, A., Duch, J., Arenas, A.: Comparing community structure identification. J. Stat. Mech. Theory Exp. **2005**(9), P09008 (2005)
10. Couzin, I.D., Krause, J., Franks, N.R., Levin, S.A.: Effective leadership and decision-making in animal groups on the move. Nature **433**(7025), 513–516 (2005)
11. Newman, M.E.: Finding community structure in networks using the eigenvectors of matrices. Phys. Rev. E **74**(3), 036104 (2006)
12. Zachary, W.W.: An information flow model for conflict and fission in small groups. J. Anthropol. Res. **33**, 452–473 (1977)
13. Lusseau, D.: The emergent properties of a dolphin social network. Proc. R. Soc. Lond. B Biol. Sci. **270**(Suppl 2), S186–S188 (2003)
14. Girvan, M., Newman, M.E.: Community structure in social and biological networks. Proc. Nat. Acad. Sci. **99**(12), 7821–7826 (2002)
15. Leskovec, J., Lang, K.J., Mahoney, M.: Empirical comparison of algorithms for network community detection. In: Proceedings of the 19th International Conference on World Wide Web, pp. 631–640. ACM (2010)
16. Rosvall, M., Axelsson, D., Bergstrom, C.T.: The map equation. Eur. Phys. J. Spec. Top. **178**(1), 13–23 (2010)
17. Clauset, A., Newman, M.E., Moore, C.: Finding community structure in very large networks. Phys. Rev. E **70**(6), 66–111 (2004)
18. Raghavan, U.N., Albert, R., Kumara, S.: Near linear time algorithm to detect community structures in large-scale networks. Phys. Rev. E **76**(3), 036106 (2007)
19. Blondel, V.D., Guillaume, J.L., Lambiotte, R., Lefebvre, E.: Fast unfolding of communities in large networks. J. Stat. Mech. Theory Exp. **2008**(10), P10008 (2008)
20. Pons, P., Latapy, M.: Computing communities in large networks using random walks. In: Yolum, I., Güngör, T., Gürgen, F., Özturan, C. (eds.) ISCIS 2005. LNCS, vol. 3733, pp. 284–293. Springer, Heidelberg (2005)

Social Web Videos Clustering
Based on Ensemble Technique

Vinath Mekthanavanh and Tianrui Li[✉]

School of Information Science and Technology, Southwest Jiaotong University,
Chengdu 611756, China
vinath.mek@gmail.com, trli@swjtu.edu.cn

Abstract. Currently, a massive amount of videos has become a challenging research area for social web videos mining. Clustering ensemble is a common approach to clustering problems, which combine a collection of clusterings into a superior solution. Textual features are widely used to describe a web video. Whereas, local and global features also have their own advantages to describe a web video as well. So we extract the local and global features as we called low-level/semantic features and high-level/visual features respectively to help to better describe a main source. In this paper, we propose a combining function of three similarity models to enhance the similarity values of videos, and then present a framework for Clustering Ensemble with the support of Must-Link constraint (CE-ML) to formulate in ensembling for clustering purposes. Experimental evaluation on the real world social web video has been performed to validate the proposed framework.

Keywords: Combining similarity · Pairwise constraint · Clustering ensemble · Social web videos mining

1 Introduction

Automatically web videos categorization is a promising direction to achieve the role to pre-defined categories of web videos. Therefore, this is a very challenging task to define the web videos within category into its category specification in an effective browsing as well as retrieving the mass amount of web videos. Several research works have been presented on this issue by utilizing the features derived from textual and visual contents. Based on the meta data (text) provided by up-loaders [1], further attempt is to extract low-level features by analyzing keyframes, audio, signal, etc., along with the textual features [2,3]. New genre or category related concepts (new technologies, domain dependent terms, etc.) appear every day. The training set should expand to incorporate all these new concepts, which makes the training very expensive. As the number of genres increases, the requirement for compound of data goes up.

Web video categorization make users to find his/her required videos. Among this motivation, many researchers try to propose any possible algorithms and

© Springer International Publishing AG 2016
V. Flores et al. (Eds.): IJCRS 2016, LNAI 9920, pp. 449–458, 2016.
DOI: 10.1007/978-3-319-47160-0_41

methods to deal with such the initial efforts in the direction of web video categorization made by [1,3,8] and then further improved by [9,10] on large scale video classification. In study of web video categorization presented in [3], semantic modalities (visual word, concept histogram) and surrounding text (title, tag) were utilized to complement low-level features. Wu et al. [4] explored techniques to boost the effectiveness of text classification for web video categorization by using contextual information associated with videos. Wang et al. [11] performed web video classification on large scale video data from 29 categories in YouTube. Recently, Mahmood et al. [13] proposed a framework for web video categorization by using the low cost of video dataset as textual features, including external information from web support like Google. Our main problem is how to enhance the text information in social media that provide very short text from up-loaders (*i.e.*, tile, tag, and description), which provide a little contextual information for clustering. The contextual resources arouse new perspectives for web video categorization. The related videos associated with a given video in social web (YouTube[1]) are usually relevant videos that may help to estimate the probability of the video categorization.

Semantic similarity is a choice with a wide range in data mining applications. The traditional TF/TF-IDF weighting schemes can not represent the semantic information of text or visual word in the keyframe of web videos. In this paper, we measure the semantic similarity from the local features (Bag of Word) by using a soft-weighting scheme to weight the significance of each visual word in the keyframe, which has been demonstrated in VIREO-374[2] to be more effective than the traditional TF/TF-IDF weighting scheme. We further include the visual similarity which plays more significantly than the other similarity that is similarity measurement of global features (ColorMoments) of web video in this research work.

Clustering ensemble offers an effective approach for aggregate multiple clustering results to improve the overall of clustering robustness and stability. Strehl et al. [5] developed a hypergraph partitioning called as ensemble methods. Wang et al. [6] presented a mixed-membership model for clustering ensemble based on Bayesian cluster ensembles. Gian et al. [7] proposed a sequential ensemble method to improve the clustering performance by using the local creation based information and clustering respectively. Tumer et al. [12] proposed a method called as voting active clusters (VACs) for combining base multiple base clusterings into a single unified ensemble clustering.

Semi-supervised method has received much attention in the last years, because it can enhance clustering quality by exploiting readily available background knowledge. By incorporating with the known prior knowledge, we here only mention the must-link since our aspect is to enforce the related videos with higher similarity with the index videos should belong to the same category. In this research work, we aim to achieve good enough similarity values between the videos in web video dataset in terms of categorization problems, by a fusion of

---

[1] http://www.youtube.com.

[2] http://vireo.cs.cityu.edu.hk/research/vireo374/.

different similarities like local similarity, global similarity and textual similarity before input to clustering ensemble methods.

The rest of this paper is organized as follows. In Sect. 2, all basic concepts will be used in the proposed framework are described. An overall of system overview, the details of feature extraction, the presentation of multimodality are shown in Sect. 3. Section 4 is an experimental framework and performance evaluation. Experimental results and corresponding comparison are presented in Sect. 5. The papers ends with conclusions and future work in Sect. 6.

2   Preliminaries

The basic concepts will be used in the proposed framework of social web videos clustering are cited in this section.

Definition 1 [14] Term Frequency-Inverse Document Frequency (TF-IDF). Suppose D is a document space, $d \in D$ and t is a term in D. The Term Frequency-Inverse Document Frequency (TF-IDF) of t to d in D is defined as follows.

$$TF - IDF(t, d, D) = TF(t, d) \times IDF(t, D) \tag{1}$$

Definition 2 [14] Original Similarity (OrS). Consider two documents $d_x = (w_{x1}, w_{x2}, ..., w_{xT})$ and $d_y = (w_{y1}, w_{y2}, ..., w_{yT})$. The original similarity between two documents can be calculated by using a normalized cosine similarity function defined as follows:

$$Sim(d_x, d_y)_{OrS} = \frac{\sum_{t=1}^{T}(w_{xt} * w_{yt})}{\sqrt{\sum_{t=1}^{T}(w_{xt})^2 * \sum_{t=1}^{T}(w_{yt})^2}} \tag{2}$$

Definition 3 Modified Cosine Similarity (MoS). Consider two documents $d_x = (w_{x1}, w_{x2}, ..., w_{xT})$ and $d_y = (w_{y1}, w_{y2}, ..., w_{yT})$. The modified cosine similarity is defined as follows:

$$Sim(d_x, d_y)_{MoS} = \frac{\sum_{t=1}^{T}(w_{xt} - \bar{w}) * (w_{yt} - \bar{w}')}{\sqrt{\sum_{t=1}^{T}(w_{xt} - \bar{w})^2 * \sum_{t=1}^{T}(w_{yt} - \bar{w}')^2}} \tag{3}$$

where T is the total number of terms, $\bar{w} = \frac{1}{T}\sum_{t=1}^{T} w_{xt}$, and $\bar{w}' = \frac{1}{T}\sum_{t=1}^{T} w_{yt}$.

Definition 4 [15]. Modified Hausdorff Distance (MHD) known as mean Hausdorff distance measures. Consider two non-empty keyframe sets X and Y of two videos. The local and global similarity between two keyframe sets can be calculated based on the following expression,

$$d_{MHD}(X, Y) = \max(d(X, Y), d(Y, X)) \tag{4}$$

$$d(X, Y) = \operatorname*{mean}_{x \in X} \min_{y \in Y} d\|x - y\|, d(Y, X) = \operatorname*{mean}_{y \in Y} \min_{x \in X} d\|y - x\| \tag{5}$$

where $\|.\|$ is the normal form, x and y are keyframes of set X and Y respectively.

3   Proposed Framework

3.1   System Overview

The framework of our research builds on text, local and global information in the documents from videos. Firstly, the bag-of-word of text information (e.g., title, tag and description) retrieval is assumed that the set of words in a document is a representative of video's content. We apply TF-IDF to find the weight of words (terms) in a document, then use the Vector Space Model with the Modified Cosine Similarity for a comparison of vectors. Secondly, we identify the local and global features from the keyframes of a video, where a set of keyframes is a representative of a video's meaning. Then, we apply the Modified Hausdorft Distance for the comparison of keyframe sets for each type of features. After that, we combine three types of similarities from the local, global and textual features to be a single similarity before input to the clustering models. This step is crucial due to the selection of weights of feature vectors value, which we aim to increase the weights of textual features. Furthermore, in the clustering purpose, two algorithms, *i.e.*, spectral clustering and graph partitioning, are selected. Finally, three graph-based cluster ensemble techniques, e.g., Clustering-based Similarity Partitioning Algorithm (CSPA), Hyper-Graph Partitioning Algorithm (HGPA) and Meta Clustering Algorithm (MCLA), are used to integrate the results as consensus functions. Pairwise constraints as ML are used to translate the related video information into the process of clustering ensembles. The proposed framework is illustrated in Fig. 1.

3.2   Feature Extraction

In the dataset, video is represented as a sequence of keyframes. We extract the local and global features based on VIREO-374 for each keyframe from these given keyframes provided by the dataset. We introduce surrounding textual information of web videos like title, tag and description which show the meaning of video content representation. The several techniques for textual features extraction are used (*i.e.*, stopword removing[3] to omit the most common word such as prepositions, articles and conjunctions; words stemming, etc.,) for getting the right as well as relevant information of videos.

3.3   Combining

The similarity of three types of features are combined together as we called Multi-Modality (MM) with a sum of similarities controlled by the weight of each feature vector.

$$Sim(d_x, d_y) = f(Sim_{local}(d_x, d_y), Sim_{global}(d_x, d_y), Sim_{textual}(d_x, d_y)) \qquad (6)$$

$$f(x, y, z) = \alpha \times (local) + \beta \times (global) + (1 - \alpha - \beta) \times (textual) \qquad (7)$$

where $0 < \alpha, \beta < 1$ are weights of feature vectors.

---

[3] www.ranks.nl/stopwords.

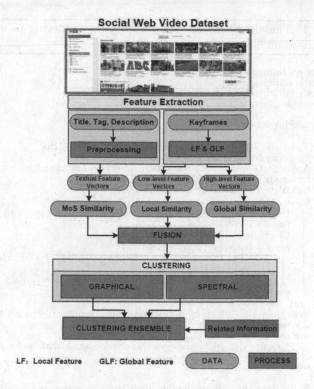

Fig. 1. The proposed framework for social web videos clustering

3.4 The Algorithm

The proposed algorithm for social web videos clustering using multimodality is shown in Algorithm 1. Here multimodality means that three types of features are extracted to use as shown in Fig. 1. Then we calculate the corresponding similarities we called local, global and textual similarity, respectively. Finally, they are fused together to get a single similarity before applying it to clustering models.

4 Experimental Framework

4.1 Datasets

We conduct our experiments using the version 2.0 of MCG web video dataset (MCG-WEBV) [16] consisting of 248,887 most viewed videos. The dataset collects the Most Viewed videos of This Month among 15 YouTube categories during December 2008 to November 2009, which are very valuable to do web videos mining for their high quality and popular contents. Meanwhile, the related videos were expanded by database, which aims to keep the original social network information on YouTube.

Algorithm 1. The algorithm for social web video clustering using a multimodality based on clustering ensembles

---

Input:
  (1) Dataset containing local features of videos (LoV).
  (2) Dataset containing global features of videos (GlV).
  (3) Dataset containing textual features of videos (TeV).
  (4) Related video information (ReV).

Output: Clustering labels.
1 **begin**
2   **for** $i \in \{LoV, GlV, TeV(DS1, DS2, DS3, DS4, DS5, DS6)\}$ **do**
3   **for** $j \in \{Keyframe\ sets\ of\ LoV\}$ **do**
4   Apply DoG and SIFT for keypoint detection and description in local feature to get 500D feature vectors.
5   Calculate the similarity matrix Sim_{local}.
6   **for** $k \in \{Keyframe\ sets\ of\ GlV\}$ **do**
7   Grid partitions based 5×5 are used in Lab color space to get 225D feature vectors.
8   Calculate the similarity matrix Sim_{global}.
9   **for** $l \in \{title, tag, description\}$ **do**
10   Text Pre-Processing is needed by applying TF-IDF scheme to find the term weights.
11   Calculate the similarity matrix with MoS $Sim_{textual}$.
12   **end**
13   **end**
14   Fuse three similarities by a function sum of similarities to get final similarity Sim.
15   **for** $m\{must\ link\}$ **do**
16   Execute Graphical and Spectral clustering algorithms for getting labels.
17   **end**
18   **end**
19   Apply different clustering ensemble algorithms to ensemble the labels with pairwise constraints.
20   **end**
21 **end**

We design our experiments on local, global and textual features part. Based on the two features from keyframes extracted from the raw videos in MCG-WEBV and the basic feature of textual information such as title, tag, description. The data is contained in separate files along the different months. We take the whole data into account in a large file and then randomly select from several months for each dataset. The number of samples in each dataset is the average number of randomly selection and sample datasets used in experiments are shown in Table 1.

4.2 Performance Evaluation

For evaluation, we use micro-precision (Micro-P) [17] to measure the accuracy of the consensus cluster with respect to true labels. Micro-P is defined as

$$Micro - P = \frac{1}{n} \sum_{k=1}^{c} a_k \tag{8}$$

where n is the number of objects and c is the number of clusters, a_k denotes the number of objects in the cluster k that is correctly assigned to the corre-

Table 1. Description of social web video datasets.

Dataset	Number of samples	Categories	Category distribution
DS1	1411	3	$4, 5, 13$
DS2	1715	6	$2, 4, 6, 8, 10, 11$
DS3	2567	8	$2, 4, 6, 8, 9, 10, 11, 13$
DS4	1999	6	$3, 4, 9, 10, 14, 15$
DS5	1018	3	$9, 10, 14$
DS6	2010	6	$2, 3, 4, 7, 8, 15$

sponding class. Note that $0 \leq Micro\text{-}P \leq 1$ with 1 indicating the best possible consensus clustering, which has to be in full agreement with the class labels.

5 Results and Discussion

5.1 Results

Using the above stated definitions and scheme, clustering labels are obtained by applying two clustering algorithms including the translation of video information with pairwise constraint as ML.

Table 2. Accuracy of the proposed algorithm in six datasets.

Dataset	Textual				MM
	Title	Tag	Description	Total	
DS1	0.88	**0.91**	0.87	0.89	**0.95**
DS2	0.71	**0.87**	0.80	0.85	**0.91**
DS3	0.79	**0.83**	0.82	0.82	**0.89**
DS4	0.85	**0.89**	0.83	0.88	**0.90**
DS5	0.73	**0.84**	0.79	0.82	**0.86**
DS6	0.74	**0.83**	0.80	0.81	**0.85**

Based on multimodality function of similarity, we incorporate the local and global similarities with each subset of textual similarity like title, tag, description and total of those for experiments. According to [18], the results are compared with ground true labels to find the accuracy. The average accuracy of six datasets is shown in Table 2, where MM refers to the proposed multimodality method. The results obtained by sub dataset like tag of each dataset show high performance of clustering result, which means that the information from tag contains more meaning of words provided by up-loaders.

According to the results above, we apply three different well-known clustering ensemble algorithms, *i.e.*, CSPA, HGPA and MCLA, to execute in each dataset to perform our proposed multimodality. Finally, we ensemble those labels with pairwise constraint of related video information. Results are shown in Table 3. The results obtained from ensembles show better than the others. Text information can show a good performance of clustering results of social web videos clustering. However, our proposed multimodality is more novelty which shows better performance than using only text information in clustering purposes shown in Fig. 2. We compare proposed Clustering Ensemble with the support of Must-Link constraint (CE-ML) with some existing state-of-the art clustering ensemble approaches, including CSPA [5], HGPA [5], MCLA [5] and SCE [19].

Table 3. Clustering ensemble results of six datasets.

Dataset	DS1		DS2		DS3		DS4		DS5		DS6	
	Text	MM	Text	MM	Text	MM	Text	MM	Text	MM	Text	MM
CSPA	0.76	**0.89**	0.74	**0.87**	0.71	**0.82**	0.76	**0.84**	0.74	**0.81**	0.75	**0.83**
HGPA	0.72	**0.78**	0.73	**0.82**	0.72	**0.76**	0.71	**0.74**	0.69	**0.74**	0.70	**0.80**
MCLA	0.75	**0.83**	0.76	**0.85**	0.74	**0.78**	0.72	**0.83**	0.70	**0.79**	0.71	**0.81**
SCE	0.73	**0.80**	0.74	**0.83**	0.71	**0.80**	0.74	**0.81**	0.72	**0.78**	0.71	**0.83**
CE-ML	0.82	**0.97**	0.80	**0.93**	0.79	**0.91**	0.76	**0.89**	0.74	**0.86**	0.72	**0.87**

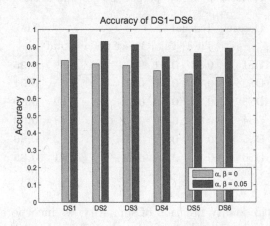

Fig. 2. Clustering performance in multimodality framework

5.2   Results Discussion

Based on the stage of experiments, we get the corresponding results that persuade to have more issues in the next stage.

(a) The idea of modified cosine similarity by adjusted weights that obtained from vector space model can produce better similarity values than original one which leads to have better clustering results.
(b) The multi-modality outperforms the single-modality of similarity.
(c) The adjusted weights and greater weights of textual features in MM model can provide us more accuracy in terms of an effective description of social web videos. In our experiments, according to the best solutions, the weight control of feature vectors α, β are 0.05.
(d) The accuracy of each dataset is high, but a bit decreases with the increasing number of categories (e.g., DS2, DS3, DS6 and DS6).
(e) There are some videos across over the categories (clusters) in final results because the name of categories has similar meaning defined by YouTube.
(f) This research work is based on ensemble technique to implement our proposed algorithms. The experimental results show that our approach is also novelty for enforcing the related videos with higher similarity to index videos should belong to the same category by the help of ML in clustering ensembles.

6   Conclusion

In this paper, we presented an approach for the construction and exploration of similarities exploited to clustering ensembles based on solving video categorization problems which containing local, global and textual information. The experimental results revealed that the proposed modified similarity and multi-modality model worked well for solving problems stated earlier. In our future work, we will emphasis on feature fusion and development of effective algorithms which may help to produce high performance of clustering accuracy. How to incorporate original information with external support information like Flickr will also be a challenge.

Acknowledgements. This work is supported by the National Science Foundation of China (No. 61573292).

References

1. Ramachandran, C., Malik, R., Jin, X., Gao, J., Nahrstedt, K., Han, J.: Videomule: a consensus learning approach to multi-label classification from noisy user-generated videos. In: Proceedings of the 17th ACM International Conference on Multimedia, pp. 721–724 (2009)
2. Ekenel, H.K., Semela, T., Stiefelhagen, R.: Content-based video genre classification using multiple cues. In: Proceedings of the 3rd International Workshop on Automated Information Extraction in Media Production, pp. 21–26 (2010)
3. Yang, L., Liu, J., Yang, X., Hua, X.-S.: Multi-modality web video categorization. In: Proceedings of the International Workshop on Multimedia Information Retrieval, pp. 265–274 (2007)

4. Wu, X., Ngo, C.-W., Zhu, Y.-M., Peng, Q.: Boosting web video categorization with contextual information from social web. World Wide Web **15**, 197–212 (2012). Springer
5. Strehl, A., Ghosh, J.: Cluster ensembles-a knowledge reuse framework for combining multiple partitions. J. Mach. Learn. Res. **3**, 583–617 (2003)
6. Wang, H., Shan, H., Banerjee, A.: Bayesian cluster ensembles. Stat. Anal. Data Mining **4**(1), 54–70 (2011)
7. Gian, T., Ching, Y.S., Tang, Y.: Sequential combination method for data clustering analysis. J. Comput. Sci. Technol. **17**(2), 118–128 (2002)
8. Zanetti, S., Zelnik-Manor, L., Perona, P.: A walk through the web's video clips. In: Proceedings of Computer Vision and Pattern Recognition Workshops, pp. 1–8. California Institute of Technology, Pasadena (2008)
9. Zhang, J.R., Song, Y., Leung, T.: Improving video classification via YouTube video co-watch data. In: Proceedings of the 2011 ACM Workshop on Social and Behavioural Networked Media Access, pp. 21–26 (2011)
10. Brezeale, D., Cook, D.J.: Automatic video classification: a survey of the literature. IEEE Trans. Syst. Man Cybern. Part C Appl. Rev. **38**, 416–430 (2008)
11. Wang, Z., Zhao, M., Song, Y., Kumar, S., Li, B.: YouTubeCat: learning to categorize wild web videos. In: Proceedings of International Conference on Computer Vision and Pattern Recognition (CVPR), pp. 879–886 (2010)
12. Tumer, K., Agogino, A.K.: Ensemble clustering with voting active clusters. Pattern Recogn. Lett. **29**, 1947–1953 (2008). Elsevier
13. Mahmood, A., Li, T., Yang, Y., Wang, H., Afzal, M.: Semi-supervised evolutionary ensembles for Web video categorization. Knowl. Based Syst. **76**, 53–66 (2015). Elsevier
14. Salton, G., Buckley, C.: Term-weighting approaches in automatic text retrieval. Inf. Process. Manag. **24**(5), 513–523 (1988)
15. Dubuisson, M.P., Jain, A.K.: A modified Hausdorff distance for object matching. In: Proceedings of the 12th International Conference on Pattern Recognition, Jerusalem, Israel, pp. 566–568 (1994)
16. Cao, J., Zhang, Y.-D., Song, Y.-C., Chen, Z.-N., Zhang, X., Li, J.-T.: MCG-WEBV: A Benchmark Dataset for Web Video Analysis, vol. 10, pp. 324–334. Institute of Computing Technology, Beijing (2009)
17. Zhou, Z.-H., Tang, W.: Clusterer ensemble. Knowl. Based Syst. **19**, 77–83 (2006). Elsevier
18. Ding, S., Jia, H., Zhang, L., Jin, F.: Research of semi-supervised spectral clustering algorithm based on pairwise constraints. Neural Comput. Appl. **24**, 211–219 (2014). Springer
19. Zhang, X., Jiao, L., Liu, F., Bo, L., Gong, M.: Spectral clustering ensemble applied to SAR image segmentation. IEEE Trans. Geosci. Remote Sens. **46**(7), 2126–2136 (2008)

Learning Latent Features for Multi-view Clustering Based on NMF

Mengjiao He, Yan Yang$^{(\boxtimes)}$, and Hongjun Wang

School of Information Science and Technology, Southwest Jiaotong University,
Chengdu 610031, People's Republic of China
yyang@swjtu.edu.cn

Abstract. Multi-view data coming from multiple ways or being presented in multiple forms, have more information than single-view data. So multi-view clustering benefits from exploiting the more information. Nonnegative matrix factorization (NMF) is an efficient method to learn low-rank approximation of nonnegative matrix of nonnegative data, but it may not be good at clustering. This paper presents a novel multi-view clustering algorithm (called MVCS) which properly combines the similarity and NMF. It aims to obtain latent features shared by multiple views with factorizations, which is a common factor matrix attained from the views and the common similarity matrix. Besides, according to the reconstruction precisions of data matrices, MVCS could adaptively learn the weight. Experiments on real-world data sets demonstrate that our approach may effectively facilitate multi-view clustering and induce superior clustering results.

Keywords: Multi-view clustering · Nonnegative matrix factorization (NMF) · Similarity matrix · Latent features

1   Introduction

Multi-view data has received widespread attention in various fields since it has more information than single-view data. Typical example are as follows: a Web document, which is depicted by its URL or described by the words on the pages; and a multilingual document, which has a representation in each language. Multi-view data with multiple descriptions contains the consistent and complementary information. Multi-view clustering is trying to exploit the latent structures to obtain more information for improving the performance of clustering [1].

Conventional machine learning method applied in multi-view data learning concatenates all multiple views into one single view, which may cause over-fitting

This work is supported by the National Science Foundation of China (Nos. 61170111 and 61572407), the Project of National Science and Technology Support Program (No. 2015BAH19F02) and the Science and Technology Planning Project of Sichuan Province (No. 2014SZ0207).

V. Flores et al. (Eds.): IJCRS 2016, LNAI 9920, pp. 459–469, 2016.
DOI: 10.1007/978-3-319-47160-0_42

in case of a small size data. And ignoring the statistical property may lead to that the process of grouping has no physically meaning. Nevertheless, multi-view clustering optimizes all the views simultaneously to improve the learning performance [2].

One of the earliest schemes of multi-view learning is co-training [3]. Since that time, a lot of successful multi-view algorithms have been proposed. They could be roughly classified into multiple kernel learning, co-training and subspace learning. The multiple kernel learning algorithms combine different views and kernels to improve performance [4,5]. Co-training algorithms aim to maximize the agreement between two distinct views [6,7]. Subspace learning algorithms obtain a latent subspace shared by multiple views [8,9]. All of them attempt to find the common consistent information, use complementary information or combine both of them to help clustering.

MVCS is proposed in this paper, inspired by NMFCSJ [10]. It combines NMF and similarity to improve the performance of clustering, as NMF is an efficient method to learn low-rank approximation of nonnegative matrix of non-negative data, and similarity among samples has information about the relationship, which help constrain the process of feature extracting. Our method makes full use of the common latent information among the different views. Moreover, it learns the weights of each view automatically. The experimental results show a better performance of MVCS compared to a number of baseline methods.

The remainder of this paper is started in Sect. 2, a brief review of some related works. In Sect. 3, we propose our MVCS approach and corresponding formula derivation. In Sect. 4, the report of experiments is shown and analyzed. In the Sect. 5, we draw a conclusion.

2   Related Work

2.1   Multi-view Clustering

Multi-view learning optimize the objective functions to exploit the whole views in order to obtain more information. Different views in multi-view data have the consistent fundamental attributes with respect to the instances and their own characteristics. So, exploiting the latent features of multi-view data could improve the learning performance.

Since Blum and Mitchell [2] classified the Web pages according to its content and linkage, they tried to maximize the common latent space of the two views to gain a better results. Muslea et al. [11] proposed an robust semi-supervised learning algorithm combing co-training with active leaning. Kumar et al. [12] developed co-training for multi-view data clustering. Multiple kernel learning (MKL) has been widely applied in multi-view data learning. It is because the kernels in MKL directly correspond to views, and be merged to improves learning performance [13]. Kloft and Blanchard [14] applied l_p-norm MKL to a tighter upper bound in multi-view clustering. Another method is subspace learning-based approaches. It aims to obtain a latent subspace shared by different views since views in one data set has something consistent. A canonical correlation analysis (CCA) based method for multi-view data was proposed by Kursun and Alpaydin [15].

2.2 NMF-Based Multi-view Learning

Nonnegative matrix factorization (NMF) is used for feature extraction in the field of data mining. It is well understood that NMF is a method to separate a non-negative matrix into two non-negative matrices, and the two non-negative matrices could rebuilt the original one. As the decomposition result is non-negative, meaningful in physics, the interpretability of NMF outputs makes it fit on large-scale and time-varying data sets.

Given an input nonnegative data matrix $X \in \mathbb{R}_+^{m*n}$, $X_{m*n} = (x_1, x_2, \ldots x_n)$, x_i is a instance of vector in m-dimensional space, and $+$ means it is non-negative. NMF aims to seek two rank-r nonnegative matrices, a basic matrix $W \in \mathbb{R}_+^{m*r}$ and a feature matrix $H \in \mathbb{R}_+^{r*n}$. They are determined by minimizing the cost function:

$$\min \|X - WH\|_F^2$$
$$s.t. W, H \geq 0. \tag{1}$$

where $W_{m*r} = (w_1, w_2, \ldots w_r)$, w_i is considered as basis vectors. $H_{r*n} = (h_1, h_2, \ldots h_n)$, h_i is a column vector with r-dimension, determined by $x_i = W * h_i$, which could be seen as the new coordinates in the new space defined by W. $r \ll m$ means a sample in a high-dimensional space can be presented by two non-negative matrices of lower rank. Lee and Seung [16] proposed the multiplicative update algorithm to find the optimal W and H. It has been demonstrated that the rule is convergent on the premise of W and H are nonnegative.

A lot of NMF-based multi-view algorithms have been proposed, such as semi-supervised NMF (SSNMF) [17] and NMFCSJ [10]. They have a shared factor matrix in collective factorization of data matrix and prior information similarity matrix. With the guidance of the prior information, their performance are shown well.

3   A Novel Multi-view Clustering Algorithm by Combining the Similarity and NMF (MVCS)

3.1   Formulation

MVCS computes similarity among samples firstly and then put it into NMF to extract the latent features shared by multiple views. According to the reconstruction precisions of data matrices it learns the weights of different views in the process of clustering. Finally, K-means is used to generate the result. In what follows, it will be introduced in detail.

Suppose given a multi-view data set with p views $X = (X^1, X^2, ..., X^p)^T$. $X \in \mathbb{R}^{M*N}$ has N samples, and x_i presents the ith sample in the data set.

$X^i \in \mathbb{R}^{m_i * N}$ means the ith view of data, in a m_i-dimensional space ($\sum m_i = M$). With NMF algorithm, the features in X^i are extracted to $\hat{H}^i$.

$$X^i = W^i \hat{H}^i \tag{2}$$

Similarity matrix of the data points are encoded in the matrix $S \in \mathbb{R}^{N*N}$ which is composed of similarities between object x_i and x_j ($i, j = 1, 2, ...N$). The larger the value between samples is, the more similar they are. As no additional information is given, it can be defined as follows:

$$S_{ij} = \frac{x_i(x_j)^T}{\|x_i\| * \|x_j\|} \tag{3}$$

S has the common information of different views, thus it constrains the NMF to extract the consistent features in different views. According to the additive fuzzy clustering model for ordinal similarity [18], the similarity S has an approximate decomposition form.

$$S = H^T * H$$
$$s.t. \sum_{j=1}^{N} H_{ij} = 1, \ \forall i = 1, 2, ...N \tag{4}$$

where H and $\hat{H}$ in (2) are both the features matrices corresponding to the original data set. They have the similar information about clustering, as they come from the same data set. MVCS utilizes the available information in H to guide the NMF to get more consistent information in every view. Therefore, the MVCS performs in (5) which combines the formulas (2) and (4).

$$\min \sum_{k=1}^{p} \omega \left\| X^k - (W^k \hat{H}^k) \right\|_F^2 + \frac{\lambda}{2} \left\| S - (H^T H) \right\|_F^2 \tag{5}$$

where ω is the weights of different views, learned in the clustering automatically according to the reconstruction precisions of data matrices [19]. λ is the tradeoff parameter of the similarity matrix, the more information the similarity matrix has, the larger the λ is.

Through establishing the relationship between $\hat{H}^k$ and H, let it be $\hat{H}^k = H^k * U^k$, to guide NMF obtain the latent features shared in different views. It means that H^k is the column-normalized matrix of $\hat{H}^k$ [17]. So U^k, a diagonal matrix is imported. It is expressed as $U_{ii}^k = \sum_i \hat{H}_{ij}^k$. We aim to find the latent features in every view, thus the MVCS model with the matrix form is

$$\min D = \sum_{k=1}^{p} \omega \left\| X^k - (W^k H^k U^k) \right\|_F^2 + \frac{\lambda}{2} \left\| S - (H^{k^T} H^k) \right\|_F^2$$
$$s.t. \sum_{j=1}^{N} H_{ij} = 1, \ \forall i = 1, 2, ...N \tag{6}$$
$$W^k, H^k, U^k \in \mathbb{R}_+$$

3.2 Algorithm

In order to get latent features H^k, an iterative optimization algorithm is proposed for the convergence of objective function.

The first step is to update W^k. We first separate the function D to get (7), which takes one view into account:

$$\min \| X - (WHU) \|_F^2$$
$$s.t. W, H, U \in \mathbb{R}_+ \tag{7}$$

It is similar with formula (1), so the multiplicative update algorithm could be applied. Matrix W in (8) can be easily derived:

$$W_{ij} \leftarrow W_{ij} \frac{(XU^T H^T)_{ij}}{(WHUU^T H^T)_{ij}} \tag{8}$$

Then, U could be derived though the Eq. (7). For its form being similar with least squares estimation, it is convenient for us to calculate:

$$U_{ii} = \frac{(WH)_i^T X}{(WH)_i^T (WH)_i} \tag{9}$$

The most important step is to calculate feature matrix H. According to [17,20], the H has the update rule:

$$H_{ij} \leftarrow H_{ij} \left(\frac{(W^T(\omega * X)U^T + \lambda * (HS))_{ij}}{(W^T W(\omega * HU)U^T)_{ij}} + \lambda * (H * (H^T H))_{ij} \right)^{1/2} \tag{10}$$

Then, updating ω. When W^k, H^k and U^k are fixed, ω is updated automatically according to the reconstruction precisions of data matrices.

$$\omega = \| X - WHU \|_F^2 \tag{11}$$

The above four formulas (8)–(11) are executed circularly, until the objective function converges. Then the common features of each view are extracted and combined together for clustering. The cluster method we choose is K-means, the simplest one of the clustering algorithms. Actually plenty of effective clustering algorithms also can be adopted. So, our MVCS algorithm may be improved further. The specific procedure of the MVCS algorithm optimization is as shown below.

4 Experimental Study

In this section, experiments on five multi-view data sets are conducted. The numerical results demonstrate the effectiveness of the MVCS algorithm. It is valid to utilize similarity matrix to obtain more consistent features of each view in multi-view data clustering.

Algorithm 1 (MVCS Algorithm)

Input:
Multi-view data set; The number of views (p); The clustering number C;
Output:
Clustering label for samples
Initial $\lambda = 0.0001$; k=1;
Generate similarity matrix by formula (4);
Repeat
    Repeat
    Initial ω, U^k, H^k, W^k
    1) update the W^k according to (8).
    2) update the U^k according to (9).
    3) update the H^k according to (10).
    4) update the ω according to (11).
    Until convergence
k++;
Until $k > p$
Combine H^k $\forall k = 1, 2, ...p$; $H = [H^1; H^2; ..., H^p]$
Put the class labels out by K-means.
End

4.1 Experiment Setting

Data Sets. The experiments are based on five real world multi-view data sets including Pendigits, BBCsports, ISetTwo, Animal and Vehicle (Table 1). The information of these data sets are as follows:

Baseline Algorithms. Seven baseline algorithms are taken to compare with MVCS. Specific arrangements are as follows:

WRMK: Weighted robust multi-view kmeans [21] integrates heterogeneous representations of large-scale data to combine these heterogeneous features for unsupervised large-scale data clustering.

Table 1. The information of datasets

Dataset	Pendigits	BBCsports	ISetTwo	Animal	Vehicle
Instances	2000	544	2100	2594	1000
Views	6	2	2	3	2
Classes	10	5	7	6	3

RMSC: It is proposed in [22] that uses transition probability matrices from views to recover a shared low-rank transition probability matrix, which is put to the standard Markov chain method for clustering.

CRMS: By co-regularizing the clustering hypotheses co-regularized multi-view spectral clustering [23] looks for clusterings that are consistent across the views.

MVKKM and MVSpec: They are distance-based and trace-based spectral iterative algorithms [24] that iterations alternate between updating the clusters and reestimating the weights.

TWkmeans: It is an automated two-level variable weighting clustering algorithm [25] for multiview data.

MultiNMF: It is developed in [26], which formulates a joint matrix factorization process with the constraint.

Quality Criteria. The normalized mutual information (NMI) [1] and Rand Index (RI) [27] are taken to evaluate the clustering performances.

4.2 Parameter Selection

The Optimum of Iteration. The values of D on different data sets with the change of iteration times in Fig. 1, proves the convergence of our algorithm.

The clustering result is associated with the iteration times. On the basis of the NMF algorithm, the large number of iteration leads to less loss of information of original data set, the more the number of iteration, the better the result is. However, a number of iterations mean a high cost. It is necessary to select the reasonable number of iteration, which would achieve better clustering results as well as less expense.

The result in Fig. 2 shows that when the number of iterations exceeds 37, the algorithm will keep a relatively good and stable status. Thus, 40 is selected as the number of iteration.

The Optimum of λ. As mentioned above, λ is the tradeoff parameter of the similarity matrix. It has an important impact on extracting the latent features. The value of λ is decided by the information of S. In another word, the more information the similarity matrix has, the larger the λ is, which has a positive effect on obtaining the latent features. On the contrary, if the quantity of information in S is small, the large λ would have a bad influence on extracting common features.

While, if the multi-view data set is large in volume, the amount of information contained in the similarity matrix S is in a very stable level. Thus, an appropriate value of λ need to be determined. The relationship between the λ and quality criteria is shown in Fig. 3.

In Fig. 3, when the initial value is greater than 0.00105, the most clustering results decrease as the increase of λ; when the initial value is less than 0.00105, they mostly maintain at a relatively stable high-level state. Based on the above analysis, our algorithm chooses 0.0001 to be the initial value of λ.

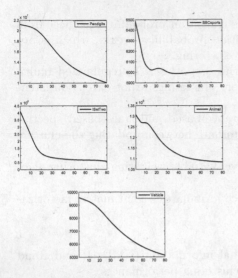

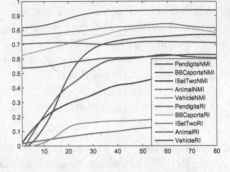

Fig. 1. The changes of D with the increase of iteration times

Fig. 2. The convergence and iteration times of MVCS

4.3   Experiment Results

As what above said, the parameters λ is 0.0001, and the iteration number is set 40. The experiment is conducted 20 times repeatedly with same conditions. The experimental results of comparison between the baseline methods and MVCS are shown via average values. The average value illustrates the clustering performance in general. Tables 2 and 3 show all the results about NMI and RI.

Table 2. The average NMI of each algorithm

Method	WRMK	RMSC	CRMS	MVKKM	MVSpec	TWkmeans	MultiNMF	MVCS
Pendigits	0.4902	0.7520	0.6799	0.4770	0.4865	0.7725	0.7287	**0.7748**
BBCsports	0.6840	0.3039	0.1942	0.0377	0.6289	0.1805	0.2451	**0.7414**
ISetTwo	0.4537	0.5826	0.5518	**0.5927**	0.6191	0.5253	0.1693	0.5357
Vehicle	0.1180	0.1926	0.1534	0.1646	0.1549	0.1838	0.1555	**0.2229**
Animal	0.1238	0.1050	0.1241	0.0971	0.0970	0.0921	0.0692	**0.1475**

It is noted that the MVCS has the highest NMI and RI in the most of data sets. It demonstrates that the MVCS algorithm has a better performance in multi-view data clustering, especially in Vehicle and Animal. It mostly because the features of two data sets are not obvious, and MVCS could extract the features better than others.

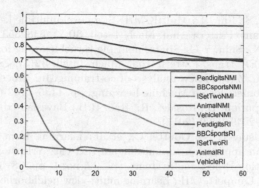

Fig. 3. Different values of λ

Table 3. The average RI of each algorithm

Method	WRMK	RMSC	CRMS	MVKKM	MVSpec	TWkmeans	MultiNMF	MVCS
Pendigits	0.7308	0.9413	0.9214	0.8659	0.8753	0.9380	0.8960	**0.9415**
BBCsports	0.8769	0.7482	0.5675	0.2602	0.8740	0.4557	0.6045	**0.9017**
ISetTwo	0.7667	0.8479	0.8235	**0.8631**	0.8558	0.6851	0.7571	0.8429
Vehicle	0.5904	0.6337	0.5905	0.5866	0.5914	0.5503	0.3793	**0.6367**
Animal	0.6635	0.7207	0.6157	0.6871	0.7175	0.5591	0.6757	**0.7258**

5 Conclusion

This paper proposes an novel multi-view clustering algorithm called MVCS. It properly combines the similarity and NMF to obtain latent features shared by multiple views to improve the clustering. Besides, the weights of different views could be learned adaptively according to the reconstruction precisions of data matrices. The experimental results also confirmed its effectiveness.

Experimental results show that in dealing with five multi-view data. And the MVCS shows a better performance than other seven baseline algorithms. As the MVCS algorithm is not good at dealing with the sparse data in high-dimensional space, we will overcome it and try to extend this algorithm with other methods of similarity measurements and clustering algorithms to improve its performance in the future work.

References

1. Liu, J., Jiang, Y., Li, Z., Zhou, Z.H., Lu, H.: Partially shared latent factor learning with multiview data. IEEE Trans. Neural Netw. Learn. Syst. **26**(6), 1233–1246 (2015)
2. Xu, C., Tao, D., Xu, C.: A survey on multi-view learning. CoRR [Online], vol. abs/1304.5634 (2013). arxiv.org/abs/1304.5634
3. Blum, A., Mitchell, T.: Combining labeled and unlabeled data with co-training. In: Proceedings of the Workshop on Computational Learning, pp. 92–100 (1998)

4. Wang, Z., Chen, S., Sun, T.: MultiK-MHKS: a novel multiple kernel learning algorithm. IEEE Trans. Pattern Anal. Mach. Intell. **30**(2), 348–353 (2008)
5. Subrahmanya, N., Shin, Y.C.: Sparse multiple kernel learning for signal processing applications. IEEE Trans. Pattern Anal. Mach. Intell. **32**(5), 788–798 (2010)
6. Wang, W., Zhou, Z.H.: A new analysis of co-training. In: Proceedings of the 27th International Conference on Machine Learning, pp. 1135–1142 (2010)
7. Yu, S., Krishnapuram, B., Rosales, R., Rao, R.B.: Bayesian co-training. J. Mach. Learn. Res. **12**, 2649–2680 (2011)
8. Amini, M.R., Usunier, N., Goutte, C., et al.: Learning from multiple partially observed viewsan application to multilingual text categorization. In: Advances in Neural Information Processing Systems, vol. 22, no. 1, pp. 28–36 (2010)
9. Quadrianto, N., Lampert, C.H.: Learning multi-view neighborhood preserving projections. In: Proceedings of the International Conference on Machine Learning, pp. 425–432 (2011)
10. Zhang, J.S., Wang, C.P., Yang, Y.Q.: Learning latent features by nonnegative matrix factorization combining similarity judgments. Neurocomputing **155**, 43–52 (2015)
11. Muslea, I., Minton, S., Knoblock, C.A.: Active+Semi-supervised Learning=Robust Multiview Learning. In: Machine Learning-international Workshop then Conference, pp. 435–442 (2002)
12. Kumar, A., Rai, P., Daume III, H.: Co-regularized multi-view spectral clustering. In: Advances in Neural Information Processing Systems, pp. 1413–1432 (2011)
13. Lanckriet, G.R.G., Cristianini, N., Bartlett, P., Ghaoui, L.E., Jordan, M.I.: Learning the kernel matrix with semidefinite programming. J. Mach. Learn. Res. **5**, 27–72 (2004)
14. Kloft, M., Blanchard, G.: The local rademacher complexity of Lp-norm multiple kernel learning. CoRR [Online], vol. abs/1304.0790 (2011). arXiv.org/abs/1304.0790
15. Kursun, O., Alpaydin, E.: Canonical correlation analysis for multiview semisupervised feature extraction. In: Rutkowski, L., Scherer, R., Tadeusiewicz, R., Zadeh, L.A., Zurada, J.M. (eds.) ICAISC 2010. LNCS, vol. 6113, pp. 430–436. Springer, Heidelberg (2010). doi:10.1007/978-3-642-13208-7_54
16. Lee, D., Seung, H.S.: Learning the parts of objects by nonnegative matrix factorization. Nature **401**, 788–791 (1999)
17. Lee, H., Yoo, J., Choi, S.: Semi-supervised nonnegative matrix factorization. IEEE Sign. Process. Lett. **17**, 4–7 (2010)
18. Sato, M., Sato, Y.: Structural model of similarity for fuzzy clustering. In: IEEE International Conference on Fuzzy Systems, vol. 2, pp. 963–968 (1997)
19. Wang, H., Nie, F., Huang, H., Yang, Y.: Learning frame relevance for video classification. In: Proceedings of the 2011 ACM Multimedia Conference and Co-located Workshops, pp. 1345–1348 (2011)
20. Miao, L.D., Qi, H.R.: Endmember extraction from highly mixed data using minimum volume constrained nonnegative matrix factorization. IEEE Trans. Geosci. Remote Sens. **45**(3), 765–777 (2007)
21. Xiao, C., Nie, F., Huang, H.: Multi-view K-means clustering on Big Data. In: Proceedings of the 23rd International Joint Conference on Artificial Intelligence, pp. 2598–2604 (2013)
22. Xia, R., Pan, Y., Du, L., Yin, J.: Robust multi-view spectral clustering via low-rank, sparse decomposition. In: Proceedings of 28th AAAI Conference on Artificial Intelligence, vol. 3, pp. 2149–2155 (2011)

23. Kumar, A., Rai, P., Daum, H.: Co-regularized multi-view spectral clustering. In: Advances in Neural Information Processing Systems (2011)
24. Tzortz, G., Likas, A.: Kernel-based weighted multi-view clustering. In: Proceedings of 12th International Conference on Data Mining, pp. 675–684 (2012)
25. Chen, X., Xu, X., Huang, J., Ye, Y.: TW-K-means: automated two-level variable weighting clustering algorithm for multiview data. IEEE Trans. Knowl. Data Eng. **25**(4), 932–944 (2013)
26. Liu, J., Wang, C., Gao, J., Han, J.: Multi-view clustering via joint nonnegative matrix factorization. In: Proceedings of the 2013 SIAN International Conference on Data Mining, pp. 252–260 (2013)
27. Rand, W.M.: Objective critera for the evaluation of clustering methods. J. Am. Stat. Assoc. **66**(336), 846–850 (1971)

A Semantic Overlapping Clustering Algorithm for Analyzing Short-Texts

Lipika Dey, Kunal Ranjan[✉], Ishan Verma, and Abir Naskar

Innovation Labs, Tata Consultancy Services, Delhi, India
{lipika.dey,k.ranjan,ishan.verma,abir.naskar}@tcs.com

Abstract. The rise in volumes of digitized short-texts like tweets or customer complaints and opinions about products and services pose new challenges to the established methods of text analytics both due to the sparseness of text and noise. In this paper we present a new semantic clustering algorithm, which first discovers frequently occurring semantic concepts within a repository, and then clusters the documents around these concepts based on concept distribution within them. The method produces overlapping clusters which generates far more accurate view of content embedded within real-life communication texts. We have compared the clustering results with LSH based clustering and show that the proposed method produces fewer overall clusters with more semantic coherence within a cluster.

Keywords: Short text clustering · Concept extraction · Overlapping clustering

1 Introduction

Most text clustering tasks treat text as bags of words. Semantics in the text is largely ignored in the process, and the results often have low interpretability. Clustering short texts become even more challenging since there is not enough content from which statistical conclusions can be drawn correctly.

In this paper, we present a clustering method that can group together semantically similar short text documents despite surface level dissimilarities. The first step is to identify conceptually related word clusters, or concept-clusters based on co-occurrence patterns, from the repository. If words are considered as atomic elements that constitute documents, frequently co-occurring words can be considered as semantic components or concepts that can represent the content of a document. Each document can be further viewed as a weighted cover of concepts present within a repository. In the second phase, document clusters are discovered as groups of documents that have similar distribution of concepts. Documents may partially overlap on contained semantic concepts. The proposed method outputs partially overlapping clusters.

The proposed method has been tried on various short text collections like News titles, customer complaints logged in a support center, and emails. It is found to be particularly useful for clustering consumer-generated short text documents where the analysis objectives are two-fold:

© Springer International Publishing AG 2016
V. Flores et al. (Eds.): IJCRS 2016, LNAI 9920, pp. 470–479, 2016.
DOI: 10.1007/978-3-319-47160-0_43

1. To find clusters of documents that contain the same problem or problem combinations, despite being worded differently.
2. To find frequency of different problems in the collection, co-occurrence of problems and temporal distribution of different types of problems.

The resulting clusters have been compared with those obtained using Latent Semantic Hashing (LSH) [13]. It is found that the proposed method provides a better semantic understanding of the content.

The rest of the paper is organized as follows. Section 2 presents a review of earlier work done in short text clustering. Section 3 presents the method for discovering concept clusters while Sect. 4 describes the clustering algorithm. Section 5 presents some results and discussions.

2 Survey of Related Work

We present a brief overview of the recent approaches to short text clustering reported in literature. In [1], it was proposed that two short segments that may not have any common words, but if terms from the first segment appear frequently with terms from the second segment in other documents, then the segments may be considered as semantically related. In order to avoid the problem of high computation time that arises while calculating correlation between all terms, this work proposed the selection of a few terms randomly, and then using these terms with the Nyström method to approximate the term-term correlation matrix. In [2] it was mentioned that tf-idf is not very efficient for short texts, since the discriminative power of the data is not captured by it, due to the sparsity. Instead their method measures term discriminability by term level instead of document level. In [3], a method was proposed to improve the accuracy of clustering short text items using Wikipedia as an additional knowledge source. In [4], a framework was proposed to improve the performance of short text clustering by exploiting the internal semantics from original text and external concepts from world knowledge like Wikipedia and WordNet. In [5] a model was proposed for document representation that captures semantic similarity between documents based on correlation measures between terms computed over a defined corpus. In [6] Affinity Propagation defined in [7] was applied to cluster related tweets and retweets that contain URLs to news stories. In [8], a method for conceptualizing short text was proposed to detect and map terms in short text to instances and attributes in a probabilistic knowledgebase. In [9] a convolutional neural networks based clustering algorithm was proposed that first embed the original keyword features into compact binary codes with a locality preserving constraint and then the word embedding's are explored and fed into convolutional neural networks to learn deep feature representations, with the output units fitting the pre-trained binary code in the training process. K-Means was used to generate clusters after obtaining the learned representations. In [10] a topic-based clustering method was proposed that mined topics from short texts data via transfer learning using a novel topic model called "Dual Latent Dirichlet Allocation (DLDA) model", which jointly learns two sets of topics on short and long texts.

While there is a large body of work concentrating on semantic clustering, none of them have addressed the problem of concept co-occurrence and overlapping clusters from analytical viewpoints, which is one of the primary contributions of our work.

3 Proposed Semantic Overlapping Clustering Algorithm

Words are basic building blocks for documents. Concepts can be thought of as groups of co-occurring words that determine its semantic content. It may be noted that different concepts may share words. Semantic clustering tries to group documents that exhibit content similarity despite surface-level dissimilarity. Finding the semantic concepts is a problem by itself, since theoretically, an intractably large number of concepts can be obtained from a document collection.

The proposed semantic clustering works in two phases.

Phase 1: *Concept discovery* - During this phase, the task is to construct a concept graph as a collection of concept clusters, where each concept cluster is a weighted connected graph of words. A concept cluster in the concept graph is a connected weighted graph of words and co-occurring word-pairs. The words constituting a single concept cluster may not always co-occur in totality in any single document, but a majority of these words co-occur frequently.

Phase 2: *Document Clustering* - Finding document clusters that are similar in terms of contained concept clusters. Conceptual similarity of documents is established as a function of the distribution of concepts across these documents.

We now introduce some basic terms and notations that are used throughout in the paper.

Definition 1: *Connectivity (κ) of a concept cluster*: A set of λ number of nodes are said to be $\kappa -$ connected to each other, provided each of them is related to at least $\kappa\lambda$ number of nodes from this set. κ lies between 0 and 1. When κ is equal to 1, the set of nodes are fully connected to each other. The connectivity parameter κ is thus used to control the complexity of concept clusters and thereafter the tightness of document clusters as well.

The process of discovering the word-clusters is unsupervised and iterative. During phase 1, concept clusters are discovered from a large collection of documents, with the more frequent concepts discovered earlier than the rare ones.

Definition 2: *Significance of word pairs or links*: The significance of a pair of words w_i and w_j, denoted by $\sigma(w_i, w_j)$, is computed as a function of their individual occurrences, their co-occurrence frequency and also the relative importance of the words in the repository.

$$\sigma(W_i, W_j) = f_{ij}/(\max(n_i, n_j))$$

where f_{ij} denotes the total number of documents in which word pair (w_i, w_j) co-occur in the whole repository,

n_i and n_j are the number of unique words co-occurring with w_i and w_j respectively across the whole repository. More common words will co-occur with a larger number of unique words than the rare ones.

All word-pairs that have a specified minimum number of occurrence, β, are sorted in decreasing order of significance. Word-pairs are then incrementally connected to each other based on co-occurrence frequency of entire groups. This way, pairs with very low frequency contribute less to the overall clustering process than the ones with higher frequency.

The above steps produce concept clusters as connected weighted word-graphs, where words are nodes and edges connect a pair of co-occurring words. Each edge is associated with its significance value. At the end of concept cluster formation process, each word or a node n in the concept graph is associated to a weight W(n) computed as follows:

$$W(n) = 1/Number\ of\ concept\ clusters\ containing\ n$$

Given parameters κ, β and a document collection D, the core task of concept cluster formation is accomplished by the function Find_concepts. Final clusters emerge through a merge and split approach as document clusters are formed iteratively.

Function **Find_concepts (D, κ, β)**

1. Compute edge-set E for D, containing only those sets of word-pairs whose frequency is greater than β.
2. Compute (w_i, w_j) for each edge $e = (w_i, w_j)$ in E. Sort E in decreasing order of (w_i, w_j).
3. Initialize graph C to NULL.
4. Repeat steps [a] to [d] until edge-set E is empty
   (a) Remove top most edge e from E.
   (b) If C is empty start a new component $C' = e$.
   (c) Otherwise
      (i) For each existing component s belonging to C,
         (1) If κ holds for s U {e}, then update s to s' such that s' = s U {e}. C gets updated automatically.
   It may be noted that this step may add e to more than one component of C.
      (ii)   If e was not added to any existing component s then start a new concept cluster $s' = \{e\}$
   (d) Update C = C U s '.
5. Return C

End Function

It is obvious, that a document may or may not contain all words in a concept cluster. Also, each document may have partial overlap with many concept clusters.

Definition 3: The strength of association of a document D to a concept cluster s, denoted by α(D, s), is computed in terms of proportional overlap of words between D and s.

$$\alpha(D, s) = \frac{\sum_{w_n \in D \cap S} W(w_n)}{\sum_{w_n \in D \cup S} W(w_n)}$$

$\alpha(D, s)$ is a value between 0 and 1.

The association value $\alpha(D, s)$ is discretized to distinguish between strong and weak associations of documents to concept clusters using a user-specified threshold δ and a binary function $M(D, s)$ computed as follows:

$$M(D, s) = \begin{array}{l} 1 \text{ if } \alpha(D, s) \geq \delta \\ 0 \text{ if } \alpha(D, s) < \delta \end{array}$$

Lower values of δ will result in weak document-concept associations and thereby later on result in bigger document clusters with less homogeneity, while higher values will result in tight clusters.

Definition 4: *Concept Cover* - A concept cluster s is said to be covered by a set of documents D_s such that $\forall d \in D_s, M(d, s) = 1$.

$M(D, s)$ is now used to generate a document-concept binary matrix.

Algorithm **Construct_Concept_Graph** presented below finds concept clusters and their corresponding covers, by calling the Find_Concept function iteratively till each document belongs to the concept cover of at least one concept cluster.

Algorithm **Construct_Concept_Graph**

1. Let D be the collection of all documents to be clustered.
2. Initialize κ to 1. Get β as input. Usually β is given in terms of percentage of documents that should contain a pair.
3. Initialize graph C to NULL. C will finally contain a set of connected independent components, where each component is a concept cluster.
4. Let D' denote a set of documents initially set equal to D.
5. Repeat until D' is empty
   (a) C_{new} = **Find_concepts** (D', κ, β)
   (b) C = C + C_{new}
   (c) For each d belonging to D'
       (i)   For each s belonging to C_{new}
             (1) Compute M(d,s)
   (d) If M(d,s) = 1 for at least one s belonging to C_{new}
       (i)   Remove d from D'

End Algorithm

We illustrate the working principle of the proposed approach using a small set of News headlines shown in Table 1. The cluster numbers in the third column reflect the grouping obtained after the process is applied on this set. It can be seen that the first two titles report the same incident, the third and fourth titles report a second incident and the fifth title refers to a third incident, and that is indeed what clustering produces.

Table 1. News titles as samples of short texts

ID	Document text	Cluster Id
D1	Tarzana-based Israeli crime leader gets 32 years in prison	C1
D2	Alleged leader of Israeli crime ring gets 32 years in prison	C1
D3	London Trader Hayes Sentenced to 14 Years	C2
D4	Former Trader Tom Hayes Sentenced to 14 Years for Libor Rigging	C2
D5	Whorton caught at country club while on home detention	C3

Figure 1 shows the concept clusters that are obtained. For the nodes that are duplicated across multiple concept clusters, the same color is assigned to the node across all clusters. The color of all non-repeated nodes is yellow. Table 2 shows the discretized association values M(d, s) computed with $\delta = 0.5$ for the news titles. It can be seen that documents D1 and D2 contain exactly same concepts, while D3 and D4 are partially overlapping. In the next section we will show how the concept covers can be exploited for document clustering. The next section will explain how these clusters are obtained.

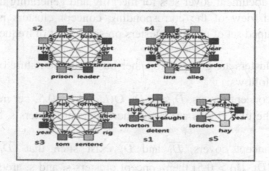

Fig. 1. Concept clusters discovered for news titles shown in Table 1 (Color figure online)

4 Generating Overlapping Document Clusters Around Concept Clusters

In this second phase document clusters are discovered by considering the distributional similarity of concept clusters within documents.

For a given concept cluster s_i let $\overline{D_i}$ be its cover. Let $O(\overline{D_i}, \overline{D_j})$ denote the similarity of the concept covers of two concept clusters, s_i and s_j. $O(\overline{D_i}, \overline{D_j})$ is computed using Jaccard similarity of concept covers as follows:

$$O(\overline{D_i}, \overline{D_j}) = \frac{\overline{D_i} \cap \overline{D_j}}{\overline{D_i} \cup \overline{D_j}}$$

We also determine the fraction of overlap between $\overline{D_i}$ and $\overline{D_j}$ as follows:

Table 2. Document-concept matrix for earlier example with $\delta = 0.5$

Doc	s1	s2	s3	s4	s5
D1	0	1	0	1	0
D2	0	1	0	1	0
D3	0	0	0	0	1
D4	0	0	1	0	1
D5	1	0	0	0	0
Concept Covers	{D5}	{D1, D2}	{D4}	{D1, D2}	{D3, D4}

$$f(\overline{D_i D_j}) \begin{cases} \dfrac{\overline{D_i} \cap \overline{D_j}}{\overline{D_i}} & when\ D_i < D_j \\[2mm] \dfrac{\overline{D_i} \cap \overline{D_j}}{\overline{D_j}} & when\ D_i < D_j \end{cases}$$

The above measures are used to decide whether to merge two document covers into a single cluster or keep them separate. While $O(\overline{D_i}, \overline{D_j})$ is a numeric value denoting the degree of overlap between the document covers, $f(\overline{D_i}, \overline{D_j})$ takes into account the nature of the overlapping document cover sets for merging and generating the clusters. In the process, a grouped view of the corresponding concept clusters is also generated. Analytically, a grouped set of concept clusters presents a set of frequently co-occurring concepts.

Two concept clusters are grouped together based on the nature of overlap of their concept covers as follows:

(a). If for two concept covers, $\overline{D_i}$ and $\overline{D_j}$, $O(\overline{D_i}, \overline{D_j}) > 0.5$, i.e. more than 50 % of the documents in the two groups are same, then concept clusters s_i and s_j are merged together.

(b). If two concept covers, $\overline{D_i}$ and $\overline{D_j}$ exist such that $|\overline{D_i}| \gg |\overline{D_j}|$, $[O(\overline{D_i}, \overline{D_j}) < 0.05]$ and $[f(\overline{D_i}, \overline{D_j}) > 0.5]$ then concept clusters s_i and s_j are grouped together. This denotes a situation where from within a large collection of similar documents, a small portion also has similarities with another small set of documents. In this case, the smaller set is merged with the bigger set.

When concept covers are merged together to create document clusters, the corresponding concept clusters are also merged to create larger concept clusters. Document overlaps are recomputed for larger concept clusters. The order of merging is chosed carefully such that larger overlaps are taken care of before smaller overlaps. Tightness of clusters are controlled through δ defined earlier. The process continues till there is no change.

Algorithm **Find_Conceptual_Clusters** repeatedly groups concept clusters based on the overlap of their concept covers and ultimately outputs the association of each document to each concept cluster or a group or clusters.

Algorithm **Find_Conceptual_Clusters**

1. Let G be the concept graph containing all concept clusters.
2. Initialize list T to NULL. T will contain the final list of concept covers for grouped or ungrouped concept clusters in G.
3. Initialize δ to 0.5.
4. Compute concept covers for all concept clusters in G.
5. Prepare ordered list L of concept cluster pairs s_i, $s_j \in$ G based on decreasing scores of $O(\overline{Dt}, \overline{Dj})$.
6. Set $S' = NULL$
7. Repeat steps 8 till L is empty OR there is no change in S'
8. For the topmost pair (s_i, s_j) in L
   (a) If $(O(\overline{Dt}, \overline{Dj}) >= 0.5)$ OR $(O(\overline{Dt}, \overline{Dj}) < 0.05$ AND $f(\overline{Dt}, \overline{Dj}) > 0.5)$ then
      (i) $s_{new} = s_i \cup s_{j.}$ /* /* Merge s_j to s_i */
      (ii) Put s_{new} in S'
      (iii) Remove s_i and s_j from G
      (iv) Remove all (s_p, s_q) from L where either s_p or s_q is same as s_i or s_j.
9. Set $G = G \cup S'$
10. If no change in G then STOP else Go to Step 4
11. Construct $\overline{D_t}$ for each $s \in$ G.
12. Output T as the final set of concept covers, $\overline{D_s}$ for each $s \in$ G.

End Algorithm

The set T contains final clusters of documents, such that documents belonging to the same cluster share a set of concepts to varying degrees. Each document has membership value $\alpha(\mathbf{D}, \mathbf{s})$ to each cluster $\mathbf{s}$ in T. Table 3 shows some sample texts picked up from a customer complaint repository and their memberships to different clusters based on concept overlaps to illustrate the overlaps. Clearly clusters C1 and C2 overlap on the concept "New car" though the actual problems are different. Similarly clusters C6 and C7 overlap on concept related to "Tread Depth" but otherwise contain documents related to kickplates and radio respectively.

Table 3. Sample text and cluster memberships using proposed algorithm.

DocID	C1	C2	C3	C5	C6	C7	Text
1	0.11	0.94	0.04	0.00	0.00	0.00	NEW VEHICLE DETAIL INSTALL DOOR EDGE GUARDS NITROGEN TIRE
2	0.46	0.18	0.00	0.00	0.00	0.00	NEW CAR DETAIL INSTALL WHEEL LOCKS
3	0.00	0.00	0.00	0.74	0.00	0.00	PERFORM THE SIRIUSXM VIGATION REFLASH
4	0.01	0.00	0.00	0.00	0.85	0.42	Tread Depth Above 6/32nds 27 PT. INSPEC.no charge PLEASE INSTALL ILLUMITED KICKPLATES
5	0.00	0.00	0.00	0.00	0.44	0.97	Tread Depth Above 6/32nds 27pt CUST STATES VI AND RADIO GLITCHES

5 Results

The proposed algorithm has been tested for multiple real repositories that include news titles, short text complaints and technical resolutions data for a vehicle manufacturer. The fourth column of Table 2 shows clustering results obtained for the News titles using LSH. Clearly the proposed algorithm is able to maintain better semantic cohesiveness.

Table 4 shows the comparison of LSH based clustering with proposed clustering method for three different datasets containing customer complaints for a specific organization. It was observed throughout that the proposed algorithm preserves better semantic similarity than LSH. Figure 2, on top left, shows a set of complaints, which are all about wheel lock and right rocker panel. Bottom left complaints are about wheel lock, touch up paint and washer solvent. Further on the right we have complaints about wheel locks and nitro. Corresponding LSH break-ups for each group is given along with some illustrative data to show how semantic similarity is maintained by the proposed algorithm despite surface-level dissimilarities. Analysis of co-occurrence of clusters reveals that the most common customer request is for installing wheel locks. Other requests like filling up tires with nitrogen or installing wheel deflectors almost always are accompanied with request for installing wheel locks. On the other hand brake and gear problems are usually unique and isolated. We are now working on mapping the concept clusters towards problem classes. Since the resulting clusters are overlapping in terms of concepts, we envisage that the sets formed with respect to constituent concepts and problem classes will be rough. Thus rough-set based analytics can be used to better estimate problem occurrence patterns

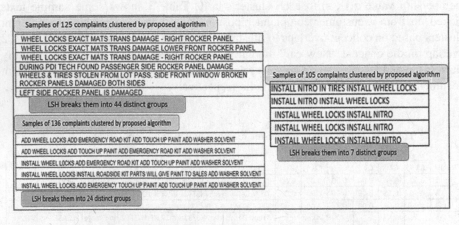

Fig. 2. Conceptually clustered customer complaints

Table 4. Results from three different datasets

Repository	Documents present	#LSH_Clusters	#Clusters (Proposed Algo)
R1	3536	200	136
R2	7629	2113	183
R3	5267	614	94

6 Conclusion

In this paper we have presented a conceptual clustering algorithm that can cluster conceptually similar short texts more effectively than existing methods. Presently we are working on developing incremental clustering algorithms around growing collections of concepts.

References

1. Seifzadeh, S., Farahat, A.K., Kamel, M.S., Karray, F.: Short-text clustering using statistical semantics. In: Proceedings of the 24th International Conference on World Wide Web Companion, pp. 805–810 (2015)
2. Yan, X., Guo, J., Liu, S., Cheng, X., Wang, Y.: Clustering short text using Ncut weighted non-negative matrix factorization. In: CIKM 12 Proceedings of the 21st ACM International Conference on Information and Knowledge Management, pp. 2259–2262 (2012)
3. Banerjee, S., Ramanathan, K., Gupta, A.: Clustering short texts using wikipedia. In: Proceedings of the 30th Annual International ACM SIGIR Conference on Research and Development in Information Retrieval, pp. 787–788 (2007)
4. Hu, X., Sun, N., Zhang, C., Chua, T.: Exploiting internal and external semantics for the clustering of short texts using world knowledge. In: Proceedings of CIKM, Hong Kong, China, pp. 919–928 (2009)
5. Farahat, A.K., Kamel, M.S.: Statistical semantics for enhancing document clustering. Knowl. Inf. Syst. 28(2), 365–393 (2010)
6. Kang, J., Lerman, K., Anon, P.: Analyzing microblogs with affinity propagation. In: Proceedings of KDD workshop on Social Media Analytics (2010)
7. Frey, B.J., Dueck, D.: Clustering by passing messages between data points. Science 312, 972–976 (2007)
8. Song, Y., Wang, H., Wang, Z., Li, H., Chen, W.: Short text conceptualization using a probabilistic knowledgebase. In: Proceedings of the Twenty-Second International Joint Conference on Artificial Intelligence, vol. 3, pp. 2330–2336. AAAI Press (2011)
9. Xu, J., Wang, P., Tian, G., Xu, B., Zhao, J., Wang, F., Hao, H.: Short text clustering via convolutional neural networks. In: Proceedings of NAACL-HLT, pp. 62–69 (2015)
10. Jin, O., Liu, N.N., Zhao, K., Yu, Y., Yang, Q.: Transferring topical knowledge from auxiliary long texts for short text clustering. In: Proceedings of the 20th ACM International Conference on Information and Knowledge Management, pp. 775–784 (2011)
11. Broder, A.Z.: On the resemblance and containment of documents. In: Compression and Complexity of Sequences, pp. 21–29. IEEE Computer Society Press, Salerno, Italy
12. Blei, D.M., Ng, A., Jordan, M.I.: Latent Dirichlet allocation. JMLR 3, 993–1022 (2003)
13. Gionis, A., Indyk, P., Motwani, R.: Similarity search in high dimensions via hashing. In: VLDB, vol. 99, No. 6, pp. 518–529 (1999)

Derivation and Application
of Rules and Trees

On Behavior of Statistical Indices in Incremental Context

Shusaku Tsumoto[(✉)] and Shoji Hirano

Faculty of Medicine, Department of Medical Informatics,
Shimane University, 89-1 Enya-cho Izumo, Matsue 693-8501, Japan
{tsumoto,hirano}@med.shimane-u.ac.jp
http://www.med.shimane-u.ac.jp/medinfo/tsumoto/index.htm

Abstract. This paper illustrates behavior of statistical indices for rule induction when an additional example is input in an incremental way by using accuracy, coverage and lift. Whereas accuracy and coverage behave monotonically, lift may behave so if some additional constraints are satisfied, of which have to be taken care for incremental rule induction.

Keywords: Rough sets · Statistical indices · Accuracy · Coverage · Lift

1 Introduction

When an additional sample will be input in an incremental way, the behavior of statistical indices is very important for rule induction. When statistical indices behave in a monotonical way, the extension of rule induction only needs some modification for evaluation of statistical indices. However, if the behavior of one index changes in a nonmonotonic way, the revision may be complicated. In this paper, the following three indices are used for illustration: accuracy, coverage and lift. Although accuracy and coverage behave monotonically, the extension of rule induction are based on the evaluation of inequalities for rule selection with the threshold fixed. However, since lift does not, rule induction should consider the classification of its behavior.

This paper is organized as follows: Sect. 2 briefly describe rough set theory and the definition of probabilistic rules based on this theory. Section 3 provides formal analysis of incremental updates of accuracy and coverage, where two important inequalities are obtained. Section 4 shows incremental updates of list, which may extend our incremental rule induction methods with these indices. Finally, Sect. 5 concludes this paper.

2 Rough Sets and Probabilistic Rules

2.1 Rough Set Theory

Rough set theory clarifies set-theoretic characteristics of the classes over combinatorial patterns of the attributes, which are precisely discussed by Pawlak [1,3].

This research is supported by Grant-in-Aid for Scientific Research (B) 15H2750 from Japan Society for the Promotion of Science (JSPS).

© Springer International Publishing AG 2016
V. Flores et al. (Eds.): IJCRS 2016, LNAI 9920, pp. 483–492, 2016.
DOI: 10.1007/978-3-319-47160-0_44

This theory can be used to acquire some sets of attributes for classification and can also evaluate how precisely the attributes of database are able to classify data. One of the main features of rough set theory is to evaluate the relationship between the conditional attributes and the decision attributes by using the hidden set-based relations. Let a conditional attribute or conjunctive formula of attributes a decision attribute be denoted by R and D. Then, a relation between R and D can be evaluated by each supporting sets ($[x]_R$ and $[x]_D$) and their overlapped region denoted by $R \wedge D$ ($[x]_R \cap [x]_D$). If $[x]_R \subset [x]_D$, then a proposition $R \to D$ will hold and R will be a part of lower approximation of D. Dually, D can be called a upper approximation of R. In this way, we can define the characteristics of classification in the set-theoretic framework. Let n_R, n_D and n_{RD} denote the cardinality of $[x]_R$, $[x]_D$ and $[x]_R \cap [x]_D$, respectively. Accuracy (true predictive value) and coverage (true positive rate) can be defined as:

$$\alpha_R(D) = \frac{n_{RD}}{n_R} \quad and \tag{1}$$

$$\kappa_R(D) = \frac{n_{RD}}{n_D}, \tag{2}$$

It is notable that $\alpha_R(D)$ measures the degree of the sufficiency of a proposition, $R \to D$, and that $\kappa_R(D)$ measures the degree of its necessity. For example, if $\alpha_R(D)$ is equal to 1.0, then $R \to D$ is true. On the other hand, if $\kappa_R(D)$ is equal to 1.0, then $D \to R$ is true. Thus, if both measures are 1.0, then $R \leftrightarrow D$.

For further information on rough set theory, readers could refer to [1–3].

2.2 Probabilistic Rules

The simplest probabilistic model is that which only uses classification rules which have high accuracy and high coverage.[1] This model is applicable when rules of high accuracy can be derived. Such rules can be defined as:

$$R \overset{\alpha,\kappa}{\to} d \quad s.t. \quad R = \vee_i R_i = \vee \wedge_j [a_j = v_k],$$
$$\alpha_{R_i}(D) > \delta_\alpha \quad and \kappa_{R_i}(D) > \delta_\kappa, \tag{3}$$

where δ_α and δ_κ denote given thresholds for accuracy and coverage, respectively. Where $|A|$ denotes the cardinality of a set A, $\alpha_R(D)$ denotes an accuracy of R as to classification of D, and $\kappa_R(D)$ denotes a coverage, or a true positive rate of R to D, respectively. We call these two inequalities *rule selection inequalities*.

It is notable that this rule is a kind of probabilistic proposition with two statistical measures, which is one kind of an extension of Ziarko's variable precision model (VPRS) [3][2].

---

[1] In this model, we assume that accuracy is dominant over coverage.

[2] In VPRS model, the two kinds of precision of accuracy is given, and the probabilistic proposition with accuracy and two precision conserves the characteristics of the ordinary proposition. Thus, our model is to introduce the probabilistic proposition not only with accuracy, but also with coverage.

3   Updates of Statistical Indices

Usually, datasets will monotonically increase. Let $n_R(t)$ and $n_D(t)$ denote cardinalities of a supporting set of a formula R in given data and a target concept d at time t.

$$n_R(t+1) = \begin{cases} n_R(t)+1 & \text{an additional example} \\ & \text{satisfies } R \\ n_R(t) & \text{otherwise} \end{cases}$$

$$n_D(t+1) = \begin{cases} n_D(t)+1 & \text{an additional example} \\ & \text{belongs to} \\ & \text{a target concept } d. \\ n_D(t) & \text{otherwise} \end{cases}$$

Let $\neg R$ and $\neg D$ be the negations of R and D, respectively. Then, the above two possibilities have the following two dual cases.

$$n_{\neg R}(t+1) = \begin{cases} n_{\neg R}(t) & \text{an additional example} \\ & \text{satisfies } R \\ n_{\neg R}(t)+1 & \text{otherwise} \end{cases}$$

$$n_{\neg D}(t+1) = \begin{cases} n_{\neg D}(t) & \text{an additional example} \\ & \text{belongs to} \\ & \text{a target concept } d. \\ n_{\neg D}(t)+1 & \text{otherwise} \end{cases}$$

Thus, from the definition of accuracy (Eq. (1) and coverage (Eq. (2)), accuracy and coverage may nonmonotonically change due to the change of the intersection of R and D, n_{RD}. Since the above classification gives four additional patterns, we will consider accuracy and coverage for each case as shown in Table 1, called incremental sampling scheme, in which 0 and +1 denote stable and increase in each value.

Table 1. Incremental sampling scheme

R	D	$\neg R$	$\neg D$	$R \wedge D$	$\neg R \wedge D$	$R \wedge \neg D$	$\neg R \wedge \neg D$
0	0	+1	+1	0	0	0	+1
0	+1	+1	0	0	+1	0	0
+1	0	0	+1	0	0	+1	0
+1	+1	0	0	+1	0	0	0

Since accuracy and coverage use only the positive sides of R and D, we will consider the following subtable for the updates of accuracy and coverage (Table 2).

Then, Table 3 is obtained as the classification of four cases of an additional example.

Table 2. Four patterns for an additional example

t:	$[x]_R(t)$	$D(t)$	$[x]_R \cap D(t)$
Original	n_R	n_D	n_{RD}
t+1	$[x]_R(t+1)$	$D(t+1)$	$[x]_R \cap D(t+1)$
Both negative (BN)	n_R	n_D	n_{RD}
R: positive (RP)	$n_R + 1$	n_D	n_{RD}
d: positive (dP)	n_R	$n_D + 1$	n_{RD}
Both positive (BP)	$n_R + 1$	$n_D + 1$	$n_{RD} + 1$

Table 3. Summary of change of accuracy and coverage

Mode				$\alpha(t+1)$	$\kappa(t+1)$
BN	n_R	n_D	n_{RD}	$\alpha(t)$	$\kappa(t)$
RP	$n_R + 1$	n_D	n_{RD}	$\frac{\alpha(t)n_R}{n_R+1}$	$\kappa(t)$
dP	n_R	$n_D + 1$	n_{RD}	$\alpha(t)$	$\frac{\kappa(t)n_D}{n_D+1}$
BP	$n_R + 1$	$n_D + 1$	$n_{RD} + 1$	$\frac{\alpha(t)n_R+1}{n_R+1}$	$\frac{\kappa(t)n_D+1}{n_D+1}$

3.1 Updates of Accuracy and Coverage

From Table 3, updates of Accuracy and Coverage can be calculated from the original datasets for each possible case. Since rules is defined as a probabilistic proposition with two inequalities, supporting sets should satisfy the following constraints:

$$\alpha(t+1) > \delta_\alpha \ \kappa(t+1) > \delta\kappa \tag{4}$$

Then, the conditions for updating can be calculated from the original datasets: when accuracy or coverage does not satisfy the constraint, the corresponding formula should be removed from the candidates. On the other hand, both accuracy and coverage satisfy both constraints, the formula should be included into the candidates. Thus, the following inequalities are important for inclusion of R into the conditions of rules for D:

$$\alpha(t+1) = \frac{\alpha(t)n_R + 1}{n_R + 1} > \delta_\alpha,$$

$$\kappa(t+1) = \frac{\kappa(t)n_D + 1}{n_D + 1} > \delta\kappa.$$

For its exclusion, the following inequalities are important:

$$\alpha(t+1) = \frac{\alpha(t)n_R}{n_R + 1} < \delta_\alpha,$$

$$\kappa(t+1) = \frac{\kappa(t)n_D}{n_D + 1} < \delta\kappa.$$

Thus, the following inequalities are obtained for accuracy and coverage.

Theorem 1. *If accuracy and coverage of a formula R to d satisfies one of the pairs of the following inequalities, then R may include into the candidates of formulae for probabilistic rules if the next dataset belongs to BP.*

$$\frac{\delta_\alpha(n_R + 1) - 1}{n_R} < \alpha_R(D)(t) \leq \delta_\alpha,$$
$$\kappa_R(D)(t) > \delta_\kappa$$

or

$$\alpha_R(D)(t) > \delta_\alpha$$
$$\frac{\delta_\kappa(n_D + 1) - 1}{n_D} < \kappa_R(D)(t) \leq \delta_\kappa.$$

or

$$\frac{\delta_\alpha(n_R + 1) - 1}{n_R} < \alpha_R(D)(t) \leq \delta_\alpha,$$
$$\frac{\delta_\kappa(n_D + 1) - 1}{n_D} < \kappa_R(D)(t) \leq \delta_\kappa.$$

A set of R which satisfies the above two constraints is called **in subrule layer**.

Theorem 2. *If accuracy and coverage of a formula R to d satisfies one of the pairs of the following inequalities, then R may exclude from the candidates of formulae for probabilistic rules.*

$$\delta_\alpha < \alpha_R(D)(t) < \frac{\delta_\alpha(n_R + 1)}{n_R},$$
$$\kappa_R(D)(t) > \delta_\kappa$$

or

$$\alpha_R(D)(t) > \delta_\alpha$$
$$\delta_\kappa < \kappa_R(D)(t) < \frac{\delta_\kappa(n_D + 1)}{n_D}.$$

or

$$\delta_\alpha < \alpha_R(D)(t) < \frac{\delta_\alpha(n_R + 1)}{n_R},$$
$$\delta_\kappa < \kappa_R(D)(t) < \frac{\delta_\kappa(n_D + 1)}{n_D}.$$

A set of R which satisfies the above two constraints is called **out subrule layer**.

It is notable that the lower and upper bounds can be calculated from the original datasets.

Select all the formulae whose accuracy and coverage satisfy the above inequalities They will be a candidate for updates. A set of formulae which satisfies the inequalities for probabilistic rules is called a *rule layer* and a set of formulae which satisfies Eq. (5) and (5) is called a *subrule layer (in)*.

Then, a space of a set of rules can be illustrated as follows. Each indice has four regions with respect to inclusion and exclusion, Table 4 show four possible cases for coverage, denoted by A, B, C and D. In the same way, Table 5 show four possible cases for coverage, denoted by $A2$, $B2$, $C2$ and $D2$.

Table 4. Four possible cases for coverage

A	$\kappa_R(D) \leq \frac{n_R-1}{n_R}\delta_\kappa$	Not included in a Set of Rules
B	$\frac{n_R-1}{n_R}\delta_\kappa < \kappa_R(D) \leq \delta_\kappa$	Included into a set of Rules in cases of BP
C	$\delta_\kappa < \kappa_R(D) \leq \frac{n_R-1}{n_R}\delta_\kappa + 1$	Removed from a set of Rules in cases of RP
D	$\kappa_R(D) > \frac{n_R-1}{n_R}\delta_\kappa + 1$	Always included

Table 5. Four possible cases for accuracy

A2	$\alpha_R(D) \leq \frac{n_R-1}{n_R}\delta_\alpha$	Not included in a Set of Rules
B2	$\frac{n_R-1}{n_R}\delta_\alpha < \alpha_R(D) \leq \delta_\alpha$	Included into a set of Rules in cases of BP
C2	$\delta_\alpha < \alpha_R(D) \leq \frac{n_R-1}{n_R}\delta_\alpha + 1$	Removed from a set of Rules in cases of dP
D2	$\alpha_R(D) > \frac{n_R-1}{n_R}\delta_\alpha + 1$	Always included

Figure 1 illustrates the area of rule layer, in which the horizontal and vertical axes show the values of coverage and accuracy. A to D, and $A2$ to $D2$ are corresponding to the regions defined in Tables 4 and 5. The region of rule layer is shown as a gray shaded region to which $(C, C2), (C, D2), (D, C2)$ and $(D, D2)$ are belonging.

Figure 2 illustrates the regions of out subrule layers: $(C, C2)$, $(C, D2)$ and $(C2, D)$. It is notable that these regions are also classified as rule layer, which can be viewed as the boundary of rule layer. On the other hand, Fig. 3 illustrates the regions of in subrule layers, into which $(B, B2),(C, B2)$, $(D, B2),(B, C2)$ and $(B, D2)$ are included.

4   Lift

Next, let us take a *lift* measure, denoted by $l_R(D)$, which is defined as:

$$l_R(D) = \frac{n \times n_{RD}}{n_R n_D},$$

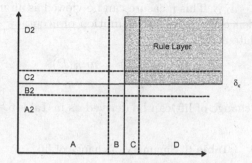

Fig. 1. Intuitive diagram of rule layers. The regions A to D and $A2$ to $D2$ are defined in Tables 4 and 5, respectively. Shaded regions correspond the regions in which rule selection inequalities are satisfied.

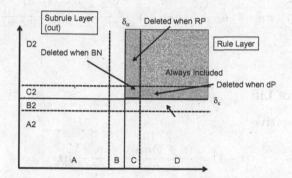

Fig. 2. Intuitive diagram of rule and out subrule layers. The regions A to D and $A2$ to $D2$ are defined in Tables 4 and 5, respectively. Thinly shaded regions correspond the regions for out subrule layers. It is notable that these regions are also included into rule layer.

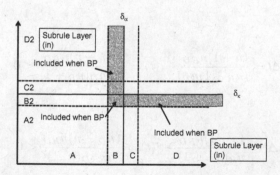

Fig. 3. Intuitive diagram of in subrule layers. The regions A to D and $A2$ to $D2$ are defined in Tables 4 and 5, respectively

where $n = n_D + n_R - n_{RD}$. This measure can be viewed as an index for degree of statistical independence. By using the definition of accuracy and coverage, the lift can be reformulate as:

$$l_R(D) = \frac{n\alpha_R(D)}{n_D} = \frac{n\kappa_R(D)}{n_R}$$

Then, summary of change of lift can be derived as in Table 6.

Table 6. Summary of change of lift

				$\alpha(t+1)$	$l(t+1)$
BN	n_R	n_D	n_{RD}	$\alpha(t)$	$\frac{(n+1)n_{RD}}{n_R n_D}$
RP	$n_R + 1$	n_D	n_{RD}	$\frac{\alpha(t)n_R}{n_R+1}$	$\frac{(n+1)n_{RD}}{(n_R+1)n_D}$
dP	n_R	$n_D + 1$	n_{RD}	$\alpha(t)$	$\frac{(n+1)n_{RD}}{n_R(n_D+1)}$
BP	$n_R + 1$	$n_D + 1$	$n_{RD} + 1$	$\frac{\alpha(t)n_R+1}{n_R+1}$	$\frac{(n+1)(n_{RD}+1)}{(n_D+1)(n_R+1)}$

4.1   Change of Lift

BN: Both Negative

$$l(t+1) = \frac{(n+1)n_{RD}}{n_R n_D} = \frac{n+1}{n}l(t)$$

Thus,

$$\Delta l = l(t+1) - l(t) = \frac{1}{n}l(t)$$

Therefore, lift will increase with an additional example.

RP

$$\Delta l = \frac{(n+1)n_{RD}}{(n_R+1)n_D} - l(t) = \frac{(n_R - n)n_{RD}}{n_R n_D(n_R + 1)} < 0$$

dP

$$\Delta l = \frac{(n+1)n_{RD}}{n_R(n_D+1)} - l(t) = \frac{(n_D - n)n_{RD}}{n_R n_D(n_+1)} < 0$$

BP

$$\Delta l = \frac{(n+1)(n_{RD}+1)}{(n_D+1)(n_R+1)} - l(t)$$

$$= \frac{n_{RD}n_R(n_D-n) + n_D n(n_R-n_{RD}) + (n_R n_D - n n_{RD})}{n_D n_R(n_D+1)(n_R+1)}$$

Thus, change of lift can be summarized as Table 7. Except for BP, the behavior of lift is monotonic. Surprisingly, lift will increase even when an additional example neither belongs to the target concept nor satisfies the formula R. In the case of BP, the denominator of the difference, $\Delta = n_{RD}n_R(n_D-n) + n_D n(n_R-n_{RD}) + (n_R n_D - n n_{RD})$ have to be larger than 0. becomes: Since $n = n_R + n_D - n_{RD}$, Δ becomes:

$$\Delta = n_{RD}^2(n_R+2) - n_{RD}(2n_D+3n_R) + n_R^2 + n_0 n_R,$$

which can be viewed as a quadratic equation. Thus, the discriminant can be used to check whether Δ is always larger than 0. The determinant, denoted by det_{lift}, is obtained as:

$$det_{lift} = (2n_D+3n_R)^2 - 4(n_R+2)(n_R^2 + n_D n_R)$$

$$= -4n_R^3 + (1-n_D)n_R^2 - 8n_D n_R + 4n_D^2$$

Table 7. Incremental sampling scheme for lift

	R	D	$R \wedge D$	$\alpha_R(D)$	$\kappa_R(D)$	$l_R(D)$
BN.	→	→	→	→	→	↑
RP.	↑	→	→	→	↓	↓
dP.	→	↑	→	↓	→	↓
BP.	↑	↑	→	↑	↑	?

Theorem 3. *If*

$$det_{lift} = -4n_R^3 + (1-n_D)n_R^2 - 8n_D n_R + 4n_D^2 < 0,$$

then the lift always increases when an additional example both belongs to a class d and satisfies a formula R.

Since the formula of det_{lift} is complex, the region where the lift always decreases with an additional example is not easy to be illustrated. However, numerical examples shown in Table 8 will show the meaning: in this case with $n_D = 20$ and $n_R = 5$, the lift will decrease when an additional example is both positive (BP).

Thus in most cases, when $n_D < n_R$, the lift measure will decrease, which should be taken care when the threshold for the lift will be set up.

Table 8. Relation between det_{lift} and n_D and n_R

n_D	20	20	20	20	20	20	20	20	20	20
n_R	1	2	3	4	5	6	7	8	9	10
det_{lift}	1417	1172	841	400	175	−908	1823	−2944	−4295	−5900

4.2   Threshold for Lift

In summary, the following two factors have to be taken care:

1. Lift will increase even when an additional example neither belongs to D nor satisfies a formula R.
2. Even when an additional example belongs to a class d and satisfies a formula R, lift may decrease.

Thus, first, the threshold of lift should be corrected as a function of n. The easiest way is to modify the original threshold into:

$$\delta_l(t+1) = \frac{n}{n+1}\delta_l(t).$$

Then, the discriminant det_{lift} should be calculated. If $det_{lift} < 0$, then a rule induction algorithm can be extended in the same way as the case where accuracy and coverage are used for rule selection. On the other hand, $det_{lift} \geq 0$, the lift value may decrease. Thus, the threshold should be smaller than the original context.

5   Conclusion

This paper applies the incremental sampling scheme to investigate behavior of statistical indices. The result shows that since accuracy and coverage change in a monotonical way, modifications of rule induction are rather simple. However, since lift may not behave monotonically, some constraints should be satisfied when simple modifications are applied, which makes incremental induction more difficult.

Thus, in general, the threshold for an statistical index should be set after it have examined whether its behavior is monotonically increasing or decreasing with an additional example.

References

1. Pawlak, Z.: Rough Sets. Kluwer Academic Publishers, Dordrecht (1991)
2. Tsumoto, S.: Automated induction of medical expert system rules from clinical databases based on rough set theory. Inf. Sci. **112**, 67–84 (1998)
3. Ziarko, W.: Variable precision rough set model. J. Comput. Syst. Sci. **46**, 39–59 (1993)

Linguistic Variables Construction
in Information Table

Shusaku Tsumoto$^{(\boxtimes)}$ and Shoji Hirano

Faculty of Medicine, Department of Medical Informatics,
Shimane University, 89-1 Enya-cho, Izumo 693-8501, Japan
{tsumoto,hirano}@med.shimane-u.ac.jp
http://www.med.shimane-u.ac.jp/med_info/tsumoto/index.htm

Abstract. Application of attribute-oriented generalization to an information often lead to inconsistent results of rule induction, which can be viewed as generation of fuzziness with partialization of attribute information. This paper focuses on fuzzy linguistic variables and proposes a solution for inconsistencies. The results show that domain ontology may play an important role in construction of linguistic variables.

Keywords: Rough sets · Multiple hierarchy · Rule induction · Linguistic varaiables

1 Introduction

Rule mining have been widely studied in machine learning [4], data mining [9] and rough set methods [3,6,8]. In the area of rough sets, supporting sets of target concepts form a partition of the universe, and rules can be obtained by subset relations of equivalence classes [3]. Furthermore, Pawlak discusses the quality of approximation of a concept by using relations between the partitioned sets. If the condition of partitioning are loosed, the extension of original rough set model is obtained such as Ziarko's variable precision rough set model [13].

All the proposed methods for rule mining have been applied in many domains, and the usefulness have been reported in many fields. However, for knowledge discovery, the only usage of rule induction is not sufficient. Interpretation of rules by domain knowledge is necessary, and its support is very important. One way is to use domain ontology acquired by domain experts. Tsumoto [8] applied a simple type of domain ontology, called a concept hierarchy, as a form of attribute-oriented generalization to rule mining, where several important knowledge was discovered.

However, domain knowledge usually cannot be captured by a simple hiearchy, but by multiple hierarchy, which causes some inconsistent problems if a concept hiearchy is used as transformation rules.

This research is supported by Grant-in-Aid for Scientific Research (B) 15H2750 from Japan Society for the Promotion of Science (JSPS).

© Springer International Publishing AG 2016
V. Flores et al. (Eds.): IJCRS 2016, LNAI 9920, pp. 493–502, 2016.
DOI: 10.1007/978-3-319-47160-0_45

In this paper, firstly, these phenomena are illustrated by a simple example: when attribute-oriented generalization is used to transform attributes in a given dataset, rule induction with generalized attributes generates inconsistent rules, which can be viewed as emergence of fuzziness.

These observations can be viewed as overgeneralization, where important information for consistent reasoning is lost from domain knowledge. The important point is that transformation rules mainly focus on *IS-A* relations, although most of the domain hiearchy needs *PART* relations. In this paper, for representation of transformation rules with PART relations, we adopt Zadeh's fuzzy linguistic variables [10–12] and introduce inductive construction of linguistic variables from a given dataset.

The paper is organized as follows. Section 2 introduces combination of rule mining and attribute-oriented generalization. Section 3 discusses the problem mentioned above and the nature of the problem. Section 4 introduces linguistic variables proposed by Zadeh and its construction algorithm. Section 5 shows an illustrative example where the algorithm is applied. Finally, Sect. 6 concludes this paper.

2   Rule Mining: Attribute-Oriented Generalization

2.1   Rough-Set Based Rule Mining

Extension of Pawlak's Decision Rule. Based on Pawlak's rough sets [3], Skowron and Grzymala-Busse reformulates the framework of rule mining as follows [5].

Let U denote a nonempty, finite set called the universe and A denote a nonempty, finite set of attributes, i.e., $a : U \rightarrow V_a$ for $a \in A$, where V_a is called the domain of a, respectively. Then, a decision table is defined as an information system, $A = (U, A \cup \{d\})$.

The atomic formulas over $B \subseteq A \cup \{d\}$ and V are expressions of the form $[a = v]$, called descriptors over B, where $a \in B$ and $v \in V_a$. The set $F(B, V)$ of formulas over B is the least set containing all atomic formulas over B and closed with respect to disjunction, conjunction and negation.

For each $f \in F(B, V)$, f_A denote the meaning of f in A, i.e., the set of all objects in U with property f, defined inductively as follows.

1. If f is of the form $[a = v]$ then, $f_A = \{s \in U | a(s) = v\}$
2. $(f \wedge g)_A = f_A \cap g_A$; $(f \vee g)_A = f_A \vee g_A$; $(\neg f)_A = U - f_a$

By the use of this formulation, confidence (classification accuracy) and support (coverage) are reformulated as:

Definition 1. *Let R and D denote a formula in $F(B, V)$ and a set of objects which belong to a decision d. Confidence $\alpha_R(D)$ and support (coverage or true positive rate) $\kappa_R(D)$ for $R \rightarrow d$ are defined as:*

$$\alpha_R(D) = \frac{|R_A \cap D|}{|R_A|}(= P(D|R)), \ and$$

$$\kappa_R(D) = \frac{|R_A \cap D|}{|D|}(= P(R|D)),$$

where $|A|$ denotes the cardinality of a set A.

Rule. By using the above two indices, we define a probabilistic rule, represented as a proposition with two inequalities:

$$R \xrightarrow{\alpha,\kappa} d \quad s.t. \quad R = \wedge_j \vee_k [a_j = v_k], \ \alpha_R(D) \geq \delta_\alpha, \ \kappa_R(D) \geq \delta_\kappa.$$

2.2   Attribute-Oriented Generalization

Rule induction methods automatically generate rules from a given table. According to Pawlak's notation, a decision rule is obtained from a decision table, which corresponds to a reduced decision table [3]. Since a reduced table only show syntactic relations between attributes and decision and do not include information about attributes, only the relations are displayed to the users: user should discover knowledge by using domain knowledge.

One way to input domain knowledge of attributes is the usage of attribute-oriented generalization [1] where a concept hiearchy is used for transforming attributes into generalized ones. Since generalized attributes will give important semantic meaning of a each attribute, rules with transformed attributes will extract important meaning of the original rules and gain higher comprehensibility. Let us illustrate this with the example discussed in [8]. For example, terolism, cornea, antimongoloid slanting of palpebral fissures, iris defects and long eyelashes are symptoms around eyes. Thus, those symptoms can be gathered into a category "eye symptoms": they will be located at lower-level of a hiearchy of symptoms. All the relations among those attributes are shown in Fig. 1, whose generation process is called attribute-oriented generalization [1].

From the viewpoint of computerization, attribute-oriented generalization can be viewed as construction of rules of transformation of attributes. A transformation rule is defined as:

$$[a_i = v_j] \rightarrow [A_k = V_l],$$

where $[A_k = V_l]$ a upper-level concept of $[a_i = v_j]$ and its supporting set of $[A_k = V_l]$ is the union of support sets of lower-level concepts:

$$[A_i = V_l]_A = \bigcup_{i,j}[a_i = v_j]_a.$$

For example, an attribute "iris defects" should be transformed into an attribute-value pair "eye symptoms = yes". On the other hand, if this system is not observed, the attribute-value pair will be changed into "eye symptoms = no". In this way, the transformation rule for iris defects is defined as:

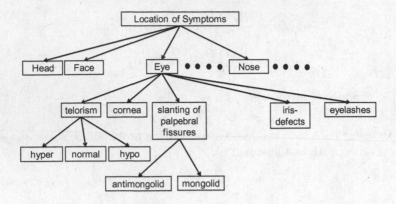

Fig. 1. An example of concept hierarchy

$$[\text{ iris-defects } = yes] \rightarrow [eye - symptoms = yes] \qquad (1)$$

$$[\text{ iris-defects } = no] \ \rightarrow [eye - symptoms = no] \qquad (2)$$

3   Background

Let us illustrate how fuzziness can be generated when transformation is applied to a given dataset. Table 1 is a small example of a data table on congenital disorders which is extracted from the database in [8].

A rule for "Aarskog" is induced as:

$$[iris\text{-}defects = yes] \rightarrow Aarskog \quad \alpha = 1.0, \kappa = 1.0.$$

After a rule of Eq. 2 and other rules obtained from Fig. 1 are applied to Table 1, Table 2 will be generated. Note that mutual exclusiveness of attributes in the original table has been lost by transformation.

Table 1. An example of datasets

U	Round-eye	Telorism	Cornea	Slanting-eye	Iris-defects	Long-eyelashes	Forward_tilted_nose	Wided_peak	Class
1	No	Normal	Megalo-cornea	Yes	Yes	Yes	Yes	Yes	Aarskog
2	Yes	Hypertelorism	Megalo-cornea	Yes	Yes	Yes	Yes	Yes	Aarskog
3	Yes	Hypotelorism	Normal-cornea	No	No	No	No	No	Down
4	Yes	Hypertelorism	Norma-corneal	No	No	No	No	No	Down
5	Yes	Hypertelorism	Large-cornea	Yes	Yes	Yes	Yes	No	Aarskog
6	No	Hypertelorism	Megalo-cornea	Yes	No	No	No	No	Cat-cry

DEFINITIONS: round: round face, slanting: antimongoloid slanting of palpebral fissures, Aarskog: Aarskog Syndrome, Down: Down Syndrome, Cat-cry: Cat Cry Syndrome.

Table 2. Transformed dataset

U	Eye-symptom	Eye-symptom	Eye-symptom	Eye-symptom	Eye-symptom	Eye-symptom	Class
1	0	0	1	1	1	1	Aarskog
2	1	1	1	1	1	1	Aarskog
3	1	0	0	0	0	0	Down
4	1	1	0	0	0	0	Down
5	1	1	1	1	1	1	Aarskog
6	0	1	1	1	0	1	Cat-cry

DEFINITIONS: 1: yes, 0: no.

By using this table, the rule shown above is transformed into:

$$[eye\text{-}symptoms = yes] \rightarrow Aarskog.$$

Since the first five attributes are changed into *eye-symptoms*, we have the following pairs of confidence (accuracy) and support (coverage): $(2/4, 2/3)$, $(2/4, 3/3)$, $(3/4, 3/3)$, $(3/4, 3/3)$, $(3/3, 3/3)$ and $(3/4, 3/3)$. Since they are not equal, this can be viewed as inconsistency.

For dealing with inconsistency, Tsumoto [7] adopted *min*-strategy, where he choose the minimum values: confidence and support are set to $2/4$ and $2/3$ from the candidates. Thus, rules can be reformulated as:

$$[eye\text{-}symptoms = yes] \rightarrow Aarskog,$$
$$\alpha = 3/4 = 0.75, \kappa = 2/3 = 0.67,$$

whose idea is applied to a database on congenital orders and several pieces of knowledge were discovered [8].

This examples show that the loss of mutual exclusiveness of attributes lead to inconsistencies of statistical indices, which can be viewed as emergence of fuzziness.

3.1   Inconsistency as Fuzziness

The above example shows that rule induction with attribute-oriented generalization easily generates rules with conflicts of statistical indices. It is because simple application of transformation violates the condition of mapping.

As shown in Sect. 2, attribute-value pair is a king of mapping from examples to values. For an attribute "$round - eye$", a set of values in "$round - eye$", $\{yes, no\}$ is equivalent to a domain of "round-eye". Since the value of $round - eye$ for the first example in a dataset, $round(1)$ is equal to *no*: $round(1) = no$. Thus, a reverse mapping can be defined as:

$$round - eye^{-1}(no) = \{1, 6\}.$$

On the other hand, the reverse mapping $[slanting - eye = yes]$ is:

$$slanting - eye^{-1}(normal) = \{3, 4\}.$$

However, simple transformation will violate this condition on mapping because transformation rules will change different attributes into the same name of generalized attributes. For example, if the following two transformation rules are applied:

$$round - eye \rightarrow \text{eye-symptoms},$$
$$slanting - eye \rightarrow \text{eye-symptoms},$$
$$normal \rightarrow no,$$
$$long \rightarrow yes,$$

then transformed relations are:

$$\text{eye-symptoms}^{-1}(no) = \{1, 6\},$$
$$\text{eye-symptoms}^{-1}(no) = \{3, 4\},$$

which are inconsistent from the viewpoint of a reverse mapping. Thus, transformed attribute-value pairs does not satisfy the original condition of mapping, where the concept of covering mapping may give some solution [2].[1]

4   Construction of Linguistic Variables

4.1   Concept Hierarchy

One solution is for us to reflect the nature of transformation. If we regard a transformation rule as a proposition in a logical sense, it represents *implication*. In other words, a supporting set the predecessor is a subset of a supporting of the antecendent: the relations should be a $IS - A$ relation. However, actually, in the above case, it is not correct from the viewpoint of domain ontology. In other words, transformation rules from a concept hierarchy are kinds of projection, which may have a risk of over-projection and inconsistencies of indices.

For example, let us consider the following three rules:

$$[Round = yes] \rightarrow [\text{Eye-symptoms} = yes],$$
$$[\text{Iris-Defects} = yes] \rightarrow [\text{Eye-symptoms} = yes],$$
$$[Telorism = hyper] \rightarrow [\text{Eye-symptoms} = yes]$$

The question is: all the symptoms are described by the characteristics of eyes' observations? The answer is in general no. Then, what kind of components are related with description of the symptoms. Table 3 summarizes the correspondence between symptoms and components. The third column also shows the values of confidence obtained from a dataset.

Incorporating the ideas of components can be viewed as transformation of a simple concept hiearchy into multiple hiearchies as shown in Fig. 2. In other

---

[1] This concept will be discussed in the near future.

Table 3. Components of symptoms

Symptoms	Components	Confidence
$[Round - eye = yes]$	: [Eye, Nose, Frontal]	$\alpha = 1/2$
$[Iris - Defects = yes]$	: [Substructure of Eye]	$\alpha = 3/3$
$[Telorism = hyper]$	: [Eye, Nose, Frontal]	$\alpha = 1/2$

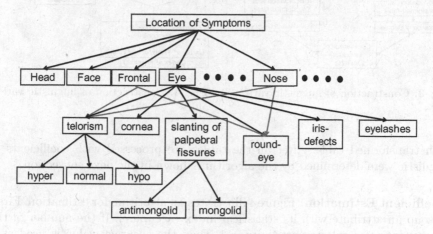

Fig. 2. Revised concept hierarchy

way, each symptoms cannot be described as a tree of $IS - A$ relation, but $PART$ relation.

Since a transformation proposition reflects a $IS - A$ relation as implications, the transforming rules should be changed in order to deal with $PART$ relation. The solution proposed here is introduction of Zadeh's linguistic variables as shown below.

4.2   Algorithm

Zadeh proposes linguistic variables to approximate human linguistic reasoning [10–12]. One of the main points in his discussion is that when human being reasons hierarchical structure, he/she implicitly estimates the degree of contribution of each components to the subject in an upper level. When concept hiearchies are provided, the following algorithm can be used to estimate the degree described in linguistic variables, as shown in Fig. 3. First, rule induction method is applied to an original data table, which calculates original accuracy and coverage for each attribute. Then, from concept hiearchy, components for each attribute are retrieved. For example, $[Round = yes]$ is composed of *Eye*, *Nose* and *Frontal*:

$$[Round - eye = yes] = \gamma[Eye] + \theta[Nose] + \eta[Frontal]$$

Then, an original table is transformed by using each component, and tables for each component are obtained. Again, rule induction method is applied to

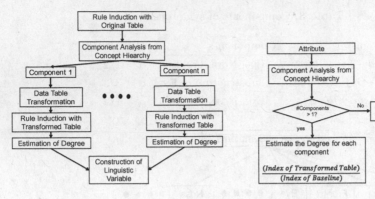

Fig. 3. Construction of linguistic variables

Fig. 4. Construction of linguistic variables

each transformed table. After all the tables are processed with coefficients of linguistic were determined by the algorithm shown in the next subsection.

Coefficient Estimation. Figure 4 illustrates an algorithm for estimation. First, pick up an attribute with its statistical index (accuracy). If the number of the components of a selected attribute is 1, then the accuracy value is used as a baseline. Otherwise, the coefficient of the attribute is obtained by:

$$\frac{(Accuracy of Transformed Table)}{(Accuracy of Baseline)}.$$

The process is repeated for all the components.

5   Experimental Evaluation

In the case of a symptom $[Round = yes]$, this symptom should be described as the combination of Eye, Nose and Frontal part of face. From the value of accuracy in Aarskog syndromes, since the contribution of Eye in [Round = yes] is equal to 0.5, the linguistic variable of $[Round = yes]$ is represented as: $[Round = yes] = 0.5 * [Eye] + \theta * [Nose] + \eta * [Frontal]$, where 0.5, θ and η are degrees of contribution of eyes and nose to this symptom, respectively.

Other two parameters θ and η can be estimated from other hiearchies. Concerning θ, wide-tilted-nose is a symptom related with a nose. Thus, similar to Table 2, Table 1 is converted into Table 4. In the same way, Table 1 is converted into Table 5 with $[round = yes]$, $[telorism = yes]$ and $[wided_peak = yes]$. Since wide-titled nose and wided-peak are composed of nose and frontal (hairline), the degrees of $round$ and $telorism$ into the nose and frontal are both 1/2. Therefore, $[Round = yes]$ can be represented as: $[Round - eye = yes] = 0.5 * [Eye] + 0.5 * [Nose] + 0.5 * [Frontal]$,

The above method was applied to the whole dataset introduced in [6] and coefficients of degree were obtained as shown in Table 6.

Table 4. Transformed dataset (nose-symptoms)

U	Nose	Nose	Nose	Class
1	0	0	1	Aarskog
2	1	1	1	Aarskog
3	1	0	0	Down
4	1	1	0	Down
5	1	1	1	Aarskog
6	0	1	0	Cat-cry

DEFINITIONS: Nose: Nose-symptoms 1: yes, 0: no

Table 5. Transformed dataset: frontal-symptoms

U	Frontal	Frontal	Frontal	Class
1	0	0	1	Aarskog
2	1	1	1	Aarskog
3	1	0	0	Down
4	1	1	0	Down
5	1	1	1	Aarskog
6	0	1	0	Cat-cry

DEFINITIONS: Frontal: frontal-symptoms, 1: yes, 0: no

Table 6. Derived linguistic variables

Symptom	Formula
$[Round - eye = yes]$	$0.5 * [Eye] + 0.5 * [Nose] + 0.5 * [Frontal]$
$[Iris - defect = yes]$	$1.0 * [Eye]$
$[Telorism = yes]$	$0.5 * [Eye] + 0.5 * [Nose] + 0.25 * [Frontal]$
$[Slanting - eye = yes]$	$0.75 * [Eye] + 0.75 * [Nose] + 0.33 * [Frontal]$
$[Long - eyelashes = yes]$	$1.0 * [Eye]$
$[Forward\_tilted\_nose = yes]$	$1.0 * [Nose]$
$[Wided\_peak = yes]$	$1.0 * [Frontal]$

6   Conclusions

This paper firstly discusses the problems of simple application of attribute-oriented generalization to rule mining for discovery of knowledge: although new interesting rules will be extracted from the original rules, inconsistencies of statistical indices will degrade the usefulness of induced rules with respect to statistical strength. This can be easily found in a simple situation as shown in Sect. 3. Even if we allow the inconsistency, we need to evaluate the strength of rules, which can be viewed as fuzzy information processing.

In this paper, we assume that these observations are caused by overgeneralization, where important information for consistent reasoning is lost from domain knowledge. The important point is that transformation rules mainly focus on *IS-A* relations, although most of the domain hiearchy needs *PART* relations. Thus in order to solve the inconsistent rule induction, transformation rules should be defined with PART relations, where Zadeh's fuzzy linguistic variables [10–12] is adopted. Then, an algorithm which generates linguistic variables from a given dataset is proposed. The method was evaluated by a dataset in [8], whose results show that linguistic variables captures the nature of attribute-oriented generalization.

References

1. Cai, Y., Cercone, N., Han, J.: Attribute-oriented induction in relational databases. In: Shapiro, G., Frawley, W. (eds.) Knowledge Discovery in Databases, pp. 213–228. AAAI press, Palo Alto (1991)
2. Munkres, J.: Topology. Featured Titles for Topology Series. Prentice Hall, Incorporated (2000). https://books.google.co.jp/books?id=XjoZAQAAIAAJ
3. Pawlak, Z.: Rough Sets. Kluwer Academic Publishers, Dordrecht (1991)
4. Shavlik, J., Dietterich, T. (eds.): Readings in Machine Learning. Morgan Kaufmann, Palo Alto (1990)
5. Skowron, A., Grzymala-Busse, J.: From rough set theory to evidence theory. In: Yager, R., Fedrizzi, M., Kacprzyk, J. (eds.) Advances in the Dempster-Shafer Theory of Evidence, pp. 193–236. Wiley, New York (1994)
6. Tsumoto, S.: Automated induction of medical expert system rules from clinical databases based on rough set theory. Inf. Sci. **112**, 67–84 (1998)
7. Tsumoto, S.: Knowledge discovery in medical databases based on rough sets and attribute-oriented generalization. In: Proceedings of IEEE-FUZZ98. IEEE Press, Anchorage (1998)
8. Tsumoto, S.: Knowledge discovery in clinical databases and evaluation of discovered knowledge in outpatient clinic. Inf. Sci. **124**, 125–137 (2000)
9. Fayyad, U.M., et al. (eds.): Advances in Knowledge Discovery and Data Mining. AAAI Press, Menlo Park (1996)
10. Zadeh, L.: The concept of linguistic variable and its application to approximate reasoning (part i). Inf. Sci. **8**, 199–249 (1975)
11. Zadeh, L.: The concept of linguistic variable and its application to approximate reasoning (part ii). Inf. Sci. **8**, 301–357 (1975)
12. Zadeh, L.: The concept of linguistic variable and its application to approximate reasoning (part iii). Inf. Sci. **9**, 43–80 (1976)
13. Ziarko, W.: Variable precision rough set model. J. Comput. Syst. Sci. **46**(1), 39–59 (1993)

Multi-objective Search for Comprehensible Rule Ensembles

Jerzy Błaszczyński[1], Bartosz Prusak[1], and Roman Słowiński[1,2(✉)]

[1] Institute of Computing Science, Poznań University of Technology,
Piotrowo 2, 60-965 Poznań, Poland
{jerzy.blaszczynski,roman.slowinski}@cs.put.poznan.pl
[2] Systems Research Institute, Polish Academy of Sciences, 01-447 Warsaw, Poland

Abstract. We present a methodology for constructing an ensemble of rule base classifiers characterized not only by a good accuracy of classification but also by a good quality of knowledge representation. The base classifiers forming the ensemble are composed of minimal sets of rules that cover training objects, while being relevant for their high support, low anti-support and high Bayesian confirmation measure. The population of base classifiers is evolving in course of a bi-objective optimization procedure that involves accuracy of classification and diversity of base classifiers. The final population constitutes an ensemble classifier enjoying some desirable properties, as shown in a computational experiment.

Keywords: Rule ensembles · Classification · Variable-consistency Dominance-based Rough Set Approach (VC-DRSA) · Regularization

1 Introduction

The interest in construction of classifier ensembles is motivated by a need to improve classification performance for the task at hand. The idea is not to rely on a single classifier while providing classification decisions. Instead, all different types of classifier ensembles involve some kind of a combination of base classifiers. Multiple combination strategies were studied in the literature [13]. When only class suggestions of base classifiers are available, the combination is performed in a way that treats each suggestion as a vote for a class (or classes). One simple solution for combining votes is by majority voting. In what concerns the ensemble, it had been observed in different studies that ensembles benefit from component classifiers being different from each other [12]. This diversity can be achieved in different ways, e.g., by using different training sets for base classifiers.

In this work, we focus on bagging type ensembles [6]. Bagging follows the idea of using bootstrap samples [9] to construct an ensemble of multiple independent base classifiers. The same algorithm is used to construct each base classifier entering the ensemble. Bagging proved to reduce the variance of base classifiers provided that they are unstable. A classifier is unstable, when small changes in the training data set cause major changes in the classifier. Independence of

V. Flores et al. (Eds.): IJCRS 2016, LNAI 9920, pp. 503–513, 2016.
DOI: 10.1007/978-3-319-47160-0_46

base classifiers is achieved by constructing each of them on a different, independently drawn sample of data. Bootstrap sample is a sample of data drawn uniformly with replacement. For more theoretical discussion on the justification "why bagging works" the reader is referred to [6,13]. Independence of bootstrap samples, which is propagated to component classifiers, is an important feature of bagging ensembles distinguishing it from different types of ensembles whose base classifiers depend on each other (i.e., arcing, boosting).

It has been shown that rough set theory can provide useful information about consistency of assignment of an object to a class [1]. Measured consistency of objects has been successfully applied to change bagging sampling strategy in variable consistency bagging (VC-bagging). In result, base classifiers are trained on bootstrap samples slightly shifted towards more consistent objects. VC-bagging proved to be able to produce more accurate ensembles than bagging [2,3]. In these ensembles, VC-DomLEM algorithm [4] has been applied to construct base classifiers being sets of "*if..., then...*" rules. VC-DomLEM is a sequential covering algorithm that is able to construct a minimal set of rules of specified consistency, which covers all objects in the training set.

Improved classification performance of ensembles comes at a cost. There are three main weaknesses of ensembles: increased storage, increased computation, and decreased comprehensibility. The first two weaknesses result from the fact that all base classifiers need to be stored and processed to obtain a classification decision. The decision provided by an ensemble is not as comprehensible as decision of a single base classifier. In case of rule classifiers, a classification decision comes with the rules that match the classified object. As each rule is supported by identifiable training objects, this permits an easy understanding on which part of the past experience, the matching rules were built. Analysis of matching rules for few dozen of base classifiers generated from different learning samples is, of course, more difficult. Similarly, it is much more difficult to read and interpret all rules coming from all base classifiers, than the rules coming from a single base classifier. Moreover, the base classifiers are trained for accuracy of classification in the ensemble, not for the quality of representation of knowledge hidden in the training set.

In this paper, we try to reduce the cost the ensembles pay in terms of knowledge representation by base classifiers, and in terms of their comprehensibility. We will construct ensembles of rule classifiers, such that the base classifiers will be composed of minimal sets of strong and confirmatory rules covering a high percentage of the training objects, and the base classifiers will be maximally diversified within the ensemble. Thus, the final ensemble should include a set of comprehensible and diversified rule classifiers without loosing too much on the accuracy of prediction. We will call such an ensemble a *comprehensible ensemble classifier*, and analogously, *comprehensible base classifiers*.

The initial set of rules for constructing a comprehensible ensemble will be the set of all different rules obtained by VC-bagging. To find a family (population) of comprehensible rule base classifiers enjoying the above mentioned good properties, we will solve a series of mixed integer linear programs (MILP) on the initial set of rules, where the goal function is a weighted sum of the following objectives: the number of rules covering a sample of the training set of objects, the support, the anti-support, and the Bayesian confirmation of the rules entering the base classifier. This weighted sum can be seen as a regularization of the minimal cover objective. The population of rule base classifiers is obtained by changing the weights (regularization parameters) and the training sample to be covered. Thus, each rule base classifier resulting from the MILP is associated with a vector of weights. The weights are tuned in an external loop, using a bi-objective evolutionary procedure.

Evolutionary procedures have been successfully applied to find an optimal trade-off between accuracy of prediction and diversity while constructing ensembles composed of neural network and SVM base classifiers [7,11], respectively. These studies, however, do not take into account the comprehensibility criterion.

The proposed methodology for constructing a comprehensible ensemble classifier is described in Sect. 2. It includes construction of the initial set of rules (Subsect. 2.1), finding comprehensible sets of rules constituting a population of rule base classifiers (Subsect. 2.2), and evolutionary bi-objective optimization of the population, that leads to a final comprehensible ensemble classifier (Subsect. 2.3). Section 3 describes a computational experiment, and Sect. 4 groups conclusions.

2 Proposed Methodology for Constructing Comprehensible Ensemble Classifier

The methodology is composed of three elements, which are considered subject to the following two data sets: the training set, and the validation set. First, a VC-bagging [2,3] rule ensemble is constructed on the training set. All the rules that compose this ensemble are integrated as one initial set of rules. Second, an evolutionary bi-objective procedure of the NSGA-II type [8] is applied to evolve a population of comprehensible sets of rules covering all objects from given training samples. The two objectives are accuracy of classification and diversity of the comprehensible set of rules, both calculated on the validation set. Third, members of this population are obtained in result of solving a series of MILP problems on the initial set of rules. This iterative procedure leads to a population that constitutes the comprehensible ensemble classifier. The above three elements will be described in the following subsections.

2.1 Construction of Initial Set of Rules

As it was shown in [5], ordinal classification problems, as well as, non-ordinal classification (and mixed ordinal/non-ordinal) problems can be analyzed within the Dominance-based Rough Set Approach (DRSA) (for a complete presentation of DRSA see [10,14]).

We construct a variable consistency bagging (VC-bagging) [2,3] ensemble on the training set. The motivation for using VC-bagging is to increase diversity of component classifiers by changing the sampling phase. The increased diversity of classifiers suits exploration of rule classifiers. We take, moreover, into account the postulate saying that base classifiers used in bagging are expected to have sufficiently high predictive accuracy apart from being diversified [6]. As we have shown in previous studies [2,3], this requirement can be satisfied by privileging consistent objects when generating bootstrap samples. To identify consistent (and inconsistent) objects we use the same consistency measures as those used to define extended lower approximations in VC-DRSA [1].

We changed the standard bootstrap sampling, where each object is sampled with the same probability, into more focused sampling, where consistent objects, are more likely to be selected than inconsistent ones. In addition, we consider consistency of objects with respect to a random subset of attributes, instead of the whole set of attributes. In this way, we introduced another level of randomization into method, which should lead to even more diversified samples, and thus also more diversified base classifiers. The base classifiers in our bagging ensemble are composed of decision rules induced from VC-bagging bootstrap samples of objects structured using VC-DRSA [1]. We use VC-DomLEM [4] sequential covering algorithm to induce rule base classifiers. The rule base classifiers are composed of sufficiently consistent and strong minimal rules that are required to cover all object included in positive regions identified by VC-DRSA.

2.2 Finding Population of Comprehensible Sets of Rules

Let $\mathbf{R}_{all}$ be the initial set of all rules constructed by VC-bagging, as described in Subsect. 2.1. Moreover, let us divide set $\mathbf{R}_{all}$ into two subsets: $\mathbf{R}_{all}^0$, and $\mathbf{R}_{all}^1$, composed of rules assigning objects to class Cl^0 or to class Cl^1, respectively. We are searching for comprehensible sets of rules $\mathbf{R}_{MC} \subset \mathbf{R}_{all}$, which can be divided, analogously, into $\mathbf{R}_{MC}^0 \subset \mathbf{R}_{all}^0$, and $\mathbf{R}_{MC}^1 \subset \mathbf{R}_{all}^1$. Rules from a comprehensible set $\mathbf{R}_{MC}$ are supposed to minimize or maximize some objectives, while covering all objects from a sample of training objects. One can find a comprehensible set of rules by solving the following multi-objective mixed integer linear programming (MO-MILP) problem:

$$\text{minimize} \quad f_1 = \sum_{r_k^0 \in \mathbf{R}_{all}^0} v(r_k^0) + \sum_{r_k^1 \in \mathbf{R}_{all}^1} v(r_k^1), \tag{1}$$

$$\text{maximize} \quad f_2 = \sum_{r_k^0 \in \mathbf{R}_{all}^0} v(r_k^0) \times sup(r_k^0) + \sum_{r_k^1 \in \mathbf{R}_{all}^1} v(r_k^1) \times sup(r_k^1), \tag{2}$$

$$\text{minimize} \quad f_3 = \sum_{r_k^0 \in \mathbf{R}_{all}^0} v(r_k^0) \times asup(r_k^0) + \sum_{r_k^1 \in \mathbf{R}_{all}^1} v(r_k^1) \times asup(r_k^1), \tag{3}$$

$$\text{maximize} \quad f_4 = \sum_{r_k^0 \in \mathbf{R}_{all}^0} v(r_k^0) \times cfir(r_k^0) + \sum_{r_k^1 \in \mathbf{R}_{all}^1} v(r_k^1) \times cfir(r_k^1), \tag{4}$$

subject to the following constraints:

$$\sum_{r_k^0 : a_i \in A(r_k^0)} v(r_k^0) \geq 1 \quad \text{for all} \quad a_i \in Cl^0 \subset A^R \tag{5}$$

$$\sum_{r_k^1 : a_i \in A(r_k^1)} v(r_k^1) \geq 1 \quad \text{for all} \quad a_i \in Cl^1 \subset A^R \tag{6}$$

$$\sum_{r_k^0 : a_i \in A(r_k^0)} v(r_k^0) \leq T_{max} \quad \text{for all} \quad a_i \in Cl^0 \subset A^R \tag{7}$$

$$\sum_{r_k^1 : a_i \in A(r_k^1)} v(r_k^1) \leq T_{max} \quad \text{for all} \quad a_i \in Cl^1 \subset A^R \tag{8}$$

$$v(r_k^0) \in \{0,1\} \text{ for all } r_k^0 \in \mathbf{R}_{all}^0, \text{ and } v(r_k^1) \in \{0,1\} \text{ for all } r_k^1 \in \mathbf{R}_{all}^1, \tag{9}$$

where r_k^0 is a rule belonging to set $\mathbf{R}_{MC}^0$, and r_k^1 is a rule belonging to set $\mathbf{R}_{MC}^1$. Rules are characterized by the following statistics: $sup(\cdot)$ is support of a rule relative to the size of the training set, $asup(\cdot)$ is anti-support of rule relative to the size of the training set, $cfir(\cdot)$ is a normalized confirmation measure of rule. A^R is a sample of training objects, where a_i is the i-th training object. The sample is a random stratified subset of the training set, i.e., it is a random subset with the same proportion of objects from both classes as the training set. Size of the sample is fixed as a percentage of the size of the training set. $A(r_k^0)$ is a set of objects $a_i \in Cl^0$ covered by rule r_k^0. Analogously, $A(r_k^1)$ is a set of objects $a_i \in Cl^1$ covered by rule r_k^1. Then, $v(r_k^0) \in \{0,1\}$ is a binary variable taking value 1 when rule r_k^0 belongs to $\mathbf{R}_{MC}^0$, and 0 otherwise. Analogously, $v(r_k^1) \in \{0,1\}$ is a binary variable taking value 1 when rule r_k^1 belongs to $\mathbf{R}_{MC}^1$, and 0 otherwise. Finally, T_{max} specifies the maximum number of times any object from A^R may be covered by rules from $\mathbf{R}_{MC}$.

In this problem, objective (1) favors minimal cover of training objects A^R by $\mathbf{R}_{MC}$. Objective (2) promotes support of rules covering A^R (either by promoting maximal sum of supports of rules in $\mathbf{R}_{MC}$ or by promoting the highest minimal support of rule belonging to $\mathbf{R}_{MC}$). Objectives (3), and (4), promote, in an analogous manner as objective (2), anti-support and confirmation of rules, respectively. Constraints (5) and (6) ensure that each object in A^R is covered by at least one rule from $\mathbf{R}_{MC}$. On the other hand, constraints (7) and (8) ensure that each object from A^R is covered at most T_{max} times by rules from $\mathbf{R}_{MC}$.

Instead of performing a multi-objective optimization, at this stage, we aggregate all objectives into one goal function, which involves regularization of objective (1):

$$\text{minimize} \quad F = \sum_{r_k^0 \in \mathbf{R}_{all}^0} v(r_k^0) + \sum_{r_k^1 \in \mathbf{R}_{all}^1} v(r_k^1) \tag{10}$$

$$- \lambda_1 \times \left(\sum_{r_k^0 \in \mathbf{R}_{all}^0} v(r_k^0) \times sup(r_k^0) + \sum_{r_k^1 \in \mathbf{R}_{all}^1} v(r_k^1) \times sup(r_k^1) \right)$$

$$+ \lambda_2 \times \left(\sum_{r_k^0 \in \mathbf{R}_{all}^0} v(r_k^0) \times asup(r_k^0) + \sum_{r_k^1 \in \mathbf{R}_{all}^1} v(r_k^1) \times asup(r_k^1) \right)$$

$$- \lambda_3 \times \left(\sum_{r_k^0 \in \mathbf{R}_{all}^0} v(r_k^0) \times cfir(r_k^0) + \sum_{r_k^1 \in \mathbf{R}_{all}^1} v(r_k^1) \times cfir(r_k^1) \right)$$

subject to constraints (5)–(9), where λ_1, λ_2, λ_3 are regularization parameters (weights) taking values from interval $[0, 1]$. Thus, an optimal solution of MILP problem (10), (5)–(9) is associated with vector of regularization parameters $\Lambda = [\lambda_1, \lambda_2, \lambda_3]$, training sample $A^{\mathbf{R}}$, and the maximal number of covers T_{max}.

To obtain a population of n comprehensible sets of rules, we solve a series of n MILP problems (10), (5)–(9) for n training samples $A^{\mathbf{R}}$ (random stratified subsets) of the same size (e.g., 90 % of the training set), associated with n vectors Λ, and n values of T_{max}. The values of regularization parameters Λ and T_{max} are tuned by a bi-objective evolutionary optimization procedure described in Subsect. 2.3. A similar regularization, but for neural network ensembles, was proposed in [7].

2.3 Evolutionary Bi-objective Search for Comprehensible Ensemble Classifier

The n solutions that form the evolving population are base rule classifiers obtained for different samples $A^{\mathbf{R}}$, vectors $\Lambda = [\lambda_1, \lambda_2, \lambda_3]$ and values of T_{max}, in the way described in Subsect. 2.2. Each base classifier i from this population is identified by its vector $\Lambda^i = [\lambda_1^i, \lambda_2^i, \lambda_3^i]$, T_{max}^i, and sample $A_i^{\mathbf{R}}$; it is evaluated with respect to two objectives, represented by the following measures calculated on the validation set:

- *G-mean$_i$* – the accuracy of prediction of base classifier i, defined as a geometric mean of its sensitivity and specificity:

$$G\text{-}mean_i = \sqrt{\frac{TP}{TP + FN} \times \frac{TN}{TN + FP}}$$

where TP, TN, FP, and FN are numbers of true positive, true negative, false positive, and false negative predictions, respectively.
- Q_i – the combination of the best and the average pairwise diversity measure of base classifier i with respect to all remaining classifiers in the current population. As pairwise diversity measure for base classifiers i and k, we use Yule's Q statistic [12]:

$$Q_{i,k} = 1 - \frac{N^{11}N^{00} - N^{01}N^{10}}{N^{11}N^{00} + N^{01}N^{10}}$$

where $N^{11}, N^{00}, N^{01}, N^{10}$ are numbers of times the two classifiers indicate the same class (11 or 00), or different classes (01 or 10), respectively. For statistically independent classifiers, the expectation of $Q_{i,k}$ is 1; $Q_{i,k} < 1$ for classifiers that recognize the same objects correctly, and $Q_{i,k} > 1$ for classifiers committing errors on different objects. $Q_{i,k} \in [0,2]$, and the higher, the better for the diversity. Thus,

$$Q_i = \max_k \{Q_{i,k}\} + \alpha \times \frac{\sum_{k=1}^{n} Q_{i,k}}{n}$$

where n is the number of base classifiers in the population, and $\alpha = 0.1$.

The aim of constructing accurate and diversified base classifiers of an ensemble can be formulated as the following bi-objective optimization problem:

$$\text{maximize } \{G\text{-mean}_i, \ Q_i\}$$

subject to MILP (10), (5)–(9), and $\lambda_1^i, \lambda_2^i, \lambda_3^i \in [0,1]$, $T_{max}^i \geq 0$, $A_i^{\mathbf{R}}$.

To achieve a set of non-dominated base classifiers with respect to $G\text{-mean}_i$ and Q_i, we adopted the elitist non-dominated sorting genetic algorithm NSGA-II [8]. The main steps of the proposed bi-objective approach to generating accurate and diverse base classifiers using NSGA-II are as follows:

Step 1: Generate an initial population $P_{t=0}$ of base classifiers for n randomly chosen vectors $\Lambda^i = [\lambda_1^i, \lambda_2^i, \lambda_3^i]$ and $T_{max}^i = \sqrt{N}$, where N is the size of training sample $A_i^{\mathbf{R}}$ randomly selected from the training set, i.e., solve n MILP problems (10), (5)–(9).

Step 2: Apply each individual i from the population P_t on the validation set, and calculate $G\text{-mean}_i$ and Q_i.

Step 3: Repeat the following steps for a fixed number of generations (e.g., 400).

3.1: Use non-dominated sorting to divide all base classifiers into fronts.

3.2: Apply binary tournament selection, recombination and mutation on vectors composed of Λ^i and T_{max}^i to generate offspring population P_t' of same size n from P_t.

3.3: Solve n MILP problems (10), (5)–(9) with n randomly selected samples $A_i^{\mathbf{R}}$ and values of Λ^i, T_{max}^i corresponding to P_t', and evaluate the resulting base classifiers in the way of Step 2.

3.4: Merge P_t and P_t' into R_t, and perform non-dominated sorting of R_t. The sorting of individuals within each front is done according to the decreasing crowding distance, with extreme individuals sorted at the top.

3.5: Create new population P_{t+1} by picking up the first n vectors $\Lambda^i = [\lambda_1^i, \lambda_2^i, \lambda_3^i]$ and T_{max}^i from R_t.

3.6: Increment the generation counter $t + 1 \rightarrow t$.

Step 4: Take the population of base classifiers from the last generation to the ensemble.

3   Experiments and Discussion

The goal of the experiment presented in this section is to compare characteristics of comprehensible ensemble classifier (CompEns) and comprehensible single classifier (CompS) with the original ensemble (Ens) and two other classifiers used as base line; these are: a random set of rules selected from the initial set of rules (Rand), and a single classifier trained on responses of the original ensemble (SoEns). The random set of rules is obtained in the following way. First a random number n_r is drawn uniformly from interval $[1, \sqrt{N_r}]$, where N_r is the number of rules in the initial set, and then, n_r rules are drawn from the initial set. The procedure is repeated for an amount of time that matches construction of CompEns. The best random subset of rules is selected as Rand. SoEns classifier is a single rule classifier that explains predictions of Ens. It is constructed on the training set with values of decision attribute replaced by decisions of Ens.

To compare the classifiers we trained them on data sets summarized in Table 1. The data sets that required binarization are marked by -b. Each of the data sets was divided into a training set and a validation set in proportion (2/3–1/3).

Table 1. Characteristics of data sets

	Data set	Objects	Attributes
1	arrhythmia-b	452	558
2	Australian	690	14
3	bank-g	1411	16
4	GermanCredit	1000	20
5	denbosch	119	8
6	Glaucoma	177	40
7	housing-b	506	13
8	windsor-b	546	10

We start our comparison with the characteristics of rule sets forming the classifiers. In Table 2, we present mean number of rules in single or base classifiers constructed by each method. Let us make a general observation that rule models constructed by CompS and CompEns are significantly smaller than the ones produced by other methods. The only method which produces smaller sets of rules is Rand. Rand is however not comparable in terms of predictive accuracy to other methods (see Table 3). Due to lack of space we do not present tables with mean numbers of conditions in rules. These results indicate that rules in models constructed by CompS and CompEns are slightly more specific than rules in other models, i.e., they are composed of slightly more elementary conditions. Also due to lack of space we do not present tables with mean support, mean anti-support and mean confirmation of rule models. We just briefly report that these results also favor CompS and CompEns. It thus turns out that CompS and CompEns construct sets of rules which, compared to rule sets constructed by other methods, are smaller, more specific, and characterized by higher support and confirmation.

Table 2. Mean number of rules constructed by each method.

	Data set	Rand	SoEns	Ens	CompS	CompEns
1	arrhythmia-b	28	79	59.1	**32**	**32.3**
2	Australian	30	96	79.9	**57**	**58.2**
3	bank-g	25	60	46.5	**30**	**29.6**
4	GermanCredit	31	212	166	**130**	**135**
5	denbosch	8	11	9.4	**8**	**7.54**
6	Glaucoma	22	37	29.2	**19**	**18.4**
7	housing-b	19	42	30.8	**19**	**18.7**
8	windsor-b	13	24	19.1	**19**	**18.7**

We follow with comparison of G-mean for ensemble and single classifiers on the validation set. First, notice that Rand is clearly the worst method in this comparison. Second, for the data sets: `Australian`, `GermanCredit`, and `denbosch` the values of G-mean of CompS and CompEns are better than results of other methods (including the original ensemble Ens). Then, for `bank-g`, the value of G-mean for CompEns is really close to the best. On the other data sets the performance of CompS and CompEns is worse than Ens. These negative differences in performance should, however, be considered in context of the benefits in terms of improved comprehensibility offered by CompS and CompEns.

Table 3. G-mean of classifiers on the validation set.

	Data set	Rand	SoEns	Ens	CompS	CompEns
1	arrhythmia-b	55.8	79.1	80.9	71.9	75.4
2	Australian	63.6	75.6	74.8	**77.8**	**80.5**
3	bank-g	81.7	80.4	89.2	85.2	**88.2**
4	GermanCredit	30.4	59.2	61.6	**62.1**	**63.2**
5	denbosch	87.2	82.2	84.9	**89.9**	**87.5**
6	Glaucoma	57.4	72.6	76.1	60.9	69.4
7	housing-b	82.1	83.8	87.8	80.5	83.1
8	windsor-b	58.5	62.6	66.3	64	63.2

4   Concluding Remarks

We presented a new methodology for constructing accurate and comprehensible ensembles of rule classifiers characterized by the following features:

- The ensemble is obtained by solving a series of n MILP problems with the objective of minimal number of rules covering all objects from a training sample, augmented by a regularization component.
- Regularization component of the MILP objective includes weighted total support, anti-support and Bayesian confirmation of rules of a base classifier.
- The parameters of MILP (weights of regularization component and allowed number of times the rules cover a single training object) are tuned in an external loop, where predictive accuracy and diversity of rule classifiers are maximized using an evolutionary bi-objective optimization procedure of the NSGA-II type on a validation set.
- In result one gets an ensemble of n rule classifiers which, compared to traditional minimal-cover rule classifiers, have significantly smaller number of rules per classifier, and a higher mean support and Bayesian confirmation, while ensuring a good predictive accuracy.

Future work will concern generalization to multi-class classification and extension of computational experiments.

References

1. Błaszczyński, J., Greco, S., Słowiński, R., Szeląg, M.: Monotonic variable consistency rough set approaches. Int. J. Approximate Reasoning **50**(7), 979–999 (2009)
2. Błaszczyński, J., Słowiński, R., Stefanowski, J.: Variable consistency bagging ensembles. In: Peters, J.F., Skowron, A. (eds.) Transactions on Rough Sets XI. LNCS, vol. 5946, pp. 40–52. Springer, Heidelberg (2010)
3. Błaszczyński, J., Słowiński, R., Stefanowski, J.: Ordinal classification with monotonicity constraints by variable consistency bagging. In: Szczuka, M., Kryszkiewicz, M., Ramanna, S., Jensen, R., Hu, Q. (eds.) RSCTC 2010. LNCS, vol. 6086, pp. 392–401. Springer, Heidelberg (2010)
4. Błaszczyński, J., Greco, S., Słowiński, R., Szeląg, M.: Sequential covering rule induction algorithm for variable consistency rough set approaches. Inf. Sci. **181**(5), 987–1002 (2011)
5. Błaszczyński, J., Greco, S., Słowiński, R.: Inductive discovery of laws using monotonic rules. Eng. Appl. Artif. Intell. **25**, 284–294 (2012)
6. Breiman, L.: Bagging predictors. Mach. Learn. **24**(2), 123–140 (1996)
7. Chen, H., Yao, X.: Multiobjective neural network ensembles based on regularized negative correlation learning. IEEE Trans. Knowl. Data Eng. **22**(12), 1738–1751 (2010)
8. Deb, K., Agrawal, S., Pratap, A., Meyarivan, T.: A fast and elitist multi-objective genetic algorithm: NSGA-II. IEEE Trans. Evol. Comput. **6**(2), 182–197 (2002)
9. Efron, B.: Nonparametric estimates of standard error: the jackknife, the bootstrap and other methods. Biometrika **68**, 589–599 (1981)

10. Greco, S., Matarazzo, B., Słowiński, R.: Rough sets theory for multicriteria decision analysis. Eur. J. Oper. Res. **129**(1), 1–47 (2001)

11. Gu, S., Jin, Y.: Generating diverse and accurate classifier ensembles using multi-objective optimization. In: Proceedings of IEEE MCDM 2014, pp. 9–15 (2015)

12. Kuncheva, L., Whitaker, C.: Measures of diversity in classifier ensembles and their relationship with the ensemble accuracy. Mach. Learn. **51**(2), 181–207 (2003)

13. Kuncheva, L.: Combining Pattern Classifiers. Methods and Algorithms. Wiley, Hoboken (2004)

14. Słowiński, R., Greco, S., Matarazzo, B.: Rough set methodology for decision aiding. In: Kacprzyk, J., Pedrycz, W. (eds.) Handbook of Computational Intelligence, pp. 349–370. Springer, Berlin (2015). Chapter 22

On NIS-Apriori Based Data Mining in SQL

Hiroshi Sakai[1(✉)], Chenxi Liu[1], Xiaoxin Zhu[2], and Michinori Nakata[3]

[1] Graduate School of Engineering, Kyushu Institute of Technology,
Tobata, Kitakyushu 804-8550, Japan
sakai@mns.kyutech.ac.jp, p350932s@mail.kyutech.jp
[2] Graduate School of Engineering, Harbin Institute of Technology,
Westdazhi Street, Nangang District, Harbin, China
XiaoxinZhu@hit.edu.cn
[3] Faculty of Management and Information Science, Josai International University,
Gumyo, Togane, Chiba 283-0002, Japan
nakatam@ieee.org

Abstract. We have proposed a framework of *Rough Non-deterministic Information Analysis* (RNIA) for tables with non-deterministic information, and applied RNIA to analyzing tables with uncertainty. We have also developed the RNIA software tool in Prolog and *getRNIA* in Python, in addition to these two tools we newly consider the RNIA software tool in SQL for handling large size data sets. This paper reports the current state of the prototype named *NIS-Apriori in SQL*, which will afford us more convenient environment for data analysis.

Keywords: Association rules · NIS-Apriori algorithm · SQL · Prototype · Uncertainty

1 Introduction

We have been coping with rough sets [7], non-deterministic information [6,7], the *Apriori* algorithm [1,12], the software tool in Prolog [9], and *getRNIA* in Python [11,15]. Recently, we are considering a software tool in SQL in order to handle large size data sets.

In rough sets, we usually employ *Deterministic Information Systems* (*DISs*) with deterministic attribute values. We can see every *DIS* is a standard table. For handling information incompleteness in tables [2,4,6,7,9], we employ *Non-deterministic Information Systems* (*NISs*) with non-deterministic values and missing values. By changing *DIS* to *NIS*, several new issues occurred, for example the *possible equivalence classes*, the *minimum* and the *maximum degrees of data dependency*, the *certain* and the *possible* rules, and so on [9]. At the same time, one computational problem occurred, namely the computational complexity may increase exponentially due to the case analysis on *NIS*. However, in rule generation, we proved some properties and escaped from the exponential order problem [10,11]. Due to this result, the rule generator in Prolog [10] and *getRNIA* in Python [15] were implemented.

© Springer International Publishing AG 2016
V. Flores et al. (Eds.): IJCRS 2016, LNAI 9920, pp. 514–524, 2016.
DOI: 10.1007/978-3-319-47160-0_47

In this paper, we focus on the rule generator in SQL, because SQL has the high versatility. Furthermore, several algorithms including *Apriori* were investigated in [12]. For handling large size data sets, we think SQL will be more suitable than the previous languages, Prolog and Python. Recently, the 'sparse' property of the data sets is considered [2,14]. The density of the important part in the data sets may not be unique, and we may ignore the meaningless part. In the sparse matrix, we may employ the special format for reducing the data size. The use of this sparse property will be another approach to large size data sets.

As for this work, we need to specify that this is not the first trial, and the first trial was done in [13]. We follow the result in [13], and consider a rule generator which we name *NIS-Apriori in SQL*.

This paper is organized as follows: Sect. 2 surveys RNIA and rule generation. Section 3 investigates *NIS-Apriori* in SQL and its prototype system. Section 4 concludes this paper.

2   RNIA and Rule Generation

At first, we clarify the rules in *DIS*. For a fixed decision attribute *Dec* and a set *CON* of attributes, we see an implication $\tau : \wedge_{A \in CON}[A, val_A] \Rightarrow [Dec, val]$ is (a candidate of) a *rule*, if τ satisfies the next two constraints.

$$For two threshold values\ 0 < \alpha, \beta \leq 1.0,$$
$$support(\tau)(= N(\tau)/|OB|) \geq \alpha,$$
$$accuracy(\tau)(= N(\tau)/N(\wedge_{A \in CON}[A, val_A])) \geq \beta, \tag{1}$$
$$Here,\ N(*)\ \text{means the number of objects satisfying}$$
$$the formula\ *,\ OB\ \text{means a set of all objects.}$$

Then, we briefly survey rule generation in RNIA. Figure 1 shows NIS Ψ_1, where we see [high,veryhigh] and nil. Here, [high,veryhigh] is non-deterministic information, namely either high or veryhigh is the actual value, and nil is missing value. Each nil may take every possible value in the attribute.

In Ψ_1, we replace each non-deterministic information and nil with a possible value, and we obtain a table with deterministic information. We named it a *derived DIS* from NIS. Let $DD(\Psi)$ be a set of all derived *DISs* from Ψ. We see an actual *DIS* ϕ^{actual} exists in $DD(\Psi)$. For Ψ_1, $DD(\Psi_1)$ consists of 4608 ($=3^2 \times 2^9$) derived *DISs*. Based on $DD(\Psi)$, we proposed the certain and the possible rules below:

Definition 1. [10] *For NIS Ψ and the decision attribute Dec, we fix the threshold values α and β ($0 < \alpha, \beta \leq 1.0$).*

(1) *We say τ is a certain rule, if τ satisfies $support(\tau) \geq \alpha$ and $accuracy(\tau) \geq \beta$ in each $\phi \in DD(\Psi)$,*
(2) *We say τ is a possible rule, if τ satisfies $support(\tau) \geq \alpha$ and $accuracy(\tau) \geq \beta$ in at least one $\phi \in DD(\Psi)$.*

object	temp(erature)	head(ache)	nau(sea)	flu
x1	high	nil	no	yes
x2	[high,veryhigh]	yes	yes	yes
x3	nil	no	no	nil
x4	high	yes	nil	nil
x5	high	nil	yes	no
x6	normal	yes	nil	nil
x7	normal	no	yes	no
x8	nil	yes	nil	yes

object	attrib	value	det
x1	temp	high	1
x1	head	yes	2
x1	head	no	2
x1	nau	no	1
x1	flu	yes	1
x2	temp	high	2
x2	temp	veryhigh	2
x2	head	yes	1

Fig. 1. An exemplary NIS Ψ_1.

Fig. 2. A part of Ψ_1 in NRDF format.

Definition 1 seems natural, but we have the computational complexity problem, because the number of elements in $DD(\Psi)$ increases exponentially. In Ψ_1, the number is 4608, and the number is more than 10^{100} in Mammographic data set in UCI machine learning repository [3]. For this computational problem, we defined two sets for a descriptor $[A, val]$ below:

$$inf([A, val]) = \{x : object \mid \text{the value of } x \text{ for A is a singleton set } \{val\}\},$$
$$sup([A, val]) = \{x : object \mid \text{the value of } x \text{ for A is a set including } val\},$$
$$inf(\wedge_{A \in CON}[A, val_A]) = \cap_{A \in CON} inf([A, val_A]),$$
$$sup(\wedge_{A \in CON}[A, val_A]) = \cap_{A \in CON} sup([A, val_A]).$$

For example, $inf([head, yes]) = \{x2, x4, x6, x8\}$ and $sup([head, yes]) = inf([head, yes]) \cup \{x1, x5\}$ hold in Ψ_1. The actual equivalence class is between two sets. For NIS Ψ, an implication τ, and $minsupp(\tau)$ and $minacc(\tau)$ defined by $min_{\phi \in DD(\Psi)}\{support(\tau) \text{ by } \phi\}$ and $min_{\phi \in DD(\Psi)}\{accuracy(\tau) \text{ by } \phi\}$, we have the following which do not depend upon the number of $DD(\Psi)$.

$$
\begin{aligned}
&\tau : \wedge_{A \in CON}[A, val_A] \Rightarrow [Dec, val], \\
&minsupp(\tau) = |inf(\wedge_{A \in CON}[A, val_A]) \cap inf([Dec, val])|/|OB|, \\
&minacc(\tau) = \frac{|inf(\wedge_{A \in CON}[A, val_A]) \cap inf([Dec, val])|}{|inf(\wedge_{A \in CON}[A, val_A])| + |OUTACC|}, \\
&OUTACC = \{sup(\wedge_{A \in CON}[A, val_A]) \setminus inf(\wedge_{A \in CON}[A, val_A])\} \\
&\quad \setminus inf([Dec, val]).
\end{aligned}
\tag{2}
$$

The $OUTACC$ means a set of objects, from which we can obtain an implication $\tau' : \wedge_{A \in CON}[A, val_A] \Rightarrow [Dec, val']$ $(val \neq val')$. Similarly, we can calculate $maxsupp(\tau)$ and $maxacc(\tau)$. We can also prove that there exists

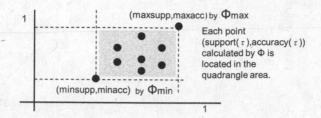

Fig. 3. The distribution of each point $(support(\tau), accuracy(\tau))$ by $\phi \in DD(\Psi)$.

$\phi_{min} \in DD(\Psi)$ which makes both $support(\tau)$ and $accuracy(\tau)$ the minimum. There exists $\phi_{max} \in DD(\Psi)$ which makes both $support(\tau)$ and $accuracy(\tau)$ the maximum. Based on these results, we have the chart in Fig. 3 and Theorem 1.

Theorem 1. *For an implication τ, we have the following.*

(1) τ is a certain rule, if and only if $minsupp(\tau) \geq \alpha$ and $minacc(\tau) \geq \beta$.
(2) τ is a possible rule, if and only if $maxsupp(\tau) \geq \alpha$ and $maxacc(\tau) \geq \beta$.
(3) Even though the certain rules and the possible rules depend upon $DD(\Psi)$, the conditions to check them do not depend upon $DD(\Psi)$.

Based on Theorem 1, we can escape from the exponential order problem. Without Theorem 1, it will be hard to handle Mammographic data set, which has more than 10^{100} derived *DISs*.

3   NIS-Apriori in SQL

3.1   NIS-Apriori Algorithm

The *Apriori* algorithm was proposed by Agrawal, and this is the representative algorithm in data mining [1,12]. This algorithm handles transaction data, and each transaction is given as a set of items. We identify each descriptor in table data with an item, then we can consider the *Apriori* algorithm in tables [10,11]. In certain rule generation, we compare the minimum point in Fig. 3 with the threshold values α and β. On the other hand, we compare the maximum point in Fig. 3 with the threshold values α and β. Since the management of the implications is almost the same as in case of the *Apriori* algorithm and the calculation does not depend upon $|DD(\Psi)|$, we figure out that the computational complexity of the *NIS-Apriori* algorithm is about twice the complexity of the *Apriori* algorithm.

3.2   The NRDF Format

In data sets, we usually have the csv format. This is very familiar, however the name of the attribute and the number of all attributes may be different in each

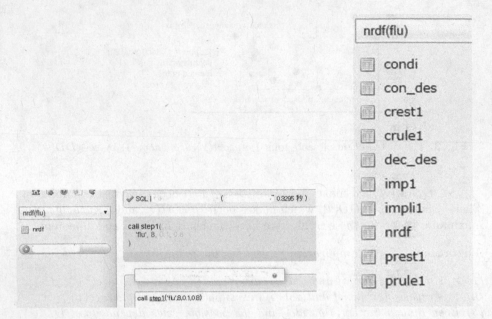

Fig. 4. SQL query execution, where Japanese characters were erased.

Fig. 5. All created tables.

data set. For handling various types of data sets, it is useful to employ another unified format. Otherwise, the program is depending upon the number of the attributes and the name of the attribute.

Based on [13], we employ the NRDF format, which is the extended RDF (resource description framework) format, for any data set. This RDF format may be called as the EAV (entity-attribute-value) format [5,14]. In [5], the KDD-related tasks of attribute selection and decision tree induction were implemented based on the EAV format.

The NRDF format employs 4 attributes, *object*, *attrib*, *value*, and *det*. Figure 2 shows a part of the NRDF format of Ψ_1. In order to specify non-deterministic information, we added the 4th column *det*. The value of *det* means the number of possible values. If *det*=1, this means that the value is deterministic. Otherwise, we know the value is non-deterministic and the number of values by *det*.

3.3 Step 1: Rule Generation in the Form of $P_1 \Rightarrow Dec$

In Step 1, the procedure *step1* generates the certain and the possible rules in the form of $P_1 \Rightarrow Dec$. This procedure consists of the following:

1. create table *condi* (the condition of the rules),
2. create table *con_des* (the descriptors for the condition),
3. create table *dec_des* (the descriptors for the decision),

4. create table *impli*1 (the implications with *inf*, *sup*, *inacc*, *outacc*),
5. create table *crule*1 (the certain rules),
6. create table *prule*1 (the possible rules),
7. create table *crest*1 (the candidates of Step 2),
8. create table *prest*1 (the candidates of Step 2).

At first, a file *nrdf* in the NRDF format in Fig. 2 is stored in the system (Fig. 4). In Fig. 4, we execute '*call step*1('*flu*', 8, 0.1, 0.8)', which means the decision attribute is '*flu*', the number of the objects is 8, the *support* value is 0.1, and the *accuracy* value is 0.8. It took about 0.33 (s) for executing the procedure *step*1 in windows PC, and all tables in Fig. 5 were generated.

In Fig. 5, two tables *con_des* and *dec_des* store the set of descriptors on the condition part and the set of descriptors on the decision part, respectively. The procedure *step*1 generates the Cartesian Products by using *con_des* and *dec_des*, and adds *inf*, *sup*, *inacc* and *outacc* to the table *impli*1 (Fig. 6).

Based on *impli*1, the procedure *step*1 calculates *minsupp* and *minacc* for each tuple and compares them with the threshold values α and β. If $minsupp \geq \alpha$ and $minacc \geq \beta$, the procedure *step*1 adds this tuple to the table *crule*1. If $minsupp \geq \alpha$ and $minacc < \beta$, the procedure adds this tuple to the table *crest*1 (Fig. 7). On the other hand, the procedure calculates *maxsupp* and *maxacc* for each tuple and compares them with the threshold values α and β. If $maxsupp \geq \alpha$ and $maxacc \geq \beta$, the procedure adds this tuple to the table *prule*1 (Fig. 8). If $maxsupp \geq \alpha$ and $maxacc < \beta$, the procedure adds this tuple to the table *prest*1. The following is the SQL procedure for generating the table *prule*1.

The procedure for *prule*1 in Step 1:

```
create table prule1 (att1 varchar,val1 varchar,deci varchar,
    deci_value varchar,maxsupp decimal,maxacc decimal)
select impli1.att1,impli1.val1,impli1.deci,impli1.deci_value,
    impli1.sup/ob as maxsupp,impli1.sup/(con_des.inf+inacc) as maxacc
from impli1,con_des
where impli1.att1=con_des.attrib and impli1.val1=con_des.value
having maxsupp >=alpha and maxacc >=beta;
```

In Step 1, the most complicated part is to generate the table *impli*1. After obtaining the Cartesian Products *imp*1, *step*1 sequentially adds *inf*, *sup*, *inacc*, and *outacc* to *impli*1. If $inf([A, val_A] \wedge [Dec, val])$ is an empty set, this tuple is not stored in the temporary table data set. Even though it is necessary to add *inf*=0 to the table *impli*1, the value NULL is added to *impli*1 in this case. Therefore, *step*1 replaces NULL with 0 after adding *inf* information to *impli*1. The same occurs for *sup*, *inacc*, and *outacc*. In the current implementation, we faithfully simulated the *NIS-Apriori* algorithm, and there are ineffective procedures including the above case. It is necessary to reduce such ineffective part.

att1	val1	deci	deci_value	inf	sup	inacc	outacc
head	no	flu	no	1.000	3.000	1.000	1.000
head	no	flu	yes	0.000	2.000	1.000	1.000
head	yes	flu	no	0.000	3.000	1.000	1.000
head	yes	flu	yes	2.000	5.000	1.000	1.000
nau	no	flu	no	0.000	3.000	2.000	3.000
nau	no	flu	yes	1.000	5.000	3.000	2.000
nau	yes	flu	no	2.000	4.000	2.000	3.000
nau	yes	flu	yes	1.000	4.000	3.000	2.000

att1	val1	deci	deci_value	inf	sup	inacc	outacc
head	no	flu	no	1.000	3.000	1.000	1.000
head	no	flu	yes	0.000	2.000	1.000	1.000
head	yes	flu	no	0.000	3.000	1.000	1.000
head	yes	flu	yes	2.000	5.000	1.000	1.000
nau	no	flu	no	0.000	3.000	2.000	3.000
nau	no	flu	yes	1.000	5.000	3.000	2.000
nau	yes	flu	no	2.000	4.000	2.000	3.000
nau	yes	flu	yes	1.000	4.000	3.000	2.000

Fig. 6. A part of $impli1$.   **Fig. 7.** Total contents in $crest1$.

3.4 Step 2: Rule Generation in the Form of $P_1 \wedge P_2 \Rightarrow Dec$

In Step 2, the procedure $step2$ generates the certain and the possible rules in the form of $P_1 \wedge P_2 \Rightarrow Dec$. Since $support(P_1 \wedge P_2 \Rightarrow Dec) \leq support(P_1 \Rightarrow Dec)$ holds, it is enough to consider the implications $P_1 \wedge P_2 \Rightarrow Dec$ satisfying $(P_1 \Rightarrow Dec), (P_2 \Rightarrow Dec) \in crest1$ in certain rule generation.

We execute '$call\ step2('flu', 8, 0.1, 0.8)$' again, and it took about 0.39 (sec) for executing the procedure $step2$. Then, all tables in Fig. 10 were generated. In Fig. 10, two tables $cimpli2$ and $pimpli2$ store the tuples with inf, sup, $inacc$ and $outacc$, respectively. Tables $crule2$ and $prule2$ store the certain rules and the possible rules in the form of $P_1 \wedge P_2 \Rightarrow Dec$. Similarly to the tables $crest1$ and $prest1$, $crest2$ and $prest2$ are generated for Step 3. In Step 2, we obtained a certain rule in Figs. 9 and 12 possible rules in $prule2$.

The rule generation in Step 3 is the same as in case of Step2. Like Step 2, we execute '$call\ step3('flu', 8, 0.1, 0.8)$', then the procedure $step3$ generates rules.

3.5 An Implementation of NIS-Apriori in SQL

This prototype system is implemented on desktop PC and note PC by using the phpMyAdmin tool [8]. Currently, we made three procedures $step1$, $step2$, and $step3$ by using SQL command procedures. The data size of this file including all procedures is about 40KB in the text format. Since SQL command procedure is familiar, we will be able to use this prototype in the most of PC with SQL. Actually, we employed both desktop PC and note PC simultaneously for this implementation. We can also handle any DIS as a special case of NIS. In the NRDF format, we specify $det=1$ in each tuple, then we have the same rules in certain rule generation and possible rule generation.

3.6 The Difference Between Two Software Tools RNIA in Prolog and NIS-Apriori in SQL

Figure 11 is the execution log for $NIS\ \Psi_1$ by RNIA in Prolog. Except the redundant case of the rules, we examined the result by RNIA in Prolog is equal to the

att1	val1	deci	deci_value	maxsupp	maxacc
head	no	flu	no	0.375	1.000
head	yes	flu	yes	0.625	1.000
nau	no	flu	yes	0.625	1.000
nau	yes	flu	no	0.500	0.800
temp	high	flu	yes	0.625	0.833
temp	normal	flu	no	0.375	1.000
temp	veryhigh	flu	no	0.125	1.000
temp	veryhigh	flu	yes	0.375	1.000

Fig. 8. Total contents in *prule1*.

a1	v1	a2	v2	deci	deci_value	minsupp	minacc
head	no	nau	yes	flu	no	0.125	1.000

Fig. 9. Total contents in *crule2*.

All created tables:
cimpli2, condi, con_des, crest1, crest2, crule1, crule2, dec_des, imp1, impli1, nrdf, pimpli2, prest1, prest2, prule1, prule2

Fig. 10. All created tables.

result by *NIS-Apriori* in SQL. This will be an assurance that two software tools were implemented correctly.

Now, we have to remark the difference between the data structures of two software tools. RNIA in Prolog employs two blocks inf and sup, and internally manages them for each calculation. On the other hand, *NIS-Apriori* in SQL does not employ them directly, and employs the total search of the data set. These two points are the big difference between two software tools. We explain these two points below.

RNIA in Prolog generates $inf([A, val_A])$ and $sup([A, val_A])$ information for each descriptor $[A, val_A]$, and $inf(\tau)$ and $sup(\tau)$ are generated for each τ. For example, the set $inf([A, val_A] \wedge [Dec, val])$ is defined by $inf([A, val_A]) \cap inf([Dec, val])$, and it is stored as a temporary set. RNIA in Prolog makes use of $inf(\tau)$ and $sup(\tau)$, and generates rules. However, the use of $inf(\tau)$ and $sup(\tau)$ for each τ may be a heavy load. Figure 12 is the beginning of the log data for Mammographic data set [3]. In Fig. 12, we see that 427 objects support this rule, and every number of the object, i.e., 3, 5, $\cdots$, 960, is stored in the list. Even

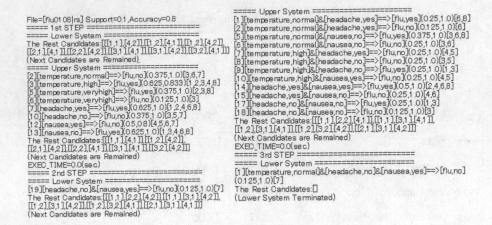

Fig. 11. The execution log by RNIA in Prolog.

though Prolog has a list processing functionality, the manipulation of such large size lists will be a heavy load.

On the other hand, *NIS-Apriori* in SQL does not store every number of the object, but stores the amount of objects (Fig. 13). For obtaining $inf(\tau)$ and $sup(\tau)$, *NIS-Apriori* in SQL executes the total search in the NRDF data set. As we have described, the most complicated part is to add inf, sup, $inacc$, and $outacc$ information to the Cartesian Products. For this part, we need to employ the total search of the NRDF data set instead of manipulating $inf(\tau)$ and $sup(\tau)$, but we can escape from the manipulation on the large size lists. In the application of *NIS-Apriori* in SQL to Mammographic data set, we obtained the same result by RNIA in Prolog. However, the execution time by the implemented *NIS-Apriori* in SQL was not good. It took about 1 (min) for Step 1. It is necessary to revise the current procedure, especially the generation of *impli*1, *cimpli*2, and *plimpli*2.

```
File=[mammo0204|rs] Support=0.2,Accuracy=0.4
===== 1st STEP ==========================
===== Lower System =====================
[1] [assessment,4]==>[severity,0](0.445, 0.763)
[3,5,6,12,13,14,15,19,22,30,33,34,35,36,39,40,41,42,43,47,50,52,53,55,56,58,62,
65,66,68,69,70,75,77,80,83,85,88,92,96,97,99,101,102,103,104,105,108,110,113,
114,115,116,117,121,122,123,125,126,127,128,133,138,141,142,143,144,149,152,
153,155,157,158,161,162,163,164,167,170,171,172,173,174,176,177,180,182,183,

909,911,913,915,916,918,919,920,922,925,927,929,930,932,933,934,936,937,938,
939,941,943,945,946,947,949,953,954,956,958,960]
```

Fig. 12. The list of the objects supporting an implication.

a1	v1	a2	v2	a3	v3	deci	deci_value	infC	sup	inacc
age	50	assess	4	density	3	severity	0	111.000	100.000	10.000
age	50	density	3	margin	4	severity	0	45.000	0.000	7.000
age	60	assess	4	density	3	severity	0	125.000	107.000	15.000
age	60	density	3	margin	4	severity	0	75.000	0.000	10.000
assess	4	density	3	margin	4	severity	1	97.000	0.000	12.000

Fig. 13. Total contents in *cimpli2* for Mammographic data set.

4   Concluding Remarks

This paper briefly described the background of RNIA for handling information incompleteness in table data, and we newly focused on SQL system for handling large size data sets. As for this prototype, we have the following consideration.

(1) Since SQL has the high versatility, *NIS-Apriori* in SQL will offer the useful environment for analyzing tables with non-deterministic values.

(2) Both RNIA in Prolog and *getRNIA* in Python internally store a list for each implication. For large size data sets, the manipulation of these lists will be a heavy load. On the other hand, *NIS-Apriori* in SQL does not employ such lists, but it employs the total search of the data sets. In two strategies, i.e., the list manipulation strategy and the total search strategy, we figure out that the list manipulation strategy will be suitable to rule generation for small size data sets, and the total search strategy will be suitable to rule generation for large size data sets.

(3) In the prototype, we faithfully simulated the *NIS-Apriori* algorithm, so the procedures in SQL might generate the meaningless tables. It is necessary to revise this point.

Acknowledgment. The authors would be grateful to the anonymous referees for their useful comments. This work is supported by JSPS (Japan Society for the Promotion of Science) KAKENHI Grant Number 26330277.

References

1. Agrawal, R., Srikant, R.: Fast algorithms for mining association rules in large databases. In: Proceedings of VLDB 1994, pp. 487–499. Morgan Kaufmann (1994)
2. Clark, P., Grzymala-Busse, J.: Mining incomplete data with many attribute-concept values and "do not care" conditions. In: Proceedings of IEEE Big Data, pp. 1597–1602 (2015)
3. Frank, A., Asuncion, A.: UCI machine learning repository. University of California, School of Information and Computer Science, Irvine, CA (2010). http://mlearn. ics.uci.edu/MLRepository.html

 4. Grzymala-Busse, J.W.: Data with missing attribute values: generalization of indiscernibility relation and rule induction. In: Peters, J.F., Skowron, A., Grzymała-Busse, J.W., Kostek, B., Świniarski, R.W., Szczuka, M.S. (eds.) Transactions on Rough Sets I. LNCS, vol. 3100, pp. 78–95. Springer, Heidelberg (2004). doi:10.1007/978-3-540-27794-1_3
 5. Kowalski, M., Stawicki, S.: SQL-based heuristics for selected KDD tasks over large data sets. In: Proceedings of FedCSIS 2012, pp. 303–310 (2012)
 6. Orłowska, E., Pawlak, Z.: Representation of nondeterministic information. Theor. Comput. Sci. **29**(1–2), 27–39 (1984)
 7. Pawlak, Z.: Systemy Informacyjne: Podstawy Teoretyczne (in Polish), WNT (1983)
 8. phpMyAdmin (2016). http://www.phpmyadmin.net/
 9. Sakai, H., Okuma, A.: Basic algorithms and tools for rough non-deterministic information analysis. In: Peters, J.F., Skowron, A., Grzymała-Busse, J.W., Kostek, B., Świniarski, R.W., Szczuka, M.S. (eds.) Transactions on Rough Sets I. LNCS, vol. 3100, pp. 209–231. Springer, Heidelberg (2004). doi:10.1007/978-3-540-27794-1_10
10. Sakai, H., Ishibashi, R., Koba, K., Nakata, M.: Rules and apriori algorithm in non-deterministic information systems. In: Peters, J.F., Skowron, A., Rybiński, H. (eds.) Transactions on Rough Sets IX. LNCS, vol. 5390, pp. 328–350. Springer, Heidelberg (2008). doi:10.1007/978-3-540-89876-4_18
11. Sakai, H., Wu, M., Nakata, M.: Apriori-based rule generation in incomplete information databases and non-deterministic information systems. Fundamenta Informaticae **130**(3), 343–376 (2014)
12. Sarawagi, S., Thomas, S., Agrawal, R.: Integrating association rule mining with relational database systems: alternatives and implications. Data Min. Knowl. Discov. **4**(2), 89–125 (2000)
13. Ślęzak, D., Sakai, H.: Automatic extraction of decision rules from non-deterministic data systems: theoretical foundations and SQL-based implementation. In: Ślęzak, D., Kim, T., Zhang, Y., Ma, J., Chung, K. (eds.) DTA 2009. CCIS, vol. 64, pp. 151–162. Springer, Heidelberg (2009). doi:10.1007/978-3-642-10583-8_18
14. Swieboda, W., Nguyen, S.: Rough set methods for large and spare data in EAV format. In: Proceedings of IEEE RIVF 2012, pp. 1–6 (2012)
15. Wu, M., Nakata, M., Sakai, H.: An overview of the getRNIA system for non-deterministic data. Procedia Comput. Sci. **22**, 615–622 (2013). http://getrnia.org

Classification for Inconsistent Decision Tables

Mohammad Azad[✉] and Mikhail Moshkov

Computer, Electrical and Mathematical Sciences and Engineering Division, King
Abdullah University of Science and Technology, Thuwal 23955-6900, Saudi Arabia
{mohammad.azad,mikhail.moshkov}@kaust.edu.sa

Abstract. Decision trees have been used widely to discover patterns
from consistent data set. But if the data set is inconsistent, where there
are groups of examples with equal values of conditional attributes but
different labels, then to discover the essential patterns or knowledge from
the data set is challenging. Three approaches (generalized, most common
and many-valued decision) have been considered to handle such inconsistency. The decision tree model has been used to compare the classification results among three approaches. Many-valued decision approach
outperforms other approaches, and M_ws_entM greedy algorithm gives
faster and better prediction accuracy.

Keywords: Decision trees · Greedy algorithms · Classifications · Many-
valued decisions · Inconsistent decision tables

1 Introduction

Often in a decision table, we have different examples with the different values
of decision and we call such table as a consistent decision table or single-valued
decision table. But it is pretty common in real life problems to have inconsistent
decision tables where there are groups of examples (objects) with equal values of
conditional attributes and different decisions (values of the decision attribute).
In this paper, we discussed three ways to discover patterns from such inconsistent
data sets.

In the rough set theory [8], generalized decision (GD) has been used to handle
inconsistency. In this case, a single example is considered from the group of equal
examples and a set of decisions has been formed consisting of different decisions
from the groups. After that, each set of decisions has been encoded by a number
(decision) such that equal sets are encoded by equal numbers and different sets
by different numbers (see Fig. 1). We have also used another approach named the
most common decision (MCD) which is derived from the concept of using most
common value in case of missing value [7]. Instead of a group of equal examples
with (probably) different decisions, we consider one example given by values of
conditional attributes and we attach to this example the most common decision
for examples from the group (see Fig. 1).

In our approach, we form a set of decisions that can be attached to the
example. We refer this approach as many-valued decision (MVD) approach (see

© Springer International Publishing AG 2016
V. Flores et al. (Eds.): IJCRS 2016, LNAI 9920, pp. 525–534, 2016.
DOI: 10.1007/978-3-319-47160-0_48

$$T^0 = \begin{array}{ccc|c} f_1 & f_2 & f_3 & \\ \hline 1 & 1 & 1 & 1 \\ 0 & 1 & 0 & 1 \\ 0 & 1 & 0 & 3 \\ 1 & 1 & 0 & 2 \\ 0 & 0 & 1 & 2 \\ 0 & 0 & 1 & 3 \\ 1 & 0 & 0 & 1 \\ 1 & 0 & 0 & 2 \end{array} \quad T^0_{MVD} = \begin{array}{ccc|c} f_1 & f_2 & f_3 & \\ \hline 1 & 1 & 1 & \{1\} \\ 0 & 1 & 0 & \{1,3\} \\ 1 & 1 & 0 & \{2\} \\ 0 & 0 & 1 & \{2,3\} \\ 1 & 0 & 0 & \{1,2\} \end{array} \quad T^0_{GD} = \begin{array}{ccc|c} f_1 & f_2 & f_3 & \\ \hline 1 & 1 & 1 & 1 \\ 0 & 1 & 0 & 2 \\ 1 & 1 & 0 & 3 \\ 0 & 0 & 1 & 4 \\ 1 & 0 & 0 & 5 \end{array} \quad T^0_{MCD} = \begin{array}{ccc|c} f_1 & f_2 & f_3 & \\ \hline 1 & 1 & 1 & 1 \\ 0 & 1 & 0 & 1 \\ 1 & 1 & 0 & 2 \\ 0 & 0 & 1 & 2 \\ 1 & 0 & 0 & 1 \end{array}$$

Fig. 1. Transformation of inconsistent decision table T^0 into decision tables T^0_{MVD}, T^0_{GD} and T^0_{MCD}

Fig. 1). Here our goal is to find a single decision from the attached set of decisions for each example. This approach is used for classical optimization problems (finding a Hamiltonian circuit with the minimum length or finding the nearest post office) where we have multiple optimal solutions but we have to give only one optimal output.

In the paper [3], one greedy algorithm using the uncertainty based on the number of boundary subtables has been addressed, but the drawback is that the uncertainty is too slow to work with medium to big ranged data sets since the running time is high order polynomial. The author only discussed about the complexity of the constructed tree but nothing is mentioned about classification accuracy.

But in this paper, we used different uncertainties which are efficient enough. We also evaluated the three approaches MVD, MCD, and GD by comparing the classification error rates by the constructed decision trees. We have presented results in the form of critical difference diagram [5] as well as average error rates using data sets from UCI ML Repository [2] and KEEL [1] repository. Finally, we found one greedy algorithm which is the fastest as well as produces good enough classification results.

2   Preliminaries

2.1   Many-Valued Decision Table

A *many-valued decision table* T is a rectangular table whose rows are filled by nonnegative integers and columns are labeled with conditional attributes $f_1, \ldots, f_n$. If we have strings as values of attributes, we have to encode the values as nonnegative integers. There are no duplicate rows, and each row is labeled with a nonempty finite set of natural numbers (set of decisions). We denote the number of examples (rows) in the table T by $N(T)$ (Table 1).

If there is a decision which belongs to all sets of decisions attached to examples of T, then we call it a *common decision* for T. We will say that T is a *degenerate* table if T does not have examples or it has a common decision.

Table 1. A many-valued decision table T'

$$T' = \begin{array}{ccc|c} f_1 & f_2 & f_3 & \\ \hline 0 & 0 & 0 & \{1\} \\ 0 & 1 & 1 & \{1,2\} \\ 1 & 0 & 1 & \{1,3\} \\ 1 & 1 & 0 & \{2,3\} \\ 0 & 0 & 1 & \{2\} \end{array}$$

A table obtained from T by removing some examples is called a subtable of T. We denote a *subtable* of T which consists of examples that at the intersection with columns $f_{i_1}, \ldots, f_{i_m}$ have values $a_1, \ldots, a_m$ by $T(f_{i_1}, a_1), \ldots, (f_{i_m}, a_m)$. Such nonempty tables (including the table T) are called *separable subtables* of T. For example, if we consider subtable $T'(f_1, 0)$ for table T', it will consist of examples 1, 2, and 5. Similarly, $T'(f_1, 0)(f_2, 0)$ subtable will consist of examples 1, and 5 (Table 2).

Table 2. Example of subtables of many-valued decision table T'

$$T'(f_1, 0) = \begin{array}{ccc|c} f_1 & f_2 & f_3 & \\ \hline 0 & 0 & 0 & \{1\} \\ 0 & 1 & 1 & \{1,2\} \\ 0 & 0 & 1 & \{2\} \end{array} \qquad T'(f_1, 0)(f_2, 0) = \begin{array}{ccc|c} f_1 & f_2 & f_3 & \\ \hline 0 & 0 & 0 & \{1\} \\ 0 & 0 & 1 & \{2\} \end{array}$$

We denote the set of attributes (columns of table T), such that each of them has different values by $E(T)$. For $f_i \in E(T)$, we denote the set of values from the attribute f_i by $E(T, f_i)$. The minimum decision which belongs to the maximum number of sets of decisions attached to examples of the table T is called the *most common decision* for T.

2.2 Decision Trees

A *decision tree* over T is a finite tree with root in which each terminal node is labeled with a decision (a natural number), and each nonterminal node is labeled with an attribute from the set $\{f_1, \ldots, f_n\}$. A number of edges start from each nonterminal node which are labeled with the values of that attribute (e.g. two edges labeled with 0 and 1 for the binary attribute).

Let Γ be a decision tree over T and v be a node of Γ. We denote $T(v)$ as a subtable of T that is mapped for a node v of decision tree Γ. If the node v is the root of Γ then $T(v) = T$, i.e. the subtable $T(v)$ is the same as T. Otherwise, $T(v)$ is the subtable $T(f_{i_1}, \delta_1) \ldots (f_{i_m}, \delta_m)$ of the table T where attributes $f_{i_1}, \ldots, f_{i_m}$ and numbers $\delta_1, \ldots, \delta_m$ are respectively node and edge labels in the path from the root to node v. We will say that Γ is a decision tree for T if Γ satisfies the following conditions:

– if $T(v)$ is degenerate then v is labeled with the common decision for $T(v)$,
– otherwise v is labeled with an attribute $f_i \in E(T(v))$. In this case, k outgoing edges from node v are labeled with $a_1, \ldots, a_k$ where $E(T(v), f_i) = \{a_1, \ldots, a_k\}$.

An example of a decision tree for the table T' can be found in Fig. 2.

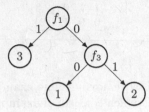

Fig. 2. Decision tree for the many-valued decision table T'

2.3 Uncertainty Measures

Uncertainty measure U is a function from the set of nonempty many-valued decision tables to the set of real numbers such that $U(T) \geq 0$, and $U(T) = 0$ if and only if T is degenerate.

Let T be a many-valued decision table having n conditional attributes, $N = N(T)$ examples (rows) and its examples be labeled with sets containing m different decisions $d_1, \ldots, d_m$. For $i = 1, \ldots, m$, let N_i be the number of examples in T that has been attached with sets of decisions containing the decision d_i, and $p_i = N_i/N$. Let $d_1, \ldots, d_m$ be ordered such that $p_1 \geq \cdots \geq p_m$, then for $i = 1, \ldots, m$, we denote by N_i' the number of examples in T such that the set of decisions attached to an example contains d_i, and if $i > 1$ then this set does not contain $d_1, \ldots, d_{i-1}$, and $p_i' = N_i'/N$. We have the following four uncertainty measures (we assume $0 \log_2 0 = 0$):

(1) Misclassification error: $me(T) = N(T) - N_{mcd}(T)$, where d_{mcd} is the most common decision for T.
(2) Sorted entropy: $entS(T) = -\sum_{i=1}^{m} p_i' \log_2 p_i'$ (see [6]).
(3) Multi-label entropy: $entM(T) = 0$, if and only if T is degenerate, otherwise, it is equal to $-\sum_{i=1}^{m}(p_i \log_2 p_i + q_i \log_2 q_i)$, where, $q_i = 1 - p_i$. (see [4]).
(4) Absent: $abs(T) = \prod_{i=1}^{m} q_i$, where $q_i = 1 - p_i$.

2.4 Impurity Functions

Let U be an uncertainty measure, $f_i \in E(T)$, and $E(T, f_i) = \{a_1, \ldots, a_t\}$. The attribute f_i divides the table T into t subtables: $T_1 = T(f_i, a_1), \ldots, T_t = T(f_i, a_t)$. We now define three types of impurity function I which gives us the impurity $I(T, f_i)$ of this partition.

(1) Weighted max (wm): $I(T, f_i) = \max_{1 \leq j \leq t} U(T_j)N(T_j)$.
(2) Weighted sum (ws): $I(T, f_i) = \sum_{j=1}^{t} U(T_j)N(T_j)$.
(3) Multiplied weighted sum (M_ws): $I(T, f_i) = (\sum_{j=1}^{t} U(T_j)N(T_j)) \times \log_2 t$.

2.5 Greedy Algorithms for Decision Tree Construction

Let I be an impurity function based on the uncertainty measure U. The greedy algorithm A_I, for a given many-valued decision table T, constructs a decision tree $A_I(T)$ for T (see Algorithm 1). We have 12 ($= 4 \times 3$) algorithms. The complexities of these algorithms are polynomially bounded above depending on the size of the tables.

Algorithm 1. Greedy algorithm A_I

Input: A many-valued decision table T with conditional attributes $f_1, \ldots, f_n$.
Output: Decision tree $A_I(T)$ for T.
  Construct the tree G consisting of a single node labeled with the table T;
  while (true) **do**
    if No one node of the tree G is labeled with a table **then**
      Denote the tree G by $A_I(T)$;
    else
      Choose a node v in G which is labeled with a subtable T' of the table T;
      if $U(T') = 0$ **then**
        Instead of T' mark the node v with the common decision for T';
      else
        For each $f_i \in E(T')$, compute the value of the impurity function $I(T', f_i)$;
        Choose the attribute $f_{i_0} \in E(T')$, where i_0 is the minimum i for which
        $I(T', f_i)$ has the minimum value; Instead of T' mark the node v with the
        attribute f_{i_0}; For each $\delta \in E(T', f_i)$, add to the tree G the node v_δ and mark
        this node with the subtable $T'(f_{i_0}, \delta)$; Draw an edge from v to v_δ and mark
        this edge with δ.
      end if
    end if
  end while

---

3 Data Sets

We consider five decision tables from UCI Machine Learning Repository [2]. For the sake of experiments, we removed from these tables more conditional attributes. As a result, we obtained inconsistent decision tables which contain equal examples with equal or different decisions. From these inconsistent tables, we derived MVD, GD and MCD tables in the way explained in the introduction. The information about obtained inconsistent (represented as many-valued decision) tables can be found in Table 3. These modified tables have been renamed as the name of initial table with an index equal to the number of removed conditional attributes. We also consider five decision tables from KEEL [1] multi-label data set repository. Note that, these tables are already in many-valued decision format. The information about these table can be found in Table 4. We can derive GD table from MVD table in a natural way by encoding sets of decisions using numbers. To obtain MCD table from MVD table, we choose for each row the smallest decision from the set of decisions attached to this row.

Table 3. Characteristics of modified UCI inconsistent data represented in MVD format

Data set T	Row	Attr	Label	lc	ld	Spectrum #1, #2, #3
CARS-1	432	5	4	1.43	0.36	258, 161, 13
FLAGS-5	171	21	6	1.07	0.18	159, 12,
LYMPH-5	122	13	4	1.07	0.27	113, 9,
NURSERY-1	4320	7	5	1.34	0.27	2858, 1460, 2
ZOO-5	42	11	7	1.14	0.16	36, 6,

Table 4. Characteristics of KEEL multi-label data

Decision table T	Row	Attr	Label	lc	ld	Spectrum								
						#1	#2	#3	#4	#5	#6	#7	#8	#9
*bibtex**	7355	1836	159	2.41	0.015	2791	1825	1302	669	399	179	87	46	18
COREL5k	4998	499	374	3.52	0.009	3	376	1559	3013	17	0	1	0	0
*enron**	1561	1001	53	3.49	0.066	179	238	441	337	200	91	51	15	3
GENBASE-1	662	1186	27	1.47	0.054	560	58	31	8	2	3	0	0	0
MEDICAL	967	1449	45	1	0.027	741	212	14	0	0	0	0	0	0

Tables 3 and 4 also contain the number of examples (column "Row"), the number of attributes (column "Attr"), the total number of decisions (column "Label"), the cardinality of decision (column "lc"), the density of decision (column "ld"), and the spectrum of this table (column "Spectrum"). The decision cardinality, lc, is the average number of decisions for each example in the table. The decision density, ld, is the average number of decisions for each example divided by the total number of decisions. If T is a many-valued decision table with N examples (x_i, D_i) where $i = 1, \ldots, N$, then $lc(T) = \frac{1}{N} \sum_{i=1}^{N} |D_i|$, where $|D_i|$ is the cardinality of decision set in i-th example, and $ld(T) = \frac{1}{|L|} lc(T)$, where $|L|$ is the total number of decisions in T. Spectrum of a many-valued decision table is a sequence #1, #2,..., where #i, $i = 1, 2, \ldots$, is the number of examples labeled with sets of decisions with the cardinality equal to i. For some tables (marked with * in Table 4), the spectrum is too long to fit in the page width.

4   Decision Tree Classifier

The examples in the many-valued decision table T have been attached with sets of decisions $D \subset L$ where L is the set of all possible decisions in the table T. We denote $D(x)$ as the set of decisions attached to the example x. If X is the domain of the examples to be classified, the goal is to find a classifier $h : X \to L$ such that $h(x) = d$, where $d \in D(x)$, that means to find a decision from the ground truth set of decisions attached to the example. To solve the problem, we construct different kinds of decision trees using various impurity functions. Here we use the common evaluation measure of classification error percentage. Please note that, we can consider decision tables with single decision as a special case of many-valued decision tables where each row contains a set of a single decision. Therefore, we can use the same technique for the MCD and GD approaches.

Let T be a many-valued decision table with conditional attributes $f_1, \ldots, f_n$. We have to divide the initial subtable into three subtables: training subtable T_1, validation subtable T_2, and test subtable T_3. The subtable T_1 is used for construction of initial classifier. The subtable T_2 is used for pruning of the initial tree. Let the initial tree Γ contain t nonterminal nodes, and Γ_i, $i = 0, \ldots, t$, be the decision trees obtained from $\Gamma_0 = \Gamma$ during the pruning.

For $i = 0, \ldots, t$, we used the decision tree Γ_i to calculate the classification error rate for the table T_2. We choose the tree Γ_i which has the minimum classification error rate (in case of tie, we choose the tree with smaller index). Now this tree can be used as the final classifier and we can evaluate the test error rate by using this tree to classify the examples in table T_3.

Table 5. Classification error rate (in %)

Data set	MVD			MCD			GD		
	M_ws_abs	M_ws_entS	M_ws_entM	M_ws_abs	M_ws_entS	M_ws_entM	M_ws_abs	M_ws_entS	M_ws_entM
BIBTEX	57.38	59.96	57.72	54.61	65.37	63.95	90.97	90.97	83.07
CARS-1	3.66	3.80	4.44	6.67	6.67	6.67	16.16	17.04	17.08
COREL5k	74.39	76.52	77.55	79.54	82.14	81.96	98.70	98.70	97.46
ENRON	35.93	27.05	30.29	32.30	30.72	32.22	85.93	85.93	86.48
FLAGS-5	61.75	63.86	63.51	60.82	63.39	63.98	68.19	72.05	65.15
GENBASE-1	4.46	5.34	4.85	6.60	6.80	6.90	13.41	13.41	16.70
LYMPH-5	25.20	24.87	24.54	29.34	29.67	29.01	29.34	40.50	30.00
MEDICAL	20.00	23.43	21.72	23.27	26.00	25.07	36.53	36.53	33.71
NURSERY-1	2.05	2.70	2.71	5.40	5.25	5.35	8.25	6.69	8.39
ZOO-DATA-5	34.76	32.38	30.48	35.24	34.76	32.86	48.10	48.10	50.00
AER	31.96	31.99	31.78	33.38	35.08	34.80	49.55	50.99	48.80

To compare the various decision tree algorithms statistically, we used Friedman test with the corresponding Nemenyi post-hoc test as suggested in [5]. Let we have k greedy algorithms $A_1, \ldots, A_k$ for constructing trees and M decision tables $T_1, \ldots, T_M$. For each decision table T_i, $i = 1, \ldots, M$, we rank the algorithms $A_1, \ldots, A_k$ on T_i based on their performance scores of classification error rates, where we assign the best performing algorithm the rank of 1, the second best rank 2, and so on. We break ties by computing the average of ranks. Let r_i^j be the rank of the j-th of k algorithms on the decision table T_i. For $j = 1, \ldots, k$, we correspond to the algorithm A_j the average rank $R_j = \frac{1}{M} \cdot \sum_{i=1}^{M} r_i^j$. For a fixed significance level α, the performance of two algorithms is significantly different if the corresponding average ranks differ by at least the critical difference $CD = q_\alpha \sqrt{\frac{k(k+1)}{6M}}$ where q_α is a critical value for the two-tailed Nemenyi test depending on α and k.

5 Experimental Results

We used 3-fold cross validation to separate test and training data set for each decision table. The data set is divided into 3 folds. At i-th ($i = 1, 2, 3$) iteration,

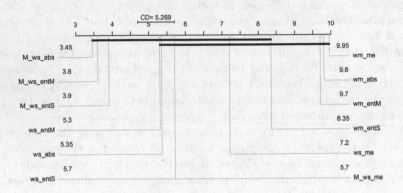

(a) MVD approach

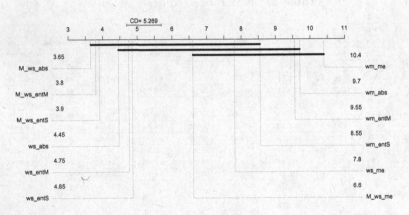

(b) MCD approach

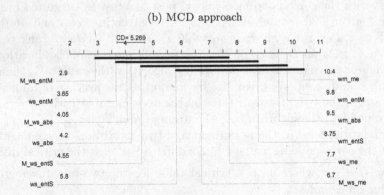

(c) GD approach

Fig. 3. Critical difference diagrams

i-th fold is used as the test subset, and the rest of data is partitioned randomly into train (70 %) and validation subset (30 %). The validation subset is used to prune the trained tree. We successively prune the nodes of the trained decision tree model based on the accuracy of the classifier from validation data set unless its accuracy is maximum. After pruning, we used trained decision tree model to predict the decisions for test data sets. For each fold, we repeat the experiment 5 times and take the average of 5 error rates.

We have four uncertainty measures (me, abs, $entS$, $entM$) and three types of impurity functions (ws, wm, M_ws). So, 12 greedy algorithms have been compared. We show the names of the algorithms as combined name of uncertainty and impurity function types separated by '_' in CDD. For example, if the algorithm name is wm_me, this means it uses wm as a type of impurity function and me as uncertainty measure. Figure 3 shows the CDD containing average rank for each algorithm on the x-axis for significance level of $\alpha = 0.05$ for the MVD, MCD and GD approaches. The best ranked algorithm are shown in the leftmost side of the figure. When Nemenyi test cannot identify significant difference between some algorithms, then those are clustered (connected).

It is clear that, M_ws_abs is the best ranked algorithm to minimize the test error for MVD and MCD approaches but M_ws_entM is the best ranked algorithm for GD approach. We have shown classification error rate for each data set and for the three best ranked algorithms for MVD approach as well as the average error rate (AER) among all the data sets in Table 5 for the three considered approaches. We have also shown the overall execution time for the three best ranked algorithm in the Table 6 for MVD, MCD, and GD approaches.

We can see that, on average, M_ws_entM gives the best error rate for MVD and GD approaches. We found that the M_ws_entM algorithm is faster on average than other two algorithms for MVD and MCD approaches. If we compare among three approaches, we find that on average MVD approach gives best classification results compared to others. But for the time, MCD approach is on average faster than others.

Table 6. Overall execution time (in sec)

Data set	MVD			MCD			GD		
	M_ws_abs	M_ws_entS	M_ws_entM	M_ws_abs	M_ws_entS	M_ws_entM	M_ws_abs	M_ws_entS	M_ws_entM
BIBTEX	375.15	1979.73	155.46	61.08	110.61	47.14	166.51	181.37	427.33
CARS-1	0.01	0.03	0.01	0.01	0.02	0.003	0.04	0.02	0.01
COREL5k	166.65	818.72	101.31	18.09	26.12	11.28	38.71	67.04	62.32
ENRON	9.21	8.27	6.44	2.52	4.26	2.56	10.06	15.48	31.32
FLAGS-5	0.02	0.01	0.02	0.01	0.02	0.02	0.03	0.05	0.03
GENBASE-1	0.36	0.32	0.23	0.25	0.25	0.31	0.54	0.82	0.56
LYMPHOGRAPHY-5	0.01	0	0.01	0.003	0.003	0.01	0.02	0.03	0.03
MEDICAL	2.03	4.57	1.61	1.84	1.7	1.29	3.18	3.38	2.65
NURSERY-1	0.05	0.06	0.05	0.06	0.09	0.06	0.21	0.16	0.23
ZOO-DATA-5	0	0.01	0.003	0.01	0.01	0.003	0.06	0.08	0.02
AVERAGE	55.35	281.17	26.51	8.39	14.31	6.27	21.94	26.84	52.45

6 Conclusion

We found that M_ws_entM gives enough good results both from the point of view of error rates of constructed classifiers and time complexity. In the future, we are planning to consider the inhibitory trees where the leaf nodes contain negation of the decision value $(d \neq v)$, and ensembles of such inhibitory trees that can be suitable for classification tasks.

References

1. Alcalá-Fdez, J., Fernández, A., Luengo, J., Derrac, J., García, S.: KEEL data-mining software tool: data set repository, integration of algorithms and experimental analysis framework (2011)
2. Asuncion, A., Newman, D.J.: UCI Machine Learning Repository (2007)
3. Azad, M., Chikalov, I., Moshkov, M.: Three approaches to deal with inconsistent decision tables - comparison of decision tree complexity. In: RSFDGrC, pp. 46–54 (2013)
4. Clare, A., King, R.D.: Knowledge discovery in multi-label phenotype data. In: Raedt, L., Siebes, A. (eds.) PKDD 2001. LNCS (LNAI), vol. 2168, pp. 42–53. Springer, Heidelberg (2001). doi:10.1007/3-540-44794-6_4
5. Demsar, J.: Statistical comparisons of classifiers over multiple data sets. J. Mach. Learn. Res. **7**, 1–30 (2006)
6. Hüllermeier, E., Beringer, J.: Learning from ambiguously labeled examples. Intell. Data Anal. **10**(5), 419–439 (2006)
7. Mingers, J.: An empirical comparison of selection measures for decision-tree induction. Mach. Learn. **3**(4), 319–342 (1989)
8. Pawlak, Z.: Rough Sets-Theoretical Aspects of Reasoning about Data. Kluwer Academic Publishers, Dordrecht (1991)

Derivation and Application
of Feature Subsets

Feature Selection in Decision Systems with Constraints

Sinh Hoa Nguyen[1,2] and Marcin Szczuka[1(✉)]

[1] Institute of Informatics, The University of Warsaw,
Banacha 2, 02-097 Warsaw, Poland
{hoa,szczuka}@mimuw.edu.pl
[2] Polish-Japanese Academy of Information Technology,
Koszykowa 86, 02-008 Warsaw, Poland

Abstract. In the paper we discuss an attribute reduction problem for a decision system with constraints. We present a new concept of decision system with constraints and a concept of constrained reduct defined for decision system with constraints. We define the problem of feature reduction for such constrained system and propose some heuristics or constrained reduct calculation and feature selection. We illustrate possible benefits of the proposed approach with an example based on the stream data obtained from sensor arrays in a coal mines.

Keywords: Rough sets · Decision system · Reduct · Sensor stream mining

1 Introduction

When dealing with high-dimensional, complicated, real-life data we are forced to hurdle several problems even before we begin constructing a model, such as classifier. The nature of data frequently makes it a challenge to represent it in a way that is at the same time computationally useful and comprehensible for the user. In real-life scenarios we are every so often facing a dichotomy between the representation used by domain expert and the representation in the form of information system or data table. The experts' representation is intuitive, understandable and convenient but may be hard or even impossible to be captured by mathematical model. Conversely, the representation in the form of information system or data table, that uses attributed selected or constructed from the measures taken, may have all of the required technical properties but may turn out to be hard to explain and to a large extend detached from the internal properties of the real-life phenomenon that it is meant to record.

In this paper we attempt to narrow the gap between the two types of representation. We introduce an extension to the traditional format of the information

The authors are partially supported by the Polish National Centre for Research and Development (NCBiR) - grant PBS2/B9/20/2013 in frame of the Applied Research Programmes.

V. Flores et al. (Eds.): IJCRS 2016, LNAI 9920, pp. 537–547, 2016.
DOI: 10.1007/978-3-319-47160-0_49

system (decision table) [11]. Albeit its simplicity and versatility, the traditional notion of information system is not best suited for data where attributes are structured or bound by relationships. In many applications attributes and their values are semantically related and domain constraints are usually enforced on them in order to stay true to the original data source. One example of such data, used further in the paper as an illustration, is a collection of measurements taken by a sensor array over a period of time. In most straightforward interpretation such data set is just a multi-dimensional time series. However, a domain expert will be well aware of relationships that constrain the range of possible sensor readings in a given setting. For example, the location of a sensor may allow for interpretation of its readings by means of readings of others.

We provide a concept called the *constrained decision system* in order to formalize the representation of data with presence of limitations (constraints) on attributes and their values. This concept is devised in a way that permits the use of various analogies with "classical" notions from the rough set toolbox. The overall goal is to have a decision system that not only records the presence of constrains but also makes it possible to apply various computational methods that use them. Of particular interest to us are the methods for feature selection and construction that work on an information system (decision table) with constrained attributes. By a constraint we will understand a subset of attributes that have to co-occur in particular way. We would like to make good use of the rough sets' concept of attribute reduction and the reduct.

Just like in the "ordinary" information system or decision table (see [13]) the task of finding the right reduct in a constrained decision system is a potentially costly undertaking. The presence of constraints may change it, although usually for the worse, as advocated in the paper. Therefore, we describe a heuristic approach that is capable of producing the semi-optimal reduction in reasonable time. We argue that the heuristic solution is sufficient for the kind of applications that we have envisioned while defining the constrained decision system.

The idea of extending the notions of information system, decision table, reduct and so on is hardly original, as are the ideas to consider subsets of attributes at once or ordering on attributes. Various attempts to do just that have been performed (see [1,2,5,9]). However, the existing extensions are rather concerned with relationships between *objects* or *attribute values* in the information system (decision table). To the best of our knowledge, the attempt at systematization of attribute reduction process in case of internal relationships between attributes (inter-attribute constraints) is novel.

2   Decision Systems with Constraints

The first step towards establishment of the apparatus for dealing with constrains in information systems and decision tables is the introduction of decision system with constraints. Traditionally, in the area of rough set theory the decision system (decision table) is a tuple $DS = (U, A \cup \{d\})$, where U is a finite set of objects and A is a finite set of attributes defined for all objects from U. As usual, we distinguish the decision attribute d that assigns the value of decision (target) to elements of the universe of objects. For any attribute $a \in A$, $a : U \to V_a$, where V_a is a domain of attribute a.

Definition 1. Decision system with constraints. *Let $DS = (U, A \cup \{d\})$ be a decision system. Constraint for DS is a subset $C_i \subseteq A$ of attributes, which are (semantically) related. A decision system with constraints is a triple $DS = (U, A \cup \{d\}, C)$, where $C = \{C_1, C_2, ..., C_k\}$ is a finite set of constraints ($C_i \subseteq A$ for $i = 1, \ldots, k$).*

One can see that a classical decision system is just a special case of a decision system with constraints. If all constraints are trivial singletons, binding only one attribute, i.e., $|C_i| = 1$ for all $1 \leq i \leq k$, then we are dealing with regular decision table.

While representation of a constraint is just a subset of attributes, the meaning of it may be much wider. The basic types of constraints that we consider are:

1. **Conjunctive**: A constraint is a set of attributes which *must appear together* in the description of target concept (decision).
2. **Disjunctive**: A constraint is a set of attributes C_i such that *at least one* of them has to occur in the description of target concept.
3. **Mutually exclusive**: A constraint is a set of attributes which *cannot occur together* in the description of target concept.

Further in this paper we will concentrate on constraints in conjunctive form. However, in *multi-faceted* classification systems we may want to use several types of constraints at the same time.

As our goal is to show the attribute reduction in decision system with constraints we need to introduce some more key notions, such as (in)discernibility and discernibility matrix.

Definition 2. Indiscernibility relation with constraints. *Let $\mathcal{CDS} = (U, A \cup d, C)$ be a decision table with a finite set of constraints C. Let $a \in A$ be an attribute. We say that x and y are indiscernible with respect to a, if*

- $a(x) = a(y)$ *and*
- $\forall_{C_i} (a \in C_i) \to (x, y) \in IND(C_i)$

Definition 3. Discernibility relation with constraints. *Let $\mathcal{CDS} = (U, A \cup d, C)$ be a decision table. For $a \in A$, a relation $DISCERN(a) \subseteq U \times U$ is defined as follows: $(x, y) \in DISCERN(a)$ if*

- $a(x) \neq a(y)$ *or*
- $\exists_{C_i} (a \in C_i) \wedge \exists_{a' \in C_i} a'(x) \neq a'(y)$

Definition 4. Discernibility matrix. *Let $\mathcal{CDS} = (U, A \cup d, C)$ be a decision table with constraints. Discernibility matrix $M[i, j]$ is defined as follows: $M[i, j] = \{a \in A : (x_i, x_j) \in DISCERN(a)\}$.*

Equipped with the definitions we are ready to show how to approach the feature selection (reduction) problem.

3   Constrained Reducts

We are given a constrained decision system $\mathcal{CDS} = (U, A \cup \{d\}, C)$ and $IND(B) \subseteq U \times U$ an indiscernibility relation defined on U as:

$$(x, y) \in IND(B) \leftrightarrow \forall_{b \in B}(x, y) \in IND(b).$$

The relation $IND(B)$ is an *equivalence relation*. Let $[x]_{IND(B)}$ denote indiscernibility class defined on x and $U/IND(B)$ denote all indiscernibility classes defined on U.

For a set of attributes $B \subseteq A$, a *positive region* of the set $X \subseteq U$ (*lower approximation*) w.r.t. B is defined by $POS_B(X) = \{x \in U : [x]_{IND(B)} \subseteq X\}$.

Definition 5. Reduct with constraints. *Let* $\mathcal{CDS} = (U, A \cup d, C)$ *be a decision table with a set of constraints* $C = \{C_1, C_2, ...C_k\}$, $C_i \subseteq A$. *Reduct of* $\mathcal{CDS}$ *is any* $R \subseteq A$, *which is satisfied the following conditions:*

- $POS_R(D) = POS_A(D)$
- *if* $C_i \cap R \neq \emptyset$ *then* $C_i \subseteq R$ $(*)$
- $\forall_{R'}(R' \subset R) \wedge$ *(R' satisfies (*))* $\rightarrow POS_{R'}(D) \neq POS_A(D)$

3.1   Searching for the Optimal Constrained Reduct

Let $\mathcal{CDS} = (U, A \cup d, C)$, $C = \{C_1, C_2, ..., C_k\}$, where $C_i \subseteq A$, for $1 \leq i \leq k$ be a decision system with constraints. We are looking for a reduct R satisfying the set of constraints C. The task of searching for a such reduct in this system can be formulated as an optimization problem. The optimization goal is to find a reduct R satisfying the set of constraints C and minimizing the cost function defined as:

$$cost(R) = F(card(R), card(C_i)), \text{ for } 1 \leq i \leq k$$

where F is some predefined function binding the quality of solution with the size of the optimization problem. The cost function may be defined, for example, as:

- The number of attributes in reduct R: $cost(R) = card(R)$
- The number of attributes appearing in constraints:

$$cost(R) = card(\bigcup C_j : C_j \subseteq R)$$

- The number of constrains and the number of unconstrained attributes:

$$cost(R) = card(\{C_j : C_j \subseteq R\}) + card(R \setminus \bigcup \{C_j : C_j \subseteq R\})$$

- The number of constrains: $cost(R) = card(\{C_j : C_j \subseteq R\})$

We say the reduction R is *optimal* if $cost(R)$ is minimal. Formally, the discrete optimization problem of searching for the best constrained reduct can be defined, following standard convention (see [3]), as follows:

PROBLEM: Optimal reduct with constraints:
GIVEN: A decision system with constraints $CDS = (U, A \cup d, C)$. An objective function $cost(.)$.
OBJECTIVE: Find a reduct with constraints R which produces the lowest $cost(R)$.

In this paper we concentrate on optimal finding constrained reducts with respect to the *minimal number of constraints* criterion, i.e., the last of previously provided possible variants of function $cost(R)$.

3.2 Computational Complexity and Heuristics

In this section we analyze the computational complexity of the problem of searching for the minimal reduct with constraints. We quickly show that since the reduction in the presence of constraints is no easier than traditional one, the problem is hard. We do that by making a series of rather simple observations, presented below as propositions with some rudimentary justification (proof).

Proposition 1. *For a decision system with constraints* $CDS = (U, A \cup d, C)$, *if all constraints are singletons then:*

– *all cost functions listed in the previous section are equal, and*
– *the problem of searching for the minimal constrained reduct becomes the problem of searching for the shortest reduct in a classical decision table [11].*

Proposition 2. *The problem of searching for the minimal constrained reduct is* NP-*hard.*

Proof. The problem of searching for the shortest (classical) reduct is an instance of the problem of searching for the minimal constrained reduct, as observer in Proposition 1. Since the former problem is NP-hard [13], so is the latter.

Given a decision system with constraints $CDS = (U, A \cup d, C)$, with $C = \{C_i : C_i \subseteq A\}$ – a set of constraints, we say that constraint C_i *covers* an attribute a if $a \in C_i$.

Proposition 3. *For any constrained decision system* CDS, *the constraint set can be extended to cover all attributes and the new constrained decision system* CDS' *is equivalent to the original one w.r.t. reduction. It means that any reduct R of CDS is a reduct for (extended) CDS' and vice versa.*

Proof. If a set of constraints C of CDS does not *cover* all attributes, we can extend it by singleton sets containing attributes which are not already covered. It is now easy to see that every reduct of CDS is a reduct of CDS' and the other way round.

Due to space limitations we do not provide the detailed definition of *constrained discernibility matrix* and *constrained discernibility function* but they are fairly easy to construct given the propositions above. The boolean constrained discernibility function is built as a conjunction of discernibility clauses. As in classical case, one can find relative constrained reducts by searching for *prime implicants* of (constrained) discernibility function. The straightforward method calculates all prime implicants by translation to Disjunctive Normal Form (DNF). Then, each conjunctive clause in DNF formula corresponds to a reduct [11].

We have established that every algorithm for finding prime implicants can be applied to the discernibility function to find constrained reducts. If we want to find exactly the shortest (minimal) one, we may face the problem of intractability. If we are prepared to sacrifice some optimality for the sake of tractability, we may resort to heuristic (approximate) methods. One of such heuristic was proposed in [13]. It is based on the greedy approach, where attributes are selected and added to candidate set in the order determined by value of *discernibility measure* and then eliminated until the set remains a reduct. The discernibility measure for an attribute is the number of pairs of objects which are discerned by it, or, equivalently, the number of its occurrences in the discernibility matrix [10]. This procedure can be applied to constrained reduct calculation with only a small modification. Namely, the discernibility measure for an attribute a in the modified algorithm is calculated as the number of object pairs which are discerned by a set attributes co-occurring with a in a constraint.

4   Feature Selection in Constrained Decision System

The reduction o attribute set based on simple rough set approach, i.e., on calculation of few reducts is usually not sufficient in either non-constrained and constrained scenario. Although decision reducts attempt to express information about data-based dependencies as compactly as possible, in practice they frequently both contain superfluous attributes and lack some others. Even the best reduct, if used "as-is" typically leads either to incomplete coverage or overfitting. To counter these problems numerous concepts, such as approximate reducts [14], were introduced. These concepts are to some extent usable in the case of constrained decision systems, but they require adjustments.

One of the more recent algorithms for feature reduction based on rough set concepts is the approach based on Minimum Redundancy Maximum Relevance framework (mRMR) [12]. The general idea of the mRMR has been adopted to the task of finding a relatively small subset of attributes. The algorithm was described in [6] and then implemented as R package RmRMR [7]. In standard mRMR-based feature selection, as shown in Algorithm 1 we first select an attribute that maximizes the difference between its dependency score $\phi(a, d)$ and its maximal dependency on already selected attributes ($\max b \in A'\phi(a, b)$). Then we stop the algorithm if the attribute selected in a given iteration does not pass the random probe test, i.e., the estimation of the probability that a

Algorithm 1. mRMR feature selection algorithm for a *classical decision system*

Input: set of attributes A and decision attribute d;
$\phi : A \times A \cup \{d\} \to R^+$ function for measuring dependency;
$N \in \mathbb{N}$; $\varepsilon \in [0, 1)$;
Output: subset of attributes $A' \subseteq A$
begin

    $A' \leftarrow \emptyset$;
    $stopFlag \leftarrow FALSE$;
    $A' \leftarrow \arg\max_{a \in A} \phi(a, d)$;
    $A \leftarrow A \setminus A'$;
    while $stopFlag == FALSE$ **do**

        $\bar{a} \leftarrow \arg\max_{a \in A} \left(\phi(a, d) - \max_{b \in A'} \phi(a, b) \right)$;
        foreach $i \in 1, \ldots, N$ **do**
          | $\bar{p}_i \leftarrow$ random permutation of A;
        end
        if $\frac{|\{i : |\phi(\bar{p}_i, d)| > |\phi(\bar{a}, d)|\}| + 1}{N + 2} > \varepsilon$ **then**
          | $stopFlag \leftarrow TRUE$;
        else
          | $A' \leftarrow A' \cup \bar{a}$;
        end
    end
end

randomly generated attribute obtains a higher score than the selected attribute exceeds an allowed threshold. Thus, we ensure compactness and a relatively high independence of the resulting feature subset.

For the decision system with constraints $\mathcal{CDS} = (U, A \cup d, C)$ one can adopt the $mRMR$ algorithm (Algorithm 1) to calculate a short constraint-preserving approximated reduct, as shown in Algorithm 2. The most essential adjustment to the classical $mRMR$ (Algorithm 1) is visible in the way the candidate attributes are considered. In the constrained version the entire constraint C_i is considered for addition to the selection, regardless of the fact that some of the attributes in the constraint would not pass the test in the "standard" version of $mRMR$ algorithm.

Note, that by providing constraints for the feature selection algorithms we are trying to kill two birds with one stone. Not only are we increasing the interpretability of prediction model, but also make the job of the attribute reduction algorithm easier since the imposition of constraints helps in reducing the size of their search space.

5   Motivational Example

To better explain how the need for defining the constrained decision systems emerged and how it helps in a real-life data analysis scenario we provide an exam-

---

Algorithm 2. mRMR feature selection algorithm for a *constrained decision system*

Input: set of attributes A and decision attribute d;
set of constraints $C = \{C_1, C_2, ..., C_k\}$, where $C_i \subseteq A$;
$\phi : 2^{A \cup \{d\}} \times 2^{A \cup \{d\}} \to R^+$ function for measuring dependency;
$N \in \mathbb{N}$; $\varepsilon \in [0, 1)$;
Output: subset of attributes $A' \subseteq A$
begin

$\quad A' \leftarrow \emptyset$;
$\quad stopFlag \leftarrow FALSE$;
$\quad A' \leftarrow A' \cup \arg\max_{C_i \in C} \phi(C_i, d)$;
$\quad A \leftarrow A \setminus A'$;
$\quad$ **while** $stopFlag == FALSE$ **do**
$\quad\quad \bar{C} \leftarrow \arg\max_{C_i \subset A} \left(\phi(C_i, d) - \max_{C_i' \subseteq A'} \phi(C_i, C_i') \right)$;
$\quad\quad$ **foreach** $i \in 1, \ldots, N$ **do**
$\quad\quad\quad | \quad \bar{P}_i \leftarrow$ random permutation of C;
$\quad\quad$ **end**
$\quad\quad$ **if** $\frac{|\{i : |\phi(\bar{P}_i, d)| > |\phi(\bar{C}, d)|\}| + 1}{N + 2} > \varepsilon$ **then**
$\quad\quad\quad | \quad stopFlag \leftarrow TRUE$;
$\quad\quad$ **else**
$\quad\quad\quad | \quad A' \leftarrow A' \cup \bar{C}$;
$\quad\quad\quad | \quad C \leftarrow C \setminus \{\bar{C}\}$;
$\quad\quad$ **end**
$\quad$ **end**
end

---

ple of application. This particular example has motivated us to consider feature selection process in the situation when attributes are bound by constraints. The data set we are dealing with came from sensor measurements collected at an active Polish coal mine provided by Research and Development Centre EMAG[1]. It is a subject of study in the R&D project aimed at identification of risks in mining [8] and has been used before in various experiments with rough set based methods as well as a basis for the IJCRS'15 Data Challenge [4].

The main data consists of multivariate time series corresponding to readings of sensors used for monitoring the conditions at the longwall. It was provided in a tabular format. In total, in the training data set there were series from 51,700 time periods, each 10 min long, with measurements taken every second (600 values in a single series for every sensor). The values for each time period were stored in a different row of the data file. Each of the rows contained readings from 28 different sensors thus, in total, the data consisted of 16,800 numerical attributes. The time periods in the training data were overlapping and given in a chronological order. Data labels indicate whether a warning threshold had been reached in a period between three and six minutes after the end of the training

---

[1] http://www.ibemag.pl/index.php?l=ang.

period, for three methane meters. If the methane concentration measured by any of the sensors reaches the alarm level, the cutter loader is switched off automatically. However, if we were able to predict ahead the warning methane concentrations, we could reduce the speed of the cutter loader and give the methane more time to spread out – before the necessity of switching off the whole production line. An important part of the background information in this problem is the position of the sensors.

The analysis of the most successful solutions from the IJCRS'15 competition [4] provided a conclusion that prediction of methane concentration levels can be achieved even when a small subset of attributes is used for constructing the model. Although the solution that obtained the highest evaluation score used nearly 5,000 features in the learning process, a few of the other top-ranked teams achieved similar results with models considering far fewer features. For instance, the model used by the second team used a total of 24 features. This observation convinced us to verify the effectiveness of our own feature selection methods in terms of the compactness and informativeness of their results.

The application of mRMR (from RmRMR [7]) provided us with very reasonable decision model that used fewer features than the best solutions in IJCRS'15 competition, sixteen in total, while retaining very reasonable level of prediction quality. However, the subset of attributes chosen by this method disregarded completely of the relationships (constraints) between attributes. From the closer examination we concluded that there is no natural, intuitive explanation for the particular choice of attributes. Our conclusion was that the methods that we apply must be adjusted to better suit the experts' demands. The application of modified mRMR algorithm (Algorithm 2) usually significantly increased the number of features being used, but provided more substantial and robust model for the decision making task.

6 Conclusions and Discussion

The proposed methodology for dealing with the task of attribute reduction in case when attributes (features) are bound by constraints is currently at the early stage. Initial implementation and experiments are currently under way. The approach promises to address some of the issues that occur in real-life scenario and can be expressed as constraints on attribute co-appearance. In particular, the preservation of constraints during reduction is intended to create sets of features that are more comprehensible and acceptable by domain experts. In decision support and/or knowledge discovery applications the intuitive interpretability of results is hard to overemphasize.

The heuristic presented in the paper is a very simple and straightforward one. It is based on the observation, that most of techniques form "classical" rough sets can be adopted with ease, provided the constraints we are dealing with are in conjunctive form. The other possibilities (disjunctive, exclusive) still require more investigation and modification of (approximate) algorithms for finding reducts. This is our plan for the future.

The particular application that spurred the development of the concept of constrained decision system is just one example. One can easily see that similar kind of challenges are posed, for example, by genetic (microarray) data [1]. In microarray data we usually have thousands if not tens of thousands attributes corresponding to gene expression levels. We are also acutely aware, that there are various constraints that bind features (genes). The proposed approach can help, at least in some situations, to simplify the task while preserving some important relationships between gene expression levels that are further used to, e.g., predict the probability of genetic disorder.

References

1. Abeel, T., Helleputte, T., Van de Peer, Y., Dupont, P., Saeys, Y.: Robust biomarker identification for cancer diagnosis with ensemble feature selection methods. Bioinformatics 26(3), 392–398 (2010)
2. Bazan, J., Skowron, A., Ślęzak, D., Wróblewski, J.: Searching for the complex decision reducts: the case study of the survival analysis. In: Zhong, N., Raś, Z.W., Tsumoto, S., Suzuki, E. (eds.) ISMIS 2003. LNCS (LNAI), vol. 2871, pp. 160–168. Springer, Heidelberg (2003). doi:10.1007/978-3-540-39592-8_22
3. Garey, M.R., Johnson, D.S.: Computers and Intractability: A Guide to the Theory of NP-Completeness. W. H. Freeman & Co., New York (1979)
4. Janusz, A., Sikora, M., Wróbel, Ł., Stawicki, S., Grzegorowski, M., Wojtas, P., Ślęzak, D.: Mining data from coal mines: IJCRS'15 data challenge. In: Yao, Y., Hu, Q., Yu, H., Grzymala-Busse, J.W. (eds.) RSFDGrC 2015. LNCS (LNAI), vol. 9437, pp. 429–438. Springer, Heidelberg (2015). doi:10.1007/978-3-319-25783-9_38
5. Janusz, A., Ślęzak, D.: Rough set methods for attribute clustering and selection. Appl. Artif. Intell. 28(3), 220–242 (2014)
6. Janusz, A., Ślęzak, D.: Computation of approximate reducts with dynamically adjusted approximation threshold. In: Esposito, F., Pivert, O., Hacid, M.-S., Raś, Z.W., Ferilli, S. (eds.) ISMIS 2015. LNCS (LNAI), vol. 9384, pp. 19–28. Springer, Heidelberg (2015). doi:10.1007/978-3-319-25252-0_3
7. Janusz, A., Stawicki, S., Ślęzak, D.: RmRMR package for R system. https://github.com/janusza/RmRMR
8. Kozielski, M., Sikora, M., Wróbel, Ł.: DISESOR - decision support system for mining industry. In: Proceedings of FedCSIS 2015, pp. 67–74. IEEE (2015)
9. Kruczyk, M., Baltzer, N., Mieczkowski, J., Dramiński, M., Koronacki, J., Komorowski, J.: Random reducts: a Monte Carlo rough set-based method for feature selection in large datasets. Fundamenta Informaticae 127(1–4), 273–288 (2013)
10. Nguyen, H.S.: Approximate Boolean reasoning: foundations and applications in data mining. In: Peters, J.F., Skowron, A. (eds.) Transactions on Rough Sets V. LNCS, vol. 4100, pp. 334–506. Springer, Heidelberg (2006). doi:10.1007/11847465_16
11. Pawlak, Z., Skowron, A.: Rudiments of rough sets. Inf. Sci. 177(1), 3–27 (2007)
12. Peng, H., Long, F., Ding, C.: Feature selection based on mutual information: criteria of max-dependency, max-relevance, and min-redundancy. IEEE Trans. Pattern Anal. Mach. Intell. 27(8), 1226–1238 (2005)

13. Skowron, A., Rauszer, C.: The discernibility matrices and functions in information systems. In: Słowiński, R. (ed.) Intelligent Decision Support: Handbook of Applications and Advances of the Rough Sets Theory, pp. 331–362. Springer, Dordrecht (1992)
14. Ślęzak, D.: Rough sets and functional dependencies in data: foundations of association reducts. In: Gavrilova, M.L., Tan, C.J.K., Wang, Y., Chan, K.C.C. (eds.) Transactions on Computational Science V: Special Issue on Cognitive Knowledge Representation. LNCS, vol. 5540, pp. 182–205. Springer, Heidelberg (2009). doi:10.1007/978-3-642-02097-1_10

Governance of the Redundancy in the Feature Selection Based on Rough Sets' Reducts

Marek Grzegorowski[✉]

Faculty of Mathematics, Informatics and Mechanics,
University of Warsaw, Banacha 2, 02-097 Warsaw, Poland
M.Grzegorowski@mimuw.edu.pl

Abstract. In this paper we introduced a novel approach to feature selection based on the theory of rough sets. We defined the concept of redundant reducts, whereby data analysts can limit the size of data and control the level of redundancy in generated subsets of attributes while maintaining the discernibility of all objects even in the case of partial data loss. What more, in the article we provide the analysis of the computational complexity and the proof of NP-hardness of the n-redundant super-reduct problem.

1   Introduction

Data exploration techniques allow analysts to discover interesting dependencies in data due to a fact that it gives the ability to efficiently verify current hypotheses about investigated phenomena and formulate new ones. In practice, this is usually done by conducting simple tests on available data and using results of those test in consecutive stages of the data exploration process. Very often, the main objective of an analyst is to define such a representation of objects described in the data, that in future will be the most useful for, e.g. constructing prediction models. Unfortunately, even though there are plenty methods for automatic feature selection that are well-described in literature, it is hard to find methods and algorithms which would take into account the risk of loss or lack of data during the long term operation of the prediction model.

One of the significant business cases where it is important that the set of extracted features have a certain level of redundancy is e.g. threat monitoring to prevent methane outbreaks in coal mines [1]. Because of harsh and extreme conditions prevailing in mines, either sensors as well as cables that transmit data are relatively often damaged. This causes gaps in the collected data and, hence, in the extracted attributes. Let us imagine the situation that the risks monitoring system is based on a very small predictive model, which is based on

This research was partially supported by Polish National Centre for Research and Development (NCBiR) grant PBS2/B9/20/2013 in frame of Applied Research Programme.

© Springer International Publishing AG 2016
V. Flores et al. (Eds.): IJCRS 2016, LNAI 9920, pp. 548–557, 2016.
DOI: 10.1007/978-3-319-47160-0_50

literally three attributes from a single sensor [2]. Indicated model is very effective, it is noticeable that achieved the second highest score in terms of AUC in the open data analysis competition [3]. However, if in the collected data the utilized attributes were missing, e.g. due to sensor failure, the entire threats monitoring system would not work properly and, hence, would become useless.

In this paper we propose a novel approach to the feature selection [4] that allows, in a controlled way, to obtain a certain level of redundancy in the generated subsets of attributes. The proposed approach is based on the concept of reducts derived from the theory of rough sets. In the article we present redundant reducts - an extended definition of reducts which allows to obtain small subsets of features, while maintaining the discernibility of every object in the data set. Moreover, redundant reducts retain the same information and knowledge [5] (in a sense of discernibility) as the entire data set even in case of the absence of part of its elements (usually called attributes or features).

The proposed approach fits in an extensive research on the use of reducts in the KDD process [6]. However, in this area, researchers mainly refer to the problem of improving the quality of classification by building ensembles of classifiers [7] or multiple neural networks [8] trained on several various reducts [9]. Ensembling based on reducts was also considered by researchers in relation to approximate reducts [10] or bireducts [11]. In real-life cases, using a variety of reducts to build ensemble of regression models turned out very successful, e.g., in solving the problem of prediction seismic bumps in coal mines [12]. In comparison to previous works, in this article we not only showed how to use many reducts to improve the reliability of the solution but also we presented explicitly the theoretical construction and definition of n-reducts.

This paper is organised as follows. In Sect. 2, we recall the basics of rough sets that are the foundation of the newly proposed concepts. In Sect. 3, we define rigid n-redundant reducts (for short: rigid n-reducts) and we discuss the pros and cons of this construction. In Sect. 4, we introduce n-redundant reducts (for short: n-reducts) and we provide the proof of NP-hardness of n-redundant super-reduct problem. Finally, in Sect. 5 we summarize the contribution of this work and describe the direction of further research.

2   Rough Sets Basics

The theory of rough sets, proposed by Zdzisław Pawlak in 1981 [13], provides a mathematical formalism for reasoning about imperfect data and knowledge [14,15]. In the rough set theory, all available information about objects $u \in U$ are represented in a structure called an *information system*. Formally, an information system $\mathbb{S}$ can be defined as a tuple: $S = (U, A)$, where U is a finite, non-empty set of objects and A is a finite, non-empty set of attributes. The most common representation of the information system is a table: rows correspond to objects from U and columns are associated with attributes from A.

However, it is possible to define an attribute, called a *decision* or *class attribute*, that can be used to define a partitioning of U into disjoint sets, e.g.

belongingness of the objects to some concept. An information system with a decision attribute is called a *decision system* and is denoted by: $\mathbb{S}_d = (U, A \cup \{d\})$, where $A \cap \{d\} = \emptyset$. A tabular representation of a decision system is sometimes called a *decision table* and the disjoint sets of objects with different values of the decision attribute are called *categories* or *decision classes*.

In many applications information about objects from a considered universe has to be reduced [16]. In the rough set theory selecting informative sets of attributes is conducted using the notion of indiscernibility, by computing reducts.

Definition 1 (Information super-reduct). *Let* $\mathbb{S} = (U, A)$ *be an information system. A subset of attributes* $ISR \subseteq A$ *will be called an information super reduct iff for any* $u \in U$ *the indiscernibility classes of* u *with regard to* ISR *and* A *are equal, i.e.* $[u]_A = [u]_{ISR}$.

Definition 2 (Information reduct). *Let* $\mathbb{S} = (U, A)$ *be an information system. A subset of attributes* $IR \subseteq A$ *will be called an information reduct iff the following two conditions are met:*

1. *IR is a super-reduct*
2. *There is no proper subset* $IR' \subsetneq IR$ *for which the first condition holds.*

Analogically, it is possible to define a decision super-reduct DSR and a decision reduct DR for a decision system $\mathbb{S}_d$:

Definition 3 (Decision super-reduct). *Let* $\mathbb{S}_d = (U, A \cup \{d\})$ *be a decision system with a decision attribute* d. *A subset of attributes* $DSR \subseteq A$ *is called a decision super reduct iff for any* $u \in U$ *if the indiscernibility class of* u *relative to* A *is a subset of some decision class, its indiscernibility class relative to* DSR *should also be a subset of that decision class, i.e.* $[u]_A \subseteq [u]_d \Rightarrow [u]_{DR} \subseteq [u]_d$.

Definition 4 (Decision reduct). *Let* $\mathbb{S}_d = (U, A \cup \{d\})$ *be a decision system with a decision attribute* d. *A subset of attributes* $DR \subseteq A$ *is called a decision reduct iff the following two conditions are met:*

1. *DR is a decision super-reduct*
2. *There is no proper subset* $DR' \subsetneq DR$ *for which the first condition holds.*

Let B be a subset of all attributes A. The core of B is the set of all indispensable attributes of B. We define the core of attributes as: $Core(B) = \bigcap Red(B)$, where Red(B) is the set off all reducts of B.

Theorem 1 *(Super-reduct problem is NP-complete). The decision problem, super-reduct, is to determine whether for a given information/decision system exists a super-reduct R containing k attributes. We want to show that vertex-cover can be reduced to super-reduct. Since we already know that vertex-cover is NP-complete, if it can be reduced to super-reduct, then super-reduct is also NP-complete.*

Proof (Super-reduct is NP). The polynomial super-reduct verification algorithm is based on the function isReduct [17] which (in polynomial time) verifies if a given set of attributes R is a super-reduct.

Proof (Super-reduct is NP-hard). Recall that a vertex-cover is a subset of vertices that covers all the edges in a graph. Formally, given an undirected graph G = (V,E), a vertex cover is a subset $V' \subset V$ such that if edge(i, j) is an edge of G, then either $i \in V'$ or $j \in V'$ (or both). The decision problem, vertex-cover, is to determine whether a graph has a vertex cover of a given size k.

Given an undirected graph G = (V,E) we can define the information system $\mathbb{S} = (U, A)$ as a table $\mathcal{T}$, where objects U are in rows, attributes A are in columns.

- Every edge $e \in E$ represents object from U and all edges are placed in rows.
- Every vertice $v \in V$ from graph G represents attribute from A and all vertices are placed in columns.
- We define the function $f(e) : E-> \mathbb{N}$ to assign unique natural number to each egde $e \in E$
- For each cell $c(i, j) \in \mathcal{T}$ we assign value $f(e)$ iff there is edge(i, j) in graph G, zero otherwise
- We create additional row in $\mathcal{T}$ which we call zero-edge and fill all of its cells with zero values (since, we do not want zeros to discern edges)

The reduction comes to create a table representation of the graph and can be performed in polynomial time. The algorithm for $VertexCover(G, k)$ by reduction to $SuperReduct(\mathbb{S}, k)$ goes as follows:

1. Given a graph G = (V,E) and an integer k
2. Create the information system $\mathbb{S} = (U, A)$ as a table $\mathcal{T}$
3. Solve the problem $SuperReduct(\mathbb{S}, k)$
4. If there is a solution, return Yes, else return No

To prove this reduction is correct, we need to show two more things: First, if there is a solution to $VertexCover(G, k)$, then there must be a solution to $SuperReduct(\mathbb{S}, k)$. Second, if there is a solution to $SuperReduct(\mathbb{S}, k)$, then there must be a solution to $VertexCover(G, k)$

First, supposing there is a solution, VC of size k, to the $VertexCover(G, k)$, we can construct the information system $\mathbb{S} = (U, A)$ as a table $\mathcal{T}$ as shown above. In graph G every edge is discernible by its unique number that is further stored in cells of $\mathcal{T}$. Because VC is a subset of vertices that covers all the edges then all unique edge numbers are in reduced set. Hence the reduced set is a super-reduct since every object (formerly edge) is discernible.

Second, the solution R to $SuperReduct(\mathbb{S}, k)$ is a subset of attributes that are sufficient do discern edges. Because each edge(i, j) has unique number stored in a cell of $\mathcal{T}$ in column labeled with vertice number, hence, to discern it at least one of columns i or j must be in reduct. Otherwise all values in columns for that edge would be zero and it would not be discernible from zero-edge.

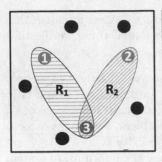 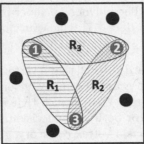 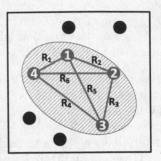

Fig. 1. In the pictures above, there are three significant examples of what rigid redundant reduct is and what is not. The convention on the left and the center figure is as follows: points represent features, ovals R_1, R_2, R_3 are grouping features in reducts. The leftmost is not a rigid redundant reduct, since after removal of the feature number 3, the remaining features: 1 and 2 are not a reduct. The middle is an example of rigid 1-redundant reduct (for short rigid 1-reduct) since after removal of any of features 1, 2 or 3 the remaining features form a reduct. The rightmost figure represents two-elemental reducts $(R_1, .., R_6)$ by the edges of the clique. In this case, the selected attributes (1, 2, 3, 4) form a rigid 2-reduct, since after removal of any two attributes the remaining features form a reduct.

3   Rigid Redundant Reducts

The comparison of reduct and super-reduct definitions, which were recalled in Sect. 2, leads to the following conclusions: The super-reduct is a structure that allows to maintain the knowledge contained in the original data set what from the point of view of accuracy and quality, is desired and expected. Unfortunately, it does not provide any estimation and limitations of data volume. On the other hand, the concept of reduct complements the definition of the super-reduct in the way allowing to minimize the amount of necessary attributes.

The concept of reduct provides tool for the selection of an irreducible set of attributes that allows to discern all of the objects in the original data set. However, it does not provide any protection against missing data. In this section, we introduce the definition of rigid n-redundant reducts (for short rigid n-reducts) that addresses both of those requirements. For the better understanding, the graphical interpretations of the introduced concept are depicted in the Fig. 1.

Definition 5 (Rigid (decision) n-reduct). *Let $\mathbb{S} = (U, A)$ be an information system. Let $\mathbb{S}_d = (U, A \cup \{d\})$ be a decision system with a decision attribute d. A subset of attributes $R \subseteq A$ is called a rigid (decision) n-reduct iff after removal of any n attributes $r_1, .., r_n$ from R the remaining set $R \setminus \{r_1, .., r_n\}$ is a (decision) reduct.*

Remark 1. *Any rigid (decision) n-reduct R is a (decision) super-reduct.*

Remark 2. *If in a given information system $\mathbb{S}$ or decision sytsem $\mathbb{S}_d$ exists any rigid (decision) n-reduct R then there is no core of attributes.*

Remark 3. *Any rigid (decision) n-reduct R of size $|R|$ is a union of $l = \binom{|R|}{|R|-n}$ overlapping reducts of size $|R| - n$.*

Definition 6 (Minimal rigid (decision) n-reduct). *We say that rigid (decision) n-reduct R is minimal in given information (decision) system iff for every rigid (decision) n-reduct R', $|R| \leq |R'|$.*

Rigid redundant reducts allow for the selection of attribute subsets which are insensitive (in the sense of preserving the discernibility of objects) to the loss of part of its features. Let s_R be the size of the subset of attributes (reduct) and s_A be the size of the set A of all attributes. Let us assume, for simplicity, that for each feature in reduct, the probability that it is missing in data is symmetric, independent and equal to $p \in (0, \frac{1}{a*s_A})$ where $a > 1$. Then, for the classical reduct the risk of objects indiscernibility is equal to $p_{cr} = p$, while for the rigid n-reduct the risk of objects indiscernibility is equal to $p_{rr} = p^{n+1}$, thus $p_{rr} < p_{cr}$.

However, the definition of rigid n-reducts has a few very clear limitations. One of them is the assumption that all features are symmetric. That is, regardless which of them is removed or missing the affect on the (in)discernibility is the same. Since a rigid n-reduct is a union of $\binom{|R|}{|R|-n}$ overlapping reducts, each of size equal to $|R| - n$, what is a very specific construction, it could be rarely available in the data set. For this reasons, in the next section we present a less restrictive definition of a redundant reduct.

4   Redundant Reducts

In real life applications as threat monitoring or recommendation systems machine learning models often have to work on incomplete data. In this section, we introduce the definition of n-redundant reducts (for short n-reducts) which extends the previously used concepts to enable the governance of the redundancy level and, hence, improving the robustness of analysis. For better intuition of readers, in Fig. 2, a graphical interpretation of n-reducts is shown. Moreover, in this section we provide the proof that the problem of finding n-redundant (decision) super-reduct of size k in given information/decision system is NP-hard.

Definition 7 (N-redundant (decision) super-reduct). *Let $\mathbb{S} = (U, A)$ be a information system. Let $\mathbb{S}_d = (U, A \cup \{d\})$ be a decision system with a decision attribute d. A subset of attributes $R \subseteq A$ is called a n-redundant (decision) super-reduct iff after the removal of any n attributes $r_1, .., r_n$ from R the remaining set $R \setminus \{r_1, .., r_n\}$ is a (decision) super-reduct.*

Definition 8 ((Decision) n-reduct). *Let $\mathbb{S} = (U, A)$ be a information system. Let $\mathbb{S}_d = (U, A \cup \{d\})$ be a decision system with a decision attribute d. A subset of attributes $R \subseteq A$ is called a (decision) n-reduct iff the following two conditions are met:*

1. *After the removal of any n attributes $r_1, .., r_n$ from R the remaining set $R \setminus \{r_1, .., r_n\}$ is a (decision) super-reduct.*

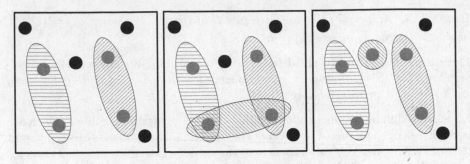

Fig. 2. In the pictures above, there are three significant examples of what redundant reduct is. The convention is as follows: points represent features, ovals are grouping features in reducts, the union of all features included in reducts on each figure forms a redundant reduct. The leftmost and the middle present an examples of 1-redundant reduct (for short 1-reduct), since after removal of any feature, the remaining features still form a super-reduct. The rightmost figure presents 2-reduct, since after removal of any two features the remaining features still form a super-reduct.

2. There is no proper subset $R' \subsetneq R$ for which the first condition holds.

Remark 4. *A 0-reduct is a classical rough sets' reduct.*

Remark 5. *If in a given information system $\mathbb{S}$ or decision system $\mathbb{S}_d$ exists any (decision) n-reduct R then there is no core of attributes.*

Remark 6. *If in a given information system $\mathbb{S}$ or decision system $\mathbb{S}_d$ exists any (decision) n-reduct R, for $n ¿ 0$. Then, $\forall$ attribute $r \in R$, $\exists$ reduct R' such as: $r \in R'$ and $R' \subsetneq R$*

Remark 7. *Any (decision) n-reduct R of size $|R|$ is a union of at least n reducts, each of size lower than or equal $|R| - n$.*

Remark 8. *After the removal of any attribute r from a n-redundant (decision) super-reduct R, the remaining set $R' = R \setminus \{r\}$ is a $(n-1)$-redundant (decision) super-reduct.*

Remark 9. *After the removal of any attribute r from a (decision) n-reduct R, the remaining set $R' = R \setminus \{r\}$ meets the following:*

1. R' is a $(n-1)$-redundant (decision) super-reduct
2. $\exists R'' \subseteq R'$, where R'' is a (decision) $(n-1)$-reduct

Definition 9 (Minimal n-redundant (decision) super-reduct). *We call a n-redundant super-reduct R minimal in information/decision system iff there is no other n-redundant super-reduct R' that $|R'| < |R|$*

Remark 10. *A minimal n-redundant (decision) super-reduct is a minimal (decision) n-reduct.*

Theorem 2 *(N-redundant super-reduct is NP-hard). The n-redundant (decision) super-reduct problem is NP-hard. The decision problem, n-redundant super-reduct, is to determine whether in given information/decision system exists n-redundant (decision) super-reduct R containing k attributes.*

Proof (N-redundant super-reduct is NP-hard). We want to show that (decision) super-reduct can be reduced to n-redundant (decision) super-reduct. Since we already know that super-reduct is NP-hard (See Theorem 1), if it can be reduced to n-redundant (decision) super-reduct, then n-redundant (decision) super-reduct is also NP-hard.

The reduction is straightforward. Given an information system $\mathbb{S} = (U, A)$ or decision system $\mathbb{S}_d = (U, A \cup \{d\})$. The reduction of data representation comes to creation of additional n unique columns-attributes (ids), where n is the number of allowed redundancy. This step can be performed in polynomial time. Hence, the whole reduction is polynomial. The algorithm for $SuperReduct(\mathbb{S}, k)$ by reduction to $RedundantSuperReduct(\mathbb{S}', k + n)$ goes as follows:

1. Given an information system $\mathbb{S} = (U, A)$ and an integer value k
2. Add n attributes with unique values (ids) to the information/decision system $\mathbb{S}' = (U, A \bigcup \{id_1, .., id_n\})$ - each id_i attribute is a reduct itself
3. Solve the problem $RedundantSuperReduct(\mathbb{S}', k + n)$
4. If there is a solution, return Yes, else return No

To prove that the above reduction is correct, we need to show two more things: First, if there is a solution to $SuperReduct(\mathbb{S}, k)$, then there must be a solution to $RedundantSuperReduct(\mathbb{S}', k + n)$. Second, if there is a solution to $RedundantSuperReduct(\mathbb{S}', k+n)$ there must be a solution to $SuperReduct(\mathbb{S}, k)$

First, if there is a solution to super-reduct R of size k, $R = SuperReduct(\mathbb{S}, k)$. Then, when we add additional n attributes $id_1, .., id_n$, each one of them being a reduct. We can easily construct a n-redundant super-reduct RR of size k+n: $RR = \bigcup_{1 \leq i \leq n} \{r_i\} \cup R$.

Second, the solution $RR = RedundantSuperReduct(\mathbb{S}', k + n)$ may contain, at most, n artificially generated unique attributes (ids). Suppose RR contain $0 \leq l \leq n$ ids attributes. When we remove all added ids attributes both from $\mathbb{S}'$ and RR, then the (k+n−l)-redundant super-reduct remains (See Remark 8). Now, directly from the definition of n-redundant super-reduct, we can remove any n-l attributes $\{r_1, .., r_{(n-l)}\}$ from RR and the remaining set $R = RR \setminus \{r_1, .., r_{(n-l)}\}$ is a super-reduct of size k in $\mathbb{S}$.

5 Summary and Future Research

In this paper we proposed a novel approach to feature selection derived from the theory of rough sets. The concept of (decision) n-reducts, allows to limit the size of the data, maintaining the information contained in the original data set in the sense of objects discernibility. Moreover, n-reducts allow to govern the level of redundancy to ensure continuity of information in case of partial data loss.

Another important contribution of this work is the proof of NP-hardness of the n-redundant super-reduct problem what opens broad possibilities for research on approximation algorithms, for which an interesting and promising approach may be pre-clustering of features [18]. Besides the fundamental properties of n-reducts presented in this work, there are still a lot of possible research fields with regards n-reducts. The proposed approach may be, in the future, adapted to other concepts of feature selection based on rough set theory [19], e.g., approximate reducts [20–22], dynamic reducts [23] or bireducts [24].

References

1. Kozielski, M., Sikora, M., Wróbel, L.: DISESOR - decision support system for mining industry. In: Ganzha, M., Maciaszek, L.A., Paprzycki, M. (eds.) 2015 Federated Conference on Computer Science and Information Systems, FedCSIS 2015, Lódz, Poland, 13–16 September 2015, pp. 67–74. IEEE (2015)

2. Grzegorowski, M., Stawicki, S.: Window-based feature engineering for prediction of methane threats in coal mines. [25], pp. 452–463

3. Janusz, A., Sikora, M., Wróbel, Ł., Stawicki, S., Grzegorowski, M., Wojtas, P., Ślęzak, D.: Mining data from coal mines: IJCRS '15 data challenge. [25], pp. 429–438

4. Janusz, A., Ślęzak, D.: Rough set methods for attribute clustering and selection. Appl. Artif. Intell. **28**(3), 220–242 (2014)

5. Wasilewski, P., Ślęzak, D.: Foundations of rough sets from vagueness perspective. In: Rough Computing: Theories, Technologies and Applications. IGI Global (2008)

6. Zhong, N., Skowron, A.: A rough set-based knowledge discovery process. Appl. Math. Comput. Sci. **11**(3), 603–620 (2001)

7. Hu, X.: Ensembles of classifiers based on rough sets theory and set-oriented database operations. In: 2006 IEEE International Conference on Granular Computing, GrC 2006, Atlanta, Georgia, USA, 10–12 May 2006, pp. 67–73. IEEE (2006)

8. Yu, D., Hu, Q., Bao, W.: Combining multiple neural networks for classification based on rough set reduction. In: 2003 Proceedings of the 2003 International Conference on Neural Networks and Signal Processing, vol. 1, pp. 543–548, December 2003

9. Pawlak, Z., Skowron, A.: Rough sets and boolean reasoning. Inf. Sci. **177**(1), 41–73 (2007)

10. Wroblewski, J.: Ensembles of classifiers based on approximate reducts. Fundam. Inf. **47**(3–4), 351–360 (2001)

11. Ślęzak, D., Janusz, A.: Ensembles of bireducts: towards robust classification and simple representation. In: Kim, T., Adeli, H., Ślęzak, D., Sandnes, F.E., Song, X., Chung, K., Arnett, K.P. (eds.) FGIT 2011. LNCS, vol. 7105, pp. 64–77. Springer, Heidelberg (2011). doi:10.1007/978-3-642-27142-7_9

12. Grzegorowski, M.: Massively parallel feature extraction framework application in predicting dangerous seismic events. In: Ganzha, M., Maciaszek, L.A., Paprzycki, M. (eds.) Proceedings of FedCSIS 2016. IEEE, September 2016 (In print)

13. Pawlak, Z.: Information systems, theoretical foundations. Inf. Syst. **3**(6), 205–218 (1981)

14. Pawlak, Z., Skowron, A.: Rough sets: some extensions. Inf. Sci. **177**(1), 28–40 (2007)

15. Pawlak, Z., Skowron, A.: Rudiments of rough sets. Inf. Sci. **177**(1), 3–27 (2007)
16. Guyon, I., Elisseeff, A.: An introduction to variable and feature selection. J. Mach. Learn. Res. **3**, 1157–1182 (2003)
17. Bazan, J.G., Nguyen, H.S., Nguyen, S.H., Synak, P., Wróblewski, J.: Rough set algorithms in classification problems. In: Polkowski, L., Lin, T.Y., Tsumoto, S. (eds.) Rough Set Methods and Applications: New Developments in Knowledge Discovery in Information Systems. Studies in Fuzziness and Soft Computing, vol. 56, pp. 49–88. Physica-Verlag GmbH, Heidelberg (2000)
18. Janusz, A., Ślęzak, D.: Utilization of attribute clustering methods for scalable computation of reducts from high-dimensional data. In: Ganzha, M., Maciaszek, L.A., Paprzycki, M. (eds.) Proceedings of Federated Conference on Computer Science and Information Systems - FedCSIS 2012, Wroclaw, Poland, 9–12 September 2012, pp. 295–302 (2012)
19. Ślęzak, D., Widz, S.: Rough-set-inspired feature subset selection, classifier construction, and rule aggregation. [26], pp. 81–88
20. Janusz, A., Ślęzak, D.: Computation of approximate reducts with dynamically adjusted approximation threshold. In: Esposito, F., Pivert, O., Hacid, M.-S., Raś, Z.W., Ferilli, S. (eds.) ISMIS 2015. LNCS (LNAI), vol. 9384, pp. 19–28. Springer, Heidelberg (2015). doi:10.1007/978-3-319-25252-0_3
21. Janusz, A., Stawicki, S.: Applications of approximate reducts to the feature selection problem. [26], pp. 45–50
22. Nguyen, H.S., Ślęzak, D.: Approximate reducts and association rules. In: Zhong, N., Skowron, A., Ohsuga, S. (eds.) RSFDGrC 1999. LNCS (LNAI), vol. 1711, pp. 137–145. Springer, Heidelberg (1999). doi:10.1007/978-3-540-48061-7_18
23. Bazan, J.G., Skowron, A., Synak, P.: Dynamic reducts as a tool for extracting laws from decisions tables. In: Raś, Z.W., Zemankova, M. (eds.) ISMIS 1994. LNCS, vol. 869, pp. 346–355. Springer, Heidelberg (1994). doi:10.1007/3-540-58495-1_35
24. Stawicki, S., Ślęzak, D.: Recent advances in decision bireducts: complexity, heuristics and streams. In: Lingras, P., Wolski, M., Cornelis, C., Mitra, S., Wasilewski, P. (eds.) RSKT 2013. LNCS (LNAI), vol. 8171, pp. 200–212. Springer, Heidelberg (2013). doi:10.1007/978-3-642-41299-8_19
25. Yao, Y., Hu, Q., Yu, H., Grzymala-Busse, J.W. (eds.): RSFDGrC 2015. LNCS (LNAI), vol. 9437. Springer, Heidelberg (2015)
26. Yao, J.T., Ramanna, S., Wang, G., Suraj, Z. (eds.): RSKT 2011. LNCS (LNAI), vol. 6954. Springer, Heidelberg (2011)

Attribute Reduction in Multi-source Decision Systems

Yanting Guo and Weihua Xu[✉]

School of Mathematics and Statistics, Chongqing University of Technology,
Chongqing 400054, People's Republic of China
1290237984@qq.com, chxuwh@gmail.com

Abstract. Data processing for information from different sources is a hot research topic in the contemporary data. Attribute reduction methods of multi-source decision systems (MSDS) are proposed in this paper. Firstly, based on the integrity of original effective information preservation, a consistent attribute reduction of the multi-source decision system is proposed. Secondly, in the case of a certain loss of original effective information, data is compressed by the fusion of conditional entropy. Then attribute reduction preserving knowledge unchanged are studied in the decision system obtained by fusion. Accordingly, examples are introduced to further elaborate the theory proposed in this paper.

Keywords: Attribute reduction · Conditional entropy fusion · Multi-source decision system

1 Introduction

Rough set theory proposed by Pawlak [6] is an important mathematical tool to deal with imprecise, inconsistent and incomplete information. In order to meet people's various requirements, many extended rough set models have been proposed, such as the fuzzy rough set and the rough fuzzy rough set, the variable precision rough set model, and other models [8,10].

Rough set theory has been widely applied in many fields, such as machine learning, knowledge discovery, data mining, decision support and analysis, information security, networking, cloud computing and biological information processing [2].

Attribute reduction is one of the core content in rough set, which has been made great development. Based on different criteria, various reduction methods are proposed in classical and generalized rough set models. According to the quantitative criteria, attribute reduction is mainly divided into two categories: qualitative reduction and quantitative reduction. From the perspective of qualitative criteria, Pawlak proposed an attribute reduction which keeps the positive region unchanged [7]. Slezak [9] provided a generalized reduction which keeps the generalized decision under the generalized decision function. Mi et al. [5] investigated the β lower distribution reduction and β upper distribution reduction in

© Springer International Publishing AG 2016
V. Flores et al. (Eds.): IJCRS 2016, LNAI 9920, pp. 558–568, 2016.
DOI: 10.1007/978-3-319-47160-0_51

the variable precision rough model. Yao and Zhao [12]investigated several different quantitative reductions in the decision-theoretic rough model. Attribute reduction is mainly to solve the problem of high-dimensional data computation complexity and accuracy.

With the development of information technology, massive data is released every day, and the volume of data is fairly large. It is important for us to efficiently acquire knowledge from information derived from different sources (namely information boxes). There is no doubt that attribute reduction can eliminate the influence of the redundant and irrelevant attributes on the calculation process and the final results. Therefore, the research of attribute reduction based on multi-source decision systems (MSDS) is of great significance.

In order to make an accurate decision, without losing any effective information is the highest requirement of data processing. Based on the consideration of integrity of information preservation, for all the source, we hope to find a common attribute reduction (namely consistent attribute reductions of MSDS) to eliminate redundant attributes. If the amount of information can not be compressed in the case of keeping the integrity of information violence, we need to appropriately reduce the standards for the preservation of the original information. Therefore, in the case of a certain loss of information, multi-source information fusion has important significance.

By integrating different sources of data, the deficiency of the single data can be made up, so as to realize the mutual complement and mutual confirmation of various data sources. In this way, the application scope of data is expanded and the accuracy of analysis can be improved. In many circumstances, integrating all information from different sources is necessary. There are some related studies on multi-source information fusion. In particular, Khan [3] based on views of membership of objects studied rough set theory and notions of approximates in Multiple-Source Approximation Systems (MSAS). Besides, Md and Khan [4] proposed a modal logic for Multiple-source Tolerance Approximation Spaces (MTAS) based on the principle of only considering the information of sources have about objects.

This paper mainly study attribute reduction of multi-source information systems (MSDS) which have the same universe and attributes and different information functions (namely Isomorphic multi-source information systems). It should be pointed out that isomorphism multi-source information systems refers to the same cardinality of the partition generated by attribute set on the universe in each information system. For heterogeneous information systems, we just need to find the ultimate goal as a middle bridge to establish the relationship between different sources. Because the more information you have on the same thing, learned knowledge should be more accurate. A higher goal may be required for the isomorphic information system in addition to finding the middle bridge. Therefore, attribute reduction of multi-source information systems (MSDS) which have the same universe and attributes and different information functions is important. Next, the most important issue is how to make full use of the information provided by each source in Multi-Source Decision Systems

(MSDS). Then in the various original information preservation requirements, attribute reduction methods of multi-source information systems (MSDS) are proposed.

This paper provides the following innovations: (1) based on the integrity of information preservation, a consistent attribute reduction of MSDS is proposed (2) in the case of a certain loss of information, data is compressed by the fusion of conditional entropy (3) research on attribute reduction keeping the knowledge unchanged in the MSDS.

The study of this paper is organized as follows. Some basic concepts in Pawlak rough set theory are briefly reviewed in Sect. 2. In Sect. 3, definitions of Multi-Source Decision Systems and consistent attribute reduction are proposed. Under the consideration of various original effective information preservation requirements, attribute reduction methods of MSDS are proposed. Section 4 concludes this paper by bringing some remarks and discussions.

2 Preliminaries

In this section, some basic concepts about rough set theory, decision systems and uncertainty measures are reviewed.

Rough set theory proposed by Pawlak is an important tool for knowledge learning. Suppose U be a nonempty and finite set of objects, which is called the universe of discourse, and R be an equivalence relation of $U \times U$. The equivalence relation R induces a partition of U, denoted by $U/R = \{[x]_R | x \in U\}$, where $[x]_R$ represents the equivalence class of x with regard to R. Then (U, R) is the Pawlak approximation space. For an arbitrary subset X of U can be characterized by a pair of upper and lower approximations which are [6]:

$$\overline{R}(X) = \{x \in U | [x]_R \cap X \neq \emptyset\},$$
$$\underline{R}(X) = \{x \in U | [x]_R \subseteq X\}.$$

And $pos(X) = \underline{R}(X), neg(X) = \sim \overline{R}(X), bnd(X) = \overline{R}(X) - \underline{R}(X)$ are called the positive region, negative region, and boundary region of X, respectively. Objects belong to positive region $pos(X)$, whose equivalence class is definitely contained in the set X. Objects belong to negative region $neg(X)$, whose equivalence class is definitely not contained in the set X. And boundary region $bnd(X)$ is composed of objects whose equivalence class may be contained in the set X.

Let $K = (U, \{R_i\}_{i \in \tau})$ be a knowledge base, where $\{R_i\}_{i \in \tau}$ is a family of equivalence relations of $U \times U$ and τ is an index set. In the knowledge base K, when some knowledge is deleted, the classification ability of knowledge base K is not weakened. In the process of knowledge processing, deleting redundant knowledge can reduce the amount of computation. When a patient visits a doctor, the doctor does not require the patient to do the whole body examination first, then gives the diagnostic conclusion. Otherwise, it will delay the patient lots of time and will greatly increase the patient's medical expenses. Therefore, the knowledge reduction is an important aspect of rough set theory. Attribute reduction is helpful to eliminate the influence of the redundant and irrelevant attributes on the calculation process and the final results.

Let $K = (U, \{R_i\}_{i \in \tau})$ be a knowledge base, $P \subseteq \widetilde{R} = \{R_i\}_{i \in \tau}$ and $r \in P$.

- If $IND(P/r) = IND(P)$, then r is not necessary or redundant in P; otherwise, r is necessary in P. It should be noted that the indiscernibility relation $IND(P)$ generated by P is the intersection of all the equivalence relations in P;
- If for arbitrary $r \in P$, r is necessary in P, then P is called independent; otherwise, P is not independent;
- If P is independent and $IND(P) = IND(\widetilde{R})$, then P is called a reduction of $\widetilde{R}$.

A decision system $I = (U, A, V, f)$ is a quadruple [1], where U is a nonempty and finite universe; $A = C \cup D$ is the set composed of condition attribute set C and decision attribute set D, and $C \cap D = \emptyset$; V is the union of attribute value domain, i.e., $V = \cup_{a \in A} V_a$; $f : U \times A \to V$ is an information function, i.e., $\forall x \in U$, $a \in A$, that $f(x, a) \in V_a$, where $f(x, a)$ is the value of the object x under attribute a. Generally, let $D = \{d\}$. Unless otherwise specified, all decision systems in this paper are defined as the shown.

Uncertainty measures can help us to analyze the essential characteristics of data. Therefore, the uncertainty measure issue is an important research direction in rough set theory. The approximation accuracy proposed by Pawlak provides the percentage of possible correct decisions when classifying objects by employing the attribute set R. Let $I = (U, A, V, f)$ be a decision system, and $U/D = \{D_1, D_2, \cdots, D_m\}$ be a classification of universe U, and R be an attribute set. Then the approximation accuracy of U/D by R is defined as

$$\alpha_R(U/D) = \frac{\sum_{D_i \in U/D} |\underline{R}(D_i)|}{\sum_{D_i \in U/D} |\overline{R}(D_i)|}.$$

Dai et al. [1] proposed a reasonable uncertainty measure for incomplete decision systems. The uncertainty measure has the property of monotonicity, namely the finer the partition of universe U generated by indiscernibility relation is, the smaller the value of uncertainty measure is. That is to say, this measure can well reflect the uncertainty of incomplete information system. And what's more, when the incomplete decision system degenerates to a complete decision system, the property of monotonicity is still true. In a complete decision system $S = (U, A, V, f)$, conditional entropy of D with respect to $B(B \subseteq C)$ is defined to be

$$H(D|B) = -\sum_{i=1}^{|U|} p([x_i]_B) \sum_{j=1}^{m} p(D_j|[x_i]_B) logp(D_j|[x_i]_B)$$

where B is the conditional attribute subset of C, $p([x_i]_B) = |[x_i]_B|/|U|$, and $p(D_j|[x_i]_B) = |[x_i]_B \cap D_j|/|[x_i]_B|$.

3   Attribute Reduction of Multi-source Decision System

With the development of information technology, there are a large amount of information is collected every day. In particular, data about the same information can be obtained from different information sources. How to make full use of the information from different sources to efficiently acquire knowledge is very important. Multi-source information fusion will become a hot spot in the field of information research. The integration of information from different sources can get more comprehensive information to help make the right decision. In this paper, we study attribute reduction under multiple information sources which have the same universe and attributes and different information functions. That is to say, the research background is the multi-source decision system. It should be pointed out that this paper studies numerical decision systems.

First of all, decision systems from different sources can form a new information system which is called the Multi-Source Decision System. Detail descriptions are as follows:

Definition 3.1. A Multi-Source Decision System (MSDS) is defined to be the structure $MI = (U, \{I_i\}_{i \in N})$, where $\forall i \in N$, each $I_i = (U, C \cup D, V_i, f_i)$ be a decision system which represents the ith source of the Multi-Source Decision System and $N = \{1, 2, 3, \cdots\cdots\}$ denotes the number of sources. And $U/D = \{D_1, D_2, \cdots, D_m\}$ for each source S_i are identical.

A Multi-Source Decision System which includes s single information sources. Let the s pieces of single-source information system overlapping together can form a information box have s levels and it comes from our previous study [11].

3.1   The First Method of Attribute Reduction in the MSDS

Next, we study attribute reduction of the Multi-Source Decision System based on requirements of original information preservation. First and foremost, based on the consideration of the integrity of original effective information preservation, we proposed a consistent attribute reduction of MSDS.

Definition 3.2. Let $MI = (U, \{I_i\}_{i \in N})$ be a Multi-Source Decision System. And for $\forall i \in N$, $I_i = (U, C \cup D, V_i, f_i)$ be a decision system. For each S_i, if $\exists A \subseteq C$ such that $IND(A) = IND(C)$ and $B \subset A$ such that $IND(B) \neq IND(C)$, then A is called a consistent attribute reduction of the Multi-Source Decision System.

It is well known that all reductions of each decision system can be obtained by discernibility matrix. If a consistent attribute reduction of the MI can be obtained, then the information box can be compressed. And the amount of calculation can be reduced.

3.2   The Second Method of Attribute Reduction in the MSDS

In the case of a certain loss of original effective information, the fusion of information from different sources is key to data compression. There is no uniform standard about information fusion from multiple sources. Mean value fusion is the most common method of information fusion. In a Multi-Source Decision System $MI = (U, \{I_i\}_{i \in N})$, for $\forall x \in U$, $a \in A$, the value of x under attribute a is equal to $\sum_{i \in N} f_i(x, a)/|N|$.

It is well known that the more accurate data is, the more precise knowledge is. In order to obtain more accurate knowledge, we evaluate the accuracy of data collected under each attribute in the multi-source decision system. Therefore, our approach is to take every condition attribute as a basic point. For each condition attribute, the reliable source is selected by conditional entropy. So the importance of arbitrary condition attribute is characterized by conditional entropy in a decision system. The conditional entropy proposed by Dai can evaluate the importance of attributes [1]. The lower conditional entropy is, the more important the condition attribute will be. According to actual requirements, other uncertainty measure can be used to select the reliable source for each condition attribute.

Definition 3.3. Let $MI = (U, \{I_i\}_{i \in N})$ be a Multi-Source Decision System. The importance of any attribute a $(\forall a \in C)$ in the Multi-Source Decision System is defined to be

$$d(a) = min_{i \in N}\{H(a|I_i)\}$$

where $H(a|I_i)$ denotes the conditional entropy of D with respect to a in the decision system I_i, which can be calculated by

$H(a|I_i) = -\sum_{k=1}^{|U|} p([x_k]_a) \sum_{h=1}^{m} p(D_h|[x_k]_a) log p(D_h|[x_k]_a).$
and

$p([x_k]_a) = |[x_k]_a|/|U|, \ p(D_h|[x_k]_a) = |[x_k]_a \cap D_h|/|[x_k]_a|.$

In a Multi-Source Decision System $MI = (U, \{I_i\}_{i \in \{1,2,\cdots,s\}})$, for $\forall a \in C$, there are r values can be calculated, namely $d(a|I_1), d(a|I_2), \cdots, d(a|I_s)$. By

Table 1. A Multi-source decision system

	1st source				2nd source				3rd source				4th source				d
	a_1	a_2	a_3	a_4	a_1	a_2	a_3	a_4	a_1	a_2	a_3	a_4	a_1	a_2	a_3	a_4	
x_1	1	2	2	1	1	2	2	1	1	2	1	1	1	2	2	1	1
x_2	1	2	1	1	1	2	2	1	1	2	1	1	1	2	1	1	1
x_3	1	1	2	1	1	1	2	1	1	1	1	1	1	1	2	1	0
x_4	0	1	1	1	1	1	1	1	0	1	2	1	0	1	2	0	1
x_5	2	1	1	2	0	1	1	1	1	1	1	1	2	2	1	1	0
x_6	0	1	1	0	0	1	2	0	0	1	1	0	1	1	2	0	1
x_7	1	1	2	1	2	2	2	1	1	2	1	1	1	2	1	1	0
x_8	1	1	1	0	2	1	1	0	1	1	1	0	1	1	1	0	1
x_9	2	1	1	0	2	1	1	1	2	1	2	1	2	1	2	1	0
x_{10}	1	1	1	0	1	1	1	1	0	1	2	1	0	1	2	0	0

comparing the size of these values, selecting the decision system with the minimum value as a reliable source of attribute a. Then a new restructuring decision system can be obtained. In the following, an example is introduced to illustrate the fusion process of multi-source information.

Example 3.2. There are four information sources about medical diagnosis, which can construct a Multi-Source Decision System with the same universe and attributes and different information functions., denoted by $MI = (U, \{I_1, I_2, I_3, I_4\})$. And $\forall i \in \{1, 2, 3, 4\}$, $I_i = (U, C \cup D, V_i, f_i)$ be a decision system, where $U = \{x_1, x_2, \cdots, x_{10}\}$ is composed of ten patients, $C = \{a_1, a_2, a_3, a_4\}$ is the conditional attribute set, and $D = \{d\}$ is the decision attribute set. Specific data information is shown in Table 1.

Through discernibility matrix method, we can easily know that there are no redundant attributes in the information sources I_1, I_2, I_4, and the information source I_3 have a unique reduction namely $\{a_1, a_2, a_4\}$. According to Definition 3.2, therefore, there is no a consistent attribute reduction of MI can be obtained. So the fusion of four information sources need to be carried out. According to Definition 3.3, conditional entropy of each attribute in each source can be obtained, and detailed results are presented in the Table 2.

Table 2. Conditional entropy

source	a_1	a_2	a_3	a_4
S_1	1.0837	1.8388	1.7020	1.2124
S_2	1.0999	1.7020	1.4614	1.8388
S_3	1.3325	1.7020	1.7020	1.8388
S_4	1.1156	1.5654	1.5654	1.3859

The smaller conditional entropy is, the more important the condition attribute will be. By comparing these values, selecting the decision system with the minimum value as a reliable source of each attribute. The reliable source of attribute a_1 is the first information source I_1, the reliable source of attribute a_2 is the fourth information source I_4, the reliable source of attribute a_3 is the second information source I_2 and the reliable source of attribute a_4 is the first information source I_1. Based on conditional entropy fusion, a new restructuring decision system can be obtained, namely Table 3.

Then attribute reduction of the new system is carried out. According the definition of knowledge reduction, we can get

$$U/C = \{\{x_1, x_2, x_7\}, \{x_3\}, \{x_4\}, \{x_5\}, \{x_6\}, \{x_8, x_{10}\}, \{x_9\}\};$$
$$U/\{C - \{a_1\}\} = \{\{x_1, x_2, x_7\}, \{x_3\}, \{x_4\}, \{x_5\}, \{x_6\}, \{x_8, x_9, x_{10}\};$$
$$U/\{C - \{a_2\}\} = \{\{x_1, x_2, x_3, x_7\}, \{x_4\}, \{x_5\}, \{x_6\}, \{x_8, x_{10}\}, \{x_9\}\};$$
$$U/\{C - \{a_3\}\} = \{\{x_1, x_2, x_7\}, \{x_3\}, \{x_4\}, \{x_5\}, \{x_6\}, \{x_8, x_{10}\}, \{x_9\}\};$$
$$U/\{C - \{a_4\}\} = \{\{x_1, x_2, x_7\}, \{x_3\}, \{x_4\}, \{x_5\}, \{x_6\}, \{x_8, x_{10}\}, \{x_9\}\}.$$

Table 3. The new system after entropy fusion

U	a_1	a_2	a_3	a_4	d
x_1	1	2	2	1	1
x_2	1	2	2	1	1
x_3	1	1	2	1	0
x_4	0	1	1	1	1
x_5	2	2	1	2	0
x_6	0	1	2	0	1
x_7	1	2	2	1	0
x_8	1	1	1	0	1
x_9	2	1	1	0	0
x_{10}	1	1	1	0	0

Therefore, a_1 and a_2 are necessary attributes of the new decision system. Further the following conclusions can be obtained, namely $IND(U/C) = IND(U/\{C - \{a_3\}\})$ and $IND(U/C) = IND(U/\{C - \{a_4\}\})$. That is to say, $\{a_1, a_2, a_3\}$ and $\{a_1, a_2, a_4\}$ are attribute reductions of the new decisiom system.

In a Multi-Source Decision System, in the case of a certain loss of original effective information, attribute reduction of the MSDS can be obtained after conditional entropy fusion.

3.3   The Effectiveness of Conditional Entropy Fusion

In order to evaluate the effectiveness of conditional entropy fusion, the approximation accuracy is used as a quantitative index to reflect the superiority of conditional entropy fusion method. The reason why we chose this uncertainty measure is that the conditional entropy has monotonicity [1]. Theoretical derivation guarantee that the proposed conditional entropy can be used as a reasonable uncertainty measure for decision system. The validity of the proposed measure is verified by experiments on some real-life data sets. The finer the partition of universe U generated by indiscernibility relation is, the smaller value of the uncertainty measure is. In the process of conditional entropy fusion, we select the information source with the minimum conditional entropy as the reliable source of each condition attribute. Therefore, the uncertainty of decision system obtained by fusion is relatively small. Next, followed by Example 3.2, classification ability of each information source, conditional entropy fusion and mean fusion is compared by approximation accuracy.

Firstly, the mean fusion information is provided in Table 4.

Let $U = \{x_1, x_2, \cdots, x_{10}\}$, $C = \{a_1, a_2, a_3, a_4\}$ and $U/d = \{D_1, D_2\}$, where $D_1 = \{x_1, x_2, x_4, x_6, x_8\}$ and $D_1 = \{x_3, x_5, x_7, x_9, x_{10}\}$. According to the information of Tables 2, 4 and 5, the lower and upper approximations of decision classes can be obtained.

Table 4. The new system after mean fusion

U	a_1	a_2	a_3	a_4	d
x_1	1	2	1.75	1	1
x_2	1	2	1.25	1	1
x_3	1	1	1.75	1	0
x_4	0.25	1	1.5	0.75	1
x_5	1.25	1.25	1	1.25	0
x_6	0.25	1	1.5	0	1
x_7	1.25	1.75	1.5	1	0
x_8	1.25	1	1	0	1
x_9	2	1	1.5	0.75	0
x_{10}	0.5	1	1.5	0.5	0

According to Table 1, the lower and upper approximation sets of decision classes and the approximation accuracy of each source can be obtained. Detailed information is shown in Table 5.

Table 5. The approximation accuracy of each source in the MSDS

$Information$	1st source	2nd source
$\overline{C}(D_1)$	$\{x_1, x_2, x_4, x_6, x_8, x_{10}\}$	$\{x_1, x_2, x_4, x_6, x_8, x_{10}\}$
$\underline{C}(D_1)$	$\{x_1, x_2, x_4, x_6\}$	$\{x_1, x_2, x_6, x_8\}$
$\overline{C}(D_2)$	$\{x_3, x_5, x_7, x_8, x_9, x_{10}\}$	$\{x_3, x_4, x_5, x_7, x_9, x_{10}\}$
$\underline{C}(D_2)$	$\{x_3, x_5, x_7, x_9\}$	$\{x_3, x_5, x_7, x_9\}$
$\alpha_C(U/d)$	$\frac{2}{3}$	$\frac{2}{3}$
$Information$	3rd source	4th source
$\overline{C}(D_1)$	$\{x_1, x_2, x_4, x_6, x_7, x_8, x_{10}\}$	$\{x_1, x_2, x_4, x_6, x_7, x_8, x_{10}\}$
$\underline{C}(D_1)$	$\{x_6, x_8\}$	$\{x_1, x_6, x_8\}$
$\overline{C}(D_2)$	$\{x_1, x_2, x_3, x_4, x_5, x_7, x_9, x_{10}\}$	$\{x_2, x_3, x_4, x_5, x_7, x_9, x_{10}\}$
$\underline{C}(D_2)$	$\{x_3, x_5, x_9\}$	$\{x_3, x_5, x_9\}$
$\alpha_C(U/d)$	$\frac{1}{3}$	$\frac{3}{7}$

According to Tables 3 and 4, the lower and upper approximation sets of decision classes and the approximation accuracy of conditional entropy fusion and mean fusion can be obtained. Detailed information is shown in Table 6.

From the perspective of approximation accuracy, compared with the mean fusion, condition entropy fusion is more objective and more close to the essential characteristics of the MSDS, such as classification ability.

Table 6. The approximation accuracy of entropy fusion and mean fusion

Information	Conditional entropy fusion	Mean fusion
$\overline{C}(D_1)$	$\{x_1, x_2, x_4, x_6, x_8, x_{10}\}$	$\{x_1, x_2, x_4, x_6, x_8\}$
$\underline{C}(D_1)$	$\{x_1, x_2, x_4, x_6\}$	$\{x_1, x_2, x_4, x_6, x_8\}$
$\overline{C}(D_2)$	$\{x_3, x_5, x_7, x_8, x_9, x_{10}\}$	$\{x_3, x_5, x_7, x_9, x_{10}\}$
$\underline{C}(D_2)$	$\{x_3, x_5, x_7, x_9\}$	$\{x_3, x_5, x_7, x_9, x_{10}\}$
$\alpha_C(U/d)$	$\frac{2}{3}$	$\frac{1}{1}$

4 Conclusions

Attribute reduction of the Multi-Source Decision System is a hot topic in data processing. Based on the consideration of original effective information preservation, two methods of attribute reduction of the MSDS are proposed. In the case of no loss of original effective information, a consistent attribute reduction of the MSDS is proposed. In the case of a certain loss of original effective information, attribute reduction of the MSDS can be obtained after conditional entropy fusion. By attribute reduction, computation complexity of high-dimensional data can be simplified and the amount of computation can be reduced effectively. Therefore, the research on attribute reduction of the Multi-Source Decision System has great significance. This paper only proposes a framework for attribute reduction of the MSDS. In the future work, there are a lot of in-depth research needs to continue, such as selection of uncertainty measurement in fusion, quantitative reduction to meet user's requirements.

Acknowledgments. This work is supported by Natural Science Foundation of China (No. 61105041, No. 61472463, No. 61402064), National Natural Science Foundation of CQ CSTC (No. cstc2015jcyjA40053), Graduate Innovation Foundation of Chongqing University of Technology (No. YCX2015227), and the Graduate Innovation Foundation of CQ (No. CYS16217).

References

1. Dai, J.H., Wang, W.T., Xu, Q.: An uncertainty measure for incomplete decision tables and its applications. IEEE Trans. Cybern. **43**, 1277–1289 (2013)
2. Peters, G., Lingras, P., Slezak, D., Yao, Y.Y. (eds.): Rough Sets: Selected Methods and Applications in Management and Engineering. Advanced Information and Knowledge Processing, pp. 95–112. Springer, London (2012)
3. Aquil, K.M.: Formal reasoning in preference-based multiple-source rough set model. Inf. Sci. **334**, 122–143 (2016)
4. Aquil, K.M., Banerjee, M.: A logic for multiple-source approximation systems with distributed knowledge base. J. Philos. Logic **40**, 663–692 (2011)
5. Mi, J.S., Wu, W.Z., Zhang, W.X.: Approaches to knowledge reduction based on variable precision rough set model. Inf. Sci. **159**, 255–272 (2004)
6. Pawlak, Z.: Rough sets. Int. J. Comput. Inform. Sci. **11**, 341–356 (1982)

7. Pawlak, Z.: Rough Sets: Theoretical Aspects of Reasoning About Data. Kluwer Academic Publishers, Dordrecht (1991)
8. Qian, Y.H., Zhang, H., Sang, Y.L., Liang, J.Y.: Multigranulation decision-theoretic rough sets. Int. J. Approximation Reasoning **55**, 225–237 (2014)
9. Slezak, D.: On generalized decision functions, reducts, networks and ensembles. In: RSFDGrC, pp. 13–23 (2015)
10. Xu, W.H., Sun, W.X., Zhang, X.Y., Zhang, W.X.: Multiple granulation rough set approach to ordered information systems. Int. J. Gen. Syst. **41**, 475–501 (2012)
11. Yu, J.H., Xu, W.H.: Information fusion in multi-source fuzzy information system with same structure. In: Proceedings of the International Conference on Machine Learning and Cybernetics, pp. 170–175 (2015)
12. Yao, Y.Y., Zhao, Y.: Attribute reductions in decision-theoretic rough set models. Inf. Sci. **178**, 3356–3373 (2008)

Matrix Algorithm for Distribution Reduction in Inconsistent Ordered Information Systems

Xiaoyan Zhang[1,2(✉)] and Ling Wei[1]

[1] School of Mathematics, Northwest University,
Xi'an 710127, People's Republic of China
zhangxyms@gmail.com, wl@nwu.edu.cn
[2] School of Mathematics and Statistics, Chongqing University of Technology,
Chongqing 400054, People's Republic of China

Abstract. As one part of some work in ordered information systems, distribution reduction is studied in inconsistent ordered information systems. The dominance matrix is restated for reduction acquisition in dominance relations based on information systems. Matrix algorithm is stepped for distribution reduction acquisition. And program is implemented by the algorithm. The approach provides an effective tool to the theoretical research and applications for ordered information systems in practices. Cases about detailed and valid illustrations are employed to explain and verify the algorithm and the program which shows the effectiveness of the algorithm in complicated information systems.

Keywords: Dominance matrix · Distribution reduction · Matrix algorithm · Ordered information systems · Rough set

1 Introduction

In Pawlak's original rough set theory [4], partition or equivalence relation (indiscernibility) is an important and primitive concept. However, partition or equivalence relation, as the indiscernibility relation in Pawlak's original rough set theory, is still restrictive for many applications. To address this issue, several interesting and meaningful extensions to equivalence relation had been proposed in the past, such as tolerance relations [5,10], neighborhood operators [16] and so on [3,6,9–11,17]. Moreover, the original rough set theory did not consider attributes with preference ordered domain, that was criteria. In many real situations, we are often faced with the problems in which the properties of ordering of considered attributes that plays a crucial role. One such type of problem is the objects of ordering. For this reason, Greco, Matarazzo, and Slowinski proposed an extension rough set theory called the dominance-based rough set approach (DRSA) to take into account the ordering properties of criteria [2]. This innovation is mainly based on substitution of the indiscernibility relation by dominance relation. Moreover, Greco, Matarazzo and Slowinski characterized the DRSA as well as decision rules induced from rough approximations, while the usefulness

© Springer International Publishing AG 2016
V. Flores et al. (Eds.): IJCRS 2016, LNAI 9920, pp. 569–579, 2016.
DOI: 10.1007/978-3-319-47160-0_52

of the DRSA and its advantages over the CRSA (classical rough set approach) are presented [3,6,10]. Several studies have been made about properties and algorithmic implementations of DRSA [1,7,10,12,14,15].

Nevertheless, only a limited number of methods using DRSA to acquire knowledge in inconsistent ordered information systems have been proposed and studied. Pioneering work on inconsistent ordered information systems with the DRSA had been proposed by Greco, Matarazzo and Słowinski [2], but they did not clearly point out the semantic explanation of unknown values. Shao and Zhang [8] further proposed an extension of the dominance relation in incomplete ordered information systems. Therefore, the purpose of this paper is to develop approaches to attribute reductions in Inconsistent Ordered Information Systems (IOIS). In this paper, theories and approaches of distribution reduction are investigated in inconsistent ordered information systems. Furthermore, algorithm of matrix computation of distribution reduction is introduced, from which we provide a new approach to attribute reductions in inconsistent ordered information systems.

As parts of these work, some other reductions have been studied and papers have been published as references [6,7,14,15]. Reductions in these literatures are different from what we study in this paper. In order to present this, we fetch the same cases as reference [14] in this paper to acquire the reductions and compare the results with those in literature [14].

The rest of this paper is organized as follows. Some preliminary concepts are briefly recalled in Sect. 2. In Sect. 3, we define the matrix algorithm for distribution reduction acquisition. The algorithm and the corresponding program we design can provide a tool to theoretical research and applications of criterion based on information systems. Cases are employed to illustrate the algorithm and the program in Sect. 4. Cases used in literature [14] shown that the algorithm and program is effective in complicated information systems. Furthermore, the results are compared with those obtained in reference [14] to show the difference of the reductions. Finally, conclusions on what we study in this paper are drawn to understand this paper briefly.

2   Distribution Reduction in Inconsistent Ordered Information Systems

The following recalls necessary concepts and preliminaries are required in the sequel of our work. Detailed description of the theory can be found in [13,17].

An information system with decisions is an ordered quadruple $\mathcal{I} = (U, A \cup D, F, G)$, where $U = \{x_1, x_2, \cdots, x_n\}$ is a non-empty finite set of objects, $A \cup D$ is a non-empty finite attributes set, $A = \{a_1, a_2, \cdots, a_p\}$ denotes the set of condition attribute $D = \{d_1, d_2, \cdots, d_q\}$ denotes the set of decision attributes, $A \cap D = \phi$; $F = \{f_k | U \to V_k, k \leq p\}$, $f_k(x)$ is the value of a_k on $x \in U$, V_k is the domain of $a_k, a_k \in A$, and $G = \{g_{k'} | U \to V_{k'}, k' \leq q\}$, $g_{k'}(x)$ is the value of $d_{k'}$ on $x \in U$, $V_{k'}$ is the domain of $d_{k'}, d_{k'} \in D$.

In an information system, if the domain of an attribute was ordered according to a decreasing or increasing preference, then the attribute is a criterion.

An information system is called an ordered information system (OIS) if all condition attributes are criterions.

Assumed that the domain of a criterion $a \in A$ is complete pre-ordered by an outranking relation $\succeq_a$, then $x \succeq_a y$ means that x is at least as good as y with respect to criterion a. And we can say that x dominates y. In the following, without any loss of generality, we consider condition and decision criterions having a numerical domain, that is, $V_a \subseteq \mathcal{R}$ ($\mathcal{R}$ denotes the set of real numbers).

We define $x \succeq y$ by $f(x,a) \geq f(y,a)$ according to increasing preference, where $a \in A$ and $x, y \in U$. For a subset of attributes $B \subseteq A$, $x \succeq_B y$ means that $x \succeq_a y$ for any $a \in B$. That is to say x dominates y with respect to all attributes in B. Furthermore, we denote $x \succeq_B y$ by $xR_B^{\succeq}y$. In general, we indicate an ordered information system with decision by $\mathcal{I}^{\succeq} = (U, A \cup D, F, G)$. Thus the following definition can be obtained.

Let $\mathcal{I}^{\succeq} = (U, A \cup D, F, G)$ be an ordered information system with decisions, for $B \subseteq A$, denote

$$R_B^{\succeq} = \{(x_i, x_j) \in U \times U | f_l(x_i) \geq f_l(x_j), \forall a_l \in B\};$$
$$R_D^{\succeq} = \{(x_i, x_j) \in U \times U | g_m(x_i) \geq g_m(x_j), \forall d_m \in D\}.$$

$R_B^{\succeq}$ and $R_D^{\succeq}$ are called dominance relations of information system $\mathcal{I}^{\succeq}$.

If we denote

$$[x_i]_B^{\succeq} = \{x_j \in U | (x_j, x_i) \in R_B^{\succeq}\}$$
$$= \{x_j \in U | f_l(x_j) \geq f_l(x_i), \forall a_l \in B\};$$
$$[x_i]_D^{\succeq} = \{x_j \in U | (x_j, x_i) \in R_D^{\succeq}\}$$
$$= \{x_j \in U | g_m(x_j) \geq g_m(x_i), \forall d_m \in D\},$$

then the following properties of a dominance relation are trivial.

Let $R_A^{\succeq}$ be a dominance relation. The following properties hold.

(1) $R_A^{\succeq}$ is reflexive, transitive, but not symmetric, so it is not an equivalent relation.
(2) If $B \subseteq A$, then $R_A^{\succeq} \subseteq R_B^{\succeq}$.
(3) If $B \subseteq A$, then $[x_i]_A^{\succeq} \subseteq [x_i]_B^{\succeq}$.
(4) If $x_j \in [x_i]_A^{\succeq}$, then $[x_j]_A^{\succeq} \subseteq [x_i]_A^{\succeq}$ and $[x_i]_A^{\succeq} = \cup\{[x_j]_A^{\succeq} | x_j \in [x_i]_A^{\succeq}\}$.
(5) $[x_j]_A^{\succeq} = [x_i]_A^{\succeq}$ iff $f(x_i, a) = f(x_j, a)$ ($\forall a \in A$).
(6) $\mathcal{J} = \cup\{[x]_A^{\succeq} | x \in U\}$ constitute a covering of U.

For any subset X of U, and A of $\mathcal{I}^{\succeq}$ define

$$\underline{R_A^{\succeq}}(X) = \{x \in U | [x]_A^{\succeq} \subseteq X\}; \qquad \overline{R_A^{\succeq}}(X) = \{x \in U | [x]_A^{\succeq} \cap X \neq \phi\}.$$

$\underline{R_A^{\succeq}}(X)$ and $\overline{R_A^{\succeq}}(x)$ are said to be the lower and upper approximation of X with respect to a dominance relation $R_A^{\succeq}$. And the approximations have also some properties which are similar to those of Pawlak approximation spaces.

For an ordered information system with decisions $\mathcal{I}^{\succeq} = (U, A \cup D, F, G)$, if $R_A^{\succeq} \subseteq R_D^{\succeq}$, then this information system is consistent, otherwise, this information system is inconsistent (IOIS).

For simple description, the following information system with decisions are based on dominance relations, i.e., ordered information systems.

Let $\mathcal{I}^{\succeq} = (U, A \cup D, F, G)$ be an information system with decisions, and $R_B^{\succeq}, R_D^{\succeq}$ be dominance relations derived from condition attributes set A and decision attributes set D respectively. For $B \subseteq A$, denote

$$U/R_B^{\succeq} = \{[x_i]_B^{\succeq} \mid x_i \in U\};$$

$$U/R_d^{\succeq} = \{D_1, D_2, \cdots, D_r\};$$

$$\mu_B^{\succeq}(x) = (\frac{|D_1 \cap [x]_B^{\succeq}|}{|U|}, \frac{|D_2 \cap [x]_B^{\succeq}|}{|U|}, \cdots, \frac{|D_r \cap [x]_B^{\succeq}|}{|U|});$$

$$\gamma_B^{\succeq}(x) = \max\{\frac{|D_1 \cap [x]_B^{\succeq}|}{|U|}, \frac{|D_2 \cap [x]_B^{\succeq}|}{|U|}, \cdots, \frac{|D_r \cap [x]_B^{\succeq}|}{|U|}\},$$

where $[x]_B^{\succeq} = \{y \in U | (x, y) \in R_B^{\succeq}\}$. Furthermore, we said $\mu_B^{\succeq}(x)$ be distribution function about attributions set B, and $\gamma_B^{\succeq}(x)$ be maximum distribution function about attributions set B.

Let $\mathcal{I}^{\succeq} = (U, A \cup D, F, G)$ be an inconsistent information system. If $\mu_B^{\succeq}(x) = \mu_A^{\succeq}(x)$, for all $x \in U$, we say that B is a distribution consistent set of $\mathcal{I}^{\succeq}$. If B is a distribution consistent set, and no proper subset of B is distribution consistent set, then B is called a distribution consistent reduction of $\mathcal{I}^{\succeq}$.

Let $\mathcal{I}^{\succeq} = (U, A \cup D, F, G)$ be an inconsistent information system. If $\gamma_B^{\succeq}(x) = \gamma_A^{\succeq}(x)$, for all $x \in U$, we say that B is a maximum distribution consistent set of $\mathcal{I}^{\succeq}$. If B is a maximum distribution set, and no proper subset of B is maximum distribution consistent set, then B is called a maximum distribution consistent reduction of $\mathcal{I}^{\succeq}$.

Theorem 2.1 (See [13]). Let $\mathcal{I}^{\succeq} = (U, A \cup D, F, G)$ be an ordered information system, and $B \subseteq A$ is a distribution consistent set of $\mathcal{I}^{\succeq}$ if and only if B is a maximum distribution consistent set of $\mathcal{I}^{\succeq}$.

Theorem 2.2 (See [13]). Let $\mathcal{I}^{\succeq} = (U, A \cup D, F, G)$ be an ordered information system.

P: $B \subseteq A$ is a distribution consistent set of $\mathcal{I}^{\succeq}$.
Q: While $\mu_A^{\succeq}(y) \not\preceq \mu_A^{\succeq}(x)$, $[y]_B^{\succeq} \not\subseteq [x]_B^{\succeq}$ holds for any $x, y \in U$.
Then we have $P \Rightarrow Q$.

The distribution consistent set requires that the classification ability of the consistent set keeps the same with the original data table. That is, $B \subset A$ is a distribution consistent set of A must satisfy that $[x]_B^{\succeq} = [x]_A^{\succeq}$ holds for any $x \in U$. This is very strict and other reductions studied in references [7, 14, 15] may not reach this special condition.

3 Matrix Algorithm for Distribution Reduction Acquisition in Inconsistent Ordered Information Systems

In this section, the dominance matrices will be put as restatement and matrices will be employed to realize the calculation of distribution reductions.

Definition 3.1. Let $\mathcal{I}^{\succeq} = (U, A \cup D, F, G)$ be an ordered information system, and $B \subset A$. Denote

$$M_B = (m_{ij})_{n \times n}, \quad \text{where} \quad m_{ij} = \begin{cases} 1, & x_j \in [x_i]_B^{\succeq}, \\ 0, & \text{otherwise.} \end{cases}$$

The matrix M_B is called dominance matrix of attributes set $B \subseteq A$. If $|B| = l$, we say that the order of M_B is l.

Definition 3.2. Let $\mathcal{I}^{\succeq} = (U, A \cup D, F, G)$ be an ordered information system, and M_B, M_C are dominance matrices of attributes sets $B, C \subseteq A$. The intersection of M_B and M_C is defined by

$$M_B \cap M_C = (m_{ij})_{n \times n} \cap (m'_{ij})_{n \times n} = (\min\{m_{ij}, m'_{ij}\})_{n \times n}.$$

The intersection defined above can be implemented by the operator '*' in Matlab platform, $M_B \cap M_C = M_B * M_C$, that is the product of elements in corresponding positions. Then the following properties are obvious.

Proposition 3.1. Let M_B, M_C be dominance matrices of attributes sets $B, C \subseteq A$, the following results always hold.

(1) $m_{ii} = 1$.
(2) $M_{B \cup C} = M_B \cap M_C$.

From the above, we can see that a dominance relation of objects has one-one correspondence to a dominance matrix. The combination of dominance relations can be realized by the corresponding matrices and the dominance relations can be compared by the corresponding matrices from the following definitions.

Definition 3.3. Let $M_A = (\alpha_1, \alpha_2, \ldots, \alpha_n)^T$ and $M_B = (\beta_1, \beta_2, \ldots, \beta_n)^T$ be matrices with $n \times n$ dimensions, α_i and β_i be row vectors respectively. If $\alpha_i \leq \beta_i$ holds for any $i \leq n$, we say that M_A is less than M_B and it is denoted by $M_A \leq M_B$.

By the definitions, dominance matrices have the following properties straightly.

Proposition 3.2. Let $\mathcal{I}^{\succeq} = (U, A \cup D, F, G)$ be an ordered information system and $B \subseteq A$. The dominance matrices with respect to A and B are, respectively, M_A and M_B. Then $M_A \leq M_B$.

In the following, we give the preparation of matrix computation for distribution reductions in ordered information systems.

Proposition 3.3. Let $\mathcal{I}^{\succeq} = (U, A \cup D, F, G)$ be an ordered information system, $U = \{x_1, x_2, \cdots, x_n\}$ and $A = \{a_1, a_2, \cdots, a_p\}$. Then

$$M_A = \bigcap_{i=1}^{p} M_{\{a_i\}} = \begin{pmatrix} a_{11} & a_{12} & \cdots & a_{1n} \\ a_{21} & a_{22} & \cdots & a_{2n} \\ \vdots & \vdots & \ddots & \vdots \\ a_{n1} & a_{n2} & \cdots & a_{nn} \end{pmatrix}.$$

and any vector $\alpha_i = (a_{i1}, a_{i2}, \cdots, a_{in})$ represents the dominance class of object x_i by the values 0 and 1, where 0 means the object not included in the class and 1 means the object included in the class.

Theorem 3.1. Let $\mathcal{I}^{\succeq} = (U, A \cup D, F, G)$ be an ordered information system and $B \subseteq A$. B is a consistent set if and only if $M_B = M_A$.

Proof. As is known, $[x]_A^{\succeq} \subseteq [x]_B^{\succeq}$ holds since $B \subseteq A$.

($\Rightarrow$) For B is a distribution consistent set, one can have that $\mu_B = \mu_A$. Then, for any x and D_j, we have that $|D_j \cap [x]_A^{\succeq}| = |D_j \cap [x]_B^{\succeq}|$. Since $[x]_A^{\succeq} \subseteq [x]_B^{\succeq}$, it is obviously that $[x]_A^{\succeq} = [x]_B^{\succeq}$. That is, the row vectors in M_B and M_A are correspondingly the same. Then $M_B = M_A$.
($\Leftarrow$) Since $M_B = M_A$, we can easily have that $[x]_A^{\succeq} = [x]_B^{\succeq}$ holds for any x and D_j. Then $|D_j \cap [x]_A^{\succeq}| = |D_j \cap [x]_B^{\succeq}|$ holds for any x and D_j. We can have that $\mu_B^{\succeq}(x) = \mu_A^{\succeq}(x)$ holds for any x. That is, B is a distribution consistent set.

　　To acquire reductions in inconsistent ordered information system, the matrices can be the only forms of storage in computing. And we illustrate the progress to calculate the reductions in the following of this section.

Algorithm 3.1. Let $\mathcal{I}^{\succeq} = (U, A \cup D, F, G)$ be an ordered information system and $B \subseteq A$. B is a consistent set if and only if $M_B = M_A$.

　　Input:　　An inconsistent ordered information system $\mathcal{I}_d^{\succeq} = (U, A \cup \{d\}, V, f)$, where $U = \{x_1, x_2, \ldots, x_n\}$ and $A = \{a_1, a_2, \ldots, a_p\}$.

　　Output:　　All distribution reductions of $\mathcal{I}_d^{\succeq}$.

　　Step1　　Load the ordered information system and simplify the system by combining the objects with same values of every attribute.

　　Step2　　Classify every single criterion and store them in separate matrices.

$$M_{a_i} = \begin{pmatrix} a_{11}^i & a_{12}^i & \cdots & a_{1n}^i \\ a_{21}^i & a_{22}^i & \cdots & a_{2n}^i \\ \vdots & \vdots & \ddots & \vdots \\ a_{n1}^i & a_{n2}^i & \cdots & a_{nn}^i \end{pmatrix}, \quad M_d = \begin{pmatrix} d_{11} & d_{12} & \cdots & d_{1n} \\ d_{21} & d_{22} & \cdots & d_{2n} \\ \vdots & \vdots & \ddots & \vdots \\ d_{n1} & d_{n2} & \cdots & d_{nn} \end{pmatrix}$$

Step3 Check the consistence of the information system.

$$M_A = \bigcap_{i=1}^{p} M_{a_i} = M_{a_1} * M_{a_2} * \cdots * M_{a_p} = \begin{pmatrix} a_{11} & a_{12} & \cdots & a_{1n} \\ a_{21} & a_{22} & \cdots & a_{2n} \\ \vdots & \vdots & \ddots & \vdots \\ a_{n1} & a_{n2} & \cdots & a_{nn} \end{pmatrix}.$$

Where '.*' is the operator in Matlab platform. If $M_A \leq M_d$, the system is consistent, terminate the algorithm. Else the system is inconsistent, go to the next step.

Step4 Acquire the consistent set. Let $B = \{b_1, b_2, \cdots, b_m\} \subset A$.

$$M_B = \bigcap_{i=1}^{m} B_i = B_1. * B_2. * \cdots. * B_m. = \begin{pmatrix} b_{11} & b_{12} & \cdots & b_{1n} \\ b_{21} & b_{22} & \cdots & b_{2n} \\ \vdots & \vdots & \ddots & \vdots \\ b_{n1} & b_{n2} & \cdots & b_{nn} \end{pmatrix}.$$

If $M_B = M_A$, B is a consistent set, store the set into the temporary storage cell. Else fetch another subset of A and repeat this step. Calculate till all subsets of A are verified, then go to the next step.

Step5 Sort the consistent sets in the storage cell and find out the minimum consistent sets which are just the reductions. Output all reductions and terminate the algorithm. □

The algorithm and the distribution reduction allow us to calculate reductions which keep the classification ability the same with the original system in a brief way. And we don't need to acquire every approximations of the decisions. It shorts the computing time and provides an effective tool to knowledge acquisition in criterion based rough set theory.

4   Experimental Computing and Case Study

We design programs and employ one case to demonstrate the effective of the method in the last section. This experimental computing program is running on a personal computer. The configuration of the computer is a bit low but the program runs well and fast. It also shows the advantage of Algorithm 3.1 and the corresponding computing program.

Case 4.1. An inconsistent ordered information system on animals sleep is presented in Table 1.

The information system is denoted by $\mathcal{I}_d^{\succeq} = (U, C \cup \{d\}, V, f)$, where C is the condition attribute set and d is the single dominance decision. There are 42 objects which represent the species of animals and 10 attributes with numerical values in the ordered information system. The interpretations and the units of attributes are represented as follows.

a_1—Bodyweight in kg;	a_6—Maximum life span (years);
a_2—Brain weight in g;	a_7—Gestation time (days);
a_3—"Non-dreaming" sleep (hrs/day);	a_8—Predation index (1–5);
a_4—"Dreaming" sleep (hrs/day);	a_9—Sleep exposure index (1–5);
a_5—Total sleep (hrs/day);	d—Overall danger index (1–5).

Table 1. $\mathcal{I}_d^{\succeq}$: An inconsistent ordered information system on animals sleep

$(U, C \cup \{d\})$		a_1	a_2	a_3	a_4	a_5	a_6	a_7	a_8	a_9	d
x_1 :	African giant pouched rat	1	6.6	6.3	2	8.3	4.5	42	3	1	3
x_2 :	Asian elephant	2547	4603	2.1	1.8	3.9	69	624	3	5	4
x_3 :	Baboon	10.55	179.5	9.1	0.7	9.8	27	180	4	4	4
x_4 :	Big brown bat	0.023	0.3	15.8	3.9	19.7	19	35	1	1	1
x_5 :	Brazilian tapir	160	169	5.2	1	6.2	30.4	392	4	5	4
x_6 :	Cat	3.3	25.6	10.9	3.6	14.5	28	63	1	2	1
x_7 :	Chimpanzee	52.16	440	8.3	1.4	9.7	50	230	1	1	1
x_8 :	Chinchilla	0.425	6.4	11	1.5	12.5	7	112	5	4	4
x_9 :	Cow	465	423	3.2	0.7	3.9	30	281	5	5	5
x_{10} :	Eastern American mole	0.075	1.2	6.3	2.1	8.4	3.5	42	1	1	1
x_{11} :	Echidna	3	25	8.6	0	8.6	50	28	2	2	2
x_{12} :	European hedgehog	0.785	3.5	6.6	4.1	10.7	6	42	2	2	2
x_{13} :	Galago	0.2	5	9.5	1.2	10.7	10.4	120	2	2	2
x_{14} :	Goat	27.66	115	3.3	0.5	3.8	20	148	5	5	5
x_{15} :	Golden hamster	0.12	1	11	3.4	14.4	3.9	16	3	1	2
x_{16} :	Gray seal	85	325	4.7	1.5	6.2	41	310	1	3	1
x_{17} :	Ground squirrel	0.101	4	10.4	3.4	13.8	9	28	5	1	3
x_{18} :	Guinea pig	1.04	5.5	7.4	0.8	8.2	7.6	68	5	3	4
x_{19} :	Horse	521	655	2.1	0.8	2.9	46	336	5	5	5
x_{20} :	Lesser short-tailed shrew	0.005	0.14	7.7	1.4	9.1	2.6	21.5	5	2	4
x_{21} :	Little brown bat	0.01	0.25	17.9	2	19.9	24	50	1	1	1
x_{22} :	Man	62	1320	6.1	1.9	8	100	267	1	1	1
x_{23} :	Mouse	0.023	0.4	11.9	1.3	13.2	3.2	19	4	1	3
x_{24} :	Musk shrew	0.048	0.33	10.8	2	12.8	2	30	4	1	3
x_{25} :	N. American opossum	1.7	6.3	13.8	5.6	19.4	5	12	2	1	1
x_{26} :	Nine-banded armadillo	3.5	10.8	14.3	3.1	17.4	6.5	120	2	1	1
x_{27} :	Owl monkey	0.48	15.5	15.2	1.8	17	12	140	2	2	2
x_{28} :	Patas monkey	10	115	10	0.9	10.9	20.2	170	4	4	4
x_{29} :	Phanlanger	1.62	11.4	11.9	1.8	13.7	13	17	2	1	2
x_{30} :	Pig	192	180	6.5	1.9	8.4	27	115	4	4	4
x_{31} :	Rabbit	2.5	12.1	7.5	0.9	8.4	18	31	5	5	5
x_{32} :	Rat	0.28	1.9	10.6	2.6	13.2	4.7	21	3	1	3
x_{33} :	Red fox	4.235	50.4	7.4	2.4	9.8	9.8	52	1	1	1
x_{34} :	Rhesus monkey	6.8	179	8.4	1.2	9.6	29	164	2	3	2
x_{35} :	Rock hyrax (Hetero.b)	0.75	12.3	5.7	0.9	6.6	7	225	2	2	2
x_{36} :	Rock hyrax (Procavia hab)	3.6	21	4.9	0.5	5.4	6	225	3	2	3
x_{37} :	Sheep	55.5	175	3.2	0.6	3.8	20	151	5	5	5
x_{38} :	Tenrec	0.9	2.6	11	2.3	13.3	4.5	60	2	1	2
x_{39} :	Tree hyrax	2	12.3	4.9	0.5	5.4	7.5	200	3	1	3
x_{40} :	Tree shrew	0.104	2.5	13.2	2.6	15.8	2.3	46	3	2	2
x_{41} :	Vervet	4.19	58	9.7	0.6	10.3	24	210	4	3	4
x_{42} :	Water opossum	3.5	3.9	12.8	6.6	19.4	3	14	2	1	1

By the experimental computing program, the distribution reductions of the system can be calculated and they are represented in the following. The operating time to compute this case is 0.158581 s.

The distribution reductions are:

$$\{a_1, a_3, a_4, a_6, a_7, a_8, a_9\}, \quad \{a_2, a_3, a_4, a_6, a_7, a_8, a_9\},$$
$$\{a_1, a_2, a_3, a_4, a_6, a_7, a_8, a_9\}, \quad \{a_1, a_3, a_4, a_5, a_6, a_7, a_8, a_9\},$$
$$\{a_2, a_3, a_4, a_5, a_6, a_7, a_8, a_9\}.$$

And it can be verified by taking computer as assistant that the above sets are reductions of the data table. Detailed progress of the verifying are not arranged here. From the results, we can easily see that the reductions studied in this paper is different with ones approached in reference [14] since those reduction are $\{a_3, a_4, a_6, a_7\}$, $\{a_4, a_5, a_6, a_7\}$, $\{a_6, a_8, a_9\}$ and $\{a_1, a_2, a_8, a_9\}$. They are different kinds of reductions in ordered information systems and can adapt to different needs in practices. From the definition of different reduction, we can also easily obtain that possible and compatible reduction are usually subsets of distribution reduction. This is not strict and should be studied and verified separately and theatrically. And the work may be taken into account as one part of the future studies in our work.

Finally, we take other inconsistent ordered information system to acquire the distribution reduction respectively. And the descriptions on the data tables are listed in the next Table 2.

From the results in Table 2, we can obtain that the algorithm and the program we studied in this paper can be effective and useful to acquire distribution reductions in practice. The numbers of objects and attributes can increase the computing time. But the matrices storage has the ability to short the memory and computing time. And it can be helpful in research theoretically and applicable.

Table 2. Descriptions on the calculations

Data name	Values	Objects	Conditions	Decisions	Reductions	Time	Operations
Body fat	Real	252	14	1	11	36.43723 s	10
Glass	Real	213	9	1	7	2.04624 s	10
Animal sleep	Real	42	9	1	5	0.13153 s	10

5   Conclusions

As is known, many information systems are data tables considering criteria for various factors in practise. Therefore, it is meaningful to study the attribute reductions in inconsistent information system on the basis of dominance relations. In this paper, distribution reduction is restated in inconsistent ordered information systems. Some properties and theorems are studied and discussed. A fact is certified that the distribution reduction is equivalent to the maximum distribution reduction in ordered information systems. Theorems on distribution reduction is implemented to create preparations for reduction acquisition and the dominance matrix is also restated to acquire distribution reductions in criterion based information systems. The Matrix algorithm for distribution reduction acquisition is stepped and programmed. Dominance matrices are the only relied parameters which need to be considered without others such as approximations and sub-information systems being brought in. Furthermore, cases are employed to illustrate the validity of the Matrix method and the program, which shows that the effectiveness of the algorithm in complicated information systems.

Acknowledgements. This work is supported by Natural Science Foundation of China (No. 61105041, No. 61472463, No. 61402064), National Natural Science Foundation of CQ CSTC (No. cstc 2013jcyjA40051).

References

1. Dembczynski, K., Pindur, R., Susmaga, R.: Generation of exhaustive set of rules within Dominance-based rough set approach. Electron. Notes Theory Comput. Sci. **82**, 96–107 (2003)
2. Greco, S., Matarazzo, B., Slowinski, R.: Rough approximation of a preference relatioin by dominance relatios. Europe J. Oper. Res. **117**, 63–83 (1999). ICS Research Report 16/96, Warsaw University of Technology (1996)
3. Hu, Q.H., Yu, D.R., Guo, M.: Fuzzy preference based rough sets. Inf. Sci. **180**, 2003–2022 (2010)
4. Pawlak, Z.: Rough sets. Int. J. Comput. Inf. Sci. **11**, 341–356 (1982)
5. Skowron, A., Stepaniuk, J.: Tolerance approxximation space. Fundam. Inf. **27**, 245–253 (1996)
6. Borkowski, M., Ślęzak, D.: Application of discernibility tables to calculation of approximate frequency based reducts. In: Ziarko, W., Yao, Y. (eds.) RSCTC 2000. LNCS (LNAI), vol. 2005, pp. 123–130. Springer, Heidelberg (2001). doi:10.1007/3-540-45554-X_14
7. Sai, Y., Yao, Y.Y., Zhong, N.: Data analysis, mining in ordered information tables.In: IEEE International Conference on Data Mining, pp. 497–509 (2011)
8. Shao, M.W., Zhang, W.X.: Dominance relation and relus in an incomplete ordered information system. Int. J. Intell. Syst. **20**, 13–27 (2005)
9. Ślęzak, D.: Searching for frequential reducts in decision tables with uncertain objects. In: Polkowski, L., Skowron, A. (eds.) RSCTC 1998. LNCS (LNAI), vol. 1424, pp. 52–59. Springer, Heidelberg (1998). doi:10.1007/3-540-69115-4_8
10. Slezak, D.: Normalized decision functions and measures for inconsistent decision tables analysis. Fundamenta Informaticae **44**, 291–319 (2000)
11. Slezak, D.: Various approaches to reasoning with frequency based decision reducts: a survey. In: Polkowski, L., Tsumoto, S., Lin, T.Y. (eds.) Rough Set Methods and Applications, pp. 235–285. Physica-Verlag HD, Heidelberg (2000)
12. Susmaga, R., Slowinski, R., Greco, S., Matarazzo, B.: Generation of reducts and rules in multi-attribute and multi-criteria classification. Control Cybern. **4**, 969–988 (2000)
13. Xu, W.H.: Rough Sets and Ordered Information Systems. Science Press, Beijing (2013)

14. Xu, W.H., Li, Y., Liao, X.W.: Approaches to attribute reductions based on rough set ad matrix computation in inconsistent ordered information systems. Knowl.-Based Syst. **27**, 78–91 (2012)
15. Xu, W.H., Zhang, W.X.: Methods for knowledge reduction in inconsistent ordered information systems. J. Appl. Math. Comput. **26**, 313–323 (2008)
16. Yu, D.R., Hu, Q.H., Wu, C.X.: Uncertainty measures for fuzzy relations and their applications. Appl. Soft Comput. **7**, 1135–1143 (2007)
17. Zhang, W.X., Wu, W.Z., Liang, J.Y., Li, D.Y.: Theory and Method of Rough Sets. Science Press, Beijing (2001)

Consistency Driven Feature Subspace Aggregating for Ordinal Classification

Jerzy Błaszczyński[1]([✉]), Jerzy Stefanowski[1], and Roman Słowiński[1,2]

[1] Institute of Computing Science, Poznań University of Technology,
Piotrowo 2, 60-965 Poznań, Poland
{jerzy.blaszczynski,jerzy.stefanowski,roman.slowinski}@cs.put.poznan.pl
[2] Systems Research Institute, Polish Academy of Sciences, 01-447 Warsaw, Poland

Abstract. We present a new method for constructing an ensemble classifier for ordinal classification with monotonicity constraints. Ordinal consistency driven feature subspace aggregating (coFeating) constructs local component classification models instead of global ones, which are more common in ensemble methods. The training classification data are first structured using Variable Consistency Dominance-based Rough Set Approach (VC-DRSA). Then, coFeating constructs local classification models in subregions of the attribute space, which is divided with respect to consistency of objects. Our empirical evaluation shows that coFeating performs significantly better than previously proposed ensemble methods on data characterized by a high number of objects and/or attributes.

Keywords: Ordinal classification · Variable-consistency Dominance-based Rough Set Approach (VC-DRSA) · Bagging · Feating

1 Introduction

Bagging ensembles considered so far in the context of rough set approach [3,4] produce multiple *global classification models* learned by the same algorithm on multiple random perturbations of the training set. Feature subspace aggregating (feating) [13] is, on the other hand an ensemble approach that differs substantially in its motivation from bagging. In feating, the idea is to divide the attribute space into not overlapping subregions and to construct multiple *local classification models* in subregions. The motivation for feating is twofold: improved predictive accuracy due to increased diversification of local models and relative ease of construction of local models, as opposed to global models.

Recall that bagging [7] follows the idea of a bootstrap sample [9], which is a sample of objects from the training set, which are drawn uniformly with replacement. Bagging ensembles were successfully extended to handle ordinal classification with monotonicity constraints problems [4]. In this type of classification problems, decision classes and attributes value sets are ordered and there exists a monotonic relationship between evaluation of an object on an attribute and its assignment to a class, such that if object a has evaluations on all considered attributes not worse than object b (i.e., a dominates b), then a is expected

© Springer International Publishing AG 2016
V. Flores et al. (Eds.): IJCRS 2016, LNAI 9920, pp. 580–589, 2016.
DOI: 10.1007/978-3-319-47160-0_53

to be assigned to a class not worse than that of b. This is called *dominance principle* [11]. Objects violating the dominance principle are called *inconsistent*.

It has been shown that rough set theory can provide useful information about consistency of object in this type of ordinal classification [2]. Measured consistency of objects has been used to change bagging sampling strategy in variable consistency bagging (VC-bagging) [3,4]. More precisely, in this more focused sampling, consistent objects are more likely to be selected than inconsistent ones. To identify consistent objects, VC-bagging is using the same consistency measures as those introduced to define extended lower approximations in VC-DRSA [2]. The supporting intuition is that decreasing a chance for selecting inconsistent objects should lead to constructing more accurate and more diversified base classifiers in the bagging scheme. VC-bagging proved to be able to produce more accurate ensembles than bagging [4].

Feating results from an observation that a local model constructed on objects similar to one we want to classify is often more accurate than a global model constructed on the entire data set [12]. Moreover, feature subspace aggregating leads to smaller samples of objects, which makes the construction of classifiers easier than in case of the complete training set. The crucial point of feating is, however, how to divide attribute space to construct local models.

Our approach to dividing attribute space is guided by an observation that consistency of objects may be useful to identify good division points from the perspective of ordinal classification with monotonicity constraints. In this way, we are going to make use of attribute values of objects selected according to degree of consistency. Following this motivation, we propose *ordinal consistency driven feature subspace aggregating* (coFeating). coFeating is going to subdivide attribute space into non-overlapping, local regions. The divisions are going to be based on objects, which are characterized by outstanding values of consistency measure. This type of feating is going to favor divisions that lead to more consistent local regions. It should also lead to diversified local classification models provided that objects that have outstanding consistency are not localized in only one region of attribute space.

We distinguish two main goals of this paper. The first is to present the methodology of construction of ordinal consistency driven feating ensembles. This part is presented in Sect. 2. The second goal is to compare experimentally coFeating with VC-bagging, and other methods proposed for ordinal classification with monotonicity constraints. This comparison will be performed in Sect. 3. We conclude the paper in Sect. 4.

2   Proposed Solution

Ordinal consistency-driven feature subspace aggregating (coFeating) involves consistency measure of objects, which was defined within Variable-Consistency Dominance-based Rough Set Approach (VC-DRSA) [2]. Before presenting coFeating, we first remind some basics of VC-DRSA in the following Subsect. 2.1. Then, we follow with a detailed description of coFeating method in Subsect. 2.2.

2.1 Basics of VC-DRSA

Dominance-based Rough Set Approach (DRSA) [10,11] concerns a finite universe U of objects described by a finite set of attributes A with ordered value sets. Attributes with preference-ordered value sets are called *criteria*, while attributes whose value sets are not preference-ordered are called *regular attributes*. Moreover, A is divided into disjoint sets of condition attributes C and decision attributes D. The value set of attribute $q \in C \cup D$ is denoted by V_q, and $V_P = \prod_{q=1}^{|P|} V_q$ is called P-evaluation space, $P \subseteq C$. For simplicity, we assume that D is a singleton, i.e., $D = \{d\}$, and values of d are ordered class labels coded by integers from 1 to n, such that the higher the number, the better the class.

When among condition attributes from C there is at least one criterion, and there exists a *monotonic relationship* between evaluation of objects on criteria and their values (class labels) on the decision attribute, then the classification problem falls into the category of ordinal classification with monotonicity constraints. In order to make a meaningful representation of classification decisions, one has to consider the *dominance relation* in the evaluation space. For each object $y \in U$, two dominance cones are defined with respect to (w.r.t.) $P \subseteq C$. The P-positive dominance cone $D_P^+(y)$ is composed of objects that for each $q_i \in P$ are not worse than y. The P-negative dominance cone $D_P^-(y)$ is composed of objects that for each $q_i \in P$ are not better than y. The decision attribute makes a partition of objects from U into ordered decision classes $X_1, X_2, \ldots, X_n$, such that if $i < j$, then class X_i is considered to be worse than X_j. The dominance-based approximations concern unions of decision classes: upward unions $X_i^{\geq} = \bigcup_{t \geq i} X_t$, where $i = 2, \ldots, n$, and downward unions $X_i^{\leq} = \bigcup_{t \leq i} X_t$, where $i = 1, \ldots, n-1$. Application of DRSA to the case of unknown monotonic relationships between condition and decision attributes has been shown in [6].

In order to simplify notation, we will use symbol X to denote a set of objects belonging to union of classes $X_i^{\geq}$ or $X_i^{\leq}$, unless it would lead to misunderstanding. Moreover, we will use $E_P(y)$ to denote any dominance cone $D_P^+(y)$ or $D_P^-(y)$, $y \in U$. If X and $E_P(y)$ are used in the same equation, then for X representing $X_i^{\geq}$ (resp. $X_i^{\leq}$), $E_P(y)$ stands for dominance cone $D_P^+(y)$ (resp. $D_P^-(y)$).

Variable-consistency rough set approaches extend lower approximation of set X by objects with sufficient evidence for membership to X. This evidence can be quantified by *object consistency measures*. In [2], we introduced gain-type and cost-type object consistency measures.

For $P \subseteq C, X \subseteq U, y \in U$, given a gain-type (resp. cost-type) object consistency measure $\Theta_X^P(y)$ and a gain-threshold (resp. cost-threshold) θ_X, the P-lower approximation of set X is defined as:

$$\underline{P}^{\theta_X}(X) = \{y \in X : \Theta_X^P(y) \propto \theta_X\}, \tag{1}$$

where $\propto$ denotes $\geq$ in case of a gain-type object consistency measure and a gain-threshold, or $\leq$ for a cost-type object consistency measure and a cost-threshold. In the above definition, $\theta_X \in [0, A_X]$ is a technical parameter controlling the degree of consistency of objects included in lower approximation of X.

In [2], we also introduced and motivated four *monotonicity properties* required from object consistency measures used in definition (1). We denoted them by $(m1)$, $(m2)$, $(m3)$, and $(m4)$.

The object consistency measure that we consider in this paper is a cost-type measure $\epsilon_X^P(y)$. For $P \subseteq C, X, \neg X \subseteq U$, it is defined as:

$$\epsilon_X^P(y) = \frac{|E_P(y) \cap \neg X|}{|\neg X|}. \tag{2}$$

This measure is an estimate of conditional probability $Pr(y \in E_p(y)|y \in \neg X)$, i.e., probability that object y belongs to $E_p(y)$ provided that it does not belong to X. As proved in [2], this measure has properties $(m1)$, $(m2)$ and $(m4)$.

Extended lower approximations of unions of decision classes are basis for induction of a set of decision rules. More precisely, rules are induced from a positive region of each union of classes X, defined as:

$$POS(X) = \bigcup_{y \in \underline{P}^\theta x(X)} E_p(y). \tag{3}$$

VC-DomLEM [5] algorithm can be applied to this end. It induces sets of rules that preserve monotonicity constraints in a degree expressed by the same consistency measure as the one used to identify sufficiently consistent objects.

A set of rules can be used to classify objects. Classification methods solve situations when the classified object is covered by multiple rules that suggest assignment to different unions of classes. In the standard DRSA classification method, an object is assigned to a class (or a set of contiguous classes) resulting from intersection of unions of decision classes suggested by the rules. Refer to [1] for more details on classification methods in DRSA.

2.2 Ordinal Consistency Driven Feature Subspace Aggregating – coFeating

coFeating makes division of attribute space into local regions identified w.r.t. object consistency measure. The division points are placed where objects characterized by outstanding values of consistency measure are located. The division is then made into non-overlapping regions in a way that maximizes their consistency. It is expected that this type of partitioning leads to local regions which are more consistent than the entire region. The local regions are then used to construct local classification models.

To illustrate this idea, let us consider the example of a division presented in Fig. 1. The ordering of values of attribute q_1 is indicated by direction of the arrow, and class "+" is better than class "−". Thus, we have two inconsistent objects in the considered region: objects having values v_1 and v_2 on q_1. In other words, these are objects with outstanding values of a consistency measure as compared to the rest of objects. Now, we can consider the following divisions of the region into local regions. First, we can assign regions $q_1 \leq v_1$ and $q_1 > v_1$. Analogously, we can assign regions $q_1 < v_2$ and $q_1 \geq v_2$. Please note that both these divisions indeed make local regions more consistent than the original region. The assigned local regions do not include any inconsistent objects. Please also note that it is possible to assign such local regions which will not improve consistency. For example, one may assign region $q_1 < v_1$ and $q_1 \geq v_1$. In result, one local region $(q_1 < v_1)$ is consistent. On the other hand, local region $(q_1 \geq v_1)$, is inconsistent (it is composed of inconsistent objects). Moreover, when using object consistency measure (2), one can observe higher inconsistency of objects in the local region than in the original region.

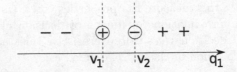

Fig. 1. Construction of a single split in coFeating.

The data structure that we use to implement coFeating is the same as the one used in the original formulation of feating [13], and it is called a *level tree*. A level tree is a restricted form of a decision tree where each node at a given level of the tree must use the same attribute. A local model is trained in each leaf of the tree. Then, to classify an object with the tree one needs to match the object to a leaf of the tree, and apply classification model found in that leaf.

The coFeating learning algorithm is presented as Algorithm 1. When in bagging mode, which is optional, for each division point a bootstrap sample is drawn. Then a number h of attributes is drawn with uniform probability. Subsequently, an object whose values are going to serve as division point coordinates is drawn according to values of object consistency measure c calculated in set D_b. Finally, the attributes are ordered according to consistency measured at the division point. More precisely, each object performance on each attribute is considered as potential division point and consistency is calculated in this point. Subsequent to that, a level tree is created basing on the selected division point and established attribute order. The whole process is repeated n times resulting in an ensemble of level trees.

Algorithm 2 shows how to construct a level tree in coFeating. It is different from the algorithm proposed for the original feating. First, in line 1, if the training set D (local region) is composed of objects from only one class d, a local model (classifier) assigning class d is constructed. Then, in line 4, if the number of objects in positive regions (see Definition (3)) constructed for classes included in D is too small, the level tree is going to abstain from assigning any class.

Algorithm 1. coFeating(D,A,n,h,o,L) - build a set of attribute space division trees based on consistency of objects

---

Input : training set D, set of attributes A, consistency measure c, number of inconsistent objects used as division points n, maximum number of subdivisions h, expected maximum number of objects at node o, base learning algorithm L.
Output: collection of division trees C.

1 $C \leftarrow \emptyset$;
2 **for** $i = 1$ *to* n **do**
3 | **if** *bagging* **then**
4 | | $D_b \leftarrow$ GetBootstrap(D)   /* This step is optional */;
5 | **else**
6 | | $D_b \leftarrow D$;
7 | $l \leftarrow$ DrawAttributesUniformly(A, h);
8 | $m \leftarrow$ DrawObjectAccordingToConsistency(D_b, c, l)   /* consistency c of objects is used as weight in the drawing process */;
9 | $l \leftarrow$ RankAttributesAccordingToConsistency(D_b, c, m, l);
10 | $C \leftarrow C \cup$ BuildDivisionLevelTree($D_b, m, l, h, o, 1, L$);

This is equivalent to constructing a model, which is assigning no class (null). On the other hand, when the maximal number of nodes in the level tree is achieved or when size of D is smaller than expected, which is checked in line 6, a local model is constructed using learning algorithm L. If neither of above conditions is met, division value (or split-value) v is selected. Finally, left and right splits are selected so that v is included in one of splits to maximize consistency, or in other words, where it fits better.

As it is specified in Algorithm 1, an ensemble of n level trees is constructed as the result of coFeating. When classifying an object, each level tree is used as a separate component classifier. Each classifier is allowed to vote for one class or for a set of contiguous classes. Then, the votes of component classifiers are aggregated as a weighted median.

3   Experiments and Discussion

The goal of the experiment is to compare ordinal consistency-driven feature subspace aggregating (coFeating) with VC-bagging and other methods previously proposed within DRSA. We previously observed in [4] that VC-bagging provides prediction accuracy at least comparable to other best methods proposed in the literature for this type of classification problem. In accord with experiments presented in [4], we measured mean absolute error (MAE) on fourteen ordinal data sets listed in Table 1 to compare the considered methods. Data sets used in this study are also the same as in other experiments with VC-bagging [4].

Analogously, to make our observations compatible with those from [4], we included in the comparison all classifiers that made use of VC-DomLEM

Algorithm 2. BuildDivisionLevelTree(D,m,l,h,o,j,L) - recursively build a feature space division tree

---

Input : training set D, split point m, list of attributes l, maximum number of nodes h, expected minimal number of objects at node o, current tree level j, base learner L.

Output: division level tree c.

1 **if** D *is composed of objects from one class d only* **then**
2 $\quad\lfloor\;$ $c.model \leftarrow$ assign d;
3 $\quad\lfloor\;$ **return** c;

4 **if** *number of objects in positive regions of D is too small* **then**
5 $\quad\lfloor\;$ **return** *null*;

6 **if** $(j = h)$ *or* (*size of D* $< o$) **then**
7 $\quad\lfloor\;$ $c.model \leftarrow$ BuildLocalModel(D, L);
8 $\quad\lfloor\;$ **return** c;

9 $a \leftarrow$ GetAttribute(l, j);
10 $v \leftarrow$ GetSplitValue(m, a);
11 $c.splitvalue \leftarrow \{a, v\}$;
12 **if** a *is ordinal* **then**
13 $\quad\mid\;$ $D_< \leftarrow$ FilterObjects($D, a < v$);
14 $\quad\mid\;$ $D_> \leftarrow$ FilterObjects($D, a > v$) /* v is included where it fits better */;
15 $\quad\mid\;$ $c.left \leftarrow$ BuildDivisionLevelTree($D_<, m, l, h, o, j + 1, L$);
16 $\quad\mid\;$ $c.right \leftarrow$ BuildDivisionLevelTree($D_>, m, l, h, o, j + 1, L$);
17 **else**
18 $\quad\mid\;$ $D_= \leftarrow$ FilterObjects($D, a = v$);
19 $\quad\mid\;$ $D_{\neq} \leftarrow$ FilterObjects($D, a \neq v$);
20 $\quad\mid\;$ $c.left \leftarrow$ BuildDivisionLevelTree($D_=, m, l, h, o, j + 1, L$);
21 $\quad\mid\;$ $c.right \leftarrow$ BuildDivisionLevelTree($D_{\neq}, m, l, h, o, j + 1, L$);

---

algorithm [5], either as a single classifier or as a component classifier in an ensemble. In other words, we included single monotonic VC-DomLEM with the standard or the new classification method. Results of these classifiers are used as a baseline for comparison. Thus, we included in the comparison: standard bagging, VC-bagging, and coFeating. Moreover, to make the results comparable, for all ensemble methods, i.e., standard bagging, VC-bagging, and coFeating, we use monotonic VC-DomLEM with the standard DRSA classification method, as component classifier or local classification model (in case of coFeating). The choice of the standard DRSA classification method in the ensembles was made due to increased computational complexity of the new classification method.

Some parameters specific to coFeating were tuned. These include the number of division points, which were selected among: 5, 10, 20, 30, 50. Then, the maximum number of nodes in the level tree was selected among: 2, 3, 5. Moreover, coFeating abstained from constructing local model when the proportion of

Table 1. Characteristics of data sets

Id	Data set	Objects	Attributes	Classes
1	balance	625	4	3
2	breast-c	286	8	2
3	breast-w	699	9	2
4	car	1296	6	4
5	cpu	209	6	4
6	bank-g	1411	16	2
7	fame	1328	10	5
8	denbosch	119	8	2
9	ERA	1000	4	9
10	ESL	488	4	9
11	housing	506	13	4
12	LEV	1000	4	5
13	SWD	1000	10	4
14	windsor	546	10	4

examples in positive regions of the local region to the size of local region was lower than a half.

All compared classifiers were carefully tuned to obtain best possible predictive accuracy. The predictive accuracy was estimated by stratified 10-fold cross-validation, repeated several times [8]. The results are shown in Table 2.

When analyzing the results, note that for two data sets: balance and breast-c, any of the ensemble methods seem not to work well. The best results are obtained with the single VC-DomLEM, regardless of the classification strategy used. The only other data set for which a single classifier turns out to be the best is windsor. For this set, however, the used classification method is important. It is also worth noting that contrary to the previous two sets, for windsor VC-bagging performs also quite good.

Observe that coFeating is the best method for the data sets which have the highest number of attributes (e.g., bank-g, housing). Moreover, it is also the best for the data sets characterized by the highest number of objects (bank-g, fame, car). coFeating is also the best method for some of the data sets with high number of attributes and objects (fame) or just attributes (breast-w). It turns out that it performs well even for the smallest data set denbosch, which can be explained by the fact that denbosch has a high number of attributes. Where it fails are artificial and highly inconsistent data sets (LEV, SWD, ERA, ESL). However, some of these data sets are also characterized by the smallest number of attributes (LEV, ERA, ESL). In case of ERA, and ESL, the results of coFeating are better than the results of single classifiers. coFeating achieved not favorable values of MAE on small data set cpu, for which VC-bagging had the best result.

Table 2. MAE resulting from repeated 10-fold cross validation

Data set	VC-DomLEM std. class.	VC-DomLEM new. class.	bagging std. class.	VC-bagging std. class.	coFeating std. class.
balance	**0.162**	**0.162**	0.201	0.197	0.216
breast-c	**0.233**	**0.233**	0.245	0.246	0.262
breast-w	0.038	0.037	0.036	0.032	**0.03**
car	0.041	0.034	0.037	0.038	**0.0257**
cpu	0.104	0.083	0.085	**0.077**	0.0861
bank-g	0.055	0.045	0.045	0.042	**0.0378**
fame	0.38	0.341	0.323	0.32	**0.308**
denbosch	0.126	0.123	0.129	**0.109**	0.118
ERA	1.39	1.39	**1.26**	1.27	1.35
ESL	0.445	0.37	0.348	**0.337**	0.37
housing	0.356	0.323	0.298	0.279	**0.273**
LEV	0.488	0.481	0.435	**0.409**	0.543
SWD	0.462	0.454	0.443	**0.43**	0.514
windsor	0.535	**0.502**	0.53	0.504	0.538

It is worth noting that, VC-bagging is the best method on data sets for which coFeating fails, except for ERA, for which bagging works best.

To conclude our observations, we would like to point out that due to the nature of the comparison that we intended to perform coFeating was tested on data sets from the previous study [4], which was made for VC-bagging. The results presented in Table 2, show that coFeating is improving classification performance on the data sets characterized by a high number of objects and/or attributes. This is concordant with observations made for original feating [13]. It would be thus interesting to extend our comparison on more data sets with these features. Especially so, since it follows the intuition that constructing local models should work better on larger data sets. Moreover, the experiments showing favorable performance of the original feating were performed on data sets significantly larger than the ones considered in this study [13].

On smaller data sets, for which coFeating performed worse, VC-bagging is the best method except for ERA data set, for which bagging is better, and windsor, for which VC-DomLEM with the new classification method is better.

4   Concluding Remarks

This paper shows advantages of using local classification models in ensembles constructed for ordinal classification with monotonicity constraints. We have proposed a new ensemble construction method, based on the idea of ordinal consistency driven feature subspace aggregating (coFeating). Object consistency measure is used in coFeating to identify division points, which allow to divide attribute space into local subregions. Local classification models are constructed in the subregions. These local models constitute coFeating ensemble. The method shows clear advantages in predictive accuracy on data sets, which are characterized by a high number of objects and/or attributes. Results of empirical evaluation show that for such data sets coFeating is more accurate than previously proposed ensemble method VC-bagging, and at least comparable to best results found in the literature.

References

1. Błaszczyński, J., Greco, S., Słowiński, R.: Multi-riteria classification - a new scheme for application of dominance-based decision rules. Eur. J. Oper. Res. **181**(3), 1030–1044 (2007)
2. Błaszczyński, J., Greco, S., Słowiński, R., Szeląg, M.: Monotonic variable consistency rough set approaches. Int. J. Approximate Reasoning **50**(7), 979–999 (2009)
3. Błaszczyński, J., Słowiński, R., Stefanowski, J.: Variable consistency bagging ensembles. In: Peters, J.F., Skowron, A. (eds.) Transactions on Rough Sets XI. LNCS, vol. 5946, pp. 40–52. Springer, Heidelberg (2010). doi:10.1007/978-3-642-11479-3_3
4. Błaszczyński, J., Słowiński, R., Stefanowski, J.: Ordinal classification with monotonicity constraints by variable consistency bagging. In: Szczuka, M., Kryszkiewicz, M., Ramanna, S., Jensen, R., Hu, Q. (eds.) RSCTC 2010. LNCS, vol. 6086, pp. 392–401. Springer, Heidelberg (2010)
5. Błaszczyński, J., Greco, S., Słowiński, R., Szeląg, M.: Sequential covering rule induction algorithm for variable consistency rough set approaches. Inform. Sci. **181**(5), 987–1002 (2011)
6. Błaszczyński, J., Greco, S., Słowiński, R.: Inductive discovery of laws using monotonic rules. Eng. Appl. Artif. Intell. **25**, 284–294 (2012)
7. Breiman, L.: Bagging predictors. Mach. Learn. **24**(2), 123–140 (1996)
8. Demsar, J.: Statistical comparisons of classifiers over multiple data sets. J. Mach. Learn. Res. **7**, 1–30 (2006)
9. Efron, B.: Nonparametric estimates of standard error. The jackknife, the bootstrap and other methods. Biometrika **68**, 589–599 (1981)
10. Greco, S., Matarazzo, B., Słowiński, R.: Rough sets theory for multicriteria decision analysis. Eur. J. Oper. Res. **129**(1), 1–47 (2001)
11. Słowiński, R., Greco, S., Matarazzo, B.: Rough set methodology for decision aiding. In: Kacprzyk, J., Pedrycz, W. (eds.) Handbook of Computational Intelligence, pp. 349–370. Springer, Berlin (2015). Chapter 22
12. Frank, E., Hall, M., Pfahringer, B.: Locally weighted naive bayes. In: Proceedings of the 19th Conference on Uncertainty in Artificial Intelligence, pp. 249–256. Morgan Kaufmann (2003)
13. Ting, K.M., Wells, J.R., Tan, S.C., Teng, S.W., Webb, G.I.: Feature-subspace aggregating: ensembles for stable and unstable learners. Mach. Learn. **82**(3), 375–397 (2010)

Author Index